ANNUAL
REPORTS IN
MEDICINAL
CHEMISTRY
Volume 29

ANNUAL REPORTS IN MEDICINAL CHEMISTRY
Volume 29

Sponsored by the Division of Medicinal Chemistry
of the American Chemical Society

EDITOR-IN-CHIEF:

JAMES A. BRISTOL

PARKE-DAVIS PHARMACEUTICAL RESEARCH
DIVISION OF WARNER-LAMBERT COMPANY
ANN ARBOR, MICHIGAN

SECTION EDITORS

WILLIAM K. HAGMANN • JOHN C. LEE • JOHN M. McCALL
JACOB J. PLATTNER • DAVID W. ROBERTSON • MICHAEL C. VENUTI

EDITORIAL ASSISTANT
LISA GREGORY

ACADEMIC PRESS, INC.
San Diego New York Boston London Sydney Tokyo Toronto

Academic Press Rapid Manuscript Reproduction

Academic Press, Inc.
A Division of Harcourt Brace & Company
525 B Street, Suite 1900, San Diego, California 92101-4495

United Kingdom Edition published by
Academic Press Limited
24-28 Oval Road, London NW1 7DX

International Standard Serial Number: 0065-7743

International Standard Book Number: 0-12-040529-6

PRINTED IN THE UNITED STATES OF AMERICA
94 95 96 97 98 99 QW 9 8 7 6 5 4 3 2 1

CONTENTS

III.   CHEMOTHERAPEUTIC AGENTS

Section Editor: Jacob J. Plattner, Abbott Laboratories, Abbott Park, Illinois

IV.   IMMUNOLOGY, ENDOCRINOLOGY AND METABOLIC DISEASES

Section Editor: William K. Hagmann, Merck Research Laboratories, Rahway, New Jersey

V.   TOPICS IN BIOLOGY

Section Editor: John C. Lee, SmithKline Beecham Pharmaceuticals, King of Prussia,
                   Pennsylvania

VI.   TOPICS IN DRUG DESIGN AND DISCOVERY

VII.   TRENDS AND PERSPECTIVES

CONTRIBUTORS

PREFACE

Annual Reports in Medicinal Chemistry continues with its objective to provide timely updates of important areas of medicinal chemistry together with an emphasis on emerging topics in biological science which are expected to provide the foundation for future therapeutic interventions.

Volume 29 retains the familiar format of previous volumes, this year with 34 chapters. As in past volumes, sections I - IV concern specific medicinal agents with annual updates on allergy, anti-infectives, and antivirals. This year we continue the trend started several years ago of reducing the number of annual updates in favor of specifically focused and mechanistically oriented chapters, where the objective is to provide the reader with the most important new results in a particular field. To this end, chapters on topics covered for the first time include neuronal cell death, animal engineering, D_3/D_4 receptor ligands, NO synthase, Ras oncogene, PDE inhibitors, HLE inhibitors and agents to control androgen action. Other topics included this year which have been occasionally reviewed in the past include antidepressants, muscarinic agents, EAAs, endothelin antagonists, thrombolytics, antithrombotics, HIV protease, HIV reverse transcriptase, opportunistic infections, IL-1, immunosuppressants, and cell adhesion.

Sections V and VI continue to emphasize important topics in medicinal chemistry, biology, and drug design as well as the important interfaces among these disciplines. In Section V, Topics in Biology, are included chapters on osteoporosis, transcription factor NHκB, protein kinases and phosphatases, translational control of gene expression and transgenics. We expect that topics of research reviewed in this section will appear in a chapter related to medicinal agents in a future volume, once sufficient time has passed to allow new compounds to be developed from a biological strategy. Chapters in Section VI, Topics in Drug Design and Discovery, reflect the current focus on mechanism-directed drug discovery and newer technologies. These include chapters on adenylate cyclase, antisense, prediction of human drug metabolism, humanized MAbs and ethnobotany.

Volume 29 concludes with chapters on NCE introductions worldwide in 1993. In addition to the chapter reviews, a comprehensive set of indices has been included to enable the reader to easily locate topics in volumes 1 - 29 of this series.

Over the past year, it has been my pleasure to work with 6 highly professional section editors and 65 authors, whose critical contributions comprise this volume.

James A. Bristol
Ann Arbor, Michigan
May, 1994

SECTION 1. CNS AGENTS

Editor: John M. McCall, The Upjohn Company
Kalamazoo, MI 49001

Chapter 1. Toward Third Generation Antidepressants

Dirk Leysen and Roger M. Pinder
Scientific Development Group, N.V. Organon and
Research Coordination, Akzo Pharma International B.V.
P.O. Box 20, 5340 BH Oss, The Netherlands

Introduction - There are no third generation antidepressants (1). The second generation era began in the 1970s when new tricyclics (TCAs) such as maprotiline and atypical agents like mianserin were introduced to reduce the cardiovascular and anticholinergic side effects of the first generation TCAs and monoamine oxidase inhibitors (MAOIs) typified by amitriptyline and isocarboxazid. It continues with selective serotonin reuptake inhibitors (SSRIs) and reversible inhibitors of monoamine oxidase A (RIMAs). New agents (1) introduced recently include the SSRIs sertraline, paroxetine and citalopram plus the RIMA moclobemide. In addition, the trazodone-like nefazodone and the monoamine reuptake inhibitor venlafaxine were approved by the FDA (U.S.). None of these drugs offers any advantages over earlier examples of the genre in terms of faster onset of action or broader efficacy. True third generation agents will retain the gains already made in minimizing side-effects and toxicity in overdosage but will additionally act in the first few days of treatment and be effective in the vast majority of patients (1). Novel targets for their design have been identified including receptor subtypes characterised by molecular cloning (2), intracellular mechanisms beyond the receptor, corticosteroids and neuropeptides (1).

RECEPTOR MULTIPLICITY

α_2-Adrenoceptors - Autoregulation of norepinephrine (NE) release in the human brain involves release-inhibiting, presynaptically located α_2-adrenoceptors, blockade of which leads to release of NE and a situation of raised synaptic NE levels akin to that following reuptake inhibition (3-5). α_2-Adrenoceptors (heteroceptors) also exist on 5-HT terminals in human brain, where their blockade facilitates 5-HT release (6, 7). Pharmacological prospects for α_2-adrenoceptor antagonists are manifold (8), including application as antidepressants in their own right or as adjuncts to inhibitors of NE and/or 5-HT reuptake (1). However, convincing evidence of antidepressant efficacy exists only for the three tetracyclic drugs mianserin, mirtazapine and setiptiline, none of which is selective for α_2-adrenoceptors (1). Two potent and selective antagonists, idazoxan and fluperoxan, have been tested in limited clinical trials as antidepressants (1, 9). Their efficacy is unconvincing, and large placebo-controlled trials in well-defined populations of patients are required to establish the value of such drugs. They are useful in challenge tests to explore the physiological role of α_2-adrenoceptors (10), which may be downregulated in depressed patients both centrally (11) and in blood platelets (12). However, many of the new selective antagonists, including idazoxan, are imidazoline derivatives which may both facilitate and inhibit NE release by interaction at presynaptic α_2-adrenergic and imidazoline-preferring receptors respectively (13).

The future applicability of α_2-adrenoceptor antagonists as antidepressants may lie in their affinity for the various receptor subtypes, of which four have been identified pharmacologically including three characterized by molecular cloning (14). At least three genes, on chromosomes 2, 4 and 10, code for α_2-adrenoceptors in man and three rat genes have also been expressed. Human blood platelets, often used as a proxy for central α_2-adrenoceptors, contain the prazosin-insensitive α_{2A}-type (14). Peripheral (15) and central (4, 5) autoreceptors in man are markedly different from the prazosin-sensitive α_{2B} and α_{2C} types, and probably belong to the α_{2A} or α_{2D} types. Cortical autoreceptors in rat and man are blocked preferentially by (+)- mianserin but are insensitive to (-)- mianserin, while the heteroceptors on 5-HT terminals in rat cortex were blocked with equal potency by the enantiomers, indicating that they belong to different subtypes (5). Given the relative affinities of the enantiomers for the four subtypes (16), the autoreceptor is α_{2D} in nature, while the heteroceptor is probably of the α_{2B} or, less likely, α_{2A} family (5). Such receptors represent attractive targets for modulation of NE or 5-HT neurotransmission by new and selective α_2-adrenoceptor antagonists (1, 17).

GABA - Multiplicity of GABA receptors is confined to two main types: GABA$_A$, associated with the chloride channel and involved in fast inhibitory signal transmission, and GABA$_B$ which operate in a modulatory fashion (18, 19). The GABA$_A$ receptor complex has additional binding sites for benzodiazepines, barbiturates and picrotoxin, and is the locus of action for anxiolytics, sedative-hypnotics, anticonvulsants and anaesthetics (20). The most intriguing ligands from the viewpoint of psychiatry are the neurosteroids, which do not interact with intracellular steroid receptors but enhance GABA-mediated chloride currents (21). GABA$_A$ agonists like fengabine are not antidepressant (22), while the GABA$_B$ agonist l-baclofen appeared to worsen the condition of some depressed patients (23). Although many antidepressants block GABA$_A$ receptors and despite advancement of a GABA$_A$-ergic predominance hypothesis for depression (24), recent attention has focussed upon antagonists of the GABA$_B$ receptor (19).

H_2N⌒⌒$\overset{\overset{\displaystyle O}{\parallel}}{\underset{\underset{\displaystyle OH}{|}}{P}}$‑R

1 : R = -CH(OC$_2$H$_5$)$_2$

2 : R = -(CH$_2$)$_3$CH$_3$

3 : R$_1$ = -H, R$_2$ = -CH(OC$_2$H$_5$)$_2$

4 : R$_1$ = -OH, R$_2$ = -cyclohexyl

GABA$_B$ receptors have been classified according to their relative sensitivity to baclofen, while the baclofen-sensitive group has been further subdivided into three types (18). As of March, 1994 no GABA$_B$ receptor had been cloned, so the structural characteristics of the proposed subtypes are unknown. Nevertheless, presynaptic GABA$_B$ receptor antagonists should enhance the release of not only GABA but also of various neurotransmitters more familiar in depression such as NE, 5-HT and dopamine. Penetration of the blood-brain barrier was first demonstrated with a series of 3-aminopropylphosphinic acids; GCP35348 (**1**), after systemic administration, antagonised the response to baclofen applied iontophoretically to rat cerebral cortex (19). The first orally active compound, GCP36742 (**2**), resembled desipramine in its ability to upregulate rat cortical GABA$_B$ receptors upon multiple dosage (25). However, both of these compounds are relatively weak antagonists whose affinity has been vastly amplified by amino substitution, as exemplified by GCP52432 (**3**) and GCP54626 (**4**). Both have nM potency in antagonizing baclofen-induced inhibition of GABA release in rat cerebral cortex and are orally effective (18, 19). Neither has been evaluated for antidepressant effects, but it is clear that antagonists selective for those GABA$_B$ receptors involved in release of NE and/or 5-HT may be interesting alternatives to α_2 or 5-HT$_1$ autoreceptor antagonists.

Serotonin - The relevance of altered 5-HT receptor subtype-mediated signal transduction to the pathophysiology and treatment of depression has attracted increasing interest (26). Electrophysiological studies indicate that chronic treatment with different antidepressant modalities enhances 5-HT neurotransmission (27). Due to the multiplicity, heterogeneous brain distribution and differential regulation of 5-HT receptor subtypes, several 5-HT signal transduction pathways are potential targets for new antidepressant treatments. Ligands for the multiple 5-HT receptor subtypes have recently been reviewed (28) although current evidence implicates involvement of 5-HT$_1$ and 5-HT$_2$ rather than 5-HT$_3$ receptors in depression.

5-HT$_{1A}$ Agonists - The 5-HT$_{1A}$ receptor seems to be a major site of antidepressant action (29). The clinical efficacy of partial agonists in treating anxiety and depression has been shown with the azapirones buspirone and the more selective gepirone, ipsapirone and tandospirone while the full agonist flesinoxan (**5**) may also be an effective antidepressant (30). A common metabolite of the azapirones is 1-(2-pyrimidinyl)piperazine, a potent α_2-adrenergic antagonist, but there is little evidence that it is responsible for the therapeutic activity of the azapirones. Antidepressant-like effects in the forced swimming test were obtained with gepirone and ipsapirone, using rats pretreated with a compound which inhibits metabolism (31). Furthermore, metabolically stable arylpiperazines have recently been described, which have longer durations of action than the corresponding azapirone (32). New 5-HT$_{1A}$ agonists based upon the tetralin skeleton of 8-OH-DPAT are the tetrahydrobenz(e)indoles **6** and **7** and the cis-(3aR)-(-)-hexahydro-benz(e)indole **8**. These compounds are highly potent and selective, with full intrinsic activity, and they seem to possess a long duration of action with improved oral availability (33-35). Stereoselectivity at the 5-HT$_{1A}$ and dopamine D$_2$ receptors is described.

Chronic antidepressant treatments produce adaptive hyporesponsivity of the 5-HT$_{1A}$ receptor-effector complex as indicated by a decrease of hypothermic and neuroendocrine responses. However, electrophysiological studies indicate enhanced postsynaptic 5-HT$_{1A}$ receptor-mediated responses in the hippocampus (36). Since there is no change in Bmax or Kd values, these findings suggest that the neuroadaptation may occur at a level beyond the receptor. Unravelling the functional alterations of components of the intracellular signal transduction pathway (G proteins, effector systems, etc.) may ultimately lead to new biochemical target mechanisms for future antidepressants.

5-HT$_{1A}$ Antagonists - Many compounds previously claimed as selective 5-HT$_{1A}$ receptor antagonists have been shown to behave as partial agonists or to have poor selectivity. Progress made toward the identification of selective and true ("silent") 5-HT$_{1A}$ antagonists has recently been reviewed (37, 38). Interesting silent 5-HT$_{1A}$ antagonists are: spiperone, despite high affinity for 5-HT$_2$ and D$_2$ receptors; the 2-aminotetralin (S)-UH-301 (**9**) which is only 8-fold selective over D$_2$ sites; and the phenylpiperazine WAY-100 135 (**10**) especially its S(+)-enantiomer (37, 38). Since antidepressants enhance 5-HT neurotransmission, a selective antagonist for the somatodendritic 1$_A$ autoreceptor is a worthwhile goal.

Terminal 5-HT Autoreceptors - A role for the terminal autoreceptor, 1$_B$ in rodents but 1$_D$ in humans, in the therapeutic action of antidepressants remains unclear in contrast to the situation with α_2-adrenoceptors. Chronic administration of SSRIs desensitizes these receptors, as shown by both in vitro release and in vivo electrophysiological experiments. However, acute or chronic treatment of rats with clomipramine, tianeptine or iprindole does not modify the density or the affinity of 5-HT$_{1B}$ receptors of the frontal cortex (39). The 5-HT$_{1B}$ agonist CGS 12066B (**11**) does not share the property of 5-HT$_{1A}$ agonists to reverse learned helplessness, indicating that 5-HT$_{1B}$ receptors do not mediate the behavioural effects of antidepressants in this model (40). It is difficult to assess the potential antidepressant activity of selective terminal 5-HT autoreceptor antagonists, as such compounds have not yet been reported. The tetrahydropyridines **12** and **13** are described as 5-HT$_{1B}$ agonists, whereas sumatriptan and the naphthylpiperazine (**14**) are potent 5-HT$_{1D}$ agonists. However, none of these compounds is really selective over 5-HT$_{1A}$ (28). Methiothepin, long regarded as the best example of an autoreceptor antagonist, has now been identified as the first inverse agonist at that receptor (41).

5-HT$_2$ Receptors - The importance of 5-HT$_2$ receptors with regard to the clinical effects of antidepressants may be related to the significant increase in the density of 5-HT$_2$ receptors found in both prefrontal cortex and amygdala of suicide victims/depressives (42). Chronic administration of virtually all antidepressants, regardless of their acute biochemical effects, results in down-regulation of 5-HT$_2$ and ß1-adrenergic receptor subtypes in rat brain, the time course of which parallels the onset of antidepressant action in patients with major depressive disorder. However, it is not clear whether receptor down-regulation is a common mechanism of action of antidepressants, or rather, an epiphenomenon (43). Both the 5-HT$_{2A}$ and 5-HT$_{2C}$ receptors typically undergo down-regulation in response to agonists but also exhibit a paradoxical down-regulation after chronic antagonist treatment. Both occurred without changes in mRNA levels, as measured in cultured cells (44). The acute stimulus properties of fluoxetine mostly resemble those of a 5-HT$_{2C}$ receptor agonist in conditioned taste aversion experiments (45). Chronic unpredictable mild stress results in an enhancement of 5-HT$_{2C}$ receptor mediated effects, whereas antidepressant treatments reduce this response (46).

No selective ligands have yet been described that can truly distinguish between the 5-HT$_2$ subtypes. Both the 5-HT$_{2C}$ antagonist **15** and the potent 5-HT$_{2A}$ agonist **16** are non-selective with regard to 5-HT$_2$ receptor subtypes (47, 48).

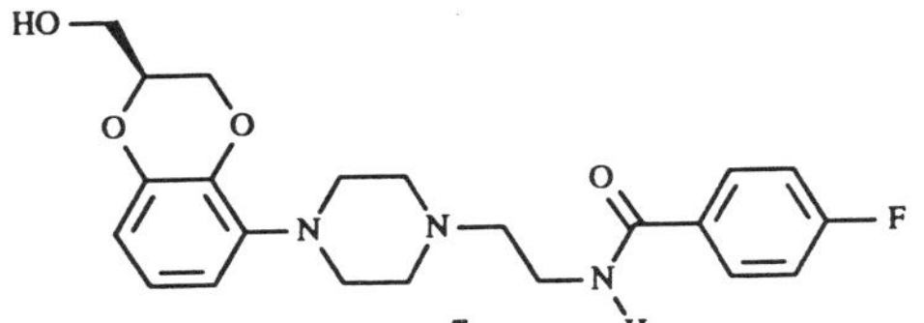

5

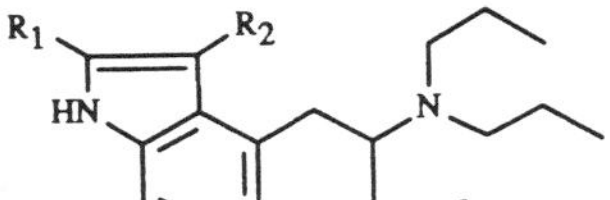

6 : R$_1$ = -H, R$_2$ = -CHO
7 : R$_1$ = -CN, R$_2$ = -H

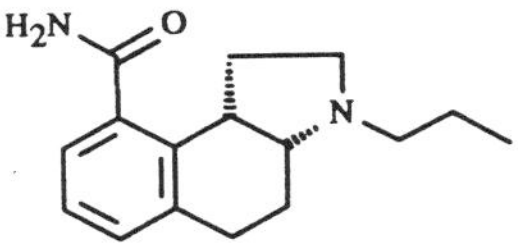

8

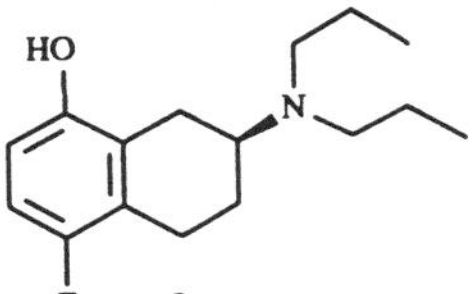

9

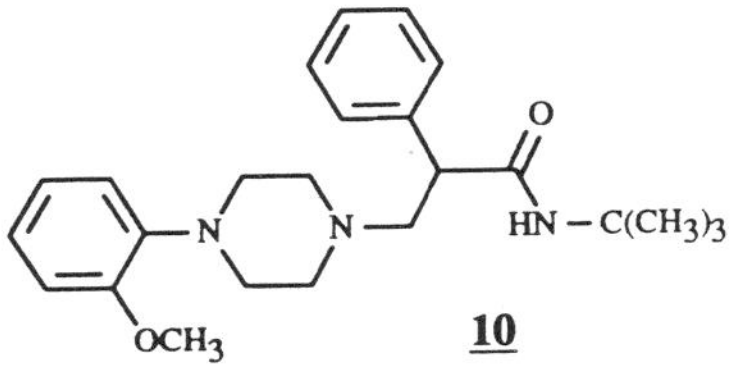

10

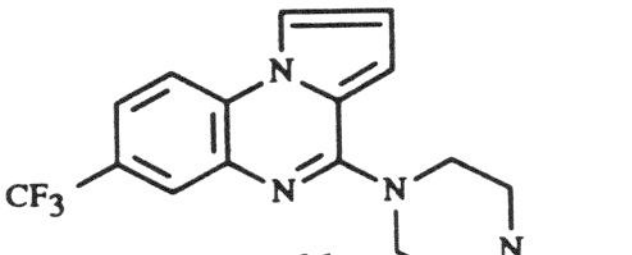

11

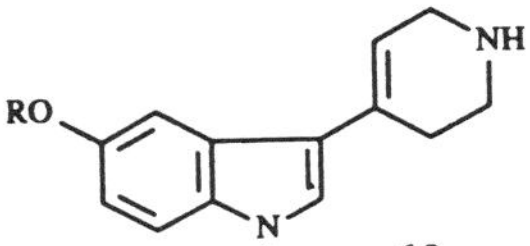

12

R = -CH$_3$, -(CH$_2$)$_2$CH$_3$

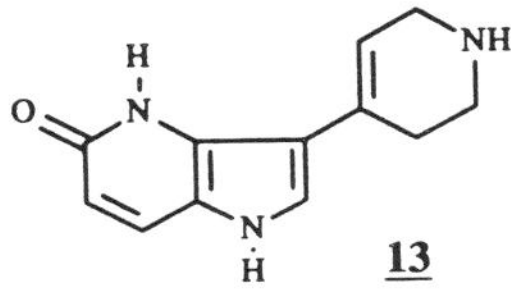

13

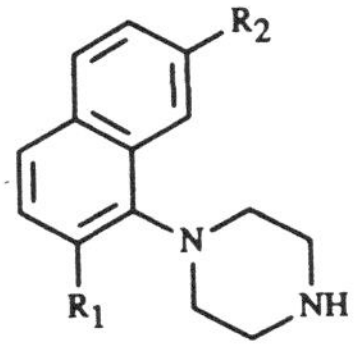

14

R$_1$, R$_2$ = -H, -OCH$_3$

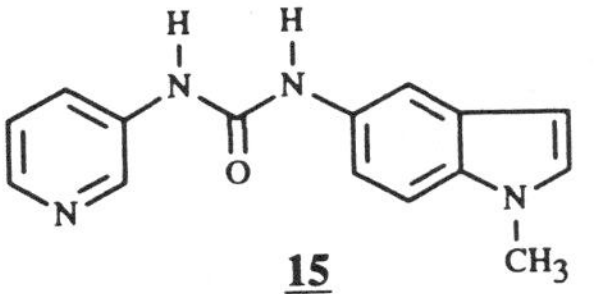

15

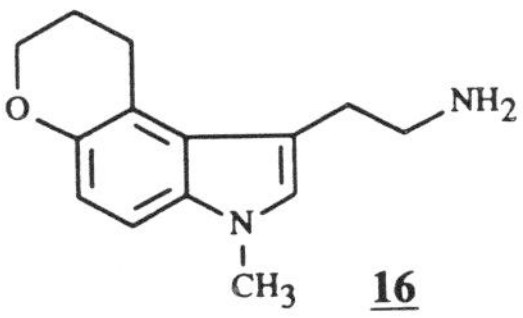

16

ALTERNATIVES TO LITHIUM

Lithium is the most commonly used drug in bipolar disorders, for both acute mania and maintenance therapy. It is also effective in the acute treatment of depression either alone or when used to augment mainline antidepressant treatment in refractory patients. However, lithium is no panacea and the need for alternatives has long been recognized. In addition to its enhancement of 5-HT transmission, lithium affects G protein-regulated phenomena such as receptor-activated phosphatidyl-inositol and cAMP-turnover, leading to altered functions of the inter-regulated cAMP-dependent protein kinase and protein kinase C (23, 49). Such mechanistic considerations may lead to the design of lithium mimetics, although rolipram, the first putative antidepressant shown to affect second messenger function by raising central cAMP levels _via_ selective inhibition of calcium-independent phosphodiesterase, was discontinued because of insufficient efficacy. Opportunistic clinical strategies have identified anticonvulsants and calcium antagonists as the leading alternatives (1).

Lithium Mimetics - The hypothesis of disturbances in G protein-mediated signal transduction in affective disorders is supported by the findings that $G_s\alpha$ levels are elevated in the cerebral cortex from postmortem brains of manic-depressive patients (50) and by the detection of lithium-sensitive hyperactive functions of G proteins in mononuclear leukocytes of untreated patients with mania (51). Furthermore, lithium completely blocked both adrenergic (ß) and cholinergic (M1) agonist-induced increases in GTP binding in rat cerebral cortex, suggesting a direct interaction at the level of receptor-G protein activation via competition with Mg^{2+} ions on magnesium low-affinity sites essential for GDP/GTP exchange on the G proteins (51). Although lithium _ex vivo_ does not seem to interfere equally with different receptor-G protein systems in various brain regions, it reduces neurotransmitter-induced GTP activation of G_s, G_i and G_q (49). A direct effect of lithium on the level of gene expression of especially G_s- and G_i -subunit mRNAs has also been reported (49). As lithium alters the function of the protein kinases, the interaction of specific nuclear transcription factors with their respective DNA sequences can thereby be modified. New biochemical targets for the design of alternatives to lithium may soon be available.

Lithium also non-competitively inhibits inositol monophosphatase, resulting in lowered cellular levels of myo-inositol and reduced agonist-induced formation of inositol polyphosphate (52).   A series of deoxy-inositol-1-phosphates (17) has been prepared as inositol monophosphatase inhibitors. These compounds suffered from low bio-availability (53), so the hydroxylated cyclohexane skeleton was replaced by the less polar phenol ring to enhance access to the brain;   the phosphate group was replaced by P-C phosphonates to avoid hydrolysis by non-specific phosphatases (18 and 19). The crystal structure of the cloned human enzyme has been determined by X-ray crystallography (54). This information might afford alternative inhibitors of inositol monophosphatase, which could further clarify whether inhibition of this enzyme relates to the therapeutic effects of lithium.

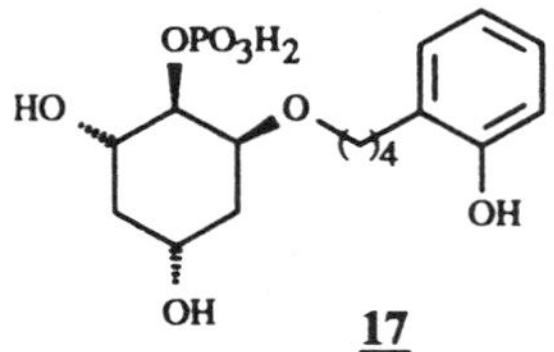

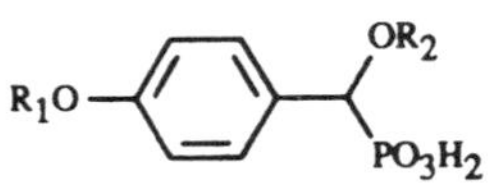

$R_1 = -Ph, R_2 = -H, -CH_3$ (for **18**) $R = cyclohexyl, menthyl, adamant-2-yl$ (for **19**)

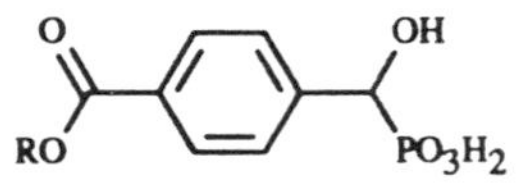

17   **18**   **19**

Anticonvulsants - Carbamazepine is now established as an alternative or adjunct to lithium (1), although both drugs, while sharing a common spectrum of clinical efficacy, may target different types of patients (23, 55). Sodium valproate also appears to be an effective antimanic agent (56, 57). For both anticonvulsants, large placebo-controlled trials in bipolar disorders are lacking to confirm their efficacy alone and in combination with lithium (1).   The acute antidepressant effects of carbamazepine and valproate are largely unproven and based almost entirely upon uncontrolled trials (1). More severely depressed patients, especially those with treatment-resistant melancholia (58), may respond better to carbamazepine than to lithium (23), whereas lithium appears to have more robust antidepressant effects than valproate (57), Neither anticonvulsant has been studied for its potential augmenting effect upon traditional antidepressants in drug-refractory patients.

In some experimental models of depression carbamazepine behaved as an atypical antidepressant drug after chronic administration, and its effects were blocked by clonidine or neuroleptics (59, 60). Carbamazepine is structurally related to TCAs, although lacking the alkylamine side chain, and has multiple pharmacological actions (1, 23). Its ability to reduce turnover of NE and DA may explain in part its antimanic action, although it also inhibits calcium function and calcium antagonists are antimanic (61). Carbamazepine also raises 5-HT function in man, suggesting a strong parallel with the action of lithium and most other antidepressant treatments (62). Carbamazepine, valproate and lithium decrease GABA turnover particularly after chronic administration, which suggests a common mechanism of antidepressant action consistent with previously mentioned notions of disturbances in GABA function in depression (23). The influence of carbamazepine upon cation pump mechanisms and ion channels, particularly its inactivation of voltage-sensitive sodium channels, may also play a role in its psychotropic action via effects upon catecholamine turnover and excitatory neurotransmission (1). Its rich panoply of central effects should provide ample mechanistic targets for the design of new antimanic and antidepressant agents.

Calcium Antagonists - The phenylalkylamine verapamil appears to be an effective antimanic agent, especially in patients responsive to but intolerant of lithium (61, 63, 64). Little information exists on the efficacy of other types of calcium antagonist except for the dihydropyridine (DHP) nimodipine which may be more effective when combined with other antimanic agents such as lithium or carbamazepine (65, 66). Even less is known about the antidepressant effects of calcium antagonists, and the only controlled clinical trial concluded that verapamil was inferior to amitriptyline (63). Nevertheless, many calcium antagonists are effective in animal models of depression and produce antidepressant-like reductions in imipramine binding and ß-adrenoceptor density upon chronic dosage (67). Conversely, chronic antidepressant treatments increase the density of cortical calcium channels (67).

Current calcium antagonists are selective for the L (long-lasting)-type of voltage-operated calcium channel and are less effective against the T(transient) and N (neuronal) types (1). Furthermore, calcium antagonists, however selective for neuronal channels involved in neurotransmitter release, will diminish monoamine availability. Calcium agonists, which activate calcium channels and increase rather than decrease calcium influx into cells, may represent alternatives to reuptake inhibitors or autoreceptor antagonists for raising synaptic monoamine levels (1). No calcium agonist has yet been developed for human use and the available DHPs have vasoconstricting and cardiostimulating properties. Behavioural changes in rodents indicative of CNS stimulation have been observed with the prototype agonist BAY K8644. The diversity of calcium channels arising from multiple genes and alternative splicing is greater than that of the T, L and N classification (68) and offers the opportunity to design new agonists and antagonists (1).

CORTICOSTEROIDS

Stressful life events can trigger the onset of depression in predisposed individuals (69). Abnormalities in the function of the hypothalamic-pituitary-adrenal (HPA)-axis, which is regulated by multiple negative feedback mechanisms, occur in depressed patients including increased levels of cortisol and corticotrophin releasing factor (CRF) but unchanged levels of adrenocorticotrophic hormone (ACTH). The number of CRF binding sites is reduced in the frontal cortex of suicide victims (70). The current hypothesis of disinhibition of the HPA axis in depression is further supported by the suppression test with the glucocorticoid dexamethasone. In addition, antidepressant treatments decrease cortisol levels. Cushing's patients, who are often depressed, have increased cortisol-, increased to normal ACTH- and normal CRF levels (69). In endogenously depressed patients and Cushing's patients, treatment with steroid synthesis inhibitors, such as metyrapone, aminoglutethimide and ketoconazole, lowers the plasma cortisol levels and reduces the severity of depression (71).

Corticosteroids influence nerve cell function <u>via</u> binding to intracellular receptors and subsequent alteration of the genomic action of the target cell. All five classes of steroid receptors: glucocorticoid (GR), mineralocorticoid (MR), estrogen-, androgen-, and progesterone receptors are ligand-dependent transcription factors (72). The GR system is widely distributed in rat brain and is involved in the feedback action of corticosteroids on stress-activated mechanisms and the HPA axis while facilitating neuroadaptation. Adrenalectomy-induced changes in 5-HT neurotransmission and behaviour can be restored by corticosterone action. The MR system is restricted predominantly to the lateral septum and hippocampus and may be involved in a tonic influence on brain function by controlling the basal activity of the HPA axis (73). Innervations of the hippocampus by midbrain serotonergic and locus ceruleus noradrenergic neurons are under control of GR and MR. These projections are inhibited by the glucocorticoid feedback at GR, leading to desensitization of serotonergic and adrenergic neurons. Chronic stress or chronic corticosterone treatment leads to a reduction in hippocampal GR and thereby reduces the negative feedback of corticosteroids on these projections. Furthermore, a reduced number of GR's has been found in the lymphocytes of depressed, but not of recovered, patients (69). In transgenic mice with GR function impaired by antisense RNA, hypothalamic GR mRNA is increased more by desipramine than in normal mice. Elevated serum ACTH and corticosterone levels in these transgenic animals were returned towards normal by desipramine (74). The SAR of 52 steroidal ligands for the glucocorticoid and progestin receptors has been described (75). Interesting GR ligands have been obtained by using the 17α-propenyl-17β-hydroxy substituents as in RU 43044 (<u>20</u>), which is a potent and selective GR antagonist but inactive <u>in vivo</u> due to very rapid metabolism. The antidepressant potential of selective GR antagonists remains hypothetical.

NEUROPEPTIDES

Some neuropeptides are intimately involved in the pathophysiology of mood disorders (1). Many depressed patients display an impaired release of thyrotropin following stimulation with thyrotropin-releasing hormone (TRH) and of growth hormone (GH) in response to α_2-agonists; their levels of CRF and ß-endorphins are increased. CRF and ß-endorphin have been shown to acutely improve mood in endogenously depressed patients, whereas TRH was

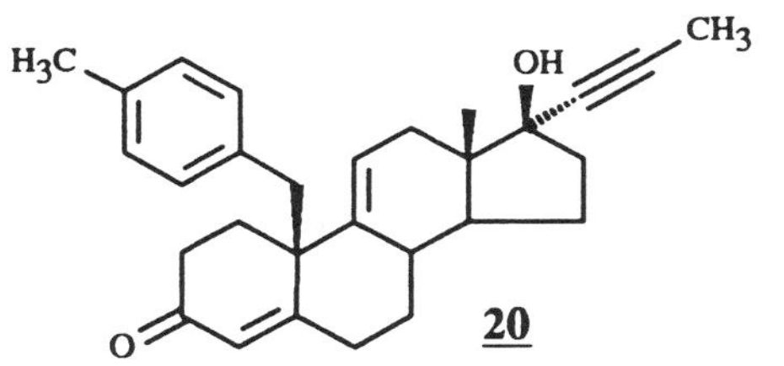

20

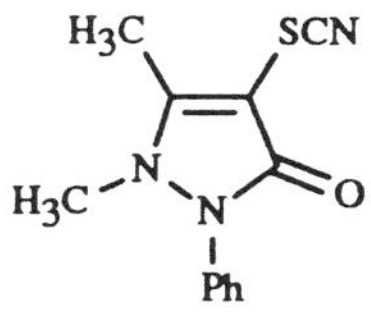

21

ineffective. The tripeptide melanocyte stimulating hormone release inhibiting factor (Pro-Leu-Gly-NH$_2$) has been claimed as being an antidepressant, equivalent in efficacy to imipramine (1).

CRF - A role for CRF in pathologic anxiety states and depression is probable. CRF is generally considered to be the primary activator of the HPA axis, as it is the major physiologic mediator of ACTH release during stress. There are clear effects of the CRF antagonists α-helical CRF$_{9-41}$ and CRF antisera on several responses in stress (76), and substantial evidence exists that CRF is hypersecreted in depression (69, 70). Direct injection of CRF into the lateral cerebral ventricles in animals produces behavioural effects related to depression and anxiety (77). Several results suggest that CRF is a neurotransmitter acting at additional CNS sites. CRF also stimulates the sympathetic tone to the heart, kidney, selected vascular beds and to the adrenal medulla resulting in epinephrine (E) secretion, whereas the cardiac parasympathetic (acetylcholine) nervous activity is inhibited (78). The secretion of CRF as well as the ACTH response to CRF levels in man is inhibited by glucocorticoids (79), while CRF secretion is activated by several classic monoamines especially 5-HT through 5-HT$_{1A}$ and 5-HT$_2$ receptors (77) and NE via α$_1$ receptors (76).

A series of oxopyrazolin thiocyanates and -disulphides (e.g. **21**), which inhibit CRF-stimulated adenylate cyclase activity in the uM range (80), as well as a series of N-terminally shortened analogues of CRF (81) have been described as CRF antagonists. The mouse pituitary and human brain CRF receptors have been cloned and functionally expressed; strong homology exists with the receptors for other small peptides such as GH releasing factor, vasoactive intestinal peptide, secretin, parathyroid hormone and calcitonin (82). This cloned human CRF receptor provides a new tool for the further development of more potent and selective CRF antagonists. Such compounds may be putative antidepressants.

Others - Neuropeptide Y (NPY) is a 36-amino acid peptide, which coexists with E and NE in some brain structures and has been claimed as a possible CSF marker for major depression (83). Acute and chronic administration of imipramine significantly lowered the level of NPY immunoreactivity in the cortex of rat brain (84) while chronic desipramine reduced the number of binding sites of NPY$_2$ receptors (85). Some enkephalins have been claimed to be antidepressant in man (1), while in mice the antidepressant action of TCAs is enhanced by inhibitors of enkephalin-degrading peptidases (86). Antidepressant-like effects were observed after systemic administration of RB 101, a mixed inhibitor prodrug of enkephalin-degrading enzymes; these effects were antagonized by the selective δ antagonist naltrindole (87). Cholecystokinin (CCK) plays an important role in anxiety and panic disorders and panic attacks can be reduced by certain CCK-B antagonists (88). Angiotensin-converting enzyme (ACE) is involved in the metabolism of several neuropeptides, and ACE inhibitors can affect central peptidergic transmission (1). As with CCK, however, ACE inhibitors, and the related angiotensin II receptor antagonists, seem to be more relevant to anxiety and cognition than to depressive disorders (1).

Conclusion - Increasing concern has been expressed about the economic burden placed upon society by depressive illness (89, 90). The overdosage risks of older TCAs have come under increasing scrutiny (91), accompanied by an enhanced awareness of the dangers of suicide by depressed patients (92). Safer alternatives like second generation antidepressants have made little impact on the risks largely because they are no more effective than TCAs which remain the mainstay of treatment because of their familiarity and cost (1, 90). Although combination therapies with different antidepressants are used to treat drug-refractory patients (1), including the now accepted practice of lithium augmentation, new agents are needed which will act more rapidly and effectively. While several targets have been identified for the design of novel antidepressants, true third generation drugs are unlikely to be introduced until the next century.

References

1.   R.M. Pinder and J.H. Wieringa, Med. Res. Rev., 13, 259 (1993).
2.   C. Gluchowski, T.A. Branchek, R.L. Weinshank and P.R. Hartig, Annu. Rep. Med. Chem., 28, 29 (1993).
3.   R.R. Ruffolo, A.J. Nichols, J.M. Stadel and J.P. Hieble, Pharmacol. Rev., 43, 475 (1991).
4.   N. Limberger, L. Späth and K. Starke, Br. J. Pharmacol., 103, 1251 (1991).
5.   M. Raiteri, G. Bonano, G. Maura, M. Pende, G.C. Andrioli and A. Ruelle, Br. J. Pharmacol., 107, 1146 (1992).
6.   M. Raiteri, G. Maura, S. Folghera, P. Cavazzini, G.C. Andrioli, E. Schlicker, R. Schalnus and M. Göthert, Naunyn-Schmiedeberg's Arch. Pharmacol., 342, 508 (1990).
7.   R. Tao and S. Hjorth, Naunyn-Schmiedeberg's Arch. Pharmacol., 345, 137 (1992).
8.   M. Berlan, J.-L. Montastruc and M. Lafontan, Trends Pharmacol. Sci., 13, 277 (1992).
9.   S.L. Dickinson, Drug News Perspect., 4, 197 (1991).
10.   N. Coupland, P. Glue and D.J. Nutt, Molec. Aspects Med., 13, 221 (1992).
11.   M. Schittecatte, G. Charles, R. Machowski, J. Garcia-Valentin, J. Mendlewicz and J. Wilmotte, Arch. Gen. Psychiat., 49, 637 (1992).
12.   I.E. Piletz, D. Chikkala, L. Khaitan, K. Jackson and Y. Qu, Neuropsychopharmacol., 9, 55 (1993).
13.   M.C. Michel and P. Ernsberger, Trends Pharmacol. Sci., 13, 369 (1992).
14.   D.B. Bylund, FASEB J., 6, 832 (1992).
15.   K. Smith and J.R. Docherty, Eur. J. Pharmacol., 219, 203 (1992).
16.   V. Simonneaux, M. Ebadi and D.B. Bylund, Molec. Pharmacol., 40, 235 (1991).
17.   R. Mongeau, P. Blier and C. de Montigny, Naunyn-Schmiedeberg's Arch. Pharmacol., 347, 266 (1993).
18.   G. Bonano and M. Raiteri, Trends Pharmacol. Sci., 14, 259 (1993).
19.   H. Bittiger, W. Froestl, S.J. Mickel and H-R. Olpe, Trends Pharmacol. Sci., 14, 391 (1993).

20.   R.B. Gammill and D.B. Carter, Annu. Rep. Med. Chem., 28, 19 (1993).
21.   Y. Hu, C.F. Zorumski and D.F. Covey, J. Med. Chem., 36, 3956 (1993).
22.   E.S. Paykel, A.E. van Woerkom, D.E. Walters, W. White and J. Mercer, Hum. Psychopharmacol. 6, 147 (1991).
23.   R.M. Post and D.M. Chuang, in "Lithium and the Cell: Pharmacology and Biochemistry" (N.J. Birch, Ed.), Academic Press, 199 (1992).
24.   R.F. Squires and E. Saederup, Neurochem. Res. 16, 1099 (1991).
25.   G.D. Pratt and N.G. Bowery, Br. J. Pharmacol. 110, 724 (1993).
26.   K. Lesch, C. Aulakh, B. Wolozin and D.L. Murphy, Pharmacol. Toxicol., 71 (Suppl. 1), 49 (1992).
27.   P. Blier, C. de Montigny and Y. Chaput, J. Clin. Psychiat., 51 (Suppl. 4), 14 (1990).
28.   A.G. Romero and R.B. McCall, Annu. Rep. Med. Chem., 27, 21 (1992).
29.   J.F.W. Deakin, F.G. Graeff and F.S. Guimaraes, Trends Pharmacol. Sci., 14, 262 (1993).
30.   M. Ansseau, W. Pichot, A. Gonzalez-Moreno, J. Wauthy and P. Papart, Hum. Psychopharmacol., 8, 279 (1993).
31.   E. Przegalinski, E. Tatarczynska and E. Chojnacka-Wojcik, J. Psychopharmacol., 4, 204 (1990).
32.   A.G. Romero, W.H. Darlington, M.F. Piercy and R.A. Lahti, Bioorg. Med. Chem. Lett., 2, 1703 (1992).
33.   P. Stjernlöf, M. Gullme, T. Elebring, B. Andersson, H. Wikstrom, S. Lagerquist, K. Svensson, A. Ekman, A. Carlsson and S. Sundell, J. Med. Chem., 36, 2059 (1993).
34.   A.G. Romero, J.A. Leiby, R.B. McCall, M.F. Piercey, M.W. Smith and F. Han, J. Med. Chem., 36, 2066 (1993).
35.   C.H. Lin, S.R. Haadsma-Svensson, G. Phillips, R.B. McCall, M.F. Piercy, M.W. Smith, K. Svensson, A. Carlsson, C.G. Chidester and P.F. Von Voigtlander, J. Med. Chem., 36, 2208 (1993).
36.   K.P. Lesch, Eur. Neuropsychopharmacol., 3, 302 (1993).
37.   I.A. Cliffe and A. Fletcher, Drugs Fut., 18, 631 (1993).
38.   A. Fletcher, I.A. Cliffe and C.T. Dourish, Trends Pharmacol. Sci., 14, 441 (1993).
39.   D. Montero, M.C. de Felipe and J. Del Rio, Eur. J. Pharmacol., 196, 327 (1991).
40.   P. Martin and A.J. Puech, Eur. J. Pharmacol., 192, 193 (1991).
41.   C. Moret and M. Briley, J. Psychopharmacol., 7, 331 (1993).
42.   P.V. Hrdina, E. Demeter, T.B. Vu, P. Sótónyi and M. Palkovits, Brain Res., 614, 37 (1993).
43.   S.M. Stahl, Prog. Neuro-Psychopharmacol. & Biol. Psychiat., 16, 655 (1992).
44.   E.L. Barker and E. Sanders-Bush, Molec. Pharmacol., 44, 725 (1993).
45.   H.G. Berendsen and C.L.E. Broekkamp, Eur. J. Pharmacol., 253, 83 (1994).
46.   J-L. Moreau, F. Jenck, J.R. Martin, S. Perrin and W.E. Haefely, Psychopharmacol., 110, 140 (1993).
47.   I.T. Forbes, G.A. Kennett, A. Gadre, P. Ham, C.J. Hayward, R.T. Martin, M. Thompson, M.D. Wood, G.S. Baxter, A. Glen, O.E. Murphy, B.A. Stewart and T.B. Blackburn, J. Med. Chem., 36, 1104 (1993).
48.   J.E. Macor, C.B. Fox, C. Johnson, B.K. Koe, L.A. Lebel and S.H. Zorn, J. Med. Chem., 35, 3625 (1992).
49.   H.K. Manji and R.H. Lenox, Synapse, 16, 11 (1994).
50.   L. Young, P. Li, S. Kish, K. Siu and J. Warsh, Brain Res., 553, 323 (1991).
51.   S. Avissar and G. Schreiber, Pharmacopsychiat., 25, 44 (1992).
52.   S. Nahorski, S. Jenkinson and R. Challiss, Pharmacol. Toxicol. 71 (Suppl. 1), 42 (1992).
53.   A. MacLeod, R. Baker, M. Hudson, K. James, M. Roe, M. Knowles and G. MacAllister, Med. Chem. Res. 2, 96 (1992).

54.   R. Bone, J. Springer and J. Attack, Proc. Natl. Acad. Sci. USA, 89, 10031 (1992).

55.   T. Okuma, Neuropsychobiol., 27, 138 (1993).

56.   T.W. Freeman, J.L. Clothier, P. Pazzaglia, M.D. Lesem and A.C. Swann, Am. J. Psychiat., 149, 108 (1992).

57.   S.L. McElroy, Neuropsychopharmacol., 9, 18S (1993).

58.   M. Cullen, P. Mitchell, H. Brodaty, P. Boyce, G. Parker, I. Hickie and K. Wilhelm, J. Clin. Psychiat., 52, 472 (1991).

59.   L. De Angelis, In Vivo, 5, 393 (1991).

60.   A. Sluzewska and A. Chodera, Pol. J. Pharmacol. Pharm., 44, 209 (1992).

61.   S.L. Dubovsky, Neuropsychobiol., 27, 184 (1993).

62.   M. Elphick, J.-D. Yang and P.J. Cowen, Arch. Gen. Psychiat., 47, 135 (1990).

63.   C. Höschl, Drugs, 42, 721 (1991).

64.   E.S. Garza-Trevino, J.E. Overall and L.E. Hollister, Am. J. Psychiat., 149, 121 (1992).

65.   R.M. Post, P.J. Pazzaglia, T.A. Ketter, M.S. George and L. Marangell, Neuropsychopharmacol. 9, 17S (1993).

66.   V. Manna, Minerva Med., 82, 757 (1991).

67.   O. Pucilowski, Psychopharmacol., 109, 12 (1992).

68.   N. Saccomano and A.H. Ganong, Annu. Rep. Med. Chem., 26, 33 (1991).

69.   S. Checkley, Br. J. Psychiatry, 160 (Suppl. 15), 7 (1992).

70.   C.B. Nemeroff, M. Owens, G. Bissette, A. Andorn and M. Stanley, Arch. Gen. Psychiatry, 45, 577 (1988).

71.   B. Murphy, J. Steroid. Biochem. Molec. Biol., 38, 537 (1991).

72.   R. Evans, Science, 240, 889 (1988).

73.   J. Reul and E. De Kloet, Endocrinology, 117, 2505 (1985).

74.   M. Pepin, F. Pothier and N. Barden, Molec. Pharmacol., 42, 991 (1992).

75.   G. Teutsch, M. Guillard-Moguilewsky, G. Lemoine, F. Nique and D. Philibert, Proc. Soc. Trans., 19, 901 (1991).

76.   A. Dunn and C. Berridge, Brain Res. Rev., 15, 71 (1990).

77.   C.B. Nemeroff, Neuropsychopharmacol., 6, 69 (1992); 9, 3S (1993).

78.   L. Fisher, Trends Pharmacol. Sci., 10, 189 (1989).

79.   D. Orth, Endocrine Rev., 13, 164 (1992).

80.   M. Abreu, W. Rzeszotarski, D. Kyle, R. Hiner and R. Elliott, US Patent 5, 063, 245 (1991).

81.   W. Kornreich, J. Hernandez, J. Rivier and W. Vale, EP 516, 450, A2 (1991).

82.   N. Vita, FEBS, 335, 1 (1993).

83.   E. Widerlöv, L. Lindström, C. Wahlestedt and R. Ekman, J. Psychiat. Res., 22, 69 (1988).

84.   M. Smialowska and B. Legutko, Neuroscience, 41, 767 (1991).

85.   P. Widdowson and A. Halaris, Brain Res., 539, 196 (1991).

86.   M. de Felipe, I. Jimenez, A. Castro and J. Fuentes, Eur. J. Pharmacol., 159, 175 (1989).

87.   A. Baamonde, V. Daugé, M. Ruiz-Gayo, I. Fulga, S. Turcaud, M.-C. Fournié-Zaluski and B. Roques, Eur. J. Pharmacol., 216, 157 (1992).

88.   J. Bradweijn, Neuropsychopharmacol., 9 (Suppl. 2), 6 (1993).

89.   P. Kind and J. Sorensen, Int. Clin. Psychopharmacol., 7, 191 (1993).

90.   D. Eccleston (Ed.), Br.J.Psychiat., 163, Suppl. 20 (1993).

91.   S. Kapur, T. Mieczkowski and J.J. Mann, JAMA, 268, 3441 (1992).

92.   ACNP Task Force, Neuropsychopharmacol., 8, 177 (1993).

Chapter 2. Neuronal Cell Death and Strategies for Neuroprotection

Christopher F. Bigge and Peter A. Boxer
Parke-Davis Pharmaceutical Research, Ann Arbor, MI

Introduction - Neuronal cell death, resulting from injury or disease, can be devastating. Currently, there are no drugs available that can prevent or delay neuronal cell death. Strategies that interrupt early events associated with acute insults and those that might preserve neuronal integrity during chronic stress are being pursued.

GENERAL MECHANISMS OF NEURONAL CELL DEATH

Cells of the central and peripheral nervous system are relatively unique in the body, since they are post mitotic and irreplaceable. Necrotic cell death, which is prominent in acute neurological conditions such as stroke, is characterized by swelling of the cell, disruption of internal and external membranes and cell lysis. Acute neuronal death can be easily modeled in primary neuronal cell culture and/or brain slices. Furthermore, experimental neuroprotective agents can be evaluated in animal models in which ischemia or trauma are the causative events. In apoptosis (programmed cell death), cells undergo nuclear condensation and fragmentation and the nuclear fragments, along with intracellular organelles, are extruded. In addition, nuclear DNA may be cleaved into fragments as a consequence of the activation of an endogenous Ca^{2+}-Mg^{2+}-dependent endonuclease. The apoptotic process is energy dependent and in some instances requires ongoing RNA and protein synthesis.

Calcium Homeostasis and the Role of Calcium in Neuronal Pathology - High levels of intracellular calcium $[Ca^{2+}]_i$ is thought to be the primary causative event in mediating necrotic neuronal death (1,2). The multiple mechanisms by which $[Ca^{2+}]_i$ is increased in neurons and some of the biochemical processes that are initiated when calcium levels exceed the buffering capacity of the neuron are shown in Figure 1. Free $[Ca^{2+}]_i$ in neurons is tightly regulated at around 100 nM and a concentration gradient of >10,000 exists compared to the extracellular concentration. Calcium enters the neuron through a variety of mechanisms including ligand-gated ion channels (most notably the N-methyl-D-aspartate (NMDA) glutamate receptor) (3), voltage-gated calcium channels (4), leak channels, and reversal of the Na^+-Ca^{++} antiporter (5). Neurons also have inositol trisphosphate receptors (IP_3) on intracellular organelles that can liberate calcium from intracellular stores (6). Since the maintenance of low free $[Ca^{2+}]_i$ is clearly energy dependent, the Na^+-Ca^{++} ATPase transporter is critical in maintaining the concentration. Both intra- and extracellular sources of calcium may operate in a positive feedback loop to increase free $[Ca^{2+}]_i$ (1,2).

Disruption of $[Ca^{2+}]_i$ homeostasis effects various protein kinase signalling pathways including calcium-calmodulin kinase II (Cam-kinase II) and protein kinase C (PKC) which in turn phosphorylate a number of proteins and activate a number of biochemical cascades.

Alterations in the mosaic of phosphorylated products with concomitant increases and decreases in the activity of cellular enzymes and ion channels influences expression of immediate early genes (IEGs) that may be involved in programmed cell death. The most likely role of IEGs is to transactivate other genes that can either contribute to cell death or be neuroprotective (7). Activation of calcium dependent proteases (calpains), at $[Ca^{2+}]_i$ in excess of 1 μM, may have the greatest impact on cell death. Calpains hydrolyze peptide bonds of structural proteins which results in the break down of the cell membrane, and ultimate destruction of the neuron.

Figure 1.   Schematic of Neuronal Processes Involved in Cell Death.

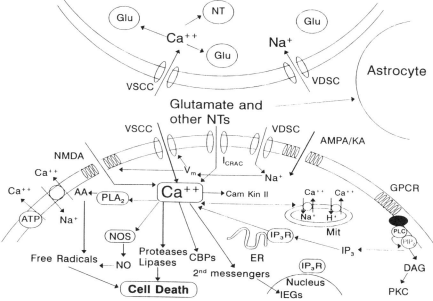

Fig. 1 Mechanisms of Ca2+ homeostasis. Abbreviations: NT=neurotransmitter; Glu=glutamate; VSCC=voltage-sensitive calcium channel; VDSC=voltage dependent sodium channel; V_m=membrane potential (depolarization); I_{CRAC}=calcium release activated channel; GPCR=G-protein coupled receptor; PLC=phospholipase C; DAG=diacylglycerol; Cam Kin II=calcium-calmodulin kinase II; CBP=calcium binding proteins; Mit=mitochondria; ER=endoplasmic reticulum; NOS=nitric oxide synthase; AA=arachadonic acid

<u>Energy Depletion and Oxidative Stress Contribute to Neuronal Death</u> - Many processes that control $[Ca^{2+}]_i$ homeostasis are interrelated and make it unlikely that blockade of a single mechanism can prevent all cellular necrosis. Neuronal cell cultures subjected to prolonged ischemia (hypoxia + hypoglycemia) ultimately die even in the presence of combinations of drugs that inhibit many of the processes illustrated in Figure 1 (8). Sporadic energy supplies can disrupt ion homeostasis as in focal ischemia where depolarization in the penumbra is initially moderate and/or intermittent although ATP is still produced (9). The resultant energy depletion does not destroy all neurons and is not a prerequisite for cell damage, but may act as a trigger in certain cells and result in irreversible damage. Oxidation of free fatty acids produces free radicals that may contribute to neuronal damage <u>via</u> inflammatory reactions or may act as downstream mediators of excitotoxicity and further enhance glutamate release

(10). Free radicals have also been implicated in lipid peroxidation that alters the structural integrity of cell membranes. Elevations in $[Ca^{2+}]_i$ may also initiate the formation of free radicals (11).

Neurotrophins and Chronic Mechanisms of Neuronal Cell Death - The molecular mechanisms responsible for neoplastic transformation may parallel those of neurodegeneration via either an effector pathway that allows permanent exit from the cell cycle or a lack of specific trophic factors (12). The pharmacology of neurotrophic factors in models of neurodegenerative disease was reviewed (13). Neurons dying from either glucose or NGF deprivation demonstrate mitochondrial dysfunction prior to morphological signs of cell damage. Misregulation of $[Ca^{2+}]_i$ seems to be a factor in cell death in both cases.

Important clues regarding apoptotic cell death have come from genetic studies in the nematode, *Caenorhabditis elegans*. Two genes, *ced-3* and *ced-4*, are required for cell death, whereas *ced-9* acts as a brake to halt cell death (14). The mammalian gene *bcl-2* is structurally and functionally homologous to *ced-9* and suppresses apoptotic cell death in many types of mammalian cells including neurons (15). Although the exact mechanism by which the protein encoded for by *bcl-2* suppresses cell death is not understood, it is clear that it exerts its action after the rise in $[Ca^{2+}]_i$ and may be related to the generation of free radicals (16). *Bcl-2* may be only one member of a gene family that regulates apoptotic cell death since a related gene (*bcl-x*) was identified recently, and a third gene, *bax*, prevents *bcl-2* from inhibiting cell death (17).

Therapeutic Indications - Excitotoxicity is involved in many neurological disorders and even some chronic neurodegenerative diseases (18). Excessive concentrations of excitatory amino acids (EAAs) kill neurons and elevated concentrations of EAAs are found in many models of neurological disorders (19). It is clear that excitoxicity contributes to neuronal loss in acute disorders such as stroke, epilepsy and head and spinal cord trauma. There is evidence that a slower form of excitoxicity may contribute to the selective neuronal loss seen in Parkinson's, Huntington's Disease, amyotrophic lateral sclerosis (ALS), AIDS dementia, and neuropathic pain. Free radical-induced neuronal death has been implicated in both ALS and Parkinson's (18). The degree to which either EAAs or free radicals contribute to Alzheimer's disease is more controversial. Alzheimer's and other chronic neurodegenerative diseases are more likely to be influenced by growth factor deprivation and/or apoptotic cell death. However, exposure to high concentrations of EAAs may lead to the death of neurons debilitated by other chronic factors.

POTENTIAL NEUROPROTECTIVE AGENTS

EAA Receptor Antagonists - EAA receptors were historically classified by agonist sensitivity into three subtypes: NMDA, α-amino-3-hydroxy-5-methylisoxazole-4-propanoic acid (AMPA), and kainate (20). Molecular cloning of the subunits which form EAA receptors has reinforced this classification (21-23). Early experiments with combinations of NMDAR1 with NMDAR2(A-D) show that they have different sensitivities to agonists, antagonists, and modulators (22,23). Ifenprodil was recently reported to have subtype selectivity in recombinant heteromeric NMDA receptors (24), and provides hope that selectivity may ultimately limit side effects. AMPA/KA receptors are also composed of subunits (GluR 1-4 for AMPA selective and GluR 5-7 and KA1 and KA2 for kainate selective receptors) (25). Glutamate can also activate metabotropic receptors; at least 7 different mGluRs have been cloned which are all linked to G-proteins and either stimulate PLC or inhibit cyclic AMP (26).

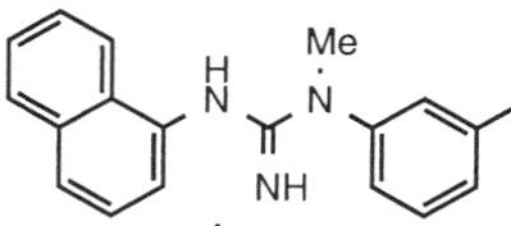

NMDA Antagonists - A unique feature of NMDA receptors is the number of sites on the receptor complex, where drugs have been shown to interact (27-31). CNS 1102 (**1**), an NMDA channel blocker in clinical trials (32), protected both the cortical and caudoputaminal regions following ischemia (33). The low efficacy partial glycine site agonists HA-966 (N-hydroxy-3-aminopyrrolidinone) and its 4-methyl analog, L-687,414, showed that neuroprotection and side effects can be separated (34). A series of 3-phenyl-4-hydroxyquinolin-2(1H)-ones were reported to be potent glycine site antagonists (35), and further modification gave 3-acyl derivatives represented by the 3-cyclopropyl ketone L-701,252 (**2**), that were anticonvulsant (36). The anticonvulsant agent, felbamate (2-phenyl-1,3-propanediol dicarbamate) is neuroprotective at clinically attainable doses in a rat pup model of hypoxia (37), and may protect _via_ an interaction with the glycine site (38). Table 1 summarizes competitive (glutamate recognition) and glycine site antagonists, channel blockers and other NMDA antagonists that have been evaluated thoroughly _in vitro_ and _in vivo_. Clinical experience thus far with NMDA antagonists indicates that the psychotomimetic side effects may limit their therapeutic potential (39).

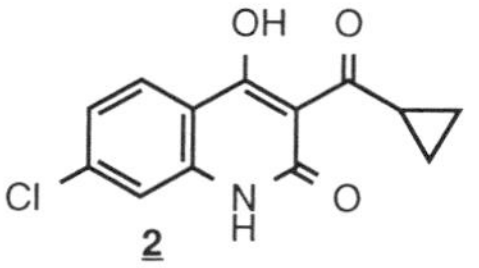

Table 1.   Selected Excitatory Amino Acid Receptor Antagonists.[a]

Competitive	Channel Blockers	Glycine Site	Other NMDA	AMPA
CGS 19755[b]	Dizocilpine[c]	HA 966	Ifenprodil[b]	GYKI 52466 (**3**)
CGP 40116[b]	CNS 1102 (**1**)[b]	L-687,414	Eliprodil[b]	NBQX (**4**)[b]
D-CPP-ene[b]	Memantine[b,d]	L-701,252 (**2**)	Nitroglycerin[d]	YM-90K (**5**)
LY 233053[b]	Remacemide[b]	Felbamate[d]		LY 215490 (**6**)
	DM[b,d,e]			NS 257 (**7**)

a. Compounds in this table are discussed in reviews (27-31), or in the text. b. Compounds are either in clinical trials or are expected to enter clinical trials. c. Compound has been in clinical trials. d. Currently approved by FDA for use in other indication. e. DM = dextromethorphan.

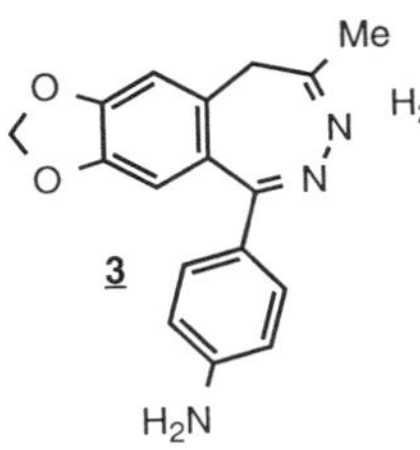

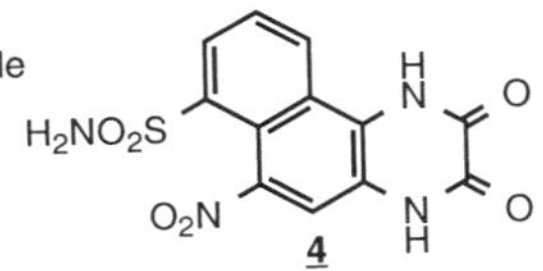

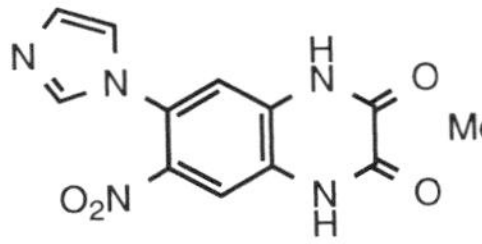

AMPA/Kainate Receptors - Activation of some AMPA/KA receptors can cause rapid neurotoxicity _via_ influx of calcium through the channels in a manner similar to that seen with NMDA receptors. Furthermore, excess sodium entry can initiate a number of sequelae that could contribute to cell death (see Figure 1). Although non-NMDA receptors appear to have fewer modulatory sites than NMDA receptors, at least two are known: 1) A 2,3-benzodiazepine site at which GYKI 52466 (**3**) inhibits AMPA-induced currents in a non-competitive manner; 2) a desensitization site at which diazoxide, cyclothiazide, aniracetam and willardines all prolong the rate of desensitization seen in response to the fast application of AMPA (40). Drugs that increase the rate of desensitization might be useful, but their feasibility is uncertain.

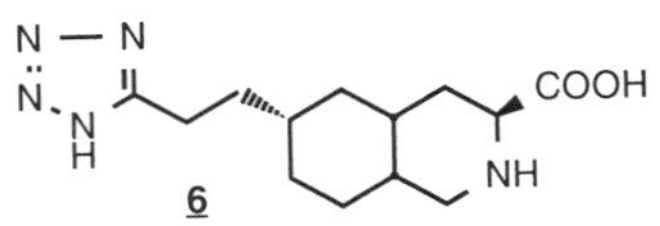

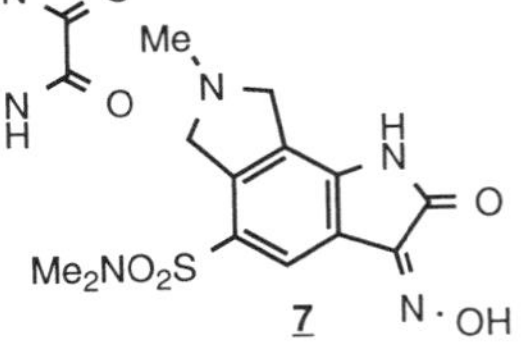

Some AMPA/KA receptor antagonists that have been evaluated as neuroprotective drugs are shown in Table 1. NBQX (**4**) and **3** have shown efficacy in both global and focal ischemia with dosing paradigms that involve significant delay in drug administration. This demonstrates that inhibition of these receptors is a productive strategy to prevent cell death, but blockade of fast synaptic neurotransmission may have significant side effects and their therapeutic potential awaits clinical trials. A few new structural types of AMPA antagonist, including YM 90K (**5**), LY215490 (**6**) and NS 257 (**7**), have been revealed that demonstrated *in vivo* activity (41-43).

Lipid Peroxidation and Free Radical Mechanisms - Free radical species of potential importance in mediating cell death include superoxide ($O_2\text{-}\bullet$) and hydroxyl ($OH\text{-}\bullet$). Hydrogen peroxide (H_2O_2) can react with iron to generate $OH\text{-}\bullet$. Several enzymes in the brain produce H_2O_2 as a normal byproduct, while others produce H_2O_2 by auto-oxidation. It has been shown that $O_2\text{-}\bullet$ is produced by NMDA receptor stimulation in cultured cerebellar granule cells using electron spin resonance techniques (44).

Free radical inhibitors such as dimethylthiourea, allopurinol and superoxide dismutase and catalase reduced stroke size (45). Barbiturates have been shown to protect brain in oxygen deprivation by a radical scavenging action, but only at high doses marked by sedation or motor depression. Vitamin E (α-tocopherol) also protects rat brain from damage due to hypoxic conditions (46,47). Known radical scavengers such as 3,5-di-tert-butyl-4-hydroxytoluene and vitamin E both contain a phenolic moiety and led to the synthesis of a 1-(acylamino)-7-hydroxyindan derivative OPC-14117 (**8**), which was cerebroprotective (48).

LY 231617 (2,6-di-*t e r t* -b u t y l -4 - (e t h y l a m i n o) m e t h y l p h e n o l) which concentrates in the CNS and significantly lowers mean arterial blood pressure, protected the hippocampus and striatum following global ischemia in rat (49).

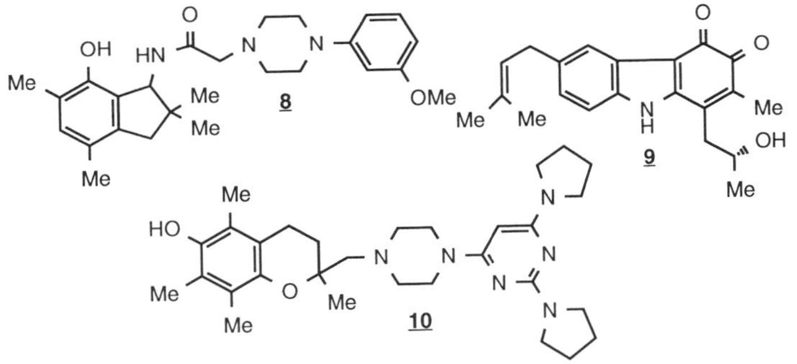

Despite the fact that xanthine oxidase (XO) concentrations are low in the brain, pretreatment with the XO inhibitor, oxypurinol, was shown to reduce ischemic neuronal injury in gerbils (50) and in rat MCAO (51,52). Carquinostatin A (**9**), isolated from *Streptomyces exfoliatus* 2419-SVT2, inhibits glutamate toxicity and has free radical scavenging activity comparable to that of vitamin E (53).

21-Aminosteroids, including tirilazad mesylate (U74006F), partially attenuated the damage induced by glucose deprivation, combined oxygen-glucose deprivation or exposure to NMDA in murine cortical cell cultures, and prevented most of the damage induced by exposure to iron (54). High doses of tirilazad mesylate reduced cortical infarct size moderately in a middle cerebral artery occlusion (MCAO) model in rat (55). U78517 (**10**), which combines the antioxidant ring portion of α-tocopherol with the piperazinyl sidechain of tirilazad, is 13-fold more potent against lipid peroxidation than tirilazad and protected cultured mouse spinal neurons against iron induced damage. __10__ has been shown to be neuroprotective in animal models of stroke and head injury (56,57).

Nitric oxide (NO), another source of free radicals, can lead to neurotoxicity by reacting with $O_2^-\bullet$ to produce the reactive peroxynitrite anion (ONOO-). ONOO- can decompose when protonated to yield highly reactive OH-• (58). NO may also contribute to excitotoxicity by facilitating glutamate release (59). An NO synthase inhibitor, L-nitroarginine methyl ester (L-NAME), reduced the volume of cortical and striatal infarct induced by MCAO in the rat (60). Nitroglycerin inhibited NMDA receptor responses by downregulation of the receptor by an action on the redox site. Once tolerance develops to its vasodilating effect, it may be useful for treatment of chronic neurological disorders (61).   The immunosuppressant, FK506, protects cultured cortical cells from NMDA neurotoxicity _via_ inhibition of calcineurin and the subsequent dephosphorylation of NO synthase which prevents its activation (62).

Calcium Channel Antagonists - Neuronal calcium channel antagonists were reviewed (63). Neuronal voltage sensitive calcium channels (VSCCs) have been classified into various subtypes (64-67). The most ubiquitous channels are those of the L-type (class C and D (66)), which are inhibited by classes of compounds represented by the dihydropyridines, diphenylalkylamines, and benzothiazepines (63,64). N-type (class B) VSCCs, which primarily inhibit catecholamine release (68,69), are selectively inhibited by the cone snail toxin ω-conotoxin GVIA (ω-CTX-GVIA) (70). The VSCC originally identified in Purkinje neurons has been called P-type and is selectively inhibited by the funnel-web spider toxin ω-agatoxin-IVA (ω-Aga-IVA).   More recently synthesis of a novel toxin, ω-conotoxin-MVIIC (SNX-230), has been shown to block VSCCs and synaptic transmission that is resistant to inhibition by either ω-CTX-GVIA or ω-Aga-IVA (tentatively called Q-type) (71).

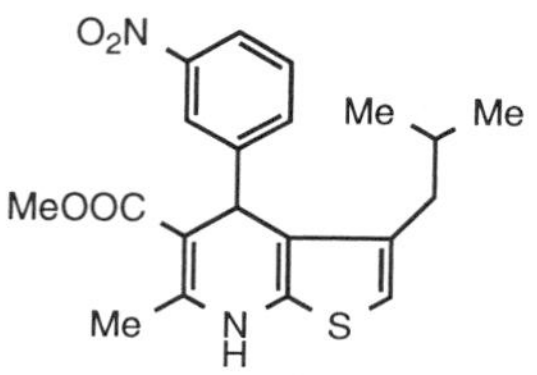

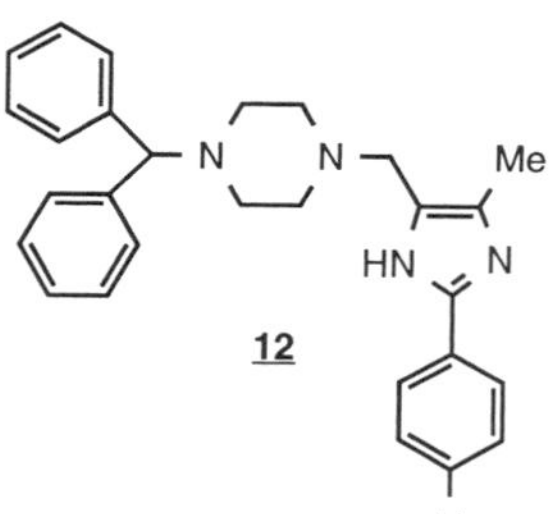

S-Emopamil, 2-isopropyl-5-(methylphenethylamino)-2-phenylvaleronitrile, attenuated the increase in hippocampal extracellular concentrations of glutamate following global cerebral ischemia while the dihydropyridine nimodipine did not (72). A novel dihydrothienopyridine calcium antagonist, S-312-d (__11__) reduced release of glutamate and taurine and protected CA1 hippocampal cells, while nimodipine did not (73). Flunarizine improved neurological recovery following experimental spinal cord injury in rabbit (74,75). Nimodipine was also protective (76). A flunarizine-like dual inhibitor of sodium and calcium channels, RS 87476 (__12__), also reduced infarct volume in a focal ischemia model (77).

The synthetic conotoxin peptide, SNX-111 (ω-conotoxin-MVIIA), which selectively blocks depolarization induced calcium fluxes through neuronal N-type VSCCs, protected hippocampal neurons from damage caused by four vessel occlusion in the rat with a single iv bolus administered up to 24 h after the ischemic insult. This suggests that the window of opportunity for therapeutic intervention after cerebral insult may be much longer than previously thought (78).

<u>Sodium Channel Antagonists</u> - Glutamate release during acute excitotoxicity coupled with energy loss does not occur <u>via</u> calcium-dependent synaptic release, but is probably due to a rundown and reversal of the energy dependent Na+-glutamate transporter (79,80). Since the transporter normally removes glutamate from the synapse, an antagonist of this transporter would not provide a good therapeutic target for neuroprotection. However, prevention of increases in [Na+]$_i$ may halt the reversal of this transporter. Blocking non-NMDA receptors is one way to reduce the increase in [Na+]$_i$. The direct blockade of voltage-dependent sodium channels would be expected to be neuroprotective, but systemic administration of potent sodium channel blockers such as tetrodotoxin is toxic.

Several anticonvulsant drugs modulate sodium channels by slowing the inactivation rate of the channel (81,82). Phenytoin is the prototypical anticonvulsant acting as a sodium channel modulator, and reduces ischemic damage in both global and focal models (83-85). Lamotrigine (6-(2,3-dichlorophenyl)1,2,4-triazine-3,5-diamine), and related analogs such as BW1003C87 (<u>13</u>) decrease glutamate release and prevent seizures with favorable therapeutic indices (86), and the 2-(N-methyl)piperazine derivative of <u>13</u>, BW619C89, inhibited veratridine-evoked release of both endogenous glutamate and aspartate from rat cerebral cortex slices (87). BW619C89 reduced cortical lesion volume (88,89). Another anticonvulsant and sodium channel modulator, riluzole (2-amino-6-(trifluoromethoxy) benzothiazole) (90-92), was also neuroprotective (93). A clinical trial indicated that riluzole slowed the progression of ALS in patients with bulbar onset (94). Interestingly, initial findings indicated that lamotrigine did not alter the progression of ALS (95).

<u>Calpain Inhibitors</u> - Calcium-activated proteolytic enzymes such as calpains have many possible biological roles (96). Calpain inhibition prevents spectrin degradation in a platelet membrane permeability assay and inhibitors may be able to attenuate excitotoxic damage in the CNS. Since calpain is only activated by high [Ca^{2+}]$_i$, interference with physiological functioning of neurons is not expected. The membrane-permeable calpain inhibitor, Cbz-Val-Phe-H, significantly protected Purkinje cells in cerebellar slices from AMPA-induced damage (97). A variety of calpain inhibitors have been reported and include transition state inhibitors, irreversible inhibitors, calmodulin antagonists and polyamines. Z-Leu-Phe-COOH, an α-ketoacid, is the most potent inhibitor of calpain I (Ki = 8.5 nM) and calpain II (5.7 nM), and is likely to be a transition state analog that forms a tetrahedral adduct with the active site cysteine and forms hydrogen bonds with the active site histidine (98).

<u>Potassium Channel Openers</u> - K$_{ATP}$ channels are abundant in hippocampal regions that are particularly vulnerable to the deleterious effects of ischemia. By hyperpolarizing neurons, K$_{ATP}$ channel openers should reduce both EAA release and activation of NMDA receptors (99,100). Three different classes of K$_{ATP}$ channel openers, (cromakalin, nicorandil and pinacidil) were found to block the ischemia induced expression of the IEGs and markedly protected neuronal cells against degeneration. Neuroprotective effects of K$_{ATP}$ channel openers were blocked by glipizide, a specific blocker of K$_{ATP}$ channels (99).

<u>Kappa agonists</u> - Kappa agonists have demonstrated neuroprotective activity due to inhibition of excitatory amino acid release, secondary vascular effects, or modulation of sodium channels (101-105). Spiradoline (U62066E) reduced hippocampal injury in a gerbil ischemia model (106). Enadoline (CI-977), an extremely potent antinociceptive agent, was

neuroprotective in animal models of cerebral ischemia (107,108).

Neurotrophins - Receptors for basic fibroblast growth factor (bFGF), nerve growth factor (NGF) and insulin-like growth factors (IGFs) are abundant in hippocampus and cerebral cortical neurons, whereas nerve growth factor (NGF) receptors are normally absent or present at low levels. bFGF, NGF and IGF-1 and IGF-II protected rat neurons against hypoglycemic damage and the damage normally caused by the mitochondrial toxins cyanide and 2,4-dinitrophenol. In addition, they prevented the elevation of $[Ca^{2+}]_i$ associated with excitotoxic and β-amyloid-induced damage (109). Cycloheximide and actinomycin D abolished the neuroprotective effect of bFGF indicating that the mechanism of action of bFGF requires protein synthesis. bFGF was neuroprotective in animal models of ischemia, and prevented kainic acid-induced neuronal cell loss, but not seizures, when administered via osmotic pumps in the hippocampus of rats (110).

The protective mechanism(s) of neurotrophins on neurons is unknown, but tyrosine kinase activity appears to be involved. Activation of neurotrophin receptors may result in the expression of IEGs (c-fos, jun, myc) that are believed to encode transcription factors. Although there are no examples of small molecules that possess the trophic actions of native neurotrophins, K-252b, a staurosporine-like tyrosine kinase inhibitor, enhances the effect of neurotrophins (NT-3 in particular) at very low concentrations. At high concentrations, K-252b acts as expected and inhibits the actions of neurotrophins (111).

Conclusions - We have reviewed some of the primary mechanisms by which neurons die, highlighting those strategies which have the greatest potential for providing neuroprotective drugs. Undoubtedly, as mechanisms of neuronal cell death become better understood, there will be many additional fruitful strategies for the development of neuroprotective agents. Given the complexity of the systems regulating neuronal cell death, it is unlikely that any single drug will be completely efficacious for any given neurological disorder. This highlights both the challenge and the promise of future research on neuroprotection.

References

1.   M. P. Mattson, Exp. Gerontol., 27, 29 (1992).
2.   B. K. Siesjö, aF. Benftsson, W. Grampp, S. Theander, Ann. N.Y. Acad. Sci., 568, 234 (1989).
3.   M. L. Mayer and R. J. Miller, Trends Pharmacol. Sci., 11, 254 (1990).
4.   R. W. Tsien, P. T. Ellinor, W. A. Horne, Trends Pharmacol. Sci., 12, 349 (1991).
5.   E. Carofoli, Ann. Rev. Biochem., 56, 395 (1987).
6.   V. Henzi and B. MacDermott, Neurosci., 46, 251 (1992).
7.   M. Vendrell, T. Curran, J. I. Morgan in "Markers of Neuronal Injury and Degeneration", V. 679, J. N. Johannessen, Ed., Ann. NY Acad Sci., New York, NY, 1993, p. 132.
8.   D. W. Choi, Prog. Brain Res., 96, 137 (1993).
9.   B. K. Siesjo, H. Minezawa, M. L. Smith, Fundam. Clin. Pharmacol., 5, 755 (1991).
10. M. Nedergaard, Acta Neurolog. Scand., 77, 81 (1988).
11. M. P. Mattson, R. E. Rydel, I. Lieberburg, V. L. Smith-Swintosky in "Markers of Neuronal Injury and Degeneration", V. 679, J. N. Johannessen, Ed., Ann. NY Acad Sci., New York, NY, 1993, p.1.
12. N. Heintz, Trends Biochem. Sci., 18, 157 (1993).
13. P. S. DiStefano, Ann. Rep. Med. Chem., 28, 11 (1993).
14. R. E. Ellis, J. Yuan, H. R. Horvitz, Annu. Rev. Cell Biol., 7, 663 (1991).
15. D. L. Vaux, I. L. Weissman, S. K. Kim, Science, 258, 1955 (1992).
16. D. J. Kane, T. A. Sarafian, R. Anton, H. Hahn, E. B. Gralla, J. S. Valentine, T. Örd, D. E. Bredesen, Science, 363, 1274 (1993).
17. D. L. Vaux, Curr. Biol. 3, 877 (1993).
18. S. A. Lipton and P. A. Rosenberg, N. Engl. J. Med., 330, 613 (1994).
19. B. Meldrum and J. Garthwaite, Trends Pharmacol. Sci., 11, 379 (1990).
20. J. Watkins, T. Honoré, P. Krogsgaard-Larsen, Trends Pharmacol. Sci., 11, 25 (1990).

21. P. H. Seeburg, Trends Neurosci., 16, 359 (1993).
22. T. Kutsuwada, N. Kashiwabuchi, H. Mori, K. Sakimura, E. Kushiya, K. Araki, H. Meguro, H. Masaki, T. Kumanishi, M. Arakawa, M. Mishina, Nature, 358, 36 (1992).
23. K. Moriyoshi, M. Masu, T. Ishii, R. Shigemoto, N. Mizuno, S. Nakanishi, Nature, 354, 31 (1991).
24. K. Williams, Mol. Pharmacol., 44, 851 (1993).
25. H. Monyer, R. Sprengel, R. Schoepfer, A. Herb, M. Higuchi, H. Lomeli, N. Burnashev, B. Sakmann, P. H. Seeburg, Science, 256, 1217 (1992).
26. S. Nakanishi, Science, 258, 597, (1992).
27. C. F. Bigge, Biochem. Pharmacol., 45, 1547 (1993).
28. C. F. Bigge and T. C. Malone, Current Drugs, 3, 951, (1993).
29. B. Scatton, Fundam. Clin. Pharmacol., 7, 389 (1993).
30. S. A. Lipton, Trends Neurosci., 16, 527 (1993).
31. M. A. Rogawski, Trends Pharmacol. Sci., 14, 325 (1993).
32. N. L. Reddy, L.-Y. Hu, R. E. Cotter, J. B. Fischer, W. J. Wong, R. N. McBurney, E. Weber, D. L. Holmes, S. T. Wong, R. Prasad, J. F. W. Keana, J. Med. Chem., 37, 260 (1994).
33. K. Minematsu, M. Fisher, L. Li, M. A. Davis, A. G. Knapp, R. E. Cotter, R. N. McBurney, K. C. H. Sota, Neurology, 43, 397 (1993).
34. R. J. Hargreaves. M. Rigby, D. Smith, R. G. Hill, Br. J. Pharmacol., 110, 36 (1993).
35. L. A. McQuaid, E. C. R. Smith, D. Lodge, E. Pralong, J. H. Wikel, D. O. Calligaro, P. J. O'Malley, J. Med. Chem. 35, 3423 (1992).
36. M. Rowley, P. D. Leeson, G. I. Stevenson, A. M. Moseley, I. Stansfield, I. Sanderson, L. Robinson, R. Baker, J. A. Kemp, G. R. Marshall, A. C. Foster, S. Grimwood, M. D. Tricklebank, K. L. Saywell, J. Med. Chem., 36, 3386 (1993).
37. C. G. Wasterlain, L. M. Adams, P. H. Schwartz, H. Hattori, R. D. Sofia, J. Wichmann, Neurology 43, 2303 (1993).
38. R. T. McCabe, C. G. Wasterlain, N. Kucharczyk, R. D. Sofia, J. R. Vogel, J. Pharmacol. Exp. Ther., 264, 1248 (1993).
39. J. Grotta, Stroke, 25, 255, (1993).
40. L. A. Wong, M. L. Mayer, Mol. Pharmacol., 44, 504 (1993).
41. J. Ohmori, S. Sakamoto, H. Kubota, M. Shimizu-Sasamata, M. Okada, S. Kawasaki, K. Hidaka, J. Togami, T. Furuya, K. Murase, J. Med. Chem., 37, 467 (1994).
42. P. L.. Ornstein, M. B. Arnold, N. K. Augenstein, D. Lodge, J. D. Leander, D. D. Schoepp, J. Med. Chem., 36, 2046 (1993).
43. F. Wätjen, C. F. Bigge, L. H. Jensen, P. A. Boxer, L. J. Lescosky, E. Ø. Nielsen, T. C. Malone, G. W. Campbell, L. L. Coughneour, D. M. Rock, J. Drejer, F. W. Marcoux, BioMed. Chem. Lett. 4, 371 (1994).
44. M. Lafon-Cazal, S. Pietri, M. Culcasi, J. Bockaert, Nature, 364, 535 (1993).
45. E. D. Hall and J. M. Braughler, Free Rad. Biol. Med., 6, 303 (1989).
46. M. Yamamoto T. Shima, T. Uozumi, T. Sogabe, K. Yamada, T. Kawasaki, Stroke, 14, 977 (1983).
47. M. A. Villalobos, J. P. de la Cruz, T. Carrasco, J. M. Smith-Agreda, F. S. de la Cuesta, Brain Res. Bull., 33, 313 (1994).
48. Y. Oshiro, Y. Sakurai, T. Tanake, T. Kikuchi, T. Hirose, K. Tottori, J. Med. Chem., 34, 2014 (1991).
49. J. A. Clemens, R. D. Saunders, P. P. Cho, L. A. Phebus, J. A. Panetta, Stroke, 24, 716 (1993).
50. J. W. Phillis, Brain Res. Bull., 23, 467 (1989).
51. Y. Lin and J. W. Phillis, Brain Res., 571, 272 (1992).
52. J. W. Phillis and S. Sen, Brain Res. 628, 309 (1993).
53. K. Shin-ya, M. Tanaka, K. Furihata, Y. Hayakawa, H. Seto, Tet. Lett., 34, 4943 (1993).
54. H. Monyer, D. M. Hartley, D. W. Choi, Neuron, 5, 121 (1990).
55. E. D. Hall, K. E. Pazara, J. M. Braughler, Stroke, 22, 361 (1991).
56. E. D. Hall, J. M. Braughler, P. A. Yonkers, S. L. Smith, K. L. Linseman, E. D. Means, H. M. Scherch, P. F. VonVoigtlander, R. A. Lahti, E. J. Jacobsen, J. Pharmacol. Exp. Ther., 258, 688 (1991).
57. E. D. Hall, K. E. Pazara, J. M. Braughler, K. L. Linseman, E. J. Jacobsen, Stroke, 21, III-83 (1990).
58. J. S. Beckman, T. W. Beckman, J. Chen, P. A. Marshall, B. A. Freeman, Proc. Natl. Acad. Sci. U S A , 87, 1620 (1990).
59. A. Kader, V. I. Frazzini, R. A. Solomon, R. R. Trifiletti, Stroke, 24, 1709 (1993).
60. A. Buisson, I. Margaill, J. Callebert, M. Plotkine, R. G. Boulu, J. Neurochem., 61, 690 (1993).
61. S. A. Lipton, Y. B. Choi, Z. H. Pan, S. Z. Lei, H. S. Chen, N. J. Sucher, J. Loscalzo, D. J. Singel, J. S. Stamler, Nature, 264, 626 (1993).
62. T. M. Dawson, J. P. Steiner, V. L. Dawson, J. L. Dinerman, G. R. Uhl, S. H. Snyder, Proc. Natl. A c a d . Sci. USA , 90, 9808 (1993).
63. S. A. Lipton, Adv. Pharmacol., 22, 271 (1991).
64. M. C. Nowycky, A. P. Fox, R. W. Tsien, Nature, 316, 440 (1985).

65.   B. P. Bean, Annu. Rev. Physiol., 51, 367 (1989).
66.   T. P.Snutch, J. P. Leonard, M. M. Gilbert, H. A. Lester, N. Davidson, Proc. Nat. Acad. Sci., 87, 3391 (1990).
67.   R. J. Miller, Science, 235, 46 (1987).
68.   D. J. Dooley, A. Lupp, G. Hertting, Naunyn-Schmiedeberg's Arch. Pharmacol., 336, 467 (1987).
69.   L. D.Hirning, A. P. Fox, E. W. McCleskey, B. M. Olivera, S. A. Thayer, R. J. Miller, R. W. Tsien, Science, 239, 57 (1988).
70.   E. W. McCleskey, A. P. Fox, D. H. Feldman, L. J.Cruz, B. M.Olivera, R. W.Tsien, D.Yoshikami, Proc. Nat. Acad. Sci., 84, 4327 (1987).
71.   D. R. Hillyard, V. D. Monje, I. M. Mintz, B. P. Bean, L. Nadasdi, J. Ramachandran, G. Miljanich, A. Azimi-Zoonooz, J. M. McIntosh, L. J. Cruz, J. S. Imperial, B. M. Olivera, Neuron, 9, 69 (1992).
72.   M. Matsumoto, M. S. Scheller, M. H. Zornow, M. A. P. Srnat, Stroke, 24, 1228 (1993).
73.   T. Gemba, M. Ninomiya, K. Matsunaga, M. Ueda, J. Pharmacol. Exp. Ther., 265, 463 (1993).
74.   S. H. Johnson, J. M. Kraimer, G. M. Graeber, Stroke, 24, 1547 (1993).
75.   L. Leybaert, G. de Ley, A. de Hemptinne, Naun.-Schmied. Arch. Pharmacol., 348, 269 (1993).
76.   H. Hara, K. Yokota, M. Shimazawa, T. Sukamoto, Jpn. J. Pharmacol., 61, 361 (1993).
77.   J. Kucharczyk, J. Mintorovitch, M. E. Moseley, H. S. Asgari, R. J. Sevick, N. Derugin, D. Norman, Radiology, 179, 221 (1991).
78.   K. Valentino, R. Newcomb, T. Gadbois, T. Singh, S. Bowersox, S. Bitner, et. al., Proc. Natl. Acad. Sci. USA, 90, 7894 (1993).
79.   D. Nicholls and D. Atwell, Trends Pharmacol. Sci., 11, 462 (1990).
80.   S. P. Burke and C. P. Taylor in "The Role of Neurotransmitters in Brain Injury", M. Globus and W. D. Dietrich, Eds., New York, NY, 1992, p. 45..
81.   M. J. McLean and R. L MacDonald, J. Pharm. Exp. Ther., 227, 779 (1983).
82.   W. A. Catterall, Trends Phrarmacol. Sci., 8, 57 (1987).
83.   J. P. Cullen, J. A. Aldrete, L. Jankovsky, F. Romo-Salas, Anesth. Analg., 58, 165 (1979).
84.   W. C. Taft, G. L. Clifton, R. E. Blair, R. J. DeLorenzo, Brain Res., 483, 143 (1989).
85.   P. A. Boxer, J. J Cordon, M. E. Mann, L. C. Rodolosi, M. G. Vartanian, D. M. Rock, C. P. Taylor, F. W. Marcoux, Stroke, 21, III-47 (1990).
86.   S. E. Smith, Z. A. Al-Zubaidy, A. G. Chapman, B. S. Meldrum, Epilepsy Res., 15, 101 (1993).
87.   M. J. Leach, J. H. Swan, D. Eisenthal, M. Dopson, M. Nobbs, Stroke, 24, 1063 (1993).
88.   H. Fujisawa, D. Dawson, S. E. Browne, K. B. Mackay, R. Bullock, J. McCulloch, Brain Res., 629, 73 (1993).
89.   D. Lekieffre, B. S. Meldrum, Neuroscience, 56, 93 (1993).
90.   J. Mizoule, B. Meldrum, M. Mazadier, M. Croucher, C. Ollat, A. Uzan, J. J. Legand, C. Gueremy, G. LeFur, Neuropharmacol., 24, 767 (1985).
91.   E. Benoit and D. Escande, Pflügers Arch., 419, 603 (1991).
92.   A. Chéramy, R. Romo, G. Godeheu, J. Glowinski, Neurosci. Lett., 147, 209 (1992).
93.   J. Pratt, L. Rataud, F. Bardot, M. Roux, J. C. Blanchard, P. M. Laduron, J. M. Stutzaman, Neurosci. Lett., 140, 225 (1992).
94.   G. Bensimmon, L. Lacomblez, V. Meininger, N. Engl. J. Med., 330, 585 (1994).
95.   A. Eisen, H. Stewart, M. Schulzer, D. Cameron, Can. J. Neurol. Sci., 20, 297 (1993).
96.   E. Melloni and S. Pontremoli, Trends Neurosci., 12, 438 (1989).
97.   H. Caner, J. L. Collins, S. M. Harris, N. F. Kassell, K. S. Lee, Brain Res. 607, 354 (1993).
98.   Z. Li, G. S. Patel, Z. E. Golubski, H. Hori, K. Tehrani, J. E. Foreman, D. D. Eveleth, R. T. Bartus, J. C. Powers, J. Med. Chem., 36, 3472 (1993).
99.   H. Schmid-Antomarchi, S. Amoroso, M. Fosset, M. Lazdunski, Proc. Natl Acad Sci USA , 87, 3489 (1990).
100. P. Ascher and L. Nowak, Trends Neurosci., 10, 284 (1987).
101. J. Hughes and G. N. Woodruff,Arzneim.-Forsch. Drug Res., 42, 250 (1992).
102. J. J. Wagner, R. M. Caudle, C. Chavkin, J. Neurosci., 12, 132 (1992).
103. S. H. Graham, J. Chen, F. R. Sharp, R. P. Simon, J. Cereb. Blood Flow Metab., 13, 88 (1993).
104. F. Wahl, M. Allit, M. Plotkine, R. G. Boulu, Eur. J. Pharmacol., 230, 209 (1993).
105. W. Fischer, R. Bodewei, P. F. Vonvigtlander, M. Mueller, J. Pharm. Exp. Tox., 267, 163 (1993).
106. E.D. Hall, K.E. Pazara, Stroke, 19, 1008 (1988).
107. P. A. Boxer, J. J. Cordon, L. C. Rodolosi, F. W. Marcoux, J. Cereb. Blood Flow Metab., 11, S429 (1991).
108. K. Kusumoto, K. B. Mackay, J. McCulloch, Brain Res., 576, 147 (1992).
109. M. P. Mattson, K. J. Tomaselli, R. E. Rydel, Brain Res., 621, 35 (1993).
110. Z. Liu, P. A. D'Amore, M. Mikati, A. Gatt, G. L. Homes, Brain Res., 626, 335 (1993).
111. B. Knüsel, D. R. Kaplan, J. W. Winslow, A. Rosenthal, L. E. Burton, K. D. Beck, S. Rabin, K. Nikolics, F. Hefti, J. Neurochem., 59, 715 (1992).

Chapter 3. Recent Advances in the Design and Characterization of Muscarinic Agonists and Antagonists

Juan C. Jaen and Robert E. Davis
Parke-Davis Pharmaceutical Research; Division of Warner-Lambert Company
Ann Arbor, MI 48105

Introduction - Muscarinic receptors are widely distributed on multiple organs and tissues and are critical to the maintenance of central and peripheral cholinergic neurotransmission. Pharmacologically, it is possible to define at least three muscarinic receptor subtypes (M_1, M_2, M_3) according to the relative affinity displayed for them by certain muscarinic antagonists (e.g. pirenzepine for M_1, AF-DX116 for M_2, hexahydrosiladifenidol for M_3) and their preferential tissue distribution (M_1 - neuronal, M_2 - cardiac, M_3 - glandular, smooth muscle) (1-3). More recently, five distinct muscarinic receptors (m1-m5) have been cloned (4-7). The regional distribution of these receptor subtypes in the brain and other organs has been documented (4,8,9). Even though some correspondence exists between both muscarinic receptor subtype classifications (M_1, M_2, and M_3 correspond to m1, m2, and m3, respectively), pharmacologically defined receptor subtypes generally represent combinations of several of the receptor subtype proteins. In light of this situation, attention should be paid to distinguishing between true muscarinic receptor subtype selectivity (at the molecular level, specific receptor subtypes expressed in cultured cell lines) and the more "real world" pharmacological muscarinic receptor subtype selectivity (intimately associated with tissue selectivity).

The receptor subtype selectivity of muscarinic agonists can be estimated by their binding selectivity, typically against a radiolabeled muscarinic antagonist (10,11), or by their functional selectivity, as measured by their activation of specific second-messenger pathways to which a given receptor subtype is linked (10,12). Multiple experimental variables can affect the estimation of subtype selectivity, such as the intrinsic subtype selectivity of the radioligand chosen for binding studies (13), or the level of expression and efficiency of second-messenger coupling of cloned receptors (14). As a result, interpretation of the literature is complicated by the lack of consistency among laboratories in the methods used to determine receptor subtype selectivity. This review concentrates on recent reports of subtype-selective muscarinic agonists and antagonists, particularly since the last report on muscarinic agents (15). The molecular biology, structure, coupling to effector systems, and other properties of muscarinic receptor subtypes have been reviewed recently (1,12,16-18).

MUSCARINIC AGONISTS

Most recent work in the design of CNS-active muscarinic agonists has centered on the search for M_1 (or m1) selective muscarinic agonists for the treatment of Alzheimer's disease (AD). Selective degeneration of basal forebrain cholinergic neurons is the earliest and most consistent neurochemical event in AD (19,20). These neurons project to cortical and hippocampal areas of the brain, where muscarinic receptors of the m1 subtype are in high density (21). Cholinomimetic therapy in AD has been aimed at compensating for the lack of cholinergic input to these areas (22,23). Selectivity for m1 (M_1) receptors should result in a reduced incidence of the side effects characteristic of earlier muscarinic agents, side-effects that were generally attributed to peripheral stimulation of non-m1 muscarinic receptors (24).

The ester group of arecoline (1), a naturally occurring muscarinic agonist, is quickly hydrolyzed *in vivo*, rendering it inactive. The search for a stable ester bioisostere lead to the synthesis of a series of aldoximes (2,3), orally active muscarinic agonists with only slight binding selectivity for M_1 receptors but displaying central cholinergic effects at doses significantly lower than those associated with peripheral side effects (25). Compound 2 (R_2=Me) is currently in clinical trials for AD (24). Arylcarbamate prodrugs of these compounds (4) display even greater central selectivity *in vivo* (26). Other heterocyclic analogues of arecoline, such as compounds 5 and 6, are potent and efficacious muscarinic agonists (27,28). Another viable ester bioisostere is the 4-substituted-1,2,5-thiadiazolyl ring, such as in 7 and 8. In particular, 7 (R= n-hexyl) is a potent M_1 agonist (rabbit vas deferens) with minimal activity at M_2 (guinea pig atrium) or M_3 (guinea pig ileum) receptors (29,30). This compound, which also displays functional m1 selectivity against cloned muscarinic receptor subtypes, is at present being studied in AD patients (31). Yet another useful ester replacement is the 3-alkoxy-2-pyrazinyl moiety, as illustrated by 9, which is also characterized as a selective M_1 agonist against isolated tissue preparations (32).

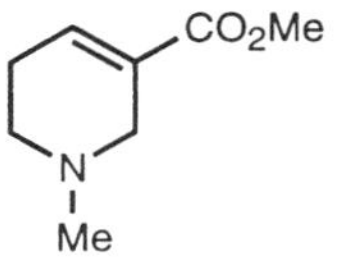

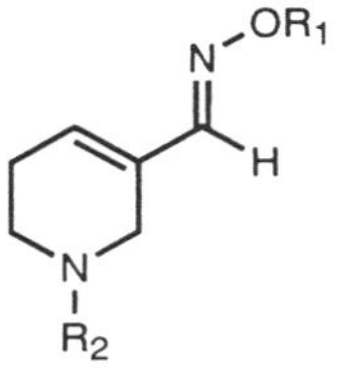

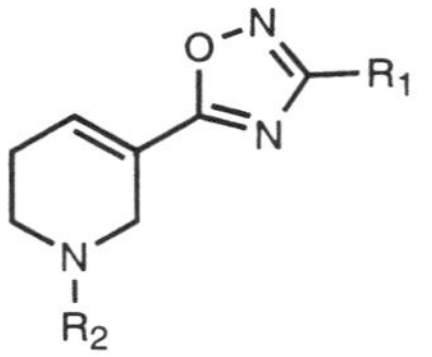

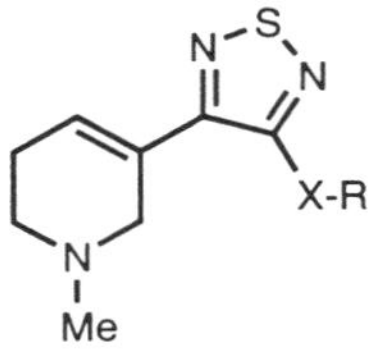

1

2; R_1= Me, R_2= H or Me
3; R_1= propargyl, R_2= H or Me
4; R_1= Me, R_2= $CO_2(C_6H_4$-p-Cl)

5; R_1= Me, R_2= H or Me
6; R_1= $^nC_4H_9$, R_2= H or Me

7; X = O, R = $^nC_4H_9$-$^nC_6H_{13}$
8; X = S, R = $^nC_4H_9$-$^nC_6H_{13}$

Bioisosteric replacement of the tetrahydropyridinyl ring of arecoline resulted in compounds 10 and 11, both with activity at central M_1 receptors (33). Compound 12, which combines bioisosteric replacement of both portions of arecoline, is a muscarinic agonist with selective effects at central receptors coupled to phosphoinositide (PI) hydrolysis, presumably M_1 receptors (34). Compound 13 is a conformationally restricted bioisostere of arecoline, displaying higher affinity but somewhat lower efficacy for muscarinic receptors than the analogous fused-piperidine compound (35).

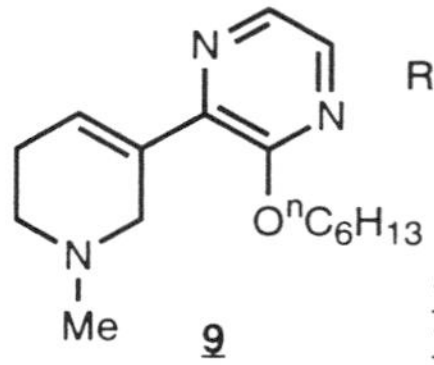

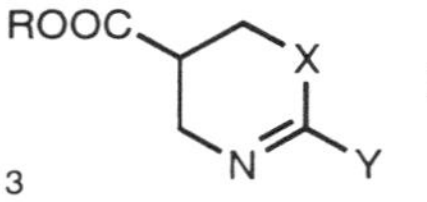

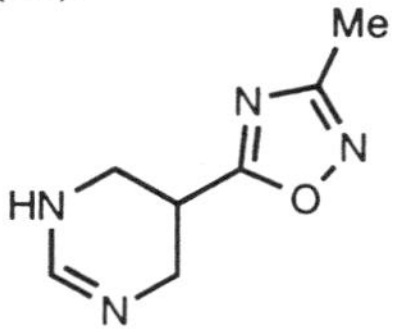

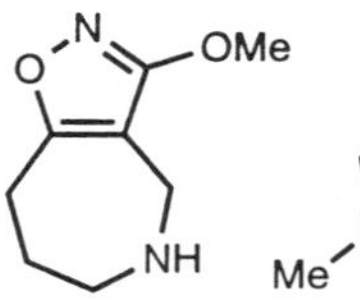

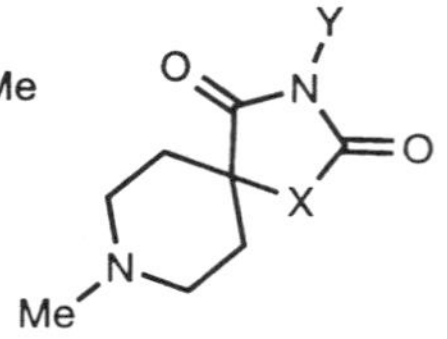

9

10; X = CH_2; Y = NH_2
11; X = NH; Y = H, Me

12

13

14; X = CH_2, Y = CH_3O
15; X = O, Y = Et

Increasing the rigidity of the muscarinic pharmacophore is a strategy that has been used occasionally to increase receptor subtype selectivity and which often affects the potency and efficacy of the compounds as well. Related to the partial muscarinic agonist RS-86, compound **14** is also a partial muscarinic agonist that reverses CO_2-induced cognitive impairment in mice (36). Compound **15**, an oxa analogue of RS-86, is a functionally-selective M_1 agonist, displaying greater separation between centrally and peripherally active doses than RS-86 (37). Other structures related to RS-86, such as **16** and **17**, are efficacious agonists at rat central M_1 receptors (38). *In vitro*, S-(-)-**17** is a partial m1 agonist with no efficacy at m2 receptors, while R-(+)-**17** is inactive against both muscarinic receptor subtypes (39). S-(-)-**17** is reported to be undergoing clinical trials in AD patients (40). Compound **18**, related to the partial muscarinic agonist AF30 (**19**), displayed greater intrinsic activity and hydrolytic stability than **19** (41).

The strategy of achieving muscarinic receptor subtype selectivity by decreasing conformational flexibility was also applied to the design of the partial muscarinic agonist (-)-**20** (some M_1 selectivity) and its enantiomer (M_2 selective), which are rigid analogues of arecoline (42). Similarly, **21**, a bicyclic analogue of **2** and **3**, is a potent partial muscarinic agonist that displays good separation between central effects (EEG) and peripheral actions (blood-pressure) (43). Related N-methoxy imidoyl chlorides and nitriles (**22**) are also centrally selective partial muscarinic agonists *in vivo* (44). The combination of 1-azabicyclic ring systems with certain heterocycles has generated potent and efficacious muscarinic agonists in recent years. Most successful in this respect have been 3-methyl- and 3-amino-1,2,4-oxadiazole derivatives **23**-**26** (45-47). The 3R-*exo* isomers (3R-**23**, 3R-**25**) are most efficacious, but display selectivity for M_2 and M_3 receptors (46). Triazole compounds, such as **27**, are also potent and highly efficacious muscarinic agonists (49,50).

18, n = 1
19; n = 2

21; R_1 = H, R_2 = propargyl
22; R_1 = CN, R_2 = Me

23; n = 1, R = Me
24; n = 2, R = Me
25; n = 1, R = NH_2
26; n = 2, R = NH_2
28; n = 1, R = cC_3H_5

27; n = 1,2

16; X = O
17; X = CH_2

20

A useful strategy towards obtaining functional M_1 selectivity has been to decrease the intrinsic efficacy of the muscarinic agonists, thus taking advantage of the potential greater central (M_1) muscarinic receptor reserves (51). It is difficult to predict how the reduced efficacy of this type of compound will affect their therapeutic utility. Compound **28** displays greatly reduced muscarinic efficacy, which translates *in vivo* into M_1 agonist activity with antagonist properties at M_2 and M_3 receptors (52). While pyrazine analogues **29** and **30** are potent and efficacious muscarinic agonists, displaying good central activity but lacking significant receptor subtype selectivity, the chloropyrazine analogue **31** possesses lower intrinsic efficacy and behaves as a full agonist at M_1 receptors, a partial agonist at M_3 receptors, and an antagonist at M_2 receptors (53,54).

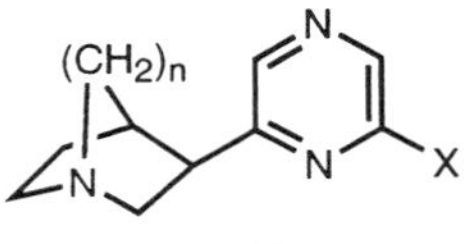

29; n = 1, X = H
30; n = 2, X = H
31; n = 2, X = Cl

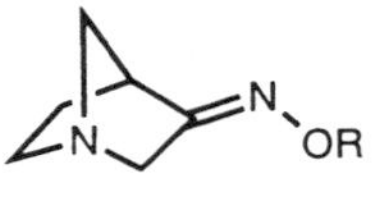

32; R= -CH₂————Ph

33; R= -CH₂

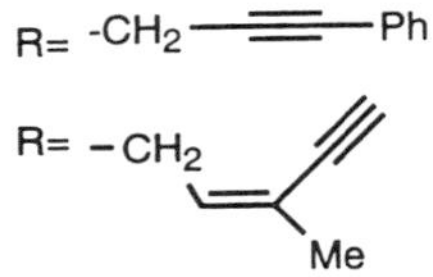

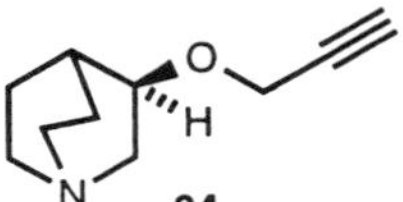

34

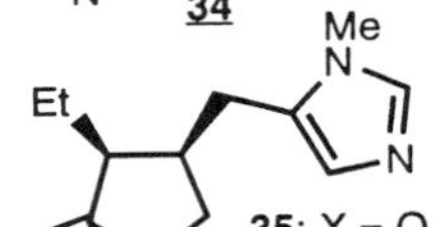

35; X = O
36; X = S

Generally, addition of even small steric bulk to muscarinic agonists leads to a loss of efficacy (55,56). An apparent exception to this observation is the discovery of bulky bicyclic oximes such as **32** and **33**, which range from partial to full muscarinic agonists, and display functional selectivity towards m1 receptors *in vitro* and good central selectivity *in vivo* (57,58). The quinuclidinyl ether **34** displays some functional selectivity for M_1 receptors in tissue preparations and *in vivo* (pithed rat) (59). The pharmacology of **35**, a thio analogue of the muscarinic agonist pilocarpine (**36**), is consistent with simultaneous activation of M_1 receptors and blockade of presynaptic M_2 receptors (60,61). The azetidine analogue **37**, related to the muscarinic agonist UH-5, is a partial m1 and m3 agonist and a full m4 agonist *in vitro* (62). A methyl-substituted derivative of oxotremorine, **38**, displays greater M_1/M_2 selectivity but somewhat reduced affinity, compared to oxotremorine (63). Oxotremorine analogues **39** and **40** are partial M_1 agonists and full M_2 agonists (64). Compound **41** is equipotent as a muscarinic agonist with BM5, of which it is a more rigid analogue (65). The oxotremorine analogue **42** is slightly less potent than oxotremorine at activating M_3 receptors in guinea pig ileum (66). An imidazolyl analogue (**43**) possesses muscarinic efficacy similar to oxotremorine (67). Addition of a methyl group to the imidazolyl ring (**44**) greatly decreased the muscarinic intrinsic activity of the compound (68), while addition of methyl groups to both the imidazole and pyrrolidinone rings produced a potent muscarinic antagonist, **45** (67).

37

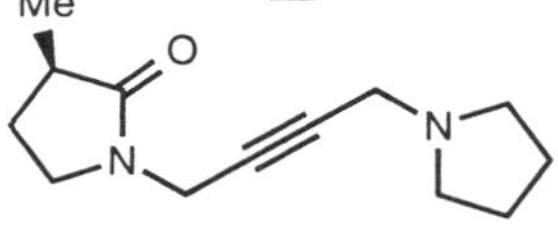

38

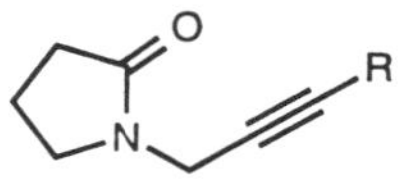

39; R = CH₂-NMe₂

40; R =

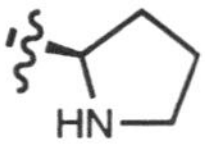

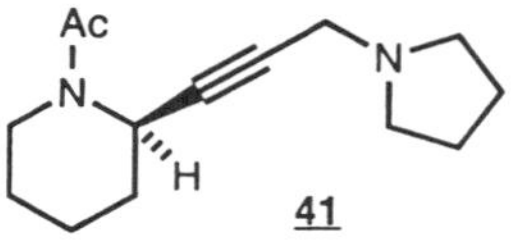

41

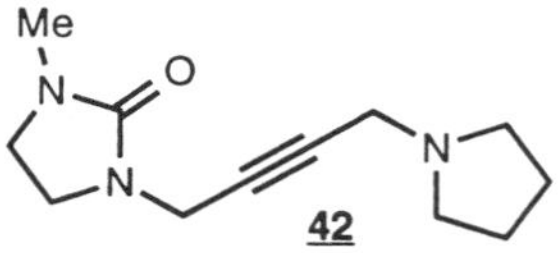

42

Quaternary carbamate **46**, structurally related to the muscarinic agonist McN-A-343, is a potent and selective M_1 agonist, displaying only weak partial M_3 agonism and competitive M_2 antagonism (69). The corresponding non-quaternized analogue **47** also displays activity at M_1 receptors with partial agonism at M_2 and M_3 receptors (70). Compound **48** is an unusual muscarinic agonist in that it does not stimulate phosphoinositide (PI) hydrolysis or inhibit adenylate cyclase, but may instead utilize phospholipase C or D (PLC, PLD) as second messenger mechanisms (71).

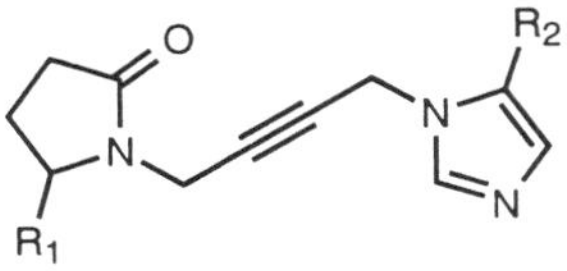

43; $R_1 = R_2 = H$
44; $R_1 = H, R_2 = Me$
45; $R_1 = R_2 = Me$

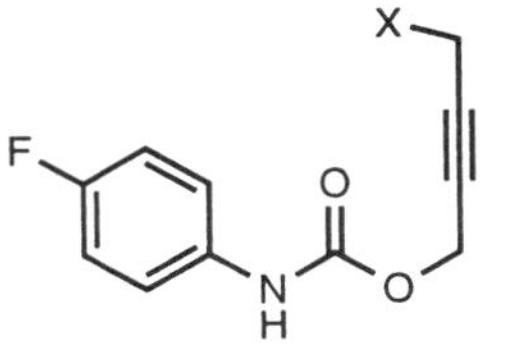

46; $X = -N^+$ **47**; $X = -N$

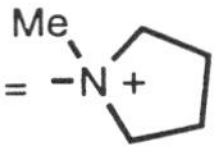

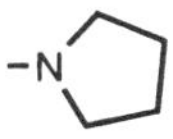

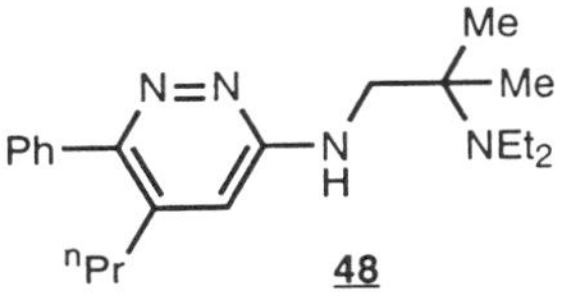

48

MUSCARINIC ANTAGONISTS

In recent years, the most common target of CNS-active muscarinic antagonists has been the m2 (M_2) receptor subtype, found in cortical and hippocampal areas of the brain, among others (72). Blockade of presynaptic inhibitory m2 receptors stimulates evoked release of acetylcholine from cholinergic neurons (73,74). Given the wide distribution of m2 receptors throughout the gastrointestinal and cardiovascular systems, implementation of m2 blockade as an indirect cholinomimetic strategy in AD will require the identification of selective m2 antagonists with preferential CNS distribution. Centrally active muscarinic antagonists with selectivity for one of the other muscarinic receptor subtypes may also possess medicinal utility. For example, selective m4 antagonists might be useful in the treatment of parkinsonism or dystonias, given the preferential expression of m4 receptors in striatum, a brain region involved in the control of movement (9). In general, subtype-selective muscarinic antagonists would be useful tools for elucidating the CNS role of each muscarinic receptor subtype.

m1 Antagonists - Hybrid structure **49** , derived from the muscarinic antagonists caramiphen and aprophen, displays moderate M_1 selectivity (75). Hydrogenation of one of the phenyl groups of aprophen lead to compound **50**, a potent antagonist with decreased affinity for M_2 receptors (76). Compound **51,** a close analogue of the non-selective muscarinic antagonist quinuclidinyl benzilate (QNB, **52**), displays about six-fold selectivity for m1 versus m2 receptors; its lipophilicity makes it a good chemical lead for the design of a SPECT imaging agent for m1 receptors *in vivo* (77). Another halogenated quinuclidinyl ester derivative (**53**) was found to be a potent (but nonselective) muscarinic antagonist, also with the potential to become a SPECT imaging agent for the mapping of cerebral muscarinic receptors (78). One of the phenyl rings in QNB can be replaced with an alkoxyalkyl group (e.g. **54**), without noticeable loss of binding affinity (79). A 64-aminoacid peptide isolated from the venom of an African snake is a potent and selective m1 antagonist (80).

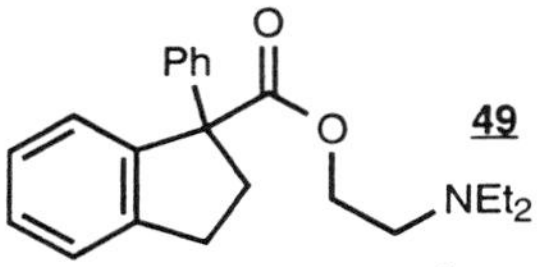

49

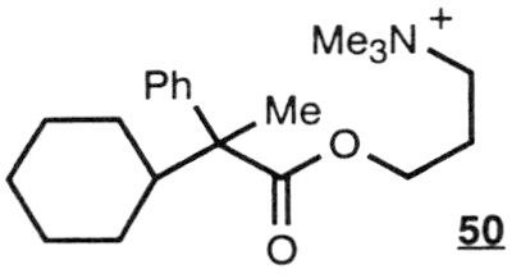

50

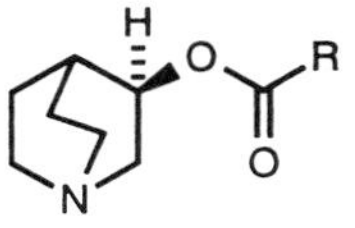

51; R =

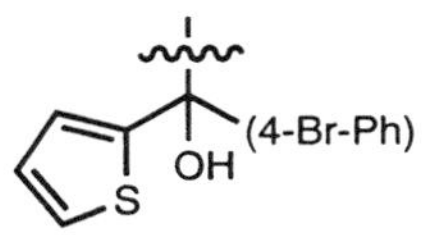

52; R =

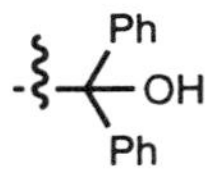

53; R =

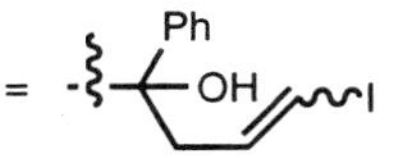

54; R =

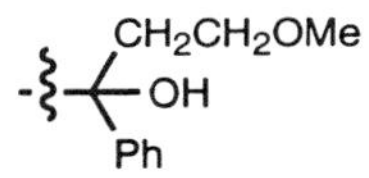

m2 Antagonists - Pirenzepine (**55**) is the prototypical M_1-selective muscarinic antagonist. However, relatively minor structural changes in the piperazine group lead to enhanced affinity for m2 receptors (e.g., **56**) (81). Replacement of the 4-methylpiperazine moiety of **55** with a 2-(diethylaminomethyl)piperidine produced AF-DX 116 (**57**), the prototypical selective M_2 antagonist. This compound enhances acetylcholine release from cortex and hippocampus (73) and possesses cognitive-enhancing properties *in vivo* (74,82). The structurally related AF-DX 384 (**58**) displays comparable selectivity but higher affinity for M_2 receptors (83). Neither compound can distinguish between m2 and m4 receptors *in vitro* (84). Systematic variation of the side chain of **58** resulted in the synthesis of AQ-RA 741 (**59**), a compound with improved affinity and selectivity for M_2 receptors (84). This line of work resulted in the synthesis of potent, lipophilic (i.e., CNS-permeable), and selective m2/M_2 antagonists (**60**, **61**), in which the side-chain basic nitrogen of earlier compounds was replaced by an amide group (85). One of these compounds (**60**) improves the performance of age-impaired rats in a water maze test and lacks significant peripheral cardiac actions (86).

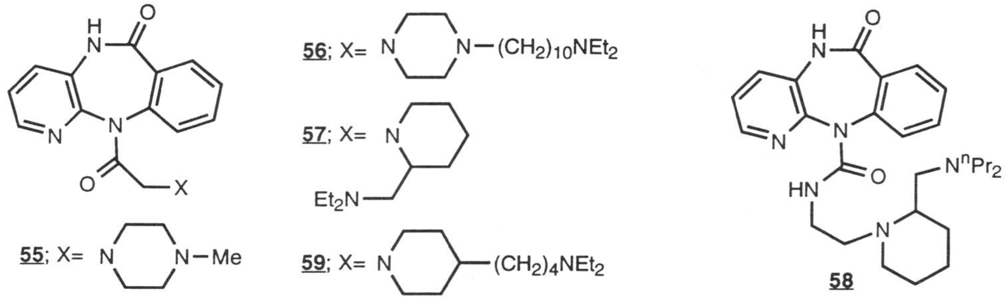

56; X= N⌒N—$(CH_2)_{10}NEt_2$

57; X= N (piperidine, Et_2N—)

55; X= N⌒N—Me

59; X= N⌒—$(CH_2)_4NEt_2$

58 HN, O, N^{n}Pr$_2$

Compound **62**, a bioisostere of **59**, is about ten times more potent than **59** at m2 receptors (sub-nanomolar affinity) and slightly more m2 selective (87). Brain permeability of this compound, like **59**, was found to be extremely low (88). A related compound, **63**, matches the affinity of **62** for m2 receptors. Although **63** does not discriminate between m2, m3 and m4 receptors *in vitro*, it appears to display significant selectivity for brain m2 receptors *in vivo* (89). A different type of M_2-selective muscarinic antagonists is represented by the polymethylene tetraamines such as methoctramine. Compound **64**, a hybrid structure between methoctramine and **59**, was found to be selective for heart M_2 receptors over M_1 and M_3 receptors. This type of compound is not expected to penetrate the CNS readily (90).

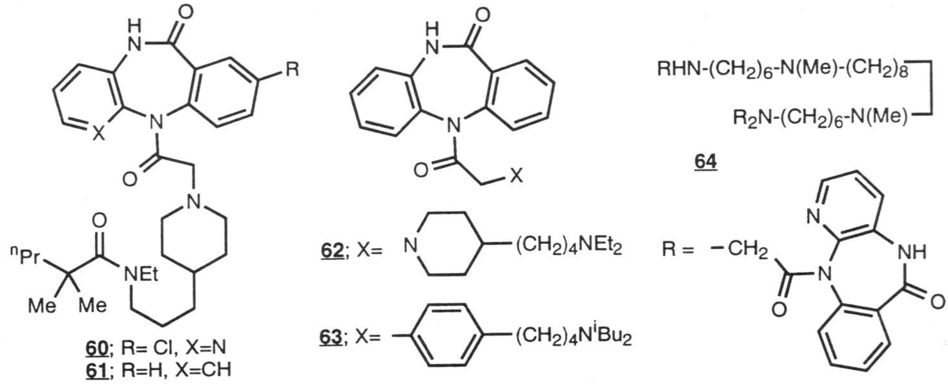

RHN-$(CH_2)_6$-N(Me)-$(CH_2)_8$⌝
R$_2$N-$(CH_2)_6$-N(Me)—⌟

64

^{n}Pr, O, Me Me, NEt

60; R= Cl, X=N
61; R=H, X=CH

62; X= N⌒—$(CH_2)_4NEt_2$

63; X= ⬡—$(CH_2)_4N^iBu_2$

R = —CH_2— (tricyclic structure)

Himbacine (**65**) and imperialine (**66**) are potent, lipophilic, m2/m4 (M_2/M_4) selective muscarinic antagonists (91,92).   Attempts to design simpler, more M_2-selective versions of himbacine, resulted in the synthesis of **67**, a compound that recognizes both M_1 and M_2 receptors with potent but equal affinity, in the low nM range (93). Compound **68**, a chlorinated derivative of imperialine, displays greater selectivity and affinity (sub-nanomolar) for m2 receptors than the parent alkaloid. Its brain penetration and long duration of action have been described (94).

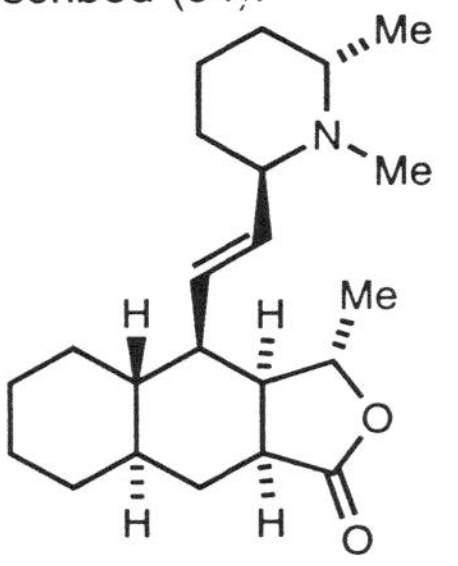

65

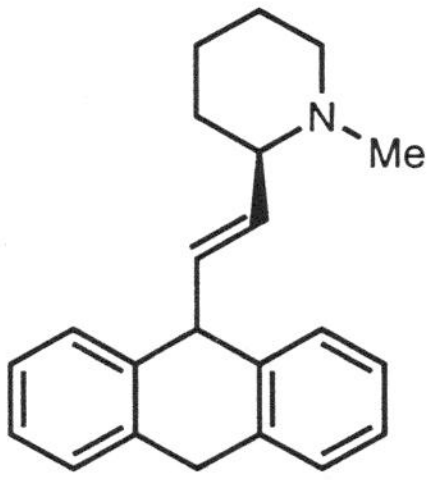

67

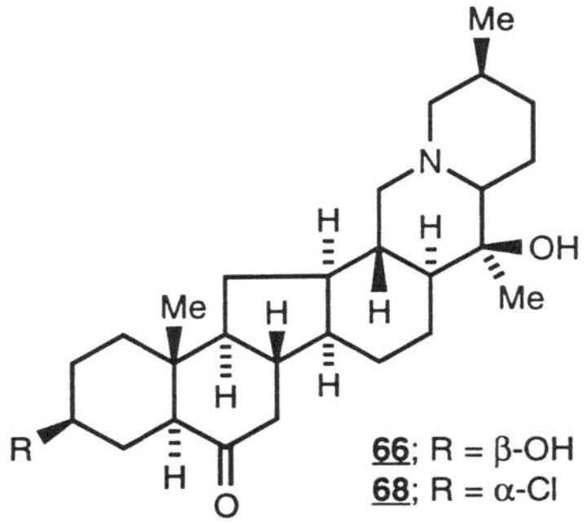

66; R = β-OH
68; R = α-Cl

Conclusion - Alzheimer's disease (AD) is at the center of most recent research in the areas of CNS-active subtype-selective muscarinic agonists and antagonists. Selective m1 agonists are being developed as direct-acting cholinomimetics and selective m2 antagonists as indirect cholinomimetics that enhance the release of acetylcholine from cholinergic presynaptic terminals. Selective m1 antagonists might be valuable as pharmacologic tools for recreating in the laboratory some of the cognitive deficits associated with AD. In recent years, significant progress has been made in the search for orally active, subtype-selective muscarinic agonists and antagonists. Some of these compounds, primarily muscarinic agonists, have already moved into clinical programs (24,31,40). Results from these trials, anxiously awaited, will be the ultimate measure of the validity of the cholinergic hypothesis of AD.

References

1.   F. Mitchelson, Pharmac. Ther., 37, 357 (1988).
2.   R.K. Goyal, N. Engl. J. Med., 321, 1022 (1989).
3.   E.C. Hulme, N.J. Birdsall, and N.J. Buckley, Annu. Rev. Pharmacol. Toxicol., 30, 633 (1990).
4.   T.I. Bonner, N.J. Buckley, A.C. Young, and M.R. Brann, Science, 237, 527 (1987).
5.   T.I. Bonner, A.C. Young, M.R. Brann, and N.J. Buckley, Neuron, 1, 403 (1988).
6.   E. Peralta, A. Ashkenazi, J. Winslow, D. Smith, J. Ramachandran, D. Capon, EMBO J., 6, 3923 (1987).
7.   C.-F. Liao, A.P. Themmen, R. Joho, C. Barberis, M. Birnbaumer, and L. Birnbaumer, J. Biol. Chem., 264, 7328 (1989).
8.   N.J. Buckley, T.I. Bonner, C.M. Buckley, and M.R. Brann, Mol. Pharmacol., 35, 469 (1989).
9.   A. I. Levey, Life Sci., 52, 441 (1993).
10.   R.D. Schwarz, R.E. Davis, J.C. Jaen, C.J. Spencer, H. Tecle, and A.J. Thomas, Life Sci., 52, 465 (1993).
11.   A. Closse, H. Bitter, D. Langenegger, and A. Wanner, Naunyn-Schmiedeberg's Arch. Pharmacol., 335, 372 (1987).
12.   M. P. Caulfield, Pharmacol. Ther., 58, 319 (1993).
13.   G. Gillard, M. Waelbroeck, and J. Christophe, Mol. Pharmacol., 32, 100 (1987).
14.   S.Z. Wang, E.E. El-Fakahany, J. Pharmacol. Exp. Ther., 266, 237 (1993).
15.   R. Baker and J. Saunders, Ann. Rep. Med. Chem., 24, 31 (1989).

16. S.V.P. Jones, A.I. Levey, D.M. Weiner, J. Ellis, E. Novotny, S.-H. Yu, F. Dörje, J. Wess, and M.R. Brann, In: Molecular Biology of G-protein Coupled Receptors, M. Brann (ed.), Birkhauser, Boston:1992, pp 170.

17. E.C. Hulme, N.J. Birdsall, and N.J. Buckley, Annu. Rev. Pharmacol. Toxicol., 30, 633 (1990).

18. J. Wess, Life Sci., 53, 1447 (1993).

19. P. Whitehouse, D. Price, R. Struble, A. Clark, J. Coyle, and M. Delong, Science, 215, 1237 (1982).

20. J.T. Coyle, D.L. Price, M.R. Delong, Science, 219, 1184 (1983).

21. D.C. Mash, D.D. Flynn, L.T. Potter, Science, 228, 1115 (1985).

22. J.C. Jaen, W.H. Moos, G. Johnson, Bioorg.Med.Chem.Letters, 2, 777 (1992).

23. J.M. Palacios and R. Spiegel, Prog. Brain Res., 70, 485 (1986).

24. R. Davis, C. Raby, M.J. Callahan, W. Lipinski, R. Schwarz, D. Dudley, D. Lauffer, P. Reece, J. Jaen, and H. Tecle, Prog. Brain Res., 98, 439 (1993).

25. E. Toja, C. Bonetti, A. Butti, P. Hunt, M. Fortin, F. Barzaghi, M.L. Formento, A. Maggioni, A. Nencioni, and G. Galliani, Eur.J.Med.Chem., 26, 853 (1991).

26. E. Toja, C. Bonetti, A. Butti, P. Hunt, M. Fortin, F. Barzaghi, M.L. Formento, A. Maggioni, A. Nencioni, and G. Galliani, Eur.J.Med.Chem., 27, 519 (1992).

27. G. A. Showell, T.L. Gibbons, C.O. Kneen, A.M. MacLeod, K. Merchant, J. Saunders, S.B. Freedman, S. Patel, and R. Baker, J.Med.Chem., 34, 1086 (1991).

28. P. Sauerberg, J.W. Kindtler, L. Nielsen, M.J. Sheardown, and T. Honoré, J.Med.Chem., 34, 687 (1991).

29. P. Sauerberg, P.H. Olesen, S. Nielsen, S. Treppendahl, M.J. Sheardown, T. Honoré, C.H. Mitch, J.S. Ward, A.J. Pike, F.P. Bymaster, B.D. Sawyer, and H.E. Shannon, J.Med.Chem., 35, 2274 (1992).

30. D. Ngur, S. Roknich, C.H. Mitch, S.J. Quimby, J.S. Ward, L. Merritt, P. Sauerberg, W.S. Messer, Jr., and W. Hoss, Biochem.Biophys.Res.Comm., 187, 1389 (1992).

31. R.A. Lucas, 3rd Internatl. Conf. on Alzheimer's and Parkinson's Diseases; Chicago, Nov.1-6, 1993.

32. J.S. Ward, L. Merritt, V.J. Klimkowski, M.L. Lamb, C.H. Mitch, F.P. Bymaster, B. Sawyer, H.E. Shannon, P.H. Olesen, T. Honoré, M.J. Sheardown, and P. Sauerberg, J.Med.Chem., 35, 4011 (1992).

33. T. Ojo, T. Rho, P.G. Dunbar, G.J. Durant, A.A. El-Assadi, S. Periyasamy, D. Ngur, Z. Fang, W. Hoss, and W.S. Messer, Abstr.Pap.Am.Chem.Soc., 206, MEDI-190 (1993).

34. P.G. Dunbar, G.J. Durant, Z. Fang, Y.F. Abuh, A.A. El-Assadi, D.O. Ngur, S. Periyasamy, W.P. Hoss, and W.S. Messer, Jr., J.Med.Chem., 36, 842 (1993).

35. A. Lagersted, E. Falch, B. Ebert, and P. Krogsgaard-Larsen, Drug Des. Discov., 9, 237 (1993).

36. Y. Ishihara, H. Yukimasa, M. Miyamoto, and G. Goto, Chem.Pharm.Bull., 40, 1177 (1992).

37. S. Tsukamoto, M. Ichihara, F. Wanibuchi, S. Usuda, K. Hidaka, M. Harada, and T. Tamura, J.Med.Chem., 36, 2292 (1993).

38. F. Wanibuchi, T. Konishi, M. Harada, M. Terai, K. Hidaka, T. Tamura, S. Tsukamoto, and S. Usuda, Eur.J.Pharmacol., 187, 479 (1990).

39. H. Wei, W.R. Roeske, J. Lai, F. Wanibuchi, K. Hidaka, S. Usuda, and H.I. Yamamura, LifeSci., 50, 355 (1991).

40. Scrip Newsletter, 1683, 11 (1992).

41. R.T. Lewis, A.M. MacLeod, L.J. Street, J.J. Kulagowski, and R. Baker, Abstr. Pap. Am. Chem. Soc., 203, MEDI-102 (1992).

42. E. Pombo-Villar, P. Supavilai, H. Weber, and H. Boddeke, Bioorg.Med.Chem.Letters, 2, 501 (1992).

43. S.M. Bromidge, F. Brown, F. Cassidy, M.S.G. Clark, S. Dabbs, M.S. Hadley, J.M. Loudon, B.S. Orlek, and G.J. Riley, Bioorg.Med.Chem.Lett., 2, 787 (1992)).

44. S.M. Bromidge, F. Brown, F. Cassidy, M.S.G. Clark, S. Dabbs, J. Hawkins, J.M. Loudon, B.S. Orlek, and G.J. Riley, Bioorg.Med.Chem.Letters, 2, 791 (1992).

45. B.S. Orlek, F.E. Blaney, F. Brown, M.S. Clark, M.S. Hadley, J. Hatcher, G.J. Riley, H.E. Rosenberg, H.J. Wadsworth, and P. Wyman, J.Med.Chem., 34, 2726 (1991).

46. E. Pombo-Villar, K.-H. Wiederhold, G. Mengod, J.M. Palacios, P. Supavilai, and H.W. Boddeke, Eur.J.Pharmacol., 226, 317 (1992).

47. R.M. Eglen, G.C. Harris, A.P. Ford, E.H. Wong, J.R. Pfister, and R.L. Whiting, Naunyn-Schmiedeberg's

Arch.Pharmacol., 345, 375 (1992).

48.   G.A. Showell, R. Baker, J. Davis, R. Hargreaves, S.B. Freedman, K. Hoogsteen, S. Patel, and R.J. Snow, J.Med.Chem., 35, 911 (1992).

49.   H.J. Wadsworth, S.M. Jenkins, B.S. Orlek, F. Cassidy, M.S.G. Clark, F. Brown, G.J. Riley, D. Graves, J. Hawkins, and C.B. Naylor, J.Med.Chem., 35, 1280 (1992).

50.   S.M. Jenkins, H.J. Wadsworth, S. Bromidge, B.S. Orlek, P.A. Wyman, G.J. Riley, and J. Hawkins, J.Med.Chem., 35, 2392 (1992).

51.   S.B. Freedman, G.R. Dawson, L.L. Iversen, R. Baker, and R.J. Hargreaves, Life Sci., 52, 489 (1993).

52.   S.B. Freedman, S. Patel, E.A. Harley, L.L. Iversen, R. Baker, G.A. Showell, J. Saunders, A. McKnight, N. Newberry, K. Scholey, and R. Hargreaves, Eur.J.Pharmacol., 215, 135 (1992).

53.   R.J. Hargreaves, A.T. McKnight, K. Scholey, N.R. Newberry, L.J. Street, P.H. Hutson, J.E. Semark, E.A. Harley, S. Patel, and S.B. Freedman, Br.J.Pharmacol., 107, 494 (1992).

54.   L.J. Street, R. Baker, T. Book, A.J. Reeve, J. Saunders, T. Willson, R.S. Marwood, S. Patel, and S.B. Freedman, J.Med.Chem., 35, 295 (1992).

55.   W.H. Moos, S.C. Bergmeier, L.L. Coughenour, R.E. Davis, F.M. Hershenson, J.A. Kester, J.S. McKee, J.G. Marriott, R.D. Schwarz, H. Tecle, and A.J. Thomas, J. Pharmac. Sci., 81, 1015 (1992).

56.   B. Ringdahl, D. Jenden, Mol. Pharmacol., 23, 17 (1983).

57.   H. Tecle, D.J. Lauffer, T. Mirzadegan, W.H. Moos, D.W. Moreland, M.R. Pavia, R.D. Schwarz, and R.E. Davis, LifeSci., 52, 505 (1993).

58.   H. Tecle, J.C. Jaen, C. Augelli-Szafran, S.D. Barrett, B.W. Caprathe, D.J. Lauffer, T. Mirzadegan, W.H. Moos, D.W. Moreland, M.R. Pavia, R.D. Schwarz, A.J. Thomas, and R.E. Davis, Abstr. Pap. Am. Chem. Soc., 206, MEDI-191 (1993).

59.   H.A. Ensinger, H.N. Doods, A.R. Immel-Sehr, F.J. Kuhn, G. Lambrecht, K.D. Mendla, R.E. Müller, E. Mutschler, A. Sagrada, G. Walther, and R. Hammer, LifeSci., 52, 473 (1993).

60.   A. Enz, H. Boddeke, A. Sauter, M. Rudin, and G. Shapiro, LifeSci., 52, 513 (1993).

61.   A. Enz, G. Shapiro, P. Supavilai, and H.W. Boddeke, Naunyn-Schmiedeberg's Arch.Pharmacol., 345, 282 (1992).

62.   B.J. Bradbury, J. Baumgold, R. Paek, U. Kammula, J. Zimmet, and K.A. Jacobson, J.Med.Chem., 34, 1073 (1991).

63.   E.J. Trybulski, J. Zhang, R.H. Kramss, and R.M. Mangano, J.Med.Chem., 36, 3533 (1993).

64.   D. Garvey, J. Wasicak, J. Chung, Y. Shue, G. Carrera, P. May, M. McKinney, D. Anderson, E. Cadman, L. Vella-Rountree, A. Nadzan, and M. Williams, J.Med.Chem., 35, 1550 (1992).

65.   U. Hacksell, J. Lundkvist, B. Nilsson, G. Nordvall, and H. Vargas, In: New Leads and Targets in Drug Research, P. Krogsgaard-Larsen, S. Christensen, H. Kofod (eds.), Munksgaard, Copenhagen:1992, p 365.

66.   B.M. Nilsson, H.M. Vargas, and U. Hacksell, J.Med.Chem., 35, 3270 (1992).

67.   M.W. Moon, C.G. Chidester, R.F. Heier, J.K. Morris, R.J. Collins, R.R. Russell, J.W. Francis, G.P. Sage, and V.H. Sethy, J.Med.Chem., 34, 2314 (1991).

68.   V. Sethy, J. Francis, D. Hyslop, G. Sage, T. Olen, A. Meyer, R. Collins, R. Russell, R. Heier, W. Hoffmann, M. Piercey, N. Nichols, P. Schreur, and M. Moon, DrugDev.Res., 24, 53 (1991).

69.   U. Moser, G. Lambrecht, C. Mellin, and E. Mutschler, Life Sci., 52, 550 (1993).

70.   G. Lambrecht, U. Moser, U. Grimm, O. Pfaff, U. Hermanni, C. Hildebrandt, M. Waelbroeck, J. Christophe, and E. Mutschler, LifeSci., 52, 481 (1993).

71.   J. Kan, R. Steinberg, F. Oury-Donat, J. Michaud, O. Thurneyssen, J. Terranova, C. Gueudet, J. Souilhac, R. Brodin, R. Boigegrain, C. Wermuth, P. Worms, P. Soubrié, and G. LeFur, Psychopharmacology, 112, 219 (1993).

72.   M.H. Richards, Br.J.Pharmacol., 99, 753 (1990).

73.   P.A. Lapchak, D.M. Araujo, R. Quirion, and B. Collier, J. Neurochem., 496, 285 (1989).

74.   M.G. Packard, W. Regenold, R. Quirion, N.W. White, Brain Res., 524, 72 (1990).

75.   R.L. Hudkins, J.F. Kachur, W.L. Dewey, D.L. DeHaven-Hudkins, and J.F. Stubbins, Med.Chem.Rev., 2, 173 (1992).

76.   H. Leader, R.K. Gordon, J. Baumgold, V.L. Boyd, A.H. Newman, R.M. Smejkal, and P.K. Chiang, J.Med.Chem., 35, 1290 (1992).

77.   V.I. Cohen, R.E. Gibson, L.H. Fan, R. De la Cruz, M.S. Gitler, E. Hariman, and R.C. Reba, J.Pharmac.Sci., 81, 326 (1992).

78.   D.W. McPherson, D.L. DeHaven-Hudkins, A.P. Callahan, and F.F. Knapp, Jr., J.Med.Chem., 36, 848 (1993).

79.   V.I. Cohen, R.E. Gibson, L.H. Fan, R. De la Cruz, M.S. Gitler, E. Hariman, and R.C. Reba, J.Med.Chem., 34, 2989 (1991).

80.   S.I. Max, J.S. Liang, L.T. Potter, J.Neurosci., 13, 4293 (1993).

81.   Y. Karton, B.J. Bradbury, J. Baumgold, R. Paek, and K.A. Jacobson, J.Med.Chem., 34, 2133 (1991).

82.   C.M. Baratti, J.W. Opezzo, S.R. Kopf, Behav.Neural. Biol., 60, 69 (1993).

83.   I. Aubert, D. Cécyre, S. Gauthier, and R. Quirion, Eur.J.Pharmacol., 217, 173 (1992).

84.   F. Dörje, J. Wess, G. Lambrecht, R. Tacke, E. Mutschler, and M.R. Brann, J.Pharm.Exp.Ther., 256, 727 (1991).

85.   H.N. Doods, R. Quirion, G. Mihm, W. Engel, K. Rudolf, M. Entzeroth, G.B. Schiavi, H. Ladinsky, W.D. Bechtel, H.A. Ensinger, K.D. Mendla, and W. Eberlein, LifeSci., 52, 497 (1993).

86.   H. Doods, M. Entzeroth, H. Ziegler, G. Schiavi, W. Engel, G. Mihm, K. Rudolf, and W. Eberlein, Eur.J.Pharmacol., 242, 23 (1993).

87.   M.S. Gitler, R.C. Reba, V.I. Cohen, W.J. Rzeszotarski, and J. Baumgold, BrainResearch, 582, 253 (1992).

88.   V.I. Cohen, J. Baumgold, B. Jin, R. De La Cruz, W.J. Rzeszotarski, and R.C. Reba, J.Med.Chem., 36, 162 (1993).

89.   M.S. Gitler, V.I. Cohen, R. De la Cruz, S.F. Boulay, B. Jin, B.R. Zeeberg, and R.C. Reba, Life Sci., 53, 1743 (1993).

90.   C. Melchiorre, M.L. Bolognesi, A. Chiarini, A. Minarini, and S. Spampinato, J.Med.Chem., 36, 3734 (1993).

91.   J.H. Miller, P.J. Aagaard, V.A. Gibson, and M. McKinney, J.Pharmacol.Exp.Ther., 263, 663 (1992).

92.   R.M. Eglen, G.C. Harris, H. Cox, A.O. Sullivan, E. Stefanich, and R.L. Whiting, Naunyn-Schmiedeberg's Arch.Pharmacol., 346, 144 (1992).

93.   M.J. Malaska, A.H. Fauq, A.P. Kozikowski, P.J. Aagaard, and M. McKinney, Bioorg.Med.Chem.Lett., 3, 1247 (1993).

94.   J. Baumgold, R.L. Pryzbyc, and R.C. Reba, Soc.Neuroscience Abstracts, 19, 194.10 (1993).

Chapter 4. Animal Engineering In Neurobiology

Donald E. Frail and Michael T. Falduto
Pharmaceutical Products Division
Abbott Laboratories, Abbott Park, IL 60064

Introduction - Transgenic animals provide a tool to study and analyze individual genes in the complex spatial and temporal array of gene expression in the mammalian brain. Moreover, heightened knowledge about the pathogenesis of neurological diseases as well as development of pharmaceutical interventions may be facilitated by transgenic animal models of human neurological diseases. Transgenic animals were last reviewed in this series in 1989, and sufficient advances have been made since then to warrant a review of the application of animal engineering in neurobiology.

The most common method for introducing a foreign gene into animals is by microinjection of DNA directly into the male pronucleus of a fertilized egg (1). After transfer of the egg into a recipient female, the progeny transgene, usually incorporated into chromosomal DNA before the first cell division, can be expressed in a developmental and tissue-specific manner. An alternative procedure, gene targeting by homologous recombination (2), requires the isolation of totipotent embryonic stem (ES) cells from blastocysts and gene transfer in vitro. Reinsertion of the ES cells into blastocysts and development of the embryo in vivo results in animals with the foreign gene, but not the targeted endogenous gene, in tissues derived from the manipulated ES cells. Chimeric animals capable of germ cell transmission of the disrupted gene can then be bred to produce homozygous gene "knockouts", which can be useful in determining the function of a normal gene.

REGULATION OF GENE EXPRESSION

Transgene expression is dependent on the regulatory sequences within promoters and enhancers. Thus, the regulation of neural genes can be studied using gene constructs containing regulatory sequences from a gene of interest linked to the coding region of a reporter protein. The reporter gene can code for any readily assayable protein, including the structural gene from which the regulatory regions are derived. In this manner, the developmental and tissue-specific regulation, physiological control, and transcriptional regulatory elements and proteins involved in gene expression can be identified and analyzed in specific cell types of the brain. This section describes studies of the regulation of neural genes in transgenic mice which have arisen in the last few years.

Calmodulin - A 362 base pair region from the rat calmodulin gene II (CaMII) promoter has been used to make transgenic mice expressing a ß-galactosidase reporter gene (3). This short fragment appeared to drive expression of the reporter in a manner similar to the endogenous CaMII gene. Transgene expression in adult mice was detected in distinct neurons in the cerebral neocortex, pyriformcortex, hippocampal regions CA1 and CA3, dentate gyrus, cerebellum, lateral vestibular nucleus of pons, and spinal cord. Two GA-rich sequence motifs found in this fragment of the rat CaMII gene are highly conserved in other species as well as in CaMI and CaMIII and may be directly involved in neuron-specific expression.

Synapsin I - Synapsin I, a synaptic vesicle protein, is thought to be involved in neurotransmitter release. Transgenic mice have been created using 4.3 kilobases (kb) of 5' flanking DNA from the synapsin I gene driving expression of the reporter gene, chloroamphenicol acetyltransferase (CAT) (4). The developmental regulation and neuronal specificity of the transgene correlated with expression of endogenous synapsin I. Furthermore, immunoreactivity of CAT was found only in neurons of the brain and spinal cord.

Catecholaminergic Proteins - The promoters for mouse, rat, and human tyrosine hydroxylase (TH), the rate limiting enzyme in catecholamine biosynthesis, have been analyzed in brains of transgenic mice (5,6,7). Lack of accurate tissue-specific expression was observed when 2.5 kb of human and 3.5 kb of mouse upstream promoter DNA was used. However, 5.0 kb of human and 4.8 kb of rat 5' flanking DNA was sufficient for expression in catecholaminergic neurons, although the human upstream DNA also confers expression of the reporter gene in other neurons and other tissues. These results demonstrate a similarity among species and suggest an evolutionary conservation in the structure of the gene for tyrosine hydroxylase. It appears that the human upstream 5 kb sequence lacks elements which suppress ectopic expression, since transgenic mice possessing the entire 11 kb tyrosine hydroxylase gene, including just 2.5 kb of upstream sequence, exhibited high level and tissue-specific expression only in the brain and adrenal glands of transgenic mice (8).

Catecholaminergic neurons of the norepinephrine subtype do not contain the epinephrine synthesizing enzyme phenylethanolamine N-methyltransferase (PNMT) which is present in epinephrine producing cells. Because the norepinephrine converting enzyme dopamine ß-hydroxylase (DBH) is expressed in norepinephrine-producing neurons of wild-type mice, two groups have utilized the promoter from human DBH, to express PNMT in catecholaminergic neurons (9,10). The aim of these studies was to perturb the neuronal phenotype and determine the functional difference between the two neurotransmitters. These experiments indicated that 4 kb of upstream regulatory sequences of the DBH gene is sufficient for elevating PNMT in noradrenergic neurons to similar levels as those found in adrenergic neurons.

Acetylcholine Receptor - The complete structural gene for the α2-subunit of the chicken neuronal acetylcholine receptor (nAChR), including 7 kb of 5' upstream and 3 kb of 3' downstream sequences, has been used to generate transgenic mice (11). Expression of the transgene was neuron-specific and was selective to subsets of CNS nuclei. The remarkable similarity of regional expression patterns between the avian gene and mammalian α-type nAChR suggests that mammalian transcription factors and the corresponding gene regulatory sequences to which they bind are at least partly conserved.

cAMP-dependent Protein Kinase - The regulatory subunit subtype Iß of cyclic AMP-dependent protein kinase (RIß) is limited to neurons in the central nervous system. A gene construct consisting of 1.5 kb of the 5' upstream regulatory region and 2 kb of the first intron of RIß driving ß-galactosidase expression was expressed in the cortex, CA1-CA3 regions of the hippocampus, and spinal cord of transgenic mice (12). The distribution and developmental regulation of the transgene is similar to the expression pattern of the endogenous RIß gene.

Olfactory Marker Protein - By using a 3 kb fragment from the regulatory region of the rat olfactory marker protein (OMP) gene, Largent and co-workers (13) were able to limit expression of the SV40 T antigen oncogene to olfactory neurons of transgenic mice. The cellular regulation was similar to that of the endogenous OMP gene.

Somatostatin - Transgenic mice were created using a 15 kb DNA fragment containing the structural gene for mouse somatostatin (14). A 30 base pair oligonucleotide sequence was inserted in the 5' untranslated region to serve as a marker of the transgene for *in situ* hybridization studies. Unlike the endogenous somatostatin gene, there was no expression of the transgene in somatostatinergic neurons, but expression was instead found in vasopressinergic magnocellular neurons of the hypothalamus. Thus, this large fragment does not contain necessary elements to activate transcription in somatostatin producing neurons. Furthermore, transcriptional factor proteins in vasopressinergic neurons may activate the somatostatin gene in the absence of all regulatory sequences contained in the endogenous gene.

Vasopressin - Ang and co-workers (15) studied the neuronal-specific expression of bovine vasopressin (BVP) by using three distinct gene constructs in transgenic mice. One construct consisted of 1.25 kb of the BVP promoter fused to the CAT reporter gene. CAT RNA was expressed in all peripheral tissues and brain regions examined. However, CAT enzyme activity was low or undetectable in most tissues. A second group of transgenic mouse lines containing the same 1.25 kb promoter but linked to the structural gene of BVP, had peripheral vasopressin expression restricted to the adrenal glands, while most brain regions expressed the transgene. The third construct contained the entire structural gene for BVP, including 9 kb of 5' upstream sequences and 1.5 kb of 3' non-coding sequence. Transgenic mice generated with this construct displayed BVP transcripts restricted to discrete groups of neurons in the hypothalamus. Transgene expression was also detected in the ovaries in three lines examined and ectopically in some brain region of one transgenic mouse line. These results suggest that silencer elements within the structural gene and the distal promoter or 3' flanking sequences may mediate the hypothalamic neuronal expression of BPV by selectively repressing gene activity in peripheral tissues and brain regions other than the hypothalamus.

Neuropeptide Y - To examine NPY gene regulation *in vivo*, transgenic mice were made with the CAT reporter gene driven by 796 base pairs of upstream promoter and 51 base pairs of exon 1 from the NPY gene. (16) Examination of mice for CAT activity revealed expression of the transgene in cerebral cortex, olfactory bulb, limbic region, adrenal gland, brain stem, spinal cord, and eyes. These data indicate that 850 base pairs from the 5' upstream regulatory region of the NPY gene is sufficient to direct expression of heterologous genes to appropriate tissues.

Preprotachykinin - Harmar and co-workers (17) created transgenic mice containing 3.3 kb of 5' flanking sequences from the rat preprotachykinin gene linked to either ß-galactosidase or thymidine kinase from herpes simplex virus. Reporter gene expression was detected in some lines of transgenic mice, but was not consistent with the expression of endogenous preprotachykinin.

Myelin Basic Protein - Oligodendrocytes express many myelin-specific components. To determine the developmental pattern and regional distribution of an oligodendrocyte protein involved in myelination, transgenic mice were created with the regulatory region of the gene for myelin basic protein (18). The timing of expression of ß-galactosidase driven by 3.2 kb of 5' flanking sequences from myelin basic protein coincided with oligodendrocyte differentiation and myelin biosynthesis. Furthermore, analysis of these mice compliments other standard techniques used to identify myelinating oligodendrocytes by revealing previously uncharacterized temporal and spatial patterns of myelination.

S100ß - The ß subunit of the dimeric protein S100, is found predominantly in astrocytes. A 17.3 kb fragment containing the human S100ß gene was used to make transgenic mice (19). There was high level and tissue-specific expression of S100ß mRNA and protein in transgenic mice that paralleled the expression of endogenous mouse S100ß. Correlation of transgene copy number with S100ß levels in the brains of the transgenic mice suggests the presence of a locus control region in the human gene.

Glial Fibrillary Acidic Protein - Another astrocyte-specific gene, glial fibrillary acidic protein (GFAP), has been used to direct expression of a transgene to the brain of mice (20). Eight lines of mice were generated with a construct consisting of 2 kb of the human GFAP promoter regulatory region linked to β-galactosidase. Six transgenic lines expressed the reporter protein in the brain, which was limited to astrocytes and inducible by gliosis.

PHENOTYPES AND ANIMAL MODELS OF NEUROLOGICAL DISEASES

The expression of proteins involved in neurological function or pathological processes can be directed to specific cell types in the mammalian nervous system of transgenic animals by using constructs containing heterologous tissue-specific promoters or relevant regulatory sequences from the structural gene of interest. Alternatively, a constituitively active promoter can be used to drive expression of a neural gene if tissue-specific expression is irrelevant to the neurological function or disease process conferred by the transgene.

Alzheimer's Disease - There have been numerous efforts in recent years attempting to reproduce the neuropathology of Alzheimer's disease using small animal models. Mechanistic and functional roles of Alzheimer's disease-related proteins might be elucidated with transgenic animals and could provide insight into the development of a true animal model for the disease. Amyloid precursor protein (APP) is proteolytically processed to generate the β-amyloid peptide deposited in senile plaques of Alzheimer's patients. Genetic mutations in the APP gene have been shown to be sufficient for the development of Alzheimer's disease in families with a hereditary, early-onset form of the disease. To date, all published reports of transgenic mice for the study of Alzheimer's disease have relied on expressing full length APP or fragments which include the β-amyloid region.

Wirak and co-workers tested the expression of APP by inserting the β-galactosidase cDNA into a 4.5 kb fragment of the 5' end of the APP gene (21). Transgenic mice with this construct demonstrated that this region of APP genomic DNA was sufficient to direct expression to neurons in a manner similar to endogenous APP. This study was followed by one in which transgenic mice were produced with a construct containing the same APP regulatory elements but substituting a cDNA encoding the 42 amino acid β-amyloid peptide for the reporter gene (22). At an age of one year, these mice had accumulations of β-amyloid in some hippocampal neurons. However, it was later discovered that non-transgenic mice of the same strain had amyloid-like structures in the hippocampus and therefore the deposits could not be attributed to the transgene expression of β-amyloid (23,24).

Transgenic mice were also created using full length APP (25). The 751 amino acid isoform of human APP was placed under the control of the promoter for rat neural-specific enolase and three lines of mice with this construct were selected for analysis. Transgene RNA was detected by polymerase chain reaction (PCR) but human APP protein in the transgenic mouse brain was made at low levels and was difficult to detect above the background of endogenous mouse APP. Immunohistochemical staining with a monoclonal antibody directed against the first 28 amino acids of β-amyloid revealed reproducibly higher levels of neuronal and extracellular amyloid containing deposits in the hippocampus and cortical regions of transgenic mice compared to controls. Another study using full length APP (695 amino acid isoform) under the control of the metallothionine IIA promoter reported a deficit in spatial learning, but no neuropathology or amyloid deposition in transgenic mice (26).

The carboxyl terminal 100 amino acids of APP, termed C100, includes the entire β-amyloid peptide sequence. A number of studies have investigated the consequences of expressing C100 in transgenic mice. Kawabata and co-workers used the human Thy-1 promoter to drive expression of C100 (27). It was reported that transgenic mice displayed the major hallmarks of Alzheimer's disease, including extracellular amyloid plaques, intracellular neurofibrillary tangles, and neurodegeneration. When the histopathological findings could not be reproduced in any of the transgenic mice, the paper was retracted (28). Another C100 transgenic mouse study used the JC viral promoter which directs expression to glial cell types (29). Transgene mRNA was expressed exclusively in the brain tissue and transgenic mice had increased ß-amyloid immunoreactivity in fixed brain sections. A later report on the same animals indicated there is no change in binding levels for excitotoxic glutamate receptor subtypes in the C100 mice (30). Kammesheidt and co-workers (31) used a brain dystrophin promoter linked to the C100 cDNA to create transgenic mice. Four to six month old mice displayed cell-body and neuropil accumulation of ß-amyloid and abnormal aggregation of the carboxy terminal APP fragment.

Mice transgenic for APP cDNA do not appear to express human APP at high levels. One possible explanation for this observation is a requirement for regulatory sequences within the APP structural gene. Two recent studies have shown that a yeast artificial chromosomes (YAC) encompassing the entire ~400 kb APP gene can elicit appropriate expression of human APP in transgenic mice (32,33). Future experiments with YACs expressing disease-linked mutant forms of human APP may lead to a better understanding of the role of APP in Alzheimer's disease. Other proteins implicated in Alzheimer's disease have not been utilized in transgenic studies

Prion Diseases - Inoculation of mice with hamster prions does not lead to scrapie. However, transgenic mice have been produced in which the hamster PrP gene is expressed in their brains and these mice do get scrapie when inoculated with hamster prions (34). Thus, synthesis of the hamster cellular prion protein is required for the synthesis of infectious prions. Furthermore, brain extracts of transgenic mice inoculated with hamster prions contained hamster prions but not mouse prions and brain extracts of transgenic mice inoculated with mouse prions contained mouse prions but not hamster prions (35). Therefore, there appears to be a species-specific interaction between the prions and the cellular prion protein present in the tissue.

A mutation at amino acid 102 of PrP^C has been implicated as the genetic defect in Gerstmann-Sträussler-Scheimnker (GSS) disease, an autosomal dominant disease that develops in adulthood (36). Transgenic mice have been established that express this mutant isoform. Adult transgenic mice developed symptoms of ataxia, lethargy, and rigidity (37). Pathological examination revealed numerous vacuoles representing spongiform degeneration, though amyloid plaques were not detected. A recent surprise is the discovery that overexpression PrP^C results in vacuolization of the CNS, as well as demyelinating polyneuropathy and skeletal muscle abnormalities in aged animals (38).

Lines of mice that are homozygous for disrupted PrP genes have been produced by homologous recombination (39). These knockout lines show no gross abnormalities and microscopic examination of the brain was normal. The mice developed and behaved normally through seven months of age (39). Most importantly, the lack of neurological problems indicate that scrapie is most likely not the result of a mere loss of function of PrP^C.

The knockout mice were resistant to infection, indicating that the cellular protein PrP^C is necessary for scrapie infection. Mice that are homozygous for disrupted PrP gene were free of scrapie symptoms for at least 13 months following inoculation with mouse scrapie prions while wild-type control mice all died within 6 months (40). Interestingly, mice that had a knockout of only one PrP^C allele showed increased resistance to scrapie, indicating that normal amounts of PrP^C are required for susceptibility. Furthermore, these mice were susceptible to hamster scrapie prions only when the PrP^C transgene was of hamster origin, thus indicating a need for host/prion PrP homology (34,35). These knockout studies indicate that sheep and cattle that are engineered to have a reduced amount of PrP^C should be resistant to scrapie and therefore of agricultural interest. The techniques to knockout genes in these animals are not currently available, but, as suggested by Büeler and co-workers (40), transgenics that express antisense PrP^C nucleic acid may accomplish the goal.

Neurotrophins and Neurotrophin Receptors - To date animals deficient in nerve growth factor (NGF) (41), brain-derived neurotrophic factor (BDNF) (42), cilliary neurotrophic factor (CNTF) (43), leukemia inhibitory factor (LIF) (44) or insulin-like growth factor I (IGF-I) (45) have been published. NGF deficient animals do not show gross abnormalities, although feeding is affected and they therefore do not thrive. They show a loss of sympathetic and sensory neurons but, interestingly, basal forebrain cholinergic neurons do not seem to be affected at the ages studied. BDNF deficient animals showed degeneration of certain sensory ganglia, especially the vestibular sensory ganglion of the inner ear, which resulted in severe deficiencies in coordination and balance. Somewhat surprisingly, sympathetic, midbrain dopaminergic, and motor neurons appeared to develop normally. In CNTF deficient animals, the morphology of motor neurons was somewhat affected postnatally, and there was a slight but significant reduction in muscle strength. Animals deficient in LIF did not show gross morphological deficits in the brain, although there did appear to be abnormal staining for the calcium binding proteins calbindin and parvalbumin. Development and growth was severely affected in animals deficient in IGF-1. Animals deficient in NT-3 and NT-4 are not yet published but are apparently being developed. Crosses between animals with different deficiencies will be an area of interest in the future.

Neurotrophin receptors have also been manipulated. In an early study, the human gene for p75 NGF receptor (low-affinity neurotrophin receptor) was expressed in transgenic mice, thereby defining the regulatory elements involved in p75 gene expression (46). The p75 gene was subsequently deleted in mice, resulting in a decrease in sensory innervation and loss of heat sensitivity (47). Further studies analyzed a variety of sympathetic target tissues and found that many, but not all, appeared quite normal (48). Striking abnormalities were found in the pineal gland and the sweat gland. Finally, the trigeminal sensory neurons of these animals also displayed an altered response to NGF but other neurotrophins tested (49). Mice that are deficient in TrkA receptor, the receptor for NGF, have severe deficiencies in sensory and sympathetic neurons (50). Mice that are deficient in the TrkB receptor, the receptor for BDNF and NT4, die postnatally as a result of lack of feeding (51). Both the CNS and the PNS have neuronal deficiencies, including smaller trigeminal and dorsal root ganglia and fewer lumbar motor neurons (51). Mice that are deficient in TrkC receptors, the receptor for NT-3, display abnormal movement due to abnormal afferent projections to the spinal motor neurons (52). Mice deficient in IGF-1 receptor are severely compromised, die at birth, and abnormalities are observed in the CNS (45).

Synaptic Mechanisms - Mice deficient in expression of α-calcium-calmodulin-dependant kinase II (CaMKII) (53,54), synapsin I (55), or protein kinase Cγ (PKCγ) (56,57) have been used to investigate synaptic mechanisms. All of the mice were viable with no gross neuroanatomical abnormalities and synaptic transmission appeared normal. Mice lacking CaMKII were deficient in their ability to produce long term potentiation (LTP) and they had impaired spatial learning. Surprisingly, LTP was normal in mice lacking synapsin I, a major substrate for CaMKII. Mice lacking PKCγ also had greatly impaired LTP and deficits in learning. An understanding of the molecular mechanisms of LTP and the relationship of LTP to learning and memory have been greatly enhanced by these studies.

Myelination and Demyelination - Myelin basic protein (MBP) and the major CNS myelin proteolipid protein (PLP) are expressed exclusively in oligodendrocytes in the CNS and therefore expression targeted to oligodendrocytes is possible. Furthermore, since these genes are expressed postnatally, the transcription units could be used to deliver secreted proteins to the CNS following development. The transcription units for these genes have been used to create transgenic animals (58,59). Signal transduction during myelination has also been analyzed. Mice that are deficient in the Fyn tyrosine kinase have about half of the myelin normally present, indicating that signaling through Fyn may play an important role in myelination (60).

Analysis of demyelinating diseases has been investigated in transgenic animals by manipulating the expression of immune system proteins. Mice lacking CD8 were generated by homologous recombination and the lack of CD8 was subsequently analyzed in a strain of mice that are susceptible to experimental allergic encephalomyelitis, an autoimmune model of multiple sclerosis (61). Although the CD8 deficient and control animals had similar disease onset and susceptibility, the mutant animals had fewer deaths but increased frequency of relapses. These results indicated that the CD8 cells are not necessary to disease induction and maintenance, though they do appear to play a supporting role. The involvement of MHC antigens in demyelination has also been examined by expressing a syngeneic MHC class I antigen specifically in oligodendrocytes using the MBP promoter (62). Hypomyelination occurred in the transgenic animals, despite the lack of obvious signs of immune infiltration and autoreactive lymphocytes.

Hypothalamic-pituitary Axis - Pituitary gonadotroph cell lines have been developed by targeting the expression of the SV40 T-antigen oncogene using the promoter/enhancer region from the human glycoprotein hormone α-subunit (63). These cell lines facilitated the analysis of several important biological questions (64), including the isolation of a clone encoding the gonadotropin-releasing hormone receptor (65). Also, targeted oncogene expression was used to develop mouse lines that develop pituitary tumors, leading to the overexpression of adrenocorticotropic hormone. This hypercorticotropism is similar to that seen in Cushing's disease (66). The promoter/enhancer elements of the growth hormone (GH) gene were used to target the expression of diphtheria toxin to pituitary somatatropes. A line of mice was developed that lacked circulating GH, resulting in dwarfism (67). Since nearly all of the somatatropes were absent, these animals also served as models to address questions concerning pituitary cell lineages.

HIV-1 gp120 - The HIV-1 coat protein gp120 has been expressed in transgenic animals under the control of the GFAP promoter to elucidate the mechanisms responsible for neuronal cell loss in HIV-1 infected individuals (68). The animals had a 40% reduction of specific neuronal subpopulations and widespread neuronal dendritic vacuolization. In addition, there was widespread reactive astrocytosis and reactive microglial involvement. Many of these neuropathological features resemble those seen in HIV-1 individuals. That gp120 alone can cause such neuropathology may explain the lack of correlation between CNS impairment and amount of detectable HIV-infected neurons.

Nitric Oxide Synthase (NOS) - Mice have been generated that are deficient in neuronal NOS, one of three NOS isoforms (69). The mice were viable and fertile. The most obvious defect was the development of enlarged stomachs. This was most likely due to abnormalities of the enteric nervous system, known to involve nitric oxide as a physiological mediator, which caused hypertrophy of the circular muscle layer of the stomach. This correlated with the human disorder infantile pyloric stenosis, in which there is a lack of NOS neurons in the circular muscle. Surprisingly, there was no gross abnormalities detected in the CNS. Very low levels of residual enzymatic activity were detected in the brain, suggesting that another, possibly novel NOS isoform may be present. The effects of the NOS deletion on long-term depression and long-term potentiation, two processes that may involve nitric oxide, have not yet been reported in these animals.

Retinoblastoma susceptibility gene (Rb) - Rb is a well-characterized tumor suppressor gene which, when mutated, leads to the development of tumors of the retina and other tissues. As a tumor suppressor gene, it is thought to play a role in mammalian development and there is evidence that it is involved in regulation of the cell cycle. Mice homozygous for functionally deleted Rb died during gestation and had severely abnormal haematopoietic and nervous systems (70,71). In the nervous system, massive cell death occurred, although ironically the development of the eye appeared to be unaffected.

FUTURE EXPECTATIONS

Transcriptional activation analyses will lead to an increasing repertoire of available promoter/enhancers for targeted gene expression in transgenic animals. Presently, for example, catecholaminergic neurons can be targeted using the regulatory sequences for tyrosine hydroxylase. Other transcriptional units may be defined in the future that allow targeting of a transgene to a very select subpopulation of catecholaminergic neurons. One use for this targeted selection will be cell ablation experiments in which a toxin gene is expressed, as was previously described for somatotrope ablation in the pituitary. Such animals may be developed as animal models or for studies of neuronal circuitry. So instead of astrocytes versus oligodendrocytes versus neurons, it may be striatal D2 dopaminergic cells versus cerebellar purkinje cells versus retina amacrine cells. In addition, important cell lines may result from targeted oncogene expression, as was described for pituitary gonadotroph cell lines.

The functions of various receptor subtypes will be analyzed using gene knockout animals. This area is just now emerging with the recent reports of knockout mice for the neurotrophin receptors. Receptor families and receptor subtypes are consistent physiological themes, however, and there is much to analyze. Analysis of subtypes of dopamine receptors, serotonin receptors and adrenergic receptors will most likely emerge. The function of cloned receptor sequences for which the ligand is not known, so-called orphan receptors, may be defined by animal engineering studies.

Other emerging areas include those of apoptosis, aging, and neuroimmunology. Programmed cell death through apoptosis is an important feature of nervous system development and is possibly involved in disease processes as well. Mice deficient in Bcl-2, a protein that can protect certain neurons from apoptosis induced by growth factor withdrawal, are viable but suffer from several abnormalities, including polycystic kidney disease (72). Neuronal abnormalities have not yet been reported. The relationship of oxidative damage to aging and neuronal cell death has been studied in one report (73), and a recent report of an extension of life-span in transgenic flies (74) may stimulate further studies. The emerging field of neuroimmunology will benefit from animal engineering studies, as exemplified by recent studies in which transgenic mice were created in which the GFAP gene was used to target the expression of the cytokine IL6 in astrocytes (75). These animals displayed tremors, ataxia, and seizures, and the degree of symptoms correlated with the expression levels of IL6. Neuropathological examination revealed neurodegeneration, astrocytosis, and the induction of acute-phase protein expression. The documented expression of IL6 in the brain during a number of CNS disorders could therefore be contributing to the neurodegenerative process. Mice deficient in IL6 have been produced, though an analysis of the nervous system was not reported (76).

Finally, important animals for neurobiological study will arise serendipitously from two directions. First, transgenic animals or knockouts that have been produced for disciplines other than neurobiology may show unpredicted phenotypic changes in the nervous system. This may occur more often than is recognized since a gross pathological examination that includes the nervous system might not always be performed. Second, the insertion of a transgene into the genome is random, and therefore the insertion site disrupts a gene with an important function. For example, a line of mice carrying an insertional mutation display lateralized circling and locomotor hyperactivity that may result from a dysfunctional basal ganglia (77). It is safe to predict that there will be surprises. Genes thought to be irrelevant in the nervous system may turn out to be essential and conversely, the deletion of genes that are predicted to be invaluable may not affect the animal at all.

References

1.   B. Hogan, F. Constantini, E. Lacy, Manipulating the Mouse Embryo: A Laboratory Manual, Cold Spring

Harbor Laboratory, Cold Spring Harbor, N.Y. (1986).
2. M.R. Capecchi, Science, 244, 1288 (1989)
3. K. Matsuo, H. Ikeshima, K. Shimoda, A. Umezawa, J. Hata, K. Maejima, H. Nojima, and T. Takano, Mol. Brain Res., 20, 9 (1993).
4. C. Hoesche, A. Sauerwald, R.W. Veh, B. Kripp, and M.W. Kilimann, J. Biol. Chem., 268, 26494 (1993).
5. W.W. Morgan and Z.D. Sharp, Neuroscience, 17, 538 (1991).
6. S. Banerjee, P. Hoppe, M. Brilliant, and D.M. Chikaraishi, J. Neuroscience, 12, 4460 (1992).
7. T. Sasaoka, K. Kobayashi, I. Nagatsu, R. Takahashi, M. Kimura, M. Yokoyama, T. Nomura, M. Katsuki, and T. Nagatsu, Mol. Brain Res., 16, 274, (1992).
8. N. Kaneda, T. Sasaoka, K. Kobayashi, K. Kiuchi, I. Nagatsu, Y. Kurasowa, K. Fujita, M. Yokoyama, T. Nomura, M. Katsuki, and T. Nagatsu, Neuron, 6, 583 (1991).
9. S. Morita, K. Kobayashi, T. Mizuguchi, K. Yamada, I. Nagatsu, K. Titani, K. Fujita, H. Hidaka, and T. Nagatsu, Mol. Brain Res., 17, 239 (1993).
10. S. G.G. Cadd, G.W. Hoyle, C.J. Quaife, B. Marck, A.M. Matsumoto, R.L. Brinster, and R.D. Palmiter, Mol. Endocrinol., 6, 1951 (1992).
11. P. Daubas, A.M. Salmon, M. Zoli, B. Geoffroy, A. Devillers-Thiery, A. Bessis, F. Medevielle, and J.P. Changeux, Proc. Nati. Acad. Sci. USA, 90, 2237 (1993).
12. K.V. Rogers, L.F. Boring, G.S. McKnight, and C.H. Clegg, Mol. Endocrinol., 6, 1756 (1992).
13. B.L. Largent, R.G. Sosnowski, and R.R. Reed, J. Neuroscience, 13, 300 (1993).
14. M. Rubinstein, B. Liu, R.H. Goodman, and M.J. Low, Mol. Cell. Neurosci., 3, 152 (1992).
15. H.L. Ang, D.A. Carter, D. Murphy, EMBO J., 12, 2397 (1993).
16. G.C. Waldbieser, C.D. Minth, C.L. Chrisman, and J.E. Dixon, Mol. Brain Res., 14, 87 (1992).
17. A.J. Harmar, P.K. Mulderry, R. Al-Shawi, V. Lyons, W.J. Sheward, J.O. Bishop, and K. Chapman, Regulatory Peptides, 46, 67 (1993).
18. D.R. Foran and A.C. Peterson, J. Neuroscience, 12, 4890 (1992).
19. W.C. Friend, S. Clapoff, C. Landry, L.E. Becker, D. O'Hanlon, R.J. Allore, I.R. Brown, A. Marks, J. Roder, and R.J. Dunn, J. Neuroscience, 121, 4337 (1992).
20. A. Messing, Y. Su, W.C. Kisseberth, F. Besnard, and M. Brenner, J. Neuropath. Exp. Neurol., 51, 349 (1992).
21. D.O. Wirak, R. Bayney, C.A. Kundel, A. Lee, G.A. Scangos, B.D. Trapp, and A.J. Unterbeck, EMBO J., 10, 289 (1991).
22. D.O. Wirak, R. Bayney, T.V. Ramabhadran, R.P. Francasso, J.T. Hart, P.E. Hauer, P. Hsiau, S.K. Pekar, G.A. Scangos, B.D. Trapp, and A.J. Unterbeck, Science, 253, 323 (1991).
23. M. Jucker, L.C. Walker, L.J. Martin, C.A. Kitt, H.K. Kleinman, D.K. Ingram, and D.L. Price, Science, 255, 1443 (1992).
24. D.O. Wirak, R. Bayney, T.V. Ramabhadran, R.P. Francasso, J.T. Hart, P.E. Hauer, P. Hsiau, S.K. Pekar, G.A. Scangos, B.D. Trapp, and A.J. Unterbeck, Science, 255, 1445 (1992).
25. D. Quon, Y. Wang, R. Catalano, J.M. Scardina, K. Murakami, and B. Cordell, Nature, 352, 239 (1991).
26. F. Yamaguchi, S.J. Richards, K. Beyreuther, M. Salbaum, G.A. Carlson, and S.B. Dunnett, Neuroreport, 2, 781 (1991).
27. S. Kawabata, G.A. Higgins, and J.W. Gordon, Nature, 354, 476 (1991).
28. S. Kawabata, G.A. Higgins, and J.W. Gordon, Nature, 356, 23 (1992).
29. F.A. Sandhu, M. Salim, and S.B. Zain, J. Biol. Chem., 266, 21331 (1991).
30. F.A. Sandhu, R.H.P. Porter, R.V. Eller, S.B. Zain, M. Salim, and J.T. Greenamyre, J. Neurochem., 61, 2286 (1993).
31. A. Kammesheidt, F.M. Boyce, A.F. Spanoyannis, B.J. Cummings, M. Ortegon, C. Cotman, J.L. Vaught, and R.L. Neve, Proc. Nati. Acad. Sci. USA, 89, 10857 (1992).
32. B.T. Lamb, S.S. Sisodia, A.M. Lawler, H.H. Slunt, C.A. Kitt, W.G. Kearns, P.L. Pearson, D.L. Price, and J.D. Gearhart, Nature Genetics, 5, 22 (1993).
33. B.E. Pearson and T.K. Choi, Proc. Nati. Acad. Sci. USA, 90, 10578 (1993).
34. M. Scott, D. Foster, C. Mirenda, D. Serban, F. Coufal, M. Wälchli, M. Torchia, D. Groth, G. Carlson, S.J. DeArmond, D. Westaway, S.B Prusiner, Cell, 59, 847 (1989).
35. S.B. Prusiner, M. Scott, D. Foster, K.-M. Pan, D. Groth, C. Mirenda, M. Torchia, S.-L. Yang, D. Serban, G.A. Carlson, P.C. Hoppe, D. Westaway, S.J. DeArmond, Cell, 63, 673 (1990).
36. K. Hsiao, S.B. Prusiner, Alzheimer Disease and Associated Disorders, 5, 155 (1991).
37. K.K. Hsiao, M. Scott, D. Foster, D.F. Groth, S.J. DeArmond, S.B. Prusiner, Science, 250, 1587 (1990).
38. D. Westaway, S.J. DeArmond, J. Cayetano-Canlas, D. Groth, D. Foster, S.-L. Yang, M. Torchia, G.A. Carlson, S.B. Prusiner, Cell, 76, 117 (1994).
39. H. Büeler, M. Fischer, Y. Lang, H. Bluethmann, H.-P. Lipp, S.J. DeArmond, S.B. Prusiner, M. Aguet, C. Weissman, C. Nature, 356, 577 (1992).
40. H. Büeler, A. Aguzzi, A. Sailer, R.-A. Greiner, P. Autenried, M. Aguet, C. Weissmann, Cell, 73, 1339 (1993).
41. C.J. Robinson, TIBTECH, 11, 496 (1993).
42. P. Ernfors, K.-F. Lee, R. Jaenisch, Nature, 368, 147 (1994).
43. Y. Masu, E. Wolf, B. Holtmann, M. Sendtner, G. Brem, H. Thoenen, Nature, 365, 27 (1993).

44. P.H. Patterson, L. Bugga, C.L. Stewart, Soc. Neurosci., 19, 1724 (1993).
45. J.-P. Liu, J. Baker, A.S. Perkins, E.J. Robertson, A. Efstratiadis, Cell, 75, 59 (1993).
46. N. Patil, E. Lacy, M.V. Chao, Neuron, 2, 437 (1990).
47. K.-F. Lee, E. Li, L.J. Huber, S.C. Landis, A.H. Sharpe, M.V. Chao, R. Jaenisch, Cell, 69, 737 (1992).
48. K.-F. Lee, K. Bachman, S. Landis, R. Jaenisch, Science, 263, 1447 (1994).
49. A.M. Davies, K.-F. Lee, R. Jaenisch, Neuron, 11, 565 (1993).
50. R.J. Smeyne, R. Klein, A. Schnapp, L.K. Long, S. Bryant, A. Lewin, S.A. Lira, M. Barbacid, Nature, 368, 246 (1994).
51. R. Klein, R.J. Smeyne, W. Wurst, L.K. Long, B.A. Auerbach, A.L. Joyner, M. Barbacid, Cell, 75, 113 (1993).
52. R. Klein, I. ilos-Santiago, R.J. Smeyne, S.A. Lira, R. Brambilla, S. Bryant, L. Zhang, W.D. Snider, M. Barbacid, Nature, 368, 249 (1994).
53. A.J. Silva, C.F. Stevens, S. Tonegawa, Y. Wang, Science, 257, 201 (1992).
54. A.J. Silva, R. Paylor, J.M. Wehner, S. Tonegawa, Science, 257, 206 (1992).
55. T.W. Rosahl, M. Geppert, D. Spillane, J. Herz, R.E. Hammer, R.C. Malenka, T.C. Südhof, Cell, 75, 661 (1993).
56. A. Abeliovich, C. Chen, Y. Goda, A.J. Silva, C.F. Stevens, S. Tonegawa, Cell, 75, 1253 (1993).
57. A. Abeliovich, R. Paylor, C. Chen, J.J. Kim, J.M. Wehner, S. Tonegawa, Cell, 75 1263 (1993).
58. M. Kimura, M. Sato, A. Akatsura, S. Nozawa-Kimura, R. Takahashi, M. Yokoyama, T. Nomura, M. Katsuki, Proc. Natl. Acad. Sci. USA , 86, 5661 (1989).
59. K.-A. Nave, C. Readhead, L. Hood, ,G. Lemke, J. Cell Biochem. Suppl., 15A, 211 (1991).
60. H. Umemori, S. Sato, T. Yagi, S. Aizawa, T. Yamamoto, Nature, 367, 572 (1994).
61. D.-R. Koh, W.-P. Fung-Leung, A. Ho, D. Gray, H. Acha-Orbea, T.-W. Mak, Science, 256, 1210 (1992).
62. T. Yoshioka, L. Feigenbaum, G. Jay, Mol. Cel. Biol., 11 5479 (1991).
63. Windle, J.J., Weiner, R.I., Mellon, P.L., Mol. Endocrinol., 4, 597 (1990).
64. F. Horn, L.M. Bilezikjian, M.H. Perrin, M.M. Bosma, J.J. Windle, K.S. Huber, A.L. Blount, B. Hille, W. Vale, P.L. Mellon, Mol. Endocrinol., 5, 347 (1991).
65. M. Tsutsumi, W. Zhou, R.P. Millar, P.L. Mellon, J.L. Roberts, C.A.. Flanagan, K. Dong, B. Gillo, S.C. Sealfon, Mol. Endocrinol., 6, 1163 (1992).
66. A. Helseth, G.P. Siegal, E. Haug, V.L. Bautch, Am. J. Pathol., 140, 1071 (1992).
67. R.R. Behringer, L.S. Mathews, R.D. Palmiter, R.L. Brinster, Genes Develop., 2, 453 (1988).
68. S.M. Toggas, E. Masliah, E.M. Rockenstein, G.F. Rall, C.R. Abraham, L. Mucke, Nature, 367, 188 (1994).
69. P.L. Huang, T.M. Dawson, D.S. Bredt, S.H. Solomon, M.C. Fishman, Cell, 75, 1273 (1993).
70. E.Y.-H.P. Lee, C.-Y. Chang, N. Hu, Y.-C.J. Wang, C.-C. Lai, K. Herrup, W.-H. Lee, A. Bradley, Nature, 359, 288 (1992).
71. T. Jacks, A. Fazeli, E.M. Schmitt, R.T. Bronson, M.A. Goodell, R.A. Weinberg, Nature, 359, 295 (1992).
72. D.J. Veis, C.M. Sorenson, J.R. Shutter, S.J. Korsmeyer, Cell, 75, 229 (1993).
73. I. Ceballos-Picot, A. Nicole, P. Briand, G. Grimber, A. Delacourte, A. Defossez, F. Javoy-Agid, M. Lafon, J.L. Blouin, P.M. Sinet, Brain Res., 552,198 (1991).
74. W.C. Orr, R.S. Sohal, Science, 263,1128 (1994).
75. I.L. Campbell, C.R. Abraham, E. Masliah, P. Kemper, J.D. Inglis, M.B.A. Oldstone, L. Mucke, Proc. Natl. Acad. Sci. USA 90, 10061 (1993).
76. M. Kopf, H. Baumann, G. Freer, M. Freudenberg, M. Lamers, T. Kishimoto, R. Zinkernagel, H. Bluethmann, G. Kohler, Nature, 368, 339 (1994).
77. L.W. Fitzgerald, K.J. Miller, A.K. Ratty, S.D. Glick, M. Teitler, K.W. Gross, Brain Res., 580, 18 (1992).

Chapter 5. Recent Advances in Dopamine D_3 and D_4 Receptor Ligands and Pharmacology

Ruth E. TenBrink, Rita M. Huff
The Upjohn Company
Kalamazoo, MI 49001

Introduction-Dopamine receptors mediate a number of central nervous system functions. Until recently, the dopaminergic system was considered to be composed of two receptor types, the D_1 and D_2 receptors. These two receptors have now been joined by the D_5 receptor, which resembles the D_1 receptor, and the D_3 and D_4 receptors, which share sequence homologies with the D_2 receptor (itself now subdivided into D_2 short and D_2 long), and whose ligands share certain pharmacological properties with D_2 ligands (1). Interest in the D_3 and D_4 receptors has been focused mainly on disease states previously linked to the D_2 receptor, such as Parkinson's disease and schizophrenia. Dopamine agonists such as pergolide and quinpirole are used in the treatment of Parkinson's disease and were reported to have a higher affinity for the D_3 receptor as compared to the D_2 receptor (2). Screening of clinical entities useful in the treatment of schizophrenia has revealed that the so-called atypical antipsychotics, which are characterized by reduced or absent extrapyramidal side effects (3, 4), typically exhibit high binding affinities for the D_3 and/or D_4 receptors, as compared to the D_2 receptor (2, 5). Seeman et al., found that dopamine D_4 receptors are elevated by 600% in postmortem schizophrenic brain tissue (6). The same investigators found about a 10% increase in the combined D_2 and D_3 receptor population of schizophrenic postmortem tissue. Results of earlier positron emission tomography studies using [^{11}C]raclopride, which has a high affinity for the D_2 receptor and low affinity for the D_4 receptor, indicated no increase in dopamine D_2 levels in schizophrenic subjects (7). In contrast, results from a study using 3-N-[^{11}C]methylspiperone, which has a high affinity for both the D_2 and D_4 receptors, showed an increase in dopamine receptors in schizophrenics as compared to controls (8). The differing results were interpreted as supporting an increase in dopamine D_4 levels in schizophrenics (9). In another attempt to measure receptor levels in schizophrenics, [^{125}I]epidepride binding to striatal D_2 and D_3 receptors in the presence of GTP was studied using autoradiography (10). The displacement of [^{125}I]epidepride by domperidone (3), a dopamine antagonist with higher affinity for D_2 than D_3 receptors, was used to visualize D_3 receptors. 7-OH-DPAT (1) was used to block the binding of [^{125}I]epidepride at the D_3 receptor in order to estimate the binding to D_2 receptors. D_2 receptors were elevated by 14% in schizophrenics compared to controls, whereas D_3 levels were found to be increased by 36%. On a different note, a loss of D_3 mRNA levels was detected in certain cortical regions of schizophrenic patients (11). Taken together, these studies indicate that there may be aberrations in the regulation of D_2, D_3, and D_4 receptors in schizophrenic brains.

Molecular Biology-Soon after the cDNA for the D_2 dopamine receptor was discovered by Civelli and collaborators in 1988 (12), the identification of related cDNA's for additional dopamine receptor proteins revealed that the long-standing classification scheme of dopamine receptors into D_1 and D_2 subtypes was inadequate (13). The cDNA's for D_2 receptors, which were first isolated from rat and then human and bovine brains, were found to be 96% identical at the amino acid level, indicating a high degree of conservation across species (14,15). There were found to be two mRNA products formed by alternate splicing of the D_2 receptor gene which differ by the inclusion (D_{2L} or D_{2A}) or absence (D_{2S} or D_{2B}) of 29 amino acids in the putative third intracytoplasmic loop (14-18). When either D_2 proteins were expressed in transfected cells, the pharmacological profile of the binding sites was similar to that expected for D_2 dopamine receptors. The mRNA for D_2 receptors is found in caudate-putamen, olfactory tubercle, nucleus accumbens, anterior and intermediate lobes of the pituitary, substantia nigra, and ventral tegmental areas of the brain (12,16,19-23). A related rat cDNA was discovered and the encoded protein was termed the D_3 dopamine receptor (24). The D_3 protein has approximately 50% overall homology but 75-80% homology within the transmembrane domains (considered to form the ligand binding pocket) to the D_2 receptor protein. Subsequently the cDNA for the human D_3 receptor was identified and found to be 97% homologous to the rat protein within the transmembrane domains (25). Expression of D_3 proteins in transfected cells revealed a similar yet distinct pharmacological profile with dopaminergic agents, indicating that D_3 receptors were a new dopamine receptor subtype. Overall RNA for D_3 receptors is much less abundant than mRNA for D_2 receptors. In addition, the distribution is not as extensive, being more predominant in mesolimbic areas including the ventral striatum, olfactory tubercle, nucleus accumbens and islands of Calleja. D_3 receptor mRNA is also found in the hypothalamus and archicerebellum (26-28).

The gene for another novel dopamine receptor, termed D_4, was discovered in human tissues and reported in 1991 (29). Subsequent cloning of the rat D_4 gene revealed 73% overall amino acid homology with the human gene but approximately 90% homology within the transmembrane domains (30). Human D_4 receptors are 41% homologous to D_2 receptors, 39% homologous to D_3 receptors and 56% homologous to either within the transmembrane domains. Expression of D_4 receptor protein in transfected cells showed the binding sites to have a pharmacological specificity distinct from D_2 and D_3 receptors. Studies of distribution of mRNA for D_4 receptors are much more limited. Information to date indicates that D_4 receptors are expressed at much lower levels than D_2 receptors (31). Brain areas which express D_4 receptors include the frontal cortex, midbrain, amygdala, medulla, and striatum (29). D_4 receptor mRNA is also found in the heart (30).

<u>D_3 Ligands</u>-The earlier appearance of cloned D_3 receptors, as compared to the D_4 receptor, for ligand and pharmacological studies is reflected in a relatively greater number of studies appearing in the recent literature. 7-Hydroxy-DPAT (<u>1</u>) is a dopamine agonist with modest selectivity for the D_3 receptor (K_i = 0.78 nM) as compared to the D_1 (K_i = 5300 nM), D_2 (K_i = 61 nM), and D_4 (K_i = 650 nM) receptors (32). Most of the affinity for the D_3 receptor resides in R-(+)-(<u>1</u>), whose IC_{50} is 6.6 nM. By comparison, the IC_{50} for the D_2 receptor was measured as 420 nM for R-(+)-(<u>1</u>) and 380 nM for the racemate (33). In another study, R-(+)-<u>1</u> and S-(-)-<u>1</u> had K_i's of 0.6 nM and 42 nM, respectively, at the D_3 receptor, whereas at the D_2 receptor the K_i's were 125 nM and 3672 nM (34). In rats, R-(+)-<u>1</u> dose-dependently inhibited both the release of striatal dopamine and induced yawning, whereas sniffing was seen only at the highest dose. In another study in rats, low doses of 7-OH-DPAT (10 µg/kg, s.c.) reduced spontaneous activity without inducing yawning. At higher doses (0.1-10 mg/kg, s.c.), 7-OH-DPAT stimulated non-stereotyped sniffing, locomotion, and chewing (35). 7-OH-DPAT was administered at varying doses to spontaneously hypertensive (SHR) and Wistar-Kyoto (WKY) rats (36). Differences were seen in the behaviors elicited by 7-OH-DPAT in SHR and WKY rats. In WKY rats, the drug caused an increase in locomotor activity at 10 mg/kg, i.p., while in SHR rats, which have higher basal

locomotor activity, a small decrease was seen at 1 mg/kg, i.p. Turning behavior in unilaterally 6-hydroxydopamine-lesioned SHR and WKY rats was increased by 7-OH-DPAT at both 0.1 mg/kg and 1 mg/kg. The dose-response curve for yawning behavior in both SHR and WKY rats was bell-shaped. [^{3}H]7-OH-DPAT has been used as a radioligand for D$_3$ receptors, although appropriate conditions must be used to prevent binding of the ligand to the high affinity state of the D$_2$ receptors, i.e., low Mg^{++} and high GTP (32). [^{3}H]7-OH-DPAT binds with high affinity at not only the D$_3$ receptor, but also at sigma sites (K$_i$ = 48 nM) in bovine caudate nucleus membranes, so caution must be taken to mask sigma sites in radioligand binding studies (37). *Trans*-7-OH-PIPAT (**2**) is an iodinated analog of 7-OH-DPAT with a 143-fold preference for the D$_3$ as compared to the D$_2$ receptor (38). R-(+)-**2** is the active isomer. [^{125}I]-**2** has higher specific activity and lower nonspecific binding than [^{3}H]-7-OH-DPAT, but it also binds with high affinity to the sigma receptor (39).

1 R$_1$, R$_2$ = n-C$_3$H$_7$
2 R$_1$ = n-C$_3$H$_7$, R$_2$ = -CH$_2$CH=CHI

The aminotetralin UH232 (**4**) binds to human D$_3$ receptors transfected in CHO cells with a K$_i$ of 10 nM and to D$_2$ receptors with a K$_i$ of 40 nM (2). In rat D$_3$ receptors expressed in CHO cells, **4** has a K$_i$ of 9 nM and at rat D$_2$ receptors a K$_i$ of 40 nM (24). Both (+)-(**4**) and (+)-**5** (AJ76) have different effects on dopamine release and metabolism as compared to haloperidol and raclopride, the classical dopamine receptor antagonists (40). Both **4** and **5** seem to prefer dopamine receptors which regulate dopamine release over the receptors which regulate dopamine metabolism. Furthermore, **5** displays a preference for the mesolimbic pathway over the nigrostriatal pathway in electrophysiological studies, at least with respect to potency. Both compounds have locomotor stimulating properties in animals. These results suggest that **4** and **5** serve as dopamine autoreceptor antagonists acting through the D$_3$ receptor, for which they have a modest (2-3-fold) preference over the D$_2$ receptor (40). Analogs of **4** and **5** in which the 5-methoxy is replaced with O-triflate (**6**), methyl ester (**7**), and nitrile (**8**) give compounds with lower affinity for the D$_3$ receptor. The Di-n-propyl analog **9** (K$_i$ of 40 nM at D$_3$) exhibited a 14-fold preference for the D$_3$ over D$_2$ receptor, vs. the 5-fold selectivity seen for **5** (41). Tetralin **9**, like **5**, also binds at the D$_4$ receptor with modest affinity (K$_i$ = 121 nM). *In vivo*, **9** and **6**, its metabolite, increase motor activity and DOPAC and HVA levels in striatum and limbic regions of the brain, indicating dopamine receptor antagonism (41).

4 R$_1$, R$_2$ = n-C$_3$H$_7$, R$_3$ = OCH$_3$
5 R$_1$ = H, R$_2$ = n-C$_3$H$_7$, R$_3$ = OCH$_3$
6 R$_1$ = H, R$_2$ = n-C$_3$H$_7$, R$_3$ = OSO$_2$CF$_3$
7 R$_1$ = H, R$_2$ = n-C$_3$H$_7$, R$_3$ = CO$_2$CH$_3$
8 R$_1$ = H, R$_2$ = n-C$_3$H$_7$, R$_3$ = CN
9 R$_1$, R$_2$ = n-C$_3$H$_7$, R$_3$ = OSO$_2$CF$_3$

10

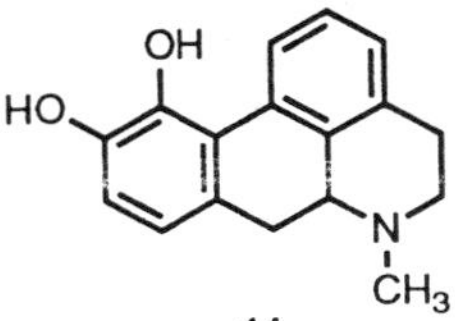

11 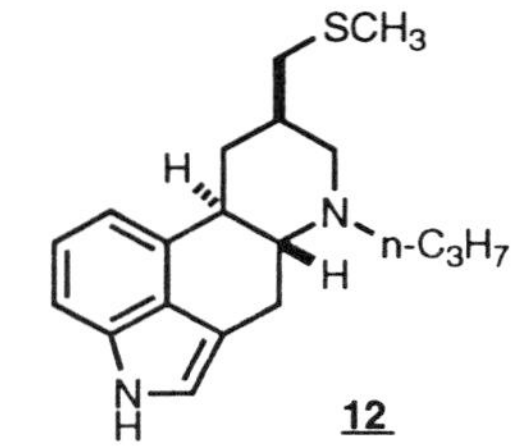   **12** 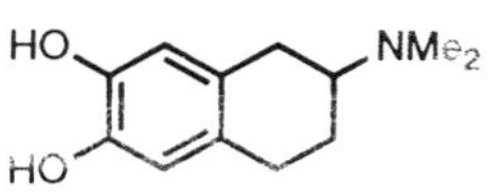   **13**

Binding affinities for a dozen new and old antipsychotics were measured against [^{125}I]iodosulpiride at sixteen receptors, including the dopamine D_3 receptor (5). Of these, chlorprothixene (**27**), clothiapine (**22**), haloperidol (**42**), risperidone (**31**), tiospirone (BMY 13,859, **32**), zotepine (**33**), and raclopride (**16**) had D_3 K_i's in the 2-30 nM range, and pipamperone (**45**), clozapine (**21**), melperone (**46**), amperozide (**34**), and remoxipride (**17**) had K_i's in the 300-1900 nM range. The dopamine agonists apomorphine (**11**), pergolide (**12**), TL 99 (**13**), quinpirole (**10**), and quinerolane (**14**) and antagonists **3**, **4**, **5**, sulpiride (**15**), remoxipride (**17**), sultopride (**18**), amisulpiride (**19**), metoclopramide (**20**), clozapine (**21**), mezilamine (**23**), pimozide (**24**), carpipramine (**25**), chlorpromazine (**26**), prochlorperazine (**28**), thioproperazine (**29**), pipotiazine (**30**), and haloperidol (**42**) were measured for their binding affinity for human dopamine D_2 and D_3 receptors in CHO transfected cell membranes (42). Of the antagonists, only **4** and **5** exhibited modest (2-3 fold) D_3 selectivity. Of the agonists, all but apomorphine had higher affinity for the D_3 receptor as compared to D_2, with selectivities ranging from 8-95-fold. However, when two site analysis is carried out on agonist competition curves at D_2 and D_3 receptors, in order to measure high affinity interactions, the large apparent D_3 selectivity for agonists is lost (43). Similarly, evaluation of the same agonists in measurements of potency at D_2 and D_3 receptor activation shows that they are relatively nonselective. Quinpirole (**10**) binds equally well to dopamine D_2, D_3, and D_4 receptors but pramipexole (SND 919CL2Y, **35**), which is under clinical investigation for the treatment of Parkinson's disease, binds preferentially to D_3 receptors (44). Both quinpirole and **35** inhibited dopamine neurons in the substantia nigra *pars compacta* and excited caudate nucleus cells. The excitation of caudate nucleus cells was thought to occur through D_3 receptor activation. In a separate study, stimulation and inhibition of substantia nigra *pars reticulata* cells by dopamine agonists such as pramipexole were concluded to be mediated indirectly via caudate D_2 and D_3 receptor subtypes, respectively (45).

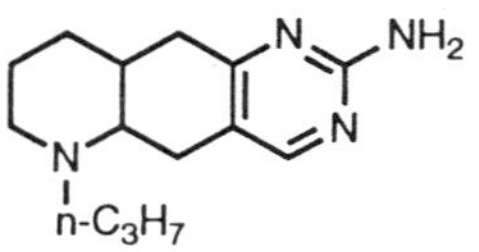

14 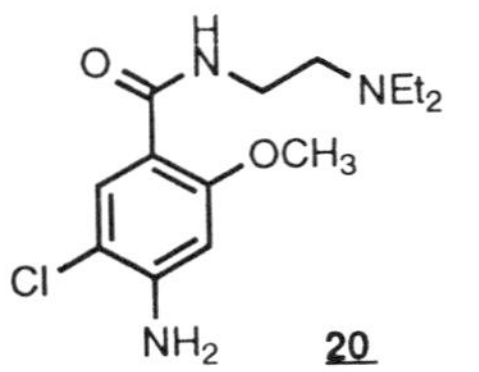   **20**

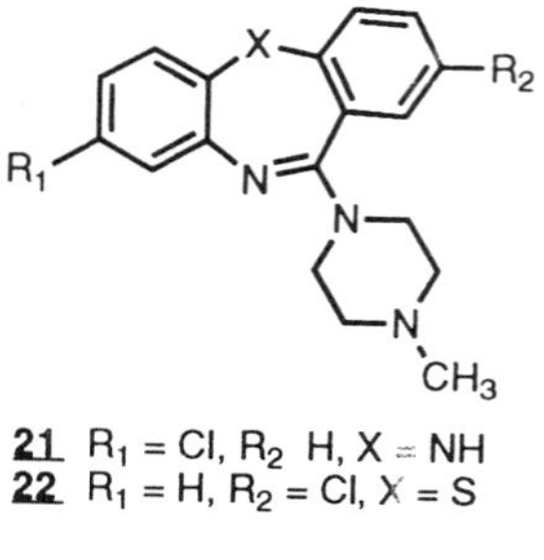

21 R_1 = Cl, R_2 H, X = NH
22 R_1 = H, R_2 = Cl, X = S

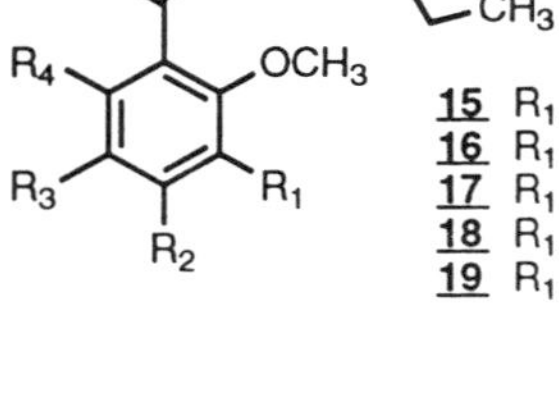

15 R_1, R_2 ,R_4 = H, R_3 = SO_2NH_2
16 R_1, R_3 = Cl, R_2 = H, R_4 = OH
17 R_1 , R_2 = H, R_3 = Br, R_4 = OCH_3
18 R_1, R_2, R_4 = H, R_3 = SO_2Et
19 R_1, R_4 = H, R_2 = NH_2, R_3 = SO_2Et

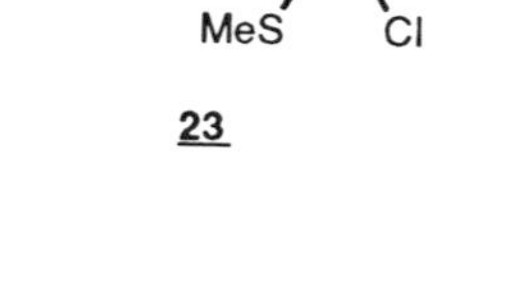

23

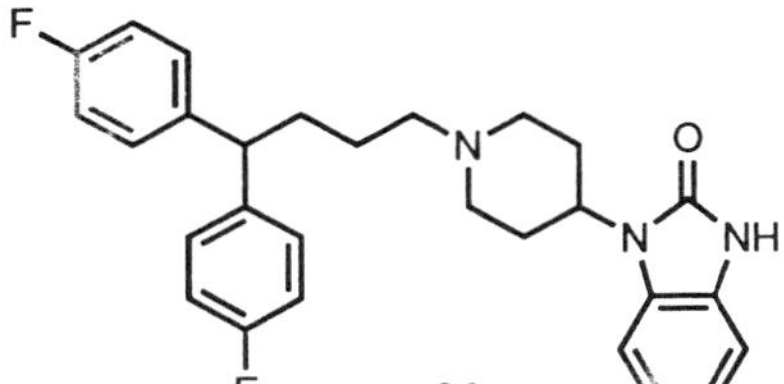

24

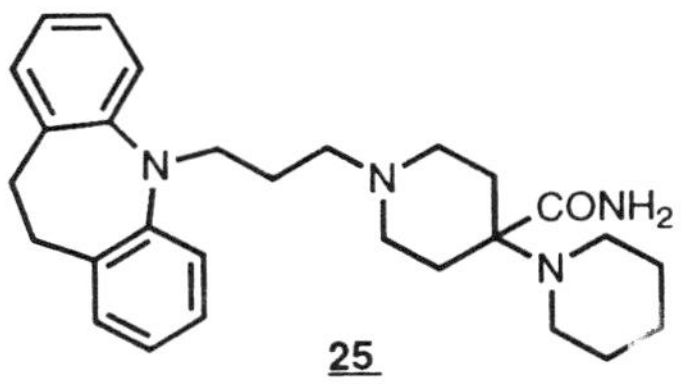

25

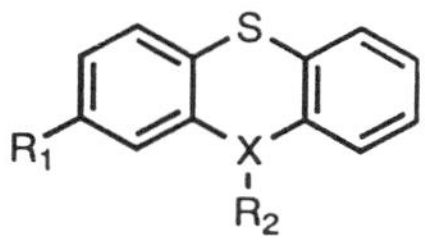

28 R$_1$ = Cl, X = N, R$_2$ = -(CH$_2$)$_3$-N⁀NMe

29 R$_1$ = SO$_2$NMe$_2$, X = N, R$_2$ = -(CH$_2$)$_3$-N⁀NMe

26 R$_1$ = Cl, X = N, R$_2$ = -(CH$_2$)$_3$NMe$_2$ **30** R$_1$ = SO$_2$NMe$_2$, X = N, R$_2$ = -(CH$_2$)$_3$-N⁀
27 R$_1$ = Cl, X-R$_2$ = C=CHCH$_2$CH$_2$NMe$_2$ ⁀OH

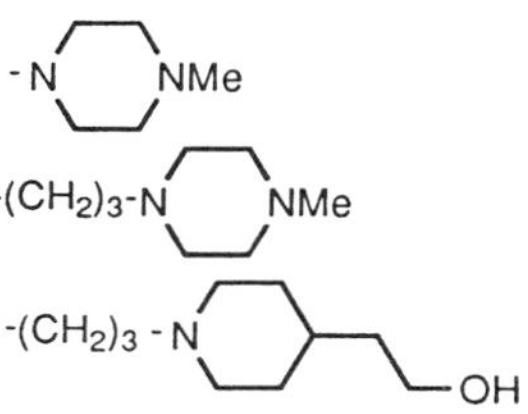

The enantiomers of 3-PPP, (-)-**36** and (+)-**36**, bind to the D$_3$ receptor with K$_i$'s of 132 nM and 217 nM, respectively, but only (-)-**36** has affinity for the D$_4$ receptor, exhibiting a K$_i$ of 130 nM (46). Substituted 3-phenylpiperidines (-)-**37** and (-)-**38** are dopamine autoreceptor antagonists with approximately three-fold higher affinity for the D$_3$ receptor over the D$_2$. Both (-)-**37** and (+)-**37** bind with high affinity (K$_i$ of 32 nM and 6 nM, respectively) to the sigma site (measured against [^{3}H]DTG). Neither **37** nor **38** has significant affinity for the D$_4$ receptor. DHX (dihydrexidine, **39**), is a dopamine D$_1$ full agonist which also exhibits some affinity for D$_2$ (K$_{0.5}$ of 1490 nM) and D$_3$ (K$_{0.5}$ of 170 nM) receptors, with an 8.8-fold preference for the D$_3$ over the D$_2$ receptor, as measured against spiperone in rat receptors expressed in C-6 glioma cells (47, 48). n-Propyl analog **40** exhibits 20-fold increased affinity at D$_3$, whereas methylated analog 4MP-DHX (**41**) exhibits high affinity for the D$_3$ receptor (K$_{0.5}$ of 2 nM) and a D$_3$/D$_2$ selectivity ratio of 110. Quinpirole and 7-OH-DPAT, as measured in this system, had D$_3$/D$_2$ ratios of 157 and 115, respectively.

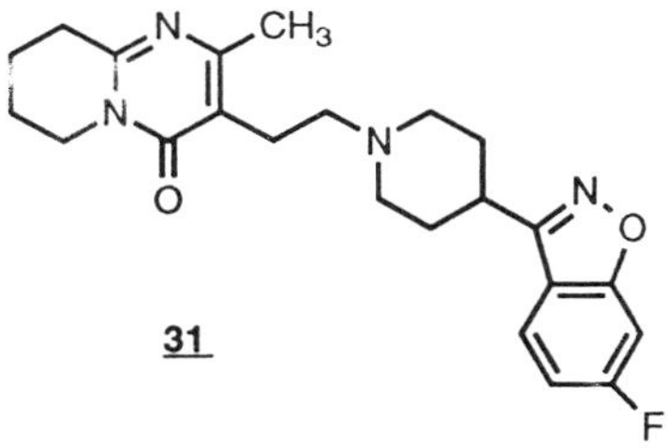

31

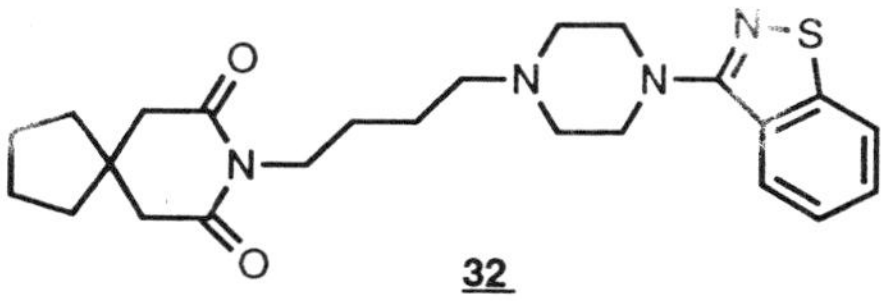

32

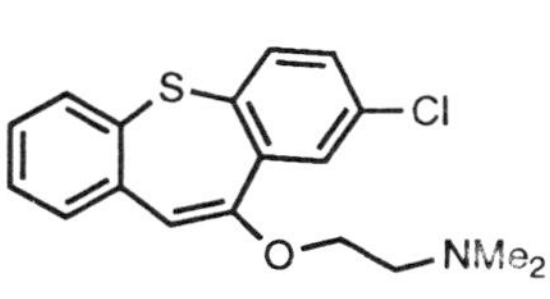

33

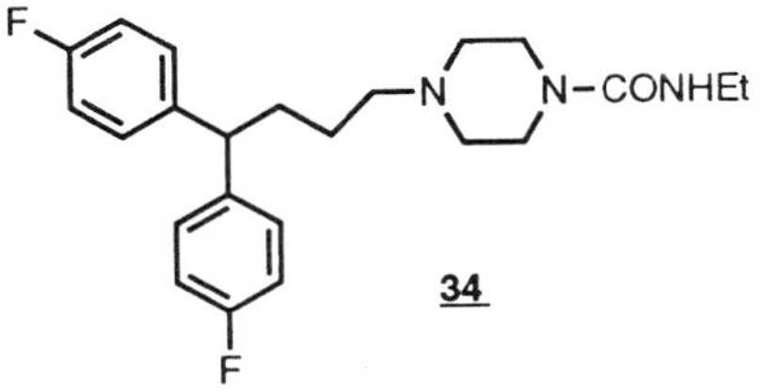

34

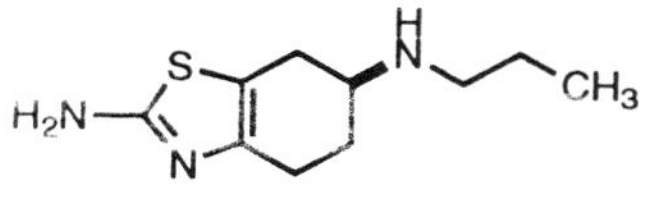

35

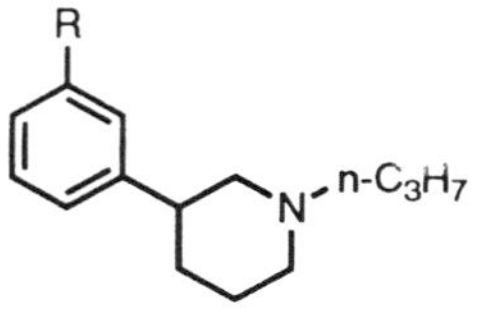

36 R = OH
37 R = OSO$_2$CF$_3$
38 R = CN

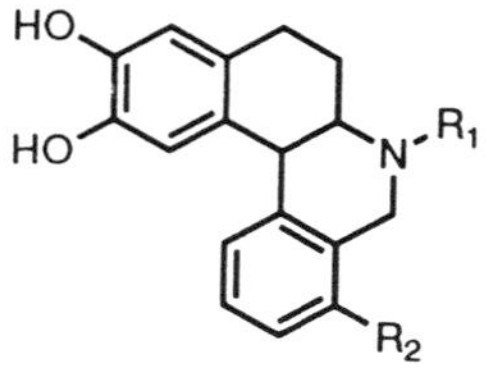

39 R$_1$, R$_2$ = H
40 R$_1$ = n-C$_3$H$_7$, R$_2$ = H
41 R$_1$ = n-C$_3$H$_7$, R$_2$ = CH$_3$

Haloperidol (**42**), a classical "typical" antipsychotic (49), exhibits subnanomolar binding to both D$_2$ and D$_3$ receptors, as measured against spiperone in cloned mammalian receptors expressed in CHO-K1 cells (50). The "reduced" haloperidol metabolites (-)-**47** and (+)-**47**, as well as the haloperidol analogs azaperone (**43**) and its reduced metabolites (+)-**48** and (-)-**48**, and BMY-14802 (**49**) exhibit reduced D$_3$ and D$_2$ affinity. Binding at sigma-1 and sigma-2 sites is much less affected. Remoxipride (**17**), an antipsychotic agent of the benzamide class with D$_2$ antagonist properties, was recently withdrawn from clinical trials due to several cases of aplastic anemia (51). Remoxipride has a K$_i$ of 125 nM at the human D$_2$ receptor, a K$_i$ of 969 nM at the human D$_3$ receptor, and no appreciable binding at D$_4$. Major differences occur in the rat and human metabolites, however, and these metabolites vary dramatically in their affinity for the D$_2$ and D$_3$ receptors (52). Rat metabolites **50** and **51** bind with nanomolar affinity at the D$_3$ receptor, with metabolites **52** and **53** exhibiting slightly diminished affinity. Binding at the D$_2$ receptor follows the same pattern. The major human metabolites (**54**, **55**, **56**, **57**, **58**), on the other hand, have no affinity for the D$_2$ or D$_3$ receptor. PD 143,188 (**59**) is a dopamine autoreceptor agonist/partial agonist which decreases dopamine synthesis in the striatum in a dose-related manner (53). It has high affinity for both D$_2$ (D$_{2L}$ and D$_{2S}$, K$_i$ ≈ 20 nM) and D$_3$ (K$_i$ = 16 nM) receptors. A series of phenyl pyrroles with high affinity for the D$_2$ and D$_3$ receptors is exemplified by **60**. The compounds are reported to be antagonists with IC$_{50}$ values ranging between 0.2-26 nM at the human D$_3$ receptor (54). RWJ-37796 (**61**) is reported to be in clinical trials for the treatment of schizophrenia (55). It exhibits nanomolar binding affinity at the D$_3$ receptor, as well as at the D$_2$, 5-HT$_{1A}$, α$_1$, and α$_2$ receptors.

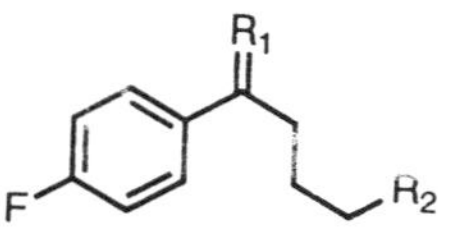

42 R$_1$ = O, R$_2$ = a
43 R$_1$ = O, R$_2$ = b
44 R$_1$ = O, R$_2$ = c
45 R$_1$ = O, R$_2$ = d
46 R$_1$ = O, R$_2$ = e
47 R$_1$ =H, OH, R$_2$ = a
48 R$_1$ =H, OH, R$_2$ = b
49 R$_1$ =H, OH, R$_2$ = c

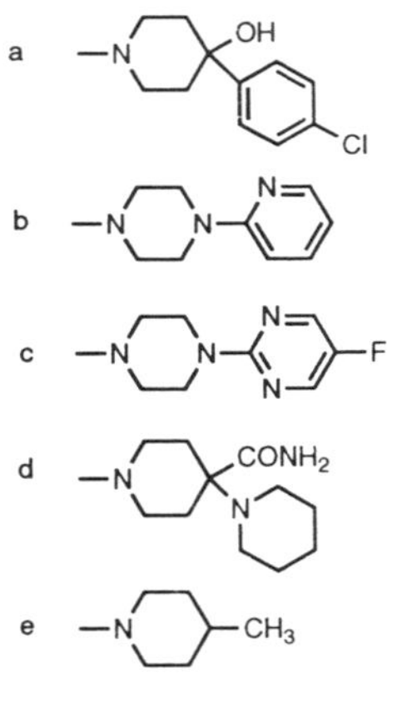

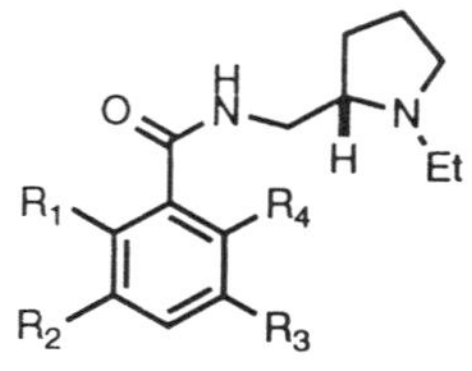

50 R$_1$= OH, R$_2$ = Br, R$_3$ = H, R$_4$ = OCH$_3$
51 R$_1$= OCH$_3$, R$_2$ = Br, R$_3$ = H, R$_4$ = OH
52 R$_1$= OCH$_3$, R$_2$ = Br, R$_3$ = OH, R$_4$ = OH
53 R$_1$= OCH$_3$, R$_2$ = Br, R$_3$ = OH, R$_4$ = OCH$_3$

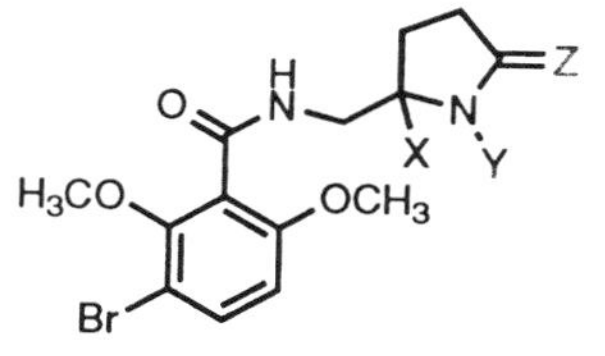

54 X = H, Y = H, Z = H, H
55 X = H, Y = C$_2$H$_5$, Z = O
56 X = H, Y = H, Z = O
57 X = OH, Y = C$_2$H$_5$, Z = O
58 X = OH, Y = H, Z = O

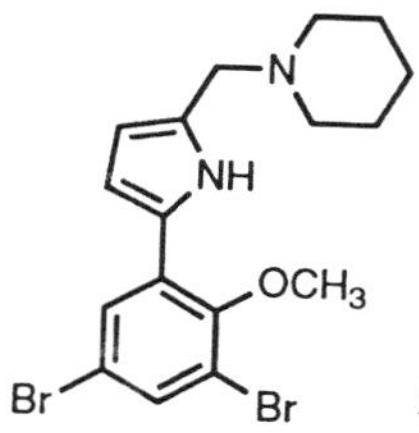

60

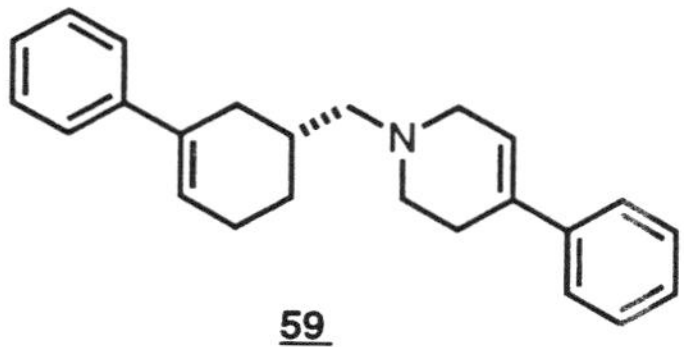

59

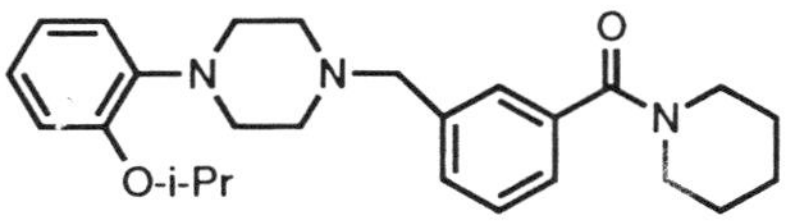

61

Ropinirole (SKF 101468, **62**), a dopamine agonist under development for the treatment of Parkinson's disease, shows 20-fold selectivity for the high affinity D$_3$ site as compared to the high affinity D$_2$ site (K$_i$'s of 70 nM and 1400 nM) as measured against iodosulpiride in cloned human receptors expressed in CHO cells (56, 57). A complicating factor in the visualization of D$_3$ receptors using autoradiography is the reported observation that dopamine apparently binds tenaciously to D$_3$ receptors (58). When preincubation of brain slices (to allow dissociation of endogenous dopamine) was omitted, no D$_3$ receptor labelling was observed, while D$_2$ receptor binding was equal in washed and unwashed brain sections. Monoamine depletion using tetrabenazine unmasked the D$_3$ receptors to [^{125}I]sulpiride binding. It is likely that the process of preparing tissues for autoradiography disrupts dopamine stores, thus allowing an abnormally high concentration of released dopamine access to the D$_3$ receptors.

D$_4$ Ligands-The therapeutic free plasma concentration of the atypical antipsychotic clozapine is about ten-fold less than its affinity constant for the D$_2$ receptor (59). This observation has been taken to mean that another receptor with higher affinity for clozapine, such as the D$_4$ receptor, might be responsible for certain of its atypical properties (29). However, clozapine was found to bind with different affinities at the D$_{2L}$ (K$_i$ = 60 nM) and D$_{2S}$ (K$_i$ = 35 nM) forms. With the D$_{2S}$ form, a biphasic curve was observed, with about 60% of the receptors exhibiting a K$_i$ of 9 nM and the remainder a Ki of 126 nM (60). At the D$_4$ receptor, clozapine binds with an affinity of 9-25 nM (29); thus the affinity of clozapine for the D$_{2S}$ and D$_4$ receptors may in fact be comparable. Olanzepine (LY-170053, **63**), a structural analog of clozapine which was recently approved for the treatment of schizophrenia, binds most strongly at muscarinic sites, but also, like clozapine, at D$_4$ sites (61). Apomorphine (**11**), NPA (**64**), and 11-OH-NPA (**65**) bind to the D$_4$ receptor (COS-7 cells, against [^{3}H]spiperone) with selectivities versus the D$_2$ receptor of 6.8 for (+)-**11** and 10.1 for (-)-**11**, 19.8 for (+)-**64** and 3.5 for (-)-**64**, and 4.2 for (+)-**65** and 5.2 for (-)-**65**. However, the (-)-aporphines cause stereotypy, hallucinations, and delusions in both animals and humans, whereas the (+)-aporphines lack pharmacological activity and block stereotyped activity induced by the (-)-aporphines (62). Lahti *et al.* (9) measure a somewhat lower affinity for (+)-**64** at the D$_4$

receptor than Seeman *et al.* (62) and thus calculate a somewhat lower D_4/D_2 selectivity ratio of 8.7. These investigators also found a 2.6-fold selectivity of clozapine for the D_4 receptor as compared to D_2, versus the 15-fold selectivity measured by van Tol *et al.* (6).

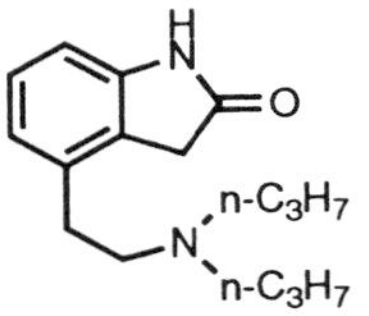

$\underline{62}$

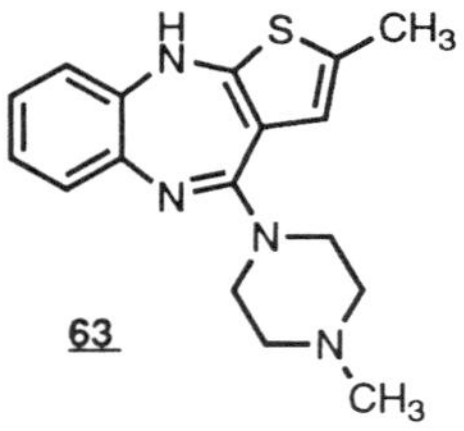

$\underline{63}$

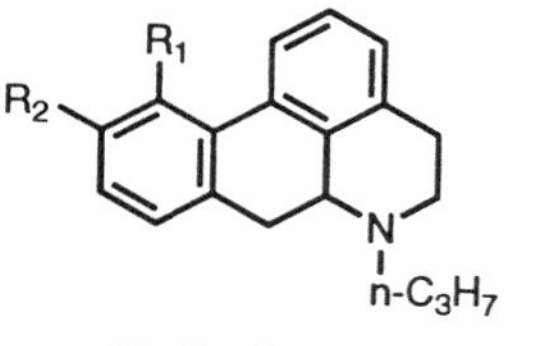

$\underline{64}$ R_1, R_2 = OH
$\underline{65}$ R_1 = OH, R_2 = H

Summary-The cloning and expression of cell lines containing the dopamine D_3 receptor has led to a re-examination of a number of both old and new compounds acting through the dopaminergic system. The more recent cloning and expression of the dopamine D_4 receptor is only now opening the door to an examination of old and new chemical entities. Full explication of the relevant pharmacology of the D_3 and D_4 receptors awaits the discovery of high affinity, selective agonists and antagonists at these receptors. Further, and most tantalizingly, their usefulness in treating CNS diseases will await human trials of these selective agents.

References

1. O. Civelli, J.R. Bunzow, D.K. Grandy, Annu. Rev. Pharmacol. Tox., 32, 281 (1993).
2. P. Sokoloff, M.-P. Martres, B. Giros, M.-L. Bouthenet, J.-C. Schwartz, Biochem. Pharmacol., 43, 659 (1992).
3. J.A. Lowe III, T.F. Seeger, F. J. Vinick, Med. Res. Rev., 8, 475 (1988).
4. M. Lader, J. Internat. Med. Res., 17, 1 (1989).
5. J. Leysen, P.M.F. Janssen, A. Schotte, W.H.M.L. Luyten, A.A.H.P. Megens, Psychopharmacology, 112, S40-S54 (1993).
6. P. Seeman, H.-C. Guan, H.H.M. Van Tol, Nature, 365, 441 (1993).
7. L. Farde, F.-A. Wiesel, S. Stone-Elander, C. Halldin, A.-L. Nordstrom, H. Hall, G. Sedvall, Arch. Gen. Psychiatry, 47, 213 (1990).
8. D.F. Wong, H.N. Wagner, C.A. Tamminga, E.P. Broussille, H.T. Ravert, A.A. Wilson, J.K.T. Toung, J. Malat, J.A. Williams, L.A. O'Tuama, S.H. Snyder, M.H. Kuhar, A. Gjedde, Science, 234, 1558 (1986).
9. R.A. Lahti, D.L. Evans, N.C. Stratman, L.M. Figur, Eur. J. Pharmacol., 236, 483 (1993).
10. E.V. Gurevich, M.-P. Kung, Y. Bordelon, J.N. Joyce, Society of Biological Psychiatry, May 19, 1994, Philadelphia, PA.
11. Schmauss, C., V. Haroutunian, K.L. Davis, and M. Davidson, Proc. Natl. Acad. Sci., 90, 8942 (1993).
12. J.R. Bunzow, H.H.M. Van Tol, D.K. Grandy, P. Albert, J. Salon, M. Chrisre, C.A. Machida, K.A. Neve, O. Civelli, Nature, 336, 783 (1988).
13. J.W. Kebabian, D.B. Calne, Nature, 227, 93 (1979).
14. C.L. Chio, G.F. Hess, R.S. Graham, R.M. Huff, Nature, 343, 266 (1990).
15. D.K. Grandy, M.A. Marchionni, H. Makam H. R.E. Stofko, M. Alfano, L. Frothingham, J.B. Fischer, K.J. Burke-Howie, J.R. Bunzow, A.C. Server, O. Civelli, Proc. Nat. Acad. Sci. USA, 86, 9762 (1989).
16. Dal Toso, R.B., B. Sommer, M. Ewert, A. Herb, D.B. Pritchett, A. Bach, B.D. Shivers, P.H. Seeburg, EMBO J., 8, 4025 (1989).

17. Monsma, F.J., L.D. McVittie, C.R. Gerfen, L.C. Mahan, D.R. Sibley, Nature, 342, 926 (1989).
18. Giros, B., Sokoloff, P., Martres, M.P., J.F. Riou, L.J. Emorine, J.C. Schwartz, Nature, 342, 923 (1989).
19. Meador-Woodruff, J.H., A. Mansour, J. R. Bunzow, H.H.M. Van Tol, S.J. Watson, Jr., O. Civelli, Proc. Natl. Acad. Sci., 86, 7625 (1989).
20. Mengod, G., M.I. Martinez-Mir, M.T. Vilaro, J.M. Palacios, Proc. Natl. Acad. Sci., 86, 8560 (1989).
21. Najlerahim, A., A.J.L. Barton, P.J. Harrison, J. Heffernan, R.C.A. Pearson, FEBS Lett., 255, 335 (1989).
22. Weiner, D.M., M.R. Brann, FEBS Lett., 253, 207, (1989).
23. Mansour, A., J.H. Meador-Woodruff, J.R. Bunzow, O. Civelli, H. Akil, S.J. Watson, J. Neurosci., 10, 2587 (1990).
24. P. Sokoloff, B. Giros, M.P. Martres, M.L. Bouthenet, J.C. Schwartz, Nature, 347, 146 (1990).
25. B. Giros, M.-P. Martres, P. Sokoloff, J.-C. Schwartz, C.R. Acad. Sci., 311, 501 (1990).
26. Bouthenet, M.-L., E. Souil, M.-P. Martres, P. Sokoloff, B. Giros, J.-C. Schwartz, Brain Res., 564, 203 (1991).
27. Landwehrmeyer, B., Mengod, G., J.M. Palacios, Mol. Brain Res., 18, 187 (1993).
28. J.-C. Schwartz, D. Levesque, M.-P. Martres, P. Sokoloff, Clin. Neuropharmacol., 16, 295 (1993).
29. H.H.M. Van Tol, J.R. Bunzow, H.-G. Guan, R.K. Sunahara, P. Seeman, H. Niznik, O. Civelli, Nature, 350, 610 (1991); H.H.M. Van Tol, C.M. Wu, H.-C. Guan, K. Ohara, J.R. Bunzow, O. Civelli, J. Kennedy, P. Seeman, H.B. Niznik, V. Jovanovic, Nature, 358, 149 (1992).
30. K.L. O'Malley, S. Harmon, L. Tang, R.D. Todd, The New Biologist, 4, 137 (1992).
31. J.A. Gingrich, M.G. Caron, Annu. Rev. Neurosci., 16, 299 (1993).
32. D. Levesque, J. Diaz, C. Pilon, M.-P. Martres, Giros, B., E. Souil, D. Schott, J.-L. Morgat, J.-C. Schwartz, P. Sokoloff, Proc. Natl. Acad. Sci. USA, 89, 8155 (1992).
33. R.J. Baldessarini, N.S. Kula, C.R. McGrath, V. Bakthavachalam, J.W. Kebabian, J.L. Neumeyer, Eur. J. Pharmacol., 239, 269 (1993).
34. G. Damsma, T. Bottema, B.H.C. Westerink, P.G. Tepper, D. Dijkstra, T.A. Pugsley, R.G. MacKenzie, T. G. Heffner, H. Wikstrom, Eur. J. Pharmacol., 249, R9-R10 (1993).
35. S.A. Daly, J.L. Waddington, Neuropharmacol., 32, 509 (1993).
36. M. Van den Buuse, Eur. J. Pharmacol., 243, 169 (1993).
37. H. Schoemaker, Eur. J. Pharmacol., 242, R1-R2 (1993).
38. C. Foulon, M.-P. Kung, H.K. Kung, J. Med. Chem., 36, 1499 (1993).
39. Kung, M.-P., Kung, H.F., Chumpradit, S., Foulon, C., Eur. J. Pharmacol., 235, 165 (1993)
40. N. Waters, S. Lagerkvist, L. Lofberg, M. Piercey, A. Carlsson, Eur. J. Pharmacol., 242, 151 (1993).
41. S.R. Haadsma-Svensson, M.W. Smith, C.-H. Lin, J.N. Duncan, C. Sonesson, H. Wikstrom, N. Waters, A. Carlsson, K. Svensson, Bioorg. Med. Chem. Lett., 4, 689 (1994).
42. P. Sokoloff, M. Andrieux, R. Besancon, C. Pilon, M.-P. Martres, B. Giros, G.-C. Schwartz, Eur. J. Pharmacol. Mol. Pharmacol. Sect., 225, 331 (1992).
43. Chio, C.L., M.E. Lajiness, R.M. Huff, Mol. Pharmacol., 45, 51 (1994).
44. D.K. Hyslop, W.E. Hoffmann, M.W. Smith, and M.F. Piercey, Society for Neurosciences Annual Meeting, Washington, DC, Nov. 7-12, 1993.
45. W.E.Hoffmann, N.F.Nichols, M.F.Piercey, Society for Neurosciences Annual Meeting, Washington, DC, Nov. 7-12, 1993.
46. C. Sonesson, N. Waters, K. Svensson, A. Carlsson, M.W. Smith, M.F. Piercey, E. Meier, H. Wikstrom, J. Med. Chem., 36, 3188 (1993).
47. W.K. Brewster, D.E. Nichols, R.M. Riggs, D.M. Mottola, T.W. Lovenberg, M.H. Lewis, R.B. Mailman, J. Med. Chem., 33, 1756 (1990).
48. V.J. Watts, C.P. Lawler, T. Knoerzer, M. A. Mayleben, K.A. Neve, D.E. Nichols, R.B. Mailman, Eur. J. Pharmacol., 239, 271 (1993).
49. P. Seeman, Pharmacol. Rev., 32, 229 (1980).
50. J.C. Jaen, B.W. Caprathe, T.A. Pugsley, J. Med. Chem., 36, 3929 (1993).
51. SCRIP, No. 1889, 25 (1994).
52. N. Mohell, M. Sallemark, S. Rosqvist, A. Malmberg, T. Hogberg, Eur. J. Pharm., 238, 121 (1993).

53. D.M. Downing, J.L. Wright, T. Mirzadegan, L.D. Wise, R.G. MacKenzie, T.A. Pugsley, T.G. Heffner, 206th Nat. Am. Chem. Soc. Meeting, Chicago, IL, Aug. 25, 1993. Medi 172.

54. G. Stemp, M.S. Hadley, D.J. Nash, C.N. Johnson, Pat. Appl. WO 9403426 (1994).

55. A.B. Reitz, E.W. Baxter, D.J. Bennett, P.S. Blum, E.E. Codd, A.D. Jordan, B.E. Maryanoff, M.E. McConnell, M.E. Ortegon, C.R. Fasmussen, M.J. Renzi, M.K. Scott, R.P. Shank, J.L. Vaught, D.J. Wustrow, 206th Nat. Am. Chem. Soc. Meeting, Chicago, IL, Aug. 25, 1993. Medi 169.

56. W.P. Bowen, M.C. Coldwell, F.R. Hicks, G.J. Riley, R. Fears, Br. J. Pharmacol., 110, 93P (1993).

57. R.B. Fears, Pat. Appl. WO 9323035 (1993).

58. A. Schotte, P.F.M. Janssen, W. Gommeren, W.H.L.M. Luyten, J.E. Leysen, Eur. J. Pharmacol., 218, 373 (1992).

59. V.M. Ackenheil, H. Brau, A. Burkhart, A. Franke, W. Pasche, Arzneim. Forsch., 26, 1156 (1976); Clozaril: Summary of Preclinical and Clinical Data (Sandoz Canada, Inc., 1990).

60. A. Malmberg, D.M. Jackson, A. Eriksson, N. Mohell, Mol. Pharmacol., 43, 749 (1993).

61. SCRIP, No. 1864, 23 (1993).

62. P. Seeman, H.H.M. Van Tol, Eur. J. Pharmacol., 233, 173 (1993).

Chapter 6. Recent Progress in Excitatory Amino Acid Research

James A. Monn and Darryle D. Schoepp
Central Nervous System Research
Eli Lilly and Company
Indianapolis, Indiana 46285

Introduction - Tremendous progress has been made in outlining various pharmacological approaches that can be used to alter excitatory amino acid (EAA) neuronal transmission. Much new information has been in the realm of EAA molecular biology. In particular, molecular studies have now given us the primary structures of multiple EAA receptors. These receptor targets can be used in novel ways to study drug-receptor interaction. However, the complexity of EAA receptors at the molecular level, including their structural features and how this relates to the *in situ* situation, is somewhat overwhelming. EAA (or glutamate) receptors fall into two distinct families based on similar molecular structure, transduction mechanisms, and pharmacological characteristics. This includes ligand-gated ion channel glutamate receptors (iGluRs) (1-5) and G-protein coupled (metabotropic) glutamate receptors (mGluRs) (4-8) (see figure). Within each family are a number of receptor subtypes. Recombinant rat and human EAA receptors can be used to study drugs at the molecular level and more clearly define sites of drug interaction, as well as further our ability to investigate the subtype selectivity of EAA receptor agonists and antagonists. Ultimately, studies with recombinant EAA receptors must be related to what is known about drug-receptor interaction at native EAA receptors, since these are the targets for potential therapeutic agents.

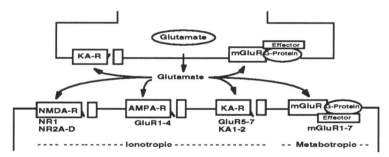

Figure. Hypothetical glutamate synapse showing the pre- and post-synaptic types of EAA receptors. Receptor proteins that have been cloned in each pharmacological class are listed.

Ionotropic Glutamate Receptors - iGluRs are intrinsic ligand-gated ion channels that when activated become permeable to cations (sodium and calcium). iGluRs fall into three pharmacologically distinct groups based on selective activation by N-methyl-D-aspartate (NMDA), α-amino-3-hydroxy-5-methyl-4-isoxazolepropanoic acid (AMPA), and kainate (figure). Receptor proteins corresponding to these pharmacological types have been cloned from rat (and in some cases mouse and human). Evidence suggests that each protein represents a subunit that is part of a heteromeric receptor complex. This is based on observations of more functional receptor channel properties when multiple subunit proteins having relatively high sequence homology are co-expressed (1-4). Both competitive and allosteric sites for drug interaction have now been described for NMDA and AMPA receptors. In addition to the competitive site at which NMDA (or glutamate) bind to the NMDA receptor complex, there are glycine, polyamine, and PCP binding sites. A number of compounds have now been described that act selectively at these different sites to alter NMDA receptor transduction. Both positive and negative allosteric interactions have also been described at AMPA receptors. A number of competitive antagonists that act at the AMPA recognition site are known. 2,3-Benzodiazepines such as GYKI 52466 act at a unique site (GYKI site) to selectively inhibit AMPA receptors by a non-competitive mechanism (9-11), and the compounds aniracetam and cyclothiazide act as positive allosteric modulators to enhance AMPA receptor transduction (10,12-15).

Metabotropic Glutamate Receptors - mGluRs are now recognized as a novel heterogenous family of G-protein coupled receptors with seven membrane spanning motifs but no homology with other such receptors (4-8). To date, seven subtypes termed mGluR1-7, and alternative splice versions of mGluR1 (α,β, c) and mGluR5 (a and b) are known. Subtypes which are coupled to phosphoinositide hydrolysis include mGluR1 and mGluR5. Other subtypes (mGluR2-3, mGluR6-7) are negatively linked to cAMP formation when expressed in non-neuronal cells. mGluRs are likely present pre- and post-synaptically where they function to modulate excitatory synaptic transmission by a variety of cellular mechanisms. This includes modulation of ion channels (K^+, Ca^{++}), enhancement of iGluR currents, induction of synaptic plasticity, and presynaptic regulation of glutamate release (4-8).

NEW PHARMACOLOGICAL AGENTS

NMDA Agonists - The preparation and pharmacological activity of D,L-(tetrazol-5-yl) glycine (1) has been disclosed (16,17). Compound 1 is the most potent NMDA agonist yet described, and is a potent convulsant following systemic administration in rodents. 4-Methylene-L-glutamic acid (2) is reported to be a potent and selective NMDA agonist, whereas 4-spirocyclopropyl-L-glutamate and 4,4-dimethyl-L-glutamate display mixed NMDA/non-NMDA receptor agonist activity in the newborn rat spinal cord (18). 2S,1'R,2'R,3'R-2,3-Dicarboxycyclopropylglycine (DCG-1/4, 3), a hybrid structure of the selective NMDA agonist, L-CCG-IV and the selective metabotropic glutamate receptor agonist, L-CCG-I, was demonstrated to be a mixed NMDA-metabotropic receptor agonist (19).

1   2   3

Competitive NMDA Antagonists - The first successful attempt to replace the α-amino acid moiety of a glutamate antagonist has been reported. A 3,4-diamino-3-cyclobutene-1,2-dione functionality was successfully utilized as the α-amino acid isostere in 4, a potent, highly selective and systemically-active NMDA receptor antagonist (20). Bicyclic derivatives (e.g., 5) display increased receptor affinity relative to 4 (21). Incorporation of a ketone functionality into the carbon backbone of 2-amino-5-phosphonovaleric acid (AP5) afforded (R)-4-oxo-5-phosphononorvaline (MDL 100,453; 6), an orally active NMDA antagonist possessing increased potency relative to AP5 (22). Introduction of a methyl substituent into the carbon chain of 6 resulted in severely diminished NMDA receptor binding (23). However, replacement of the ketone with an oxime group afforded analogs 7 and 8 which possess appreciable NMDA receptor binding affinity (23). Conformational restriction of 6 into a cis-2,3-disubstituted piperidine nucleus also resulted in potent NMDA receptor affinity for MDL

100,925 (**9**) (24,25). A conformation-stabilizing internal hydrogen bond between the protonated amino functionality and the ketone oxygen of **9** has been suggested as a possible explanation for the enhanced binding affinity of this analog (25). Introduction of a *trans*-4-methyl substituent into **9** results in significantly enhanced receptor affinity for **10** (26). The enzymatic asymmetric synthesis of NPC 17742 (**11**) was reported (27). Compound **11** is a potent, selective and systemically-active NMDA antagonist (27-30).

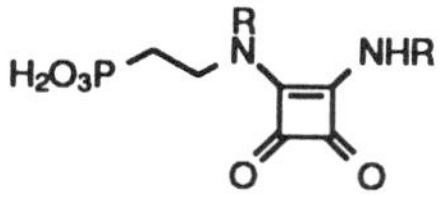

4: R = H
5: R = -(CH₂)₃-

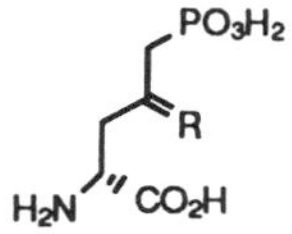

6: R = O
7: R = NOCH₃
8: R = NOCH₂Ph

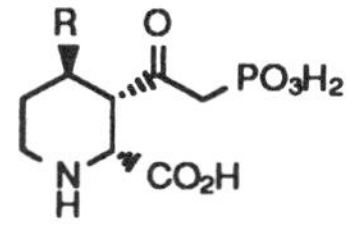

9: R = H
10: R = CH₃

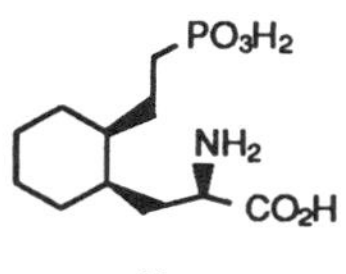

11

A series of quinoxaline-spaced phosphono α-amino acids have been reported (31). From this group, the 7-chloro derivative **12** was identified as a potent and selective competitive NMDA antagonist. Interestingly, substitution by chlorine at the 6-position as in **13** resulted in appreciable affinity for this analog at the glycine co-agonist site (31). Potential photoaffinity labeling agents (e.g., **14**) based on CGP37849 were prepared; they demonstrated potent binding affinity at the NMDA recognition site in spite of the sterically imposing alkylbenzamide-containing side chain (32). Similarly, incorporation of a phenyl substituent on the aromatic ring of **15** (33,34) as in SDZ EAB 515 (**16**) resulted in increased NMDA receptor binding affininty for this compound (35).

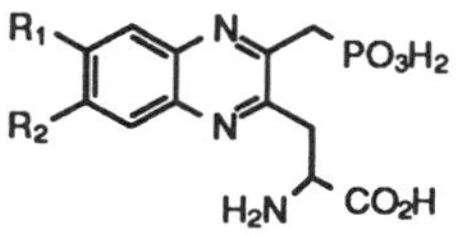

12: R₁ = H, R₂ = Cl
13: R₁ = Cl, R₂ = H

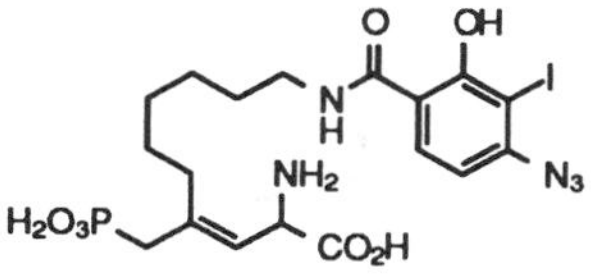

14

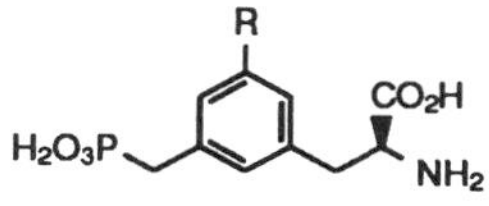

15: R = H
16: R = C₆H₅

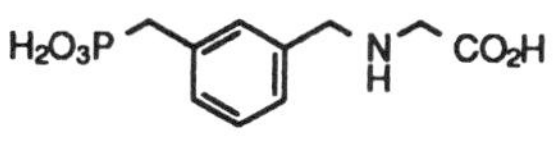

17

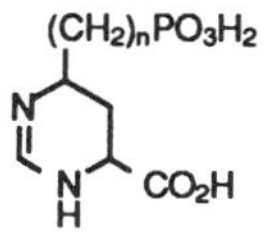

18: n = 3
19: n = 1

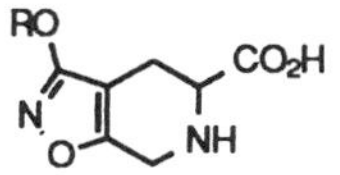

20: R = H
21: R = CH₂CO₂H

ω-Phosphono α-amino acids in which the phosphonic acid moiety is linked to the amino acid via the nitrogen atom have been prepared in an attempt to further define the glutamate recognition site. From this study, several moderately active NMDA antagonists (e.g., **17**) were identified and a pharmacophore model for antagonist binding was proposed (36). A tetrahydropyrimidine nucleus was employed as a surrogate for the six membered rings present in CGS19755 and CPP (37,38). In contrast to the similar NMDA receptor affinities observed for these agents, tetrahydropyrimidine-based ω-phosphono α-amino acid **18** was an order of magnitude more potent than **19**. The poor activity of **19** was postulated to be due to an intramolecular hydrogen bond between the phosphonic acid and the imino nitrogen of the pyrimidine ring which holds the phosphonate side chain in a non-active conformation (38). In an attempt to convert the potent AMPA receptor agonist 5-HPCA (**20**) into an AMPA antagonist by addition of an acetic acid chain to the hydroxyl group of the isoxazole ring, the moderately active, selective NMDA antagonist **21** was obtained (39). Common conformations for competitive agonists (40) as well as for competitive antagonists (36,40,41) have been addressed by molecular modeling techniques, and models for the complementary glutamate recognition site have been proposed. D-CPP-ene was evaluted as an anticonvulsant in eight patients with intractable complex partial seizures; however, all of the patients withdrew from treatment due to severe side effects produced by this compound (42).

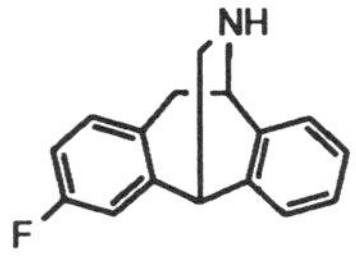

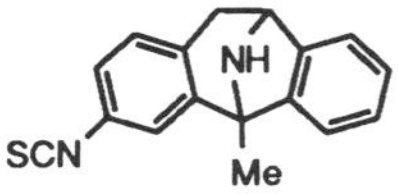

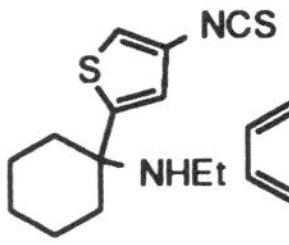

<u>Noncompetitive NMDA Antagonists</u> - A structure-activity relationship study for iminomethanobenzocycloheptenes related to MK801 was performed. Compound **22** was identified as the most potent analog from this series (43). MK801 3-isothiocyanate **23** was characterized as the most potent and highly selective site-directed irreversible agent for the MK-801 binding site reported to date (44). Isothiocyanate **24** has also been prepared (45), but is considerably less potent than **23**. A tetrahydroisoquinoline variant of MK801, FR115427 (**25**), was shown to be a relatively potent and selective NMDA antagonist (46). Related spirotetrahydroisoquinoline **26** (47) and spiroisoindoline **27** (48) were also reported. An extensive structure-activity relationship study of tricyclic (e.g., PD134365, **28**) (49) and tetracyclic (e.g., **29**) (50) 4a-phenanthrenamine derivatives was undertaken. Neuroprotective diarylalkylamines (e.g., **30**) (51) and the diarylalkylamine prodrug remacemide (**31**) (52) which interact at the NMDA channel have been disclosed. High affinity binding of [^{3}H]dextrorphan to the NMDA receptor channel was characterized (53). A series of imipramine analogs were also shown to interact with this site, though with low affinity (54). The antiparkinsonian agent memantine was found to potentiate the anticonvulsant activity of valproate and to protect against NMDA-receptor mediated neurotoxicity (55,56). Another antiparkinsonian agent, budipine, was also characterized as a low affinity non-competitive NMDA antagonist (57).

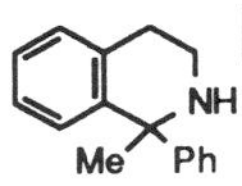

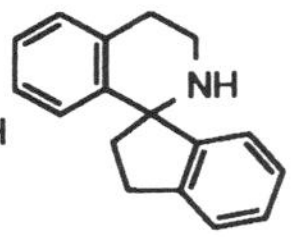

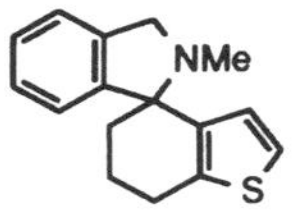

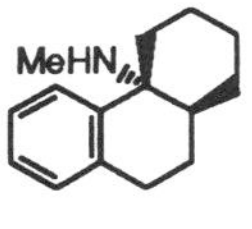

<u>Glycine Antagonists</u> - The 5-nitro derivative of 6,7-dichloroquinoxalinedione acid (ACEA 1021, **32**) has been disclosed as a highly potent, systemically-active glycine antagonist with efficacy in both maximal electroshock and focal ischemia models (58,59). A number of 3-substituted indole-2-carboxylates (e.g., **33**, **34**) have been claimed as potent glycine-site antagonists (60-65). It seems clear that a large lipophilic pocket exists at the glycine recognition site adjacent to the 3-postion of the indole nucleus. Reduction of the indole nucleus completely abolishes affinity for the glycine site (66), while 4,6-dichlorobenzimidazole-2-carboxylate retains good receptor affinity (67). *trans*-4-Substituted tetrahydroquinoline-2-carboxylates (e.g., L-689,560, **35**) are exceptionally potent glycine antagonists (68,69). A radioligand based on **35** has been characterized (70) and utilized in whole brain autoradiography (71). Benzylamine- and phenylalanine-containing derivatives

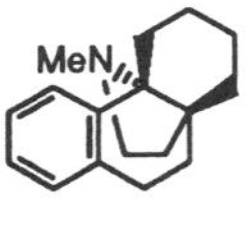

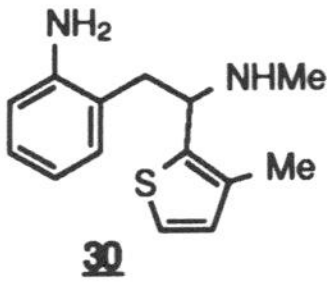

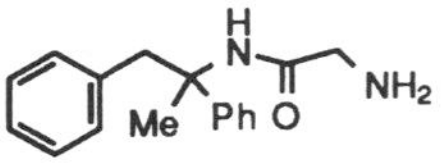

(**36** and **37**, respectively) retain high binding affinity and are systemically active in the DBA/2 mouse seizure model (72).

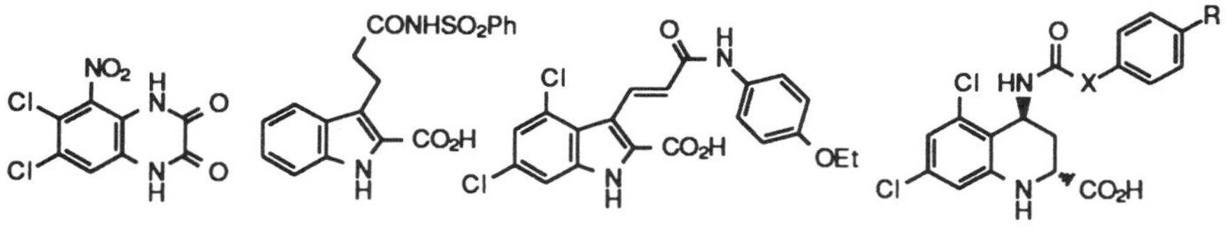

32   **33**   **34**   **35**: X = NH, R = H
                                                36: X = CH$_2$, R = CH$_2$NH$_2$
                                                37: X = CH$_2$, R = CH$_2$CH(NH$_2$)CO$_2$H

Systemic activity was also conferred on 5,7-dichlorokynurenic acid (**38**) by replacement of the 2-carboxylic acid with various dialkylaminoalkyl esters (e.g., **39**) (73). Substitution on the 3-position of **38** with a carboxymethyl side chain resulted in increased receptor binding affinity for **40** (74). 3-Substituted-4-hydroxyquinolones (e.g., **41**, **42**) bind with high affinity to the glycine site (75-77), and **42** shows good anticonvulsant activity following systemic administration (77). Isosteric pyrrolo-pyridazinone derivatives (e.g., **43**) have been claimed (78). Potent non-selective glycine/AMPA antagonist activity was observed for the systemically-active 3-nitro-7-chlorotetrahydroquinolone suggesting the importance of C4-substitution on EAA receptor selectivity (79). Similar non-selective antagonist activity was observed for acylsulfonamide **44** (80), and for carboxymethyl-substituted quinoxalinedione **45** (81). Substituted monocyclic and bicyclic derivatives of the low efficacy glycine partial agonist HA-966 have been reported (82). None are more potent than the *cis*-4-methyl derivative L-687,414, which was shown in a separate study to be anticonvulsant in baboons (83).

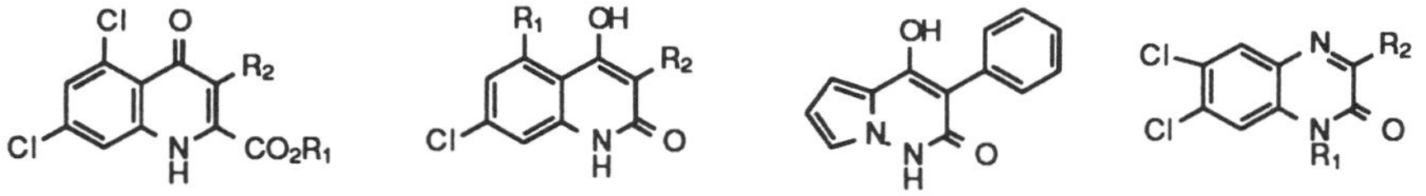

38: R$_1$ = R$_2$ = H   **41**: R$_1$ = Cl, R$_2$ = Ph   **43**   **44**: R$_1$ = H, R$_2$ = CONHSO$_2$Ph
39: R$_1$ = (CH$_2$)$_2$N(CH$_2$CH$_3$)$_2$, **42**: R$_1$ = H, R$_2$ = C(O)C$_3$H$_5$ **45**: R$_1$ = CH$_2$CO$_2$H, R$_2$ = OH
          R$_2$ = H
40: R$_1$ = H, R$_2$ = CH$_2$CO$_2$H

Polyamine Antagonists - The putative polyamine antagonist eliprodil (SL82.0715, **46**) was found to protect cultured hippocampal neurons from NMDA toxicity (84) and was active in a fluid percussion brain trauma model in rats when administered systemically up to 18 hours after injury (85). Furthermore, doses of **46** threefold higher than those protective in the traumatic injury model did not produce cytoplasmic vacuolization in neruons of the retrosplenial and posterior cingulate cortices (86). Oxindole variants (e.g., **47**) of **46** have been prepared (87). Compound **47** shows enhanced potency compared to **46** in protection of cultured rat cortical neurons toward glutamate toxicity, as well as improved selectivity for this activity when compared to its α_1-adrenergic receptor binding affinity (87). Other variations of **46** have appeared in the patent literature (88,89), as has the indole-based polyamine antagonist **48** (90). The constrained polyamine **49** appears to act as a polyamine antagonist *in vitro* (91). N-(3-Aminopropyl)1,10-diaminodecane has been reported to act as a mixed partial agonist/antagonist at the polyamine site and to alter the properties of the glutamate recognition site (92).

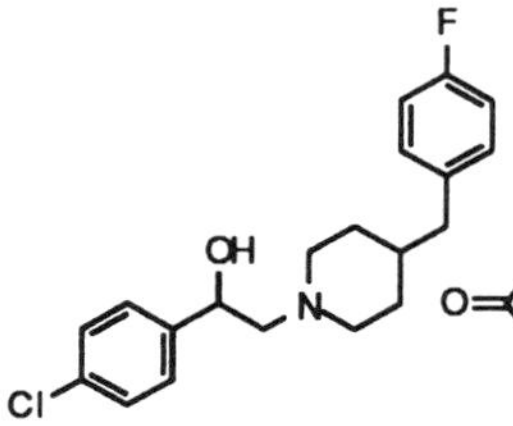

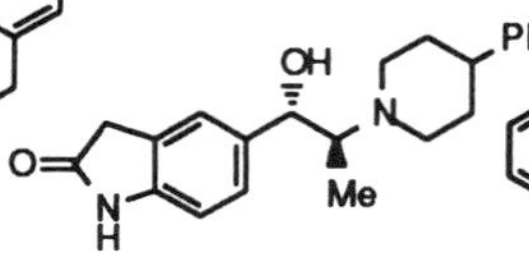

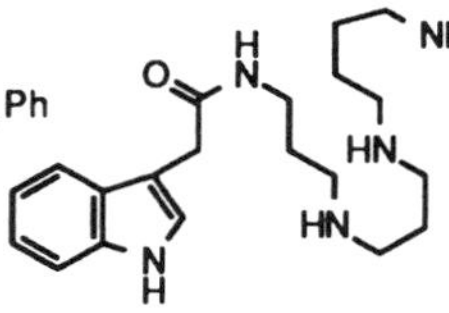

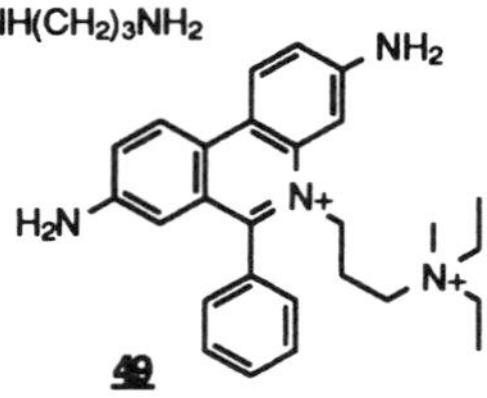

46   47   48   49

AMPA Agonists - Trifluoro-AMPA (**50**) was prepared and shown to be a potent AMPA agonist in rat cortical slice neurons *in vitro* , though slightly less potent than AMPA itself (93). Replacement of the hydroxyl functionality present on the isoxazole ring of AMPA by a chlorine atom as in **51** resulted in a complete loss of binding affinity for this compound at AMPA receptors, underscoring the requirement for an acidic group at this position (93). In accord with this hypothesis, substitution of the hydroxyl functionality by a carboxylic acid group as in ACPA (**52**) resulted in an increase in AMPA receptor affinity and agonist potency for this compound relative to AMPA itself (94). A number of isoxazolones related to the selective AMPA agonist TAN-950 A (**53**) were evaluated for their binding affinity and agonist potency at AMPA, kainate and NMDA receptors. None were as potent as **53** in their receptor affinity, though **54** was slightly more potent than **53** as an agonist in rat hippocampal neurons (95). Tetrazolylproline **55** was demonstrated to be a relatively weak, though highly selective and systemically active AMPA agonist *in vivo* (96).

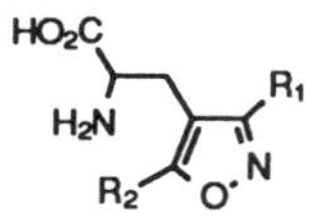

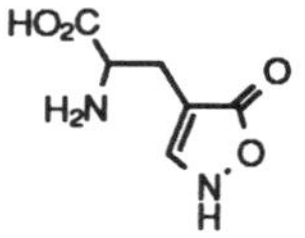

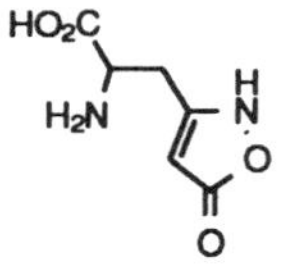

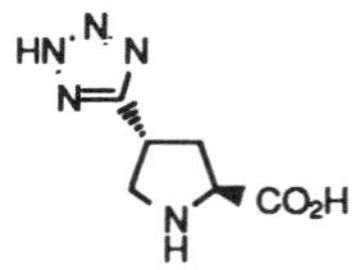

50: R_1 = OH, R_2 = CF_3 **53**   **54**   **55**
51: R_1 = Cl, R_2 = CH_3
52: R_1 = CO_2H, R_2 = CH_3

AMPA Positive Allosteric Modulators - Aniracetam (**56**) was found to reversibly potentiate responses to L-glutamate, potentially through the reduction of receptor desensitizaiton (97). Its effects on long term potentiation have been studied (98). Cyclothiazide (**57**) was similarly shown to reduce AMPA receptor desensitization, though at much lower concentrations than those required for aniracetam (99). This effect appears to be specific for AMPA receptors, as cyclothiazide does not modulate kainate receptor mediated currents in hippocampal neurons (15).

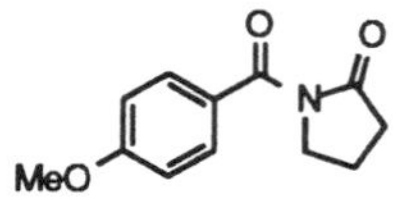

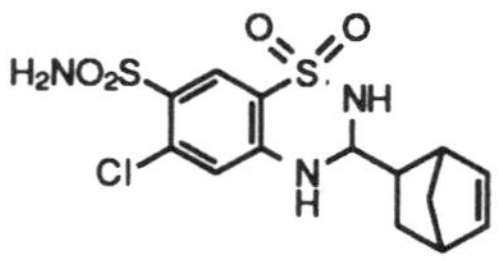

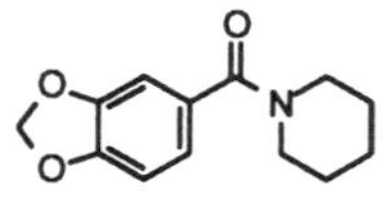

56   57   58

Cyclothiazide decreases [3H]AMPA binding *in vitro* (100), reverses AMPA receptor antagonism by the allosteric antagonist GYKI53655 (12, *vide infra*) and was shown to unmask AMPA excitotoxicity in cultured hippocampal neurons (101). The so-named AMPAkines (e.g., **58**) have been shown to increase glutamatergic transmission in the hippocampus and to improve the memory encoding in rats following systemic administration (102,103).

AMPA Negative Allosteric Modulators - The 2,3-benzodiazepine GYKI 52466 (**59**) was further characterized as a novel, selective noncompetitive antagonist of AMPA receptors (9,10,104-110). GYKI 52466 was more potent than NBQX in blocking both electroshock- and chemically-induced seizures in mice (104). In addition, **59** was found to protect hippocampal neurons from kainate toxicity *in vitro* (105) and showed efficacy in reducing the

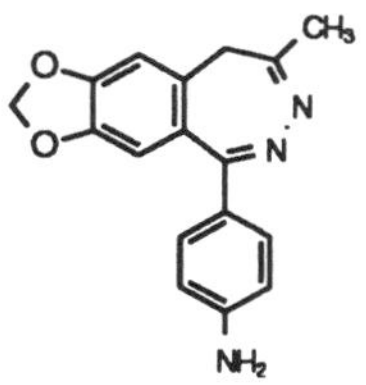

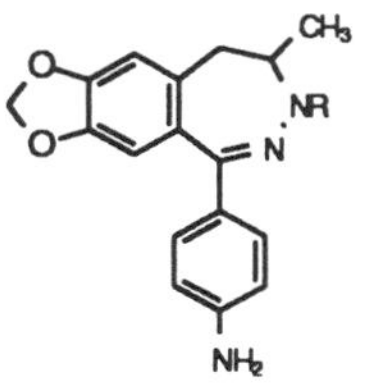

60: R = COCH$_3$
61: R = CONHCH$_3$

infarction volume following focal ischemia in rats (106,107). There is some evidence to suggest that the cerebroprotective effect of **59** may be due in part to an inhibition of glutamate release (108), though this is has been disputed (109). Antagonism of AMPA receptors by **59** also blocked MK801-induced locomotor stimulation as well as MK801-induced reversal of haloperidol catalepsy in rats (110). A structure-activity relationship study for 2,3-benzodiazepines has been performed (111), resulting in the identification of more potent analogs (e.g., **60**, **61**). Specific binding of [^{3}H] **60** has been observed in *Xenopus* brain membranes (112).

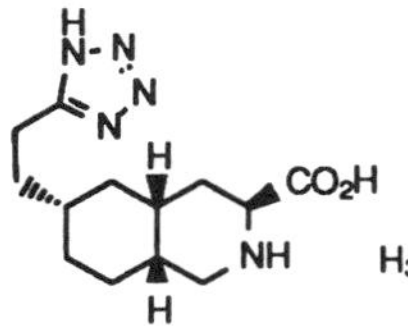

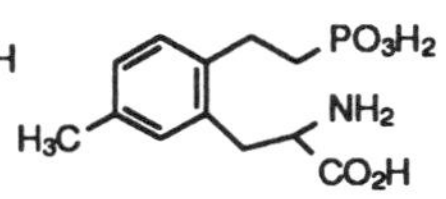

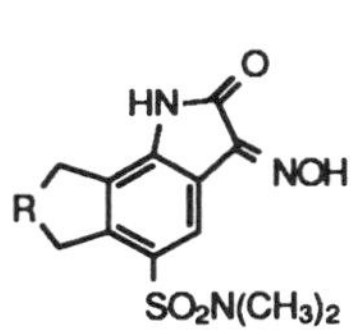

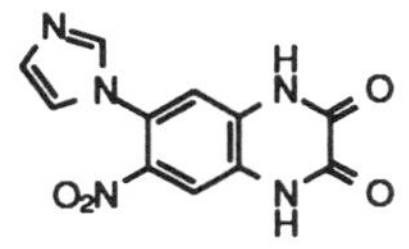

64: R = -CH$_2$CH$_2$-
65: R = N-CH$_3$

AMPA Competitive Antagonists - Decahydroisoquinoline derivative **62** was disclosed as a potent, selective and systemically-active AMPA antagonist capable of blocking ATPA-induced rigidity in mice (113). A structure-activity relationship study was performed on 2-phosphonoethylphenylalanines from which the systemically-active 5-methyl analog **63** was identified as the most potent member of this series (114). Compound **63** was effective in blocking PTZ-induced seizures in mice and kainate-induced striatal toxicity in rats. Isatin oximes (e.g., **64**, **65**) were also identified as potent, systemically active AMPA antagonists (115,116). Notably, antagonist activity in the AMPA-induced seizure model was observed following oral administration of **64**. Imidazole-substituted quinoxalinedione **66** (YM-90K) has been disclosed as a potent and selective AMPA antagonist with systemic activity against audiogenic seizures in mice (117) and ischemia-induced neuronal death in gerbils (118) and rats (119). Centrally-administered CNQX reduced knee joint inflammation and behavioral manifestations of arthritis in rats, suggesting a role for central AMPA receptors in the development of peripheral joint inflammation (120). NBQX has been extensively studied as an anticonvulsant (104,121) and neruoprotectant (122-124). Additionally, NBQX has been demonstrated to decrease mechanical hypersensitivity in rats arising from transient spinal cord ischemia (125), supporting the hypothesis that central AMPA receptors may be involved in the central processing of peripheral nociceptive stimulii.

<u>Kainate Agonists</u> - Pyridine derivatives (e.g., <u>**67**</u>) of acromelic acid have been prepared and were shown to be potent agonists in isolated rat spinal neurons (126). Methoxyphenyl and hydroxyphenyl derivatives (e.g., <u>**68**</u>, <u>**69**</u>) of kainic acid were found to be highly selective and potent ligands for the kainate receptor (127). Configurational variants of <u>**69**</u> have been prepared (128). Isosteric substitution of the ω-carboxylate functionality of kainic acid by a phosphonic acid as in <u>**70**</u> severely reduced the affinity of this analog for kainate receptors *in vitro* (129). The effect of dose of the potent kainate agonist, domoic acid, on hippocampal degeneration in monkeys was examined (130). Amnesic shellfish poisoning of 107 humans who injested mussels containing domoic acid has been reviewed (131). Concanavalin A was shown to selectively inhibit desensitization of kainate-preferring glutamate receptors (99).

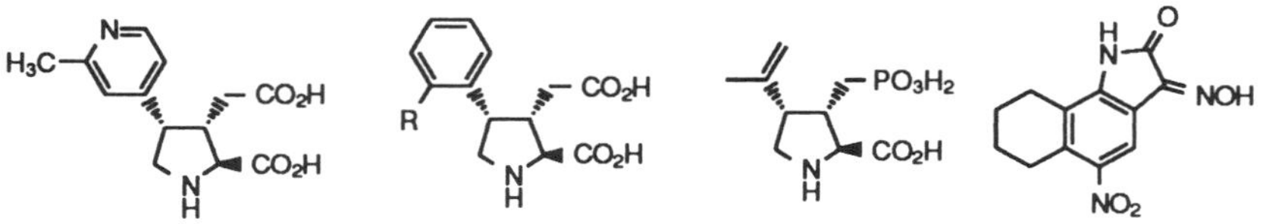

67	**68**: R = OCH₃ **69**: R = OH	**70**	**71**

<u>Kainate Antagonists</u> - The isatin oxime NS-102 (<u>**71**</u>) was shown to be a highly selective and potent ligand for the low-affinity kainate binding site *in vitro*, and to block domoate-induced depolarizations and Ca^{2+} influx in rat cortical neurons (132).

<u>Metabotropic Agonists</u> - 1*S*,3*R*-ACPD (<u>**72**</u>) was demonstrated to be the optimal configurational isomer for the stimulation of phosphoinositide hydrolysis (133,134) in rat brain slices, while for inhibition of forskolin-stimulated c-AMP formation, both the 1*S*,3*R*- and 1*S*,3*S*-isomers display similar agonist potency (134,135). A number of accounts describing the behavioral effects evinced by <u>**72**</u> upon direct administration into the brain have appeared (136-140). Like <u>**72**</u>, L-CCG-I (<u>**73**</u>) is a nonspecific metabotropic glutamate receptor agonist (141-143), although it shows a high degree of selectivity for mGluR2 over mGluR1 and mGluR4 receptors (143). In rat striatal slices, <u>**73**</u> produced a reduction of K^{+} induced efflux of D-[^{3}H]-aspartate and an inhibition of forskolin-stimulated c-AMP accumulation at similar concentrations (144). A pharmacophore model of metabotropic glutamate receptors based on superimposition of <u>**72**</u> and <u>**73**</u> has been proposed (145). *trans*-Azetidine-2,4-dicarboxylate (<u>**74**</u>) is a selective metabotropic agonist which stimulates phosphotidyl inositol (PI) hydrolysis in rat cerebellar granual cells, but not does not activate mGluR1 (146). 3,5-Dihydroxyphenylglycine (<u>**75**</u>) is a potent metabotropic agonist which is selective for PI-linked mGluRs (147,148). Serine-O-phosphate (SOP) and N-acetylaspartylglutamate (NAAG) have each been shown to selectively activate metabotropic receptors which are negatively coupled to adenylyl cyclase (149,150). In the case of SOP, selective activation of rat mGluRIV has been demonstrated (149).

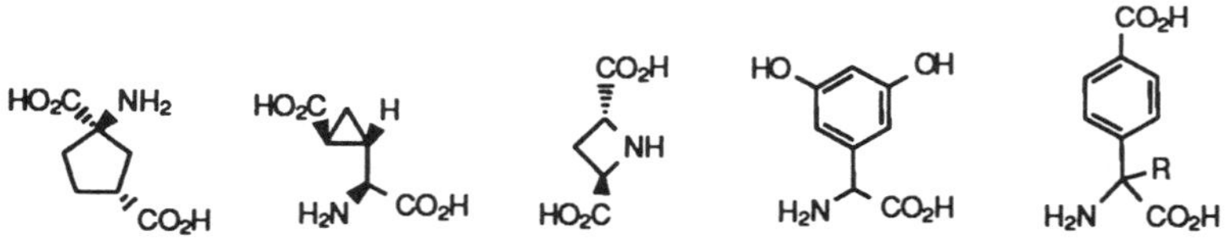

72	**73**	**74**	**75**	**76**: R = H **77**: R = CH₃

<u>Metabotropic Antagonsts</u> - S-4-Carboxyphenylglycine (<u>7 6</u>) and RS-α-methyl-4-carboxyphenylglycine (<u>77</u>) competitively antagonize the PI stimulant and neuroexcitatory effects of <u>72</u> (151-153). Compound <u>77</u> also reduced the transient inward Ca^{2+} current in CA1 pyramidal neurons (154), suggesting that metabotropic receptors play a role in neuronal signalling within the hippocampus, although this compound did not prevent the induction of LTP nor did it antagonize ACPD-induced currents in these cells (155). However, <u>77</u> was effective in preventing the LTP-induced increased post-synaptic AMPA sensitivity in hippocampal slices following tetanic stimulation, suggesting a role for metabotropic receptors in the maintenance of LTP (156). L-Aspartate-β-hydroxamate was shown to be a relatively potent and selective antagonist of ibotenate-stimulated PI hydroysis, although this effect does not appear to be competitive or reversible (157).

REFERENCES

1. G.P. Gasic and M. Hollman, Annu. Rev. Physiol., <u>54</u>, 507 (1992).
2. B. Sommer and P.H. Seeburg, Trends in Pharmacol. Sci., <u>13</u>, 291 (1992).
3. P. Seeburg, Trends in Neurosci., <u>16</u>, 359 (1993).
4. S. Nakanishi, Science, <u>258</u>, 597 (1992).
5. J.M. Barnes and J.M. Henley, Prog. Neurobiol., <u>39</u>, 113 (1992).
6. P.J. Conn and M.A. Desai, Drug Development Res., <u>24</u>, 207 (1991).
7. D.D. Schoepp and P.J. Conn, Trends in Pharmacol. Sci., <u>14</u>, 13 (1993).
8. J. Bockaert, J. Pin, and L. Fagni, Fundam. Clin. Pharmacol., <u>7</u>, 473 (1993).
9. S.D. Donevan and M.A. Rogawski, Neuron, <u>10</u>, 51 (1993).
10. C.F. Zorumski, K.A. Yamada, M.T. Price, and J.W. Olney, Neuron, <u>10</u>, 61 (1993).
11. I. Tarnawa, S. Farkas, P. Bersenyi, A. Pataki, and F. Andrasi, Eur. J. Pharmacol., <u>167</u>, 193 (1990).
12. A.J. Palmer and D. Lodge, Eur. J. Pharmacol.-Mol. Pharmacol. Section, <u>244</u>, 193 (1993).
13. K. Tsuzuki, T. Takeuchi, and S. Ozawa, Mol. Brain Res., <u>16</u>, 105 (1992).
14. K.M. Partin, D.K. Patneau, C.A. Winters, M.L. Mayer, and A. Buonanno, Neuron, <u>11</u>, 1069 (1993).
15. L.A. Wong and M.L. Mayer, Mol. Pharmacol., <u>44</u>, 504 (1993).
16. W. H. W. Lunn, D. D. Schoepp, D. O. Calligaro, R. T. Vasileff, L. J. Heinz, C. R. Salhoff and P. J. O'Malley, J. Med. Chem., <u>35</u>, 4608 (1992).
17. D. D. Schoepp, C. L. Smith, D. Lodge, J. D. Millar, J. D. Leander, A. I. Sacaan and W. H. W. Lunn, Eur. J. Pharmacol., <u>203</u>, 237 (1993).
18. O. Ouerfelli, M. Ishida, H. Shinozaki, K. Nakanishi and Y. Ohfune, Syn. Lett., 409 (1993).
19. Y. Ohfune, K. Shimamoto, M. Ishida and H. Shinozaki, BioMed. Chem. Lett., <u>3</u>, 15 (1993).
20. W. A. Kinney, N. E. Lee, D. T. Garrison, E. J. Podlesny Jr., J. T. Simmonds, D. Bramlett, R. R. Notvest, D. M. Kowal and R. P. Tasse, J. Med. Chem., <u>35</u>, 4720 (1992).
21. W. A. Kinney and D. C. Garrison, Eur. Patent Application 496,561 A2 (1992).
22. J. P. Whitten, B. M. Baron, D. Muench, F. Miller, H. S. White and I. A. McDonald, J. Med. Chem., <u>33</u>, 2961 (1990).
23. J. P. Whitten, B. M. Baron and I. A. McDonald, BioMed. Chem. Lett., <u>3</u>, 23 (1993).
24. J. P. Whitten, D. Muench, R. V. Cube, P. L. Nyce, B. M. Baron and I. A. McDonald, Biomed. Chem. Lett., <u>1</u>, 441 (1991).
25. A. Claesson, B.-M. Swahn, K. M. Edvinsson, H. Molin and M. Sandberg, BioMed. Chem. Lett., <u>2</u>, 1247 (1992).
26. J. P. Whitten, R. V. Cube, B. M. Baron and I. A. McDonald, BioMed. Chem. Lett., <u>3</u>, 19 (1993).
27. G. S. Hamilton, Z. Huang, R. J. Patch, B. A. Narayanan and J. W. Ferkany, BioMed. Chem. Lett., <u>3</u>, 27 (1993).
28. J. W. Ferkany, J. Willetts, S. A. Borosky, D. B. Clissold, E. W. Karbon and G. S. Hamilton, BioMed. Chem. Lett., <u>3</u>, 33 (1993).
29. J. W. Ferkany, G. S. Hamilton, R. J. Patch, Z. Huang, S. A. Borosky, D. L. Bednar, B. E. Jones, R. Zubrowski, J. Willetts and E. W. Karbon, J. Pharmacol. Exp. Ther., <u>264</u>, 256 (1993).
30. J. Willetts, D. B. Clissold, T. L. Hartman, R. R. Brandsgaard, G. S. Hamilton and J. W. Ferkany, J. Pharmacol. Exp. Ther., <u>265</u>, 1055 (1993).
31. R. B. Baudy, L. P. Greenblatt, I. L. Jirkovsky, M. Conklin, R. J. Russo, D. R. Bramlett, T. A. Emrey, J. T. Simmonds, D. M. Kowal, R. P. Stein and R. P. Tasse, J. Med. Chem., <u>36</u>, 331 (1993).
32. R. Heckendorn, H. Allgeier, J. Baud, W. Gunzenhauser and C. Angst, J. Med. Chem. <u>36</u>, 3721 (1993).
33. C. F. Bigge, J. T. Drummond, G. Johnson, T. Malone, A. W. Probert, Jr., F. W. Marcoux, L. L. Coughenour and L. J. Brahce, J. Med. Chem., <u>32</u>, 1580 (1989).

34.   C. F. Bigge, J. T. Drummond, G. Johnson, T. Malone, A. W. Probert, Jr., F. W. Marcoux, L. L. Coughenour and L. J. Brahce, J. Med. Chem., 32, 2583 (1989).

35.   W. Müller, D. A. Lowe, H. Neijt, S. Urwyler, P. L. Herrling, D. Blaser and D. Seebach, Helv. Chim. Acta, 75, 855 (1992).

36.   C. F. Bigge, G. Johnson, D. F. Ortwine, J. T. Drummond, D. M. Retz, L. J. Brahce, L. L. Coughenour, F. W. Marcoux and A. W. Probert, Jr., J. Med. Chem., 35, 1371 (1992).

37.   C. F. Bigge, J.-P. Wu and J. R. Drummond, Tetrahedron Lett., 32, 7659 (1991).

38.   C. F. Bigge, J.-P. Wu, J. T. Drummond, L. L. Coughenour and C. M. Hanchin, BioMed. Chem. Lett., 2, 207 (1992).

39.   U. Madsen, L. Andresen, G. A. Poulsen, T. B. Rasmussen, B. Ebert, P. Krogsgaard-Larsen and L. Brehm, BioMed. Chem. Lett., 3, 1649 (1993).

40.   D. F. Ortwine, T. C. Malone, C. F. Bigge, J. T. Drummond, C. Humblet, G. Johnson and G. W. Pinter, J. Med. Chem., 35, 1345 (1992).

41.   J. P. Whitten, B. L. Harrison, H. J. R. Weintraub and I. A. McDonald, J. Med. Chem., 35, 1509 (1992).

42.   S. Sveinbjornsdottir, J. W. A. S. Sander, D. Upton, P. J. Thompson, P. N, Patsalos, D. Hirt, M. Emre, D. Lowe and J. S. Duncan, Epilepsy Res., 16, 165 (1993).

43.   K. R. Gee, P. Barmettler, M. R. Rhodes, R. N. McBurney, N. L. Reddy, L-Y. Hu, R. E. Cotter, P. N. Hamilton, E. Weber and J. F. W. Keana, J. Med. Chem., 36, 1938 (1993).

44.   J. T. M. Linders, J. A. Monn, M. V. Mattson, C. George, A. E. Jacobson and K. C. Rice, J. Med. Chem., 36, 2499 (1993).

45.   S. O. Casalotti, A. P. Kozikowski, A. Fauq, W. Tückmantel and K. E. Krueger, J. Pharmacol. Exp. Ther., 260, 21 (1992).

46.   J. P. Hodgkiss, H. J. Sherriffs, D. A. Cottrell, K. Shirakawa, J. S. Kelly, A. Kuno, M. Ohkubo, S. P. Butcher and H. J. Olverman, Eur. J. Pharmacol., 240, 219 (1993).

47.   R. C. Griffith, J. R. Matz and J. J. Napier, WO Patent 9203420 (1992).

48.   T. M. Bare, C. W. Draper, C. D. McLaren, L. M. Pullan, J. Patel and J. B. Patel, BioMed. Chem. Lett., 3, 55 (1993).

49.   C. F. Bigge, T. C. Malone, S. J. Hays, G. Johnson, P. M. Novak, L. J. Lescosky, D. M. Retz, D. F. Ortwine, A. W. Probert Jr., L. L. Coughenour, P. A. Boxer, L. J. Robichaud, L. J. Brahce and J. L. Shillis, J. Med. Chem., 36, 1977 (1993).

50.   T. C. Malone, D. F. Ortwine, G. Johnson and A. W. Probert, Jr., BioMed. Chem. Lett., 3, 49 (1993).

51.   U. Elben, H. Anagnostopulus and K. Rudolphi, Eur. Pat. 0516978 (1992).

52.   R. C. Griffith and J. J. Napier Eur. Pat. 0427427 A (1991).

53.   P. H. Franklin and T. F. Murray, Mol. Pharmacol., 41, 134 (1992).

54.   L. A. McQuaid, J. D. Leander, L. G. Mendelsohn, E. C. R. Smith, R. R. Lawson and N. R. Mason, Res. Commun. Chem. Pathol. Pharmacol., 77, 171 (1992).

55.   E. Urbanska, M. Dziki, S. J. Czuczwar, Z. Kleinrok and W. A. Turski, Neuropharmacol., 31, 1021 (1992).

56.   H.-S. Vincent Chen, J. W. Pellegrini, S. K. Aggarwal, S. Z. Lei, S. Warach, F. E. Jensen and S. A. Lipton, J. Neurosci., 12, 4427 (1992).

57.   T. Klockgether, P. Jacobsen, P.-A. Löschmann and L. Turski, J. Neural Transm. [P-D Sect], 5, 101 (1993).

58.   S. X. Cai, C. Dinsmore, K. R. Gee, A. G. Glenn, I. C. Huang, B. L. Johnson, S. M. Kher, Y. Lu, P. L. Oldfield, P. Marek, H. Zheng, E. Weber and J. F. W. Keana, Soc. Neurosci. Abst., 718: 296.11 (1993).

59.   R. M. Wookward, J. E. Huettner, J. Guastella and E. Weber, Soc. Neurosci. Abst., 718: 296.14 (1993).

60.   F. G. Salituro, B. L. Harrison, B. M. Baron, P. L. Nyce, K. T. Stewart, J. H. Kehne, H. S. White and I. A. McDonald, J. Med. Chem., 35, 1791 (1992).

61.   W. F. Hood, N. M. Gray, M. S. Dappen, G. B. Watson, R. P. Compton, A. A. Cordi, T. H. Lanthorn and J. B. Monahan, J. Pharmacol. Exp. Ther., 262, 654 (1992).

62.   T. S. Rao, N. M. Gray, M. S. Dappen, J. A. Cler, S. J. Mick, M. R. Emmett, S. Iyengar, J. B. Monahan, A. A. Cordi and P. L. Wood, Neuropharmacol., 32, 139 (1993).

63.   A. Cugola, G. Gavira and S. Giacobbe, WO Patent 9321153 (1993).

64.   A. Cugola, G. Gaviraghi and S. Giacobbe, Eur. Patent 0568136 A (1993).

65.   C. F. Bigge, G. Johnson and P.-W. Yuen, WO Patent 9216205 (1992).

66.   M. Rowley, P. D. Leeson, S. Grimwood, A. Foster and K. Saywell, BioMed. Chem. Lett., 2, 1627 (1992).

67.   P. Louvet, G. Lallement, I. Pernot-Marino, C. Luu-Duc and G. Blanchet, Eur. J. Med. Chem., 28, 71 (1993).

68.   R. W. Carling, P. D. Leeson, A. M. Moseley, R. Baker, A. C. Foster, S. Grimwood, J. A. Kemp and G. R. Marshall, J. Med. Chem., 35, 1942 (1992).

69.   P. D. Leeson, R. W. Carling, K. W. Moore, A. M. Moseley, J. D. Smith, G. Stevenson, T. Chan, R. Baker, A. C. Foster, S. Grimwood, J. A. Kemp, G. R. Marshall and K. Hoogsteen, J. Med. Chem., 35, 1954 (1992).

70.   S. Grimwood, A. M. Moseley, R. W. Carling, P. D. Leeson and A. C. Foster, Mol. Pharmacol., 41, 923 (1992).

71.   S. Grimwood, L. Struthers and A. C. Foster, Br. J. Pharmacol., 107, 342 (1992).
72.   R. W. Carling, P. D. Leeson, A. M. Moseley, J. D. Smith, K. Saywell, M. D. Tricklebank, J. A. Kemp, G. R. Marshall, A. C. Foster and S. Grimwood, BioMed. Chem. Lett., 3, 65 (1993).
73.   K. W. Moore, P. D. Leeson, R. W. Carling, M. D. Tricklebank and L. Singh, BioMed. Chem. Lett., 3, 61 (1993).
74.   E. C. R. Smith, L. A. McQuaid, D. O. Calligaro and P. J. O'Malley, BioMed. Chem. Lett., 3, 81 (1993).
75.   L. A. McQuaid, E. C. R. Smith, D. Lodge, E. Pralong, J. H. Wikel, D. O. Calligaro and P. J. O'Malley, J. Med. Chem., 35, 3423 (1992).
76.   P. D. Leeson, R. Baker, R. W. Carling, J. J. Kulagowski, I. M. Mawer, M. P. Ridgill, M. Rowley, J. D. Smith, I. Stansfield, G. I. Stevenson, A. C. Foster and J. A. Kemp, BioMed. Chem. Lett., 3, 299 (1993).
77.   M. Rowley, P. D. Leeson, G. I. Stevenson, A. M. Moseley, I. Stansfield, I. Sanderson, L. Robinson, R. Baker, J. A. Kemp, G. R. Marshall, A. C. Foster, S. Grimwood, M. D. Tricklebank and K. L. Saywell, J. Med. Chem., 36, 3386 (1993).
78.   J. J. Kulagowski and P. D. Leeson, GB Patent 2265372 A (1993).
79.   R. W. Carling, P. D. Leeson, K. W. Moore, J. D. Smith, C. R. Moyes, I. M. Mawer, S. Thomas, T. Chan, R. Baker, A. C. Foster, S. Grimwood, J. A. Kemp and G. R. Marshall, J. Med. Chem., 36, 3397 (1993).
80.   S. J. Hays, P. A. Boxer, C. P. Taylor, M. G. Vartanian, L. J. Robichaud and E. Ø. Nielsen, BioMed. Chem. Lett., 3, 77 (1993).
81.   J. R. Epperson, P. Hewawasam, N. A. Meanwell, C. G. Boissard, V. K. Gribkoff and D. Post-Munson, BioMed. Chem. Lett., 3, 2801(1993).
82.   P. D. Leeson, B. J. Williams, M. Rowley, K. W. Moore, R. Baker, J. A. Kemp, T. Priestley, A. C. Foster and A. E. Donald, BioMed. Chem. Lett., 3, 71 (1993).
83.   S. E. Smith and B. S. Meldrum, Eur. J. Pharmacol., 211, 109 (1992).
84.   I. A. Shalaby, B. L. Chenard, M. A. Prochniak and T. W. Butler, J. Pharmacol. Exp. Ther., 260, 925 (1992).
85.   S. Toulmond, A. Serrano, J. Benavides and B. Scatton, Brain Res., 620, 32 (1993).
86.   D. Duval, N. Roome, C. Gauffeny, J. P. Nowicki and B. Scatton, Neurosci. Lett., 137, 193 (1992).
87.   B. L. Chenard, T. W. Butler, I. A. Shalaby, M. A. Prochniak, B. K. Koe and C. B. Fox, BioMed. Chem. Lett., 3, 91 (1993).
88.   J. L. Roba, C. I. Gillet, M. F. Rafferty, B. Jarrot and P. M. Beart, WO Patent 9203131 (1992).
89.   B. L. Chenard, US Patent 5185343 (1993).
90.   N. A. Saccomano and R. A. Volkmann, US Patent 5185369 (1993).
91.   I. J. Reynolds, K. D. Rothermund and S. Rajdev, BioMed. Chem. Lett., 3, 85 (1993).
92.   K. Williams, L. M. Pullan, C. Romano, R. J. Powel, A. I. Salama and P. B. Molinoff, J. Pharmacol. Exp. Ther., 262, 539 (1992).
93.   U. Madsen, B. Ebert, P. Krogsgaard-Larsen and E. H. F. Wong, Eur. J. Med. Chem., 27, 479 (1992).
94.   U. Madsen and E. H. F. Wong, J. Med. Chem., 35, 107 (1992).
95.   N. Tamura, T. Iwama and K. Itoh, Chem. Pharm. Bull., 40, 381 (1992).
96.   J. A. Monn, M. J. Valli, R. A. True, D. D. Schoepp, J. D. Leander and D. Lodge, BioMed. Chem. Lett., 3, 95 (1993).
97.   L. Vyklicky Jr., D. K. Patneau and M. L. Mayer, Neuron, 7, 971 (1991).
98.   U. Staubli, J. Ambros-Ingerson and G. Lynch, Hippocampus, 2, 49 (1992).
99.   D. K. Patneau, L. Vyklicky Jr. and M. L. Mayer, J. Neurosci., 13, 3496 (1993).
100.   R. A. Hall, M. Kessler, A. Quan, J. Ambrose-Ingerson and G. Lynch, Brain Res., 628, 345 (1993).
101.   P. C. May and P. M. Robison, J. Neurochem., 60, 1171 (1993).
102.   R. Granger, U. Staubli, M. Davis, Y. Perez, L. Nilsson, G. A. Rogers and G. Lynch, Synapse, 15, 326 (1993).
103.   U. Staubli, G. Rogers and G. Lynch, Proc. Natl. Acad. Sci. USA, 91, 777 (1994).
104.   S.-i. Yamaguchi, S. D. Donevan and M. A. Rogawski, Epilepsy Res., 15, 179 (1993).
105.   P. C. May and P. M. Robison, Neurosci. Lett., 152, 169 (1993).
106.   S. E. Smith an d B. S. Meldrum, Stroke, 23, 861 (1992).
107.   E. LePeillet, B. Arvin, C. Moncada and B. S. Meldrum, Brain Res., 571, 115 (1992).
108.   B Arvin, C. Moncada, E. LePeillet, A. Chapman and B. S. Meldrum, NeuroReport, 3, 235 (1992).
109.   J. W. Phillis, M. Smith-Barbour, L. M. Perkins and M. H. O'Regan, NeuroReport, 4, 109 (1993).
110.   W. Hauber and R. Andersen, N.-S. Arch. Pharmacol., 348, 486 (1993).
111.   I. Tarnawa, P. Berzsenyi, F. Andrási, P. Botka, T. Hámori, I. Ling and J. Körösi, BioMed. Chem. Lett., 3, 99 (1993).
112.   G. Szabó and J. M. Henley, NeuroReport, 5, 93 (1993).
113.   P. L. Ornstein, M. B. Arnold, N. K. Augenstein, D. Lodge, J. D. Leander and D. D. Schoepp, J. Med. Chem., 36, 2046 (1993).
114.   G. S. Hamilton, Z. Huang, J. Wu, S. A. Borosky, D. L. Bednar, J. W. Ferkany and E. W. Karbon, BioMed. Chem. Lett., 2, 1269 (1992).

115. F. Wätjen, E. Ø. Nielsen, J. Drejer and L. H. Jensen, BioMed. Chem. Lett., 3, 105 (1993).
116. F. Wätjen, C. F. Bigge, L. H. Jensen, P. A. Boxer, L. J. Lescosky, E. Ø. Nielsen, T. C. Malone, G. W. Cambell, L. L. Coughenour, D. M. Rock, J. Drejer and F. W. Marcoux, BioMed. Chem. Lett., 4, 371 (1994).
117. J. Ohmori, S. Sakamoto, H. Kubota, M. Shimizu-Sasamata, M. Okada, S. Kawasaki, K. Hidaka, J. Togami, T. Furuya and K. Murase, J. Med. Chem., 37, 467 (1994).
118. M. Shimizu-Sasamata, S. Kawasaki, S. Yatsugi, J. Ohmori, S. Sakamoto, K. Koshiya, S. Usuda and K. Murase, Soc. Neurosci. Abst., 44.14 (1992).
119. M. Shimizu-Sasamata, S. Kawasaki, S. Yatsugi, J. Ohmori, S. Sakamoto, K. Koshiya, S. Usuda and K. Murase, J. Cereb. Blood Flow Metab., 13 (Suppl. 1), S664 (1993).
120. K. A. Sluka and K. N. Westlund, Pain, 55, 217 (1993).
121. T. Zarnowski, Z. Kleinrok, W. A. Truski and S. J. Czuczwar, Neuropharmacol., 32, 895 (1993).
122. I. Westergren and B. B. Johansson, Brain Res., 573, 324 (1992).
123. H. Li and A. M. Buchan, J. Cerebr. Blood Flow, 13, 933 (1993).
124. M. J. Sheardown, P. D. Suzdak and L. Nordholm, Eur. J. Pharmacol., 236, 347 (1993).
125. X.-J. Xu, J.-X. Hao, A. Seiger and Z. Wiesenfeld-Hallin, J. Pharmacol. Exp. Ther., 267, 140 (1993).
126. K. Konno, K. Hashimoto and H. Shirahama, Heterocycles, 33, 303 (1992).
127. L. Kwak, H. Aizawa, M. Ishida and H. Shinozaki, Neurosci. Lett., 139, 114 (1992).
128. K. Hashimoto, M. Horikawa, M. Ishida, H. Shinozaki and H. Shirahama BioMed. Chem. Lett., 2, 743 (1992).
129. I. Jefferies, BioMed. Chem. Lett., 2, 1519 (1992).
130. A. C. Scallet, Z. Binienda, F. A. Caputo, S. Hall, M. G. Paule, R. L. Rountree, L. Schmued, T. Sobotka and W. Slikker Jr., Brain Res., 627, 307 (1993).
131. E. C. D. Todd, J. Food. Protection, 56, 69 (1993).
132. T. H. Johansen J. Drejer, F. Wätjen and E. Ø. Nielsen, Eur. J. Pharmacol. [Mol. Pharmacol. Sect.], 246, 195 (1993).
133. D. D. Schoepp, B. G. Johnson, R. A. True and J. A. Monn, Eur. J. Pharmacol. [Mol. Pharmacol. Sect.], 207, 351 (1991).
134. J. Cartmell, A. R. Curtis, J. A. Kemp, D. A. Kendall and S. P. H. Alexander, Neruosci. Lett., 153, 107 (1993).
135. D. D. Schoepp, B. G. Johnson and J. A. Monn, J. Neurochem., 58, 1184 (1992).
136. A. I. Sacaan, J. A. Monn and D. D. Schoepp, J. Pharmacol. Exp. Ther., 259, 1366 (1991).
137. M. Koch, Brain Res., 629, 176 (1993).
138. J. P. Tizzano, K. I. Griffey, J. A. Johnson, A. S. Fix, D. R. Helton and D. D. Schoepp, Neurosci. Lett., 162, 12 (1993).
139. P. Laudrup and H. Klitgaard, Eur. J. Pharmacol., 250, 15 (1993).
140. H. Klitgaard and P. Laudrup, Eur. J. Pharmacol., 250, 9 (1993).
141. M. Ishida, H. Akagi, K. Shimamoto, Y. Ohfune and H. Shinozaki, Brain Res., 537, 311, (1990).
142. Y. Nakagawa, K. Saitoh, T. Ishihara, M. Ishida and H. Shinozaki, Eur. J. Pharmacol. 184, 205 (1990).
143. Y. Hayashi, Y. Tanabe, I. Aramori, M. Masu, K. Shimamoto, Y. Ohfune and S. Nakanishi, Br. J. Pharmacol., 107, 539 (1992).
144. G. Lombardi, M. Alesiani, P. Leonardi, G. Cherici, R. Pellicciari and F. Moroni, Br. J. Pharmacol., 110, 1407 (1993).
145. G. Costantino, B. Natalini and R. Pellicciari, BioMed. Chem., 1, 259 (1993).
146. M. Favaron, R. M. Manev, P. Candeo, R. Arban, N. Gabellini, A. Kozikowski and H. Manev, Neuroreport, 4, 967 (1993).
147. I. Ito, A. Kohda, S. Tanabe, E. Hirose, M. Hayashi, S. Mitsunaga and H. Sugiyama, NeuroReport, 3, 1013 (1992).
148. E. F. Birse, S. A. Eaton, D. E. Jane, P. L. St. J. Jones, R. H. P. Porter, P. C.-K. Pook, D. C. Sunter, P. M. Udvarhelyi, B. Wharton, P. J. Roberts, T. E. Salt and J. C. Watkins, Neurosci., 52, 481 (1993).
149. C. Thomsen and P. D. Suzdak, NeuroReport, 4, 1099 (1993).
150. B. Wroblewska, J. T. Wroblewski, O. H. Saab and J. H. Neale, J. Neurochem., 61, 943 (1993).
151. S. A. Eaton, D. E. Jane, P. L. St. J. Jones, R. H. P. Porter, P. C.-K. Pook, D. C. Sunter, P. M. Udvarhelyi, P. J. Roberts, T. E. Salt and J. C. Watkins, Eur. J. Pharmacol. Mol. Pharmacol. Sect., 244, 195 (1993).
152. D. E. Jane, P. L. St. J. Jones, P. C.-K. Pook, T. E. Salt, D. C. Sunter and J. C. Watkins, Neuropharmacol., 32, 725 (1993).
153. K. Lingenhohl, H.-R. Olpe, N. Bendali and T. Knopfel, Neurosci. Res., 18, 229 (1993).
154. B. G. Frenguelli, B. Potier, N. T. Slater, S. Alford and G. L. Collingridge, Neuropharmacol., 32, 1229 (1993).
155. P. Chinestra, L. Aniksztejn, D. Diabira and Y. Ben-Ari, J. Neurophysiol., 70, 2684 (1993).
156. O. A. Sergueeva, N. B. Fedorov and K. G. Reymann, Neuropharmacol., 32, 933 (1993).
157. G. O. Ormandy, Brain Res., 572, 103 (1992).

SECTION II. CARDIOVASCULAR AND PULMONARY AGENTS

Editor: David W. Robertson, Ligand Pharmaceuticals
San Diego, CA 92121

Chapter 7. Small Molecule Endothelin Receptor Antagonists

David C. Spellmeyer
Chiron Corporation, Emeryville, CA 94608

Introduction - An aggressive effort to identify small molecule receptor antagonists has followed the 1988 discovery of endothelin (ET), the most potent known vasoconstrictor (1,2). It has taken only six years from the discovery of ET to the identification of bioavailable small molecule ET receptor antagonists which should be useful in elucidating pathological effects and disease states involving ET-1.

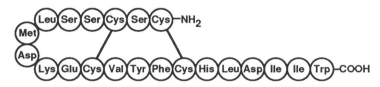

Endothelin-1

The ET family of 21-amino acid peptides (ET-1, ET-2, and ET-3) possess an unusual chemical structure containing 2 disulfide bonds (1). At least two different ET converting enzymes (ECEs), one a metalloproteinase and one an aspartyl proteinase, are thought to be responsible for converting the biological precursor peptides (big-ET-1, big-ET-2, and big-ET-3) to the biologically active ETs (3,4). While ECEs remain targets for therapeutic intervention, the identification, purification, and expression of ECEs has been difficult; consequently, the discovery of small molecule receptor antagonists has outpaced the discovery of ECE inhibitors.

Three unique receptor subtypes have been identified and cloned (5-7)— all belonging to the seven-transmembrane G-protein coupled receptor superfamily (8). The ET_A receptor subtype shows higher affinity for ET-1 and ET-2 than for ET-3 and is known to mediate vasoconstriction (5,6). The ET_B receptor subtype binds ET-1, ET-2, and ET-3 with equal affinities (9,10), and has been shown to mediate both vasoconstriction and vasodilation (10). Recently, the ET_C receptor subtype, which shows higher affinity for ET-3 than for ET-1, has been identified in *Xenopus laevis* (7), but not in higher species. Progress in the discovery of the ET_A and ET_B receptor antagonists has been summarized in previous volumes of this series (11-13) and elsewhere (14).

Biological Activities of ET-1 - Elevated levels of ET-1 have been found in many disease states, including cardiovascular (15-18), respiratory (19), gastric (20,21), renal (18,22-26), and urological diseases (27). The earliest reported peptidic receptor antagonists suggested a pathophysiologic role of ET-1 in several disease states. These include cyclosporin-induced nephrotoxicity (23,24) and cerebral vasospasm (28,29). Several general reviews of the biological functions of the ETs have appeared (14,30-34).

Peptidic Receptor Ligands - Several classes of peptidic ligands have been reported. ET$_A$-selective ligands include the cyclic pentapeptides such as BQ-123 (1 , cyclo[D-Trp-D-Asp-Pro-D-Val-Leu]) (35-37), the acyl tripeptides such as 2 (FR-139317) (28,38), hexapeptides (39-44), and depsipeptides (45). A series of hexapeptide analogues, such as 3 (PD-145065), are the first reported ET$_A$/ET$_B$ non-selective antagonists (46). Monocyclic and linear analogues of ET-1 have been reported to be ET$_B$-selective antagonists (47-49). NMR structural information of several of these peptidic receptor ligands has been reported (50-56).

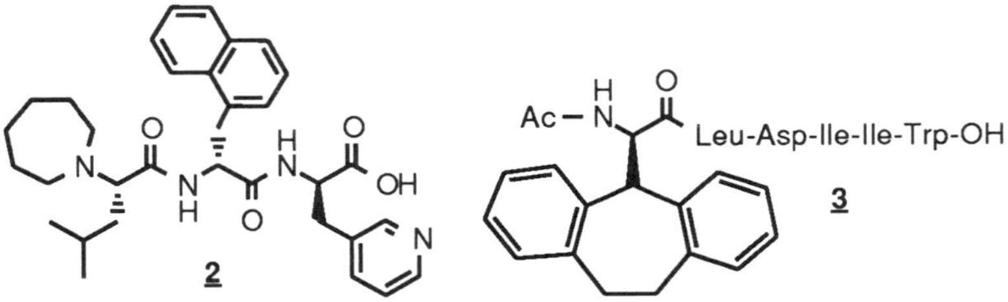

SMALL MOLECULE ENDOTHELIN RECEPTOR ANTAGONISTS

Only a few non-peptidic ligands had been reported until 1993. These include a series of anthraquinones (57), asterric acid (58), and myricerone caffeoyl esters (59,60). Recently, several series of small molecule ET antagonists have been reported, including diketopiperazines (61), N -pyrimidinylbenzenesulfonamides (62), N – isoxazolylnaphthalenesulfonamides (63), dihydropyridine anhydrides (64), and indanecarboxylic acids (65).

Myricerone Caffeoyl Analogues - Compound 4 (Shionogi code number 97-139) is a synthetic derivative of a non-peptide identified through natural-product screening (59) that binds selectively to the ET$_A$ receptor subtype (IC$_{50}$ = 1.0 ± 0.2 nM and 1000 ± 200 nM for ET$_A$ and ET$_B$, respectively) (66). Compound 4 is an antagonist in second messenger assays (ED$_{50}$ = 2.6 ± 0.4 nM in Ca^{2+}-flux assays in A7r5 cells), and antagonizes ET-1-induced contractions on rat thoracic aorta (pA$_2$ = 8.8 ± 0.4). In pithed rats, an i.v. dose of 0.3 mg/kg of 4 inhibited the ET-1-induced hypertension for 60 minutes.

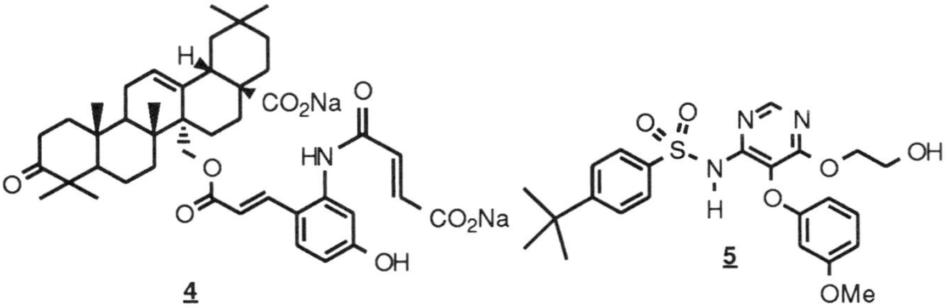

N-pyrimidinylbenzenesulfonamides - A non-selective ET$_A$/ET$_B$ compound (Ro 46-2005, 5) is the first reported orally bioavailable ET receptor antagonist (62). This molecule was optimized from a pyrimidinyl sulfonamide identified as a weak ET_A ligand through random screening of a corporate database. Compound **5** shows modest affinity for both the ET_A and ET_B receptors, with a four-fold preference for the ET_A receptor (62,67). Table 1 shows the receptor binding affinities of **5** in several species.

Table 1: Receptor binding affinities for compound **5** in several species (67).

ET_A Assays		ET_B Assays	
Tissue	IC_{50} (nM)	Tissue	IC_{50} (nM)
human smooth muscle	220 ± 60	human placenta	160 ± 77
rat mesangial cells	430 ± 140	porcine cerebellum	227 ± 92
		rat endothelial cells	1140 ± 340
recombinant ET_A	360 ± 160	recombinant ET_B	530 ± 150

Compound **5** inhibits ET-1-induced and sarafotoxin S6b-induced contractions *in vitro*, behaving as a competitive antagonist on both the ET_A ($pA_2 = 6.50 \pm 0.04$ on rat aorta rings) and ET_B ($pA_2 = 6.47 \pm 0.16$ on rat small mesenteric arteries) receptors (62,67). Ro 46-2005 shows no agonist activity and does not block the contraction induced by angiotensin II, serotonin, noradrenaline, potassium chloride, nor prostaglandin $F_{2\alpha}$, demonstrating that **5** is specific for ET-1-induced contractions (62). Bosentan (Ro 47-0203), an analogue of **5** has recently been reported (68,69).

A 3 mg/kg i.v. dose of compound **5** reduced renal vasoconstriction in a renal ischemia model and prevented decreased blood flow in models of subarrachnoid hemorrhage in rats (62). This compound has also been shown to be a potent, orally active hypotensive agent when dosed at 10 - 100 mg/kg in sodium-depleted squirrel monkeys (62).

Further *in vivo* work shows that i.v. dosing of compound **5** leads to a prolonged two-fold increase in the circulating levels of ET-1 and a slight increase in circulating levels of ET-3, without affecting circulating levels of big-ET-1 (70). In contrast, the ET_A-selective ligands BQ-123 (**1**) and FR-139317 (**2**) did not lead to increased levels of ET-1 or ET-3. This led the authors to speculate that the prolonged increased levels of ET-1 may be due to displacement of ET-1 from ET_B receptors (70).

N-isoxazolylnaphthalenesulfonamides - A second class of sulfonamide-derived ET receptor antagonists has been reported (63). This class of compounds was also identified through screening a proprietary compound collection. In contrast to **5**, this class is highly specific for the ET_A receptor.

Structure-activity relationship data have been reported for this class of compounds. The sulfonamide nitrogen must not be alkylated. Compound **6** shows good affinity for the ET_A receptor, while compound **7** shows no affinity at concentrations as high as 32 μM.

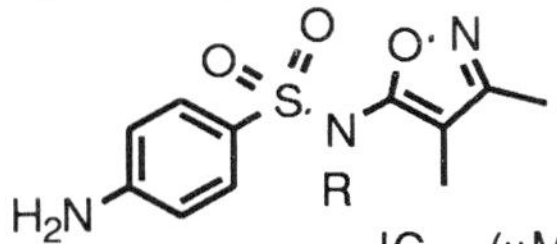

Chemical optimization led to compound **10** (BMS-182874), which has modest affinity for the ET_A receptor, and high selectivity (63). Compound **10** is orally active in a hypertension model using one-kidney rats (63). These rats were implanted with deoxycorticosterone acetate pellet and given saline to drink, inducing hypertension. When **10** was dosed at 100 μmol/kg, mean arterial pressure fell by 25% within 1 hour and remained depressed for up to 24 hours (63).

	IC_{50} (μM)
6 R = H	0.8
7 R = CH$_3$	> 32

Table 2: Receptor Binding Affinities and Functional Activities of a Variety of 5-substituted *N*-(3,4-Dimethyl-5-isoxazolyl)-1-naphthalenesulfonamides (**63**).

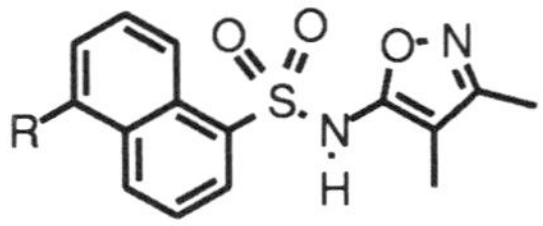

Compound	R	Receptor Binding IC_{50} (μM)	Functional Activity K_B (μM)
8	NH2	4.0 ± 0.8	
9	NHCOCH3	0.88 ± 0.12	11
10	N(CH3)2	0.15 ± 0.01	0.5
11	CH2N(CH3)2	2.8 ± 0.7	
12	OH	7.8 ± 0.1	

Several of these compounds are reasonably good ligands, showing micromolar or better affinity for the ET_A receptor. Table 2 shows compound **10** displays antagonism of ET-1-induced contractions of rabbit carotid artery rings (**63**). Interestingly, **9** shows significantly reduced functional antagonism relative to **10**, but has a similar receptor binding affinity. The reason for this difference remains unclear.

Dihydropyridine Anhydrides - The symmetric anhydride CGS 27830 (**13**) binds to both the ET_A and ET_B receptors subtypes with IC_{50}s of 15.9 ± 1.3 nM and 295 ± 19 nM, respectively (64). The R,R-isomer of **13** shows modest affinity for the ET receptors ($IC_{50} = 422 \pm 56$ nM and 2.7 ± 0.5 μM for ET_A and ET_B, respectively), and the S,S-isomer is inactive. Compound **13** has a half-life of less than 60 minutes when dosed i.v. at 10 mg/kg (64). Pretreatment of **13** on isolated rabbit aorta produced an insurmountable inhibition of ET-1-induced contractions. Compound **13** is specific to ET-1-induced contractions, as it does not inhibit phenylephrine-induced or KCl-induced contractions. The crystal structure of **13** has been reported (64).

Indane Carboxylic Acids - Compound **14** (SB 209670) is the first published example of the indane carboxylic acid class of non-selective ET_A/ET_B receptor antagonists (65). Compound **14** binds to cloned human ET_A and ET_B receptors with IC_{50}s of 0.43 ± 0.09 nM and 14.7 ± 3.0 nM, respectively. Thus, this is the first small molecule with subnanomolar affinity for an ET receptor. In addition, compound **14** shows potent *in vitro* functional activity in contractile responses to ET-1 on either rat aorta ($K_B = 0.4 \pm 0.04$ nM, mediated by ET_A receptors) or on rabbit pulmonary artery ($K_B = 199 \pm 9$ nM, mediated by ET_B receptors).

In contrast to the sulfonylureas, this class of compounds was discovered through screening compounds similar to known antagonists of other seven-transmembrane G-protein coupled receptors (65). Comparison of the original lead compound (**15**) with NMR structures of ET-1 suggested that the aromatic rings and carboxylic acid might overlap with the sidechains of the Tyr-13, Phe-14, and Asp-18 residues. Comparison of NMR structures of ET-1 and model structures of BQ-123 suggested further modifications that ultimately led to the discovery of **14**. This approach demonstrates the powerful combination of synthetic chemistry, database methods, molecular modeling, and NMR structural information of peptides and proteins.

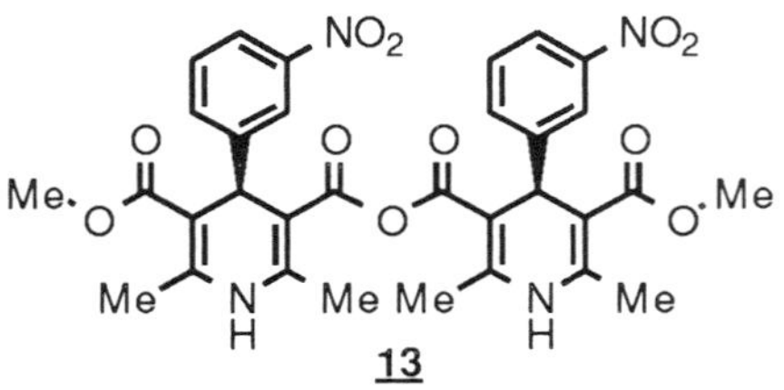

13

Conclusions - Advances in molecular biology have led to the discovery of the ET family of peptides, their biological precursors, their receptors, and their processing enzymes. The discovery — by a variety of techniques — of several classes of potent small molecule ET receptor antagonists only six years after the discovery of ET-1 is noteworthy. Natural product screening has produced peptides, depsipeptides, and small molecules. Both random screening and directed screening of chemical libraries have led to the discovery of several classes of small molecules. Chemical optimization of many of these leads has produced potent and bioavailable small molecule ET receptor antagonists. Although several of these compounds have been shown to be effective in specific models of disease, it is still unclear what therapeutic benefit any ET receptor antagonist might have. The development of bioavailable small molecule ET antagonists should be useful in unraveling the role of ET in disease states and in elucidating which unmet medical needs might be treated with ET antagonists.

References

1. M. Yanagisawa, H. Kurihara, S. Kimura, Y. Tomobe, M. Kobayashi, Y. Mitsui, Y. Yazaki, K. Goto and T.A. Masaki, Nature, 332, 411 (1988).
2. M. Yanagisawa, A. Inouye, T. Ishikawa, Y. Kasuya, S. Kamura, S. Kumagaye, K. Nakajima, T. Watanabe, S. Sakakibara, K. Goto and T. Masaki, Proc. Natl. Acad. Sci., USA, 85, 6964 (1988).
3. K. Shiosaki, A.S. Tasker, G.M. Sullivan, B.K. Sorensen, T.W. von Geldern, J.R. Wu-Wong, C.A. Marselle and T.J. Opgenorth, J. Med. Chem., 36, 468 (1993).
4. A.J. Turner, Biochem. Soc. Trans., 21, 697 (1993).
5. H.Y. Lin, E.H. Kaji, G.K. Winkel, H.E. Ives and H.F. Lodish, Proc. Natl. Acad. Sci., USA, 88, 3185 (1991).
6. H. Arai, S. Hori, I. Aramori, H. Ohkubo and S. Nakanishi, Nature, 348, 730 (1990).
7. S. Karne, C.K. Jayawickreme and M.R. Lerner, J. Biol. Chem., 268, 19126 (1993).
8. W.C. Probst, L.A. Snyder, D.I. Schuster, J. Brosius and S.C. Sealfon, DNA Cell Biol., 11, 1 (1992).
9. T. Sakurai, M. Yanagisawa, Y. Takuwa, H. Miyazaki, S. Kimura, K. Goto and T. Masaki, Nature, 348, 732 (1990).
10. M. Clozel, G.A. Gray, V. Breu, B.-M. Löffler and R. Osterwalder, Biochem. Biophys. Res. Comm., 186, 867 (1992).
11. A.M. Doherty and R.E. Weishaar, Annu. Rep. Med. Chem., 25, 89 (1990).
12. A.M. Doherty, Annu. Rep. Med. Chem., 26, 83 (1991).
13. A.M. Doherty, Annu. Rep. Med. Chem., 27, 79 (1992).
14. A.M. Doherty, J. Med. Chem., 35, 1493 (1992).
15. A. Lerman, J. Fredric L. Hildebrand, K.B. Margulies, B. O'murchu, M.A. Perrella, D.M. Heublein, T.R. Schwab and J. John C. Burnett, Mayo Clin. Proc., 65, 1441 (1990).
16. I. Ziv, G. Fleminger, R. Djaldetti, A. Achiron, E. Melamed and M. Sokolovsky, Stroke, 23, 1014 (1992).
17. K. Ide, K. Yamakawa, T. Nakagomi, T. Sasaki, I. Saito, H. Kurihara, M. Yoshizumi, Y. Yazaki and K. Takakura, Neurol. Res., 11, 101 (1989).
18. G. Remuzzi and A. Benigni, Lancet, 342, 589 (1993).
19. D.W.P. Hay, P.J. Henry and R.G. Goldie, Trends Pharmacol. Sci., 14, 29 (1993).
20. J.L. Wallace, C.M. Keenan, W.K. MacNaughton and G.W. McKnight, Eur. J. Pharmacol., 167, 41 (1989).

21. J.L. Wallace, G. Cirino, G. De Nucci, W. McKnight and W.K. MacNaughton, Am. J. Physiol., 256, G661 (1989).
22. E.P. Nord, Kidney Int., 44, 451 (1993).
23. I.T.M. Bloom, F.R. Bentley and R.N. Garrison, Surgery, 114, 480 (1993).
24. D.M. Lanese and J.D. Conger, J. Clin. Invest., 91, 2144 (1993).
25. N. Perico and G. Remuzzi, Kidney Int., 43, S76 (1993).
26. M.S. Simonson and M.J. Dunn, Annu. Rev. Physiol., 55, 249 (1993).
27. S. Kobayashi, R. Tang, B. Wang, T. Opgenorth, P. Langenstroer, E. Shapiro and H. Lepor, Mol. Pharmacol., 45, 306 (1994).
28. S. Itoh, T. Sasaki, K. Ide, K. Ishikawa, M. Nishikibe and M. Yano, Biochem. Biophys. Res. Comm., 195, 969 (1993).
29. M. Clozel and H. Watanabe, Life Sci., 52, 825 (1993).
30. W.G. Haynes, A.P. Davenport and D.J. Webb, Trends Pharmacol. Sci., 14, 225 (1993).
31. W.G. Haynes and D.J. Webb, Clin. Sci., 84, 485 (1993).
32. A. Inoue, M. Yanagisawa, S. Kimura, Y. Kasuya, T. Miysuchi, K. Goto and T. Masaki, Proc. Natl. Acad. Sci., USA, 86, 2863 (1989).
33. J. Leppäluoto and H. Ruskaoho, Ann. Med., 24, 153 (1992).
34. Y. Takuwa, M. Yanagisawa, N. Takuwa and T. Masaki, Prog. Gro. Fact. Res., 1, 195 (1989).
35. K. Ishikawa, T. Fukami, T. Nagase, K. Fujita, T. Hayama, K. Niiyama, T. Mase, M. Ihara and M. Yano, J. Med. Chem., 35, 2139 (1992).
36. M. Ihara, K. Noguchi, T. Saeki, T. Fukuroda, S. Tsuchida, S. Kimura, T. Fukami, K. Ishikawa, M. Nishikibe and M. Yano, Life Sci., 50, 247 (1992).
37. M. Ihara, T. Fukuroda, T. Saeki, M. Nishikibe, K. Kojiri, H. Suda and M. Yano, Biochem. Biophys. Res. Commun., 178, 132 (1991).
38. I. Aramori, H. Nirei, M. Shoubo, K. Sogabe, K. Nakamura, H. Kojo, Y. Notsu, T. Ono and S. Nakanishi, Mol. Pharmacol., 43, 127 (1993).
39. A.M. Doherty, W.L. Cody, N.L. Leitz, P.L. DePue, M.D. Taylor, S.T. Rapundalo, G.P. Hingorani, · T.C. Major, R.L. Panek and D.G. Taylor, J. Cardiovasc. Pharmacol., 17, S59 (1991).
40. W.L. Cody, A.M. Doherty, J.X. He, P.L. DePue, S.T. Rapundalo, G.A. Hingorani, T.C. Major, R.L. Panek, D.T. Dudley, S.J. Haleen, D. LaDouceur, K.E. Hill, M.A. Flynn and E.E. Reynolds, J. Med. Chem., 35, 3301 (1992).
41. A.M. Doherty, W.L. Cody, X. He, P.L. DePue, D.M. Leonard, J.B.J. Dunbar, K.E. Hill, M.A. Flynn and E.E. Reynolds, Bioorg. Med. Chem. Lett., 3, 497 (1993).
42. A.M. Doherty, W.L. Cody, P.L. DePue, J.X. He, L.A. Waite, D.M. Leonard, N.L. Leitz, D.T. Dudley, S.T. Rapundalo, G.P. Hingorani, S.J. Haleen, D.M. LaDouceur, K.E. Hill, M.A. Flynn and E.E. Reynolds, J. Med. Chem., 36, 2585 (1993).
43. D.C. Spellmeyer, S. Brown, G.S. Stauber, H.M. Geysen and R. Valerio, Bioorg. Med. Chem. Lett., 3, 519 (1993).
44. D.C. Spellmeyer, S. Brown, G.S. Stauber, H.M. Geysen and R. Valerio, Bioorg. Med. Chem. Lett., 3, 1253 (1993).
45. Y.K.T. Lam, O.D. Hensens, J.M. Liesch, D.L. Zink, L. Huang, D.L.J. Williams and O.R. Genilloud, Eur. Pat. Appl. 0496452 A1 (1992).
46. W.L. Cody, A.M. Doherty, J.X. He, P.L. DePue, L.A. Waite, J.G. Topliss, S.J. Haleen, D. LaDouceur, M.A. Flynn, K.E. Hill and E.E. Reynolds, Med. Chem. Res., 3, 154 (1993).
47. M. Takai, I. Umemura, K. Yamasaki, T. Watakabe, Y. Fujitani, K. Oda, Y. Urade, T. Inui, T. Yamamura and T. Okada, Biochem. Biophys. Res. Comm., 184, 953 (1992).
48. H. Urade, Y. Fujitani, K. Oda, T. Watakabe, I. Umemura, M. Takai, T. Okada, K. Sakata and H. Karaki, FEBS Lett., 311, 12 (1992).
49. W.L. Cody, J.X. He, P.L. DePue, S.T. Rapundalo, G.P. Hingorani, D.T. Dudley, K.E. Hill, E.E. Reynolds and A.M. Doherty, Bioorg. Med. Chem. Lett., 4, 567 (1994).
50. H. Tamaoki, Y. Kobayashi, S. Nishimura, T. Ohkubo, Y. Kyogoku, K. Nakajima, S.-I. Kumagaye, T. Kimura and S. Sakakibara, Prot. Eng., 4, 509 (1991).
51. M.D. Reily and J. James B. Dunbar, Biochem. Biophys. Res. Comm., 178, 570 (1991).
52. M.D. Reily, V. Thanabal, D.O. Omecinsky, J.B. Dunbar, A.M. Doherty and P.L. DePue, FEBS Lett., 300, 136 (1992).
53. S.R. Krystek, D.A. Bassolino, R.E. Bruccoleri, J.T. Hunt, M.A. Porubcan, C.A. Wandler and N.H. Anderson, FEBS Lett., 299, 255 (1992).
54. R.A. Atkinson and J.T. Pelton, FEBS Lett., 296, 1 (1992).
55. E.K. Bradley, S.C. Ng, R.J. Simon and D.C. Spellmeyer, Bioorg. Med. Chem., in press (1994).
56. N.H. Andersen, C. Chen, T.M. Marschner, J. Stanley R. Krystek and D.A. Bassolino, Biochemistry, 31, 1280 (1992).
57. S. Miyata, N. Fukami, M. Neya, S. Takase and S. Kiyoto, J. Antibiot., 45, 788 (1992).
58. H. Ohashi, H. Akiyama, K. Nishikori and J.-I. Mochizuki, J. Antibiot., 45, 1684 (1992).
59. M. Fujimoto, S.-I. Mihara, S. Nakajima, M. Ueda, M. Nakamura and K.-S. Sakurai, FEBS Lett., 305, 41 (1992).

60. S.-I. Mihara, K. Sakurai, M. Nakamura, T. Konoike and M. Fujimoto, Eur. J. Pharmacol., 247, 219 (1993).
61. M.F. Chan and V.N. Balaji, WO 9323404 (November 25,1993).
62. M. Clozel, V. Breu, K. Burri, J.-M. Cassal, W. Fischli, G.A. Gray, G. Hirth, B.-M. Löffler, M. Müller, W. Neidhart and H. Ramuz, Nature, 365, 759 (1993).
63. P.D. Stein, J.T. Hunt, D.M. Floyd, S. Moreland, K.E.J. Dickinson, C. Mitchell, E.C.-K. Liu, M.L. Webb, N. Murugesan, J. Dickey, D. McMullen, R. Zhang, V.G. Lee, R. Serafino, C. Delany, T.R. Schaeffer and M. Kozlowski, J. Med. Chem., 37, 329 (1994).
64. B. Mugrage, J. Moliterni, L. Robinson, R.L. Webb, S.S. Shetty, K.E. Lipson, M.H. Chin, R. Neale and C. Cioffi, Bioorg. Med. Chem. Lett., 3, 2099 (1993).
65. J.D. Elliot, M.A. Lago, R.D. Cousins, A. Gao, J.D. Leber, K.F. Erhard, P. Nambi, N.A. Elshourbagy, C. Kumar, J.A. Lee, J.W. Bean, C.W. DeBrosse, D.S. Eggleston, D.P. Brooks, G. Feuerstein, J. Robert R. Ruffolo, J. Weinstock, J.G. Gleason, C.E. Peishoff and E.H. Ohlstein, J. Med. Chem., in press (1994).
66. S.-I. Mihara, S. Nakajima, S. Matumura, T. Kohnoike and M. Fujimoto, J. Pharmacol. Exp. Ther., 268, (1994).
67. V. Breu, B.-M. Löffler and M. Clozel, FEBS Lett., 334, 210 (1993).
68. S.P. Roux, M. Clozel, U. Sprecher, G. Gray and J.P. Clozel, Circulation, 88, I-170 (1993).
69. B. Seo, B.S. Oemar, R. Siebenmann, L. von Segesser and T.F. Luscher, Circulation, 89, 1320 (1994).
70. B.-M. Löffler, V. Breu and M. Clozel, FEBS Lett., 333, 108 (1993).

Chapter 8. Emerging Opportunities in the Treatment of Asthma and Allergy

Allen J. Duplantier and John B. Cheng
Central Research Division, Pfizer Inc
Groton, CT 06340

Introduction - Over the last decade the pharmaceutical industry has made a significant effort to discover leukotriene (LT) modulators for asthma therapy. Many potent LTD_4 antagonists and 5-lipoxygenase inhibitors have been reported and some are entering anti-asthma clinical trials (1, 2). This chapter focuses on two novel molecular targets: modulators of intracellular cAMP [i.e., phosphodiesterase type IV (PDE IV) inhibitors and cAMP-dependent protein kinase (cA-PKA) agonists] and inhibitors of lymphocyte-derived cytokines [e.g., anti-IgE, and interleukin 4 and 5 (IL-4 and IL-5) inhibitors]. These classes of compounds are expected to block the inflammatory process prior to the production of chemical mediators (e.g., LTs) and activation of inflammatory cells (e.g., mast cells, basophils, eosinophils and lymphocytes). Thus, they may have a broad spectrum of anti-inflammatory actions. Both PDE IV inhibitors and cA-PKA agonists can modulate the cAMP signaling pathway (see Fig. 1). Current asthma therapies such as theophylline and steroids are thought to stimulate cAMP pathways (3, 4), but due to the nonspecific nature of their effects one cannot conclude that this is their sole mechanism of action (MOA). It is well known that IgE is the underlying cause of allergic asthma. Recent advances in molecular biology and immunology have identified the cellular mechanism that controls IgE biosynthesis in response to allergen challenge: A certain subset of helper T lymphocytes, Th-2 cells, is now thought to be a key triggering cell in the initiation of IgE production and pulmonary hypereosinophilia. IL-4 and IL-5 are signaling molecules produced from the Th-2 cells. The therapeutic implication of this theory is that inhibitors of IL-4 and/or IL-5 biosynthesis should reduce serum IgE and pulmonary inflammation in patients with asthma.

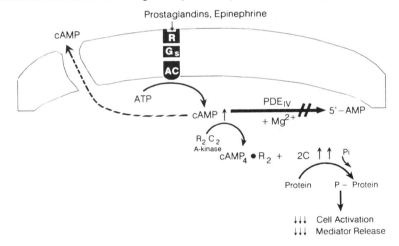

Figure 1. cAMP as a secondary messenger.

MODULATORS OF INTRACELLULAR cAMP

PDE IV Inhibitors - Cyclic nucleotide PDEs are viewed as attractive molecular targets for the treatment of asthma and inflammation. PDE isozymes are classified into five families (PDE I - V) based on their specificity for either cAMP or cGMP, inhibitor sensitivity and calcium-calmodulin dependence (5, 6). All five of the PDE isozyme families are found in human tracheal smooth muscle (7). PDE IV (cAMP-specific), however, is the primary cAMP PDE isozyme present in the inflammatory cells most clearly associated with asthma: neutrophils, macrophages, mast cells, eosinophils, basophils and T-lymphocytes (8, 9). In addition, PDE IV and PDE III/IV inhibitors possess both bronchodilatory and anti-inflammatory activity in animal models. The assumed MOA of a PDE IV inhibitor is to increase the concentration of cAMP within the target cell. cAMP, an intracellular second messenger, is hydrolyzed specifically and catalytically by PDE IV to 5' AMP in many inflammatory cells and in airway smooth muscle (Figure 1). Thus, generalized regulation of leukocyte activity via this second messenger system has a theoretical advantage over single mediator inhibition (9). However, a major obstacle in the development of PDE IV inhibitors is the elimination of emetic side effects (8). Several comprehensive reviews on PDE inhibition have recently appeared in the literature (8, 9, 10, 11, 12).

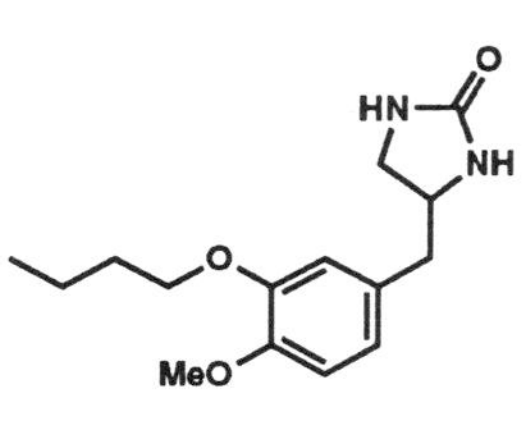

$\underline{\mathbf{1}}$ 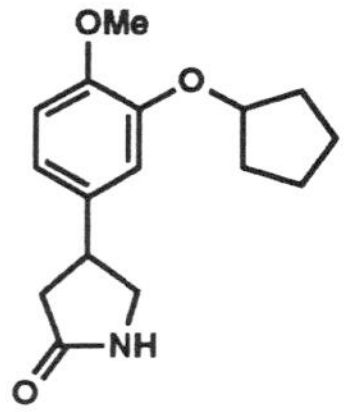 $\underline{\mathbf{2}}$ 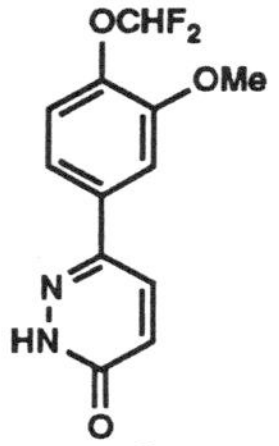 $\underline{\mathbf{3}}$ 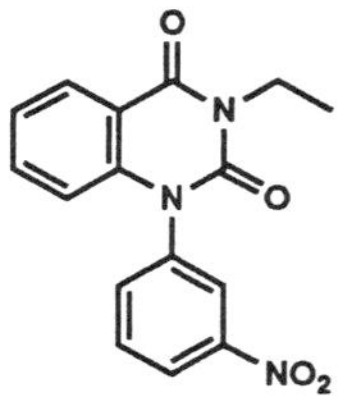 $\underline{\mathbf{4}}$

Anti-inflammatory Properties of PDE IV Inhibitors - Cytokines such as IL-4 (see IL-4 section of this chapter) and IFN-γ can modulate human monocyte PDE activity, and at low doses, IL-4 and IFN-γ have a synergistic effect (13). PDE IV inhibitor **1** (Ro 20-1724) reduces abnormal levels of IL-4 production, suggesting a possible means for controlling defective immune function in atopic dermatitis (14). Compound **2** (rolipram, ZK-62,711), the most widely studied selective PDE IV inhibitor, and **3** (zardaverine, B-842-92), a mixed PDE III/IV inhibitor, suppress lipopolysacharide (LPS)-induced TNF synthesis in mouse peritoneal macrophages (15). The enantiomers of **2** have been prepared by asymmetric synthesis (16,17) and the absolute stereochemistry has been assigned (18). Although R-(-) and S-(+)-**2** are equipotent PDE-IV inhibitors, (-)-**2** suppresses LPS-induced TNF production in human mononuclear cells 5-fold more effectively than (+)-**2** (19). Compound **4** (nitraquazone, TVX-2706) inhibits TNFα (>80% at 10μM) and IL-1β (50% at 10μM) secretion from LPS-stimulated human monocytes (20). Compounds **5** (PDA-641) and **2** significantly inhibit the secretion of TNFα from human donors, but are weak inhibitors of IL-1β (21, 22). Thus, PDE IV inhibitors may be effective against diseases involving the production of TNF α such as rheumatoid arthritis and septic shock (23).

In intact eosinophils, (-)-**2** is 10-fold more potent than (+)-**2** in enhancing isoprenaline-stimulated cAMP accumulation (24). Racemic **2** inhibits human neutrophil production of leukotriene B$_4$ (LTB$_4$) (25). Eosinophil counts in conjunctival epithelium in LT-challenged guinea pigs that receive (-)-**2** are significantly reduced compared to LT-challenged controls (26). Compared to placebo pre-treated, LPS exposed rats, compound **3** (30μmol/kg) significantly inhibits LPS induced neutrophil increase, increase in elastase activity and TNF α release in bronchoalveolar lavage fluid (27). In addition, compound **2** inhibits mouse ear edema and inflammatory cell infiltration as shown by myeloperoxidase activity (25).

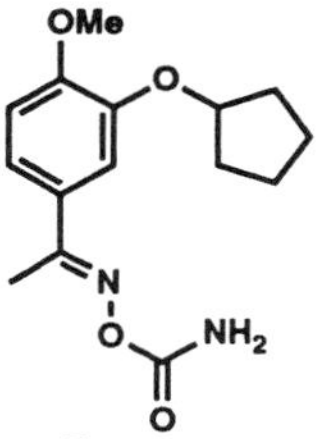

5 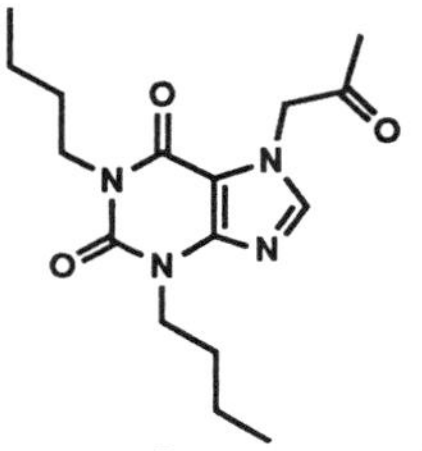   **6** 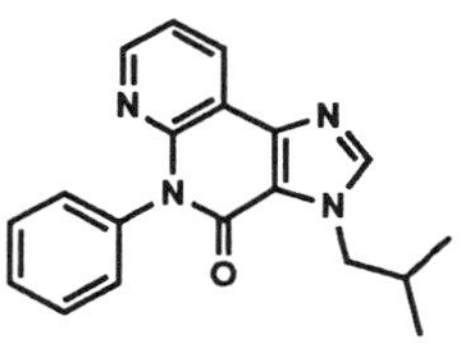   **7**

Bronchodilatory Activity of PDE IV Inhibitors - The pulmonary antiallergic activity of theophylline results in part from inhibition of PDE IV, and the bronchodilator activity of theophylline results in part from inhibition of PDE III (28). Siguazodan (PDE III inhibitor) and **2** produce only a modest reversal of agonist-induced (carbachol, 10μM) tone in intact human bronchi (7). However, when used in combination these two PDE inhibitors act in at least an additive fashion to induce relaxation, suggesting that PDE III and PDE IV co-regulate cAMP content in human airway smooth muscle. However, **3** (PDE III/IV inhibitor) demontrates only modest and short-lasting bronchodilatory activity after intranasal administration (29). Compounds **2** and **6** (denbufylline, BRL-30892) exhibit bronchodilatory activity suggesting that selective PDE IV inhibitors could act as bronchodilators in the clinic (30). The bronchodilator effects of a series of imidazo-[4,5-*c*]-naphthyridine-2-ones (*e. g.* **7**) do not correlate well with their PDE IV inhibitory potencies (31). Possible explanations might be differences in their membrane permeabilities and/or in their effects on other PDE isozymes. It is interesting that some of these compounds exhibit more potent antibronchospastic activity *in vitro* and *in vivo* than theophylline. Other examples of selective PDE IV inhibitors acting as bronchodilatory agents *in vivo* include **2** (32), **5** (33), **8** (AH-21-132) (34) and **9** (ORG-20241) (35).

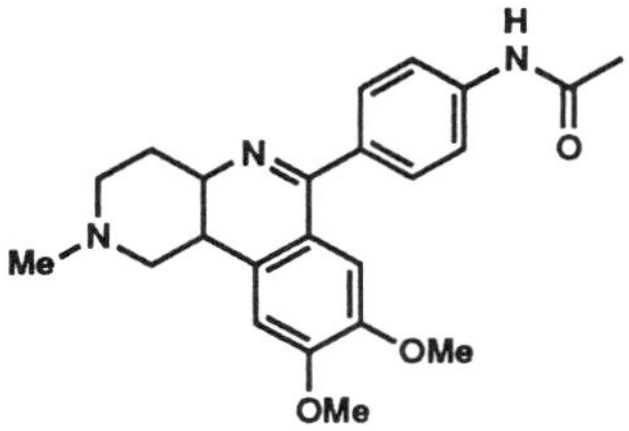

8 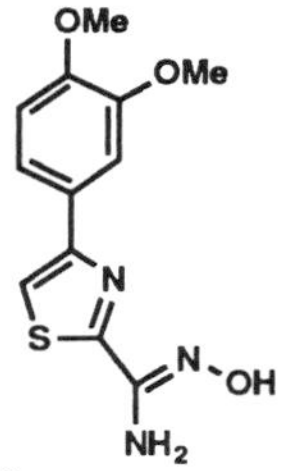   **9** 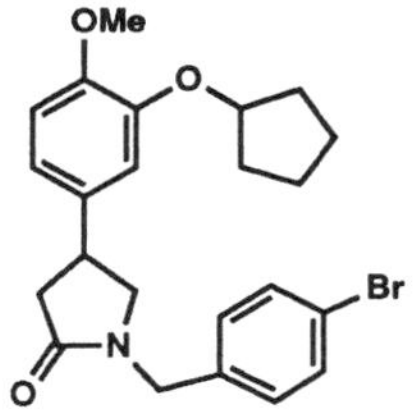   **10**

SAR of PDE IV Inhibitors - The SAR of PDE IV inhibitors has been reviewed (9, 10, 36). More recently, SAR development has been centered around the pyrrolidinone moiety of **2**. For example, the oxime carbamate derivative **5** is equipollent to **2** and is currently in development for the prophylactic treatment of asthma (37). The optically active analog *S*-**10**, a 4-bromobenzyl analog of *S* -**2**, is 10 times more potent than **2** and 400 times more potent than its *R*-enantiomer (18). Benzopyran derivative **11** displays much reduced potency on the high affinity form of PDE IV (38). Moreover, the 3,5-dichloropyridyl amide analog **12** (RP-73,401) is 3000-fold more potent than **2** (PDE IV, pig aorta) (39). In a different structural series, adenosine receptor-blocking xanthines (40,41), the PDE IV potencies are comparable to **2** [*e.g.* theophylline and **13** (BRL-61063)].

11   **12**   **13**

<u>PDE IV Subtypes</u> - The molecular biology of the PDE family isozymes has been reviewed (9). Several isoforms of the PDE IV isozyme have been cloned from rat and from man, and access to these isoforms should lead to even more selective PDE inhibitors. The degree of protein sequence identity is much greater between human brain PDE IV (*h*PDE IV-B) and a homolog derived from rat brain (92% over 562 amino acids) than between *h*PDE IV-B and human monocyte PDE IV, *h*PDE IV-A (76% over 538 amino acids), suggesting a greater subtype-specific versus species-specific conservation of protein sequence (42). Several cDNA clones from alternatively spliced mRNAs that encode *h*PDE IV-B were isolated by homology to a rat PDE IV-B clone and characterized (43). At least two alternatively spiced forms of the rat homologs of *h*PDE IV-D are present in rat Sertoli and FRTL-5 thyroid cells, but not in the brain (44). These findings demonstrate that the cAMP-specific PDE genes have a complex structure and that cAMP PDE proteins with different amino termini are derived from these genes. However, the significance of alternatively spliced forms is not clear. Three genes were expressed in *Saccharomyces cerevisiae* and encode cAMP-specific PDEs (45). The products of the expressed genes display the pattern of sensitivity to **2** and **6**. Each of the genes demonstrate a distinctive pattern of expression in RNA from human cell lines. cDNA clones that encode *h*PDE IV-D, a novel human cAMP-specific PDE of 604 amino acids, were isolated by homology to a rat PDE IV$_D$ clone and characterized (46). *h*PDE IV-D produced in *E. coli* was inhibited by **2**. Transcription of *h*PDE IV-D was most abundant in skeletal muscle.

<u>cA-PKA Agonists</u> - cA-PKAs are likely responsible for all known biological events of cAMP in eukaryotic cells (47,48,49). Growing evidence suggests that increased levels of cAMP within inflammatory cells block cell activation and inhibit release of chemical mediators (9, 10, 50). Thus, a cA-PKA agonist, similar to a PDE IV inhibitor, should manifest anti-inflammatory activities (51). Figure 1 shows the possible action and metabolism of cAMP following adenylate cyclase stimulation seen in inflammatory cells. Elevated cAMP binds two regulatory (R$_2$) subunits of the PKA holoenzyme, promoting dissociation of two monomeric catalytic (2C) subunits. The free C-subunit can phosphorylate preferred substrates in the cytoplasm and/or participate in the expression of cAMP-regulated genes resulting in decreased cell activation and mediator release (52,53,54). The anti-inflammatory properties of the cAMP analogs, dibutyryl cAMP and 8-bromo-cAMP, are classical examples of actions by cA-PKA agonists. In addition, cA-PKA regulates pulmonary receptor activity (55) and is defective in certain inflammatory diseases (56,57,58). For example, cA-PKA may be involved in the development of human lung β-adrenergic receptors (42). In psoriatic patients, cA-PKA activity in skin fibroblasts and erythrocytes correlates well with the severity of the disease (43,44). Recent molecular biology studies indicate that the enzyme has evolved into a cA-PKA family (59), which consists of 4 different R-subunit isozymes (i.e. RIα, RIβ, RIIα and RIIβ) and 3 C-subunit isozymes with splice variants (i.e. Cα, Cβ, Cγ, Cα2 and Cβ2). Two major cA-PKA holoenzymes are generally considered as Type I cA-PKA and Type II cA-PKA depending upon whether the RI- or RII-subunit is present in the holoenzyme structure. The structure function analysis of the catalytic subunits of cA-PKA has recently been reviewed (60).

A cAMP-mediated biological response for a given cell is presumably dictated by an individualized cA-PKA isozyme and its subcellular location (61,62). Thus, isozyme specificity and intracellular compartmentalization offer discrete biological responses to the stimulus of the multiple functional ligand, cAMP. While these features present opportunities to discover drugs selective for cA-PKA in inflammatory cells, the cellular distribution and functional role of the isozyme is unknown. Type II cA-PKA appears to be ubiquitous, whereas the tissue distribution of Type I cA-PKA is more restricted (63,64). RI isozymes are primarily cytoplasmic, and most of RII forms are bound to cell particulates, possibly via a specific A-kinase anchor protein (65). Recent publications suggest that activation of Type I cA-PKA may be responsible for cAMP-mediated proliferation response of human lymphocytes and human neoplastic B cells (66,67).

INHIBITORS OF LYMPHOCYTE-DERIVED CYTOKINES

Anti IgE - IgE is a likely cause of human atopic diseases. Inhibition of IgE production is well-recognized as a target for anti-allergy/anti-asthma drugs. Several small molecules [**14** (TEI-1338) and **15** (KSU-2178)] have recently been shown to prevent allergen-dependent IgE production *in vitro* or *in vivo* (68, 69). While precise mechanisms of the action remain unknown, these compounds may serve as tools to uncover a rate-limiting step of IgE production in lymphocytes. Without a defined mechanism or an established SAR for a given compound, it is difficult to design a drug with improved intrinsic activity and selectivity for IgE antibodies. Recent advances in immunology and molecular biology have shown that both T-cells and T-cell derived cytokines are required for IgE synthesis (70,71). These participants include T-cell surface molecules [gp39 (72,73), a ligand for CD40 on B cells, and CD21 (74,75)], IL-4 and IL-13 (76). A low-affinity IgE receptor (CD23) on B cells may also play a role in promoting IgE production by interacting specifically with surface CD21 on T cells (70,71). The potential role of these molecules allows the design of new anti-IgE therapies (chemicals or proteins) targeting T and B cell interactions. Interfering with allergic reactions prior to antibody antigen coupling may offer a broader spectrum of therapeutic activity than symptom-relief agents (77).

14   **15**

IL-4 modulators (IL-4 Blockers and IL-4 Biosynthesis Inhibitors) - IL-4 is a B and T cell stimulatory factor that is synthesized by Th-2 cells (78), mast cells (79), and basophils (80). IL-4 is a key mediator that initiates IgE synthesis in human lymphocytes. IL-4 also induces expression of vascular cell adhesion molecule on endothelial cells, thus promoting blood eosinophils migrating into pulmonary tissues (81). The results of animal and IL-4 transgenic mice studies (82,83,84) support the *in vitro* findings. Overexpression of IL-4 in Th-2-like cells has been postulated to be the cause of elevated serum IgE in asthmatic patients (85,86). A drug that interferes with the action and biosynthesis of IL-4 may be potentially useful for the treatment of allergy and asthma. The biological action of IL-4 is species-dependent. To evaluate potential therapeutic effect of IL-4 blockers in man, an immunodeficient murine model with engrafted human peripheral mononuclear leukocytes has recently been developed (87,88). *In vitro*, several compounds inhibit IgE production by mixed human lymphocytes after IL-4 challenge. Cromolyn (**16**), an inhaled anti-asthmatic drug, partially blocks IL-4 induced IgE production in normal subjects (89,90). The blocking effect of **16** is within concentration ranges of 10 nM - 1 μM, which is 10^2-10^3 more potent than its inhibitory effect on the degranulatory response in mast cells. Besides IgE, **16** has no effect on baseline levels of IgG, IgM and IgA *in vitro*. The potent anti-IgE property of **16** can be seen with another chromanone analog, **17** (nedocromil) (91). It appears that both compounds have no effect on the induction of germline ε transcript but inhibit Sμ --> Sε deletional switch recombination (92). In addition, compounds known to modulate transmembrane signaling pathways such as inhibitors of either tyrosine kinase (herbimycin A), phosphatase (okadaic acid, microcystin-LR, calyculin A), or PDEs and inducers of cAMP (PGE_2, forskolin) are reported to prevent IL-4 mediated IgE response (93). IFN-γ and non-T cell derived proteins such as vasoactive intestinal peptide, somatostatin (94), and IL-8 (95) are also blockers of the IL-4 response. None of the aforementioned agents are antagonists of IL-4 receptors, and they may block the IL-4 effect at an IL-4 signaling pathway level or at a post-transcriptional level. Efforts are needed to dissect the MOA of these blockers. Use of biological agents such as anti-IL-4 antibodies, soluble IL-4 receptors and IL-4 peptides to inhibit the IL-4 action have been reviewed (96).

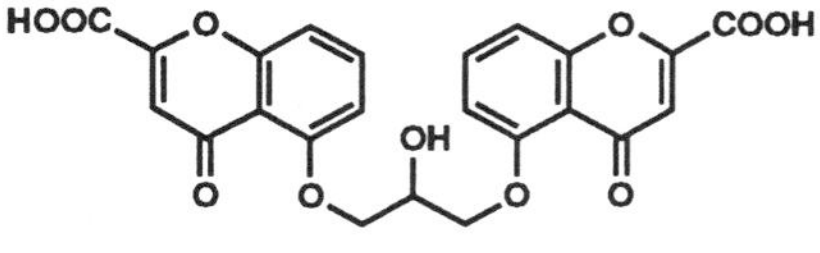

16 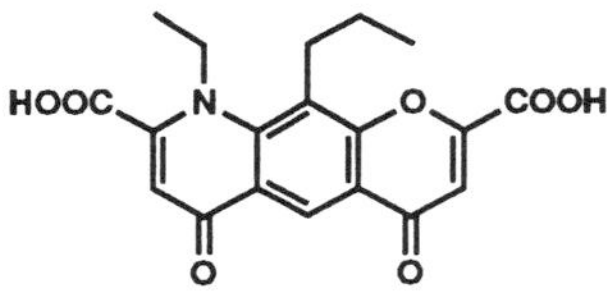   17

To date no small molecules have been shown to antagonize IL-4 receptors. IL-4 binds to IL-4 receptors with a binding constant (Kd) of ~0.05 nM (97,98,99,100). IL-4 dissociates from its binding site with a $t_{1/2}$ of 2-4 hours at 4°C. Site-directed mutagenesis, anti-peptide antibody, and synthetic peptide studies reveal the importance of certain amino acids (e.g. the disulfide bridge, cys[46]-cys[99], tyr[124]) of the IL-4 molecule involved in the binding to its receptors (101,102,103). Based on the crystal and solution structures of recombinant human IL-4 and the hypothesis that IL-4/IL-4 receptor binding is topologically analogous to growth hormone/growth hormone receptor interactions (104,105), critical receptor binding regions of IL-4 have been proposed.

IL-13 is a recently discovered cytokine that also induces germline ε transcript expression and IgE biosynthesis. Since the biological activities of IL-4 and IL-13 are very similar, receptors for both ligands may share a common subunit. One likely candidate is the IL-2 γ chain, which is a functional component of the IL-4 receptor complex (106,107,108).

Besides blocking the action of IL-4, inhibitors of IL-4 biosynthesis should be anti-IgE agents. A case study revealed that after 1.5-5 weeks treatment with the IL-4 biosynthesis inhibitor, Intron A (IFNα), a patient showed an improvement in symptoms (eczema) and a decrease in serum IgE, supporting the potential usefulness of IL-4 inhibitor in man (109). Compound 18 (IPD-1151T) (110) inhibits allergen-mediated IL-4 production from helper T cells. A phase III clinical study showed that serum levels of total and allergen-specific IgE were significantly reduced by 18 (300 mg/day for 6-13 wks).

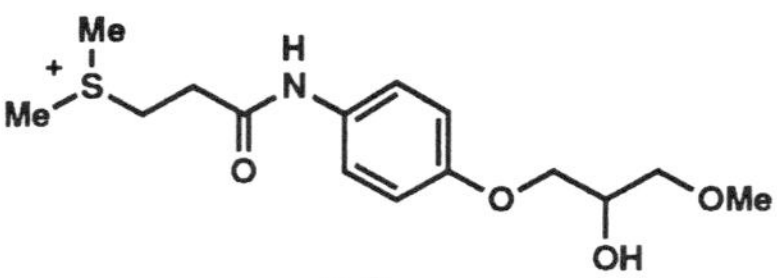

18

IL-5 Modulators - IL-5 is a Th-2 cell derived stimulatory factor that is selective for eosinophil proliferation, differentiation, and activation (111,112). IL-5 prolongs survival of eosinophils in cell cultures (113). Involvement of IL-5 or overexpression of cytokines from Th2-like cells in asthma has been recently reported. IL-5 induces pulmonary eosinophilia and bronchial hyperresponsiveness in guinea pigs (114,115). Intraperitoneal administration of prednisolone or anti-IL-5 antibody prevents the pulmonary eosinophil response induced by IL-5 or antigen. When given orally to mice, roxithromycin, a macrolide antibiotic, blocks Con A-induced IL-5 production from splenic cells *ex vivo* (116). Again, the mode of actions of these IL-5 modulators needs to be studied. The solution structure of IL-5 has been solved (117).

References

1.   N. Chanarin and S. L. Johnston, Drugs, 47, 12 (1994).
2.   J. S. Sawyer and D. L. Saussy, Jr., Drug News & Perspectives, 6, 139 (1993).
3.   J. B. Cheng, A. Goldfien, P. L. Ballard and J. M. Roberts, Endocrinology, 107, 1646 (1980).
4.   P. J. Barnes and R. A. Pauwels, Eur. Respir. J., 7, 579 (1994).
5.   J. A. Beavo and D. H. Reifsnyder, Trends Pharmacol. Sci., 11, 150 (1990).
6.   C. D. Nicholson, R. A. J. Challiss and M. Shahid, Trends Pharmacol. Sci., 12, 19 (1991).
7.   T. J. Torphy, B. J. Undem, L. B. Cieslinski, M. A. Luttmann, M. L. Reeves and D. W. P. Hay, J. Pharmacol. Exp. Ther., 265, 1213 (1993).
8.   C. D. Nicholson and M. Shahid, Pulmonary Pharmacol., 7, 1 (1994).
9.   T. J. Torphy, G. P. Livi and S. B. Christensen, Drug News & Perspectives, 6, 203 (1993).
10. J. A. Lowe, III and J. B. Cheng, Drugs of the Future, 17, 799 (1992).
11. M. A. Giembycz and G. Dent, Clinical Exp. Allergy, 22, 337 (1992).
12. M. Giembycz, Biochem. Pharmacol., 43, 2041 (1992).
13. S-H. Li, S. C. Chan, A. Toshitani, D. Y. M. Leung and J. M. Hanifin, J. Invest. Dermatol., 99, 65 (1992).
14. S. C. Chan, S-H. Li and J. M. Hanifin, J. Invest. Dermatol., 100, 681 (1993).
15. F. U. Schade and C. Schudt, Eur. J. Pharm., 230, 9 (1993).
16. J. Mulzer, R. Zuhse and R. Schmiechen, Angew. Chem. Int. Ed. Engl., 31, 870 (1992).
17. A. I. Meyers and L. Snyder, J. Org. Chem., 58, 36 (1993).
18. P. W. Baures, D. S. Eggleston, K. F. Erhard, L. B. Cieslinski, T. J. Torphy and S. B. Christensen, J. Med. Chem., 36, 3274 (1993).
19. J. Semmler, H. Wachtel and S. Endres, Int. J. Immunopharmacol., 15, 409 (1993).
20. K. Molnar-Kimber, L. Yonno, R. Heaslip and B. Weichman, Agents and Actions, 39, Special Conference Issue, C77 (1993).
21. K. L. Molnar-Kimber, L. Yonno, A. Rhoad, B. M. Weichman and R. J. Heaslip, J. Immunol, 150, Part 2, 295A (1993).
22. D. A. Hartman, S. J. Ochalski and R. P. Carlson, Agents and Actions, 39, Special Conference Issue, C70 (1993).
23. W. Fiers, FEBS Lett., 285, 199 (1991).
24. J. E. Souness and L. C. Scott, Biochem. J., 291, 389 (1993).
25. D. E. Griswold, E. F. Webb, J. Breton, J. R. White, P. J. Marshall and T. J. Torphy, Inflammation, 17, 333 (1993).
26. S. J. Newsholme and L. Schwartz, Inflammation, 17, 25 (1993).
27. J. C. Kips, G. F. Joos, R. A. Peleman and R. A. Pauwels, Clin. Exp. Allergy, 23, 518 (1993).
28. R. E. Howell, B. D. Sickels and S. L. Woeppel, J. Pharmacol. Exp. Ther., 264, 609 (1993).
29. T. Brunnee, R. Engelstatter, V. W. Steinijans and G. Kunkel, Eur. Respir. J., 5, 982 (1992).
30. J. Cortijo, J. Bou, J. Beleta, I. Cardelus, J. Llenas, E. Morcillo and R. W. Gristwood, Br. J. Pharmacol., 108, 562 (1993).
31. F. Suzuki, T. Kuroda, T. Kawakita, H. Manabe, S. Kitamura, K. Ohmori, M. Ichimura, H. Kase and S. Ichikawa, J. Med. Chem., 35, 4866 (1992).
32. A. Tomkinson, J-A. Karlsson and D. Raeburn, Br. J. Pharmacol., 108, 57 (1993).
33. R. J. Heaslip, L. J. Lombardo, J. M. Golankiewicz, B. A. Ilsemann, D. Y. Evans, B. D. Sickels, J. K. Mudrick, J. Bagli and B. M. Weichman, J. Pharmacol. Exp. Ther., 268, 888 (1994).
34. R. W. Foster, K. Rakshi, J. R. Carpenter and R. C. Small, Br. J. Pharmacol., 34, 527 (1992).
35. C. D. Nicholson, M. Shahid, J. Bruin, J. de Boer, R. G. M. van Amsterdam, J. Zaagsma, G. Dent, M. A. Giembycz and P. J. Barnes, Br. J. Pharmacol., 107, 252P (1992).
36. P. W. Erhardt in "Cyclic Nucleotide Phosphodiesterases: Structure, Regulation and Drug Action," J. Beavo and M. D. Housley, Eds., Wiley, Chichester, 1990, pp. 320-328.
37. L. J. Lombardo, J. F. Bagli, J. M. Golankiewicz, R. J. Heaslip and J. K. Mudrick, 205th ACS National Meeting, Denver, MEDI-128 (1993).
38. I. L. Pinto, D. R. Buckle, S. A. Readshaw and D. G. Smith, Bioorg. Med. Chem. Lett., 3, 1743 (1993).
39. J. F. Souness, S. E. Webber, K. Pollock, M. N. Palfreyman, M. J. Ashton, D. Raeburn and J-A. Karlsson, FASEB J., 8, A371 (1994).
40. D. Ukena, C. Schudt and G. W. Sybrecht, Biochem. Pharmacol., 45, 847 (1993).

41.   D. R. Buckle, J. R. S. Arch, B. J. Connolly, A. E. Fenwick, K. A. Foster, K. J. Murray, S. A. Readshaw, M. Smallridge and D. G. Smith, J. Med. Chem, <u>37</u>, 476 (1994).

42.   M. M. McLaughlin, L. B. Cieslinski, M. Burman, T. J. Torphy and G. P. Livi, J. Biol. Chem., <u>268</u>, 6470 (1993).

43.   R. Obernolte, S. Bhakta, R. Alvarez, C. Bach, P. Zuppan, M. Mulkins, K. Jarnagin and E. R. Shelton, Gene, <u>129</u>, 239 (1993).

44.   L. Monaco, E. Vicini and M. Conti, J. Biol. Chem., <u>269</u>, 347 (1994).

45.   G. Bolger, T. Michaeli, T. Martins, T. St. John, B. Steiner, L. Rodgers, M. Riggs, M. Wigler and K. Ferguson, Mol. Cell. Biol., <u>13</u>, 6558 (1993).

46.   P. A. Baecker, R. Obernolte, C. Bach, C. Yee and E. R. Shelton, Gene, <u>138</u>, 253 (1994).

47.   D. A. Wash, J. P. Perkins, and E. G. Krebs, J. Biol. Chem., <u>243</u>, 3763 (1968).

48.   E. G. Krebs, and J. A. Bevo, Ann. Rev. Biochem., <u>48</u>, 923 (1979).

49.   S. J. Beebe, and J. D. Corbin, The enzymes: Control by Phosphorylation, Part A ed. by E.G. Crebs and P. D. Boyer) 17, 43 (1986).

50.   T. J. Torphy, and B. J. Umdem, Throax, 46, 512 (1991).

51.   P. N. Christopher, J. J. Crowley, M. E. Morgan, and R. E. Vestal, Am. Rev. Resp. Dis. 137, 25, (1988)

52.   J. L. Meinkoth, Y. Ji, S.S. Taylor, and J.R. Feramisco, Proc. Natl. Acad. Sci. (USA) <u>87</u>, 9595 (1990).

53.   S.R. Adams, A.T. Harootunian, Y.J. Buechler, S. S. Taylor, and R.Y. Tsien, Nature, <u>349</u>, 694 (1991).

54.   J.C. Chrivla, R. P.S. Kwok, N. Lamb, M. Hagiwara, M.R. Montminy and R.H. Goodman, Nature, <u>365</u>, 865 (1993).

55.   D. M. Duffy, P.L. Ballard, A. Goldfien, and J.M. Roberts, Endocrinol., <u>131</u>, 841 (1992).

56.   D. E. Brion, F. Raynaud, A. Plet, A. P. Laurent, B. Leduc, and W. Anderson, Proc. Natl. Acad. Sci. (USA) 83 (1986).

57.   F. Raynaud, P. Gerbaud, O. Enjolras, I. Gorin, M. Donnadieu, W.B. Anerson, and D. Evain-Brion, Lancet, <u>1</u>, 1153 (1989).

58.   F. Raynaud, P. Gerbaud, A. Bouloc, I. Gorin, W.B. Anderson, and D. Evain-Brion, J. Invest. Dermatol., <u>100</u>, 77,(1993).

59.   S.S. Taylor, J.A. Buechler, and W. Yonemoto, Ann. Rev. Biochem., <u>59</u>, 971 (1990).

60.   S.S. Taylor, D.R. Knighton, J. Zheng, J.M. Sowadski, C.S. Gibbs, and M.J. Zoller, TIBS, <u>18</u>, 84 (1993).

61.   S.mmmm. Lohmann, P. DeCamilli, I. Einig, and U. Walter, Proc. Natl. Acad. Sci. (USA) <u>82</u>, 6723 (1984).

62.   J.D. Scott. Pharmacol. Therap., <u>50</u>, 123 (1991).

63.   C.S. Rubin, J. Erlichman, and O.M. Rosen, J. Biol. Chem., <u>247</u>, 6135 (1972).

64.   J.D. Corbin, P.H. Sugden, L. West, D.A. Flockhart, T.M. Lincoln, and D. McCathy, J. Biol. Chem., <u>253</u>, 3997 (1978).

65.   V.M. Coghlan, L.K. Langeberg, A. Fernandez, N.J.C. Lamb, and J. D. Scott, J. Biol. Chem., <u>269</u>, 7658 (1994).

66.   B.S. Skalhegg, K. Tasken, V. Hansson, H.S. Huitfeidt, T. Jahnsen and T. Lea, Science, <u>263</u>, 84 (1994).

67.   K. Tasken, B. S. Skalhegg, R. Solberg, K.B. Andersson, S.S. Taylor, T. Lea, H.D. Blomhoff, T. Jahnsen, and V. Hansson, J. Biol. Chem., <u>268</u>, 21276 (1993).

68.   H. Ohmori, A. Hazato, Y. Kato, and S. Kurozumi, Int. J. Immunopharmacol., <u>12</u>, 333 (1990).

69.   Y. Kurashina, and Y. Tanioka, Jap. J. Pharmacol., <u>49</u> (suppl.), 165P (1989).

70.   W.E. Paul, Blood, <u>77</u>, 1859 (1991).

71.   R.L. Coffman, J. Ohara, M.W. Bond, J. Carty, E. Zlotnick, and W.E. Paul, J. Immunol., <u>136</u>, 4538 (1986).

72.   R.J. Noelle, M. Roy, D.M. Shepherd, I. Stamenkovic, J.A, Ledbetter, and A. Aruffo, Proc. Natl. Acad. Sci. (USA), <u>89</u>, 6550 (1992).

73.   R.J. Armitage, W.C. Fanslow, L. Strockbine, T.A. Sato, K.N. Clifford, B.M. Macduff, D.M. Anderson, S.D. Gimpel, T. Davis-Smith, C.R. Malixzewski, E.A. Clark, C.S. Smith, K.H. Grabstein, D. Cosman, and M.K. Spriggs, Nature, <u>357</u>, 80 (1992).

74.   L. Flores-Romo, J. Shields, Y. Humbert, P. Graber, J-P. Aubry, J-F. Gauchat, G. Ayala, B. Allet, M. Chavez, H. Bazin, M. Capron, and J-Y. Bonnefoy, Science, <u>261</u>, 1038 (1993).

75.   J-P. Aubry, S. Pochon, P. Graber, K.U. Jansen, and J-Y. Bonnefoy, Nature, <u>358</u> (1992).

76.   T. Defrance, P. Carayon, G. Billian, J-C. Guillemot, A. Minty, D. Caput, and P. Ferrara, J. Exp. Med., <u>179</u>, 135 (1994).

77.   T. A. E. Platts-Mills, Am. Rev. Respir. Dis., <u>145</u>, S44 (1992).

78.   M. Howard, J. Farrar, M. Hilfiker, B. Johnson, K. Takatsu, T. Hamaoka, and W.E. Paul, J. Exp. Med., <u>155</u>, 914, (1982).

79.   M. Plaut, J.H. Pierce, C.J. Watson, J. Hanley-Hyde, R.P. Nordan, and W.E. Paul, Nature, <u>339</u>, 64 (1989).

80. D.C. Seder, W.E. Paul, A.M. Dvorak, S.J. Sharkis, A. Kagey-Sobotka, Y. Niv, F.D. Finkelman, S.A. Barbieri, S.J. Galli, and M. Plaut, Proc. Natl. Acad. Sci. (U.S.A.), 88, 2835 (1991).

81. R. P. Schleimer, S.A. Sterbinsky. J. Kaiser, C.A. Bickel. D.A. Klunk, K. Tomioka, W. Newman, F.W. Luscinskas, M.A. Gimbrone, Jr., B. W. McIntyre, and B.S. Bochner, J. Immunol., 148, 1066 (1992).

82. R.I. Tepper, D.A. Levinson, B. Z. Stanger, J. Campos-Torres, A.K. Abbas, and P. Leder, Cell, 62, 457 (1990).

83. G.G. Brusselle, J.C. Kips, J.H. Tavernier, J.G. van der Heyden, C.A. Cuvelier, R.A. Pauwels, and H. Bluethmann, Clin. Exptl. Allergy, 24, 73 (1994).

84. R. Kuhn, K. Rajewsky and W. Muller, Science, 254, 707 (1991).

85. D.S. Robinson, Q.Hamid, S. Ying, A. Tsicopoulos, J. Barkans, A.M. Bentley, C. Corrigan, S. R. Durham, and A.B. Kay, New Engl. J. Med. 326, 298 (1992).

86. M.L. Kapsenberg, E.A. Wiwewnga, J.D. Bos, and H.M. Jansen, Eur. Resp. J., 4, (Suppl. 13), 27s (1991).

87. M. Ito, G. Matsuzaki, Uno, S., Katakai, Y., Suko, M., Endo, S. and Okudaira, Int. Arch. Allergy Immunol., 99, 373 (1992).

88. H.L.Spiegelbery, L. Beck, H.P. Kocher, W.C. Fanslow, and A.H. Lucas, J. Clin. Invest., 93, 711(1994).

89. H. Kimata, A. Yoshida, C. ishioka, and H. Mikawa, Clin. Exp. Immunol., 84, 395 (1991).

90. R.K.S. Loh, H.H. Jabara, and R.S. Geha, J. Allergy Clin. Immunol., 93, 219 (1994).

91. H. Kimata, and H. Mikawa, J. Immunol., 151, 6723 (1993).

92. R. K. S. Loh, H. H. Jabara and R. S. Geha, J. Allergy Clin. Immunol., 93, A219 (1994).

93. D. Armerding, and A. Hren, Int. Arch. Allergy Immunol., 101, 143 (1993).

94. H. Kimata, A. Yoshida, M. Fujimoto, and H. Mikawa, J. Immunol., 150, 4630 (1993).

95. H. Kimata, A. Yoshida, C. Ishioka, I. Lindley, and H. Mikawa, J. Exp. Med., 176, 1227 (1992).

96. K. Cooper and H. Masamune, Ann. Rep. Med. Chem., 27, 209 (1992).

97. J. Ohara and W.E. Paul, Nature, 325, (1987).

98. L.S. Park, D. Friend, H.M. Sassenfeld and D.L. Urdal, J. Exp. Med., 166, 476 (1987).

99. H. Cabrillat, J-P. Galizzi, O. Djossou, N. Arai, T. Yokota, K. Arai, and J. Banchereau, Biochem. Biophys. Res. Comm., 149, 995 (1987).

100. P. Valent, J. Besemer, K. Kishi, F. D. Padava, K. Geissler, K. Lechner, and P.Bettlheim, Blood, 76, 1734 (1990).

101. N. Kruse, T. Lehrnbecher, and W. Sebald, FEBS Lett., 286, 58 (1991).

102. A.E. Postlethwaite, and J.M. Seyer, J. Clin. Invest., 87, 2147 (1990).

103. S.C. Bischoff, A.de Weck, and C.A. Dahinden, Lymphokine Cytokine Res., 11, 33 (1992).

104. R.Powers, D.S. Garrett, C.J. March, E.A. Frieden, A.M. Gronenborn, and G.M. Clore, Science, 256, 1673 (1992).

105. M. R. Walter, W. J. Cook, B. G. Zhao, R. P. Cameron, Jr., S. E. Ealick, R. L. Walter, Jr., P. Reichert, T.L. Nagabhushan, P. P. Trotta, and C. E. Bugg, J. Biol. Chem., 267, 20371 (1992).

106. M. Kondo, T. Takeshita, N. Ishii, M. Nakamura, S. Watanabe, K-i. Arai, K. Sugamura, Science, 262, 1874 (1993).

107. S.M. Russell, A.D. Keegan, N. Harada, Y. Nakamura, M. Noguchi, P. Leland, M.C. Griedmann, A. Miyajima, R.K.Puri, W.E. Paul, and W.J. Leonard, Science, 262, 1880 (1993).

108. A.D. Keegan, K. Nelms, M. White, L-M. Wang, J.H. Pierce, and W.E. Paul, Cell, 76, 811 (1994).

109. J-F. Gauchat, H. Gascan, M-G. Roncarolo, F. Rousset, J. Pene, and J.E. de Vries, Eur. Resp.J. 4 (Suppl 13), 31s (1991).

110. Y. Yanagihara, M. Kiniwa, T. Shida, and A. Koda, Eur. J. Pharmacol., 183, P626 (1990).

111. H.D. Campbell, W.Q.J. Tucker, Y. Hort, M.E. Martinson, G. Mayo, E.J. Clutterbuck, C.J. Sanderson, and I.G. Young, Proc. Natl. Acad. Sci. (USA), 84, 6629 (1987).

112. C.J. Sanderson, H.D. Campbell, and I.G. Young, Immunol. Rev., 102, 29 (1988).

113. G. J. Gleich, C. R. Adolphson and K. M. Leiferman, Ann. Rev. Med., 44, 85 (1993).

114. T. Iwama, H. Nagai, H. Suda, N. Tsuruoka, and A. Koda, Brit. J. Pharmacol., 105, 19 (1992).

115. T. Iwama, H. Nagai, N. Tsuruoka, and A. Koda, Clin. Exp. Allergy, 23, 32 (1993).

116. S-I. K, M. Adachi, K. Asano, K-I. Okamoto and T. Takahashi, Life Sci., 52, 25 (1990).

117. M. V. Milburn, A. M. Hassell, M. H. Lambert, S. R. Jordan, A. E. I. Proudfoot, P. Graber and T. N. C. Wells, Nature, 363, 172 (1993).

Chapter 9. The Enzymology and Manipulation of Nitric Oxide Synthase

Jon M. Fukuto[1] and Yumiko Komori[2]

[1]Department of Pharmacology, UCLA School of Medicine, Center for the Health Sciences, Los Angeles, California 900-24-1735. [2]Faculty of Pharmacy, Meijo University, 150 Yagotoyama, Tenpaku, Nagoya, 468 Japan.

Introduction - Until several years ago, nitric oxide (NO) was considered to be a unique and curious molecule of significant interest only to inorganic, organometallic and environmental chemists. However, with the discovery of its biosynthesis in mammalian cells, NO has become a molecule of extreme interest to biologists, pharmacologists, physiologists and biochemists as well. In mammalian systems, NO is synthesized via an enzymatic oxidation of the amino acid arginine by a class of enzymes known collectively as the nitric oxide synthases (NOS). The details of this enzymatic process will be discussed in some detail later in the text. The physiological functions of NO are diverse and ubiquitous and are the subject of several recent reviews (1-4). As a testimony to its importance and diversity in biological systems, NO was recently chosen as "1992 Molecule of the Year" by the editors of Science (5). Although endogenous NO generation has only recently been demonstrated, the pharmacology of NO has been of interest for awhile. NO and NO-donor compounds have been shown to cause vasorelaxation by activating the enzyme guanylate cyclase which catalyzes the conversion of GTP to cGMP (for example, see 6). Elevated intracellular cGMP levels, through a series of events, results in smooth muscle relaxation (7). Thus, NO-generating compounds, such as nitroglycerin, have been used as vasorelaxants for many years. However, it is only since 1987 that it became apparent that these compounds were eliciting their biological response by utilizing an endogenous NO system. One of the first observations of a biological response mediated by endogenous NO generation was the report that acetylcholine treatment of endothelial cells resulted in the release of a short-lived species capable of causing smooth muscle relaxation (8). At that time, this as yet unidentified short-lived species was termed endothelium-derived relaxing factor (EDRF). Seven years later, the identity of EDRF was proposed to be NO by several groups (9,10).

Along with its function as a messenger molecule in the vascular system, NO has also been shown to be an effector molecule used in immune system response. Prior to the discovery of mammalian NO biosynthesis, it was observed that macrophages are "activated" to produce nitrate and nitrite when exposed to infectious agents (11). Later it was found that nitrate and nitrite were produced as a result of initial NO generation by macrophages (the primary oxidative decomposition products of NO in a biological system are nitrate and nitrite) (12). Thus, NO appears to be utilized by macrophages as a chemical agent to combat infection. It has also been proposed that NO can disrupt cellular metabolism by ligating crucial iron ions involved in, for example, mitochondrial respiration or DNA biosynthesis (13). Interestingly, in the presence of superoxide, (O_2^-), another species released by activated macrophages, NO can be oxidized to peroxynitrite (^-OONO) which is a potent oxidant and is capable of destroying or modifying biological molecules. Thus, peroxynitrite may be responsible for some of the observed cytotoxicity caused by activated macrophages (14).

The apparent biological utility of NO also extends into the central nervous system. For example, it has been known since 1974 that the excitatory amino acid neurotransmitter glutamate causes an elevation of cGMP in cerebellar slices and that this phenomenon is dependent on Ca^{+2} (15). Further work on this system led to the proposal that glutamate binding to the postsynaptic receptor allows an influx of Ca^{+2} into the neurons which, in turn, stimulates the biosynthesis of NO. As in the vascular system, NO is capable of activating guanylate cyclase which results in elevated cGMP levels. Although the implications of these events in the brain are incompletely understood, NO biosynthesis appears to be integral to the excitatory amino acid neurotransmitter signal transduction pathway and in cell to cell signaling (16,17).

In the peripheral nervous system, NO has been implicated as a neurotransmitter in non-adrenergic, non-cholinergic (NANC) neurons. That is, electrical stimulation of these neurons results in the release of a substance which causes an increase in cGMP in smooth muscle tissues which then results in relaxation. The physical properties of the released species are similar to those of NO (18). NANC neurons innervate a variety of smooth muscle tissues including intestine, trachea, corpus cavernosum and esophagus. Although, it has not been unequivocally demonstrated that NO is the neurotransmitter in all these systems, it appears to be involved somewhere in the signal transduction pathway.

NO plays a role in a variety of other physiological functions as well. For example, it has been postulated to be involved in long-term potentiation (thus involved in memory and learning) and platelet adhesion and aggregation [1,2]. Moreover, various disease states or physiological disorders have been attributed to abnormal NO biosynthesis. For example, endotoxin shock, insulin-dependent diabetes and reperfusion toxicity can apparently result from an overproduction of NO whereas angina pectoris, male impotence or hypertension could be due to a lack of NO biosynthesis[1,2] Clearly NO is an important messenger and/or effector molecule in biological systems and it is likely that many other NO-mediated physiological events will be discovered in the years to come.

It is quite evident that pharmacological manipulation of NO levels or NO biosynthesis may prove to be an important strategy for the treatment of a variety of disorders related to abnormal NO production. However, in order to fully understand the factors governing endogenous NO generation, it is imperative to have an appreciation of the enzymology of NO biosynthesis. Therefore, the purpose of this short review is to provide an introduction to the current understanding of the nature of the NOS catalyzed oxidation of arginine. For a more comprehensive treatment of the enzymology of NO biosynthesis, the reader is referred to the recent review by Stuehr and Griffith (19).

NOS Isoforms and Regulation - As mentioned previously, NO is biosynthesized from the enzymatic oxidation of a terminal guanidinium nitrogen on the amino acid arginine. The other product of the reaction is citrulline which is formed in equimolar amounts (20) (Figure 1).

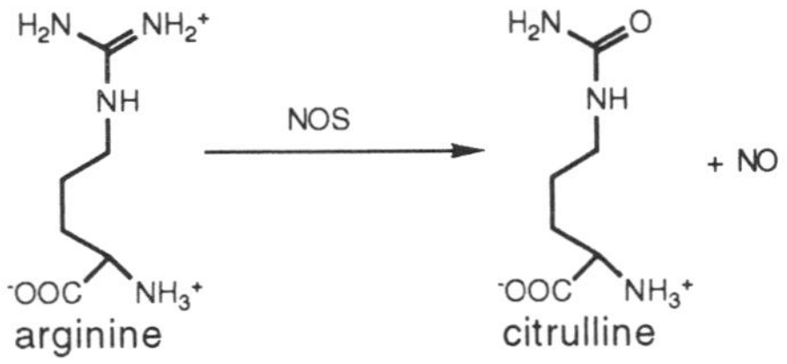

Figure 1: Reaction catalyzed by NOS.

There are a variety of NOS isoforms which can be distinguished on the basis of their method of regulation, physical properties and subcellular distribution [19]. In general, the NOS enzymes can be divided into two distinct classes, an inducible (iNOS) and a constitutive (cNOS) class. The constitutive isoform can be found in a variety of tissues and cells including brain, heart, endothelial cells, epithelial cells and platelets. One of the outstanding characteristics of the cNOS isoforms is that they are all subject to regulation by calcium (Ca^{+2}) *via* the Ca^{+2} binding protein, calmodulin (CaM). That is, increases in local or intracellular Ca^{+2} levels results in cNOS activation through an interaction with a Ca^{+2}/CaM complex. Thus, NO biosynthesis by cNOS is primarily regulated by physiological control of intracellular Ca^{+2} levels. In contrast to cNOS, the primary method of iNOS regulation appears to be the induction of *de novo* iNOS protein synthesis. Thus, macrophages, hepatocytes or tumor cells, for example, can be induced to synthesize iNOS by exposure to a variety of cytokines or bacterial-derived agents such as lipopolysaccharides, gamma-interferon or tumor necrosis factor. Currently it is not known how or if NO production from expressed iNOS is physiologically regulated. Interestingly, iNOS does bind CaM, but this binding is sufficiently strong that it can be considered to be more like a subunit as opposed to a regulatory protein (21). Although it is not known if Ca^{+2} is involved in the binding of CaM to iNOS, it is evident that Ca^{+2} does not have a regulatory role. Clearly, the differences in iNOS and cNOS regulation reflect the differences in their physiological roles or function. For example, the use of NO from cNOS for the maintenance of vascular tone requires careful regulation whereas NO produced from macrophages for use as a cytotoxic agent may not require the same degree of physiological control. It should also be pointed out that brain, endothelial cell and macrophage derived NOS have consensus sites for cAMP-dependent phosphorylation which may play a role in their regulation as well (22).

Along with the distinctions associated with the regulation of activity, NOS can be further subdivided into cytosolic and particulate isoforms. For example, the cNOS in endothelial cells was shown to be primarily associated with membranes as a result of N-myrisoylation (23,24). cNOS from brain and iNOS from macrophages, on the other hand, are primarily cytosolic proteins (although a particulate isoform from rat brain has been reported (25)). To give some indication of the diversity of NOS proteins, Table 1 provides a partial listing of some NOS isoforms, their method of regulation and some of their physical properties. Surely this list will be expanded and modified as more is learned about NOS and new isoforms are discovered in other tissues and cells.

Table 1: Representative NOS Isoforms

Source	Type/Regulation	Properties
macrophages (72,73)	Inducible/*de novo* synthesis	Soluble, K_m^1 = 2.8-16 µM 250-260 kDa (native)
rat liver (74)	Inducible/*de novo* synthesis	Soluble, K_m^1 = 11 µM 135 kDa (denatured)
bovine endo. cells (75)	Constitutive/Ca^{+2}-CaM	Particulate, K_m^1 = 2.9 µM 135 kDa (denatured)
rat cerebellum (25)	Constitutive/Ca^{+2}-CaM	Particulate, K_m^1 = 15.5 µM 150 kDa (denatured)
rat cerebellum (27)	Constitutive/Ca^{+2}-CaM	Soluble, K_m^1 = 2.2 µM 279 kDa (native)
porcine cerebellum (76)	Constitutive/Ca^{+2}-CaM	Soluble, K_m not reported 200 kDa (native)

[1] K_m for L-arginine.

Many of the mechanistic or biophysical studies on NOS have focused on iNOS from macrophages and cNOS from rat brain. In depth studies on iNOS from macrophage indicate that the enzyme is active only as a homodimer (26). It has also been shown that the dimerization of the monomers is promoted by substrate and several of the enzyme cofactors and/or prosthetic groups (26). cNOS from rat cerebellum also exists as a homodimer in the native form (27). It has yet to be demonstrated whether or not the cNOS monomers are active. However, if cNOS is analogous to iNOS it might be expected that only the dimeric protein is active. It is not known if a monomer-dimer equilibrium has any regulatory role in physiological NO production. However, this phenomenon must be considered when interpreting experiments with purified or semi-purified NOS since *in vitro* conditions may cause the dissociation of the monomers with consequent loss of activity.

Cofactor and Prosthetic Group Requirements - Both iNOS and cNOS (at least those isoforms which have been well characterized) have identical cofactor and prosthetic group requirements. Both utilize NADPH and O_2 and require FAD, FMN, iron-heme and tetrahydrobiopterin (BH_4) for catalytic activity (for example, see 19 and references therein). Thiols also appear to be required for maximum activity (28). Based on cofactor and prosthetic group requirements, NOS bears a striking similarity to the well-studied cytochrome P450 monooxygenase system (P450). The P450 system is composed of two proteins, cytochrome P450 reductase and an ironheme containing cytochrome P450. In this system, the reductase protein is responsible for binding NADPH and, through a conduit consisting of FAD and FMN, transfers electrons to the ironheme component of cytochrome P450. The reduced ironheme then binds and activates molecular oxygen which results in the eventual incorporation of one oxygen atom into the substrate and one into water (eqn. 1) (29).

1) $NADPH + O_2 + S\text{-}H \rightarrow NADP^+ + S\text{-}OH + H_2O$ (S-H = substrate)

NOS has significant sequence homology with cytochrome P450 reductase (30) and incorporation of oxygen atoms from molecular oxygen (as opposed to water) into both citrulline and NO has been demonstrated (31). Moreover, it has been shown that CaM regulates cNOS activity by allowing electron transfer to occur from NADPH to the ironheme group (32) which is required for NO generation. (Since iNOS has CaM tightly bound to it, Ca^{+2} does not regulates its activity). Thus, based on the above observations, it appears that NOS is a self-sufficient NADPH-dependent oxygenase with mechanistic similarity to P450.

The analogy between P450 and NOS is not, however, absolute. A significant difference between the two systems is that NOS, unlike P450, has a requirement for BH_4. The possible role of BH_4 in NOS catalysis is, as yet, unclear. Other non-heme oxygenase enzyme systems, such as tyrosine or phenylalanine hydroxylase, utilize BH_4 as a redox active cofactor for the direct activation of molecular oxygen (33). In these systems BH_4 is oxidized to a dihydro-form (BH_2) during catalysis which is then recycled back to BH_4 by separate NAD(P)H-dependent reductase enzymes. Clearly, NOS is not utilizing BH_4 in this way since NOS is a single protein system (although redox-cycling of BH_4 while bound to the NOS protein is certainly possible and has been proposed (34)). It has been postulated that BH_4 may have an allosteric role in NOS activity or simply act to keep a crucial catalytic component in a reduced state (35). Consistent with an allosteric role for BH_4, dimerization of inactive monomeric protein to the active dimer is promoted by the presence of BH_4 (26). Other workers have postulated a redox role for BH_4 (36). The exact nature of the BH_4 requirement by NOS is a matter of considerable current interest and speculation and it remains a mechanistic enigma.

The Mechanism of NOS Catalyzed NO Generation - The generation of NO from the NOS catalyzed oxygenation of L-arginine is formally an overall 5e⁻ oxidation of a guanidinium nitrogen. This is a remarkable process in that this presumed multi-step reaction is performed by a single protein and it constitutes an overall odd-electron oxidation. It represents a truly unique transformation as there is little chemical or biochemical precedence for this type of reaction. The first step in the transformation appears to be an N-hydroxylation since it has been shown that N-hydroxy-L-arginine is a biosynthetic intermediate (37-41). This step constitutes an overall 2e⁻ oxidation of the guanidinium nitrogen of arginine. Further oxidation of N-hydroxy-L-arginine to give citrulline and NO requires an odd, 3e⁻ oxidation process. The NADPH stoichiometry has been determined to be 1.5 NADPH used per NO formed from L-arginine and 0.5 NADPH used per NO formed from N-hydroxy-L-arginine (37). However, the validity of these stoichiometric measurements has been questioned (42). These experimental observations are summarized in Figure 2.

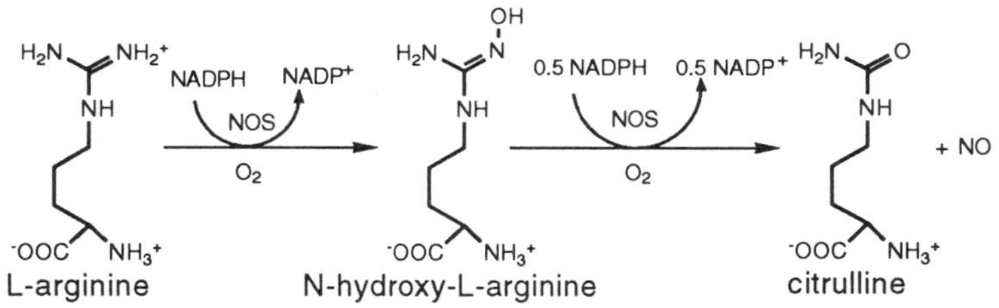

Figure 2: Pathway for NOS catalyzed NO formation from arginine.

The mechanistic intricacies of the reaction pathway depicted in Figure 2 are not fully understood. The first step in the reaction, N-hydroxylation of arginine, has biochemical precedence and is likely to be analogous to other N-hydroxylations performed by P450 (43). A recent study indicates that N-hydroxylation of a guanidine compound by P450 is possible and that, in this system, further oxidation to generate the citrulline equivalent urea product occurs by a superoxide (O_2^-) dependent mechanism (44). In other related studies, the oxidation of N-hydroxyguanidine compounds has been examined using chemical model systems and it was found that HNO and not NO was the preferred reaction product (45,46). The possibility that HNO is a precursor or biosynthetic intermediate to NO generation has been postulated (47,48, 19). That is, 4e⁻ oxidation of a guanidinium nitrogen on arginine to give citrulline and HNO followed by a one electron oxidation of HNO would generate NO (Figure 3).

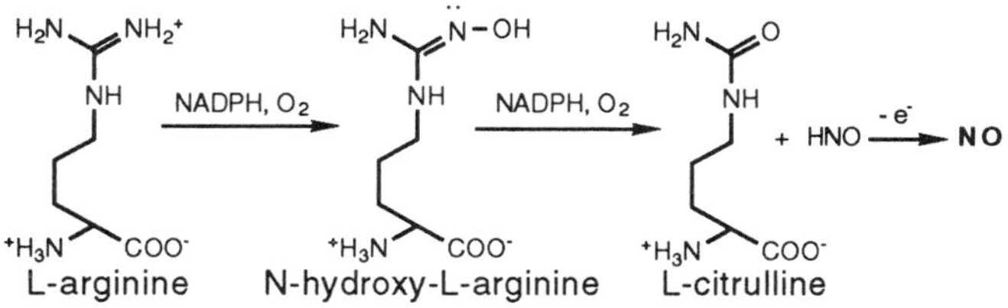

Figure 3: Possible NO biosynthesis from initial HNO generation

Since NO is a radical species generated from a non-radical, even-electron precursor, arginine, there is a mechanistic requirement for an odd-electron step. That is, an odd or one electron oxidation of a biosynthetic intermediate must occur in order to generate the requisite radical product, NO. In the above scheme (Figure 3), the odd electron step occurs at the end as a one electron oxidation of HNO. A variety of other possible and provocative mechanisms for NO generation from NOS catalyzed oxidation of arginine have been proposed as well (49,42,19,41). All of these current hypothesis accommodate the odd-electron requirement by proposing a single electron oxidation of a biosynthetic intermediate by the enzyme. Below is a schematic which depicts the fundamental aspects of the currently hypothesized mechanisms (Figure 4).

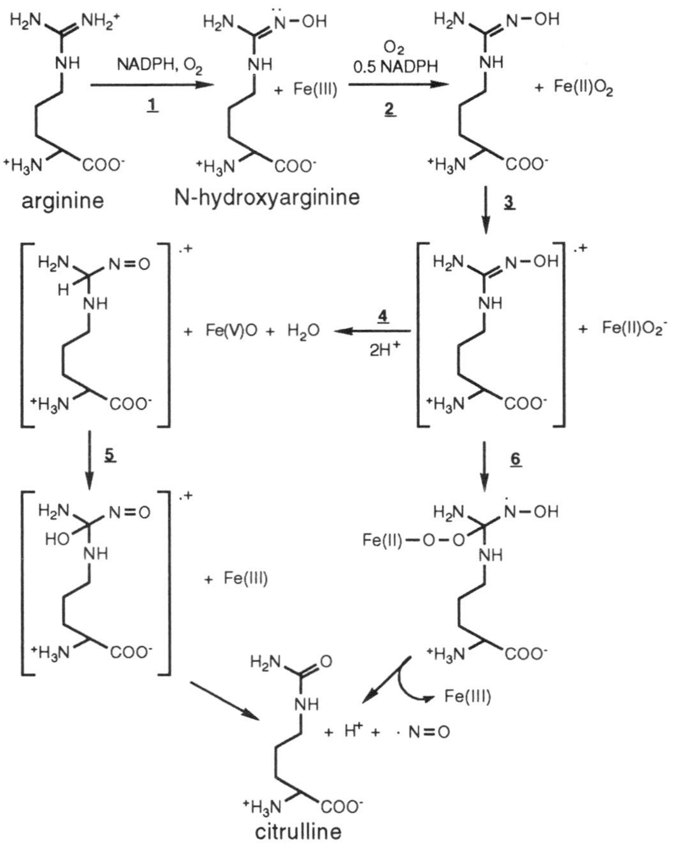

Figure 4. Proposed mechanisms for NOS catalyzed NO biosynthesis.

Step 1 in Figure 4 represents a P450-like N-oxidation of arginine to give N-hydroxyarginine. Step 2 depicts a single electron transfer from NADPH to the ferric heme to give a ferrous heme. The reduced iron then binds O_2 and is further reduced by the electron rich N-hydroxyguanidine intermediate (step 3). This oxygen-bound, reduced-iron species can then nucleophilically attack the oxidized N-hydroxy intermediate (step 6) to give a metal peroxo species which should decompose to citrulline and NO. Alternatively, generation of an iron-oxo heme through the proton assisted loss of water (step 4) and subsequent hydroxylation of the tautomerized N-hydroxy intermediate (step 5) would result in an unstable hydroxy-nitroso radical cation species which should also decompose to release citrulline and NO. Although all the above mentioned mechanisms are highly speculative, they are useful as a basis for discussion and they illustrate the likely complex and multi-step nature of the NO biosynthetic process. Certainly other, reasonable mechanisms will be proposed in the future as the field progresses.

Manipulation of Enzyme Activity - Due to the role of NO in a variety of important biological functions, manipulation of physiological NO levels may prove to be of tremendous therapeutic benefit. Certainly, a desired decrease in physiological NO levels can be accomplished by inhibition of NOS activity. This has been accomplished in *in vitro* systems by a wide variety of agents. For example, cNOS activity is decreased in the presence of calmodulin inhibitors such as calmidazolium (50). Expectedly, iNOS is not affected by calmodulin inhibitors. Flavoprotein inhibitors such as diphenyleneiodonium are capable of inhibiting both cNOS from endothelial cells and iNOS from macrophages (51). Imidazole and phenylimidazole compounds have been reported to inhibit cNOS by binding the sixth coordination site on the heme and thus blocking oxygen activation (52). Other imidazole agents appear to inhibit by binding both the heme and calmodulin sites of cNOS (53). Rat cerebellar cNOS is inhibited competitively by 7-nitro indazole and related compounds, although the mechanism of inhibition has not been elucidated (54). 3-Amino-1,2,4-triazole and 3-amino-1,2,4-triazine both inhibited iNOS from macrophages presumably by interacting with the heme component (55). Chlorpromazine has recently been reported to inhibit both cNOS from brain and the induction of iNOS in lung (56). Methylene blue was found to inhibit NO biosynthesis in cultured endothelial cells, albeit at fairly high concentrations (57). Significantly, selective inhibition of iNOS has been observed for aminoguanidine (58). It thus represents a potential compound for the treatment of pathological overproduction of NO. Somewhat surprisingly, it has been reported that a compound structurally unrelated to arginine, phencyclidine, is a suicide inhibitor of rat brain NOS (59). This finding may have significant implications in explaining drug side effects since it shows that NOS may interact with arginine-unrelated substrates.

The majority of the above mentioned agents are useful tools for studying NOS and the role of NO in isolated preparations. However, since they predominantly act by interfering with common regulatory systems (i.e. calmodulin) or cofactor/prosthetic group action (ironheme or flavins) ,*in vivo* they might be expected to affect other physiological functions which also utilize these factors. For example, heme binding agents may also inhibit other heme requiring enzymes such as P450 or catalase and calmodulin inhibitors would have an even greater global effect on a variety of signal transduction systems. Therefore, it is unlikely that many of these relatively nonspecific agents will be therapeutically useful. Also, since iNOS and cNOS have identical cofactor and prosthetic group requirements, specific inhibition on the basis of these components may be difficult (however, as mentioned above, aminoguanidine has been reported to be fairly selective for iNOS, although the mechanism of inhibition has not yet been elucidated). Therefore, in order to obtain specificity for NOS inhibition in *in vivo* situations, it may be worthwhile to try to first take advantage of the specificity of the enzyme for its substrate, arginine, rather than rely on manipulation of its cofactors or prosthetic groups.

Arginine analogs represent the largest class of compounds used as NOS inhibitors. The structures of the commonly utilized arginine analogs are given in Figure 5. N-methyl-L-arginine is one of the most commonly used and fully studied of the arginine analogs. It is both a competitive and irreversible inhibitor of NOS (60). The irreversible component of N-methyl-L-arginine inhibition appears to be mechanism based in that it is initially metabolized in a manner similar to L-arginine itself to give an N-methyl-N-hydroxy intermediate. Further oxidation of the N-hydroxylated intermediate results in the apparent generation of an

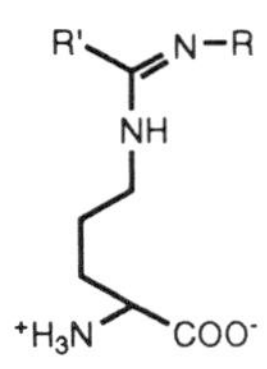

$$R = H, R' = NH_2; \text{ L-arginine}$$
$$R = CH_3, R' = NH_2; N^G\text{-methyl-L-arginine}$$
$$R = NO_2, R' = NH_2; N^G\text{-nitro-L-arginine}$$
$$R = NH_2, R' = NH_2; N^G\text{-amino-L-arginine}$$
$$R = cyclopropyl, R' = NH_2; N^G\text{-cyclopropyl-L-arginine}$$
$$R = allyl, R' = NH_2; N^G\text{-allyl-L-arginine}$$
$$R = H, R' = CH_3; \text{ N-iminoethyl-L-ornithine}$$

Figure 5: The structures of commonly used arginine analogs used as NOS inhibitors.

irreversibly binding species (49,61). N-Allyl-L-arginine has also been found to be a potent reversible and irreversible inhibitor of iNOS and, as with the N-methyl analog, the irreversible component of the inhibition is likely to be due to enzyme activation of the substrate to form a covalently binding species (62). Irreversible inhibition of NOS from phagocytic cells by N-iminoethyl-L-ornithine (63) and of NOS from rat brain by N-nitro-L-arginine (64) have also been described (although the mechanisms of inhibition have not been studied in depth). N-Nitro-L-arginine inhibition of endothelial cell NOS exhibits an initial reversible phase followed by enzyme inactivation after prolonged exposure (65). A comparison of the kinetics of N-nitro-L-arginine inhibition with that of N-methyl-L-arginine of endothelial cell NOS suggest that these two analogs differ dramatically in their mechanisms of inhibition (66).

As mentioned previously, NO generation in the various systems, at one level, differs on the basis of the NOS isoform involved (iNOS vs. cNOS, etc.). Therefore, in order to achieve some degree of system or tissue selectivity, it would be desirable to develop specific or selective agents for the various NOS isoforms. Selectivity for NOS isoform inhibition by a variety of arginine analogs has been reported. For example, N-cyclopropyl-L-arginine exhibited clear selectivity for cNOS from brain over iNOS from macrophages or smooth muscle cells whereas N-amino-L-arginine was relatively indiscriminate (67). N-Nitro-L-arginine has also been reported to be fairly selective for cNOS compared to iNOS (68). A comparison of N-amino-L-arginine, N-methyl-L-arginine and N-nitro-L-arginine revealed that the N-amino and N-nitro analogs were significantly better inhibitors of cNOS from endothelial cells than was N-methyl-L-arginine. However, in iNOS from macrophage, N-amino and N-methyl-L-arginine were more potent than N-nitro-L-arginine (69). A study comparing the inhibition of NO production in three cell types by N-nitro-, N-amino- and N-methyl-L-arginine, reported a striking selectivity for cNOS over iNOS by the N-nitro analog (70). It should, however, be realized that in *in vivo* or cellular *in vitro* preparations, the inhibition of NO production is the result of not only a direct action on NOS. For example, it has been demonstrated that amino acid transport into the cell is an important consideration as well (71).

Conclusion - The diversity and ubiquity of NO function in physiology makes it an interesting and important species for investigation (1,2,4). However, these properties also make specific, therapeutic targeting of NO-related phenomenon a challenging task. A complicating aspect of NO physiology which must be addressed when developing an NO-related therapeutic strategy is that NO is involved in so many seemingly unrelated physiological events. That is, manipulation of one NO-mediated event may impact other important and unrelated physiological functions as well. NO and NO-donors used to elevate physiological NO levels are already being used or developed as treatment for disorders related to inadequate smooth muscle dilation (i.e. coronary artery disease, male impotence or respiratory distress). Pharmacological agents which decrease endogenous NO levels by specifically alterating NOS activity will undoubtedly be developed as well for the treatment of a variety of diseases or disorders resulting from an overproduction of NO (i.e. migraine headaches, reperfusion toxicity or septic shock). The development of NOS inhibitors for the treament of a specific NO-mediated malady is somewhat hindered by the physiological ubiquity of NO and NOS. Luckily, endogenous NO production is the result of the actions of related, yet distinct isozymes which differ not only in their mechanisms of regulation and biophysical properties but also in their substrate binding specificities. Exploitation of these differences may prove useful in the development of specific drugs.

References

1.   S. Moncada, R. M. J. Palmer and E. A. Higgs, Pharmacol. Rev., 43, 109 (1991).
2.   C. Nathan, FASEB J., 6, 3051 (1992).
3.   P. L. Feldman, O. W. Griffith and D. J. Stuehr, Chem. Eng. News, 71, 26 (1993).
4.   M. Gibaldi, J. Clin. Pharmacol., 33, 488 (1993).
5.   E. Culotta and D. E. Koshland, Science, 258, 1862 (1992).
6.   H. H. H. W. Schmidt, S. M. Lohman and U. Walter, Biochim. Biophys. Acta, 1178,153 (1993).
7.   L. J. Ignarro, Biochem. Pharmacol., 41, 485 (1991).
8.   R. F. Furchgott, J. V. Zawadzki, Nature (London), 288, 373 (1980).
9.   R. M. J. Palmer, A. G. Ferrige and S. Moncada, Nature (London), 327, 524 (1987).
10.   L. J. Ignarro, G. M. Buga, K. S. Wood, R. E. Byrns and G. Chaudhuri, Proc. Natl. Acad. Sci., USA, 84 ,9265 (1987)
11.   D. J. Stuehr and M. A. Marletta, Proc. Natl. Acad. Sci., USA, 82, 7738 (1985)
12.   M. A. Marletta, P. S. Yoon, R. Iyengar, C. D. Leaf and J. S. Wishnok, Biochemistry, 27, 8706 (1988).
13.   J. B. Hibbs, Jr., R. R. Taintor, Z. Vavrin and E. M. Rachlin, Biochem. Biophys. Res. Commun., 157, 87 (1988).
14.   J. S. Beckman and J. P. Crow, Biochem. Soc. Trans., 21, 330 (1993).
15.   J. A. Ferrendelli, M. M. Chang and D. A. Kinscherf, Brain Res., 22, 535 (1974).
16.   J. Garthwaite, Trends in Neuroscience, 14, 60 (1991).
17.   T. M. Dawson, V. L. Dawson and S. H. Snyder, Ann. Neurology, 32, 297 (1992).
18.   H. Bult, G. E. Boeckxstaens, P. A. Pelckmans, F. H. Jordaens, Y. M. Maercke and A. G. Herman, Nature (London), 345, 346 (1990).
19.   D. J. Stuehr and O. W. Griffith, Adv. Enzymol., A. Meister, ed., Wiley and Sons, New York, 287 (1992).
20.   P. A. Bush, N. E. Gonzalez, J. M. Griscavage and L. J. Ignarro, Biochem. Biophys. Res. Commun., 185, 960 (1992).
21.   H. J. Cho, Q. Xie, J. Calaycay, R. A. Mumford, K. M. Swiderek, T. D. Lee and C. Nathan, J. Exp. Med., 176, 599 (1992).
22.   D. S. Bredt, C. D. Ferris and S. H. Snyder, J. Biol. Chem., 267(, 10976 (1992).
23.   W. C. Sessa, C. M. Barber and K. R. Lynch, Circ. Res., 72, 921 (1993).
24.   L. Busconi and T. Michel, J. Biol. Chem., 268, 8410 (1993).
25.   K. Hiki, R. Hattori, C. Kawai, C. and Y. Yui, J. Biochem., 111, 556 (1992).
26.   K. J. Baek, B. A. Thiel, S. Lucas and D. J. Stuehr, J. Biol. Chem., 268, 21120 (1993).
27.   H. H. H. W. Schmidt, J. S. Pollock, M. Nakane, L. D. Gorsky, U. Forstermann and F. Murad, Proc. Natl. Acad. Sci., USA, 88, 365 (1991).
28.   D. J. Stuehr, N. S. Kwon and C. F. Nathan, Biochem. Biophys. Res. Commun., 168, 558 (1990).
29.   Y. Watanabe and J. T. Groves, "The Enzymes, 3rd edition", D. S. Sigman ed., vol. 20: Mechanisms of Catalysis, Academic Press, San Diego, pp. 405 (1992).
30.   D. S. Bredt, P. M. Hwang, C. E., Glatt, C. Lowenstein, R. R. Reed and S. H. Snyder, Nature, 351, 714 (1991).
31.   A. M. Leone, R. M. J. Palmer, R. G. Knowles, P. L. Francis, D. S, Ashton, and S. Moncada, J. Biol. Chem., 266, 23790 (1991).
32.   H. M. Abu-Soud and D. J. Stuehr, Proc. Natl. Acad. Sci., USA, 90, 10769 (1993).
33.   T. A. Dix and S. J. Benkovic, Acc. Chem. Res., 21, 101 (1988).
34.   B. Mayer, M. John, B. Heinzel, E. R. Werner, H. Wachter, G. Schultz, G. and E. Böhme, FEBS Lett., 288, 187 (1991).
35.   J. Giovanelli, K. L. Campos, and S. Kaufman, Proc. Natl. Acad. Sci., USA, 88, 7091 (1991).
36.   J. M. Hevel and M. A. Marletta, Biochemistry, 31, 7160 (1992).
37.   D. J. Stuehr, N. S. Kwon, C. F. Nathan, O. W. Griffith, P. L. Feldman and J. Wiseman, J. Biol. Chem., 266, 6259 (1991).
38.   G. C. Wallace and J. M Fukuto, J. Med. Chem., 34, 1746 (1991).
39.   A. Zembowicz, M. Hecker, H. MacArthur, W. C. Sessa and J. R. Vane, Proc. Natl. Acad. Sci., USA, 88, 11172 (1991).

40.   R. A. Pufahl, P. G. Nanjappan, R. W. Woodard and M. A. Marletta, Biochemistry, 31, 6822
       (1992).
41.   P. Klatt, K. Schmidt, G. Uray and B. Mayer, J. Biol. Chem., 268, 14781 (1993).
42.   M. A. Marletta, J. Biol. Chem., 268, 12231 (1993).
43.   A. K. Cho and B. Lindeke, vol. eds., "Progress in Basic and Clinical Pharmacology", Lomax,
P.   and Vesell, series eds., Karger, Basel, (1988).
44.   B. Clement, M.-H. Schultze-Mosgau and H. Wohlers, Biochem. Pharmacol., 46, 2249
       (1993).
45.   J. M. Fukuto, G. C. Wallace, R. Hszieh and G. Chaudhuri, Biochem. Pharmacol., 43, 607
       (1992).
46.   J. M. Fukuto, D. J. Stuehr, P. L. Feldman, M. P. Bova and P. Wong, J. Med. Chem., 36,
       2666 (1993).
47.   J. M. Fukuto, K. Chiang, R. Hszieh, P. Wong and G. Chaudhuri, G., J. Pharmacol. Exp. Ther.,
       263, 546 (1992).
48.   J. M. Fukuto, A. J. Hobbs and L. J. Ignarro, Biochem. Biophys. Res. Commun., 196, 707
       (1993).
49.   P. L. Feldman, O. W. Griffith, H. Hong and D. J Stuehr, J. Med. Chem., 36, 491 (1993).
50.   V. B. Schini and P. M. Vanhoutte, J. Pharmacol. Exp. Ther., 261, 553 (1992).
51.   D. J. Stuehr, O. A. Fasehun, N. S. Kwon, S. S. Gross, J. A. Gonzalez, R. Levi and C. F.
       Nathan, FASEB J., 5, 98 (1991).
52.   D. J. Wolff, G. A. Datto, R. A. Samatovicz and R. A. Tempsick, R. A., J. Biol. Chem., 268,
       9425 (1993).
53.   D. J. Wolff, G. A. Datto and R. A. Samatovicz, J. Biol. Chem., 268, 9430 (1993).
54.   R. C. Babbedge, P. A. Bland-Ward, S. L. Hart and P. K. Moore, Br. J. Pharmacol., 110, 225
       (1993).
55.   Y. Buchmüller-Rouller, P. Schneider, S. Betz-Corradin, J. Smith and J. Mauël, Biochem.
       Biophys. Res. Commun., 183, 150 (1992).
56.   M. Palacios, J. Padron, L. Glaria, A. Rojas, R. Delgado, R. Knowles and S. Moncada,
       Biochem. Biophys. Res. Commun., 196, 280 (1993).
57.   S. Shimizu, T. Yamamoto and K. Momose, Res. Commun. Chem. Path. Pharmacol., 82,
       35 (1993).
58.   T. P. Misko, W. M. Moore, T. P. Kasten, G. A. Nickols, J. A. Corbett, R. G. Tilton, M. L.
       McDaniel, J. R. Williamson and M. Currie, Eur. J. Pharmacol., 233, 119 (1993).
59.   Y. Osawa and J. C. Davila, Biochem. Biophys. Res. Commun., 194, 1435 (1993).
60.   N. M. Olken, K. M. Rusche, M. K. Richards and M. A. Marletta, Biochem. Biophys. Res.
       Commun., 177, 828 (1991).
61.   R. A.Pufahl, P. G. Nanjappan, R. W Woodard and M. A. Marletta, Biochemistry, 31, 6822
       (1992).
62.   N. M. Olken and M. A. Marletta, J. Med. Chem., 35, 1137 (1992).
63.   T. B. McCall, M. Feelisch, R. M. J. Palmer and S. Moncada, Br. J. Pharmacol., 102, 234
       (1991).
64.   M. A. Dwyer, D. S. Bredt and S. H. Snyder, Biochem. Biophys. Res. Commun., 176, 1136
       (1991).
65.   B. Mayer, M. Schmid, P. Klatt and K. Schmidt, FEBS Lett., 333, 203 (1993).
66.   P. Klatt, K. Schmidt, F. Brunner and B. Mayer, J. Biol. Chem., 269, 1674 (1994).
67.   L. E. Lambert, J. F. French, J. P. Whitten, B. M. Baron and I. A. McDonald, Eur. J.
       Pharmacol., 216, 131 (1992).
68.   E. S. Furfine, M. F. Harmon, J. E. Paith and E. P. Garvey, Biochemistry, 32, 8512 (1993).
69.   S. S. Gross, D. J. Stuehr, K. Aisaka, E. A. Jaffe, R. Levi and O. W. Griffith, Biochem. Biophys.
       Res. Commun., 170, 96 (1990).
70.   L. E. Lambert, J. P. Whitten, B. M. Baron, H. C. Cheng, N. S. Doherty and I. A. McDonald,
       Life Sci., 48, 69 (1991).
71.   K. Schmidt, P. Klatt and B. Mayer, Mol. Pharmacol., 44, 615 (1993).
72.   J. M. Hevel, K. A. White and M. A. Marletta, J. Biol. Chem., 266, 22789 (1991).
73.   D. J. Stuehr, H. J. Cho, N. S. Kwon, M. F. Weise and C. F. Nathan, Proc. Natl. Acad. Sci.,
       USA, 88, 7773 (1991).
74.   T. Evans, A. Carpenter and J. Cohen, Proc. Natl. Acad. Sci., USA, 89, 5361 (1992).
75.   J. S. Pollock, U. Förstermann, J. A. Mitchell, T. D. Warner, H. H. H. W. Schmidt, M. Nakane
       and F. Murad, Proc. Natl. Acad. Sci., USA, 88, 10480 (1991).
76.   B. Mayer, M. John and E. Böhme, FEBS Lett., 277, 215 (1990).

Chapter 10. Thrombolytic Agents

Jack Henkin[a] and William D. Haire[b]
[a]Abbott Laboratories, Abbott Park, IL 60064
and [b]University of Nebraska Medical Center
Omaha, NB 68124-3330

Introduction - Intravascular thrombi leading to myocardial infarction, stroke, pulmonary embolism or peripheral occlusion are the major cause of mortality and of much morbidity in economically advanced countries. Administration of thrombolytic agents, all of which are currently plasminogen activators, can be an efficient means to restore blood flow, often preserving life and limb (1). Plasminogen is an abundant 90 kDa proenzyme, about 2μM in plasma which incorporates into clots. Enzymatic cleavage of a unique Arg560-Val561 peptide bond in plasminogen converts it into active plasmin, a trypsin-like enzyme whose primary targets in plasma are fibrin, the matrix protein of thrombi, and its soluble precursor fibrinogen. Thrombolytic therapy and the plasminogen/plasmin system have been extensively reviewed (2,3). This chapter will focus on new molecular and clinical progress, new activators under study, and problems remaining in thrombolysis.

CLOT SPECIFICITY

Fibrin Specificity - Fibrin-specificity of a plasminogen activator refers to the relative activation of clot-bound versus free plasminogen and depends on many factors. Free plasminogen adopts a compact conformation resistant to activation while clot-bound plasminogen exists in a more open accessible structure like that of lys-plasminogen (truncated by 76 residues at the N-terminus). Free plasmin is rapidly quenched by circulating α_2-antiplasmin (α_2-AP) while plasmin formed on a clot is much more stable since fibrin and α_2-AP compete for binding sites on plasminogen. Indeed, α_2-AP has been shown to be the most important quencher of free plasmin (4), and α_2-AP addition to rabbits treated with tissue plasminogen activator (tPA) enhanced clot specificity and reduced lysis of fibrinogen (5). The above facts make all plasminogen activators sufficiently clot specific to lyse at least small thrombi while preserving some circulating plasminogen and fibrinogen. The most specific plasminogen activators in use are tPA and prourokinase (proUK). The least specific is streptokinase (SK). Urokinase (UK) and anistreplase (APSAC), are somewhat more specific than SK. Fibrin-specificity does not lead to greater safety but enhances efficacy via preservation of the circulating plasminogen needed to sustain activity of systemically administered plasminogen activators (6,7). This should have greatest impact on large or resistant thrombi and is mirrored in the greater efficacy of local plasminogen activator delivery via catheter. Beyond efficacy, specificity can be doubled edged sword. While large fluxes of plasmin can be anticoagulating, low fluxes can be prothrombotic via plasmin's activation of factors V and VIII (3). Partial degradation of fibrinogen by plasminogen leads to accumulation of a clottable truncated fibrinogen known as X-fragment (XF). Plasminogen has higher affinity for polymerized XF than for fibrin, binding to their end-to-end junctions (8). Since XF levels may be maximal with low plasmin fluxes, and since clots rebuild dynamically, XF could convert protective "good" clots into more unstable ones, leading to bleeding at a later time. Resistance to lysis is affected by the presence of natural inhibitors. These include PAI-1 and protein C inhibitor (PCI,PAI-3). PAI-1, a 50kDa serpin, reacts very rapidly with both UK and tPA but is low in plasma compared with therapeutic doses of these plasminogen activators. However it is abundant in platelets, a significant mass within thrombi, and may block tPA action locally, adding to other platelet resistance mechanisms (9). PCI, a heparin-dependent serpin, while slower reacting than PAI-1, is much more abundant in plasma and may represent a major buffer capacity against UK and possibly tPA activities when these are given gradually (10). The unmet needs of thrombolysis span safety, efficacy and cost. Bleeding at puncture sites and hematuria are common. Potentially lethal intracranial bleeding occurs at a rate of about 1% with all agents. The greatest need in this area is an agent which can discriminate in favor of lysing occlusive clots rather that protective ones.

ESTABLISHED AGENTS

SK and APSAC - SK, a 47 kDa protein isolated from *streptococcus,* is an indirect plasminogen activator. It binds in 1:1 complex with plasminogen, forming a pseudo-plasmin active site which can activate other plasminogen molecules, yielding 2-chain plasmin. The complex does not attack fibrinogen or react with α_2-AP. The lowest cost agent, SK, displays some properties which limit its usefulness. A foreign protein, SK elicits a specific antibody response, and fear of acute clinical allergy precludes its reuse for at least several months. SK causes allergic reaction, occasionally severe (11), in about 5% of patients presumed to have had streptococcal exposure. Others are simply refractory to SK until its dose exceeds their antibody titre. As the least specific agent, SK consumes considerable plasminogen before complete lysis is achieved, which may explain lower patency rates. While all plasminogen activators may be prothrombotic via the action of plasmin, SK has recently been shown to to lead to thrombin formation by an unknown plasmin-independent mechanism (12). This clot building propensity surely reduces lytic efficacy. Anisoylated plasminogen-SK activator complex (APSAC, anistreplase) is a prodrug form of SK in which the essential active site serine of complexed SK:lys-plasminogen is blocked via a hydrolytically labile group. The block decomposes at a rate comparable to that of SK clearance, with a net effect of simulating a continuous steady state level after a single bolus injection (13). This offers convenience compared to other agents which must be delivered systemically by iv infusion. It was also hoped that lys-plasminogen component of APSAC would deliver the activity more specifically to clots. However significantly greater clot specificity than that of SK has not been demonstrated in man.

Tissue Plasminogen Activator - tPA (alteplase), a 72kDa enzyme secreted into plasma by endothelial cells is probably the body's main endogenous mechanism for clearing thrombi. It is unusual in that its single chain form is nearly as active as the 2-chain form; the latter is formed by plasmin. The enzyme has high affinity for fibrin, and forms a ternary complex with fibrin and plasminogen. In this complex tPA is many times more active than when free, yielding clot specificity. Very low concentrations of tPA can initiate clot lysis. Fibrin binding involves the D region and has been further dissected into peptides within this region having both lysine-dependent and independent interactions with tPA (14). As lysis proceeds, fibrin fragment E becomes exposed and new C-terminal lysines, produced via plasmin action, become new binding sites for plasminogen and tPA which then accelerate lysis (15). tPA is comprised of a catalytic C-terminal domain preceded by several other protein domains: The N-terminal fibronectin-finger like region contributes to fibrin binding. This is followed by a triple disulfide looped growth factor domain (about 40 aa), then two 80 aa triple disulfide looped kringles, K1 and K2. Epitope mapping with MABs suggests that fibrin contacts tPA at several sites and closely surrounds the enzyme (16). Analysis of mutants has led to new knowledge of domain function. Binding to terminal lysines on fibrin is mediated by K2 and is competed by ω-amino acids such as EACA. Lys33 within K2 (tPA residue #212) is prominent in this interaction (17). Clustered mutation of charged groups to Ala residues has shown that this K2 site is not required for fibrin binding or fibrin stimulation of tPA (18), while the finger and growth factor regions were important for these fibrin interactions. The catalytic domain was also suggested to interact with fibrin. A KHRR sequence at position 296-299 was critical for PAI-1 interaction, and some mutations on the catalytic domain could lead to greater zymogen-like behavior of the 1-chain form, enhancing clot- specificity. Clearance of tPA via a specific hepatic receptor is rapid ($t_{1/2}$ 3-6 min) and dependent on a key determinant, Tyr67, in the growth factor domain (19). Several studies have shown that tPA is cleared by the liver receptor in complex with its natural serpin inhibitor PAI-1. The complex is internalized for degradation via the LDL-receptor-related protein (LRP) which binds a number of other ligands including Apo-E and α_2-macroglobulin (20, 21). tPA contains 3 glycosylation sites which normally carry one high mannose-type and two complex-type oligosaccharides. The mannose receptor of liver endothelial and Kupffer cells binds the former and enhances tPA clearance (22). tPA has recently been reported to be a mitogen for human aortic smooth muscle cells in culture (23); this stimulation depends an intact tPA active site. Thus tPA may play a role in intimal hyperplasia following injury.

Urinary PA - uPA exists in an active 2-chain form, urokinase (UK,tcu-PA) which is found in both full size (54 kDa) and truncated (32kDa) versions. UK with a growth factor domain and one kringle, does not bind to fibrin. Its intrinsic Vm is about 5-fold greater than that of fully stimulated tPA. r-UK may be expected to replace some of the natural urine or kidney cell-derived material in the near future. The 1-chain precursor, prourokinase (proUK, scu-PA) also can exist in 32 and 54 kDa forms, each with slightly different kinetic behavior (24). All clinical work thus far has been with 54 kDa proUK, both natural and recombinant. R-proUKs have been expressed both in *E. coli* and in mammalian cells. The bacterial products require refolding of 12 bonds and lack the oligosaccharide normally found at Asn302. Significant kinetic differences have been noted between glycosylated and nonglycosylated proUK (25). ProUK is a true zymogen with very little activity in free solution. It is the only plasminogen activator which is essentially inert in plasma in the absence of a clot. In plasma α_2-AP contributes to proUK stability, preventing its conversion to UK (26). Presumably PAI-1 and PAI-3 also stabilize proUK by scavenging trace UK. Thrombi are rapidly attacked by proUK with local evolution of UK. Clots previously exposed to mild plasmin action are activated with respect to proUK lysis, leading to a notion of synergism between tPA or UK and proUK. At fixed levels of clot specificity proUK could lyse thrombi about twice as completely as tPA suggesting that different types of clot-bound plasminogen were accessible to the two agents, a smaller subset being activatable by tPA (27). The mechanism of proUK clot specificity has been difficult to unravel because of the expanding activation cycle with plasminogen/plasmin. Newer studies indicate that conversion to 2-chain UK is not required for at least initial clot lysis by proUK. Work with plasminogen and proUK mutants showed that proUK can attack plasminogen on a partially degraded fibrin matrix (28). Thrombin-inactivated proUK (29), and proUK itself (30) could activate plasminogen bound to fibrin fragment E-2, the latter being more available after partial clot lysis. Kinetic studies also suggested that free plasminogen binds to circulating proUK at a second negative regulatory site suppressing its off-clot reactivity (31).

Endothelial cells and smooth muscle cells, monocytes and many malignant tumor cells display on their surface a high affinity glycolipid-anchored uPA receptor (uPAR) which binds to the growth factor domain of uPA (32). uPAR has been shown to enhance plasminogen activation, to be essential to a variety of invasive processes and to be involved in signal transduction in some cells (33). Endocytosis of the UK:PAI-1 complex appears also to involve the LRP protein (34) analogous to tPA. Rapid liver clearance of both HMW and LMW forms of proUK appears to be mediated by yet another receptor on parenchymal cells, and is not competed by tPA or other glycoproteins (35). Platelets have recently been shown to take up proUK, possibly acting as a reservoir (36). It is unclear whether this involves receptors like those described above. As platelets appear to have a major role in rethrombosis, proUK may be expected to show lower rates of reocclusion.

ESTABLISHED CLINICAL APPLICATIONS

Myocardial Infarction (MI) - The use of and precise indications for thrombolytic therapy in MI has been the subject of recent in-depth reviews (37, 38). That use of thrombolytic agents could result in reperfusion of occluded infarct-related arteries was well established by the mid-1980's. Early, complete lysis of coronary obstruction with maintenance of persistent patency of the infarct-related artery is important in determining the amount of myocardium damaged, the long-term left ventricular function, the short- and long-term frequency of significant arrhythmias and both short- and long-term mortality (39, 40). A recent clinical trial comparing the use of (SK) followed by IV heparin, SK followed by subcutaneous heparin, t-PA in a novel dosing strategy followed by IV heparin and a combination of both SK and t-PA followed by IV heparin in over 41,000 patients has shown that the combination of t-PA and IV heparin provides a survival benefit compared to the other arms of the study, with mortality in the t-PA group of 6.3% compared to the 7.2% and 7.4% in the SK groups (p=0.001, 14% risk reduction, 95% CI 5.9-21.3%). However, it does so with more strokes and greater cost (41). The cost-effectiveness of this improvement in survival and the quality of life issues in patients with non-fatal strokes have been discussed at length with little, if any, consensus on their impact on the decision to use either SK or t-PA in patients with MI. The relative roles of other effective thrombolytic agents, either FDA approved (APSAC) or those undergoing clinical trials (pro-UK) compared to tPA and SK remain to be determined.

Achieving complete reperfusion of the infarct-related artery is required to maximally impact on outcome. This occurs in only 75% of patients treated with thrombolytic therapy. Persistent occlusion after lytic therapy is often felt to be non-thrombus related (atherosclerotic) and treated with angioplasty or surgery. Newer thrombolytic agents might improve this rate of incomplete reperfusion and result in improved outcomes of MI. Even without improving the effectiveness of current agents, the number of lives saved with thrombolytic therapy could be increased by expanding the number of patients treated and/or providing therapy earlier (42). Earlier therapy, often provided prior to arriving at a hospital, decreases infarct size, improves left ventricular function and decreases mortality (43). Expanding the window of treatment, from the currently recommended 6 hours to 12 hours after the onset of chest pain, may be able to improve the mortality rate of MI (44). This hypothesis has been recently born out (45). Patients presenting 6 to 12 hours after the onset of pain were randomly allocated to receive either t-PA or placebo. The relative mortality rates were 8.90% and 11.97%, respectively (risk reduction 25.6%, p=0.023, 95% CI 6.3-45%).

Thrombolysis is not the only treatment modality that can establish early reperfusion of occluded coronary arteries. Direct angioplasty can achieve the same endpoint, albeit invasively. The relative roles of lytic therapy and angioplasty are difficult to assess due to conflicting results of published clinical trials (46, 47). At present the use of one over another currently probably reflects the bias of the treating physician.

Deep Vein Thrombosis (DVT) - The precise role of thrombolytic therapy relative to anticoagulation for DVT of the leg is not well defined with less than twenty studies published between 1967 and 1982 comparing the outcomes of these two treatments. The results of these studies have been critically reviewed with diametrically opposed results (48, 49). There is no doubt that thrombolytic agents can dissolve thrombi, as their use clearly results in smaller thrombi and a greater likelihood of venous patency than anticoagulation. However, the issue remains whether this results in any improvement of patient status. Recent work has provided insight into the immediate outcome of thrombolysis in DVT in various patient populations. Combining heparin with three boluses of UK over 24 hours for treatment of DVT of both arm and leg veins resulted in roughly half of the patients achieving some degree of clot lysis, with arm veins more likely to achieve complete lysis than leg veins (50). In 174 DVT's of the leg treated with IV SK, thrombi totally occluding the vein were less likely to achieve complete lysis than non-occlusive thrombi (14 vs 60%, p<0.001) (51). For the completely occluded veins, using a catheter to infuse thromboytic agents directly into the thrombus, either by continuous infusion (52) or pulse spray (53), may result in more frequent clot dissolution.

The use of thrombolytic therapy for thrombosis of the veins of the arm/thorax (axillary/subclavian system) has been critically reviewed recently (54). In these patients, treatment decisions are often based on the observation that patients with spontaneous thoracic vein thrombosis have a high incidence of incapacitating long-term sequelae (55). Recent investigators have employed thrombolytic agents generally infused directly into the thrombus, combined with angioplasty and/or surgery to correct any underlying anatomic abnormality felt to predispose to recurrent thrombosis (55, 56). This approach has yielded good long-term resolution of symptoms, but in a small number of patients. For patients receiving lytic therapy but no surgery to correct underlying stenotic lesions the results were less impressive (55). Thrombolytic approaches to subclavian catheter-induced thoracic vein thrombosis have been described, using both local and systemic infusions of thrombolytic agents (57). Short-term vessel patency and long-term symptom relief is the rule after such therapy.

Pulmonary Embolism (PE) - The data supporting the use of thrombolysis in PE have been recently reviewed with divergence of opinion on its clinical utility (58, 59). As with DVT of the leg, this is not due to questions about the ability of currently available agents to dissolve thrombi. Indeed, recent studies have demonstrated that short infusions (2 hour) of high doses of both UK and t-PA can rapidly dissolve thrombi and improve microvascular perfusion (60). This degree of improvement may also be achieved with a 10 minute bolus of t-PA (61). The uncertainty is due to a lack of data regarding the long term outcome of patients treated with thrombolytic agents compared to anticoagulant therapy. Two recent studies have addressed this issue. The first showed a marked improvement in right ventricular function and a trend to fewer recurrent emboli in patients treated with t-PA compared to those treated with anticoagulation alone (62). There were 5 recurrent PE's, 2 fatal, in this study of 101 patients - all in the group treated with anticoagulation alone. All of these patients had impaired right ventricular wall motion. This study suggests that thrombolytic therapy might improve mortality in the subpopulation of PE patients with right ventricular dysfunction as well as the group of patients with overt shock (63) that have generally been acknowledged to be likely to benefit from thrombolytic therapy. Patients who do not die of PE are at risk for long term disability from loss of functional cardiopulmonary reserve. In a seven year follow-up of patients randomly assigned to thrombolysis or anticoagulation, those who received thrombolytics were less likely to have basal or exercise-induced pulmonary hypertension (64). Other functional differences between these two groups may become evident when the data from this abstract is published in its entirety. These studies should form the basis of future large-scale clinical trials designed to determine what functional advantages thrombolysis of PE offers to patients compared to standard anticoagulation.

Peripheral Arterial Occlusion (PAO) - Thrombolytic therapy of PAO has been the subject of recent reviews (65, 66), and entire issues of vascular journals (67, 68). Thrombolytic agents are generally infused directly into the occluding lesion via a variety of catheters, either as a continuous infusion or pulse spray, with no clear advantage of one method over the other (69). UK is generally felt to be the agent of choice because of its greater likelihood of achieving complete thrombolysis and its better safety profile compared to SK. Beyond this, "the optimal patient, dose, method of delivery, duration of therapy, and outcome remain to be established" (65). In practice, lytic therapy alone is often inadequate. Approximately two thirds of patients require surgery to relieve focal lesions uncovered by thrombolytic therapy. Thus, complications of operative intervention must be considered in determining the best approach to PAO in any given patient. In this regard, it was been hypothesized that thrombolytic therapy would result in shorter, less dangerous operations than would be required if surgery were the first line of therapy. This has been born out in a randomized trial comparing primary surgery to thrombolysis followed by surgery (when necessary) in 114 patients (70). One year mortality was higher in the surgery group (42% vs 16%, p = 0.01), mainly due to cardiopulmonary complications. Studies of this caliber may establish thrombolysis as the preferred initial therapy of most patients with PAO.

INVESTIGATIONAL AGENTS

Hybrids - Under examination as possible new generation agents have been numerous conjugates or mutants of tPA and uPA, including truncations, point mutations, deletions and fusions (71). ProUK was conjugated to a bispecific Mab directed against fibrin and against proUK and tested in baboons (72). It cleared about 4-fold slower than free proUK and also had 5-fold greater thrombolytic potency, partly due to increased specificity. A chimeric anti-fibrin Mab fused with 33kDa proUK gave slower clearance and higher clot specificity with an 11-fold net increase in potency (73). A series of fused chimeras between uPA and various combinations of plasminogen kringles was examined. While the plasminogen domains retained their binding of ε-aminocaproic acid they did not confer fibrin binding to the fused enzyme (74). Another fusion of anti-fibrin with the catalytic domain of uPA was linked via a region mutated so as to be activated into UK by thrombin rather than by plasmin. As expected the activation of this plasminogen activator was inhibited by heparin and hirudin (75). Since clearance of plasminogen (a-chain)-tPA hybrids was known to be mediated largely by α_2-AP, it should be possible to decrease clearance by blocking the active site as in APSAC. This was probed in a hybrid using various blocking groups (76) generating half-lives from 7-35 fold longer than unblocked hybrid. A truncated tPA, BM 06.022, in which only the K2 and catalytic domains remain, displays a 4-5 fold prolonged half-life at the price of a 4-fold reduction in maximal fibrin stimulation (77). It was hoped that this mutant could be delivered by bolus injection, and a double bolus regimen was found to be optimal (78). Another promising variant was constructed with multiple mutations that simultaneously slowed clearance, enhancing fibrin stimulation and reducing PAI interaction (79). This mutant gave excellent lysis of platelet rich clots after bolus injection in rabbits. A recent follow-up report with it described how 6-fold less mutant was required to achieve equal patency compared with tPA in rabbit arterial clots. Mean time to patency was cut by more than half. Duration of patency was nearly doubled and the mutant gave less bleeding than tPA (80). A new strategy is to mutate tPA so that the 1-chain form has lower activity. A more zymogenic tPA might be resistant to PAI, and be more inert in the circulation. Modeling considerations based on the mechanisms of serine protease activation were used to design point mutants of tPA, yielding one with 140-fold greater activity in the 2-chain form (81).

Staphylokinase (STA) - The high cost of human plasminogen activators has prompted a search for an agent more like SK in source, but with lower immunogenicity and greater specificity for fibrin. Recent work has focused on STA from *S. aureus,* and its recombinant form (STAR) expressed in *E. coli.* In a hamster model of clot lysis STA was comparable to SK in potency against simple clots, and was several-fold more active against platelet-enriched or retracted clots; yet it showed less induction of neutralizing antibody than SK upon repeated injection in dogs (82). In baboons, STAR gave lytic potency similar to SK but without significant appearance of neutralizing antibody or resistance to repeated treatments (83). The hypotension often seen with SK was also not observed with STAR. STA appears to be considerably more clot specific than SK. Their plasminogen activation mechanisms are both indirect (84). A lag seen with STA (85) which is prolonged by plasmin inhibitors, precedes 2-chain STA:plasmin, the active plasminogen activator complex. This form reacts with α_2-AP, STA dissociates and fully recycles (86). Fibrin competes with α_2-AP for STA:plasmin, thus explaining its inertness until reaching the clot. Fibrin also stimulates the action of STAR through interactions with the kringles of plasminogen within the complex (87). It remains to be seen whether the promising preclinical properties of STAR translate into clinically significant improvements over SK.

Bat Plasminogen Activator - Recent studies have been reported on a recombinant form of a PA isolated from vampire bat saliva (Bat PA). This foreign protein displays remarkably low immunogenicity and has about 80% homology with human tPA. Bat PA is much more stimulated by polymeric fibrin than by soluble fibrin, fibronectin, or x-fragment, compared with tPA (88). Since the Fibrin-specificity of tPA degrades as a clot dissolves and releases soluble stimulatory fragments, it is possible that Bat PA may retain specificity longer during the lytic process.

NEEDS IN THROMBOLYTIC THERAPY

The inadequacies remaining in thrombolytic therapy are: 1) failure to achieve reflow, 2) rethrombosis following initial patency (89), and 3) internal (especially intracranial) bleeding.

The first 2 are related since rethrombosis may be viewed as a late extension of the dynamic clot rebuilding in response to lysis. Paradoxically, plasminogen activator treatment induces new thrombin formation (90). Platelet activation is also seen after treatment with tPA or SK (91, 92). Some new mechanical approaches have been examined in local settings. Ultrasound enhanced lysis (93) in one study, but a second study showed that this was confounded by higher reocclusion (94). Pulsed-spray techniques (95) and restriction of lysis to an isolated limb by "lysis block" (96) have been described. Both inadequate patency and rethrombosis can be attacked by increased lytic potency in combination with anticoagulant and/or antiplatelet agents. Antithrombin adjunctive agents used in combination with thrombolytic therapy have included anti-Xa inhibitors such as a tick peptide (97), direct thrombin inhibitors such as hirudin (98) or combinations of both (99). A variety of antiplatelet strategies has also been attempted and reviewed (100). One comparison of thrombin inhibition and platelet glycoprotein IIB/IIIa antagonism favored the former, ostensibly because thrombin itself is a major platelet stimulator (101). While virtually all the above have succeeded to enhance stable clot lysis, the extra homeostatic defect incurred will likely increase bleeding. Unfortunately no design thus far has convincingly discriminated between protective and occlusive clots and no animal or other model has been shown to predict bleeding safety for a particular thrombolytic therapy regimen. These remain the challenges for future work.

References

1. C.R. Benedict, S.Mueller, H.V. Anderson and J.T. Willerson, Hospital Practice, 61 (1992).
2. E.G. Bovill, R. Becker and R.P. Tracy, Prog in Cardiovasc Dis 34, 4, 279 (1992).
3. J. Henkin, P. Marcotte, and H. Yang, Prog. Cardiovasc Dis 34, 2, 135 (1991)
4. E. Angles-Cano, D. Rouy and H.R. Lijnen, Biochim Biophys Acta, 1156, 34 (1992).
5. J.I. Weitz, B. Leslie, J. Hirsh and P. Clement, J Clin Invest, 91, 1343 (1993).
6. P.T. Önundarson, H. Haraldsson, L. Bergmann, C.W. Francis and V.J. Marder, Thromb Haemost, 70, 998 (1993).
7. S.R. Torr, D.A. Nachowiak, S. Fuji and B.E. Sobel, JACC 19, 1085 (1992).
8. J.A. Weisel, C. Nagaswami, B. Korsholm, L.C. Petersen and E. Suenson, J Mol Biol., 235, 1117 (1994).
9. W.P. Fay, D.T. Eitzman, A.D. Shapiro, E.L. Madison and D. Ginsberg, Blood, 83, 351 (1994).
10. F. Espana, A. Estelles, P.J. Fernandez, J. Gilabert, J. Sanchez-Cuenca and J.H. Griffin, Thromb Haemost, 70, 989 (1993).
11. K.A. Davies, P. Mathieson, C.G. Winearls, A.J. Rees and M.J. Walport, Clin Exp Immunol 80, 83 (1990).
12. E.J.P. Brommer and P. Meijer, Thromb Haemost 70, 6, 995 (1993).
13. H. Silber, M.J. Hausmann, A. Katz, H. Gilutz, N. Zueker and I. Ovsyshcher, Angiol, 43, 572 (1992).
14. P. Grailhe, W. Nieuwenhuizen and E. Angeles-Cano, Eur J Biochem, 219, 961 (1994).
15. V. Fleury, S. Loyau, H.R. Lijnen, W. Nieuwenhuizen and E. Angles-Cano, Eur J Biochem, 216, 549 (1993).
16. U. Zacharias, B. Fischer, F. Noll and H. Will, Thromb Haemost, 67, 88 (1992).
17. V.S. De Serrano, L.C. Sehl and F.J. Castellino, Arch Biochem Biophys, 292, 206 (1992).
18. W.F. Bennett, N.F. Paoni, B.A. Keyt, D. Botstein, A.J.S. Jones, L. Presta, F.M. Wurm and M.J. Zoller, J Biol Chem, 266, 5191 (1991).
19. R.Bassel-Duby, N.Y.Jiang, T. Bittick, E. Madison, D. McGookey, K. Orth, R. Schohet, J. Sambrook and M-J. Gething, J Biol Chem, 267, 9668 (1992).
20. K. Orth, E.L. Madison, M.J. Gething, J.F. Sambrook and J.Herz, PNAS, USA, 89, 7422 (1992).
21. G. Bu, E.A. Maksymovitch and A.L. Schwartz, J Biol Chem, 268, 13002 (1993).
22. M. Otter, M.M. Barrett-Bergshoeff and D.C. Rijken, J Biol Chem, 266, 13931 (1991).
23. J.M. Herbert, I.Lamarche, V. Prabonnaud, F. Dol and T. Gautheir, J Biol Chem, 269, 3076 (1994).
24. G.A.W. de Munk, E. Groeneveld and D.C. Rijken, Thromb Haemost, 70 481 (1993).
25. C. Lenich, R. Pannell, J. Henkin and V. Gurewich, Thromb Haemost, 68, 539 (1992).

26.   P.J. Declerck, H.R.. Lijnen, M. Verstreken and D. Collen, Thromb Haemost, 65, 4 (1991).
27.   R. Pannell and V. Gurewich, Fibrinolysis 6, 1 (1992).
28.   V. Fleury, H. R. Lijnen, E.Anglés-Cano, J Biol Chem 268 (25), 18554 (1993).
29.   J. Liu and V. Gurewich, Blood, 81, 980 (1993).
30.   J. Liu and V. Gurewich, J Biol Chem, 267, 15289 (1992).
31.   C. Longstaff, A.M. Clough and P.J. Gaffney, J Biol Chem, 267, 1 (1992).
32.   L.B. Møller, Blood Coagulation and Fibrinolysis, 4, 293, (1993).
33.   S.A. Rabbani, A.P. Mazar, S.M. Bernier, M.Haq, I. Bolivar, J. Henkin and D. Goltzman, J Biol
      Chem, 267, 14151 (1992).
34.   P.A. Andreasen, L. Sottrup-Jensen, L. Kjøller, an. Nykjær, S.K. Moestrup, C.M. Petersen and J.
      Gliemann, FEBS Letters, 338, 239 (1994).
35.   J. Kuiper, D.C. Rijken, G.A.W. deMunk and T.J.C.van Berkel, J Biol Chem, 267, 1589 (1992).
36.   V. Gurewich, M. Johhstone, J-P. Loza and R. Pannell, Fed Eur Bioch Soc 318, 317 (1993).
37.   H.V. Anderson, and J.T. Willerson, N Eng J Med, 329,703 (1994).
38.   N.W., Shammas, R. Zeitler, and P. Fitzpatrick, Clin Cardiol, 16, 283 (1993).
39.   H.D. White, D.B. Cross, J.M. Elliot, R.N. Norris, and T.W. Yee, Circulation, 89, 61 (1994).
40.   R.F.E. Pedretti, E. Colombo, S.S. Braga, and B.J. Caru, JACC, 23, 19 (1994).
41.   The GUSTO Investigators, N Eng J Med, 329,673 (1993).
42.   A.J. Doorey, E.L. Michelson, and E.J. Topol, J Am Med Assoc. 268, 3108 (1992).
43.   W.D. Weaver, M. Cerqueira, A.P. Hallstrom, P.E. Litwin, J.S. Martin, P.J. Kudenchuck and M.E.
      Eisenberg, JAMA, 270, 1211 (1993).
44.   Fibrinolytic Therapy Trialists' Collaborative Group, Lancet, 343, 311 (1994).
45.   LATE Study Group, Lancet, 342, 759 (1993).
46.   F. Zijlstra, M. Jan de Boer, J.C.A. Hoorntje, S. Reiffers, J.H.C. Reibner, and H. Suryapranata, N
      Eng J Med, 328, 680 (1993).
47.   E.E. Ribiero, L.A. Silva, R. Carneiro, L.G.D. D'Olivera, A. Gasquez, J.G. Amino, J.R. Tavares, A.
      Petrizzo, S. Torossian, R. Duprat, E. Buffolo and S.G. Ellis, J Am Coll Cardiol, 22, 376 (1993).
48.   L.Q. Rogers, and C.L. Lutcher, Am J Med, 88, 389 (1990).
49.   Y. Sidorov, Arch Intern Med, 149, 1841 (1989).
50.   S.Z. Goldhaber, J.F. Polak, M.L. Feldstein,M.F. Meyerovitz, M.F. and M.A. Creager, Am J
      Cardiol, 73, 75 (1994).
51.   C. Thery, J.J. Bauchart, M. Lesenne, P. Asseman, J.G. Flajollet, R. Legghe and P. Marache,
      Am J Cardiol, 69, 117 (1992).
52.   A.J. Comerota, and S.C. Aldridge, Can J Surgery, 36, 359 (1993).
53.   D.L. Robinson, G.P. and Teitelbaum, AJR, 160, 1288 (1993).
54.   D.M. Becker, J.T. Philbrick, and F.B. Walker, Arch Intern Med, 151, 1934 (1991).
55.   J. Malcynski, T.F.O'Donnell, W.C. Mackey, and Milan, V.A. Can J Surgery, 36, 365 (1993).
56.   H.I.Machleder, Seminars Vascular Surgery, 5, 82 (1992).
57.   E.L. Seigel, A.C. Jew, R. Delcore, J.I. Iliopoulos, and H.J.H. Thomas, Am J Surgery, 166, 716
      (1993).
58.   J.W. ten Cate, Lancet, 342, 1315 (1993).
59.   S.Z. Goldhaber, Prog Cardiovasc Dis, 34, 113 (1991).
60.   S.Z. Goldhaber, C.M. Kessler, J.A. Heit, C.G. Elliott, W.R. Friedenberg, D.E. Heiselman, D.B.
      Wilson, J.A. Parker, D. Bennett, M.L. Feldstein, A.P. Selwin, D. Kim, G.V.R.K. Sharma, J.S.
      Nagel and M.F. Meyerovitz, JACC, 20, 24 (1992).
61.   J.L. Diehl, G. Meyer, J Igual, M.A. Collignon, M. Giselbrecht, P. Even and H. Sors, Am J Cardiol,
      70, 1477 (1992).
62.   S.Z. Goldhaber, W.D. Haire, M.L. Feldstein, M. Miller, R. Toltzis, J.L. Smith, A.M. Taviera da-
      Silva, P.C. Come, R.T. Lee, J.A. Parker, A. Mogtader, T.J. McDonough and E. Braunwald,
      Lancet, 341, 507 (1993).
63.   J.P. Mitchell, E.P. and Trulock, Am J Med, 90, 255 (1991).
64.   G.V.R.K. Sharma, E.E. Folland, K.M. McIntyre, and A.A. Sasahara, JACC, 15, 65A (1990).
65.   J.D. Durham, and R.B. Rutherford, Seminars Interventional Radiology, 9, 166 (1992).
66.   J.A. Kaufman, M.A. and Bettmen, Seminars Interventional Radiology, 9, 159 (1992).
67.   E.C. Martin, Circulation, 83, suppl 2, I-1 (1991).
68.   A. J. Comerota, Seminars Vascular Surgery, 5, 67 (1992).
69.   K. Kandarpa, P.S. Chopra, J.E. Aruny, J.F. Polak, M.C. Donaldson, A.D. Whittemore, J.A.
      Mannick, S.Z. Goldhaber, M.F. Meyerovitz, Radiology, 188, 861 (1993).

70.   K. Ouriel, C.K. Shortell, J.A. DeWeese, R.M. Green, C.W. Francis, M.V.U. Azodo, O.H. Gutierrez, J.V. Manzione, c. Cox and V.J. Marder, J Vasc Surgery, in press, (1994).
71.   W. Markland, D. Pollock and D.J. Livingston, Protein Engineering, 3, 117 (1989).
72.   Y. Imura, J.M. Stassen, T. Kurokawa, S. Iwasa, H.R. Lijnen and D. Collen, Blood, 79, 9, 2322 (1992).
73.   P. Holvoet, Y. Laroche, J.M. Stassen, H. R. Lijnen, B. Van Hoef, F. De Cock, A. Van Houtven, Y. Gansemans, G. Matthyssens and D. Collen, Blood, 81, 696 (1993).
74.   A. Boutaud and F.J. Castellino, Arch Biochem Biophys, 303, 222 (1993).
75.   W-P. Yang, J. Goldstein, R. Procyk, G.R. Matsueda and S-Y. Shaw, Biochem, 33, 2306 (1994).
76.   S. Wilson, D.W. Cronk, I. Dodd, A.F. Esmail, S.B. Kalindjian, L. McMurdo, M.J. Browne, R.A.G. Smith and J.H. Robinson, Thromb Haemost, 70, 984 (1993).
77.   U. Martin, R. Bader, E. Böhm, U. Kohnert, E. von Möllendorf, S. Fischer and G. Sponer, Cardiovascular Drug Reviews, 11, 299 (1993).
78.   U. Tebbe, R. von Essen, A. Smolarz, P. Limbourg, J. Rox, J. Rustige, A. Vogt, J. Wagner, W. Meyer-Sabellek and K-L. Neuhaus, Am J Cardiol, 72, 518 (1993)
79.   C.J. Refino, N.F. Paoni, B.A. Keyt, C.S. Pater, J.M. Badillo, F.M. Wurm, J. Ogez and W.F. Bennett, Thromb Haemost, 70, 313 (1993).
80.   C.R. Benedict, C. Refino, B. Keyt, N. Paoni, J. Todd, C. Pater, J. Badillo, R. Thomas and W. Bennett, JACC, (spec. issue, 43rd Annual Scientific Session, ACC, Atlanta, GA, 3/13-17/94), 314A (Feb 1994).
81.   E.L. Madison, A. Kobe, M-J. Gething, J.F. Sambrook and E.J. Goldsmith, Science, 262, 419 (1993).
82.   D. Collen, F. De Cock, I. Vanlinthout, P.J. Declerck, H.R. Lijnen and J.M. Stassen, Fibrinolysis, 6, 232 (1992).
83.   D. Collen, F. De Cock and J-M. Stassen, Circulation, 87, 996 (1992).
84.   H.R. Lijnen, B. Van Hoef and D. Collen, Eur J Biochem, 211, 91 (1992).
85.   D. Collen, B. Schlott, Y. Engelborghs, B. Van Hoef, M. Hartmann, H.R. Lijnen, and D. Behnke, J Biol Chem, 268, 8284 (1993)
86.   K. Silence, D. Collen and H.R. Lijnen, J Biol Chem, 268, 9811 (1993).
87.   K. Silence, D. Collen and H.R. Lijnen, Blood, 82, 1175 (1993).
88.   P.W. Bergum and J. Gardell, J Biol Chem, 267, 17726 (1992).
89.   R.C. Becker, Cardiology, 82, 265 (1993).
90.   R. Seitz, H. Pelzer, A. Immel and R. Egbring, Fibrinolysis, 7, 109 (1993).
91.   P. Chang, D.L. Aronson, J. Scott and C.M. Kessler, Amer J Cardiol, 70, 406 (1992).
92.   G. Rasmanis, O. Vesterqvist, K. Gréen, O. Edhag and P. Henriksson, Br Heart J, 68, 374 (1992).
93.   D. Harpaz, X. Chen, C.W. Francis, F.J. Marder and R.S. Meltzer, JACC, 21, 1507 (1993).
94.   R. Kornowski, R.S. Meltzer, A. Chernine, Z. Vered and A. Battler, Circulation, 89, 339 (1994).
95.   K. Valji, A.,C. Roberts, G.B. Davis, and J.J. Bookstein, AJR, 156, 617 (1991).
96.   M. martin, T. Heimig, B.J. Othmar Fieback, L. Magnus, and C. Riedel, Angiol, 45, 143 (1994).
97.   G.R. Sitko, D.R. Ramjit, I.I. Stabilito, D. Lehman, J.J. Lynch and G.P. Vlasuk, Circulation, 85, 805 (1992).
98.   W.E.Rote, D-X. Mu, E.R. Bates, M.A. Nedelman and B.R. Lucchesi, J Cardiovasc Pharmacol, 23, 203 (1994).
99.   J.J. Lynch, G.R. Sitko, M.J. Mellott, E.M. Nutt, E.D. Lehman, P.A. Friedman, C.T. Dunwiddie and G.P. Vlasuk, Cardiovasc Res, 28, 78 (1994).
100. R.N. Puri and R.W. Colman, Blood Coagulation and Fibrinolysis, 4, 465 (1993).
101. E.J. Haskel, N.A. Prager, B.E. Sobel, D.R. Abendschein, Circulation, 83, 1048 (1991).

Chapter 11. Antithrombotic And Anti-Inflammatory Agents Of The Protein C Anticoagulant Pathway

S. Betty Yan and Brian W. Grinnell
Eli Lilly and Company
Indianapolis, Indiana 46285-1543

Introduction - Thrombosis is a principal cause of mortality that plays a key role in a number of disease processes including stroke and inflammation, and is a major complication of the atherosclerotic process resulting in myocardial infarction. In recent years, significant advances in the treatment of thrombotic disease and in the understanding of underlying mechanisms has increased the awareness of the liabilities of conventional anticoagulants and the need for better antithrombotic agents. The widespread uses of thrombolytics in the treatment of myocardial infarction, and the growing use of various angioplasty procedures in the past decade both have helped to propel the research on antithrombotics, as well as on agents with anti-ischemic properties. Because of the key role of thrombin in the control of coagulation and thrombosis, the ability to modulate thrombin generation and activity has been a central theme in the development of new therapeutic agents. Two previous chapters in this series (1, 2) in 1991 and 1992 reviewed specific aspects in the research and development of better antithrombotics. The 1991 chapter gave a broad overview of different mechanism of actions with a brief description of each class of antithrombotics, anti-platelets, and anticoagulants. The 1992 chapter gave an in-depth discussion of the development of short peptide and low molecular weight non-peptide organics as anti-platelets and anticoagulant agents. The aim of this chapter is to discuss therapeutic potentials and issues/challenges in the development of antithrombotics from the Protein C anticoagulant pathway, namely recombinant Protein C, Protein S and soluble thrombomodulin.

Many excellent recent review articles have been written about the Protein C anticoagulation pathway and its role in hemostasis and thrombosis (3). The intent of this article is not to review the detailed physiology of this pathway, but rather to describe some of the data and rationale relevant to the development of human Protein C, Protein S, and thrombomodulin as potential antithrombotic agents. Clearly, the fact that thrombosis and hemostasis *in vivo* is a membrane surface phenomenon, not a solution phase event, points to the potential advantage of developing therapeutics from this pathway, as each protein functions through interaction with components of the coagulation cascade on the platelet and endothelial surfaces. Because thrombin has biological activities other than catalyzing fibrin clot formation, such as those mediated through the thrombin receptor (4), therapeutic targets amenable to therapy with Protein C, Protein S and thrombomodulin may be quite broad. In this regard, a potential advantage to the use of protein therapeutics from this pathway, as compared to low molecular organic or peptidyl agents, is the potential for multiple biological activities in one structure. We will review the preclinical and clinical (limited) pharmacology of Protein C, revealing both antithrombotic and anti-inflammatory properties. Emphasis will be placed on Protein C since it is the more advanced of the three proteins in drug development. In the last part of this chapter, clinical trial data on several antithrombotics that were published in 1993 and 1994 will be reviewed briefly.

Background on the Protein C Anticoagulant System - The Protein C anticoagulant pathway plays a key role in maintaining normal hemostasis. This pathway very efficiently inhibits the conversion of prothrombin to thrombin, the serine protease that catalyses the generation of fibrin and thus clot formation. In addition, thrombin plays a major role in platelet activation, especially in arterial thrombosis. Until very recently, the Protein C pathway was thought to consist of the three glycoproteins, Protein C, Protein S, and thrombomodulin. Recent studies have demonstrated that Factor V plays an important cofactor role (5, 6) and can be considered part of this pathway. Protein C itself is a serine protease that circulates in the blood as a zymogen. The major physiological enzyme complex that converts Protein C to its active form is thrombin in complex with an endothelial surface membrane protein, thrombomodulin (3). Activated Protein C, along with cofactor Protein S, functions to block thrombin generation primarily by inactivating the activated forms of Factors V and VIII, thus inhibiting the prothrombinase and Factor Xase enzyme complexes, respectively. Because of these activities, the Protein C pathway serves as a controlling point for both the extrinsic and intrinsic coagulation pathways in preventing thrombin generation. As is evident from this brief description, homeostasis of blood depends on an intricate balance between the procoagulant and anticoagulation factors.

PROTEIN C AND PROTEIN C DERIVATIVES

Therapeutic Targets and Rationale - The importance of the Protein C pathway is clearly demonstrated by homozygous Protein C deficient individuals who generally suffer life threatening neonatal purpura fulminans. While heterozygous Protein C deficiency is generally not life threatening, many of these patients suffer from recurrent thrombotic episodes and, in those treated with coumadin, commonly present with coumadin-induced skin necrosis (7). Recent studies on treatment of both of these genetically deficient patient groups, with plasma purified human Protein C as a replacement therapy, have been very successful (8).

Because of the important role of Protein C in maintaining hemostasis, activated Protein C (both plasma-derived and recombinant) have been tested in most of the major available animal thrombotic models. For example, in baboon arterial thrombotic models, activated Protein C is effective as an antithrombotic agent (9, 10) and as an adjunctive therapy to thrombolytics (11). It is also effective in a stroke model in the rabbit (12) and in a model of ischemia in the porcine heart (13). Activated Protein C was also efficacious in venous thrombosis models in the dog and non-human primate models (14). From these data in many animal models of thrombosis, activated Protein C would be predicted to be effective in both venous and arterial thrombosis.

The potential use of activated Protein C in the treatment of microvascular thrombosis and sepsis merits a more detail discussion. Activated Protein C has recently been shown to be effective in the treatment of microvascular thrombosis presenting as disseminated intravascular coagulopathy (DIC) in cancer patients (15, 16). Recently the rationale for the use of an antithrombotic in the treatment of sepsis, another pathological condition complicated by underlying microvascular thrombosis, was reviewed (17). Most current anti-sepsis compounds have been aimed at preventing the formation of cytokines induced by gram-negative bacterial lipopolysaccharide or as antagonists to the action of cytokines, such as interleukin-1. However, these direct anti-inflammatory approaches have been disappointing in clinical trials to date (18, 19). Several studies have suggested a role of the Protein C pathway in the inflammatory response to endotoxin or bacteria in vivo or in vitro (20-22). Protein C levels are depressed (50% or lower) in patients suffering from meningococcemia (23, 24), especially for those that developed purpura fulminans. The mortality rate for patients suffering from meningococcemia with purpura fulminans is over 50%. In several recent studies, reduced Protein C levels have been shown to be correlated with mortality in both sepsis and septic shock (25-28). This has suggested that Protein C or activated Protein C could be effective during sepsis as a replacement therapy to correct declining levels due to consumptive coagulopathy. In support of this, Protein C has demonstrated efficacy in preventing shock associated with lethal endotoxemia in a baboon model (29). Of even greater interest, plasma-derived Protein C was recently successfully used to treat 4 patients with severe meningococcal sepsis with purpura fulminans (30) and a patient with gram positive sepsis with purpura fulminans and associated disseminated intravascular coagulation (31).

Direct antithrombotic activity, preventing microvascular thrombosis, is likely a major mechanism of action for the anti-sepsis activity of Protein C. However, it has been suggested that Protein C has anti-inflammatory properties in addition to its antithrombotic properties (32). The potential direct anti-inflammatory effects of Protein C have not been established clearly, but recent research has begun to elucidate several possibilities. The ability of Protein C to inhibit thrombin generation could, in and of itself, reduce thrombin-mediated inflammatory effects from decreased thrombin-receptor cleavage, and subsequently reduce platelet and endothelial cell activation (4). Because activated Protein C is generated in complexes on the membrane, this direct inhibition of thrombin generation at the site of thrombin receptor activation may help explain why thrombin inhibition through heparin has not been effective in septic shock. Activated Protein C has also been shown to have possible direct anti-inflammatory activity by inhibiting the generation of cytokines such as tissue necrosis factor (TNF) and interleukin-1 (IL-1), and to uncouple lipopolysaccharide interactions with monocyte CD14, through as yet unknown mechanisms (29, 33, 34). Protein C is also linked to inflammatory mediators and the complement cascade via its interaction with Protein S, and the carrier protein for C4BP.

A critical event following inflammatory insult at the endothelium is the recruitment, adhesion and activation of leukocytes. In recent studies, it has been found that human Protein C can inhibit E-selectin-mediated cell adhesion, an effect primarily mediated through glycosyl residues shown to be ligands for this selectin (35). The selectin cell adhesion receptors appear to play a role in septic shock and ischemic heart injury (36, 37). These specific Asn-linked carbohydrates on Protein C will be discussed in more detail in a later section. While these *in vitro* data on the action of protein C are suggestive, the amounts of Protein C required for these effects are greater than those required for antithrombotic effects. Further studies will be required to fully assess the potential direct anti-inflammatory properties of Protein C *in vivo*.

Structure of Human Protein C and Activated Protein C - The cDNA sequence of human Protein C (HPC) codes for a protein of 461 amino acids (38). The first 42 amino acids are a leader sequence consisting of a signal peptide (residues -42 to -25), and a pro-peptide (residues -24 to -1) that contains the recognition site for a vitamin K-dependent carboxylase (39). There are two proteolytic cleavage sites resulting in the removal of the propeptide and an internal Lys 156-Arg 157 dipeptide during cell secretion. 90-95% of the zymogen circulates as a heterodimer consisting of a light chain (residues 1-155), disulfide linked, to the heavy chain serine protease domain (residues 157-419). The remaining 5-10% of the secreted mature HPC zymogen circulates in a single chain form, lacking the removal of the internal dibasic dipeptide Lys-Arg. The first 9 glutamic acid residues in the light chain are sites of post-translational modification to γ-carboxyglutamates(Gla) by the vitamin K-dependent carboxylase. The light chain also contains one residue (Asp 71) post-translationally modified to ε−β-hydroxyaspartate (40). There are also four Asn-linked glycosylation sites for HPC (Asn 97, Asn 248, Asn 313, Asn 329) (41), but there is no evidence of any O-linked glycosylation in the two EGF-like domains in the light chain (42). From the study of natural and recombinant Protein C by a variety of cell biology, biochemical and molecular genetic approaches, the role of each of these post-translational modifications to the function of the protein has been extensively studied (35, 41, 43-45). The complete γ-carboxylation of the light-chain, β-hydroxylation of Asp 71 and correct pro-peptide processing are all required for full functional anticoagulant activity. Zymogen Protein C can be converted to aPC *in vitro* by either thrombin alone (in the absence of calcium ions) or by thrombomodulin-thrombin complex (in the presence of calcium ions), both conditions resulting in the removal of a 12 amino acid activation peptide (residues 158-169) from the amino-terminus of the heavy chain of HPC.

<u>Development of Recombinant Human Protein C</u> - While Protein C can be isolated from human plasma, the amount of available blood-bank plasma and the potential for contaminating infections agents hinders the development and availability of natural Protein C for wide therapeutic application. This section will be devoted to the issues, challenges and experiences related to the development of the recombinant version of the native sequence of Protein C. Many of the experiences encountered for recombinant HPC will also apply to Protein S and thrombomodulin, and other highly post-translationally modified proteins.

It has been difficult to identify mammalian cell lines that can express functionally active and correctly modified rHPC, especially at a commercially viable level, because of the number and complexity of the post-translational modifications described above. There have been substantial efforts both in academic and industrial laboratories to express and secrete fully active vitamin K-dependent proteins. However, with the exception of an engineered, adenovirus-transformed human kidney cell line, 293, all cell lines thus far engineered secrete only partially active material (46). Because of the now known versatility of the HK293 cell line, it has been used for expressing other vitamin K-dependent proteins such as human Protein S (47), and many other complex proteins such as thrombomodulin (48) and tissue factor pathway inhibitor (49). Unique vectors have been developed for the efficient expression of rHPC in the HK293 cell line (50), in addition to methods for the development of genetically stable HK293 production cell lines. Through efforts to purify and clone the enzymes involved in the complex modification of Protein C (51-53) it may be possible to engineer any cell to express fully biologically active rHPC at high levels. However, recent results reported for such efforts with another vitamin K-dependent protein, structurally very similar to HPC, have been quite disappointing (54, 55). While the HK293 cell line secretes primarily fully processed protein, we have found that the rate-limiting step in secretion is not due to the processing enzymes above, but instead is due to specific glycosyl processing in the endoplasmic reticulum (44). Many processing enzymes are involved in glycosylation and little is known about their regulation and control. Therefore, it is not anticipated that the molecular engineering of the glycosylation pathway will be a feasible approach in the near future. However, in the HK293 cell line, it has been possible to increase secretion of HPC by clonal selection and re-integration of strong expression vectors (46).

Another important aspect in the development of recombinant Protein C as a therapeutic has been the creation of an efficient and selective purification system. This procedure, termed pseudo-affinity chromatography , takes advantage of the conformational changes in HPC upon the addition or removal of calcium divalent ions (43). The concept of conformationally specific purification was first achieved using monoclonal antibodies that only recognized Protein C in the presence of calcium ions (56) When immobilized, these antibodies could be used to purify HPC from human plasma (57). The novel pseudo-affinity method, however, eliminates the inherent difficulties of scale-up with immunoaffinity procedures. In addition, the purification step is economical, efficient and highly selective, resulting in the recovery of only those molecules with full γ-carboxylation.

<u>Characterization of Recombinant Protein C from 293 Cells</u> - The HK293 cell-derived rHPC has been extensively characterized (43, 58). The amidolytic (protease) and anticoagulant activities of recombinant Protein C produced from the engineered HK293 cells displays high functional activity, comparable to or higher than that of plasma-derived HPC. Plasma-derived HPC and the purified HK293 cell-produced rHPCs both display apparent molecular weights from 62 to 66 kD as analysed by SDS-PAGE. Like plasma-derived Protein C, the 293 cell rHPC is greater than 90% two chain, and contains a similar pattern of the three heavy chain α,β,γ glycoforms. We have found that the respective N-terminal sequences of the heavy and light chains of rHPC are identical to those previously reported for plasma-derived HPC (59). The purified rHPC from HK293 cells has complete γ-carboxylation and a similar degree of β-OH-Asp when compared to plasma-derived HPC. Thus, with respect to

the proteolytic processing and amino acid modification, the rHPC from the HK293 cell line is processed identically to plasma-derived material, if not more completely in the case of γ-carboxylation.

The only structural difference between plasma-derived and recombinant HPC resides in the carbohydrate component. Oligosaccharides found on a glycoprotein are determined to a large extent by the cell line used (60), and glycosylation can influence parameters that are important to drug design, such as circulatory clearance, tissue targeting, immunogenicity, and efficacy (60). Therefore, the oligosaccharide structures on the rHPC from HK293 cells merit a more detail discussion. While little has been published about the carbohydrate on plasma derived HPC (57), the major oligosaccharides on rHPC from HK293 cell line, on the other hand, have been fully elucidated (58). They consist of some novel and rare Asn-linked oligosaccharides. The following structures are some examples that illustrate these novel and rare epitopes on the non-reducing ends of these oligosaccharides.

$$\begin{array}{l}
\text{Fuc}\alpha(1\rightarrow3)\backslash \\
\text{GalNAc}\beta(1\rightarrow4)\text{GlcNAc}\beta(1\rightarrow2)\text{Man}\alpha(1\rightarrow6)\backslash \qquad\qquad\qquad \text{Fuc}\alpha(1\rightarrow6)\backslash \\
\qquad\qquad\qquad\qquad\qquad\qquad\qquad\qquad\qquad\qquad \text{Man}\beta(1\rightarrow4)\text{GlcNAc}\beta(1\rightarrow4)\text{GlcNAc} \\
\text{GalNAc}\beta(1\rightarrow4)\text{GlcNAc}\beta(1\rightarrow2)\text{Man}\alpha(1\rightarrow3)^{/} \\
\qquad \text{Fuc}\alpha(1\rightarrow3)^{/}
\end{array}$$

$$\begin{array}{l}
\qquad\qquad\qquad \text{Fuc}\alpha(1\rightarrow3)\backslash \\
\qquad \text{GalNAc}\beta(1\rightarrow4)\text{GlcNAc}\beta(1\rightarrow2)\text{Man}\alpha(1\rightarrow6)\backslash \qquad\qquad \text{Fuc}\alpha(1\rightarrow6)\backslash \\
\qquad\qquad\qquad\qquad\qquad\qquad\qquad\qquad\qquad\qquad \text{Man}\beta(1\rightarrow4)\text{GlcNAc}\beta(1\rightarrow4)\text{GlcNAc} \\
\text{NeuAc}\alpha(2\rightarrow6)\text{GalNAc}\beta(1\rightarrow4)\text{GlcNAc}\beta(1\rightarrow2)\text{Man}\alpha(1\rightarrow3)^{/}
\end{array}$$

$$\begin{array}{l}
\text{NeuAc}\alpha(2\rightarrow3)\text{Gal}\beta(1\rightarrow4)\text{GlcNAc}\beta(1\rightarrow2)\text{Man}\alpha(1\rightarrow6)\backslash \qquad\qquad \text{Fuc}\alpha(1\rightarrow6)\backslash \\
\qquad\qquad\qquad\qquad\qquad\qquad\qquad\qquad\qquad\qquad \text{Man}\beta(1\rightarrow4)\text{GlcNAc}\beta(1\rightarrow4)\text{GlcNAc} \\
\qquad \text{GalNAc}\beta(1\rightarrow4)\text{GlcNAc}\beta(1\rightarrow2)\text{Man}\alpha(1\rightarrow3)^{/} \\
\qquad\qquad \text{Fuc}\alpha(1\rightarrow3)^{/}
\end{array}$$

The novel trisaccharide epitope, GalNAcβ(1→4)[Fucα(1→3)]GlcNAcβ(1→•), on rHPC was named PC293 determinant. The PC293 determinant and the two rare disaccharide epitopes, GalNAcβ(1→4)GlcNAcβ(1→2) and NeuAcα(2→6)GalNAcβ(1→4), have been found in naturally occurring human glycoproteins (58), suggesting that these novel and rare oligosaccharides on HK293-produced glycoproteins probably would not pose an immunogenicity problem in humans, although actual clinical trial data will be needed to confirm this. From glycosyl composition analysis, these GalNAc-containing rare oligosaccharides are not present in plasma derived HPC. Studies in a primate thrombotic model show these rare Asn-linked oligosaccharides do not appear to significantly affect the circulatory half-life and efficacy of recombinant activated HPC (r-aPC) when compared with plasma aPC (9). Because of the carbohydrate sequence homology of the PC293 determinant to the Lewis X determinant, Asn-linked oligosaccharides containing the PC293 determinants were tested for binding to E-selectin. Surprisingly, oligosaccharides containing the PC293 determinants had a higher affinity for the E-selectin than the natural ligand sialyl Le X (35). Thus, the carbohydrate portion of rHPC may impart some of the observed anti-inflammatory activity discussed earlier. While the structures of the carbohydrates on plasma-derived Protein C are not known, this material does contain fucosylated oligosaccharide (43), suggesting that it could contain ligands for the selectin family. This is consistent with the fact that the site of normal biosynthesis for HPC is the liver, and it is known that fucosylated oligosaccharides, including sialyl Le X, are produced on glycoproteins synthesized by the liver (61, 62).

<u>Protein C Derivatives as Second Generation Antithrombotics</u> - Several approaches have been taken to improve either the process for activated Protein C (aPC) production or the molecule itself by the creation of site-specific changes in the amino acid backbone. An effective method (63) in expressing r-aPC directly from HK293 cells by replacing the thrombin-cleaved activation peptide with a sequence in the insulin receptor precursor (PRPSRKRR), a substrate for the cellular insulin receptor processing protease, has been reported. This derivative was secreted from the cell as aPC, giving the potential advantage of eliminating the costly activation and purification steps needed with the secreted zymogen. It has been possible to increase the functional anticoagulant activity of Protein C by alterations in the sites for glycosylation (41). For example, elimination of partial glycosylation at the unusual consensus sequence Asn329-X-Cys resulted in a high-activity β-form Protein C molecule that could result in improved therapeutic efficacy. Forms of Protein C with increased activity through improved catalysis and reduced inhibitor binding activity have been described (64). By altering amino acids near the thrombin cleavage site, several Protein C derivatives have been created that are much more sensitive to thrombin, less dependent on thrombomodulin and less sensitive to inhibition of activation by calcium ion (65-69). These could be more effective antithrombotics when administered as a zymogen, i.e., as a "pro-drug" form of Protein C at a potentially reduced dose due to longer half-life. These Protein C derivatives would have the clinical advantage of being activated at the site of arterial thrombin generation, i.e., activation would not be confined to the microvasculature where sufficient thrombomodulin is present. Further, since such derivatives are essentially thrombomodulin-independent, activation could occur in prothrombotic inflammatory disease states such as sepsis where thrombomodulin is down-regulated (22, 70).

RECOMBINANT THROMBOMODULIN AS AN ANTITHROMBOTIC

<u>Therapeutic Potential</u> - There are excellent review articles on the biochemistry, structure-function relationships and potential cardiovascular uses of thrombomodulin (71, 72). Thrombomodulin may also have a physiological role outside of hemostasis. Thrombomodulin was independently discovered as a cyclic-AMP responsive cell surface glycoprotein in murine teratocarcinoma F9 cells (73) and was named fetomodulin. Subsequent cDNA cloning of fetomodulin confirmed the identity of fetomodulin as thrombomodulin. The role of thrombomodulin in embryogenesis and in developmental biology is not clear and is currently being investigated (74). Thrombin and thrombomodulin may be involved in tissue remodeling and wound healing, a hypothesis supported by the discovery that thrombomodulin is synthesized by nonvascular surfaces of body cavities (75), and unexpected tissues or cells such as synovial fluid cells (76) and osteoblasts (77). While the specific substrates for thrombin-thrombomodulin complex outside of coagulation are yet to be identified, it is possible that thrombomodulin could have a yet unknown therapeutic potential outside of thrombosis.

<u>Development of Recombinant soluble Thrombomodulin</u> - Since thrombomodulin is a membrane protein, the strategy for its development as a potential therapeutic has been to express a recombinant truncated molecule without the transmembrane and cytoplasmic domains (78, 79). In the cell lines used to date, two forms of recombinant soluble thrombomodulin are secreted which differ in the degree of glycosaminoglycan modification. The presence of the two glycoforms in the recombinant thrombomodulin is similar to that observed in human urine, suggesting that natural endothelial surface thrombomodulin may also exist in two glycoforms (80). Of significance, the two glycoforms have different affinities for thrombin and different cofactor activity for Protein C (81, 82). As mentioned above in the section on recombinant Protein C expression, the degree and structure of oligosaccharides on a recombinant glycoprotein is determined to a large extent by the cell line used for expression. Studies to date have found the glycosaminoglycan on human and rabbit thrombomodulin to be of the chondroitin-sulfate type, but detailed structural analysis has not been reported to determine how similar or dissimilar the structures are to the

natural wild type thrombomodulin. Future studies on the structural elucidation of the oligosaccharides on natural and recombinant thrombomodulin and the effect of the glycosaminoglycan moiety on pharmacokinetics will be very helpful in guiding the development of thrombomodulin and other recombinant glycoproteins with glycosaminoglycan type oligosaccharides.

RECOMBINANT PROTEIN S AS AN ANTITHROMBOTIC

The therapeutic potential for recombinant Protein S is the least clear of the three major proteins in the Protein C anticoagulant pathway. Protein S is also a vitamin K-dependent protein with a very similar arrangement of structural domains and post-translational modifications as Protein C. However, unlike Protein C, Protein S does not have the serine protease domain, and has no enzymatic activity. In place of the protease domain, Protein S has a domain with sequence homology to sex hormone binding protein (83, 84). Even though no direct steroid binding activity has been demonstrated with Protein S, there are indications that the level of Protein S in circulation is influenced by hormonal levels (85-88), but also to be decreased in male smokers or in Type I diabetics, conditions that have no obvious link to sex hormonal levels (89, 90). Protein S deficiency, familial or acquired, has been reported to be associated with increase risk for arterial thrombosis (91, 92).

While Protein S is clearly an important risk factor for thrombosis, the exact mechanism of action is not clear. Protein S was discovered as a cofactor for activated Protein C (aPC). However, unlike other cofactor molecules in the coagulation cascade, which typically accelerate activities by several orders of magnitude (93), Protein S only accelerates the aPC activity by several fold (94, 95). A recent report has suggested that in addition to a cofactor role, Protein S has a direct anticoagulant activity rendering Factor Va susceptible to aPC (96). Protein S circulates in the blood either free or in complex with C4b-binding protein (97, 98), and when in complex with C4b-binding protein is devoid of cofactor activity to aPC. C4b-binding protein level increases during pathological conditions such as infection when the complement cascade is activated, resulting in a decrease in free Protein S and thus a decrease in the function of the Protein C pathway. However, the amount of total Protein S in circulation is estimated at around 25-30 μg/ml, suggesting that the amount of exogenously added recombinant Protein S needed to shift the equilibrium to more free Protein S could be prohibitively high and costly in terms of drug development. Furthermore, recent data suggest that there may be as yet unidentified factors that regulate the binding of Protein S to C4b-binding protein (92). For these reasons, there has been very little pharmacological testing of recombinant human Protein S in animal thrombotic models, although expression of fully active recombinant Protein S has been reported (47). There have been studies to identify the peptide residues that are involved in the interaction between Protein S and C4b-binding protein (98-100). Perhaps a small peptide or a peptide mimetic may be more feasible as a way to develop an antithrombotic targeted to increase free Protein S in circulation.

Conclusion - It is indeed an exciting time for the design and development of new antithrombotics. The results of clinical studies have become available for several of the antithrombotic compounds mentioned in the two earlier chapters in this series (1, 2). The clinical data with a monoclonal antibody against platelet glycoprotein GPIIbIIIa, 7E3, (101-106), hirulog (107-111) and hirudin (112-114), are very encouraging and have shed additional insight into the understanding of treatments for thrombotic conditions. Of importance, cessation of these three antithrombotics did not appear to result in the rebound coagulation phenomenon previously observed with heparin (115) and another thrombin inhibitor, agatroban (116). These early data have suggested that hirudin and hirulog could be used safely in conjunction with an anti-platelet agent, such as aspirin. Phase III clinical data have just been reported for c7E3 (a chimeric Fab fragment version of 7E3) (105, 106). Ischemic complications of coronary angioplasty and atherectomy were reduced in patients given c7E3, although the risk of bleeding was increased. In the next several years, as more Phase III data become available, the utility of these agents in the treatment of specific thrombotic diseases will become more clear.

We have reviewed the major components in the Protein C anticoagulant pathway and the rationale for the importance of this pathway in the design and development of antithrombotics. The first generation therapeutic compounds based on this important regulatory pathway will likely be recombinant versions of the natural proteins. Future therapeutics may be those designed to enhance the levels of thrombomodulin, to increase free Protein S or to increase the efficiency of zymogen conversion to activated Protein C. While only well designed human clinical trials on antithrombotics based on this important pathway will allow us to understand potential therapeutic applications and limitations, the unique mechanism of action, potential for dual antithrombotic and anti-inflammatory activities, and natural ability of proteins in this pathway to bind to cell membrane, may offer agents with a wide therapeutic utility .

References

1.   R.J. Shebuski, Ann. Reports in Med. Chem. 26, 93 (1991).
2.   J.A. Jakubowski, G.F. Smith and D.J. Sall, Ann. Reports in Med. Chem. 27, 99 (1992).
3.   C.T. Esmon, Thromb. Haemost. 70, 29 (1993).
4.   S.R. Coughlin, T.-K.H. Vu, D.T. Hung and V.I. Wheaton, J. Clin. Invest. 89, 351 (1992).
5.   B. Dahlback and B. Hildebrand, Proc. Natl. Acad. Sci. 91, 1396 (1994).
6.   P.J. Svensson and B. Dalback, N. Engl. J. Med. 330, 517 (1994).
7.   L.H. Clouse and P.C. Comp, N. Engl. J. Med. 314, 1298 (1986).
8.   H.P. Schwarz, K. Nelson, M. Dreyfus, M. Materson, R. Montogomery, G. Rivard, M. David, W. Kreuz, L. Parapia, A. Minford, L. Tillyer, J. Allgrove, J. Conrad and K. Bauer, Thromb. Haemost. 69, Abstract no. 1698 (1993).
9.   A. Gruber, S.R. Hanson, A.W. Kelly, S.B. Yan, N. Bang, J.H. Griffin and L.A. Harker, Circulation 82, 578 (1990).
10.   S.R. Hanson, J.H. Griffin, L.A. Harker, A.B. Kelly, C.T. Esmon and A. Gruber, J. Clin. Invest. 92, 2003 (1993).
11.   A. Gruber, L.A. Harker, S.R. Hanson, A.B. Kelly and J.H. Griffin, Circulation 84, 2454 (1991).
12.   Y. Oda, K. Tsumoto and H. Mizokami, Thromb. Haemost. 69, Abstract no. 665 (1993).
13.   T.R. Snow, M.T. Deal, D.T. Dickey and C.T. Esmon, Circulation 84, 293 (1991).
14.   T.W. Wakefield, S.K. Wrobleski, M.S. Sarpa, F.B. Taylor, C.T. Esmon, A. Cheng and L.J. Greenfield, J. Vasc. Surg. 14, 588 (1991).
15.   K. Okajima, H. Imamura, S. Koga, I. Masayasu, K. Takatsuki and N. Aoki, Am. J. Hematol. 33, 277 (1990).
16.   K. Okajima, S. Koga, M. Kaji, M. Inoue, T. Nakagaki, A. Funastsu, H. Okabe, K. Takatsuki and N. Aoki, Thromb. Haemost. 63, 48 (1990).
17.   M. Levi, H. ten Cate, T. van der Poll and S.J.H. van Deventer, JAMA 270, 975 (1993).
18.   M. Gibaldi, Pharmacotherapy 13, 302 (1993).
19.   Z.M. Quezado, C. Natanson, D.W. Alling, S. Banks M., C.A. Koev, R.J. Elin, J.M. Hosseini, J.P. Bacher, R.L. Danner and W.D. Hoffman, JAMA 269, 2221 (1993).
20.   W.W. Hancock, K. Tanaka, H.H. Salem, N.L. Tilney, R.C. Atkins and J.W. Kupiec-Weglinski, Transplant. Proc. 23, 235 (1991).
21.   F.B. Taylor, A. Chang, G. Ferrell, T. Mather, R. Catlett, K. Blick and C.T. Esmon, Blood 78, 357 (1991).
22.   T.A. Drake, J. Cheng, A. Chang and F.B. Taylor, Am. J. Pathology 142, 1458 (1993).
23.   D. Powars, R. Larsen, J. Johnson, T. Hulbert, T. Sun, M. Patch, R. Francis and L. Chan, Clin. Infectious Disease 17, 254 (1993).
24.   P. Brandtzaeg, P.M. Sandset, G.B. Joo, R. Ovstebo, U. Abildgaard and P. Kieruef, Thromb. Res. 55, 459 (1989).
25.   J. Roman, F. Velasco, F. Fernandez, M. Fernandez, R. Villalba, V. Rubio and A. Torres, J. Perinat. Med. 20, 111 (1992).
26.   F. Fourrier, C. Chopin, J. Goudemand, S. Hendrycx, C. Caron, A. Rime, A. Marey and P. Lestavel, Chest 101, 816 (1992).
27.   J.F. Hesselvik, J. Malm, B. Dahlback and M. Blomback, Thromb. Haemost. 65, 126 (1991).
28.   F. Leclerc, J. Hazelzet, B. Jude, W. Hofhuis, V. Hue, A. Martinot and E. Van der Voort, Intensive Care Med. 18, 202 (1992).
29.   F.B. Taylor, A. Chank, C.T. Esmon, A. D'Angelo, S. Vigano-D'Angelo and f.E. Blick, J. Clin. Invest. 79, 918 (1987).

30. G.E. Rivard, M. David, C. Farrell, W. Gerson, J.D. Dickerman, E.G. Bovil and H.P. Schwarz, Thrombo. Haemost. 96, A2339 (1993).
31. W.T. Gerson, J.D. Dickerman, E.G. Bovill and E. Golden, Pediatrics 91, 418 (1993).
32. C.T. Esmon, F.B. Taylor and T.R. Snow, Thromb. Haemost. 66, 160 (1991).
33. W.W. Hancock, A. Tsuchida, H. Hau, N.M. Thomson and H.H. Salem, Transplant. Proc. 24, 2302 (1992).
34. S. Grey, H. Hau, H. Salem and W.W. Hancock, Transplant. Proc. 25, 2913 (1993).
35. B.W. Grinnell, R.B. Hermann and S.B. Yan, Glycobiology, 4, In press , (1994).
36. M.P. Bevilacqua and R.M. Nelson, J. Clin. Invest. 91, 379 (1993).
37. W. Newman, L.D. Beall, C.W. Carson, G.G. Hunder, N. Graben, Z.I. Randhawa, T.V. Gopal, J. Wiener-Kronish and M.A. Matthay, J. Immunol. 150, 644 (1993).
38. R.J. Beckmann, R.J. Schmidt, R.F. Santerre, J. Plutzsky, G.R. Crabtree and G.L. Long, Nucleic Acids Res. 13, 5233 (1985).
39. D.C. Foster, M.S. Rudinski, B.G. Schach, K.L. Berkner, A.A. Kumar, F.S. Hagen, C.A. Sprecher, M.Y. Insley and E.W. Davie, Biochem. 26, 7003 (1987).
40. T. Drakenberg, P. Fernlund, P. Roepstorff and J. Stenflo, Proc. Natl. Acad. Sci. 80, 1802 (1983).
41. B.W. Grinnell, J.D. Walls and B. Gerlitz, J. Biol. Chem. 226, 9778 (1991).
42. R.J. Harris and M.W. Spellman, Glycobiology 3, 219 (1993).
43. S.B. Yan, P. Razzano, Y.B. Chao, J.D. Walls, D.T. Berg, D.B. McClure and B.W. Grinnell, Bio/Technology 8, 655 (1990).
44. D.B. McClure, J.D. Walls and B.W. Grinnell, J. Biol. Chem. 267, 19710 (1992).
45. E.G. Bovill, J.A. Tomczak, B. Grant, F. Bhushan, E. Pillemer, I.R. Rainville and G.L. Long, Blood 79, 1456 (1992).
46. B.W. Grinnell, J.D. Walls, B. Gerlitz, D.T. Berg, D.B. McClure, H. Ehrlich, N.U. Bang and S.B. Yan. in Protein C and Related Anticoagulants (eds. Bruley, D. and Drohan, W.) 13 (Gulf Publishing Co., Houston, 1990).
47. B.W. Grinnell, J.D. Walls, C. Marks, GlasebrookA.L., D.T. Berg, S.B. Yan and N.U. Bang, Blood 76, 2546 (1990).
48. J.F. Parkinson, B.W. Grinnell, R.E. Moore, J. Hoskins, C.J. Vlahos and N.U. Bang, J. Biol. Chem. 265, 12602 (1990).
49. P.L. Smith, T.P. Skelton, D. Fiete, S.M. Dharmesh, M.C. Beranek, L. MacPhail, G.J. Broze and J.U. Baenziger, J. Biol. Chem. 267, 19140 (1992).
50. D.T. Berg, D.B. McClure and B.W. Grinnell, Biotechnique. 6, 972 (1993).
51. C.K. Derian, W. VanDusen, C.T. Przysieck, P.N. Walsh, K.L. Berkner, R.J. Kaufman and P.A. Friedman, J. Biol. Chem. 264, 6615 (1989).
52. D.P. Morris, B.A.M. Soute, C. Vermeer and D.W. Stafford, J. Biol. Chem. 268, 8735 (1993).
53. D.C. Foster, R.D. Holly, C.A. Sprecher, K.M. Walker and A.A. Kumar, Biochem. 30, 367 (1991).
54. A. Rehemtulla, D.A. Roth, L.C. Wasley, A. Kuliopulos, C.T. Walsh, B. Furie, B.C. Furie and R.J. Kaufman, Proc. Natl. Acad. Sci. 90, 4611 (1993).
55. L.C. Wasley, A. Rehemtulla, J.A. Bristol and R.J. Kaufman, J. Biol. Chem. 268, 8458 (1993).
56. S. Vigano-D'Angelo, P.D. Comp, C.T. Esmon and A. D'Angelo, J. Clin. Invest. 77, 416 (1986).
57. B.W. Grinnell, D.T. Berg, J. Walls and S.B. Yan, Bio/Technology 5, 1189 (1987).
58. S.B. Yan, Y.B. Chao and H. vanHalbeek, Glycobiology 3, 597 (1993).
59. W. Kisiel, J. Clin. Invest. 64, 761 (1979).
60. D.A. Cumming, Glycobiology 1, 115 (1991).
61. T.W. De Graaf, M.E. Vander Stelt, M.G. Anbergen and W. van Dijk, J. Exp. Med. 177, 657 (1993).
62. G. Walz, A. Aruffo, W. Kolanus, M. Bevilacqua and B. Seed, Science 250, 1132 (1990).
63. H.J. Ehrlich, S.R. Jaskunas, B.W. Grinnell, S.B. Yan and N.U. Bang, J. Biol. Chem. 264, 14298 (1989).
64. S.B. Yan and B.W. Grinnell, Perspective. Drug Discovery. Design In Press, (1994).
65. H.J. Ehrlich, B.W. Grinnell, S.R. Jaskunas, C.T. Esmon, S.B. Yan and N.U. Bang, EMBO J. 9, 2367 (1990).
66. M.A. Richardson, B. Gerlitz and B.W. Grinnell, Nature 360, 261 (1992).
67. A.R. Rezaie, T. Mather, F. Sussman and C.T. Esmon, J. Biol. Chem. 269, 3151 (1994).
68. A.R. Rezaie and C.T. Esmon, J. Biol. Chem. 267, 26104 (1992).
69. M.A. Richardson, B. Gerlitz and B.W. Grinnell, Protein Sci. 3, 711 (1994).
70. C.T. Esmon, Arteriosclerosis and Thromb. 12, 135 (1992).
71. W.A. Dittman, Trends in Cardiovascular Medicine 1, 331 (1991).
72. J.E. Sadler, S.R. Lentz, J.P. Sheehan, M. Tsiang and Q. Wu, Haemost. 23, 183 (1993).
73. M. Imada, S. Imada, H. Iwasaki, A. Kume, H. Yamaguchi and E.E. Moore, Dev. Biol. 122, 483 (1987).
74. V.A. Ford, J.E. Wilkinson and S.J. Kennel, Roux's Arch. Dev. Biol. 202, 364 (1993).
75. M. Boffa, B. Burke and A. Haudenschild, J. Histochem. Cytochem. 35, 1267 (1987).
76. E.M. Conway and B. Nowakowski, Blood 81, 726 (1993).

77. C. Maillard, M. Berruyer, C.M. Serre, J. Amiral, M. Dechavanne and P.D. Delmas, Endocrinology 133, 668 (1993).
78. M. Ogata, K. Wakita, K. Kimura, Y. Marumoto, K. Oh-i and S. Shimizu, Appl. Microbiol. and Biotech. 38, 520 (1993).
79. J.F. Parkinson, C.J. Vlahos, S.B. Yan and N.U. Bang, Biochem. J. 283, 151 (1992).
80. S. Yamamoto, T. Mizoguchi, T. Tamaki, M. Ohkuchi, S. Kimura and N. Aoki, J. Biochem. 113, 433 (1993).
81. J. Ye, C.T. Esmon and A.E. Johnson, J. Biol. Chem. 268, 2373 (1993).
82. B. Gerlitz, T. Hassell, C. Vlahos, J.F. Parkinson, N.U. Bang and B.W. Grinnell, Biochem. J. 295, 131 (1993).
83. M.E. Baker, F.S. French and D.R. Joseph, Biochem. J. 243, 293 (1987).
84. S. Gershagen, P. Fernlund and A. Lundwall, Febs. Lett. 220, 129 (1987).
85. P.C. Comp, G.R. Thurnau, J. Welsh and C.T. Esmon, Blood 68, 881 (1986).
86. L.M. Boerger, P.C. Morris, G.R. Thurnau, C.T. Esmon and P.C. Comp, Blood 69, 692 (1987).
87. J. Malm, M. Laurell and B. Dahlback, Br. J. Haematol. 68, 437 (1988).
88. A. D'Angelo, S. Vigano-D'Angelo, C.T. Esmon and P.C. Comp, J. Clin. Invest. 81, 1445 (1988).
89. B.D. Scott, C.T. Esmon and P.C. Comp, Am. Heart J. 122, 76 (1991).
90. H.P. Schwarz, G. Schernthaner and J.H. Griffin, Thromb. Haemost. 57, 240 (1987).
91. P.A. Taheri, B.A. Eagel, H. Karamanoukian, E.L. Hoover and G. Logue, Am. Surgeon 58, 496 (1992).
92. M.E. Carr and S.L. Zekert, Haemost. 23, 159 (1993).
93. K.G. Mann and L. Lorand, Meth. Enzymol. 222, 1 (1993).
94. S. Solymoss, M.M. Tucker and P.B. Tracy, J. Biol. Chem 263, 14884 (1988).
95. G. Tans, J. Rosing, M.C.L.G.D. Thomassen, M.J. Heeb, R.F.A. Zwaal and J.H. Griffin, Blood 77, 2641 (1991).
96. M.J. Heeb, R.M. Mesters, G. Tans, J. Resing and J.H. Griffin, J. Biol. Chem. 268, 2872 (1993).
97. G. Dahlback, Thromb. Haemost. 66, 49 (1991).
98. Y. Hardig, A. Razaie and B. Dahlback, J. Biol. Chem. 268, 3033 (1993).
99. J.A. Fernandez, M.J. Heeb and J.H. Griffin, J. Biol. Chem. 268, 16788 (1993).
100. R.E. Weinstein and F.J. Walker, J. Clin. Invest. 86, 1928 (1990).
101. N.S. Kleiman, M. Ohman, R.M. Califf, B.S. George, D. Kereiakes, F.V. Aquirre, H. Weisman, T. Schaible and E.J. Topol, J. Am. Coll. Cardiol. 22, 381 (1993).
102. E.M. Faioni, F. Franchi, D. Asti, E. Sacchi, F. Bernardi and P.M. Mannucci, Thromb. Haemost. 70, 1067 (1993).
103. S.G. Ellis, J.E. Tcheng, F.I. Navetta, D.W.M. Muller, H.F. Weisman, C. Smith, K.M. Anderson, R.M. Califf and E.J. Topol, Coronary Artery Disease 4, 167 (1993).
104. M.L. Simoons, M. Jan de Boer, M. van den Brand, A.J.M. van Miltenburg, J.C.A. Hoorntje, G.R. Heyndrickx, L.R. van der Wilken, D. de Bono, W. Rutsch, T.F. Schaible, H.F. Weisman, R. Klootwijk, K.M. Nijssen, J. Stibbe and P.J. Feyter, Circulation 89, 596 (1994).
105. The EPIC Investigators, N. Engl. J. Med. 330, 956 (1994).
106. E.J. Topol, R.M. Califf, H.F. Weisman, G.E. Stephen, J.E. Tcheng, S. Worley, R. Ivanhoe, B.S. George, D. Fintel, M. Weston, K. Sigmon, K.M. Anderson, K.L. Lee, J.T. Willerson and o.b.o.t.E. Investigators, Lancet 343, 881 (1994).
107. E.J. Topol, R. Bonan, D. Jewitt, U. Sigwart, V.V. Kakkar, M. Rothman, D. de Bono, J. Ferguson, J.T. Willerson, J. Strony, P. Ganz, M.D. Cohen, R. Raymond, I. Fox, J. Maraganore and B. Adelman, Circulation 87, 1622 (1993).
108. R. Lidon, P. Theroux, M. Juneau, B. Adelman and J. Maraganore, Circulation 88, 1495 (1993).
109. C.P. Cannon, J.M. Maraganore, J. Loscalzo, A. McAllister, K. Eddings, D. George, A.P. Selwyn, B. Adelman, I. Fox, E. Braunwald and P. Ganz, Am. J. Cardiol. 71, 778 (1993).
110. G.V.R.K. Sharma, D. Lapsley, J.A. Vita, S. Sharma, E. Coccio, B. Adelman and J. Loscalzo, Am. J. Cardiol. 72, 1357 (1993).
111. I. Fox, A. Dawson, P. Loynds, J. Eisner, K. Findlen, E. Levin, D. Hanson, T. Mant, J. Wagner and J. Maraganore, Thromb. Haemost. 69, 157 (1993).
112. P. Zoldhelyi, M.W.I. Webster, V. Fuster, D.E. Grill, D. Gasper, S.J. Edwards, C.F. Cabot and J.H. Chesebro, Circulation 88, 2015 (1993).
113. M. Verstraete, M. Nurmohamed, J. Kienast, M. Siebeck, G. Silling-Engelhardt, H. Buller, B. Hoet, J. Bichler and P. Close, J. Am. Coll. Cardiol. 22, 1080 (1993).
114. A. van den Bos, J.W. Deckers, G.R. Heyndrickx, G. Laarman, H. Suryapranata, F. Zijlstra, P. Close, J.J.M.M. Rijnierse, H.R. Buller and P.W. Serruys, Circulation 88, 2058 (1993).
115. P. Theroux, D. Waters, J. Lana, M. Juneau and J. McCans, N. Engl. J. Med. 327, 141 (1992).
116. H.K. Gold, F.W. Torres, H.D. Garabedian, W. Werne, I. Jamng, A. Khan, J.N. Hagstrom, T. Yasuda, R.C. Leinbach, J.B. Newell, E.G. Bovill, D.C. Stump and D. Collen, J. Am. Coll. Cardiol. 21, 1039 (1993).

SECTION III. CHEMOTHERAPEUTIC AGENTS

Editor: Jacob J. Plattner
Abbott Laboratories, Abbott Park, IL 60064

Chapter 12. Antibacterial Agents

Martin R. Jefson*, Scott J. Hecker+, and John P. Dirlam*
*Pfizer Inc, Central Research Division, Groton, CT 06340 and
+Microcide Pharmaceuticals Inc., 850 Maude Ave., Mountain View, CA 94043

Introduction - This review describes developments of the past year in the major areas of antibacterial research, including ß-lactams, quinolones, and macrolides. Key objectives in this field are the search for more effective agents against methicillin-resistant *S. aureus* (MRSA), other resistant organisms and opportunistic infections. Research in new areas that offer the potential for the discovery of novel agents is also discussed.

ß-Lactams - The use of third-generation cephalosporins in clinical practice was reviewed (1). Multicenter comparisons of the activity of a variety of ß-lactams against clinical isolates have appeared (2-4), as well as a comparison of the pharmacokinetics of new oral cephalosporins (5).

In the cephalosporin area, significant new information has appeared on compounds previously described in Ann. Rep. Med. Chem. A summary of the *in vitro* and *in vivo* activity of parenteral agent E1077 (**1**) has appeared (6), as well as a study comparing this agent to others against clinical isolates (7). Although **1** shows potent activity against *E. coli* and *H. influenza*, and moderate activity against *S. pneumoniae* and methicillin-susceptible *S. aureus* (MIC$_{90}$s of ≤0.015, 0.06, 0.5 and 1 µg/ml, respectively), it has poor activity against ceftazidime-resistant *P. aeruginosa* (MIC$_{90}$ >64 µg/ml) and MRSA (MIC$_{90}$ = 32 µg/ml). A large body of information has appeared on parenteral agent FK037 (**2**), including two comparative studies with other cephalosporins of *in vitro* activity (8,9). This agent, as well as Ro 40-6890 (**3**, the active metabolite of ester prodrug Ro 41-3399) share similar spectrum strengths and weaknesses with **1** (10). Reports describing the pharmacokinetics (11) and efficacy against experimental infections in mice (12) of the injectable agent cefozopran (SCE-2787, **4**) have appeared.

A number of new cephalosporins have been reported in the last year, most of which appear to have similar spectrum deficiencies to the agents described above. E1101, the isopropoxycarbonyloxyethyl ester prodrug of E1100 (**5**), exhibits good oral absorption (13), and has potent activity against susceptible strains (14). The pharmacokinetics in humans of oral cephem S-1090 (**6**) were reported (15), as well as discussions of SAR in this series (16-18). The synthesis and *in vitro* activity of some C-3-triazolylthiomethyl cephems including SYN-454 (**7**) was described (19).

Some improvements in activity against MRSA have been observed with cephalosporins. The discovery and SAR of a series of 3-heteroarylthiocarbapenems was described (20,21). Impressive activity against MRSA was found (MICs of 2-4 µg/ml), although early leads exhibited extremely high protein-binding in human plasma. Further work led to the discovery of LY-206763 (**8**), which demonstrates lower protein-binding while retaining good activity against MRSA. PBP studies revealed good affinity for PBP2a, the mediator of methicillin resistance. The hydrolysis of **8** by staphylococcal ß-lactamase was 168% relative to cephaloridine (100%), although this apparently does not compromise its antimicrobial activity. Improved MRSA activity was also observed (22) in a series including

KA-154 (**9**, MIC$_{90}$s of 0.78-6.25 µg/ml), as well as with new compounds TOC-39 and TOC-50 (MIC$_{90}$s of 3.13 µg/ml) (23).

Regarding the use of a catechol substituent in order to take advantage of the *tonB*-dependent iron transport system of bacteria, two new reports have appeared. Placement of the catechol in the oxime functionality, with inclusion of a 3-vinyl group, as in **10**, affords good activity against *S. aureus* in addition to anti-*Pseudomonas* activity (24). On the other hand, placement of the catechol in the C-3 side chain, as in **11**, broadened the spectrum of Gram-negative activity (25).

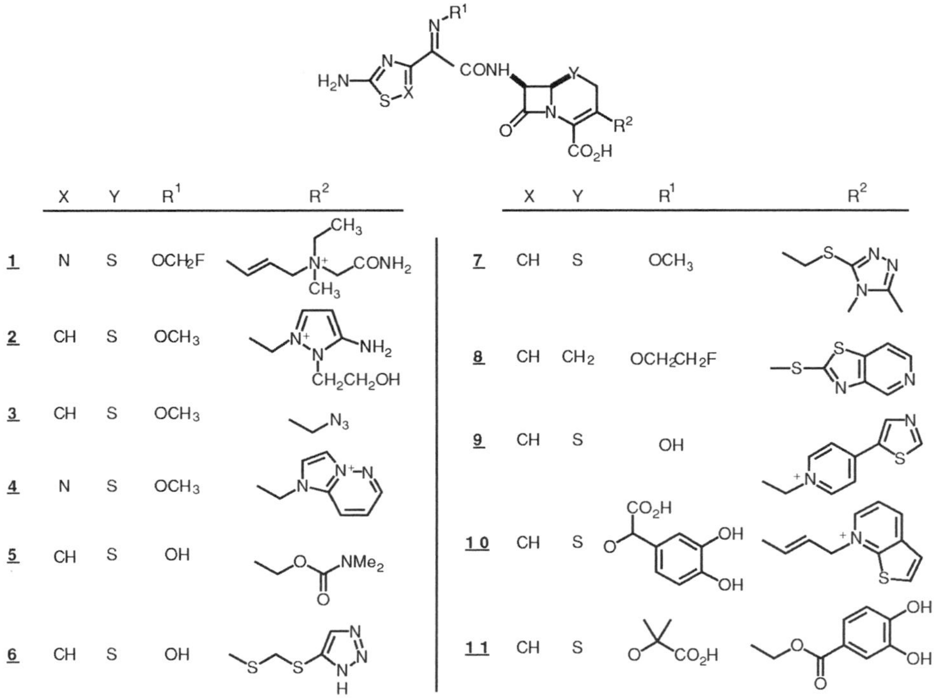

Among carbapenems, most new agents contain the 1ß-methyl group which has been found to impart good stability to renal dehydropeptidase I (26). Of compounds previously described in Ann. Rep. Med. Chem., biapenem (L-627, LJC 10,627, **12**) has been the subject of a number of new publications, including a summary of its activity against clinical isolates (27). Overall, this compound offers a broad spectrum of potent activity, but has deficiencies against MRSA and *P. aeruginosa*. Similar spectrum attributes are found with sulopenem (**13**, CP-70,429), which, while lacking the 1ß-methyl group, displays good stability to renal dehydropeptidase I as well as ß-lactamases (28). DX-8739 (**14**) was found to be superior to a number of other agents against clinical isolates of *P. aeruginosa*, although efficacy against resistant organisms was poor (29).

New members of the carbapenem class include BO-2727 (**15**), a potent parenteral agent with particularly good activity (30,31) against imipenem-resistant *P. aeruginosa* (MIC$_{90}$ = 4-6.25 µg/ml). A related compound, B2502A (**16**), selected from a series of 2-(5-substituted pyrrolidin-3-ylthio) carbapenems (32), displays even better activity against this organism (MIC_{90} = 3.13 µg/ml). The *in vitro* activity of a series of C-2 triazolylthio and pyridinylmethylthio carbapenems, of which SYN-513 (**17**) was the most active, was reported (33), as was the antibacterial activities (34) of new oral penem TMA-230, the acetoxymethyl ester prodrug of AMA-3176 (**18**).

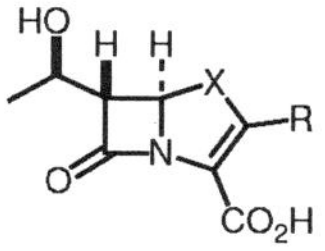

	X	R		X	R
12	CH-β-CH$_3$				
13	S		**16**	CH-β-CH$_3$	
14	CH-β-CH$_3$		**17**	CH$_2$	
15	CH-β-CH$_3$		**18**	S	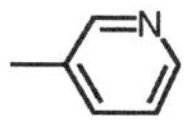

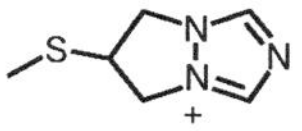

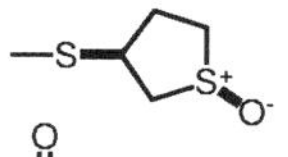

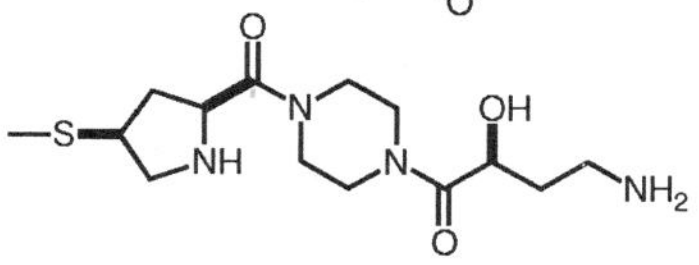

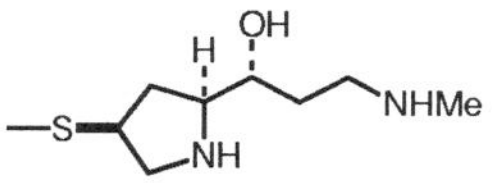

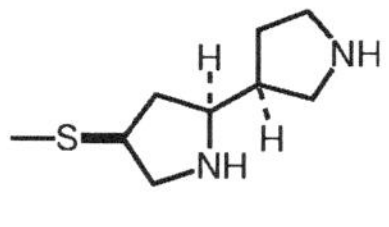

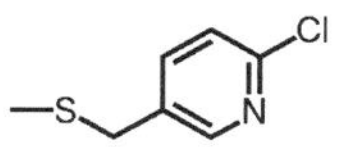

A number of reports appeared describing ß-lactams with more unusual structures. Particularly promising are a series of 2-oxaisocephems and 2-isocephems including OPC-20000 (**19**) and BOF-12013 (**20**). Compound **19** displays potent activity against MRSA (MICs of 0.2-6.25 µg/ml), and is five times as effective as vancomycin in a mouse multiply-resistant MRSA systemic infection model (35). Compound **20** shows excellent activity (36) against two strains of *P. aeruginosa*, both *in vitro* (MIC = 0.05, 0.1 µg/ml) and *in vivo* in mice (ED_{50} = 3.2, 21.8 mg/kg). The previously described oral agent GV-118819 (**21**) shows excellent activity in rodent models of bacterial respiratory infection (37). The synthesis and SAR of C-7 aminoimidazole-substituted cephalosporins such as **22**, as well as 6ß-acryl-amido penicillins such as **23**, were reported; both series display good broad-spectrum activity (38,39).

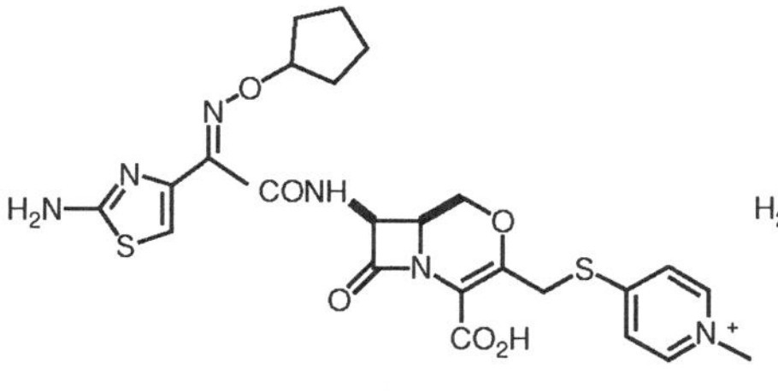

19 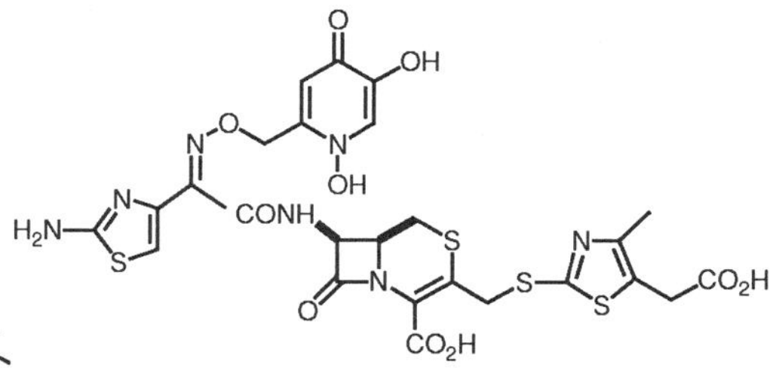   **20**

In the arena of ß-lactamase inhibitors, BRL42715 (**24**) compared favorably with clavulanic acid, sulbactam and tazobactam in synergy with known ß-lactams against several penicillinase- and cephalosporinase-producing bacteria (40-42). The synthesis and activity of a series of potent new ß-lactamase inhibitory 6-substituted penicillanic acid sulfones, including **25**, was described (43).

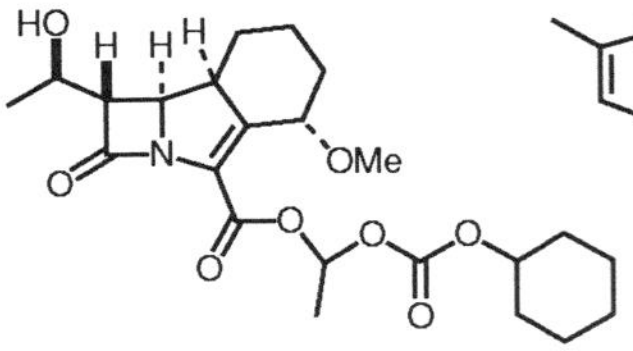

Quinolones - The majority of the currently marketed quinolones, such as ciprofloxacin (**26**), contain a piperazine or substituted piperazine at the 7-position. As summarized in the proceedings of the 4th International Symposium on New Quinolones (44), these agents are used for a broad variety of clinical indications including respiratory tract, urinary tract, skin and soft tissue, bone, and sexually transmitted diseases. However, their activity against important Gram-positive pathogens such as MRSA, enterococci, and hemolyic streptococci is only modest. Furthermore, an increasing number of *Enterobacteriaceae* and *P. aeruginosa* isolates have become resistant to these agents (45). Reviews on the mechanism of resistance to quinolones have appeared (46,47), and recent studies have examined their mechanism of action against *E. coli* DNA gyrases (48,49), efflux-mediated resistance in *S. aureus* (50), and drug features that contribute towards activity against mammalian topoisomerase II and cultured cells (51,52).

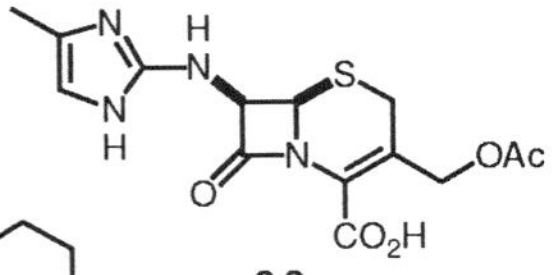

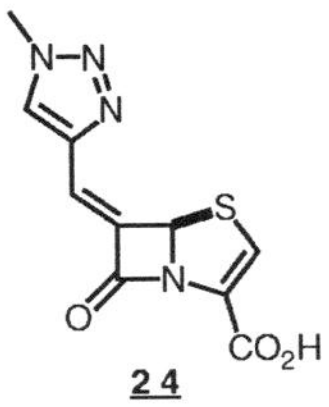

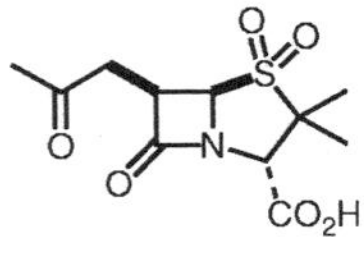

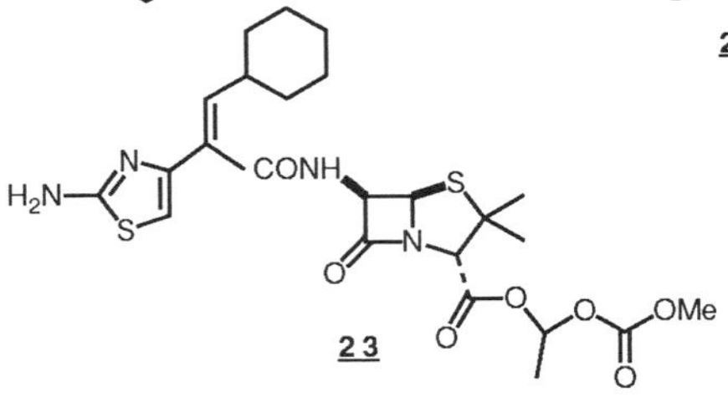

New information has appeared on previously described compounds that are currently under development. DU-6859a (**27**), which has the addition of a fluorine atom to the N-1 cyclopropyl group, has been extensively evaluated *in vitro* and in laboratory animal models. It has potent, broad-spectrum antibacterial activity, and greater activity against MRSA and quinolone-resistant *S. aureus* strains than any other earlier introduced quinolone antibacterials (53). Bay y 3118 (**28**) has shown promising results in a wide-range of tests (54), and the *in vitro* activity of this quinolone has been compared with that of ciprofloxacin, sparfloxacin, and other antibiotics against a host of clinical isolates including *Enterobacteriaceae*, *Staphylococcus* spp. and strains of *P. aeruginosa* (55,56). Reports have also appeared on a number of other quinolones under development: CP-99,219 (**29**) (57-60), AM-1155 (**30**) (61,62), NM-394 (**31**) (63,64), T-3761 (**32**) (65,66), PD-138312 (**33**) (67,68), and PD-140248 (**34**) (68,69).

A key structural feature of some of the above mentioned quinolones under development is the replacement of the C-7 piperazine by 3-amino (or aminomethyl)-1-pyrrolidine or other amino-substituted heterocycles, which provide a dramatic improvement in Gram-positive activity. The good potency observed against Gram-negative pathogens for the piperazinyl substituted quinolones is also maintained. However, the *in vivo* activity is not as impressive as the *in vitro* data would suggest, presumably due to the relatively lower solubility of the pyrrolidinyl-substituted compounds (70). Methyl substitution on the pyrrolidine ring was found to improve the solubility and therefore provide a better pharmacokinetic profile (e.g., the peak plasma concentration is higher than in the desmethyl analog). Recent studies describe the SAR observed for a series of C-7 pyrrolidines (71,72), azetidines (73), morpholines (74), and some carbon isosteres of the piperazine and pyrrolidine side chains (75). QSAR of various substituted azetidine, pyrrolidine, piperazine, and piperidine derivatives were also examined (76), and a 3-amino-4-methoxypyrrolidine moiety was found the most promising C-7 substituent group (77).

Some new quinolones that are under development include Q35 (**35**) and U-95376A (**36**). Compound **35** has reduced phototoxicity owing to a methoxy group at the 8-position (78), and its spectrum includes good antimycoplasmal activity (79,80). Compound **36**, which also contains an 8-methoxy group, is under development as a veterinary agent (81).

	R^1	R^7	X
26	CP[a]		C-H
28	CP[a]		C-Cl
29	DFP[b]		N
30	CP[a]		C-OMe
33	CP[a]		N
34	DFP[b]		N
35	CP[a]		C-OMe
36	CP[a]		C-OMe

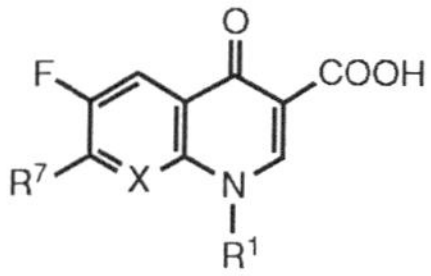

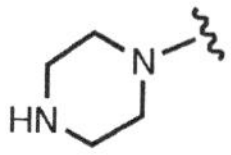

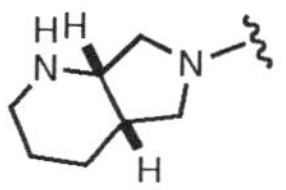

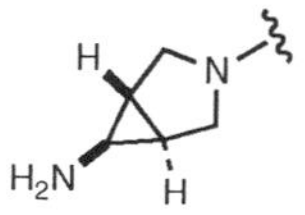

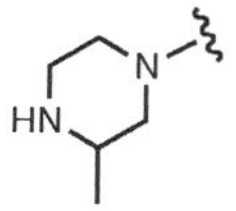

[a] cyclopropyl; [b]2,4-difluorophenyl

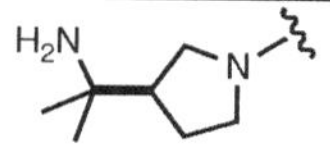

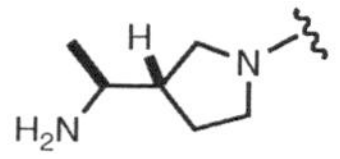

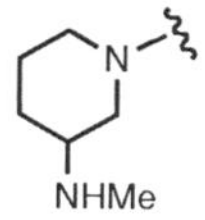

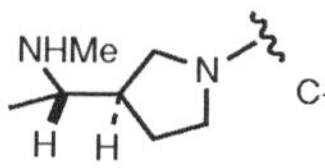

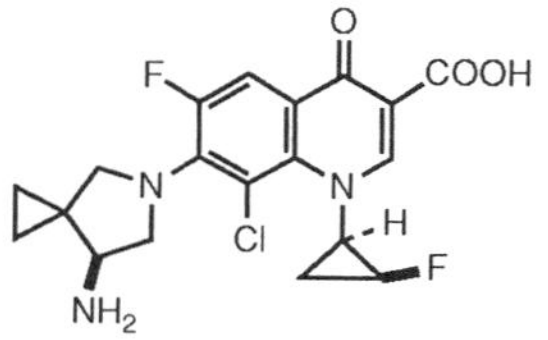

27

31

32

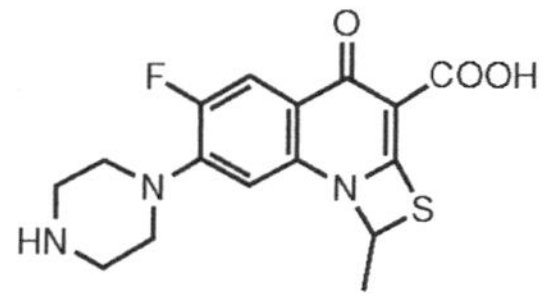

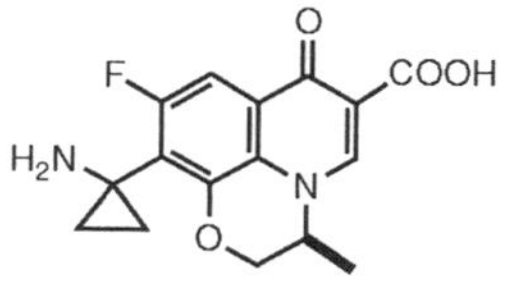

New ß-lactam-quinolone hybrids were reported at the 33rd ICAAC (82). Ro 25-0534, composed of a cephalosporin moiety linked at the 3'-position to ciprofloxacin, displays a dual action mechanism. However, this hybrid was poorly tolerated in rats when administered IV and has been dropped from development because of toxicity. Mechanism of action studies against *E. coli* suggest that while intact cephalosporin 3'-quinolone esters act as cephalosporins, carbamates and amines may possess both cephalosporin and quinolone activity in the intact molecule (83). A novel 'tetrazole-tethered' cephalosporin-quinolone hybrid was synthesized that affords cephalosporin-like activity; however, quinolone-like activity was not observed (84).

<u>Macrolides</u> - Voluminous new information pertaining to the now well described macrolides azithromycin (**37**), clarithromycin (**38**), dirithromycin (**39**) and roxithromycin (**40**) continues to appear in the literature. An international symposium dealing almost exclusively with the clinical utility of these agents was recently convened (85), and comprehensive reviews of the antimicrobial activity, pharmacokinetics, clinical efficacy and safety of the former three were published (86-89). Reports describing the *in vitro,* intracellular and *in vivo* activity of these agents against atypical mycobacteria (90-95), their uptake into and activity inside polymorphonuclear leucocytes, monocytes and macrophages (96-101), their effects on the immune system (102,103) and activity against *Borrelia burgdorferi* (104,105), *Toxoplasma gondii* (106,107) and *Chlamydia* (108) have appeared in significant numbers.

The properties of several newer macrolides have also been described in the recent literature. The *in vitro* and *in vivo* activities, pharmacokinetics and tissue distribution of a new series of 9-deoxo-8a-aza-8a-homoerythromycin derivatives (8a-azalides) have been reported (109-112). 8a-Alkyl analogs of the series, typified by L-701677 (**41**), the positional isomer of **37**, exhibit *in vitro* potency and spectrum comparable to that of **37**. Replacement of the 4"-hydroxy substituent with an amino group (e.g., L-709479 (**42**)) generally results in a 2-8 fold increase in activity against Gram-negative organisms due to improved outer membrane penetration, and a similar decrease in activity against Gram-positive organisms. The 8a-azalides exhibit improved acid stability and comparable activity in mouse infection models to **37**. The serum and tissue concentrations of the best derivative, L-709936 (**43**), in the mouse are slightly better than those of **37** with the exception of lung tissue. 11-Deoxy-**41** (**44**) and 11-deoxy-10-epi-**41** (**45**) retain much of the *in vitro* potency of **41** itself (113). 9a,11-Cyclic carbamates of **37** exhibited substantially decreased antibacterial activities *in vitro* (114). The *in vitro* antibacterial potency of a series of 14-membered azalides parallels the SAR previously reported for analogs of **37**, however this series exhibits reduced activity relative to the 15-membered azalides (115).

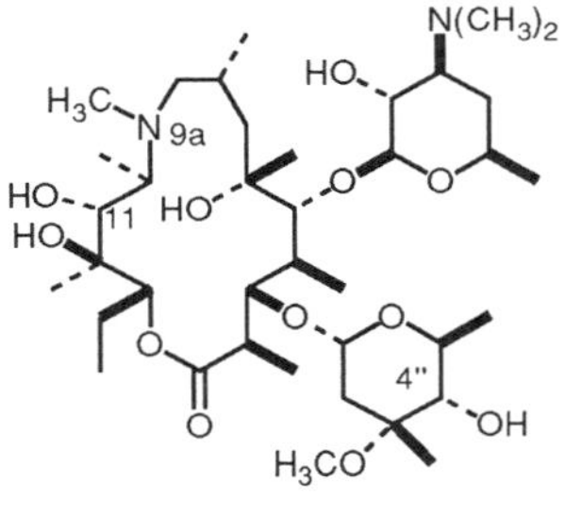

37

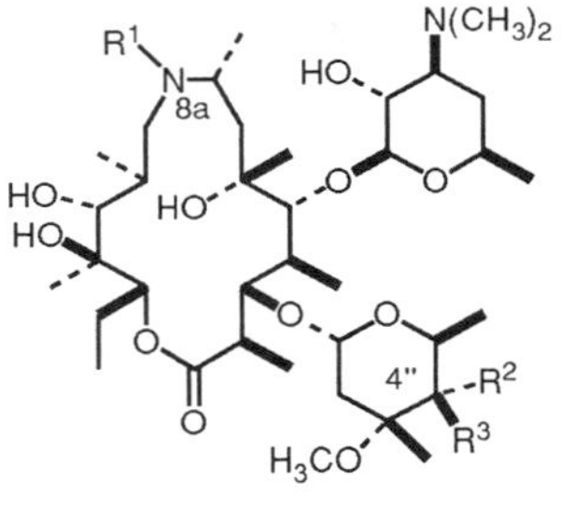

4 1: R^1=CH_3, R^2=OH, R^3=H
4 2: R^1=CH_3, R^2=NH_2, R^3=H
4 3: R^1=CH_2CH=CH_2, R^2, R^3=H,NH_2

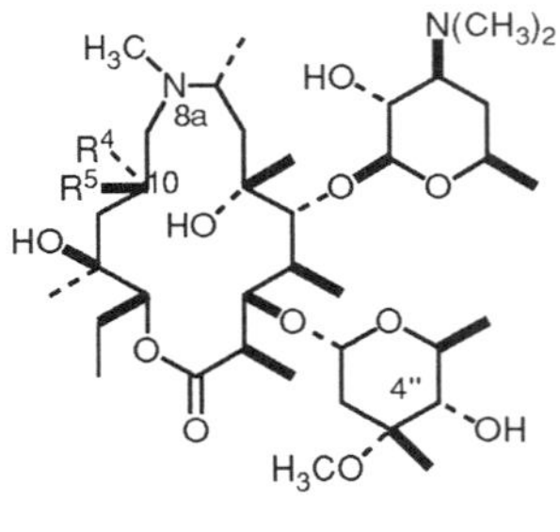

4 4: R^4=CH_3, R^5=H
4 5: R^4=H, R^5=CH_3

3,4'-Dideoxy-5-*O*-mycaminosyltylonolide (**46**) showed potent activity against Gram-positive and Gram-negative bacteria. Its antimicrobial spectrum has been compared to that of **37**, **38** and **40** (116). Compound **46** is well absorbed orally, highly distributed to various tissues including lung and has superior acid stability relative to **38** (117). Some 3-deoxy-3,4-didehydro derivatives of 5-*O*-mycaminosyltylonolide are reported to have reduced antibacterial activity relative to their parent compound (118). Both 3"-*epi*-erythromycin A (**47**) and (9S)-11-dehydroxy-9-deoxo-9-hydroxy-11-oxoerythromycin A (**48**) are reported to have *in vitro* antibacterial activity that is comparable to that of erythromycin A itself (119).

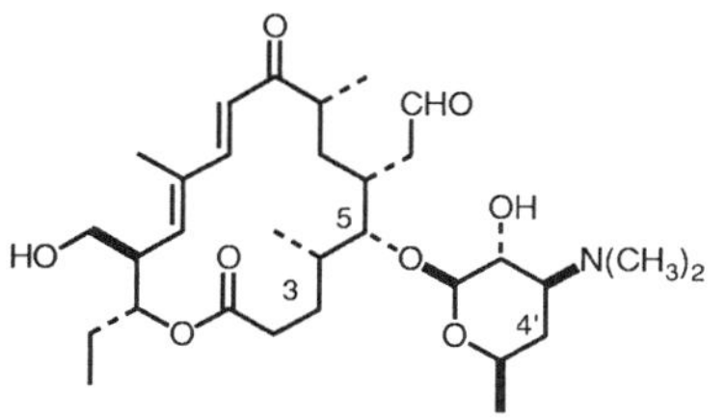

46 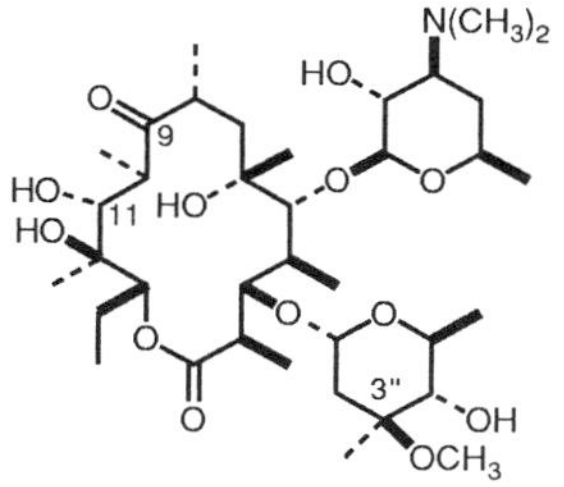   **47** 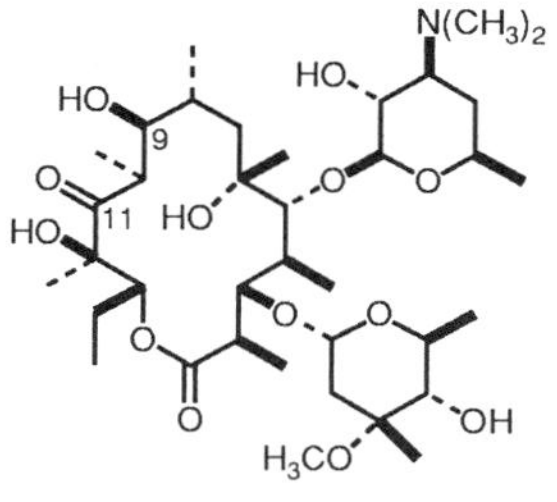   **48**

Miscellaneous - Several recent reports have been devoted to the glycylcyclines, a new series of tetracycline derivatives found to have a broad spectrum of activity against tetracycline susceptible and resistant bacteria, including strains containing ribosomal protection (*tet M*) and efflux (*tet A,B,C* and *D*) resistance determinants (120-122). Some of these analogs also exhibited good *in vitro* activity against streptococcal and enterococcal strains that were resistant to vancomycin. Two analogs, DMG-MINO (**49**) and DMG-DMDOT (**50**), were shown to have good *in vivo* efficacy (IV and SC, less active PO) against murine systemic infections caused by strains having characterized tetracycline resistance.

The synthesis and activity against *S. hyodysenteriae* of a wide variety of analogs of hygromycin A (**51**) was reported (123-128). Interestingly, replacement of the dehydro-fucose sugar moiety with an allyl group, as in **52**, affords activity comparable to that of the natural product (129). A study of the cell-free protein synthesis inhibitory activity of key analogs, in comparison with MICs, has also appeared (130).

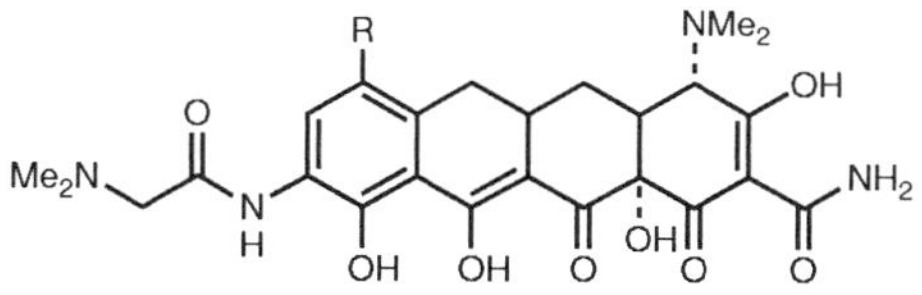

49: R = NMe$_2$
50: R = H

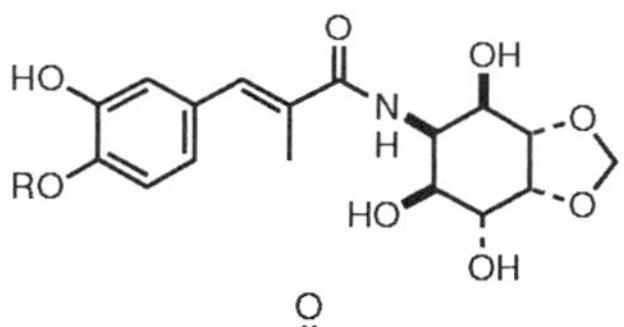

51: R =

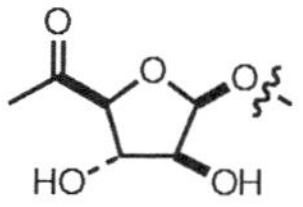

52: R = allyl

References

1.   B.A. Cunha, Clin. Ther., 14, 616 (1992).
2.   P.R. Murray, R.N. Jones, S.D. Allen, M.E. Erwin, P.C. Fuchs and E.H. Gerlach, Diagn. Microbiol. Infect. Dis., 16, 191 (1993).
3.   S.K. Spangler, M.R. Jacobs, G.A. Pankuch and P.C. Appelbaum, J. Antimicrob. Chemother., 31, 273 (1993).
4.   C.C. Johnson, Clin. Inf. Dis., 16, Suppl. 4, S371 (1993).
5.   M. Fassbender, H. Lode, T. Schaberg, K. Borner and P. Koeppe, Clin. Inf. Dis., 16, 646 (1993).
6.   T. Toyosawa, S. Miyazaki, A. Tsuji, K. Yamaguchi and S. Goto, Antimicrob. Agents Chemother., 37, 60 (1993).
7.   H.B. Huang, N.X. Chin, J.F. Wu and H.C. Neu, 33rd ICAAC, 894 (1993).
8.   J.A. Washington, R.N. Jones, E.H. Gerlach, P.R. Murray, S.D. Allenand and C.C. Knapp, Antimicrob. Agents Chemother., 37, 1696 (1993).
9.   H.C. Neu, N.-X. Chin and H.-B. Huang, Antimicrob. Agents Chemother., 37, 566 (1993).
10.   M.A. Pfaller, A.L. Barry and P.C. Fuchs, Antimicrob. Agents Chemother., 37, 893 (1993).

11. W. Paulfeuerborn, H.-J. Muller, K. Borner, P. Koeppe and H. Lode, Antimicrob. Agents Chemother., 37, 1835 (1993).
12. Y. Iizawa, K. Okonogi, R. Hayashi, T. Iwahi, T. Yamazaki and A. Imada, Antimicrob. Agents Chemother., 37, 100 (1993).
13. R. Hiruma, M. Moriyama, J. Ueno, T. Toyosawa, K. Hata, Y. Uemura and K. Katsu, 33rd ICAAC, 896 (1993).
14. R. Hiruma, J. Ueno, M. Moriyama, S. Negi and K. Katsu, 33rd ICAAC, 895 (1993).
15. M. Nakashima, H. Imoto, Y. Kimura, S. Sasaki, T. Oguma and S. Yamamoto, 33rd ICAAC, 416 (1993)
16. M. Kume, T. Kubota, Y. Kimura, H. Nakashimizu, K. Motokawa and M. Nakano, J. Antibiot., 46, 177 (1993).
17. M. Kume, T. Kubota, Y. Kimura, H. Nakashimizu and K. Motokawa, J. Antibiot., 46, 316 (1993).
18. M. Kume, T. Kubota, Y. Kimura, H. Nakashimizu and K. Motokawa, Chem. Pharm. Bull., 41, 758 (1993).
19. R. Singh, C. Fiakpui, M.P. Singh and R.G. Micetich, 33rd ICAAC, 885 (1993).
20. R.J. Ternansky, S.E. Draheim, A.J. Pike, F.W. Bell, S.J. West, C.L. Jordan, C.Y.E. Wu, D.A. Preston, W. Alborn, Jr., J.S. Kasher and B.L. Hawkins, J. Med. Chem., 36, 1971 (1993).
21. W.E. Alborn, Jr., J.E. Flokowitsch, C.Y.E. Wu, D.A. Preston, R.J. Ternansky, 33rd ICAAC, 926 (1993).
22. M. Okunishi, Y. Kurita, A. Mizutani and S. Hayashi, 33rd ICAAC, 891 (1993).
23. H. Hanaki, H. Akagi, C. Shimizu, A. Hyodo, N. Unemi, M. Yasui, 33rd ICAAC, 889 (1993).
24. J. Aszodi, A. Bonnefoy, J.F. Chantot, S.G. D'Ambrieres, P. Fauveau and P. Mauvais, 33rd ICAAC, 890 (1993).
25. T. Okita, K. Imae, T. Hasegawa, S. Iimura, S. Masuyoshi, H. Kamachi and H. Kamei, J. Antibiot., 46, 833 (1993).
26. D.H. Shih, F. Baker, L. Cama and B.G. Christensen, Heterocycles, 21, 29 (1984).
27. G.J. Malanoski, L. Collins, C. Wennersten, R.C. Moellering, Jr. and G.M. Eliopoulos, Antimicrob. Agents Chemother., 37, 2009 (1993).
28. M. Minamimura, Y. Taniyama, E. Inoue and S. Mitsuhashi, Antimicrob. Agents Chemother., 37, 1547 (1993).
29. H. Giamarellou, P. Grecka, E.J. Giamarel-Los-Bourboulis, A. Spyridaki and M. Grammatikou, 33rd ICAAC, 922 (1993).
30. B.A. Pelak, L.S. Gerckens, K. Dorso, C. Pacholok, L. Lynch, J. Kahan, D. Shungu and H. Kropp, 33rd ICAAC, 899 (1993).
31. N. Hazumi, Y. Hioki, T. Hashizume, K. Matsuda, M. Sanada, S. Nakagawa and N. Tanaka, 33rd ICAAC, 900 (1993).
32. N. Ohtake, E. Mano, R. Ushijima, M. Sanada, S. Nakagawa and N. Tanaka, 33rd ICAAC, 921 (1993).
33. C. Fiakpui, R. Singh, M.P. Singh and R.G. Micetich, 33rd ICAAC 924 (1993).
34. K. Okonogi, T. Iwahi, M. Nakao and Y. Noji, 33rd ICAAC, 927 (1993).
35. M. Matsumoto, K. Yokomi, K. Yasumura, H. Tubouchi, K. Tsuji, H. Tamaoka and M. Kikuchi, 33rd ICAAC, 886 (1993).
36. M. Matsumoto, H. Horimoto, T. Shizuta, H. Ishikawa and M. Kikuchi, 33rd ICAAC, 887 (1993).
37. E. DiModugno, S.M. Hammond, J. Lowther, L. Piccoli, D. Sabatini and L. Xerri, 33rd ICAAC, 923, (1993).
38. F. Jung, D. Boucherot, C. Delvare and A. Olivier, J. Antibiot., 46, 992 (1993).
39. R.K. Anderson, P.C. Chapman, S.C. Cosham, J.S. Davies, T.J. Grinter, M.A. Harris, D.J. Merrikin, C.A. Mitchell, R.J. Ponsford, C.F. Smith and A.V. Stachulski, J. Antibiot., 46, 331 (1993).
40. T. Muratani, E. Yokota, T. Nakane, E. Inoue and S. Mitsuhashi, J. Antimicrob. Chemother. 32, 421 (1993).
41. X.Y. Zhou, M.D. Kitzis, J.F. Acar and L. Gutmann, J. Antimicrob. Chemother., 31, 473 (1993).
42. N.X. Chin and H.C. Neu, 33rd ICAAC, 135 (1993).
43. S. Adam, R. Then and P. Angehrn, J. Antibiot., 46, 641 (1993).
44. Drugs, 45, Suppl. 3, 1 (1993).
45. R. R. Muder, C. Brennen, A. M. Goetz, M. M. Wagener, and J. D. Rihs, Antimicrob. Agents Chemother., 35, 256 (1991).
46. E. Cambau and L. Gutmann, Drugs, 45, Suppl. 3, 15 (1993).
47. D. C. Hooper and J. S. Wolfson in "Quinolone Antimicrobial Agents," 2nd Edition, D. C. Hooper and J. S. Wolfson Eds., American Society for Microbiology, Washington, D. C. , 1993, p. 97.

48.   H. Yoshida, M. Nakamura, M. Bogaki, H. Ito, T. Kojima, H. Hattori and S. Nakamura, Antimicrob. Agents Chemother., 37, 839 (1993).

49.   E. Cambau, F. Bordon, E. Collatz and L. Gutmann, Antimicrob. Agents Chemother., 37, 1247 (1993).

50.   G. W. Kaatz, S. M. Seo and C. A. Ruble, Antimicrob. Agents Chemother., 37, 1086 (1993).

51.   S. J. Froelich-Ammon, P. R. McGuirk, T. D. Gootz, M. R. Jefson and N. Osheroff, Antimicrob. Agents Chemother., 37, 646 (1993).

52.   S. H. Elsea, P. R. McGuirk, T. D. Gootz, M. Moynihan and N. Osheroff, Antimicrob. Agents Chemother., 37, 2179 (1993).

53.   33rd ICAAC (1993), Abs. Nos. 79, 80, 975-1004, 1188, and 1189.

54.   33rd ICAAC (1993), Abs. Nos. 75, 76, 235, 999, 1021A, 1197, 1465-1488.

55.   R. Wise, J. M. Andrews and N. Brenwald, J. Antimicrob. Chemother., 31, 73 (1993).

56.   A. Bauernfeind, J. Antimicrob. Chemother., 31, 505 (1993).

57.   33rd ICAAC (1993), Abs. Nos. 1509-1512.

58.   B. B. Gooding and R. N. Jones, Antimicrob. Agents Chemother., 37, 349 (1993).

59.   G. M. Eliopoulos, K. Klimm, C. T. Eliopoulos, M. J. Ferraro and R. C. Moellering, Jr., Antimicrob. Agents Chemother., 37, 366 (1993).

60.   R. Teng, N. B. Khosla, D. N. Renouf, D. Girard, B. M. Silber, G. Foulds and T. E. Liston, Pharm. Res., 10, S336 (1993).

61.   33rd ICAAC (1993), Abs. Nos. 235, and 1007-1009.

62.   H. Tomioka, H. Saito and K. Sato, Antimicrob. Agents Chemother., 37, 1259 (1993).

63.   K. Totsuka and K. Shimizu, 33rd ICAAC, 1005 (1993).

64.   T. Yoshida and S. Mitsuhashi, Antimicrob. Agents Chemother., 37, 793 (1993).

65.   33rd ICAAC (1993), Abs. Nos. 235, and 1010-1013.

66.   Y. Fukuoka, Y. Ikeda, Y. Yamashiro, M. Takahata, Y. Todo and H. Narita, Antimicrob. Agents Chemother., 37, 384 (1993).

67.   M. A. Shapiro, J. A. Dever, E. T. Joannides and J. C. Sesnie, 33rd ICAAC, 1021 (1993).

68.   M. D. Huband, M. A. Cohen, M. A. Meservey, G. E. Roland, S. L. Yoder, M. E. Dazer and J. M. Domagala, Antimicrob. Agents Chemother., 37, 2563 (1993).

69.   J. Sesnie, T. Desaty, J. Dever, E. Joannides, M. Shapiro and S. Vanderroest, 33rd ICAAC, 1020 (1993).

70.   T. Rosen, D. T. W. Chu, I. M. Lico, P. B. Fernandes, K. Marsh, L. Shen, V. G. Cepa and A. G. Pernet, J. Med. Chem., 31, 1598 (1988).

71.   J. M. Domagala, S. E. Hagen, T. Joannides, J. S. Kiely, E. Laborde, M. C. Schroeder, J. A. Sesnie, M. A. Shapiro, M. J. Suto and S. Vanderroest, J. Med. Chem., 36, 871 (1993).

72.   R. A. Bucsh, J. M. Domagala, E. Laborde and J. C. Sesnie, J. Med. Chem., 36, 4139(1993).

73.   J. Frigola, J. Pares, J. Corbera, D. Vano, R. Merce, A. Torrens, J. Mas and E. Valenti, J. Med. Chem., 36, 801 (1993).

74.   K. Araki, T. Kuroda, S. Uemori, A. Moriguchi, Y. Ikeda, F. Hirayama, Y. Yokoyama, E. Iwao and T. Yakushiji, J. Med. Chem., 36, 1356 (1993).

75.   E. Laborde, J. S. Kiely, T. P. Culbertson and L. E. Lesheski, J. Med. Chem., 36, 1964 (1993).

76.   T. Okada, K. Ezumi, M. Yamakawa, H. Sato, T. Tsuji, T. Tsushima, K. Motokawa and Y. Komatsu, Chem. Pharm. Bull., 41, 126 (1993).

77.   T. Okada, H. Sato, T. Tsuji, T. Tsushima, H. Nakai, T. Yoshida and S. Matsuura, Chem. Pharm. Bull., 41, 132 (1993).

78.   K. Marutani, M. Matsumoto, Y. Otabe, M. Nagamuta, K. Tanaka, A. Miyoshi, T. Hasegawa, H. Nagano, S. Matsubara, R. Kamide, T. Yokota, F. Matsumoto and Y. Ueda, Antimicrob. Agents Chemother., 37, 2217 (1993).

79.   Y. Gohara, S. Arai, A. Akashi, K. Kuwano, C.-C. Tseng, S. Matsubara, M. Matumoto and T., Furudera, Antimicrob. Agents Chemother., 37, 1826 (1993).

80.   33rd ICAAC (1993), Abs. Nos. 1500-1505.

81.   33rd ICAAC (1993), Abs. Nos. 1190-1195.

82.   33rd ICAAC (1993), Abs. Nos. 1489-1495.

83.   N. H. Georgopapadakou and A. Bertasso, Antimicrob. Agents Chemother., 37, 559 (1993).

84.   S.L. Dax, D.L. Pruess, P.L. Rossman and C.-C. Wei, BioMed. Chem. Lett., 3, 209 (1993).

85.   Program and Abstracts of The Second International Conference on the Macrolides, Azalides and Streptogramins, Venice, Italy, January 19-22, 1994.

86.   D. H. Peters, H. A. Friedel and D. McTavish, Drugs, 44, 750 (1992).

87.   C. H. Ballow and G. W. Amsden, Ann. Pharmacotherapy, 26, 1253 (1992).

88. M. G. Sturgill and R. P. Rapp, Ann. Pharmacotherapy, 26, 1099 (1992).
89. R. G. Finch, J. M. T. Hamilton-Miller and A. M. Lovering, Editors, J. Antimicrob. Chemother., 31, Suppl. C, 1 (1993).
90. N. Rastogi, K. S. Goh and A. Bryskier, Antimicrob. Agents Chemother., 37, 1560 (1993).
91. L. Heifets, N. Mor and J. Vanderkolk, Antimicrob. Agents Chemother., 37, 2364 (1993).
92. N. Mor and L. Heifets, Antimicrob. Agents Chemother., 37, 111 (1993).
93. N. Mor and L. Heifets, Antimicrob. Agents Chemother., 37, 1380 (1993).
94. T. Lazard, C. Perronne, Y. Cohen, J. Grosset, J-L. Vilde and J-J. Pocidalo, Antimicrob. Agents Chemother., 37, 692 (1993).
95. S. T. Brown, F. F. Edwards, E. M. Bernard, W. Tong and D. Armstrong, Antimicrob. Agents Chemother., 37, 398 (1993).
96. A. Wildfeuer, I. Reisert and H. Laufen, Arzneim.-Forsch./Drug Res., 43, 484 (1993).
97. A. P. Meyer, C. Bril-Bazuin, H. Mattie and P. J. Van Den Broek, Antimicrob. Agents Chemother., 37, 2318 (1993).
98. W. L. Hand and D. L. Hand, Antimicrob. Agents Chemother., 37, 2557 (1993).
99. G.K. Joone, C.E.J. Van Rensburg and R. Anderson, J. Antimicrob. Chemother., 30, 509 (1993).
100. M. Bonnet and P. Van Der Auwera, Antimicrob. Agents Chemother., 37, 1015 (1993).
101. A. Fietta, P. Boeri, M. L. Colombo, C. Merlini and G. Gialdroni Grassi, Chemotherapy, 39, 48 (1993).
102. P. J. McDonald and H. Pruul, Scand. J. Infect. Dis. - Suppl. 83, 34 (1992).
103. J. Tomazic, V. Kotnik and B. Wraber, Antimicrob. Agents Chemother., 37, 1786 (1993).
104. L. L. Dever, J. H. Jorgensen and A. G. Barbour, Antimicrob. Agents Chemother., 37, 1704 (1993).
105. J. Alder, M. Mitten, K. Jarvis, P. Gupta and J. Clement, Antimicrob. Agents Chemother., 37, 1329 (1993).
106. J. Blais, V. Garneau and S. Chamberland, Antimicrob. Agents Chemother., 37, 1701 (1993).
107. L. Cantin and S. Chamberland, Antimicrob. Agents Chemother., 37, 1993 (1993).
108. A. Agacfidan, J. Moncada and J. Schachter, Antimicrob. Agents Chemother., 37, 1746 (1993).
109. R. R. Wilkening, R. W. Ratcliffe, M. J. Szymonifka, K. Shankaran, A. M. May, T. A. Blizzard, J. V. Heck, C. M. Herbert, A. C. Graham and K. Bartizal, 33rd ICAAC, 426 (1993).
110. C. J. Gill, G. K. Abruzzo, A. Flattery, J. Smith, H. Kropp and K. Bartizal, 33rd ICAAC, 427 (1993).
111. B. A. Pelak, L. S. Gerckens and H. Kropp, 33rd ICAAC, 428 (1993).
112. R. R. Wilkening, R. W. Ratcliffe, G. A. Doss, K. F. Bartizal, A. C. Graham and C. M. Herbert, BioMed. Chem. Lett., 3, 1287 (1993).
113. A. B. Jones and C. M. Herbert, BioMed. Chem. Lett., 3, 1999 (1993).
114. G. Kobrehel, G. Lazarevski, Z. Kelneric and S. Dokic, J. Antibiot., 46, 1239 (1993).
115. A. B. Jones and C. M. Herbert, J. Antibiot., 45, 1785 (1992).
116. S. Hirano, K. Inouye and S. Mitsuhashi, 33rd ICAAC, 429 (1993).
117. K. Mori, Y. Kikuchi, N. Kojima, C. Ikeda, H. Tsuchiyama, S. Kageyama, K. Tomioka, T. Shibanuma and T. Tsuchiya, 33rd ICAAC, 430 (1993).
118. S. Kageyama, T. Tsuchiya and S. Umezawa, J. Antibiot., 46, 1265 (1993).
119. M. Nakata, T. Tamai, Y. Miura, M. Kinoshita and K. Tatsuta, J. Antibiot., 46, 813 (1993).
120. 33rd ICAAC (1993), Abs. Nos. 431, 432, 433, and 442.
121. R. T. Testa, P. J. Petersen, N. V. Jacobus, P.-E. Sum, V. J. Lee and F. P. Tally, Antimicrob. Agents Chemother., 37, 2270 (1993).
122. P.-E. Sum, V. J. Lee, R. T. Testa, J. J. Hlavka, G. A. Ellestad, J. D. Bloom, Y. Gluzman and F. P. Tally, J. Med. Chem., 37, 84 (1994).
123. S.J. Hecker, M.L. Minich and K.M. Werner, BioMed. Chem. Lett., 2, 533 (1992).
124 S.J. Hecker, S.C. Lilley, M.L. Minich and K.M. Werner, BioMed. Chem. Lett., 2, 1015 (1992).
125. S.J. Hecker, S.C. Lilley and K.M. Werner, BioMed. Chem. Lett., 2, 1043 (1992).
126. S.J. Hecker, C.B. Cooper, K.T. Blair, S.C. Lilley, M.L. Minich and K.M. Werner, BioMed. Chem. Lett., 3, 289 (1993).
127. S.J. Hecker, S.C. Lilley, M.L. Minich and K.M. Werner, BioMed. Chem. Lett., 3, 295 (1993).
128. B.H. Jaynes, C.B. Cooper, S.J. Hecker, K.T. Blair, N.C. Elliott, S.C. Lilley, M.L. Minich, D.L. Schicho and K.M. Werner, BioMed. Chem. Lett., 3, 1531 (1993).
129. B.H. Jaynes, N.C. Elliott and D.L. Schicho, J. Antibiot., 45, 1705 (1992).
130. S.F. Hayashi, L.J.L. Norcia, S.B. Seibel, 33rd ICAAC, 471 (1993).

Chapter 13. Advances in the Development of HIV Reverse Transcriptase Inhibitors

Donna L. Romero
Upjohn Laboratories, Kalamazoo, MI 49001

Introduction - Since the start of the AIDS pandemic, approximately 13 million people world-wide have been infected with human immunodeficiency virus (HIV). Two million of these people have already progressed to full-blown AIDS and most have died (1). The magnitude of the AIDS problem has continued to grow since its inception over a decade ago. Efforts to eradicate the disease have intensified and, as a result, tremendous resources have been focussed on the study of the retrovirus HIV, the etiological agent. The delineation of the lifecycle of HIV has shown that the virus requires the catalytic activity of many unique enzymes thereby providing attractive targets for therapeutic intervention (2-4). In terms of drug development activity, researchers have most actively targeted inhibition of the enzymes reverse transcriptase (RT) and protease (5). This chapter will discuss recent results in the development of novel inhibitors of HIV-1 RT.

Enzyme Structure and Function - RT is a heterodimeric enzyme comprised of two polypeptides of 51-kDa and 66-kDa. These peptides have an identical N-terminus, and in fact, the smaller p51 peptide is derived from the larger p66 peptide via proteolytic cleavage. This multi-purpose enzyme has three distinct catalytic functions: an RNA-dependent DNA polymerase activity, an RNAse H activity which degrades the RNA template to make way for the final function, a DNA-dependent DNA polymerase activity. RT is responsible for the synthesis of double-stranded viral DNA from proviral RNA for subsequent incorporation into the host cell chromosome (6), thus its function is essential for the replication of HIV making this enzyme ideal for chemotherapeutic intervention. Since HIV infection is a chronic process which involves high rates of viral replication and employs the highly error prone enzyme HIV RT (7, 8), the virus is highly mutable allowing for its survival (9). Thus, one limitation which might impact the ultimate efficacy of antiviral RT inhibitors (RTIs) is the development of drug resistance. Three-dimensional structural studies of HIV-1 RT have provided an enhanced understanding of the structure-function relationships and may aid in future drug design (10-12). Thus far the RTIs AZT, ddI, and ddC are the only drugs approved for the treatment of AIDS. Furthermore, these three drugs all belong to the dideoxynucleoside class of inhibitors which operate by simulating the natural dideoxynucleotide substrates that are incorporated into the growing DNA chain by RT. Therefore, these drugs require sequential phosphorylation to the 5'-triphosphate level before they become substrates or competitive inhibitors of RT. Since these compounds do not contain the C3'-hydroxyl group necessary for further chain synthesis, their incorporation into the DNA chain results in chain termination (2-4, 13). Reviews covering the synthesis (14), and the biology and development (13) of nucleoside RTIs have appeared.

Due to the serious side effects (15) and the emergence of resistant viral strains (16-19) which result from treatment with nucleoside drugs, many researchers have attempted to identify non-nucleoside RT inhibitors usually via strategies involving broad screening of chemical inventories. As a consequence of these efforts, several non-nucleoside reverse transcriptase inhibitors (NNRTIs) of disparate structures have been discovered. These NNRTIs, which act as non-competitive inhibitors with respect to deoxynucleoside triphosphates (dNTPs) at an allosteric site on the enzyme, share many common properties. For example, phosphorylation is not required for RT inhibitory activity, they are highly specific for HIV-1 relative to HIV-2 or other cellular polymerases, and they are effective against HIV strains resistant to AZT. In fact, a number of competition experiments suggest that the NNRTIs bind at the same or overlapping sites on HIV-1 RT (20-22). A crystal structure of RT complexed with the non-nucleoside nevirapine (*vide infra*, **11**) suggests that it binds in a hydrophobic pocket close to the polymerase active site (23). Several of the residues critical for conferring resistance to other NNRTIs are also located in or near this pocket (7, 8) supporting the suggestion that these inhibitors are binding at similar sites. Historical descriptions of the discovery and development of many of these NNRTIs have appeared (24-27).

1 X = N$_3$ base = Thymine
2 X = H   base = Hypoxanthine
3 X = H   base = Cytosine

Competitive Inhibitors of RT - The debate concerning the timepoint at which to begin AZT (zidovudine, **1**) therapy in the treatment of HIV-1 infection continues (28). Preliminary results from the Concorde trial, wherein 1749 asymptomatic patients were randomly given AZT (1000 mg, q.i.d.) or placebo, indicated that early treatment did not offer significant benefit when death or progression to AIDS or AIDS-related complex were used as endpoints (29). On the other hand, initial results from another large European-Australian collaborative study suggested that early AZT therapy of asymptomatic patients with CD4 counts between 500-749 per cubic millimeter delayed progression to symptomatic HIV (30). However, the duration of such an effect may be limited (31). Recent studies which attempt to probe the reasons for loss of AZT effectiveness over time indicate that AZT is not very effective against a syncytium inducing viral phenotype, a phenotype that has been correlated with a decline in CD4 cells and rapid disease progression (32-34). Resistance to AZT also may contribute to the loss of efficacy of AZT during the timecourse of treatment, and such AZT-resistant HIV-1 may still be virulent enough to contribute to disease progression. Support for sustained virulence is garnered by a recent report which shows that AZT-resistant HIV-1 can be sexually transmitted (35).

Ongoing clinical trials of ddI (didanosine, **2**) have demonstrated that a modest decrease in sensitivity occurred in viral isolates from 10 of 15 patients after one year of therapy. Fortunately, this decrease in ddI sensitivity was accompanied by an increase in AZT susceptibility (36). The clinical significance of these findings is unknown. A novel mutation at site 184 has been implicated in conferring resistance to both ddI and ddC (zalcitibine, **3**). Cross-resistance to AZT was not observed, indicating some rationale for the use of combination nucleoside therapy (9).

Accounts of the SAR of oxathiolanyl nucleosides such as (-)-3TC (**4**) and (-)-FTC (**5**) have been published (37, 38). The apparent differential antiviral activity of the enantiomers of FTC is due to differences in cellular metabolism, not inherent differences in RT inhibitory activity (39). A Met 184 to Val or Ile mutation caused resistance to (-)-3TC and (-)-FTC in cell culture (40).

4 X = H
5 X = F

Non-nucleoside Inhibitors of RT - A series of pyrrolo-[1,2-d]-[1,4]-benzodiazepin-6-ones (e.g. **6**), analogs of the potent TIBO inhibitors **7** (R-82913) and **8** (R-82150) were synthesized and antiviral testing revealed that the potency of **6** and **7** were similar (41). Converting the seven-membered diazepine ring of **8** into open chain congeners, such as the 2-mercaptobenzimidazoles **9** and **10**, provided compounds with only slight RT inhibitory activity (42).

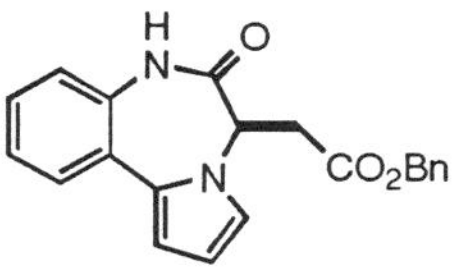

| **6** | **7** X = Cl | **9** R = H |
| | **8** X = H | **10** R = CH$_3$ |

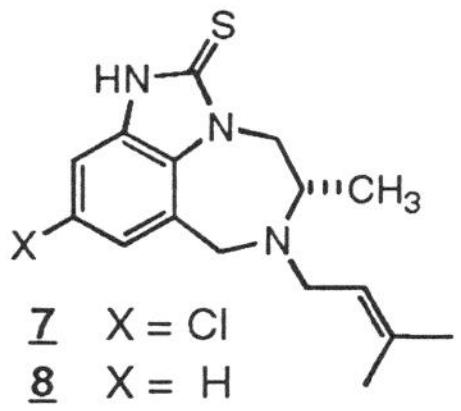

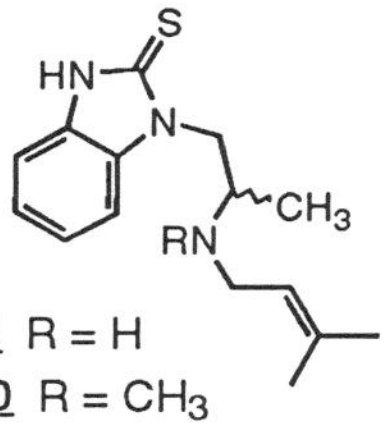

Initial clinical evaluation of nevirapine (**11**) as a monotherapy has resulted in the rapid emergence of resistant viral strains (43). Therefore, more recent trials have been conducted with nevirapine both at high doses and in combination with AZT. Analogs of nevirapine in which the lactam bridge was replaced with fused heterocycles, such as imidazole (**12**), demonstrated increased RT inhibitory activity (44). This increase was also reflected in cell culture data where a greater than 10-fold improvement in antiviral activity was observed in an assay measuring p24 protein production (**12**, EC$_{90}$ = 0.0023 µM; nevirapine, EC$_{90}$ = 0.078 µM; H9, IIIb).

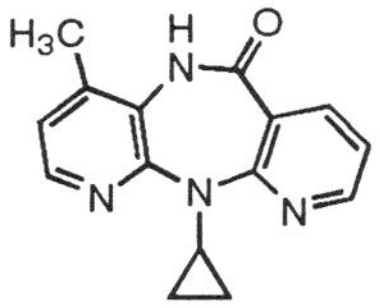

| **11** | **12** | **13** X = OCH$_3$ Y = Et | **14** X = NHSO$_2$CH$_3$ Y = *i*-Pr |

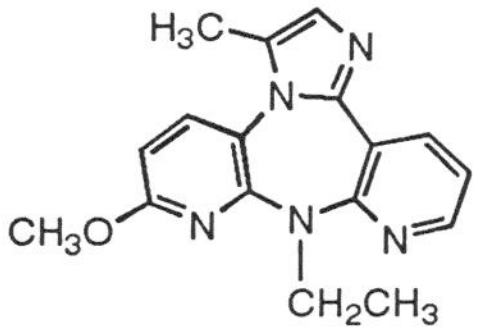

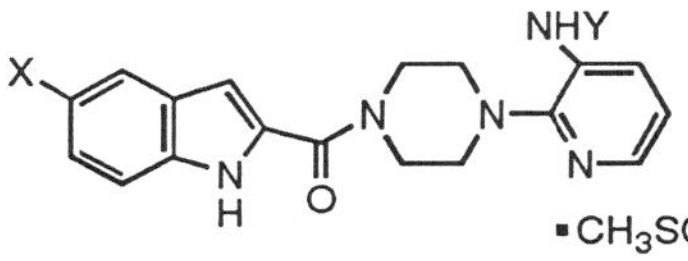

The bis(heteroaryl)piperazine (BHAP), atevirdine mesylate (U-87201E, ATV, **13**) continues to undergo clinical evaluation (45). Co-administration of ATV and AZT (ATV at 1,800 mg/day and AZT at 600 mg/day) in a Phase I study demonstrated that the combination was well tolerated and trough levels of ATV were maintained at 5-10 µM. More importantly, no phenotypic or genotypic resistance was detected up to 12 weeks of therapy (46). Pre-therapy isolates had a median ATV EC$_{50}$ of 1.1 µM, and post-therapy isolates (6-12 weeks) had a median ATV EC$_{50}$ of 1.9 µM. One patient who received 20 weeks of combination therapy developed some resistance (ATV EC$_{50}$ = 23 µM) (47). Structural modifications, such as inclusion of certain 5- or 6-substituents on the BHAP indole nucleus demonstrated that slight structural modifications can cause dramatic increases in potency (48). These studies led to the selection of a second BHAP drug candidate, delavirdine mesylate (U-90152S, **14**), which recently entered clinical trials (49). Delavirdine demonstrates good absolute bioavailability in preclinical studies in laboratory rats and dogs. Serum drug levels in excess of those needed for *in vitro* antiviral activity can be safely maintained for prolonged periods of time. Delavirdine prevented the spread of HIV-1 in human lymphocytes significantly longer than AZT at the same concentration (50). Serial passage of HIV-1$_{JR-CSF}$ and HIV-1$_{MF}$ in the presence of ATV or delavirdine selects for a novel Pro to Leu mutation at amino acid 236 in the RT. Viral strains containing this mutation were more vulnerable to other NNRTIs than was wild-type (51). Although delavirdine was less effective against the known non-nucleoside-resistant forms of RT (Tyr181Cys and Lys103Asp) as compared to wild-type RT, it still demonstrated significant activity against these mutant RTs *in vitro* (IC_{50}s ~8 µM).

L-697,661 (**15**) entered clinical trials early in 1991. Publication of the initial clinical trials indicated that blood levels of 1 µM were obtained after a single oral dose of 500 mg (52). Treatment of HIV-positive patients

with **15** at doses of 100 mg t.i.d. or 500 mg b.i.d. caused an increase in CD4 levels at week 1 which subsequently declined to base-line by week 6. Analysis of pre-treatment and post-treatment virus from these patients for *in vitro* susceptibility to **15** demonstrated that resistant HIV-variants were rapidly selected (53). Phenotypic evaluation revealed mutations at Lys103Ala and Tyr181Cys, or both, in the drug resistant viral isolates, consistent with mutations previously observed in cell culture passage experiments (54). Development of this drug is no longer being pursued (55). Further delineation of the SAR of such pyridinones revealed that substitution of the benzoxazole ring with a pyridine ring (L-702,007, **16**) enhanced the potency (56, 57). Pyridinones **15** and **16** possessed equivalent RT inhibitory activities (**15, 16**, IC_{50} = 0.019 µM). However, **16** demonstrated a EC_{95} = 0.013 µM in cell culture assays (MT4 cells, IIIb) whereas **15** had a EC_{95} = 0.050 µM. This difference most likely reflects an improvement in the cellular penetration of **16**. Since **16** exhibited the best oral bioavailability in rats and monkeys of the compounds synthesized in this study, it was selected for further development (56, 57).

Some 3,4-dihydro-2-alkoxy-6-benzyl-4-oxopyrimidines (DABOs), such as **17** (EC_{50} = 6.2 µM, MT-4, IIIb) and **18** (EC_{50} = 2.9 µM) possess potencies similar to HEPT analogs, such as **19** (EC_{50} = 23 µM) and **20** (EC_{50} = 2.2 µM) to which they are structurally related (58, 59). Recently, combinations of 6-benzyl-1-(ethoxymethyl)-5-isopropyluracil, a HEPT analog, and AZT were synergistic in their inhibition of HIV-1 replication in infected MT-4 cells (60). Purine analogs of pyrimidine containing TSAO-T (**21**) possess antiviral activities three times lower (61). However, introducing an N-1 methyl group on the purines, as in **22** (TSAO-m¹Hx),

R		R	R¹	R²
17 butyl		**19** CH_2OH	CH_3	H
18 s-butyl		**20** CH_2CH_3	CH_2CH_3	CH_3

significantly decreases cytotoxicity without sacrificing antiviral potency, resulting in increased cytotherapeutic ratios (62). Cross resistance was observed between a viral strain selected for resistance to **22** and other TSAO-purine and -pyrimidine analogs. A novel mutatation (Glu138Lys) was responsible for conferring resistance in five independently generated TSAO-resistant strains (63). This mutation is still sensitive to other NNRTIs such as L-697,661 (**15**) and TIBO (**7**).

Several novel classes of NNRTIs discovered via broad screening and subsequent SAR development have been disclosed within the past year. For example, screening led to the discovery of **23**, a member of the α-anilinophenylacetamide (α-APA) series of compounds (64). Subsequent evaluation of similar structures already present in the chemical library led to **24** (R-18893, EC_{50} = 0.088 μM, IIIb, MT-4 cells). Chemical optimization resulted in the synthesis of **25** (R-89439, EC_{50} = 0.013 μM) which provided a seven-fold increase in potency. RT mutations at Tyr181Cys or Tyr181Ile, or Tyr188Leu displayed reduced

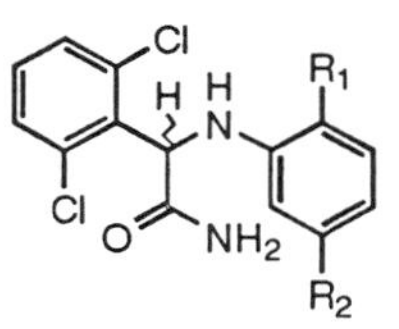

23	R_1 = OCH_3, R_2 = H
24	R_1 = NO_2, R_2 = H
25	R_1 = $COCH_3$, R_2 = CH_3

sensitivity to **25**. However, a TIBO resistant strain containing a Lys100Ile mutation was still sensitive to **25**. Clinical studies have been initiated with both drugs as racemates, although activity resides primarily in the (-)-enantiomers. Plasma levels (10-20 ng/mL) of the active enantiomer in the range of the EC_{50} were obtained upon dosing **24** in seven healthy but HIV-infected volunteers (65, 66). No resistant virus was observed after three months of therapy, although this dose was sufficient to cause a significant rise in CD4 counts (67). After four months of therapy, virus containing a Tyr181Cys mutation was obtained from one patient who had high plasma levels (200 ng/mL) of **24** due to liver dysfunction (65). A multiple dose, phase I, safety study of **25** obtained trough plasma concentration levels of the active enantiomer one hundred times the EC_{50} (68). In addition, surrogate markers appeared to improve and a good safety profile was obtained.

Molecular modeling studies comparing TIBO (**8**), nevirapine (**11**) and the novel 2,3-dihydrothiazolo[2,3-a]isoindol-5(9bH)-oneR-(+)-**26** (IC_{50} = 0.28 μM), led to the proposal of a butterfly model which describes the relative orientation of the two aromatic rings in each of these inhibitors (69). Based on this model, a novel inhibitor (R-(+)-**27**) was designed,

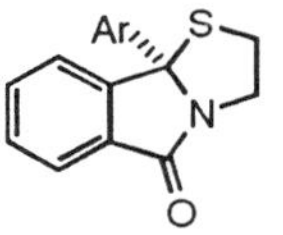

26	Ar = Ph
28	Ar = 1-napthyl
29	Ar = 3,5-dimethylphenyl

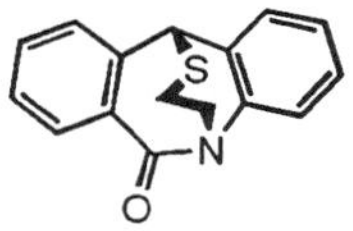

27

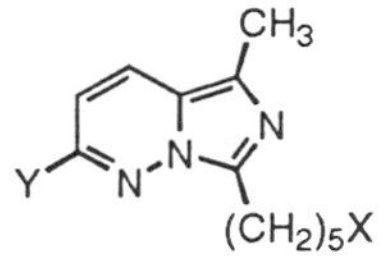

synthesized, and shown to have activity similar to **26** (IC_{50} = 0.83 μM). Optimization of the aryl substituent of **26** resulted in the synthesis of more potent compounds (R-(+)-**28**, IC_{50} = 0.15 μM; R-(+)-**29**, IC_{50} = 0.016 μM) (70). An HIV-1 mutant resistant to R-(+)-**29** was generated in cell culture via serial passage of HIV-1 in the presence of **29**. It was shown to possess the previously described Tyr181Cys substitution in RT. Various other mutations such as Lys101Ala, Lys103Asn, or Tyr188Leu also conferred resistance to **29** (71).

30 X = OCOPh, Y = Cl	**32** n = 0 X = H, Z = $CH_2S(O)Ph$	**34**
31 X = COPh, Y = 1-imidazole	**33** n = 2 X = Cl, Z = $CONH_2$	

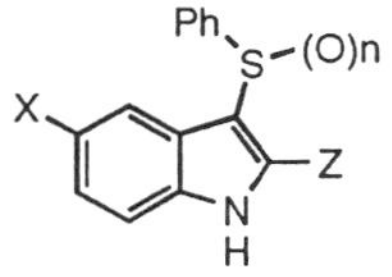

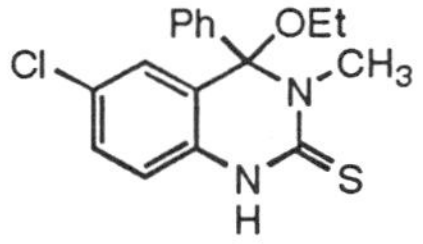

Traditional SAR optimization of lead compound **30** (IC_{50} = 1.34 μM) led to more potent imidazo[1,5-b]pyridazines such as **31** (GR142086X, IC_{50} = 0.00065 μM). The imidazo[1,5-b]pyridazines appear to bind to the same site as other NNRTIs. Pyridazine **31** had reduced sensitivity to A17, a mutant virus containing the Tyr181Cys mutation (72). 5-Chloro-3-(phenylsulfonyl)indole-2-carboxamide **33** (IC_{50} = 0.003 μM; EC_{95} = 0.003 μM, MT-

4, IIIb) was derived from initial lead **32** (IC_{50} = 0.063 µM; EC_{95} = 0.4 µM) (73).   Sulfone **33** was 45% orally bioavailable and a peak plasma level of 2.0 µM was obtained 2 hours after a 10 mg/kg dose in rhesus monkeys. RTs containing Tyr181Cys and Lys103Asn mutations were less potently inhibited by **33** than wild-type RT. Chemical modification of **34** (IC_{50} = 0.056 µM; EC_{95} = 0.1 µM, MT-4, IIIb) led to the resolved **35** (L-732,801, IC_{50} = 0.007 µM; EC_{95} = 0.025-0.050 µM). Quinazolinone **35** was less active against RT mutants (Tyr181Cys, Lys103Asn, or both) as compared to wild-type. However, it still retained significant activities against such mutants, i.e. IC_{50}s in the micromolar range (74-76). Unfortunately, **35** has pronounced first pass metabolism, most likely via demethylation. Substituting a 2-pyridylalkynyl moiety for the propyl group afforded a compound (**36**, L-738,372) with good potency and good oral bioavailability in rhesus monkeys (77). Compound **36** also retained significant activity against singly mutated RTs, but a much higher level of resistance was observed against doubly mutated RTs.

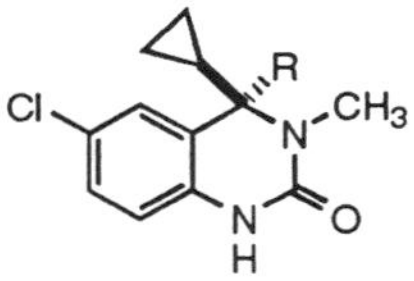

35   R = *n*-propyl

36   R = ═══〈pyridyl〉

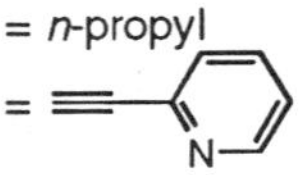

3,3-Dialkyl-3,4-dihydroquinoxalin-2(1H)-thiones, such as **37** (S-2720) are potent inhibitors of RT (78). Quinoxaline **37** has potencies 10-fold lower than L-697,661 (**15**), nevirapine (**11**) and TIBO (**8**) in cell culture against the viral isolates HIV-1 MN and D34. As in other cases, **37** is less active against the mutants selected for by other NNRTIs when compared to wild-type. For example, there is a 107-fold increase in the EC_{50} against a TIBO resistant viral strain containing the Lys100Ile mutation as compared to wild-type virus. Even with the substantial decrease in activity of **37** against this mutant, its IC_{50} remains in the micromolar range (EC_{50} = 0.32 µM). Generation of virus resistant to **37** resulted in a Gly190Glu mutation in RT. The enzymatic activity of the RT containing this mutation was only 3 % relative to the enzymatic activity of wild-type RT. The corresponding virus resistant to **37** also appears to display slower growth in tissue culture (78). Previously, it has been shown that virus containing the Tyr181Cys mutation displayed unchanged growth characteristics relative to wild-type virus (79).

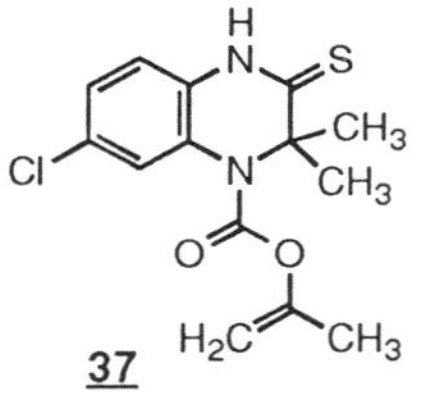

37 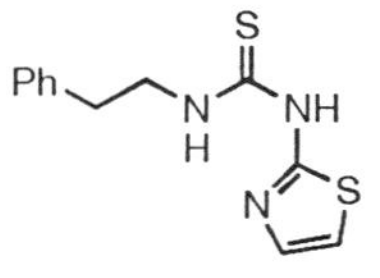   **38** 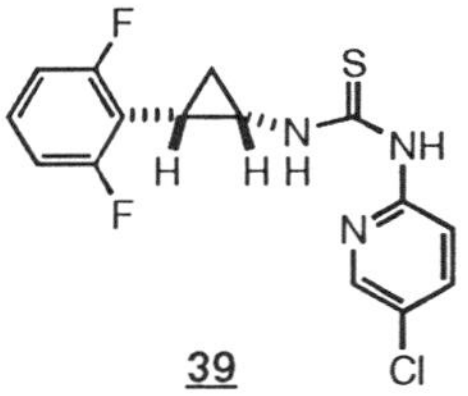   **39**

A strategy which involved iterative molecular conformational modeling, synthesis, and testing beginning with lead compound **38** (ED_{50} = 1.3 µM, MT-4, IIIb) resulted in the design and synthesis of compounds with improved activities (80, 81). One of these, **39** ((-)-MSC-127, ED_{50} = 1.1 nM), conformationally restricts the phenethyl moiety by incorporating a phenyl substituted *cis*-cyclopropane ring. Such PETT compounds (for Phenyl Ethyl Thiazolyl Thiourea) appear to bind at the same site as other NNRTIs, for they too are sensitive to RT mutations at Tyr 181 (82).

Resistance Issues - Because of the likelihood of resistance development to anti-HIV drugs, understanding the mechanism by which HIV acquires resistance may be useful in the development of effective strategies for the treatment of HIV infection. For example, the aquisition of resistance to some drugs such as ddI (**2**) and delavirdine (**14**) affords RTs that are more vulnerable to other RTIs (51, 83).   If clinical results parallel cell culture experiments, initial treatment with delavirdine may result in a virus population sensitized to inhibition by other NNRTIs. Another potential strategy could involve treatment with combinations of drugs which select for mutations that engender the RT enzyme non-functional (84, 85).

Broad screening against a panel of mutant HIV-RTs containing mutations known to confer resistance to existing classes of NNRTIs may be one approach towards solving the resistance dilemma. Promising results were observed when a broad array of pyridinones were screened for activity against enzymes containing alterations at Lys103Asn, Tyr181Cys or both mutations (86). One analog, **40** (L-702,019), was only three-fold less active against the doubly mutated RT whereas clinical candidates L-697,661 (**15**) and L-696,229 were 4000-fold less inhibitory. Two amino acid substitutions in RT were shown to be required for the generation of significant resistance to **40**.

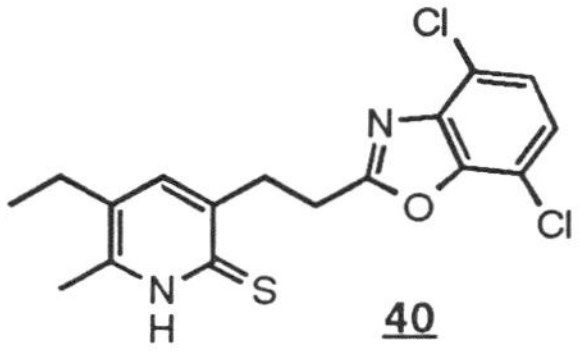

The observation that some amino acids around the nevirapine binding site are highly conserved among the RTs of other retroviruses, and the premise that mutations at these residues may produce non-functional enzymes has prompted some researchers to adopt an alternative approach to combating resistance (87). This strategy involves the use of X-ray crystallography and molecular modeling to design inhibitors which possess favorable interactions with non-mutable amino acid residues such as the catalytic aspartic acids (Asp110, Asp185, Asp186) and Tyr183.

Summary - Clinical trials of nucleoside drugs continue to demonstrate that these agents are modestly effective in slowing the progression of the disease. In order to discover less toxic, more effective chemotherapeutics, research strategies have emphasized the discovery and development of NNRTIs as evidenced by the number of recent reports of novel structural types. Strategies which target alternative points in the life-cycle of the virus, such as protease, TAT and RNAse H, are also becoming more popular. The successful discovery of a broad variety of structurally distinct, low molecular weight, NNRTIs has been due in large part to the broad screening of chemical libraries. Several of the initially reported NNRTIs rapidly entered clinical trials as monotherapies. Unfortunately, some of these clinical trials were discontinued due to the rapid emergence of resistance. In retrospect, it appears as though the exquisite selectivity of the NNRTIs for HIV-1 RT, which was desirable in terms of lowering toxicity, might actually be undesirable in the sense that the specificity may reflect a lack of tolerance of the NNRTIs to mutations in their binding pocket on RT. It is hoped that combination therapy (multiple RTIs; RTI and protease inhibitor; etc.) will provide a solution to this problem. Other approaches, such as screening against panels of mutant RTs or designing inhibitors which interact with non-mutable amino acids, have been initiated.

References

1. M. H. Merson, Science, 260, 1266 (1993).
2. H. Mitsuya, S. Broder, Nature, 325, 773 (1987).
3. H. Mitsuya, R. Yarchoan, S. Broder, Science, 249, 1533 (1990).
4. M. Nasr, J. Cradock, M. Johnston, Drug News Perspect., 5(6), 338 (1992).
5. S. Thaisrivongs, Ann. Rep. Med. Chem. 29, Chapter 15 (1994).
6. J. Saunders, R. Storer, Drug News Perspect., 5(3), 153 (1992).

7. B. D. Preston, B. J. Poiesz, L. A. Loeb, Science, 242, 1168 (1988).
8. J. D. Roberts, K. Bebebek, T. A. Kunkel, Science, 242, 1171 (1988).
9. M. A. Wainberg, Z. Gu, Q. Gao, E. Arts, R. Geleziunas, S. Bour, R. Bealieu, C. Tsoukas, J. Singer, J. Montaner, J. Acq. Immun. Defic. Syndrome, 6(S1), S36 (1993).
10. R. G. Nanni, J. Ding, A. Jacobo-Molina, S. H. Hughes, E. Arnold, Perspect. Drug Discov. Des., 1, 129 (1993).
11. L. A. Kohlstaedt, J. Wang, J. M. Friedman, P. A. Rice, T. A. Steitz, Science, 256, 1783 (1992).
12. J. F. Davies, Z. Hostomska, Z. Hostomsky, S. Jordan, D. A. Matthews, Science, 252, 88 (1991).
13. R. F. Schinazi, Perspect. Drug Discov. Des., 1, 151 (1993).
14. D. Huryn, M. Okabe, Chem. Rev., 92, 1745 (1992).
15. D. D. Richman, M. A. Fischl, M. H. Grieco, M. S. Gottlieb, P. A. Volberding, O. L. Laskin, J. M. Leedom, J. E. Groopman, D. Mildvan, M. S. Hirsch, G. G. Jackson, D. T. Durack, D. Phil, S. Nusinoff-Lehrman, & The AZT Collaborative Working Group, N. Engl. J. Med., 317, 192 (1987).
16. B. A. Larder, S. D. Kemp, Science, 246, 1155 (1989).
17. B. A. Larder, G. Darby, D. D. Richman, D.D. Science, 243, 1731(1989).
18. P. Kellam, C. A. B. Boucher, B. A. Larder, Proc. Natl. Acad. Sci. U.S.A., 89, 1934 (1992).
19. M. H. St. Clair, J. L. Martin, G. Tudor-Williams, M. C. Bach, C. L. Vavro, D. M. King, P. Kellam, S. D. Kemp, B. A. Larder, Science, 253, 1557 (1991).
20. J. C. Wu, T. C. Warren, J. Adams, J. Proudfoot, J. Skiles, P. Rhagavan, C. Perry, I. Potocki, P. R. Farina, P. M. Grob, Biochemistry, 30, 2022 (1991).
21. T. J. Dueweke, F. J. Kezdey, G. A. Waszak, M. R. Deibel, W. G. Tarpley, J. Biol. Chem. 267, 27 (1992).
22. M. Goldman, J. H. Nunberg, J. A. O'Brien, J. C. Quintero, W. A. Schleif, K. F. Freund, S. L. Gaul, W. S. Saari, J. S. Wai, J. M. Hoffman, P. S. Anderson, D. J. Hupe, E. A. Emini, A. M. Stern, Proc. Natl. Acad. Sci. U.S.A., 88, 6863 (1991).
23. L. A. Kohlstaedt, J. Wang, J. M. Friedman, P. A. Rice, T. A. Steitz, Science, 256, 1783 (1992).
24. S. D. Young, Perspect. Drug Discov. Des., 1, 181(1993).
25. J. Saunders, Drug Des. Discov., 8, 255 (1992).
26. E. De Clercq, Med. Res. Rev., 13, 229 (1993).
27. "The Search for Antiviral Drugs", J. Adams, V. J. Merluzzi, Eds., Birkhauser, Boston, 1993.
28. J. G. Bartlett, N. Engl. J. Med., 329, 351 (1993).
29. J.-P. Aboulker, A. M. Swart, Lancet, 341, 889 (1993).
30. D. A. Cooper, J. M. Gatell, S. Kroon, N. Clumeck, J. Millard, F. D. Goebel, J. N. Bruun, G. Stingl, R. L. Melville, J. Gonzalez-Lahox, J. W. Stevens, A.P. Fiddian & the European-Australian Collaborative Group, N. Eng. J. Med., 329, 297 (1993).
31. J. P. Aboulker, A. M. Swart, Lancet, 341, 980 (1993).
32. M. Tersmette, M. Koot, C. A. B. Boucher, J. W. Mulder, J. M. A. Lange, R. A. Coutinho, F. Miedema, J. Cell Biochem., S17E, 24 (1993).
33. C. A. Boucher, J. M. A. Lange, F. Miedema, G. J. Weverling, M. Koot, J. W. Mulder, J. Goudsmit, P. Kellum, B. A. Larder, M. Tersmette, AIDS, 6, 1259 (1992).
34. R. I. Connor, H. Mohri, Y. Cao, D. D. Ho, J. Virol., 67, 1772 (1993).
35. A. Erice, D. L. Mayers, D. G. Strike, K. J. Sannerud, F. E. McCutchan, K. Henry, H. H. Balfour, N. Engl. J. Med. 328, 1163 (1993).
36. R. C. Reichman, N. Tejani, J. L. Lambert, J. Strussenberg, W. Bonnez, B. Blumberg, L. Epstien, R. Dolin, Antiviral Res., 20, 267 (1993).
37. L. S. Jeong, R. F. Schinazi, J. W. Beach, H. O. Kim, K. Shanmuganathan, S. Nampalli, M. W. Chun, W.-K. Chung, B. G. Choi, C. K. Chu, J. Med. Chem., 36, 2627 (1993).
38. L. S. Jeong, R. F. Schinazi, J. W. Beach, H. O. Kim, S. Nampalli, K. Shanmuganathan, A. J. Alves, A. McMillan, C. K. Chu, R. Mathis, J. Med. Chem. 36, 181 (1993).
39. J. E. Wilson, J. Louise Martin, K. Borroto-Esoda, S. Hopkins, G. Painter, D. C. Liotta, P. A. Furman, Antimicrob. Agents Chemother., 37, 1720 (1993).

40.   R. F. Schinazi, R. M. Lloyd, M.-H. Nguyen, D. L. Cannon, A. McMillan, N. Ilksoy, C. K. Chu, D. C. Liotta, H. Z. Bazmi, J. W. Mellors, Antimicrob. Agents Chemother., 37, 875 (1993).

41.   G. V. De Lucca, M. J. Otto, Biorg. Med. Chem. Lett., 2, 1639 (1992).

42.   E. E. Swayze, S. M. Peiris, L. S. Kucera, E. L. White, D. S. Wise, J. C. Drach, L. B. Townsend, Biorganic and Med. Chem. Lett., 3, 543 (1993).

43.   J. Adams, K. D. Hargrave, Spec. Publ. R. Soc. Chem. 119, 282 (1993).

44.   N. K. Terrett, D. Bojanic, J. R. Merson, P T. Stephenson, Biorg. Med. Chem. Lett., 2, 1745 (1992).

45.   D. L. Romero, Drugs Future, 19, 7 (1994).

46.   Reichman, R.; Fischl, M.; Para, M.; Powderly, W.; Timpone, J.; Bassiakos, Y.; and the ACTG 199 team, IXth Int. Conf. AIDS/IVth STD World Congress; Berlin, Po-B26-2055 (1993).

47.   Demeter, L. M.; Resnick, L.; Tarpley, W. G.; Fischl, M.; Para, M.; Reichman, R. C. and the ACTG 199 team, IXth Int. Conf. AIDS/IVth STD World Congress; Berlin, Po-A26-0643, (1993).

48.   D. L. Romero, R. A. Morge, M. J. Genin, C. Biles, M. Busso, L. Resnick, I. W. Althaus, F. Reusser, R. C. Thomas, W. G. Tarpley, J. Med. Chem., 36, 1505 (1994).

49.   D. L. Romero, Drugs Future, 19, no. 3, (1994).

50.   T. J. Dueweke, S. M. Poppe, D. L. Romero, S. M. Swaney, A. G. So, K. M. Downey, I. W. Althaus, F. Reusser, M. Busso, L. Resnick, D. L. Mayers, J. Lane, P. A. Aristoff, R. C. Thomas, W. G. Tarpley, Antimicrob. Agents Chemother., 37, 1127 (1993).

51.   T. J. Dueweke, T. Pushkarskaya, S. M. Poppe, S. M. Swaney, J. Q. Zhao, I. S. Y. Chen, M. Stevenson, W. G. Tarpley, Proc. Natl. Acad. Sci. U. S. A., 90, 4713 (1993).

52.   M. S. Saag, E. A. Emini, O. L. Laskin, J. Douglas, W. I. Lapidus, W. A. Schleif, R. J. Whitley, C. Hildebrand, V. W. Byrnes, J. C. Kappes, K. W. Anderson, F. E. Massari, G. M. Shaw, N. Engl. J. Med., 329, 1065 (1993).

53.   R. T. Davey, R. L. Dewar, G. F. Reed, M. B. Vasudevachari, M. A. Polis, J. A. Kovacs, J. Fallon, R. E. Walker, H. Masur, S. E. Haneiwich, D. G. O'Neill, M. R. Decker, J. A. Metcalf, M. A. Deloria, O. L. Laskin, N. Salzman, H. C. Lane, Proc. Natl. Acad. Sci. U.S.A., 90, 5608 (1993).

54.   J. H. Nunberg, W. A. Schleif, E. J. Boots, J. A. O'Brien, J. C. Quintero, J. M. Hoffman, E. A. Emini, M. E. Goldman, J. Virol., 65, 4887 (1991).

55.   SCRIP, 1858, 21(7/24/93).

56.   J. S. Wai, T. M. Williams, D. L. Bamberger, T. E. Fisher, J. M. Hoffman, R. J. Hudcosky, S. C. MacTough, C. S. Rooney, W. S. Saari, C. M. Thomas, M. E. Goldman, J. A. O'Brien, E. A. Emini, J. H. Nunberg, J. C. Quintero, W. A. Schleif, P. S. Anderson, J. Med. Chem., 36, 249 (1993).

57.   J. M. Hoffman, A. M. Smith, C. S. Rooney, T. E. Fisher, J. S. Wai, C. M. Thomas, D. L. Bamberger, J. L. Barnes, T. M. Williams, J. H. Jones, B. D. Olson, J. A. O'Brien, M. E. Goldman, J. H. Nunberg, J. C. Quintero, W. A. Schleif, E. A. Emini, P. S. Anderson, J. Med. Chem., 36, 953 (1993).

58.   M. Artico, S. Massa, A. Mai, M. E. Marongiu, G. Piras, E. Tramontano, P. La Colla, Antivir. Chem. Chemother., 4(6), 361 (1993).

59.   M. Baba, Z. Debyser, S. Shigeta, E. De Clercq, Drugs Future, 17, 891 (1992).

60.   S. Yuase, Y. Sadakata, H. Takashima, K. Sekiya, N. Inouye, M. Ubasawa, M. Baba, Mol. Pharm., 44, 895 (1993).

61.   J. Balzarini, M.-J. Camarase, A. Karlsson, Drugs Future, 18(11), 1043 (1988).

62.   J. Balzarini, S. Velazquez, A. San-Felix, A. Karlsson, M.-J. Perez-Perez, M.-J. Camarasa, E. De Clercq, Mol. Pharm., 43, 109 (1993).

63.   J. Balzarini, A. Karlsson, A.-M. Vandamme, M.-J. Perez-Perez, H. Zhang, L. Vrang, B. Oberg, K. Backbro, T. Unge, A. San-Felix, S. Velazquez, M.-J. Camarasa, Proc. Natl. Acad. Sci. U.S.A., 90, 6952 (1993).

64.   R. Pauwels, K. Andries, Z. Debyser, P. Van Daele, D. Schols, P. Stoffels, K. De Vreese, R. Woestenborghs, A.-M. Vandamme, C. G. M. Janssen, J. Anne, G. Cauwenbergh, J. Desmyter, J. Heykants, M. A. C. Janssen, E. De Clercq, P. A. J. Janssen Proc. Natl. Acad. Sci, U.S.A., 90, 1711 (1993).

65.   N. Clumeck, M. Gerard, L. Kestens, M. Peeters, M. Vandenbruane, R. Colebunders, J. Delescluse, J.D. Van der Endt, R. Soete, M. Farber, J. Decree, M. De Brabander, R. Van den Broeck, P. Stoffels, P. A. J. Janssen, IX Int. Conf. AIDS/IVSTD World Congress; Berlin, PoB26-1999 (1993).

66.   M. P. de Bethune, R. Pauwels, M. Peeters, J. Desmyter, E. De Clercq, K. Andries, IX Int. Conf. AIDS/IVSTD World Congress; Berlin, PoA26-0631 (1993).

67.   M. De Brabander, L. Kestens, R. Colebunders, J. De Cree, R. Van den Broeck, P. A. J. Janssen, IX Int. Conf. AIDS/IVSTD World Congress; Berlin, PoB26-2062 (1993).

68.   J. De Cree, M. Vandenbruaene, R. Woestenborghs, V. Van de Velde, R. Van Den Broeck, M. De Brabander, P. Stoffels, G. Cauwenbergh, B. Colebunders, H. Verhaegen, P. A. J. Janssen, IX Int. Conf. AIDS/IVSTD World Congress; Berlin, PoB26-2073 (1993).

69.   W. Schafer, W.-G. Friebe, H. Leinert, A. Mertens, T. Poll, W. von der Saal, H. Zilch, B. Nuber, M. L. Ziegler J. Med. Chem., 36, 726 (1993).

70.   A Mertens, H. Zilch, B. Konig, W. Schafer, T. Poll, W. Kampe, H. Seidel, U. Leser, H. Leinert J. Med. Chem., 36, 2526 (1993).

71.   G. Maas, U. Immendoerfer, B. Koenig, U. Leser, B. Mueller, R. Goody, E. Pfaff, Antimicrob. Agents Chemother., 37, 2612 (1993).

72.   D. G. H. Livermore, R. C. Bethell, N. Cammack, A. P. Hancock, M. M. Hann, D. V. S. Green, R. Brian Lamont, S. A. Noble, D. C. Orr, J. J. Payne, M. V. J. Ramsey, A. H. Shingler, C. Smith, R. Storer, C. Williamson, T. Willson, J. Med. Chem. 36, 3784 (1993).

73.   T. M. Williams, T. M. Ciccarone, S. C. MacTough, C. S. Rooney, S. K. Balani, J. H. Condra, E. A. Emini, M. E. Goldman, W. J. Greenlee, L. R. Kauffman, J. A. O'Brien, V. V. Sardana, W. A. Schleif, A. D. Theoharides, P. S. Anderson J. Med. Chem., 36, 1291 (1993).

74.   T. A. Lyle, S. D. Young, P. S. Anderson, S. K. Balani, S. F. Britcher, S. S. Carroll, C. Culberson, E. A. Emini, M. E. Goldman, C. F. Homnick, J. R. Huff, W. C. Lumma, J. A. O'Brien, D. B. Olsen, L. S. Payne, D. J. Pettibone, J. C. Quintero, W. M. Sanders, P. E. J. Sanderson, W. A. Schleif, S. J. Smith, M. Stahlhut, A. M. Theorides, C. M. Thomas, L. O. Tran, T. J. Tucker, C. M. Wiscount, 206th ACS National Meeting, Med. Chem. Abstr. no. 135, Chicago, IL, August 22-27 (1993).

75.   S. F. Britcher, W. C. Lumma, M. E. Goldman, T. A. Lyle, J. R. Huff, L. S. Payne, M. L. Quesada, S. D. Young, W. M. Sanders, P. E. Sanderson, T. J. Tucker, EP 0 530 994 A1, 10Mar1993.

76.   T.A. Lyle, T. J. Tucker, C. M. Wiscount, EP 0 569 083 A, 29Apr93.

77.   T. A. Lyle, S. D. Young, P. S. Anderson, S. F. Britcher, E. A. Emini, M. E. Goldman, C. F. Homnick, J. R. Huff, W. C. Lumma, J. A. O'Brien, L. S. Payne, D. J. Pettibone, J. C. Quintero, W. M. Sanders, W. A. Schleif, S. J. Smith, A. M. Theoharides, T. J. Tucker, C. M. Wiscount, J. Cell. Biochem., S18B, 164 (1994).

78.   J. P. Kleim, R. Bender, U.-M. Billhardt, C. Meichsner, G. Riess, M. Rosner, I. Winkler, A. Paessens, Antimicrob. Agents Chemother., 37, 1659 (1993).

79.   D. Richman, C.-K. Shih, I. Lowy, J. Rose, P. Prodanovich, S. Goff, J. Griffin Proc. Natl. Acad. Sci. U.S.A., 88, 11241 (1991).

80.   R. Noreen, P. Engelhardt, J. Kangametsa, C. Sahlberg, L. Vrang, J. M. Morin, R. J. Ternansky, H. Zhang, Antiviral Res. 20(S1), 68 (1993).

81.   P. T. Lind, J. M. Morin, R. Noreen, R. J. Ternansky, PCT INT. Appl. WO 93 03,022, 18Feb1993.

82.   R. F. Abdulla, J. M. Morin, R. J. Ternansky, M. D. Kinnick J. B. Deeter, Antiviral Res. 20(S1), 68 (1993).

83.   M. H. St. Clair, J. L., Marin, G. Tudor-Williams, M. C. Bach, C. L. Vavro, S. M. King, P. Kellam, S. D. Kemp, B. A. Larder, Science, 263, 1557 (1991).

84.   D. D. Richman, Nature, 361, 588 (1993).

85.   E. De Clercq, Biochem. Pharm., 47, 155 (1994).

86.   M. E. Goldman, J. A. O'Brien, T. L. Ruffing, W. A. Schleif, V. V. Sardana, V. W. Byrnes, J. H. Condra, J. M. Hoffman, E. A. Emini, Antimicrob. Agents Chemother., 37, 947 (1993).

87.   L. Tong, M. Cardozo, P.-J. Jones, J. Adams, Biorg. Med. Chem. Lett., 3, 721 (1993).

Chapter 14. HIV Protease Inhibitors

Suvit Thaisrivongs

Upjohn Laboratories, Kalamazoo, MI 49001

Introduction - The protease of the Human Immunodeficiency Virus (HIV) is essential for cleavage of the viral *gag* and *gag/pol* polypeptide precursors into individual structural proteins and enzymes during the final stage of viral maturation. It has been well established that when the protease is catalytically defective (1), viral maturation in HIV-infected cell culture is blocked, and consequently, infection is arrested. Inhibitors of HIV protease have continued to receive much attention as potential therapeutic agents for the treatment of HIV infection (2-10). The resulting immature and non-infectious viral particles, as a consequence of inhibition of HIV protease, may also serve as available antigen for potential stimulant to the immune system. The previous review on the subject of HIV protease inhibitors in the *Annual Reports in Medicinal Chemistry* was in 1991 (11). At that time, a clinical study of one protease inhibitor Ro 31-8959 (**3**) was being initiated. Although many of the peptidomimetic inhibitors possess potent antiviral activity *in vitro*, the low oral bioavailability and rapid biliary excretion have limited their potential utility as successful pharmaceutical agents. High molecular weight, low solubility and high lipophilicity contribute significant obstacles to the development of peptidic inhibitors (12). Recent advances, however, have resulted in compounds with much reduced peptidic character and also non-peptidic inhibitors that are more orally bioavailable; consequently, an increasing number of protease inhibitors are currently undergoing clinical evaluation.

Peptide-derived HIV Protease Inhibitors - Successful design of potent inhibitors continued to use mimics of the transition-state of the amide bond hydrolysis. A series of 2-hetero-substituted statine analogues were shown to result in effective inhibitors such as **1** (K_i = 13 nM) with antiviral activity (ED_{50} = 0.14 μM) (13-14). Phenylnorstatine containing **2** (where R = CH_3 and H) showed IC_{50} values of 2.3 and 6.5 nM (15), and K_i values of 2.3 and 5.5 pM, respectively (16). In HIV-1 infected ATH8 cells, they showed antiviral activity (ED_{50} = <0.1 μM); and the oral bioavailability in rats was 5.9% and 42.3%, respectively (17). For the hydroxyethylamine based inhibitors, **3** (18) continued to be extensively studied, and a number of modified inhibitors were reported. Replacement of asparagine with 3'(R)-tetrahydrofuranylglycine gave **4** with improved inhibitory activity (IC_{50} = 0.054 nM) and antiviral activity (CIC_{95} = 8 nM) in preventing the spread of HIV-1 in MT4 cells (19). The N-terminal truncated tetrahydrofuran and tetrahydropyran urethanes, such as **5** (IC_{50} = 160 nM and CIC_{95} = 800 nM) were less potent (20). The corresponding cyclic sulfolane **6** (IC_{50} = 3 nM and CIC_{95} = 50 nM), however, was much more effective (21). The N-terminal quinoline was replaced by a number of heterocycles; **7** (IC_{50} = 0.07 nM and CIC_{95} = 12 nM) showed good activity (22). The phenylalanine-derived **8** (IC_{50} = 5.4 nM and CIC_{95} = 200 nM) was an effective inhibitor (23). The hydroxyethylurea SC-52151 (**9**, IC_{50} = 6.3 nM and antiviral ED_{50} = 10 nM) showed oral bioavailability of 1% in both rats and dogs; with improved formulation, oral bioavailability in rats was 17% (24-27). The trifluoromethylproline-containing **10** (IC_{50} = 0.05 nM) showed reduced antiviral activity in HIV-1 infected PBL (ED_{50} = 0.87 µM) (28).

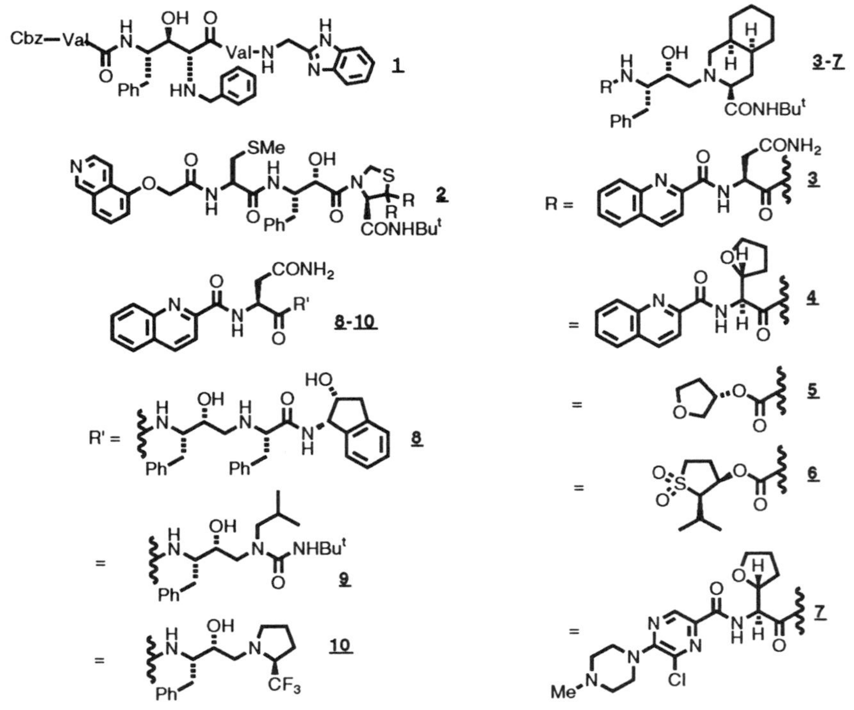

The hydroxyethylene isostere-containing inhibitors have continued to yield new series of peptidic inhibitors: Ala-Ala-PheOHXxx-Val-Val-OMe (Xxx = Gly, Ala, Nval, Phe; K_i = 4.0, 3.0, 1.2, 0.6 nM, respectively) (29). As exemplified by **11** (IC_{50} = 0.45 nM and ClC_{95} = 12 nM) with the aminohydroxyindane C-terminus (30), the *para*-position of the phenylmethyl sidechain at P_1 and P_1' sites could be substituted with water-solubilizing groups extending out of the enzyme active site (30-32). The tetrahydrofuran and tetrahydropyran urethanes, such as **12** (IC_{50} = <0.03 nM and ClC_{95} = 3 nM), were also effective inhibitors (20). After 120 mg/kg oral dosing in rats, plasma concentration of **13** (K_i = 0.2 nM and ED_{90} = 0.1 µM in HIV infected MT2 cells) above ED_{90} value was maintained for several hours (33-34). Water soluble phosphate prodrug **14** was reported to have potent antiviral activity (ED_{90} = 47 nM in HIV-1 infected human PBMC); area-under-the-curve after iv administration to rats was greatly improved over that of the parent non-phosphorylated compound (35). The dihydroxyethylene isostere-containing **15** (K_i = <1 nM) showed potent antiviral activity (ED_{50} = 3 nM) in HIV-1 infected PBL (36). The benzoic acid analogue **16** (IC_{50} = 1 nM and ED_{50} = 14 ng/mL) of the proline-containing hydroxyethylamine isostere was also shown to be a potent inhibitor (38). Compound **17** (IC_{50} 9 nM) incorporated the hydroxyethyl hydrazine dipeptide isostere (39-40). The hydroxyethylamine isostere was combined with the hydroxyethylene isostere and resulted in L-735,524 (**18**), which was found to be a very potent inhibitor (IC_{50} = 0.36 nM and ClC_{95} = 25-50 nM). The oral bioavailability was 22%, 70%, and 12% in rats, dogs, and monkeys, respectively (41).

11

12

13

Noa—Ser(OPO$_3$K$_2$)—Cha$\overset{OH}{—}$Val—Ile—Amp **14**

15

16

17

18

19

20

21

22

23

24

25

26

27

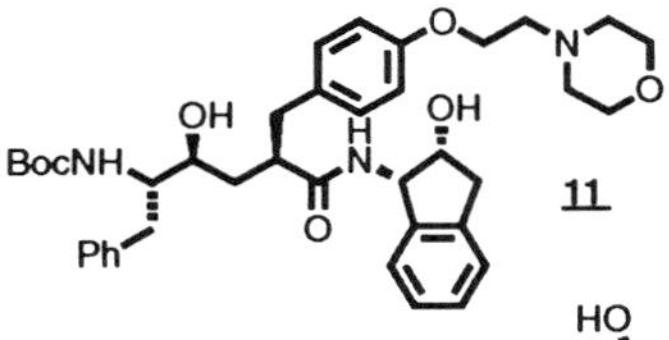

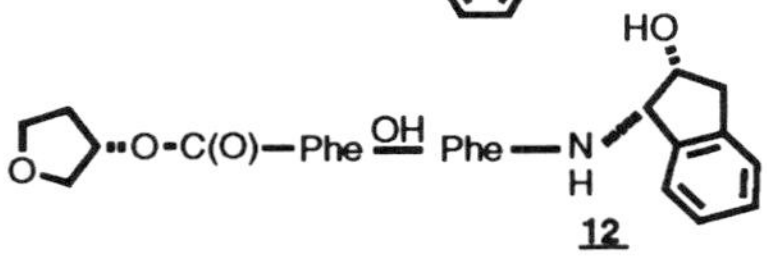

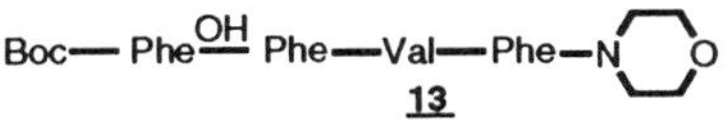

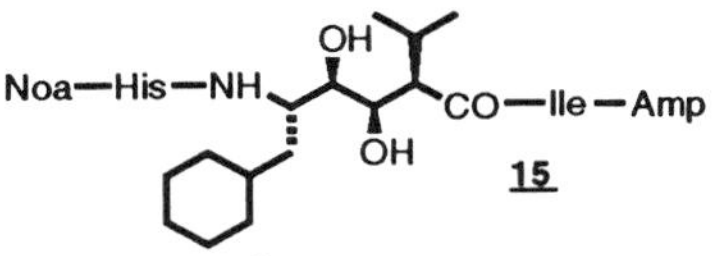

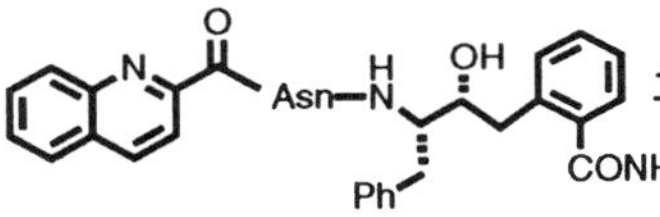

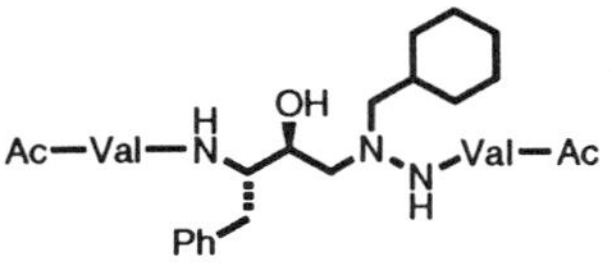

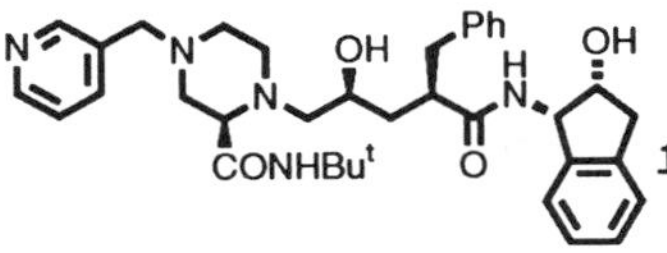

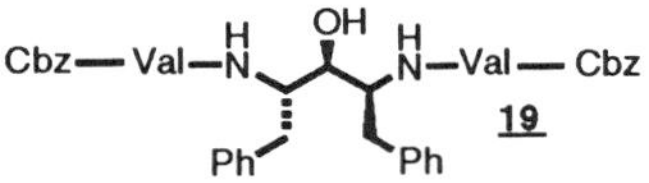

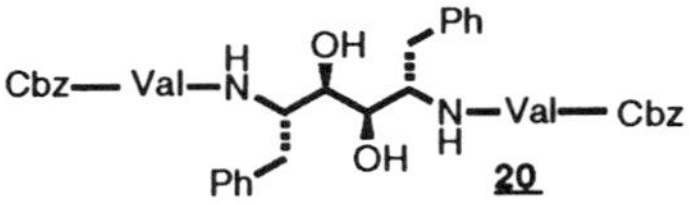

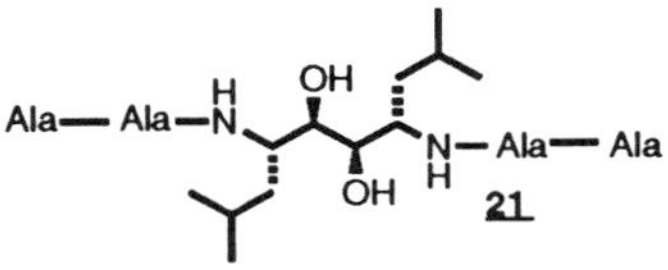

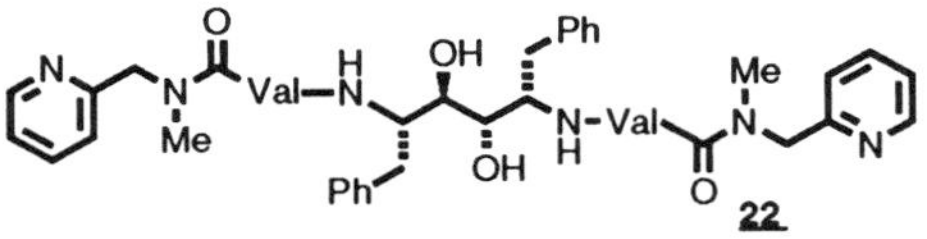

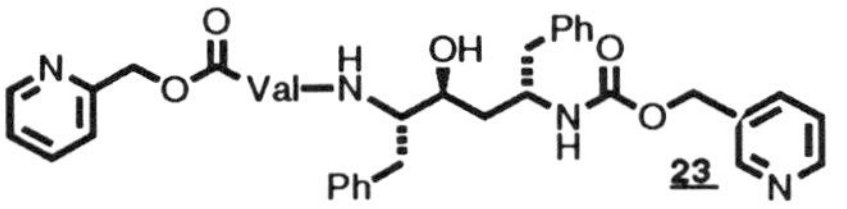

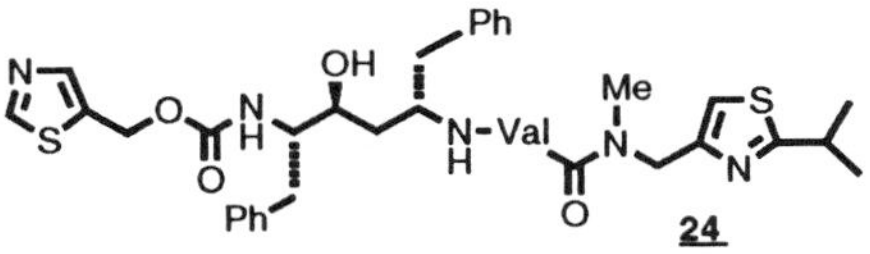

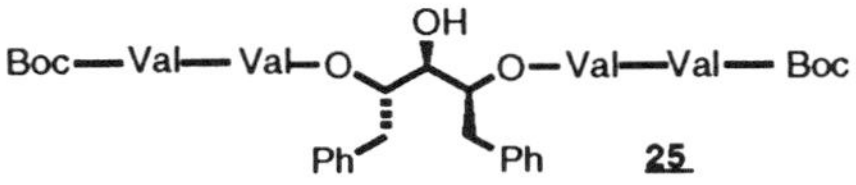

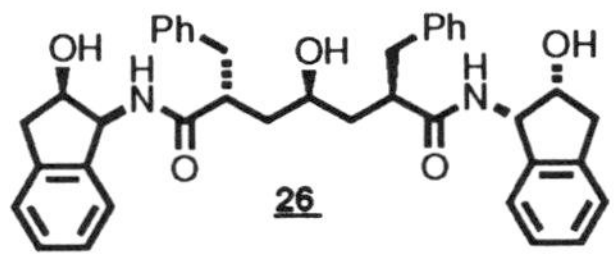

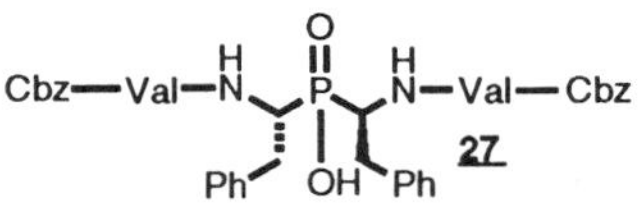

A number of structure-based approaches based upon complementarity to the C_2 symmetry of HIV protease have given rise to potent C_2 symmetric or quasi-symmetric inhibitors. C_2-symmetric **19** and **20** were shown to be highly effective inhibitors (42-43). A related C_2 symmetric **21** (K_i = 0.8 μM), with isobutyl sidechains, was found to be a much weaker inhibitor (44). Modifications of **20** to evaluate the effect of polar, heterocyclic end groups led to the non-C_2 symmetric A-77003 (**22**, IC_{50} = <1 nM and ED_{50} = 0.2 μM in HIV infected MT4 cells) with solubility of 197 μg/mL at pH 7.4, and half-lives after iv administrations to rats, dogs, and monkeys of 0.5, 1.1, 3.2 hrs, respectively (45-48). Effort to further improve oral bioavailability led to A-80987 (**23**), a non-symmetric inhibitor, with potent activity (K_i = 0.25 nM and ED_{50} = 0.13 μM in MT4 cells) and good oral bioavailability of 13-26% in three species (49). Further improvement was found in the non-symmetric A-84538 (**24**, K_i = 15 pM and ED_{50} = 30 nM) with >70% oral bioavailability in three species (50). Compound **25**, an oxygen analogue of **19**, showed good inhibitory activity (IC_{50} = 5 nM) (51). The C_2 symmetric dicarboxylic acid **26** exhibited potent activity (IC_{50} = 0.67 nM and MIC_{100} = 100 nM) (52). The C_2 symmetric phosphinate **27** was found to be a good inhibitor (K_i = 2.8 nM), although with only weak antiviral activity (53). From a screening program, penicillin-derived C_2 symmetric dimers were found to be good lead structures for analogue optimization (54-58). The dimeric structure **28** (R = CH_2Ph) was found to be an active inhibitor (IC_{50} = 0.9 nM and ED_{50} = 25 nM in HIV-1 infected H9 cells) (54). Structure **28** (R = Et, IC_{50} = 4.8 nM and ED_{50} = 5 µM in MT4 cells) was found to have rapid plasma clearance in rats, with an oral bioavailability of 5% (57). A hybrid of the penicillin-derived structure and a statine insert, which was derived from a D-phenylalanine, led to **29** (K_i = 0.25 nM and ED_{50} = 4.7 µM in HIV-1 infected MT4 cells) (58).

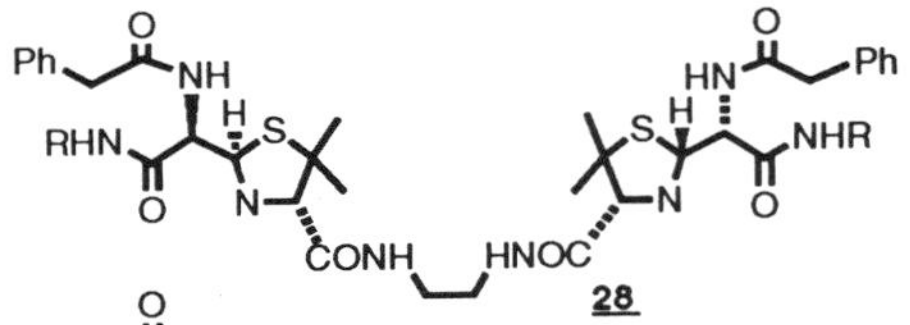

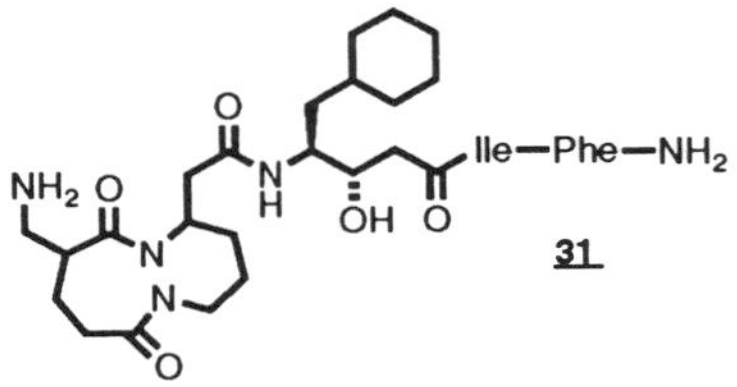

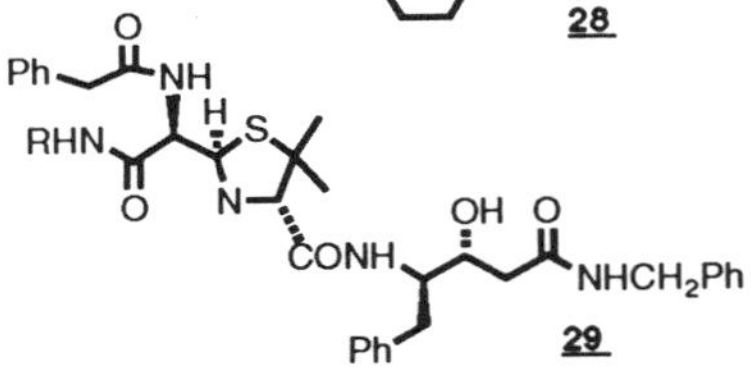

Incorporation of a γ-turn mimetic into HIV protease inhibitor led to the preparation of inhibitors with a constrained reduced-bond isostere, such as **30** (K_i = 0.43 µM) (59). Application of a molecular modelling study of crystal structures of enzyme/inhibitor complexes led to the preparation of **31** (IC_{50} = 26 nM) (60). Conformational constraint at the P_1' site was exemplified by **32** with potent activity (IC_{50} = 0.07 nM and CIC_{95} = 10 nM). It was found to be 15-20% orally bioavailable in dogs, and plasma levels exceeding 10 times the CIC_{95} value could be maintained for several hrs (61). Using the 3,5-linked pyrrolin-4-one unit to stabilize the β-strand and reduce solvation, compound **33** (IC_{50} = 1.3 nM and CIC_{95} = 0.8 µM) was shown to be an effective inhibitor (62).

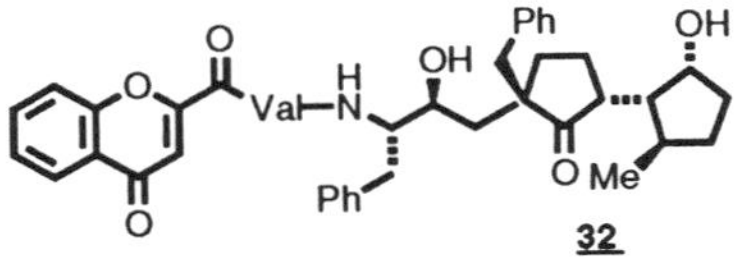

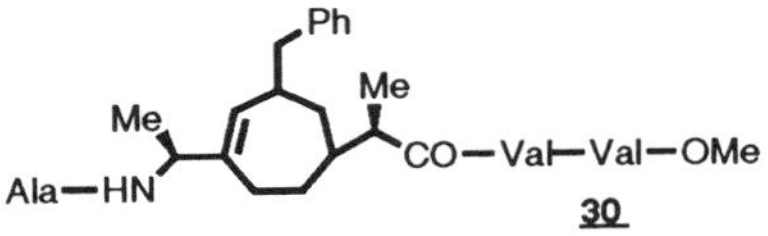

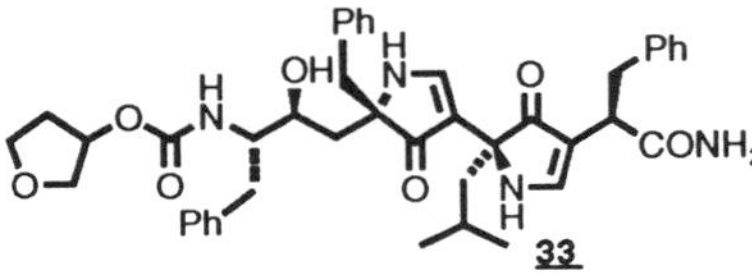

Non-peptidic HIV Protease Inhibitors - There are increasing numbers of reported non-peptidic HIV protease inhibitors which have been discovered through screening (of natural products and libraries of synthetic compounds) or from structural-based design based upon crystal structures of HIV protease/inhibitor complexes. From *Hypoxylon fragiforme*, a bark-inhabiting Ascomycete, was isolated 18-dehydroxy cytochalasin H (**34**) which was shown to be a competitive inhibitor (K_i = 1 µM) (63-65). From the magenta ascidian *Didemnum* sp. were isolated, didemnaketals A and B (**35**, R = $COCH_3$ and $C(CH_3)=CHCH_2CH_2CH(CH_3)CH_2CO_2CH_3$), IC_{50} = 2 and 10 µM, respectively) (66). A number of brominated polyacetylenic acids (IC_{50} = 6 to 12 µM) were isolated from the marine sponge *Xestospongia muta*, such as the acid **36** (IC_{50} = 7 µM) (67). Various cation complexes of cycloprazonic acid (CPA, **37**) were found to be competitive inhibitors (68). A CPA_2Tb^{3+} complex was shown to have very high inhibitory activity (K_i = 6 pM). From random screening of a sample collection, three classes of compounds were identified as lead inhibitors (69): the 3-substituted-4-hydroxycoumarin **38** (K_i = 0.97 µM), the 3,6-disubstituted-4-hydroxy-2-pyrone **39** (K_i = 2.2 µM), and the biphenylcarboxylic acid **40** (K_i = 0.14 µM). Carboxyl carborane ester derivatives of porphyrins, previously evaluated as experimental neutron-capture therapeutics, were found to be HIV protease inhibitors. The tetrakiscarborane carboxylate ester of 2,4-bis-(α,β-dihydroxyethyl)deuteroporphyrin IX (**41**), where $R_{1,2}$ = $O(CO)B_{10}H_{11}C_2$ and R_3 = H_2, exhibited inhibitory activity (IC_{50} = 185 nM) (70). The bis(phenethylamino-succinate)C_{60} (**42**), where X = $HO(CO)(CH_2)_2C(O)NH(CH_2)_2$-, was shown to be a competitive inhibitor (K_i 5.3 µM) with antiviral activity (EC_{50} = 7 µM) in PBMC. It was hypothesized that a C_{60} fullerene had approximately the same radius as the cylinder that described the active site of HIV protease (71-72).

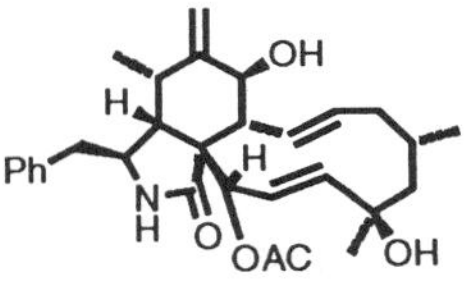

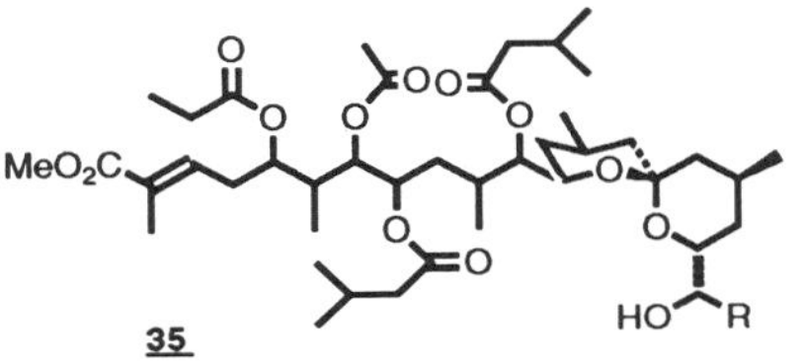

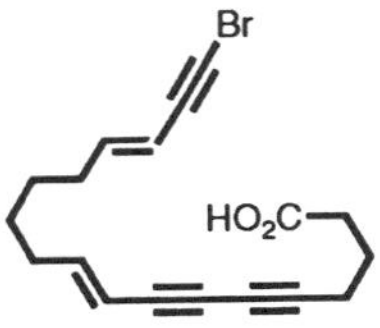

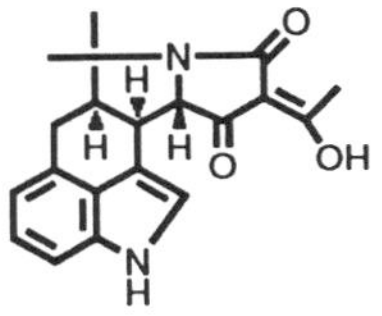

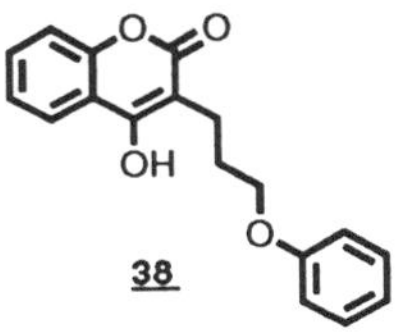

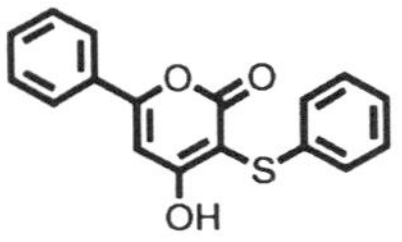

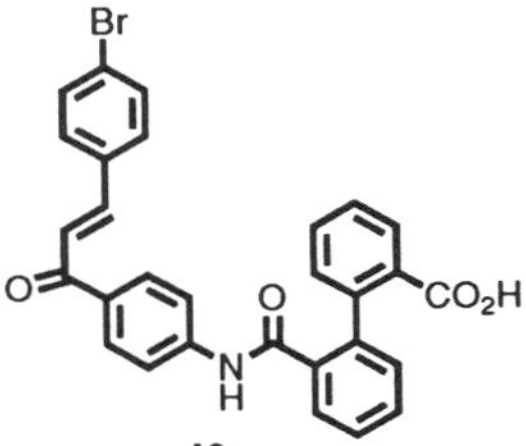

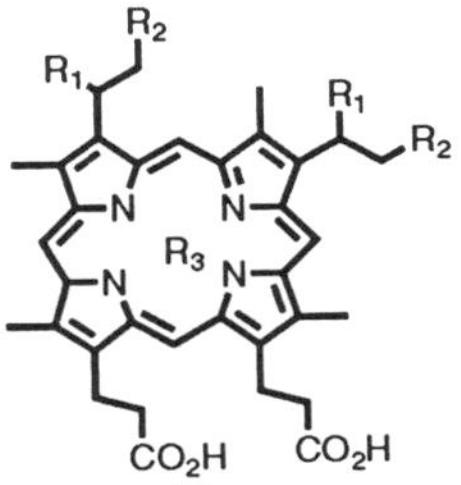

A computational search of the Cambridge Structural Database based on shape complementarity with the HIV-1 protease active site led to the antipsychotic agent haloperidol (**43**, X = O) as a weak inhibitor (K_i = 100 µM). The 1,3-dithiolane derivative (X = -SCH_2CH_2S-) was found to be a more active inhibitor (K_i = 15 µM) (73-74). In another study (75), a computer-assisted substructure search was based upon the crystal structure of HIV-1 protease/**19** complex, and **44** was found to have inhibitory activity (IC_{50} = 11 µM). Searching compound databases for potential molecules with one oxygen displacing the flap-associated water and a second oxygen interacting with the two catalytic aspartic acid residues, *trans*-1,4-cyclohexanediols and hydroquinones were suggested. Compounds **45** (R = OH, X = C(O); and R = CH_2OH, X = S(O)) were found to have K_i values of 10 and 7 µM, respectively (76). A novel core structure was developed by a computer-assisted design and led to the dibenzocycloheptadienone derivative **46** which was found to inhibit HIV-1 protease non-competitively (IC_{50} = 5 nM) (77). The related seven-membered cyclic urea XM-323 (**47**) showed potent activity (K_i = 0.27 nM and ED_{90} = 0.06 µM in HIV-1 infected MT-2 cells) (78-79). The oral bioavailability was 27% in rats and 37% in dogs. Another structure-based design of non-peptidic inhibitors, based upon the structure of **17** led to **48**, which was shown to be an active inhibitor (K_i = 1 nM and ED_{50} = 0.5 µg/mL in HIV-1 infected MT-2 cells). It also showed >30% oral bioavailability in rats, dogs and monkeys (80).

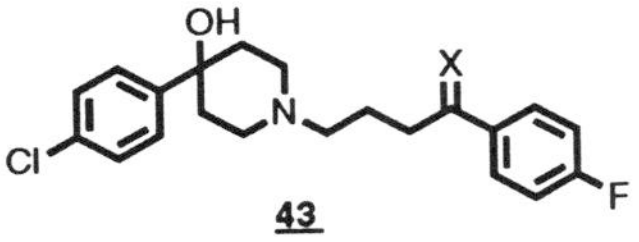

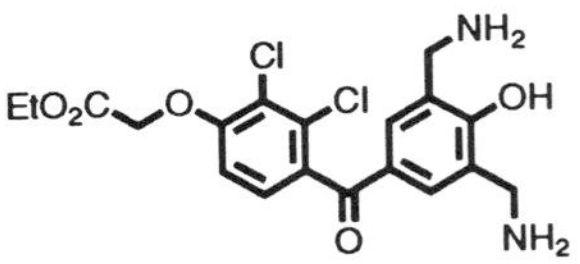

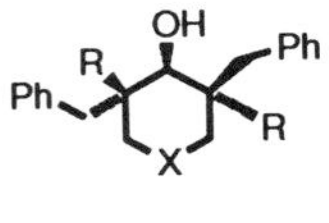

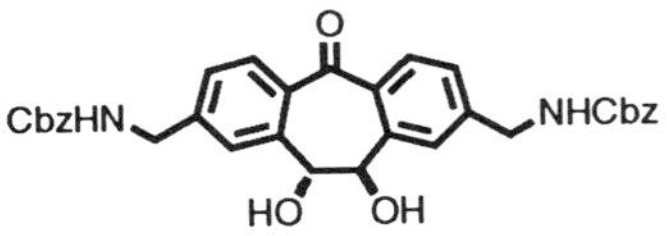

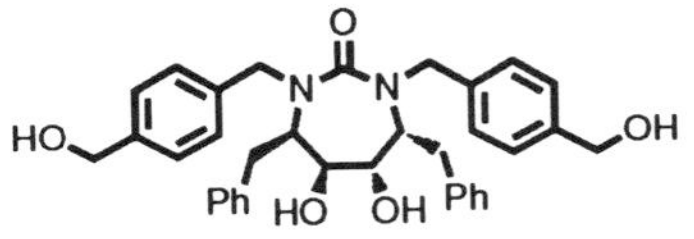

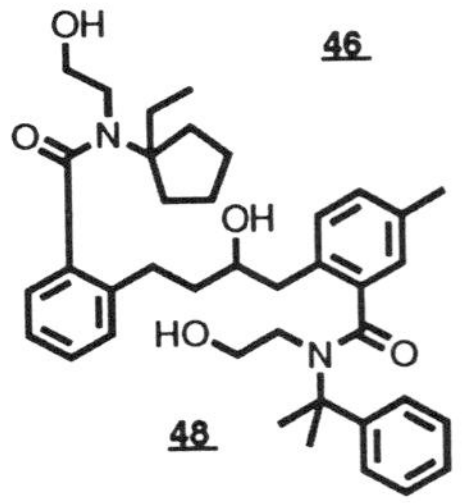

Crystal Structures of Enzyme/ligand Complexes - The availability of X-ray crystallographic structures of HIV protease/inhibitor complexes has provided researchers with an opportunity for in-depth evaluation of enzyme/inhibitor interaction and a practical effort toward the iterative cycle of structure-based design of inhibitors. Even though only a fraction of the determined crystal structures of these complexes have been disclosed, these reported structures constituted an unprecedently large number of enzyme/inhibitor complexes (81-84). Crystal structures of HIV protease complexes with many of the inhibitors, in essentially all different classes of molecules, described in this review have been determined. For the peptide-derived inhibitors, the complexes with HIV-1 protease include: substituted statine derivative **1** (14), hydroxyethylamine **3** (18), hydroxyethylamine urea **9** (26), hydroxyethylene Ala-Ala-PheOHXxx-Val-Val-OMe (Xxx = Gly, Ala, Nval, Phe) (29, 85), modified P_1' of hydroxyethylene **11** (30), dihydroxyethylene **15** (86), and hydroxyethyl hydrazine **17** (39). For the C_2 symmetric inhibitors, the complexes with HIV-1 protease include: the diamine monohydroxy **19** (43), the diamine dihydroxy **21** (44), the dicarboxylic acid **26** (52), the phosphinate **27** (53), the penicillin-derived dimer **28** (56). One key common feature of all of these complexes is a trapped water molecule which bridge the two carbonyl groups, flanking the transition-state analogue dipeptide site, to the NH groups of isoleucine 50 from each of the enzyme flap. This structure is remarkably constant even though various dipeptide transition-state analogues are of different sizes and shapes. Except for inhibitor **28**, the hydroxyl functionality of the transition-state analogue inserts is within hydrogen bonding distance of the two catalytic aspartic acid residues. Since the enzyme dimer is C_2 symmetric, many inhibitors can be found in two orientations related by a 2-fold axis. C_2 symmetric inhibitors, however, may bind asymmetrically, as was observed with inhibitor **21** (44), for example, which was found in two orientations related by an almost perfect 2-fold axis. X-ray crystallographic structures of Phe-Val-Pheψ[CH$_2$NH]Leu-Glu-Ile-NH$_2$(87) and compound **15** (88) in HIV-2 protease and of inhibitor Ala-Ala-Phe<u>OH</u>Gly-Val-Val-OMe in SIV protease (89) were also reported. The latter two inhibitors are found in these complexes in a conformation which is nearly the same as that found for the same inhibitors bound to HIV-1 protease, supporting the fact that the active sites for these related enzymes are highly conserved. Additionally, a crystal structure of an SIV protease with a covalently bound 1,2-epoxy-3-(*p*-nitrophenoxy)-propane was also described (90).

For the non-peptidic structures, the crystal structures of complexes with HIV-1 protease include the symmetric (CPA)$_2$Tb^{3+} and (CPA)$_2$(Fe^{3+})$_2$ complexes (68). The crystal structure of the HIV-1 protease complexed with an analogue of the C_2 symmetric cyclic urea **47**, in which the 4'-hydroxymethylbenzyl sidechains were replaced by 2'-naphthylmethyl groups, revealed that the carbonyl oxygen forms hydrogen bonds with the backbone amides of the two Ile50 residues on the flaps. This carbonyl oxygen replaces the bound water molecule which was commonly found in all of the complexes of linear peptidomimetic inhibitors (78). In the crystal structure of HIV-1 protease complex with the 4-hydroxycoumarin **38**, the hydroxyl group at C-4 is positioned to hydrogen bond with the two catalytic aspartic acid residues, while the carbonyl oxygen at C-2 forms hydrogen bonds with the backbone amides of the two Ile50 residues on the flaps. This carbonyl oxygen also replaces the commonly found trapped water molecule (69). In the HIV-1 protease complex of the haloperidol derivative **43** where X = -SCH$_2$CH$_2$S-, the conformation of the flaps is intermediate between that of the free enzyme and that of commonly found inhibitor-bound enzyme. There is a Cl$^-$ ion in between the protonated nitrogen atom of the inhibitor and the two amides of the two Ile50 residues on the flaps. The binding site is on the opposite surface of the active site relative to that of a peptidic inhibitor (73). The crystal structure of the bound inhibitor was significantly translated and rotated from the predicted orientation as was used in the computational search.

In Vitro Combination of Protease and RT Inhibitors - The rationale of combination therapy for treatment of HIV infection includes increased efficacy, reduced toxicity, dose reduction, and delay of emergence of resistant strains (91-92). A number of reports described *in vitro* experiments in which HIV protease inhibitors were shown to act synergistically with nucleoside and non-nucleoside reverse transcriptase inhibitors using various viral strains in different cell types. Compound **3** is synergistic with AZT or ddC in CEM-T4 cells infected with HIV-1 strain GB8 (93-94). It was shown in HIV-1 infected PBMC to demonstrate additive to synergistic effect with AZT, ddC, or recombinant interferon-αA (95). A divergent combination of **3**, AZT and interferon-α was also studied in adherent monocyte cultures infected with HIV-1 Ba-L monocytotropic strain (96). Compound **9** was reported to act synergistically with AZT or ddI (27). Z-Ala-Phe<u>OH</u>Ala-Val-Valinol, was shown to be synergistic with AZT in acutely infected H9 cells, and also in co-cultivation of chronically infected with non-infected cells (97). Compound **15** was shown to have synergistic effect with the non-nucleoside reverse transcriptase inhibitor U-90152 (98). Compound **22** showed strong trend toward synergy with AZT and additivity to ddI (99). The inhibitor **47** was shown to act synergistically with AZT or ddI to inhibit viral replication (100).

In Vitro Selection of HIV Resistance - There are an increasing number of reports on the _in vitro_ selection of HIV resistance to protease inhibitors by serial passages of viruses in gradually increasing concentrations of inhibitors (101-104). Two months of passages led to the mutant G48V and L90M for **3** (105). HIV-1 variants that showed a 35-fold increase in ED_{50} value for **22**, with no decrease in replicative capacity, were consistently identified as R8Q and M46I (106-107). In a separate study, mutations V32I, M46L, V82I and M46F were found as early as 8 passages (108). In another study with **22**, mutations identified, in acutely infected cells with a 10-fold increase in ED_{50} value, were V32I, V82I, and M46L; R8Q and M46F were found in chronically infected cells (109-111). The mutant V32I, M46L and I47V was reported after serial passage of HIV in the presence of **23** (112-113). Virus with 4-fold resistance was identified for **47** with two changes, V82F and L97V (114). The conclusion to be drawn from these studies is that HIV protease is able to mutate at various sites to become resistant to its inhibitors. It should be noted that V32I, M46I, I47V, and V82I are changes from HIV-1 to HIV-2 protease.

Compounds in Clinical Trials - A number of HIV protease inhibitors have been reported to undergo clinical evaluation. Compound **3** was the first protease inhibitor to enter clinical trial which started in early 1991. Although the oral bioavailability is low, the mean plasma concentration at 600 mg TID was reported to be 70 nM, which is higher than the ED_{90} value _in vitro_ (94); the bioavailability was reported to be markedly increased by administration with food. A 16-week study in HIV-infected patients (CD4 counts of 50-250) showed that the drug was well tolerated and there was a trend towards an increase in CD4 counts and a decrease in p24 levels, with maximal effect at 4 weeks (115). With combination of 600 mg TID of **3** and 200 mg TID of AZT (CD4 counts of <300), there was a trend towards increasing CD4 to a greater extent than monotherapies. HIV-infected asymtomatic volunteers received a single escalating dose of **9**; peak plasma concentrations occurred at 1.5 hr, and plasma levels declined with a half-life of 2 hr. At doses of 100, 250, 500, and 1000 mg, C_{max} values were 61, 144, 294, and 827 ng/mL, respectively. No adverse reactions were noted. It was projected that 600 mg TID would result in C_{max} of 0.35 µM and C_{min} of 0.01 µM, the ED_{50} value (116). The sulfate salt of **18** gave more consistent blood levels than the free base (oral doses from 20 to 1000 mg), and higher blood levels were observed in fasted individuals. At 400 mg dose, the C_{max} was above 1 µM, and blood levels were above 0.1 µM ($CIC_{95} < 0.1$ µM) for approximately 6 hr. At 400 mg QID in HIV-positive patients, C_{min} value was 0.1-0.2 µM. HIV-infected patients (CD4 counts of 67-665) were dosed for 12 days without significantly adverse effects. There was approximately 40% and 70% reduction of p24 on day 2 and day 12, respectively; p24 gradually rose after drug was discontinued on day 12. There was no significant change in CD4 (117). In asymptomatic HIV-infected patients (CD4 counts of 200-500), **22** (0.35 hr half-life) was given by continuous IV administration over 28 days at 0.035 to 0.28 µg/kg/hr doses, showing C_{max} of 0.6 µg/mL at the highest dose. There was no serious side effects, although there were a number of reports of local vein irritation and phlebitis. There was no change in CD4 or p24 antigen level (118). In a phase I clinical trial, **23** was given orally at 500, 750 and 1000 mg single doses, and resulted in blood levels of 0.5, 2 and 5 µg/mL, respectively (119). Phase I clinical trial showed **47** to have low and highly variable oral bioavailability. The poor water solubility characteristic and high metabolism profile of this compound are among the likely factors that contributed to the observed poor oral bioavailability (120).

Prospects - The efficacy of HIV protease inhibitors for the treatment of HIV infection is yet to be established; however, early indication of positive results in HIV-infected patients after treatment with protease inhibitors holds promise for these compounds as an important addition to anti-AIDS therapy. Recent evidence of the emergence of resistance to a number of non-nucleoside RT inhibitors in clinical studies has been demonstrated for the chronic treatment with these agents. Indeed, a number of experimental non-nucleoside RT inhibitors have been abandoned as potential single-agent therapeutic agents due to very early emergence of resistant clinical isolates. The reduced efficacy of long-term treatment with the nucleoside RT inhibitor AZT is due in part to the development of resistance. Most of the protease inhibitors bind to the active site of the enzyme, and many are effective against both HIV-1 and HIV-2 proteases which contain conserved binding sites. In vitro resistance to many protease inhibitors has been reported, although the degree of resistance is notably less when compared to that of the RT inhibitors. Convergent or divergent multiple-drug therapy, or sequential administration of several agents might enhance the efficacy, reduce adverse side effects, and significantly delay the emergence of resistant clinical HIV strains. Some of the effective combinations demonstrated in vitro need to be further characterized in a clinical setting.

References

1. N.E. Kohl, E.A. Emini, W.A. Schleif, L.J. Davis, J.C. Heimbach, R.A.F. Dixon, E.M. Scolnick and I.S. Sigal, Proc. Nat. Acad. Sci. U.S.A., 85, 4686 (1988).
2. T. Robins and J. Plattner, J. AIDS, 6, 162 (1993).
3. S.K. Grant, T.D. Meek, B.W. Metcalf and S.R. Petteway, Jr., Biomed. Appl. Biotechnol., 1, 325 (1993).
4. S.C. Pettit, S.F. Michael and R. Swanstrom, Perspect. Drug Discovery Des., 1, 69 (1993).
5. T.D. Meek, J. Enzyme Inhib., 6, 65 (1992).
6. H. Mitsuya, J. Enzyme Inhib., 6, 1 (1992).
7. J.A. Martin, Antiviral Res., 17, 265 (1992).
8. S.R. Petteway, Jr., G.B. Dreyer, T.D. Meek, B.W. Metcalf and D.M. Lambert, AIDS Res. Rev., 1, 267, (1991).
9. J.R. Huff, J.Med.Chem., 34, 2305 (1991).
10. A.G. Tomasselli, W.J. Howe, T.K. Sawyer, A. Wlodawer and R.L. Heinrikson, Chimica Oggi, 5, 1 (1991).
11. D.W. Norbeck and D.J. Kempf, Annu. Rep. Med. Chem., 26, 141, (1991).
12. J.J. Plattner and D.W. Norbeck in "Drug Discovery Technologies", R. Clark and W.H. Moos, Ed., Ellis Horwood Ltd., Chichester, 1990, p. 92.
13. A. Billich, A. Aziz, P. Lehr, B. Charpiot, H. Gstach and D. Scholz, J. Enzyme Inhib., 7, 213 (1993).
14. A. Billich, D. Scholz, H. Retscher, B. Charpiot, H. Gstach, P. Lehr and B. Rosenwirth, 1st Natl. Conf. on Human Retroviruses and Related Infections, Washington, D.C., December 12-16, 1993, Abstract #262.
15. T. Mimoto, J. Imai, S. Kisanuki, H. Enomoto, N. Hattori, K. Akaji and Y. Kiso, Chem. Pharm. Bull., 40, 2251 (1992).
16. S. Kageyama, T. Mimoto, Y. Murakawa, M. Nomizu, F. Motoyoshi, H. Ford, Jr., T. Shirasaka, S. Gulnik, J. Erickson, K. Takada, H. Hayashi, S. Broder, Y. Kiso and H. Mitsuya, Antimicrob. Agents Chemother., 37, 810 (1993).
17. A. Kiriyama, T. Mimoto, S. Kisanuki, Y. Kiso and K. Takada, Biopharm. Drug Dispos., 14, 697 (1993).
18. A. Krohn, S. Redshaw, J.C. Ritchie, B.J. Graves and M.H. Hatada, J. Med. Chem., 34, 3340 (1991).
19. W.J. Thompson, A.K. Ghosh, M.K. Holloway, H.Y. Lee, P.M. Munson, J.E. Schwering, J. Wai, P.L. Darke, J. Zugay, E.A. Emini, W.A. Schleif, J.R. Huff and P.S. Anderson, J. Am. Chem. Soc., 115, 801 (1993).
20. A.K. Ghosh, W.J. Thompson, S.P. McKee, T.T. Duong, T.A. Lyle, J.C. Chen, P.L.. Darke, J.A. Zugay, E.A. Emini, W.A. Schleif, J.R. Huff and P.S. Anderson, J. Med. Chem., 36, 292 (1993).
21. A.K. Ghosh, W.J. Thompson, H.Y. Lee, S.P. McKee, P.M. Munson, T.T. Duong, P.L. Darke, J.A. Zugay, E.A. Emini, W.A. Schleif, J.R. Huff and P.S. Anderson, J. Med. Chem., 36, 924 (1993).
22. A.K. Ghosh, W.J. Thompson, M.K. Holloway, S.P. McKee, T.T. Duong, H.Y. Lee, P.M. Munson, A.M. Smith, J.M. Wai, P.L. Darke, J.A. Zugay, E.A. Emini, W.A. Schleif, J.R. Huff and P.S. Anderson, J. Med. Chem., 36, 2300 (1993).
23. T.J. Tucker, W.C. Lumma, Jr.; L.S. Payne, J.M. Wai, S.J. De Solms, E.A. Giuliani, P.L. Darke, J.C. Heimbach, J.A. Zugay, W.A. Schleif, J.C. Quintero, E.A. Emini, J.R. Huff and P.S. Anderson, J. Med. Chem., 35, 2525 (1992).
24. D.P. Getman, D E. Bertenshaw, G.A. DeCrescenzo, J.N. Freskos, R.M. Heintz and K.C. Lin, 206th Amer. Chem. Soc. Natl. Meeting, Chicago, Illinois, August 22-27, 1993, MEDI 137.
25. J.N. Freskos, D.P. Getman, M.L. Bryant, K.P. Houseman and R.A. Muller, 206th Amer. Chem. Soc. Natl. Meeting, Chicago, Illinois, August 22-27, 1993, MEDI 136.

26. D.P. Getman, G.A. DeCrescenzo, R.M. Heintz, K.L. Reed, J.J. Talley, M.L. Bryant, M. Clare, K.A. Houseman, J.J. Marr, R.A. Mueller, M.L. Vazquez, H.-S. Shieh, W.C. Stallings and R.A. Stegeman, J. Med. Chem., 36, 288 (1993).

27. M. Bryant, M. Smidt, D. Getman, J. Talley, M. Vazquez, G. Decrescenzo, R. Mueller, A. Roy, J. Ng, J. Stolzenbach, S. Snook, R. Cavalier, M. Herin, M. Cole, A. Karim and J. Sherman, 1st Natl. Conf. on Human Retroviruses and Related Infections, Washington, D.C., December 12-16, 1993, Abstract #261.

28. D. Häbich, J. Hansen, A. Paessens, W. Röben and G. Streissle, 9th Internatl. Conf. on AIDS, Berlin, June 7-11, 1993, PO-A25-0616.

29. G.B. Dreyer, D.M. Lambert, T.D. Meek, T.J. Carr, T.A. Tomaszek, Jr., A.V. Fernandez, H. Bartus, E. Cacciavillani, A.M. Hassell, M. Minnich, S.R. Petteway, Jr. and B.W. Metcalf, Biochemistry, 31, 6646 (1992).

30. W.J. Thompson, P.M.D. Fitzgerald, M.K. Holloway, E.A. Emini, P.L. Darke, B.M. McKeever, W.A. Schleif, J.C. Quintero, J.A. Zugay, T.J. Tucker, J.E. Schwering, C.F. Homnick, J. Nunberg, J.P. Springer and J.R. Huff, J. Med. Chem., 35, 1685 (1992).

31. R.B. Lingham, B.H. Arison, L.F. Colwell, A. Hsu, G. Dezeny, W.J. Thompson, G.M. Garrity, M.M. Gagliardi, F.W. Hartner, P.L. Darke, S.K. Balani, S.M. Pitzenberger, J.S. Murphy, H.G. Ramjit, E.S. Inamine and L.R. Treiber, Biochem. Biophys. Res. Commun., 181, 1456 (1991).

32. S.D. Young, L.S. Payne, W.J. Thompson, N. Gaffin, T.A. Lyle, S.F. Britcher, S.L. Graham, T.H. Schultz, A.A. Deana, P.L. Darke, J. Zugay, W.A. Schleif, J.C. Quintero, E.A. Emini, P.S. Anderson and J.R. Huff, J. Med. Chem., 35, 1702 (1992).

33. E. Alteri, G. Bold, R. Cozens, A. Faessler, T. Klimkait, M. Lang, J. Lazdins, B. Poncioni, J.L. Roesel, P. Schneider, M. Walker and K. Woods-Cook, Antimicrob. Agents Chemother., 37, 2087 (1993).

34. R.M. Cozens, E. Alteri, G. Bold, F. Cumin, A. Faessler and T. Klimkait, 33rd ICAAC, New Orleans, Louisiana, U.S.A., October 17-20, 1993, Abstract 396.

35. K.T. Chong, M.J. Ruwart, R.R. Hinshaw, K.F. Wilkinson, R.D. Rush, M.F. Yancey, J.W. Strohbach and S. Thaisrivongs, J.Med. Chem., 36, 2575 (1993).

36. S. Thaisrivongs, A.G. Tomasselli, J.B. Moon, J. Hui, T.J. McQuade, S.R. Turner, J.W. Strohbach, W.J. Howe, W.G.Tarpley and R.L. Heinrikson, J. Med. Chem., 34, 2344 (1991).

37. S. Thaisrivongs, S.R. Turner, J.W. Strohbach, R.E. TenBrink, W.G. Tarpley, T.J. McQuade, R.L. Heinrikson, A.G. Tomasselli, J.O. Hui and W.J. Howe, J. Med. Chem., 36, 941 (1993).

38. S.W. Kaldor, M. Hammond, B.A. Dressman, J.E. Fritz and T.A. Crowell, 206th Amer. Chem. Soc. Natl. Meeting, Chicago, Illinois, August 22-27, 1993, MEDI 138.

39. A. Fässler, J. Rösel, M. Grütter, M. Tintelnot-Blomley, E. Alteri, G. Bold and M. Lang, Bioorg. Med. Chem. Lett., 3, 2837 (1993).

40. H.L. Sham, D.A. Betebenner, C. Zhao, N.E. Wideburg, A. Saldivar, D.J. Kempf, J.J. Plattner and D.W. Norbeck, J. Chem. Soc., Chem. Commun., 1052 (1993).

41. J. Vacca, 33rd ICAAC, New Orleans, Louisiana, U.S.A., October 17-20, 1993, Symposium #82.

42. D.J. Kempf, D.W. Norbeck, L. Codacovi, X.C. Wang, W.E. Kohlbrenner, N.E. Wideburg, D.A. Paul, M.F. Knigge, S. Vasavanonda, A. Craig-Kennard, A. Saldivar, W. Rosenbrook, Jr., J.J. Clement, J.J. Plattner and J. Erickson, J. Med. Chem., 33, 2687 (1990).

43. J. Erickson, D.J. Neidhart, J. VanDrie, D.J. Kempf, X.C. Wang, D.W. Norbeck, J.J. Plattner, J.W. Rittenhouse, M. Turon, N. Wideburg, W.E. Kohlbrenner, R. Simmer, R. Helfrich, D.A. Paul and M. Knigge, Science, 249, 527 (1990).

44. G.B. Dreyer, J. C.Boehm, B. Chenera, R.L. DesJarlais, A.M. Hassell, T.D. Meek, T.A.Tomaszek, Jr. and M. Lewis, Biochemistry, 32, 93 (1993).

45. D.J. Kempf, L. Codacovi, X.C. Wang, W.E. Kohlbrenner, N.E. Wideburg, A. Saldivar, S. Vasavanonda, K.C. Marsh, P. Bryant, H.L. Sham, B.E. Green, D.A. Betebenner, J. Erickson and D.W. Norbeck, J. Med. Chem., 36, 320 (1993).

46. A.H. Kaplan, J.A. Zack, M. Knigge, D.A. Paul, D.J. Kempf, D.W. Norbeck and R. Swanstrom, J. Virol., 67, 4050 (1993).

47. J.J. Kort, J.A. Bilello, G. Bauer and G.L. Drusano, Antimicrob. Agents Chemother., 37, 115 (1993).

48. D.J. Kempf, K.C. Marsh, D.A. Paul, M.F. Knigge, D.W. Norbeck, W.E. Kohlbrenner, L. Codacovi, S. Vasavanonda, P. Bryant, X.C. Wang, N.E. Wideburg, J.J. Clement, J.J. Plattner and J. Erickson, Antimicrob. Agents Chemother., 35, 2209 (1991).

49. D.J. Kempf, D.W. Norbeck, K.C. Marsh and J. Erickson, 206th Amer. Chem. Soc. Natl. Meeting, Chicago, Illinois, August 22-27, 1993, ORGN 119.

50. D.J. Kempf, C. Flentge, K.C. Marsh, E. McDonald, T. Robins, S. Vasavanonda, C.-M. Chen, N.E. Wideburg, C. Park, X. Kong, J. Denissen, D.D. Ho, M. Markowitz, T. Toyoshima, M.K. Singh, J. Erickson, H.L. Sham, K. Stewart, B.E. Green, M. Turon, A. Saldivar and D.W. Norbeck, Keystone Symposia, Santa Fe, New Mexico, March 5-12, 1994, Abstract S038, p. 130.

51. R.E. Babine, N. Zhang, A.R. Jurgens, S.R. Schow, P.R. Desai, J.C. James and M.F. Semmelhack, J. Bioorg. Med. Chem. Lett., 2, 541 (1992).

52. R. Bone, J.P. Vacca, P.S. Anderson and M.K. Holloway, J. Am. Chem. Soc., 113, 9382 (1991).

53. S.S. Abdel-Meguid, B. Zhao, K.H.M. Murthy, E. Winborne, J.K. Choi, R.L. DesJarlais, M.D. Minnich, J.S. Culp,

C. Debouck, T.A. Tomaszek, Jr., T.D. Meek and G.B. Dreyer, Biochemistry, 32, 7972 (1993).
54. D.C. Humber, N. Cammack, J.A.V. Coates, K.N. Cobley, D.C. Orr, R. Storer, G.G. Weingarten and M.P. Weir, J. Med. Chem., 35, 3080 (1992).
55. D.S. Holmes, I.R. Clemens, K.N. Cobley, D.C. Humber, J. Kitchin, D.C. Orr, B. Patel, I.L. Paternoster and R. Storer, Bioorg. Med. Chem. Lett., 3, 503 (1993).
56. A. Wonacott, R. Cooke, F.R. Hayes, M.M. Hann, H. Jhoti, P. McMeekin, A. Mistry, P. Murray-Rust, O.M.P. Singh and M.P. Weir, J. Med. Chem., 36, 3113 (1993).
57. D.C. Humber, M.J. Bamford, R.C. Bethell, N. Cammack, K. Cobley, D.N. Evans, N.M. Gray, M.M. Hann, D.C. Orr, J. Saunders, B.E.V. Shenoy, R. Storer, G.G. Weingarten and P.G. Wyatt, J. Med. Chem., 36, 3120 (1993).
58. D.S. Holmes, R.C. Bethell, N. Cammack, I.R. Clemens, J. Kitchin, P. McMeekin, C.L. Mo, D.C. Orr, B. Patel, I.L. Paternoster and R. Storer, J. Med. Chem., 36, 3129 (1993).
59. K.A. Newlander, J.F. Callahan, M.L. Moore, T.A. Tomaszek, Jr. and W.F. Huffman, J. Med. Chem., 36, 2321 (1993).
60. M. Kahn, H. Nakanishi, R.A. Chrusciel, D. Fitzpatrick and M.E. Johnson, J. Med. Chem., 34, 3395 (1991).
61. R.W. Hungate, J.L. Chen, K.E. Starbuck, S.L. McDaniel, R. B. Levin and B.D. Dorsey, 206th Amer. Chem. Soc. Natl. Meeting, Chicago, Illinois, August 22-27, 1993, MEDI 143.
62. A.B. Smith III, R. Hirschmann, A. Pasternak, R. Akaishi, M.C. Guzman, D.R. Jones, T.P. Keenan, P.A. Sprengeler, P.L. Darke, E.A. Emini, M.K. Holloway and W.A. Schleif, J. Med. Chem., 37, 215 (1994).
63. J. Ondeyka, O.D. Hensens, D. Zink, R. Ball, R.B. Lingham, G. Bills, A. Dombrowski and M. Goetz, J. Antibiot., 45, 679 (1992).
64. A.W. Dombrowski, G.F. Bills, G. Sabnis, L.R. Koupal, R. Meyer, J.G. Ondeyka, R.A. Giacobbe, R.L. Monaghan and R.B. Lingham, J. Antibiot., 45, 671 (1992).
65. R.B. Lingham, A. Hsu, K.C. Silverman, G.F. Bills, A. Dombrowski, M.E. Goldman, P.L. Darke, L. Huang, G. Koch, J.G. Ondeylka and M.A. Goetz, J. Antibiot., 45, 686 (1992).
66. B.C.M. Potts, D.J. Faulkner, J.A. Chan, G.C. Simolike, P. Offen, M.E. Hemling and T.A. Francis, J. Am. Chem. Soc., 113, 6321 (1991).
67. A.D. Ashok, W.C. Kokke, S. Cochran, T.A. Francis, T. Tomszek and J.W. Westley, J. Nat. Prod., 55, 1170 (1992)
68. P.L. Darke, P.M.D. Fitzgerald, J.A. Zugay, J. Ondeyka, M. Goetz and L.C. Kuo, 1st Natl. Conf. on Human Retroviruses and Related Infection, Washington, D.C., December 12-16, 1993, Abstract #263.
69. P.J. Tummino, D. Ferguson, L. Hupe, A. Heldsinger and D. Hupe, 1st Natl. Conf. on Human Retroviruses and Related Infection, Washington, D.C., December 12-16, 1993, Session #77, L7.
70. D.L. DeCamp, L.M. Babe, R. Salto, J.L. Lucich, M.S. Koo, S.B. Kahl and C.S. Craik, J. Med. Chem., 35, 3426 (1992).
71. S.H. Friedman, D.L. DeCamp, R.P. Sijbesma, G. Srdanov, F. Wudl and G.L. Kenyon, J. Am. Chem. Soc., 115, 6506 (1993).
72. R. Sijbesma, G. Srdanov, F. Wudl, J.A. Castoro, C. Wilkins, S.H. Friedman, D.L. DeCamp and G.L. Kenyon J. Am. Chem. Soc., 115, 6510 (1993).
73. E. Rutenber, E.B. Fauman, R.J. Keenan, S. Fong, P.S. Furth, P.R. Oritz de Montellano, E. Meng, I.D. Kuntz and D.L. DeCamp, J. Biol. Chem., 268, 15343 (1993).
74. Z. Sui, J.J. DeVoss, D.L. DeCamp, J. Li, C.S. Craik and P.R. Ortiz de Montellano, Synthesis, 803, (1993).
75. M.G. Bures, C.W. Hutchins, M. Maus, W. Kohlbrenner, S. Kadam and J.W. Erickson, Tetrahedron Comput. Methodol., 3, 673 (1990).
76. B. Chenera, R.L. DesJarlais, J.A. Finkelstein, D.S. Eggleston, T.D. Meek, T.A. Tomaszek, Jr. and G.B. Dreyer, Bioorg. Med. Chem. Lett., 3, 2717 (1993).
77. H. Wild, J. Hansen and J. Lautz, 9th Internatl. Conf. on AIDS, Berlin, June 7-11, 1993.
78. P.Y.-S. Lam, P.K. Jadhav, C.J. Eyermann, C.N. Hodge, Y. Ru, L.T. Bacheler, J.L. Meek, M.J. Otto, M.M. Rayner, Y.N. Wong, C.-H. Chang, P.C. Weber, D.A. Jackson, T.R. Sharpe and S. Erickson-Viitanen, Science, 263, 380 (1994).
79. M.J. Otto, C.D. Reid, S. Garber, P.Y.-S. Lam, H. Scarnati, L.T. Bacheler, M.M. Rayner and D.L. Winslow, Antimicrob. Agents Chemother., 37, 2606 (1993).
80. K. Appelt, 33rd ICAAC, New Orleans, Louisiana, U.S.A., October 17-20, 1993, Symposium #82.
81. P.M.D. Fitzgerald, Curr. Opinion. Struc. Biol., 3, 868 (1993).
82. A. Wlodawer and J.W. Erickson, Annu. Rev. Biochem., 62, 543 (1993).
83. K. Appelt, Perspect. Drug Discovery Des., 1, 23 (1993).
84. C.L. Waller, T.I. Oprea, A. Giolitti and G.R. Marshall, J. Med. Chem., 36, 4152 (1993).
85. K.H.M. Murthy, E.L. Winborne, M.D. Minnich, J.S. Culp and C. Debouck, J. Biol. Chem., 267, 22770 (1992).
86. N. Thanki, J.K.M. Rao, S.T. Foundling, J.W. Howe, J.B. Moon, J.O. Hui, A.G. Tomasselli, R.L. Heinrikson, S. Thaisrivongs and A. Wlodawer, Protein Sci., 1, 1061 (1992).
87. L. Tong, S. Pav, C. Dargellis, F. Do, D. LaMarre and P.C. Anderson, Proc. Natl. Acad. Sci. USA, 90, 8387 (1993).
88. A.M. Mulichak, J.O. Hui, A.G. Tomasselli, R.L. Heinrikson, K.A. Curry, C.S. Tomich, S. Thaisrivongs, T.K. Sawyer and K.D. Watenpaugh, J. Biol. Chem., 268, 13103 (1993).

89. B. Zhao, E. Winborne, M.D. Minnich, J.S. Culp, C. Debouck and S.S. Abdel-Meguid, Biochemistry, 32, 13054 (1993).
90. R.B. Rose, J.R. Rosé, R. Salto, C.S. Craik and R.M. Stroud, Biochem., 32, 12498 (1993).
91. A.S. Fauci, Ann. Intern. Med., 116, 85 (1992).
92. M.A. Fischl, Hospital Practice, 43 (1994)
93. J.C. Craig, I.B. Duncan, L. Whittaker and N.A. Roberts, Antiviral Chem. Chemother., 4, 161 (1993).
94. N.A. Roberts, J.C. Craig and I.B. Duncan, Biochem. Soc. Trans., 20, 513 (1992).
95. V.A. Johnson, D.P. Merrill, T.C. Chou and M.S. Hirsch, J. Infect. Dis., 166, 1143 (1992).
96. S. Rusconi, Y.K. Chow, D.P. Merrill and M.S. Hirsch, 33rd ICAAC, New Orleans, Louisiana, U.S.A., October 17-20, 1993, Abstract #676.
97. D.M. Lambert, H. Bartus, A.V. Fernandez, C. Bratby-Anders, J.J. Leary, G.B. Dreyer, B.W. Metcalf and S.R. Petteway, Jr., Antiviral Res., 21, 327 (1993).
98. K.T. Chong, P.J. Pagano and R.R. Hinshaw, 9th Internatl. Conf. on AIDS, Berlin, June 7-11, 1993, PO-A25-0557.
99. S. Kageyama, J.N. Weinstein, T. Shirasaka, D.J. Kempf, D.W. Norbeck, J.J. Plattner, J. Erickson and H. Mitsuya, Antimicrob. Agents Chemother., 36, 926 (1992).
100. M.J. Otto, S. Garber, C.A. Baytop and C.D. Reid, 33rd ICAAC, New Orleans, Louisiana, U.S.A., October 17-20, 1993, Abstract #37.
101. F. Dianzani, G. Antonelli, O. Turriziani, E. Riva, G. Dong and D. Bellarosa, Antiviral Chem. and Chemother., 4, 329 (1993).
102. J.C. Craig, L. Whittaker, I.B. Duncan and N.A. Roberts, Antiviral Chem. Chemother., 4, 335 (1993).
103. M.A. El-Farrash, M.J. Kuroda, T. Kitazaki, T. Masuda, K. Kato, M. Hatanaka and S. Harada, J. of Virology, 68, 233 (1994).
104. M.J. Otto, S. Garber, D.L. Winslow, C.D. Reid, P. Aldrich, P.K. Jadhav, C.E. Patterson, C.N. Hodge and Y.-S.E. Cheng, Proc. Natl. Acad. Sci. USA, 90, 7543 (1993).
105. J.C. Craig, I.B. Duncan, H. Jacobsen, J. Mous, N.A. Roberts and L.N. Whittaker, 2nd Internatl. Workshop on HIV Drug Resistance, Noordwijk, The Netherlands, June 3-5, 1993, Abstract book page 22.
106. D. Ho, 2nd Internatl. Workshop on HIV Drug Resistance, Noordwijk, The Netherlands, June 3-5, 1993, Abstract book page 18.
107. D.D. Ho, T. Toyoshima, H. Mo, D.J. Kempf, D. Norbeck, C.-M. Chen, N.E. Wideburg, S.K. Burt, J.W. Erickson and M.K. Singh, J. Virology, 68, 2016 (1994).
108. S.F. Michael, A. Kaplan, D. Kempf, M. Knigge, D. Paul, D. Norbeck, J. Erickson, L. Everitt and R. Swanstrom, 2nd Internatl. Workshop on HIV Drug Resistance, Noordwijk, The Netherlands, June 3-5, 1993, Abstract book page 19.
109. R. Wehbie, S. Pettit, S. Michael, A. Kaplan and R. Swanstrom, 2nd Internatl. Workshop on HIV Drug Resistance, Noordwijk, The Netherlands, June 3-5, 1993, Abstract book page 20.
110. R. Swanstrom, A.H. Kaplan, S.F. Michael, R.S. Wehbie, S.C. Pettit, M.F. Knigge, D.A. Paul, L. Everitt, D.J. Kempf, D.W. Norbeck and J. Erickson, 1st Natl. Conf. on Human Retroviruses and Related Infections, Washington, D.C., December 12-16, 1993.
111. C.-M. Chen, A. Saldivar, M. Turon, N.E. Wideburg, T. Robins, S. Vasavanonda, K. Stewart, D.J. Kempf, D. Ho and D.W. Norbeck, 1st Natl. Conf. on Human Retroviruses and Related Infections, Washington, D.C., December 12-16, 1993, Abstract #265.
112. D. Kempf, 5th Internatl. Conf. of the NCDDG-HIV, Washington, D.C., July 11-16, 1993.
113. T. Robins, S. Vasavanonda, S. Blohm, C.-M. Chen, N. Wideburg, M. Turon, A. Saldivar, D. Kempf, D. Norbeck and J. Clement, 1st Natl. Conf. on Human Retroviruses and Related Infections, Washington, D.C., December 12-16, 1993, Abstract #267.
114. M. Otto, S. Garber, S. Stack and D. Winslow, 2nd Internatl. Workshop on HIV Drug Resistance, Noordwijk, The Netherlands, June 3-5, (1993).
115. J.F. Delfraissy, D. Sereni, F. Brun-Vézinet, E. Dussaix, A. Krivine, J. Dormant and K. Bragman, 9th Internatl. Conf. on AIDS, Berlin, June 7-11, 1993, WS-B26-3.
116. M. Cole, A. Karim, C. Wallemark, S. Bondy and J. Sherman, 1st Natl. Conf. on Human Retroviruses and Related Infections, Washington, D.C., December 12-16, 1993, Abstract #572.
117. H. Teppler, R. Pomerantz, T. Bjornsson, J. Pientka, B. Osborne, E. Woolfe, K. Yeh, P. Duetsch, E. Emini, K. Squires, M. Saag and S. Waldman, 1st Natl. Conf. on Human Retroviruses and Related Infections, Washington, D.C., December 12-16, 1993, Session 77, L8.
118. S. Danner, M. Reedjik, C.A.B. Boucher, K.M. Mayer, J.M. Leonard and T.B. Tzeng, 9th Internatl. Conf. on AIDS, Berlin, June 7-11, 1993, WS-B26-6.
119. D.W. Norbeck, 1st Natl. Conf. on Human Retroviruses and Related Infections, Washington, D.C., December 12-16, 1993, Session #71.
120. M.M. Rayner, B.C. Cordova, R.P. Meade, P.Y.Lam and S. Erickson-Viitanen, 1st Natl. Conf. on Human Retroviruses and Related Infections, Washington, D.C., December 12-16, 1993, Abstract #264.

Chapter 15. Antiviral Agents

Richard E. Boehme, Alan D. Borthwick and Paul G. Wyatt
Glaxo Research and Development Limited
Greenford, Middlesex, UB6 OHE, UK

Introduction - This chapter focuses on developments in the antiviral field with the exclusion of HIV. HIV reverse transcriptase inhibitors and HIV protease inhibitors are reviewed in chapters 13 and 14, respectively. We will review important advances with particular reference to new chemical entities, clinical trials and molecular targets.

Herpesviruses - Antivirals for the treatment of herpes infections have been reviewed (1,2). The virus-specific anti-herpes agents such as acyclovir, ganciclovir and foscarnet have had a significant impact on the management of herpesvirus infections. Concomitant with the use of these agents has been an increase in emergence of drug-resistant virus strains (3).

Acyclovir (**1**) continues to be the major treatment of choice for herpes simplex infections including cold sores (HSV-1) and genital herpes (HSV-2). A recent five-year study extends the safety and efficacy profile of oral acyclovir for suppression of genital herpes (4). It has also been licensed recently for over-the-counter treatment of HSV-1 infections in the UK. Research leading to the clinical development of its L-valyl ester prodrug, valaciclovir hydrochloride (**2**) which has an oral bioavailability of 63%, has been reviewed (5). Penciclovir's (**3**) overall spectrum of activity and selectivity against HSV-1, HSV-2, VZV and EBV is comparable to that of acyclovir, and the majority of acyclovir-resistant HSV and VZV clinical strains are cross-resistant to penciclovir (6). Famciclovir (**4**) the diacetyl, 6-deoxy oral prodrug of penciclovir has been licensed for herpes zoster (VZV) in the UK (7).

Several compounds with enhanced activity against VZV and HCMV are in clinical development. Research leading to clinical development of PYaraU (**5**) as a selective inhibitor of VZV has been reviewed (4).

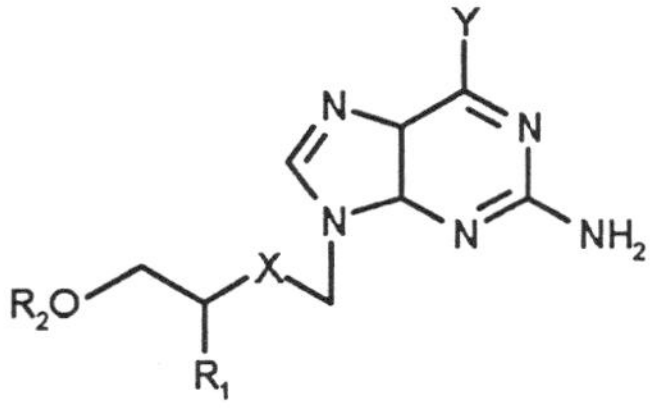

	X	Y	R₁	R₂
1	O	OH	H	H
2	O	OH	H	L-valyl
3	CH₂	OH	CH₂OH	H
4	CH₂	H	CH₂OAc	Ac

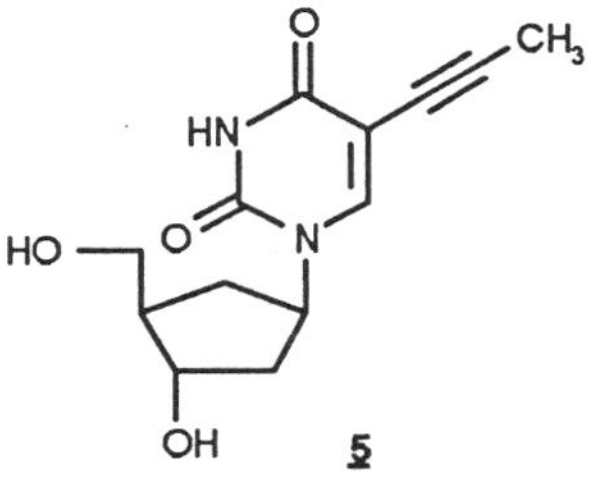

Phosphonate isosteres of nucleotides continue to receive attention as agents active against a wide range of herpes viruses. The phosphonylmethoxyalkyl derivatives of purines and pyrimidines have been reviewed (8). HPMPC (**6**) is a derivative of this class of compound and is a potent and selective inhibitor of HCMV in vivo. However it showed reversible nephrotoxicity which was dose-limiting after two doses at 10mg/kg (9). The therapeutic potential of HPMPC has been reviewed (10). A novel acyclic analogue of ganciclovir (**7**) (SR3727A) is reported to have potent and selective activity against a range

ANNUAL REPORTS IN MEDICINAL CHEMISTRY—29

of HCMV strains (IC_{50} = 6-17μM) *in vitro* (11). It has a similar activity to ganciclovir against murine CMV *in vivo*, but is much less toxic to dividing cells.

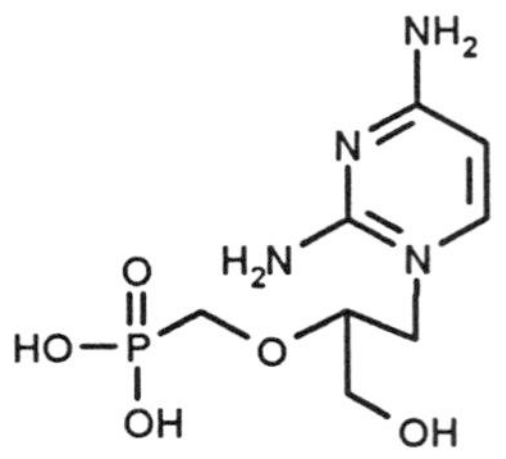

Studies on several new nucleoside analogues have been reported. Evaluation of hexose nucleosides, where the sugar ring has been expanded to a six-membered ring led to the discovery of a novel class of 1,5-anhydrohexitol nucleosides which are active against a range of human herpes viruses *in vitro* (12). The 5-iodouracil analogue (**8**) has highly selective activity against HSV-1 and HSV-2 (IC_{50} = 0.07μg/ml) which depends on specific phosphorylation by the virus-encoded thymidine kinase (TK). The isonucleosides are another class of nucleosides with a different sugar template surrogate. The guanine analogue (**9**) was shown to be more active than ACV *in vivo* against HSV-1(IC_{50} = 84 mg/kg/day) and HSV-2(IC_{50} = 52 mg/kg/day) with no overt toxicity at 200mg/kg for 5 days in mice (13).

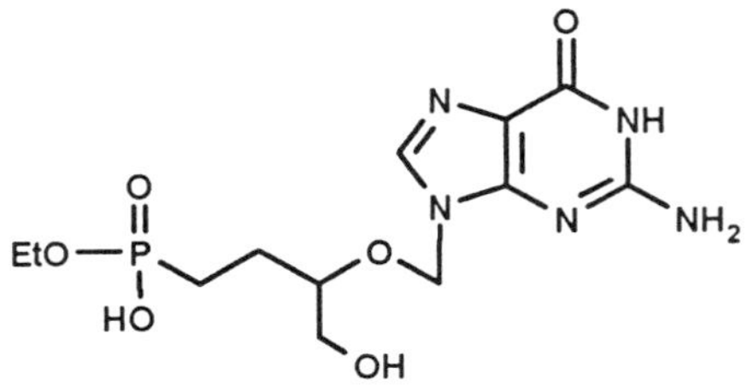

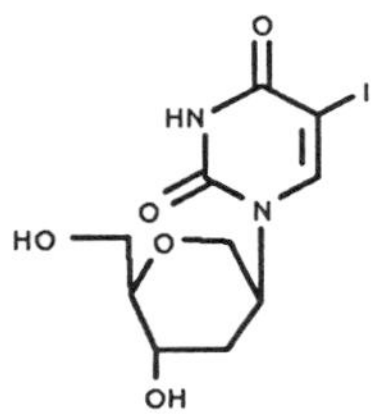

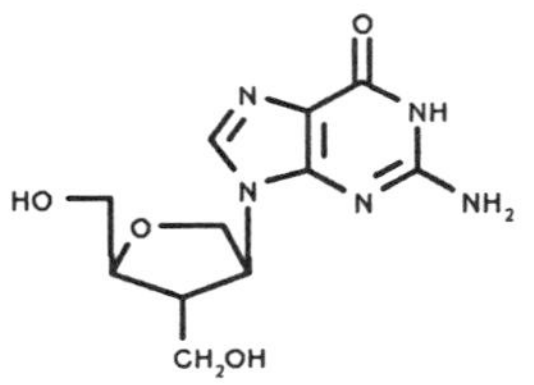

The oxetanocin analogue (**10**) A-73209 was reported to be potent and selectively active against a panel of strains of VZV (IC_{50} = 0.01μg/ml), HSV-1 (IC_{50} = 0.03μg/ml) and HSV-2 (IC_{50} = 2.2μg/ml) *in vitro* and to be more active than ACV orally against HSV-1 and equipotent to ACV against HSV-2 via the i.p. route (14). Carbocyclic nucleosides continue to be exploited as novel, metabolically stable antiherpes compounds. The chiral 4'-hydroxy carbocyclic 2'-deoxyguanosine (**11**) was shown to be more potent than ACV against HSV-1(IC_{50} = 0.15μg/ml) and HSV-2 (IC_{50} = 0.26μg/ml) *in vitro* and less toxic to Vero cells than carbocyclic 2'-deoxyguanosine (15).

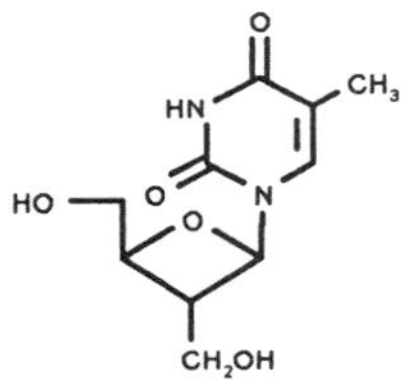

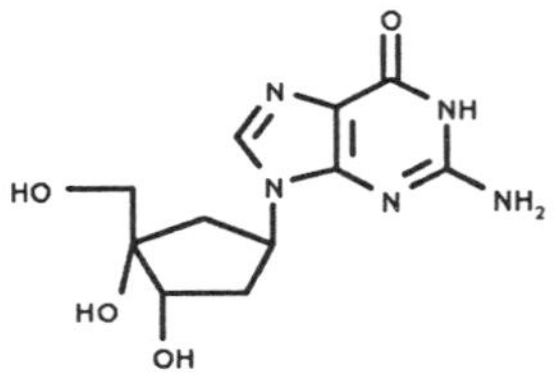

A series of novel 3-quinolinecarboxamides that are structurally similar to the quinolone class of antibacterials possess good antiherpetic activity (16). The 4-fluorophenyl derivative (**12**) was five-fold more potent than ACV *in vitro* against HSV-2, and in a multiple-dose regimen in mice exhibited half the potency of ACV, however it showed a smaller separation between activity and toxicity than ACV. It did not inhibit topoisomerase-2, and showed no antibacterial activity *in vitro*. Mappicine ketone **(13)** is an analogue of camptothecin and inhibits a broad range of herpesviruses including HSV-1, HSV-2 and CMV with a potency comparable to that of ACV (17). It is inactive against other DNA and RNA viruses, is not an inhibitor of mammalian topoisomerase-1, and inhibits a different viral target than ACV or camptothecin.

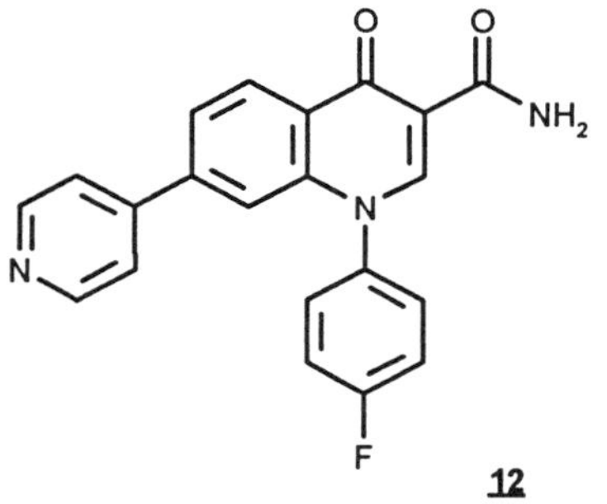

12 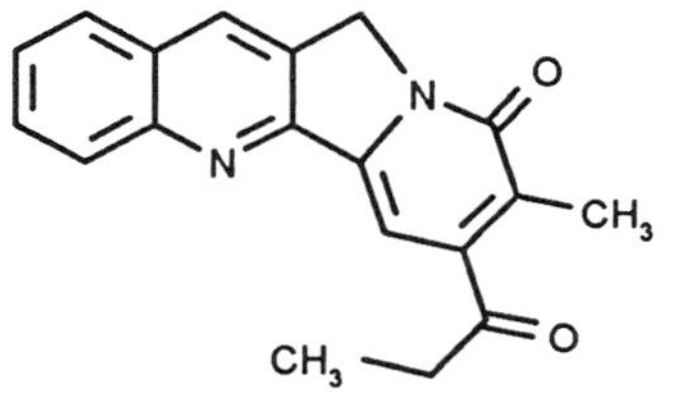   **13**

Other Targets - The need for new classes of anti-HSV compounds with novel mechanisms of viral inhibition is becoming increasingly acute as mutants resistant to conventional nucleoside and pyrophosphate analogues emerge, especially in the vulnerable immunocompromised patients.

Inhibition of the subunit association required for catalytic activity of HSV ribonucleotide reductase (RR) has been investigated as a potential strategy for antiviral therapy (18). A potent substituted tetrapeptide (IC_{50} = 0.18μM) has been developed from an SAR investigation of the minimum structural requirement for inhibition of HSV RR subunit association (19).

Uracil-DNA glycosylase is involved in post-replicative DNA repair processes. Recent work to isolatate and characterise this enzyme from HSV-1 infected cells has enabled the first specific inhibitors of HSV-1 uracil-DNA glycosylase to be designed (20). The most potent and selective was found to be 6-(para-n-octylanalino)uracil (IC_{50} = 8μM).

Recent research has discovered that all human herpesviruses encode a serine protease. There is substantial homology among the human herpesviruses with respect to both the amino acid sequence of their proteases and the proteolytic cleavage sites favored by these enzymes (21). The properties and function of this protease have been best described for HSV-1 where the enzyme plays a crucial role in cleaving the scaffold proteins required for construction of viral capsids (22). A protease-deficient HSV-1 mutant does not cleave the scaffold protein, and DNA entry and capsid maturation are blocked (23). The obligate requirement for this proteolytic processing to produce mature virus makes this an exciting new antiviral target. Mutation and inhibition studies for CMV (21,24,25) and HSV-1 (26,27) indicate that the herpesvirus protease is a member of the serine protease super family and its conserved substrate cleavage sites V/LXA↓S/V differ from that of chymotrypsin and subtilisin. Recently P5-P8' was indentified as the minimal substrate peptide for the Ala[610]/Ser[611] cleavage site in HSV-1, and a requirement was demonstrated for residues flanking the conserved core P4-LVNA/S-P1' in substrate recognition and hydrolysis (28).

Antisense oligodeoxynucleotides are a promising new class of antiviral agents. Because they bind in a sequence-specific manner to complementary regions of mRNA, oligos can inhibit gene expression highly selectively. The 'antisense' approach has been used successfully to block cellular expression and replication of several viruses including HSV-1 and HSV-2 (29). ISIS 2922 is an antisense phosphorothioate oligonucleotide 21 nucleotides in length that is complementary to the coding region of the mRNA of the CMV major immediate early gene 2 (IE2). The IE2 region encodes multiple proteins which regulate CMV gene expression and are required for replication. ISIS 2922 is 16-30 fold more potent *in vitro* than ganciclovir against HCMV and it specifically reduces IE protein synthesis in HCMV-infected cells in a dose dependant manner that correlates with antiviral activity (30). A new class of anti-EBV agents has been developed based on 28mer phosphorothioate oligonucleotides, where the mechanism of action appears to be inhibition of EBV DNA synthesis. The 28 mer ($IC_{90} = 0.5\mu M$) was shown to be more potent against EBV *in vitro* than ACV, DHPG or PFA and to have a better therapeutic index than these compounds (31).

The inhibition of S-adenosylhomocysteine (AdoHcy) as the basis of the antiviral activity of certain adenosine analogues active against HCMV has been reviewed (32). Finally, the structure and function of HSV DNA replicating proteins and their potential as antiviral targets have been reviewed (33).

Hepatitis Viruses - Three nucleoside analogues intended for treatment of hepatitis B virus (HBV) infections were tested in human phase 2 studies in 1993. Lamivudine (**14**) and FIAU (**15**) were administered to patients for periods ranging from 2 weeks to several months (34,35). The results of famciclovir (**4**) clinical trials have not yet been reported, however the compound is very active in animal models of HBV (36). In addition to HBV, FIAU has broad-spectrum activity *in vitro* against herpesviruses and HIV. Previous human trials of FIAU involving herpesviruses were unsuccessful due to toxicity (37), and HIV trials failed to demonstrate sufficient efficacy (38). Although short-term FIAU therapy was effective and well-tolerated in chronic HBV patients, extended therapy caused irreversible hepatic and pancreatic toxicity, culminating in several deaths (35). The compound was subsequently withdrawn from development. These unfortunate results prompted ongoing investigations to determine the mechanism of toxicity of FIAU and assess the potential of other nucleoside analogues to elicit similar toxicity.

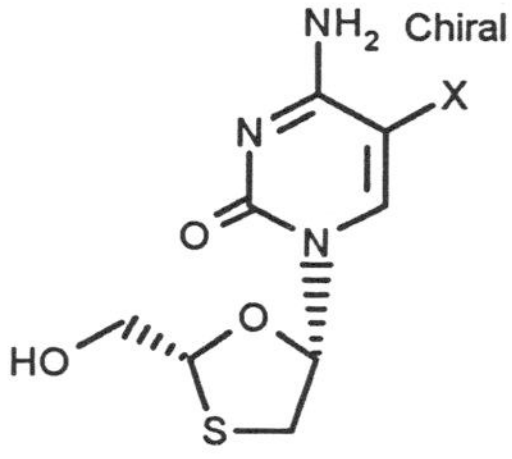

14 X=H
16 X=F

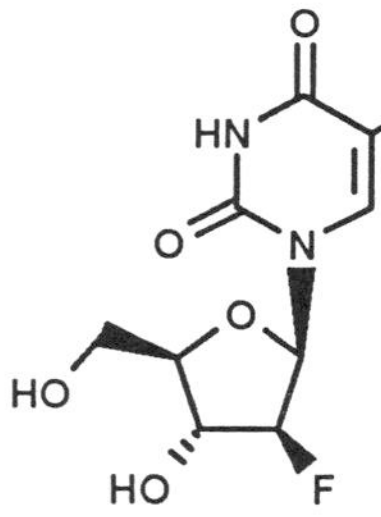

15

Lamivudine produced >90% reductions of viraemia in chronic HBV patients during 4 weeks of therapy, although viraemia rapidly returned to pre-treatment levels at the conclusion of therapy (34). No significant toxicity was observed, and further studies to assess the efficacy of lamivudine for treatment of HBV-associated liver disease are planned. Lamivudine and the related compound FTC (**16**) (39) are unique in that their unnatural enantiomers are more active and less toxic than their natural (+)-enantiomers. The unnatural enantiomers of other antiviral nucleoside analogues may provide safer and more potent therapies in the future. Famciclovir has been licensed in the UK for treatment of varicella zoster virus infections, and ongoing clinical studies are evaluating its efficacy for treatment of chronic HBV. Other nucleoside analogues have been evaluated for efficacy against HBV, including azidothymidine and PMEA, however none appears to be as promising as lamivudine and famciclovir (40, 41). All of these nucleoside analogues, including those in clinical development, appear to act through interference with the HBV-encoded DNA polymerase. At present no other virus-encoded molecular targets suitable for drug discovery work have been identified.

Immune stimulation by interferon-α is effective in some HBV patients, and other immune modulators are being evaluated for this therapeutic application (42). Thymosin α-1, a thymic polypeptide, has shown efficacy in early studies and is being evaluated further for both HBV and hepatitis C virus (HCV; 43). Granulocyte-macrophage colony stimulating factor (GM-CSF) has similar efficacy to that of interferon-α for reduction of hepatitis markers. This cytokine may have some clinical utility, but efficacy will probably be limited, like that of interferon-α, to a minority of HBV patients. Improved strategies for immunotherapy of HBV infections, currently in early development, include induction of HLA class 1-mediated immunity with peptide antigens. This approach has the potential to stimulate a useful immune response in chronic HBV patients. About one-third of chronic HCV patients respond to interferon-α therapy, however the effect is often transient (44). Ribavirin is undergoing clinical trials presently for treatment of HCV, but the difficulty of working with this virus has slowed progress in identifying new antiviral compounds (45).

HCV encodes a serine protease, a metalloprotease, an RNA helicase and an RNA polymerase (46). One or more of these may be a good target for drug discovery, and work to investigate this area is underway.

Influenza viruses - Rimantidine HCl (**17**) (to be marketed as flumadine) has been approved for treatment and prophylaxis in influenza A infections almost 7 years after the original filing of the New Drug Application. Rimantidine exhibits similar efficacy to amantidine against influenza A. When taken before or within 48 hours of onset of symptoms, rimantidine reduces and shortens the fever and systemic symptoms of infection. Like amantidine, rimantidine is inactive against influenza B (47).

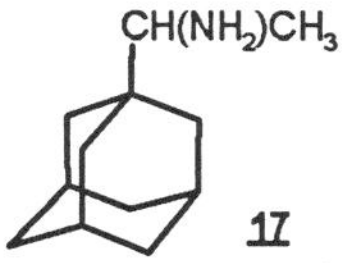

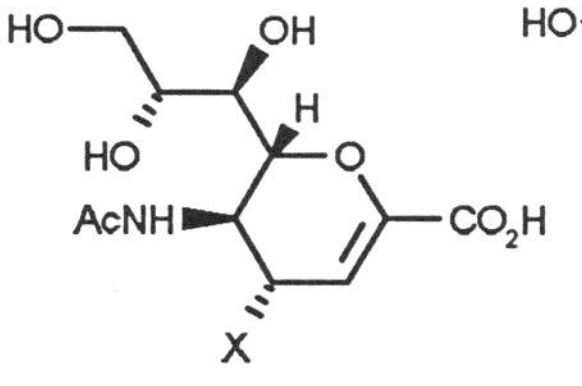

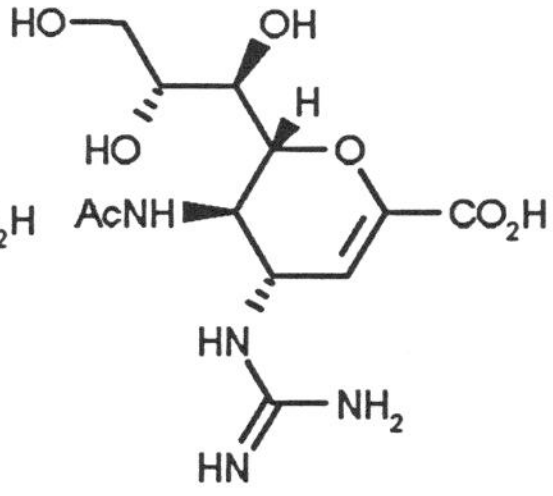

Two potent and selective inhibitors of influenza neuraminidase, 4-amino Neu 5Ac2en (**18**) and 4-guanidino Neu 5Ac2en (**20**) were first reported last year (48). The compounds were rationally designed after computational investigation of the sialic acid binding site of a crystal structure of fluA NA. Replacement of the 4-hydroxyl of Neu 5Ac2en (**21**) with amino and the more basic guanidino group gave 100-fold and 10,000-fold increases in activity, respectively. Compound (**19**) inhibited a range of flu A and B NA's (IC$_{50}$'s between 0.64 and 7.9 nM). *In vitro*, (**19**) inhibited a number of laboratory strains of flu A and B (IC$_{50}$'s 5 - 14 nM) and a range of clinical isolates (IC$_{50}$'s 0.02 - 16 μM). In all cases it was more active than amantadine, rimantidine and ribavirin. No *in vitro* cytotoxicity was seen with concentrations up to 10 mM. Compound **(19)** is inactive against other viruses tested and exhibits a 10^6-fold selectivity over human sialidases tested . **(19)** is highly active when given intranasally to both mice and ferrets and is ~1000-fold more active than amantidine (49,50). Modification of the 4-guanidino group resulted in substantial loss of activity. The selectivity of **(19)** was related to the unique size and polar nature of the 4-binding position of flu A and B when compared to that of other sialidases (51).

Three series of 2'-deoxy-2'-fluororibosides containing pyrimidines, 2-unsubstituted purines or 2-aminopurines have been reported to inhibit influenza viruses. The cytosine derivative **(21)**, although active *in vitro* (IC_{50} = 2.5 µM vs. fluA), is inactive *in vivo* (flu A in mice) presumably due to deamination to the inactive uracil derivative. The *in vitro* activity of the purine derivatives is dependent on cell type and has been correlated to intracellular triphosphate levels of the derivatives. No differences in triphosphate levels have been found between infected and uninfected cells. The most active compound reported, 2' - deoxy - 2' fluoroguanosine **(22)**, is a substrate for calf thymus deoxycytidine kinase. **(22)** exhibits IC_{50}'s against fluA of 18 and 0.55 µM in MCDK cells and CEF cells, respectively and toxicity (IC_{50}) of 437 and 17 µM in the respective cell lines. The compound is particularly active in human respiratory epithelial cells (IC_{90} vs. fluA < 0.01 µg/ml and fluB ~ 0.1 µg/ml) with therapeutic indices of greater than 1000. Selection of partially resistant mutants (IC_{50}'s 5-fold greater than wild type) indicated the mode of action was at least partially virus specific. In a mouse model vs. fluA or B commencing treatment (40 mg kg^{-1}) after virus challenge, **(22)** was significantly more active than either amatidine or ribavirin. No signs of toxicity were observed with any of the compounds tested (52-54).

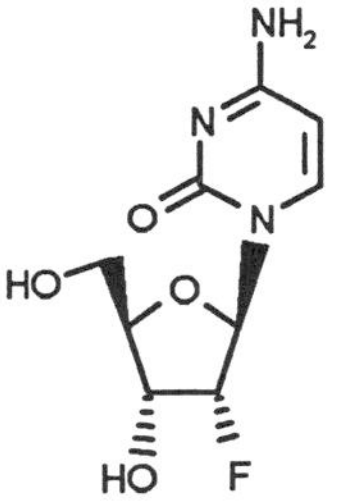

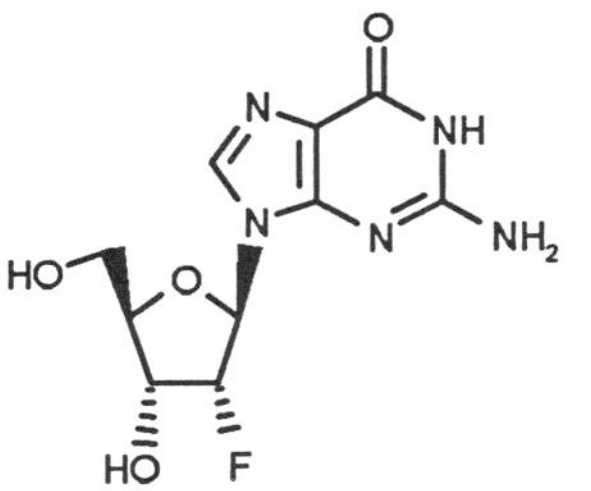

Influenza is an enveloped virus and must fuse with the host cell to allow virus entry. This fusion process between the virus membrane and the endosomal membrane of the host cell is mediated by the viral surface protein hemagglutinin (HA). This process involves a pH dependent conformational change of HA to expose a hydrophobic segment normally buried in the trimer interface of the native protein (55). The structure of this lipophilic region (56), the mechanism of fusion (57-59) and the effect of the conformational change on the lateral mobility of HA (60) have been investigated. A study to identify small molecule inhibitors of HA conformational change used computational techniques to identify putative binding sites in the hydrophobic region of HA. Compounds were selected for test using further computational techniques. This process identified compounds that inhibited the conformational change of HA, virus induced cell-cell fusion and viral replication, with the most active compound **(23)** exhibiting IC_{50} values in the µM range (55).

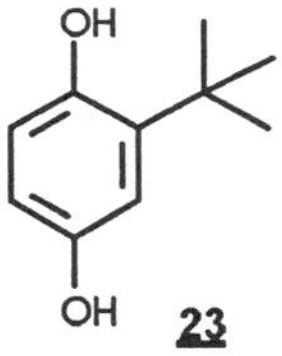

Phosphorothioate (61) or unmodified (62) oligonucleotides complementary to conserved regions of influenza virus RNA exhibited anti-influenza activity at nontoxic concentrations *in vitro* (IC_{50} down to $0.1\mu M$). Studies in mice (i.p. dosing once a day) using modified phosphorothioate nucleotides have demonstrated antiviral efficacy (62).

The posttranslational proteolytic cleavage of influenza HA by host cell proteases is essential for infectivity and spread of the virus in the host organism. Studies using known protease inhibitors identified two compounds, camostat mesilate (**24**) (IC_{50} vs. fluA and B ~7µM) and nafamostat mesilate (**25**) (IC_{50} vs fluA and B ~1.5 µM).

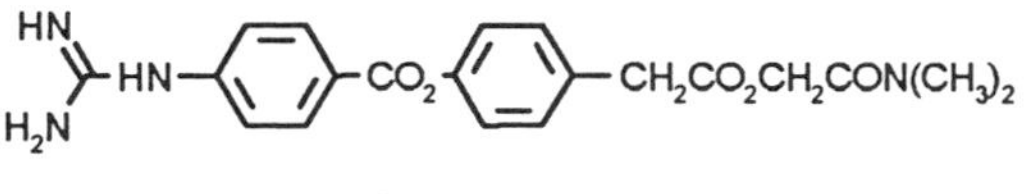

Both compounds protected chick embryos from virus-induced mortality with selectivity indices of 280 and 28, respectively (cf. ribavirin has a selective index of 55 in this test). As (**24**) is widely used for the treatment of pancreatitis, it is an interesting compound for further investigation (63).

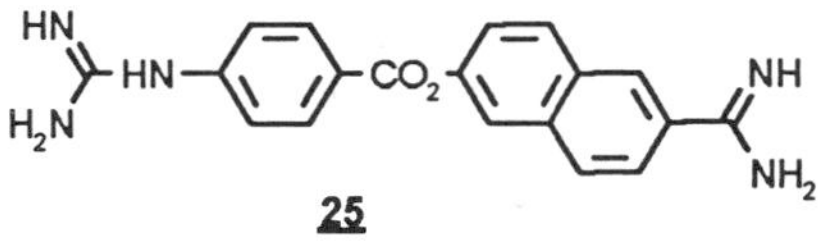

Respiratory Syncial Virus - RespGam (formerly Respivir), a preparation of immune globulins from patients with high levels of antibodies against RSV, has been reported to be safe and efficacious in prevention of RSV infections in high risk infants and children. However, an FDA advisory committee voted against approval of RespiGam after concluding that the Phase 3 efficacy trials were flawed. The decision from the FDA is pending (64).

Provir (SP-303) is a natural product polyphenolic polymer of average molecular weight of 2100 daltons. The compound exhibits activity against a range of virus serotypes (IC_{50} vs. RSV 2-10µM, therapeutic ratio 2-7). Provir inhibits viral penetration (IC_{50} 0.48µM) indicating the latter is at least one of the modes of action. Provir inhibited RSV in cotton rats either by the i.p. route or by small-particle aerosol treatment but the therapeutic ratio was approximately 10 (65,66). Despite this low therapeutic ratio in animal models, Provir has been reported to be safe in Phase 1 clinical trials (67).

Rhinoviruses - Human rhinoviruses (HRV's) are a major cause of the common cold. A series of anti-HRV agents act by binding to a specific hydrophobic pocket within the virion capsid protein VP1 (68). A number of these compounds possess potent *in vitro* activity against a range of HRV serotypes, e.g. pirodavir (**26**) (69) and WIN 54954 (**27**) (70,71).

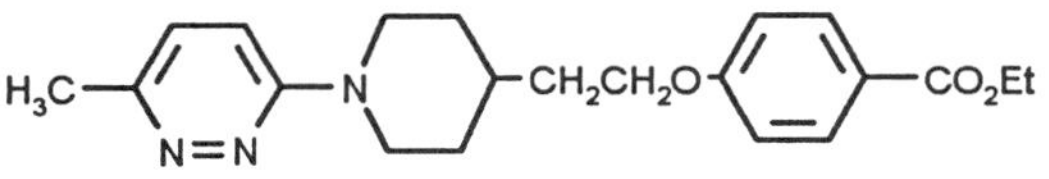

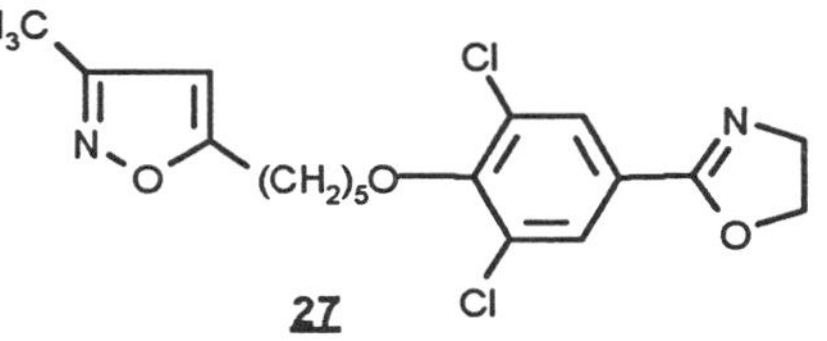

Pirodavir is a potent inhibitor (IC_{50} ~0.064µg/ml) of a range of HRV serotypes, and intranasal prodavir is protective against experimental RSV illness. However, in a randomised double blind, placebo-controlled trial in naturally occurring colds, prodavir reduced virus shedding but did not offer any clinical benefits (72). WIN 54954 is one of a class of oxazolines derivatives and has activity against a broad range of rhinovirus (MIC ~0.28µg/ml) and enterovirus (MIC ~0.06µg/ml) serotypes. However, two clinical trials using oral WIN 54954 in prophylaxis of experimental rhinovirus infection failed to demonstrate significant antiviral or clinical effect. Although trough plasma levels of WIN 54954 were above the MIC's for the serotypes used, drug levels in nasal washes and saliva were below 1 - 2 % of those in plasma. The lack of activity was attributed to insufficient drug at the site of infection. (70) Further examples of this type of compound have been reported (73,74). In a study to find water-soluble binders of picornavirus VP1, SCH 38057 (**28**) was shown to bind in the same pocket as the WIN compounds and to inhibit HRV 14 growth (IC_{50} 27.6µM). However, this compound was shown act at a later stage of viral replication than the WIN compounds (75).

28

Papillomavirus - Papillomaviruses cause a range of benign and malignant diseases in both animals and humans. Of the human papillomaviruses (HPV), certain types are frequently associated with anogenital warts and cancers. However, no specific antiviral agents exist and current treatment consists primarily of surgical removal or antimitotic agents. Until recently the search for anti-papilloma agents has been hampered by the lack of *in vitro* systems that duplicate epithelial differentiation. However, recent advances have allowed the propagation of HPVs *in vitro* (76). ISIS 2105 is a 20-residue phosphorothioate oligonucleotide targeted to the translational initiation of both HPV types 6 and 11 E2 mRNA. The compound inhibits E2-dependent transactivation by HPV type 11 E2 expressed from a surrogate marker (IC_{50} 5-7µM) (77). ISIS 2105 is currently undergoing clinical trails for treatment of HPV-associated genital warts (78).

The papillomavirus E1 protein has recently been identified as an ATP-dependent DNA helicase (79). This is the only papillomavirus gene product with a known catalytic function, and may have potential as a target for future drug discovery.

References

1.   E. De Clercq, Antimicrob. Agents Chemother., 32, Suppl. A, 121 (1993).
2.   A. Lapucci, M. Macchia and A. Parkin, Farmaco.,48, 871 (1993).
3.   P. Collins, Annals of Medicine, 25, 441 (1993).
4   L.H. Golberg, R. Kaufman, T.O. Kurtz, M.A. Conant, J.L. Eron, R.L. Batenhorst and G.S. Boone, Arch. Dermatol., 129, 582 (1993).
5.   D.J.M. Purifoy, L.M. Beauchamp, P. Demiranda, P. Ertl, S. Lacey, G. Roberts, S.G. Rahim, G. 6. Darby, T.A. Krenitsky and K.L. Powell, J. Med. Virol., 93, (Suppl. 1), 139 (1993).
6.   M.R. Boyd, S. Safrin and E.R. Kern, Antiviral Chem. Chemother., 4, (Suppl. 1), 3 (1993).
7.   Marketletter, Dec 20, 1993.
8.   A. Holy in "Isopolar Phosphorus-Modified Nucleotide Analogues in: Advances in Antiviral Drug Design," Vol. 1, E. De Clercq, Ed., JAI Press Inc., Greenwich, C.T., 1993, p. 179 .
9.   L.W. Drew, J.P. Lalezari, E. Glutzer, J. Flaherty, J.C. Martin, J.P. Fisher and H.S. Jaffe, Antiviral Research, 20 (Suppl. 1), 55 (1993).
10.   E. De Clercq, Reviews in Med.Virol., 3, 85, (1993).
11.   D.L. Barnard, J.H. Huffman, R.W. Sidwell and E.J. Reist, Antiviral Research, 22, 77 (1993).
12.   I.Verheggen, A.V. Aerschot, S.Toppet, R. Snoeck, G. Janssen, J. Balzarini, E. De Clercq and P. Herdewijn, J. Med. Chem., 36, 2033 (1993).
13.   J.A. Tino, J.M. Clark, A.K. Field, G.A. Jacobs, K.A. Lis, T.L. Michalik, B. McGeever-Rubin, W.A. Slusarchyk, S.H. Spergel, J.E. Sundeen, A.V. Tuomari, E.R. Weaver, M.G. Young and R. Zahler, J. Med. Chem., 36, 2033 (1993).

14.   J. Clement, J. Alder, K. Marsh, D. Norbeck, W. Rosenbrook, T. Herrin, C. Hartline and E. Kern, Antiviral Research, 20 (Suppl. 1),108 (1993).

15.   A. D. Borthwick, K. Biggadike, I.L. Paternoster, J.A.V. Coates and D.J. Knight, Bioorg. Med. Chem. Lett., 2577 (1993).

16.   M.P. Wentland, R.B. Perni, P.H. Dorff, R.P. Brundage, M.J. Castaldim, T.R. Bailey, P.M. Carabateas, E.R. Bacon, D.C. Young, M.G. Woods, D. Rosi, M.L. Drozd, R.K. Kullnig and F.J. Dutko, J. Med. Chem., 36, 1580 (1993).

17.   D.M. Lambert, S. Barney, R. Wittrock, H.S. Allaundeen, M. Massare, W. Kinsbury, D. Berges, G. Gallagher, J. Taggart, G. Hofmann, M. Mattern, R. Hertzberg, R. Johnson and S.R. Petteway Jr., Antiviral Research, 20 (Suppl. 1),140 (1993).

18.   M. Liuzzi and R. Deziel in "The Search for Aniviral Drugs: Case Histories from Concept to Clinic," J. Adams and V.J. Merluzzi, Eds., Birkhauser, Boston MA, 1993 p. 225.

19.   N. Moss, R. Deziel, J. Adams, N. Aubry, M. Bailey, M. Baillet, P. Beaulieu, J. Dimaio, J.S. Duceppe, J.M. Ferland, J. Gauthier, E. Ghiro, S. Goulet, L. Grenier, P. Lavallee, C. Lepinefrenette, R. Plante, S. Rakhit, F. Soucy, D. Wernic, and Y. Guindon, J. Med. Chem., 36, 3005 (1993).

20.   F. Focher, A. Verria, S. Spadari, R. Manservigi, J. Gambino and G.E. Wright, Biochem. Journal, 292, 883 (1993).

21.   A.R. Welch, A.S. Woods, L.M. McNally, R.J. Cotter and W. Gibson, Proc. Natl. Acad. Sci. U.S.A., 88, 10792 (1991).

22.   F.J. Rixon, Seminars in Virology, 4, 135 (1993).

23.   V.G. Preston, J.A.V. Coates and F.J. Rixon, J. Virology, 45, 1056 (1983).

24.   A.R. Welch, L.M. McNally, R.T. Hall and W. Gibson, J. Virology, 67, 7360 (1993).

25.   W. Gibson, L.M. McNally, A.R. Welch, M.R.T. Hall, D. Smith, M. Wakulchick and E.C. Villarreal, in "Multidisciplinary Approach to Understanding Cytomegalovirus Disease," S. Michelson and S.A. Plotkin, Eds., Elsevier Science Publishers B.V.; Amsterdam, 1993 p. 21.

26.   F. Liu and B. Roizman, Proc. Natl. Acad. Sci. U.S.A., 89, 2076 (1992).

27.   F. Liu and B. Roizman, J. Virology, 67, 1300 (1993).

28.   C.L. Dilanni, D.A. Drier, I.C. Deckman, P.J. McCann, F. Liu, B. Roizman, R.J. Colonno and M.G. Cordingley, J.Biol.Chem., 268, 25449 (1993).

29.   J.L. Tonkinson and C.A. Stein, Antiviral Chem. Chemother., 4, 193 (1993).

30.   R.F. Azad, V.B. Driver, K. Tanaka, R.M. Crooke and K.P. Anderson, Antimicrob. Agents Chemother., 37, 1945 (1993).

31.   G-Q. Yao, S. Grill, W. Egan and Y-C. Cheng, Antimicrob. Agents Chemother., 37, 1420 (1993).

32.   R. Snoeck, G. Andrei, J. Neyts, D. Schols, M. Cools, J. Balzarini and E. De Clercq, Antiviral Research, 21, 197 (1993).

33.   J.T. Matthews, B.J. Terry and A.K. Field, Antiviral Research, 20, 89 (1993).

34.   D.L.J. Tyrrell, M.C. Mitchell, R.A. DeMan, S.W. Schalm, J. Main, H.C. Thomas, J. Fevery, F. Nevens, P. Beranek and C. Vicary, Hepatology, 18, 112A (1993).

35.   Clin-Pharm, 12, 715 (1993).

36.   K. Tsiquaye, D. Sutton, M. Maung and M.R. Boyd, Proc. 33rd ICAAC, October 1993, p. 1594.

37.   C.W. Young, R. Schneider, F.S. Leyland-Jones, D. Armstrong, C.T.C. Tan, C. Lopez, K.A. Watanabe, J.J. Fox and F.S. Philips, Cancer Res., 34, 5006 (1983).

38.   T.M. Hooton, L. Dejarnette, T. Tartaglione, D. Paar, T. Jones, J. Santangelo, K.A. Smiles, S. Straus, D.D. Richman and L. Corey, Proc. 31st ICAAC, October 1991,p. 302.

39.   P.A. Furman, M. Davis, D.C. Liotta, M. Paff, L.W. Frick, D.J. Nelson, R.E. Dornsife, J.A. Wurster, L.J. Wilson and J.A. Fyfe, Antimicrob. Agents Chemother., 36, 2686 (1992).

40.   H.L. Janssen, L. Berk, R.A. Heijtink, F.J. ten Kate and S.W. Schalm, Heptology, 17, 383 (1993).

41.   R.A. Heijtink, G.A. De Wilde, J. Kruining, L. Berk, J. Balzarini, E. DeClercq, A. Holy and S.W. Schalm, Antiviral Res., 21, 141 (1993).

42   G.M. Dusheiko and A.J. Zuckerman, Antimicrob. Chemother., 32 (Suppl A), 107 (1993).

43   I. Rezakovic, C. Zavaglia, R. Bottelli and G. Ideo, Hepatology, 18, 252A (1993).

44   E. DeAlava, J. Camps, J. Pardo-Mindan, M. Garcia-Granero, J. Sola, M. Munoz, M.P. Civeira, F. Contreras, J. Vazquez, A. Castilla and J. Prieto, Liver, 13, 73 (1993).

45.   S. Kakumu, K. Yoshioka, T. Wakita, T. Ishikawa, M. Takayanagi and Y. Higashi, Hepatology, 18, 258 (1993).

46.   J.I. Esteban, J. Genesca and H.J. Alter, Prog. Liver Dis., 10, 253 (1992).

47   Antiviral Agents Bulletin, 6, 257 (1993).
48   C.R. Penn, D.M. Ryan, J.M. Woods, R.C. Bethell, V.J. Hotham, P. Colman, W.-Y. Wu and M. von Itzstein, 32nd ICAAC, Abst. 1325 (1992).
49   M. von Itzstein, W.-Y. Wu, G.B. Kok, M.S. Pegg, J.C. Dyason, B. Jin, T.v. Phan, M.L. Smythe, H.F. White, S.W. Oliver, P.M Colman, J.N. Varghese, D.M. Ryan, J.M. Woods, R.C. Bethell, V.J. Hotham, J.M. Cameron and C.R. Penn, Nature, 363, 418 (1993).
50   J.M. Woods, R.C. Bethell, J.A.V Coates, N. Healy, S.A. Hiscox, B.A. Pearson, D.M. Ryan, J. Ticehurst, J. Tilling, S.M. Walcott and C.R. Penn, Anitmicrob. Agents Chemother., 37, 1473 (1993).
51   C.T. Holzer, M. von Itzstein, B. Jin, M.S. Pegg, W.P. Stewart and W.-Y. Wu, Glycoconjugate J., 10, 41 (1993).
52   J.V. Tuttle, M. Tisdale and T.A. Krenitsky, J. Med. Chem., 36, 119 (1993).
53   M. Tisdale, G. Appleyard, J.V. Tuttle, D.J. Nelson, S. Nusinoff-Lehrman, W. Al Nakib, J.N. Stables, D.J.M. Purifoy, K.L. Powell and G. Darby, Antiviral Chem. Chemother., 4, 281 (1993).
54   B.S. Rollins, A.H.A. Elkhatieb and F.G. Hayden, Antiviral Res., 21, 357 (1993).
55   D.L. Bodian, R.B. Yamasaki, R.L. Buswell, J.F. Stearns, J.M. White and I.D. Kuntz, Biochemistry, 32, 2967 (1993) and refs. within.
56   C.M. Carr and P.S. Kim, Cell, 23 (1993).
57   R. Bron, A.P. Kendal, H.D. Klenk and J. Wilschut, Virology, 195, 808 (1993).
58   F.W. Tse, A. Iwata and W. Almers, J. Cell Biol., 121, 543 (1993).
59   C. Schoch and R. Blumenthal, J. Biol. Chem., 268, 9267 (1993).
60   O. Gutman, T. Danieli, J.M. White and Y.I. Henis, Biochemistry, 32, 101 (1993).
61   J.H. Huffman, R.W. Sidwell, A.P. Gessman, B.J. Moscone, J.Y. Tang and S. Agrawal, Antiviral Res., 20 (suppl 1), 152 (1993).
62   N.B. Ledovskikh, L.V. Yurchenko, G.A. Nevinskii, E.I. Frolova, E.M. Ivanova, A.A. Koshkin, N.A. Bulychev, V.F. Zarytova and V.V. Vlasov, Mol. Biol., 26, 490 (1992).
63   M. Hosoya, S. Shigeta, T. Ishii, H. Suzuki and E. De Clercq, J. Infect. Diseases, 168, 641 (1993).
64.   Antiviral Agents Bulletin 6, 353 (1993).
65.   D.L. Barnard, J.H. Huffman, L.R. Meyerson and R.W. Sidwell, Chemotherapy, 39, 212 (1993).
66.   B.E. Gilbert, P.R. Wyde, S.Z. Wilson and L.R. Meyerson, Antiviral Res., 21, 37 (1993).
67.   Scrip 1861, 24 (1993).
68.   D.H. Sheppard, B.A. Heinz and R.R. Rueckert, J. Virology, 67, 2254 (1993).
69.   K. Andries, B. Dewindt, J. Snoeks, R. Willebrords, K. van Eemeren, R. Stokbroekx and P.A.J. Janssen, Anitmicrob. Agents Chemother., 36, 100 (1992).
70.   R.B. Turner, F.J. Dutko, N.H. Goldstein, G. Lockwood and F.G. Hayden, Anitmicrob. Agents Chemother., 37, 297 (1993).
71.   G.D. Diana, T.J. Nitz, J.P. Mallamo and A. Treasurywala, Antiviral Chem. Chemother., 4, 1 (1993).
72.   F.G. Hayden, G. Eisen, M. Janssens, K. Andries and P. Janssen, Antiviral Res., 20 (suppl 1), 160 (1993).
73.   G.D. Diana, D.M. Volkots, D. Cutcliffe, R.C. Oglesby, T.R. Bailey, J.P. Mallamo, T.J. Nitz and D.C. Pevear, Antiviral Res., 20 (suppl 1), 100 (1993).
74.   G.D. Diana, US Patent 5,242,924 (1993).
75.   E. Rozhon, S. Cox, P. Buontempo, J. O'Connell, W. Slater, J. De Martino, J. Schwartz, G. Miller, E. Arnold, A. Zhang, C. Morrow, S. Jablonski, P. Pinto, R Versace, T. Duelfer and V. Girijavallabhan, Antiviral Res., 21 15 (1993).
76.   L.A. Laimins, Infect. Agents Dis., 2, 74 (1993).
77.   L.M. Cowsert, M.C. Fox, G. Zon and C.K. Mirabelli, Anitmicrob. Agents Chemother., 37, 171 (1993).
78.   Antiviral Agents Bulletin, 6 42 (1993).
79.   L. Yang, I. Mohn, E. Fonts, D.A. Lim, M. Nohaile and M. Botcham, Proc. Natl. Acad. Sci. U.S.A., 90, 5086 (1993).

Chapter 16. Problems and Progress in Opportunistic Infections

Robert C. Goldman and Larry L. Klein
Pharmaceutical Research Division
Abbott Laboratories
Abbott Park, IL 60064-3500

Introduction - The term 'opportunistic infection' is usually used to delineate virulent microorganisms, capable of causing infection in a normal host, from less virulent microorganisms, which usually require an immune deficit in order to establish infection. Although this serves to focus attention on a defined opportunity, it eliminates the equally important parameters of altered patterns of contact between man and animals, land use, travel, behavior, infrastructure deterioration, population growth and density, hospital settings, pathogen emergence, and even global warming as factors defining opportunities for infection (1). This chapter will none-the-less adhere primarily to the classical definition of opportunistic infection in order to focus on the class of microorganisms which usually do not cause disease, or at least not severe disease, in the normal host, but rather require an immune deficit in the host (e.g. inherited disease, aging, premature infancy, HIV infection, radiation and chemotherapy, immunesuppressive drugs during transplantation, malnutrition, pregnancy, severe trauma, burns, concurrent infection or malignancy). It will further more focus on infections due to fungi, bacteria of the genus *Mycobacterium*, and the protozoan *Toxoplasma gondii*, as each serves as a timely and representative example of opportunists from diverse evolutionary backgrounds, and exemplifies problems and progress in discovery and development of new therapeutic regimes.

FUNGAL INFECTIONS

Infections due to fungi are a major cause of morbidity and mortality in immunocompromised hosts, and efforts in drug discovery and development have continued.

Cispentacin - Cispentacin ($\underline{1}$, where X = CH_2) is weakly active *in vitro*, but was quite effective in animal models of *C. albicans* infection (2, 3); it was inactive against *Aspergillus fumigatus in vivo,* but active against *Cryptococcus neoformans*. This agent exhibited excellent oral activity with low toxicity, which stimulated further interest in determining its mode of action. Intracellular concentration reached mM levels due to active transport by the proline permease, and to a lesser extent other amino acid permeases in *C. albicans*, and at these high concentrations charging of prolyl tRNA was inhibited (4). Since expression of amino acid permeases is regulated, composition of test media could influence uptake and thus activity. Recently, a set of patents (5) have described the tetrahydrofuran (X=O), tetrahydrothiophene (X=S), and pyrrolidine (X=NCO_2Me) analogs along with a dihydropyran compound, none of which were shown to exhibit any biological or pharmacological advantages over cispentacin. No data are yet available on the clinical status of cispentacin or these related structures.

Pradimicins - The mechanism of action of another series of promising anti-fungal drugs, the pradimicins, was also elucidated (6). Binding to mannoprotein in the fungal cell wall appears to the primary event leading to severe morphological alterations and fungicidal action, and such binding might inhibit the action of 'morphogenetic' mannoproteins involved in cell wall assembly. Pradimicins are broad spectrum, and are active against *Candida sp.*, *Aspergillus fumigatus*, and *Cryptococcus neoformans* in animal models (7, 8), and thus are being considered for clinical development. Along with the isolation and characterization of new naturally occuring analogs (9, 10), several synthetic programs are underway with the goal of producing analogs of pradimicin A (**2**) with improved potency and/or water solubility. Studies regarding structure-activity relationships at the C-15 amino acid sidechain (11, 12), the sugar portion (13), the C-4' amine (14), such as the n-cyano analog (**3**) and the C-11 position (15, 16), have led to analogs having equal or improved activity to pradimicin A. Data from a recent study replacing the C-4' amino group with a hydroxyl group has established that the amine function is not required for activity (17). In fact, the hydroxyl analog, BMS-181184 (**4**), shows lower toxicity, good antifungal activity and a similar spectrum to pradamycin A in addition to exhibiting greater water solubility .

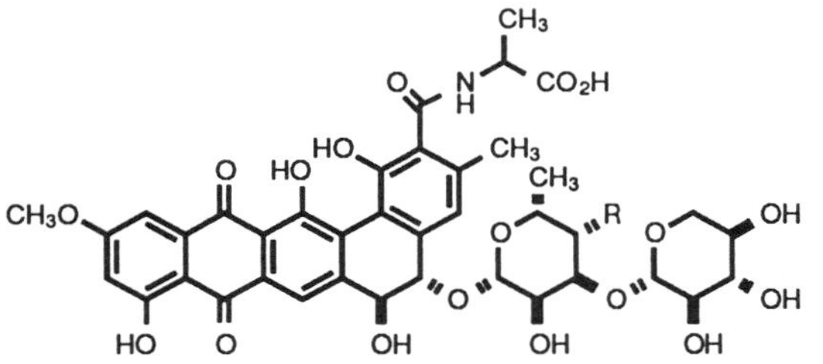

2 R = NHCH₃
3 R = NHCN
4 R = OH

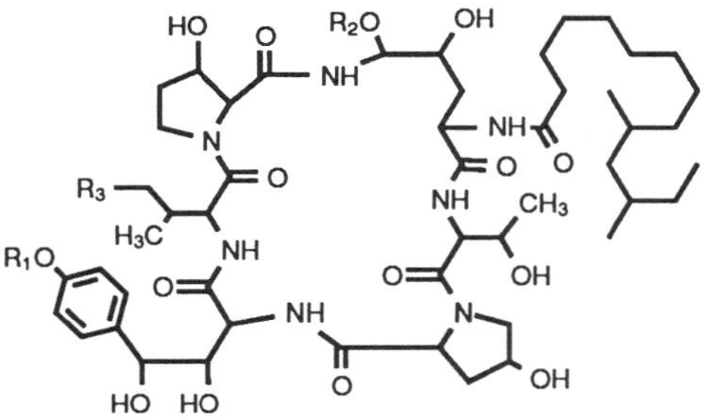

5 R₁ = PO₃H⁻ Na⁺; R₂ = H; R₃ = CONH₂
6 R₁ = H; R₂ = CH₂CH₂NH₂; R₃ = CH₂NH₂
7
n-C₅H₁₁O

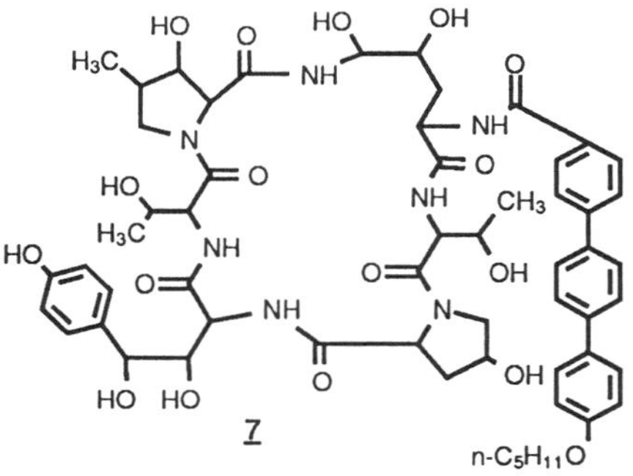

Lipopeptides - The antifungal activity of one class of cyclic lipopeptides results from inhibition of ß-1,3 glucan synthesis, followed by a tighlty coupled shut down of protein synthesis (18). Although the first clinical candidate, cilofungin, was withdrawn from clinical trials, efforts have continued to develop this class of agent (19). Structure-activity data concerning the homotyrosine residue were reported (20) and showed that the phenolic oxygen was required for antifungal activity. Phosphorylation of this hydroxyl group afforded prodrug **5** having improved water solubility, and **5** was shown to be readily cleaved *in vivo* to the parent compound (21). Another SAR study was reported which replaced various amino acids in the echinocandin system with simplified residues via a solid-phase approach, and several active compounds were prepared (22). In addition, water soluble cationic derivatives of the pneumocandins were prepared by reducing the unique glutamate residue to an amine. In combination with this modification, the hemiaminal function was also stabilized via formation of an aminoethyl aminal (**6**) at this position (23). Both modifications resulted in increased solubility and equal or improved *in vitro* and *in vivo* activities; in addition, the aminals also possessed excellent activity againat *Aspergillus*. An SAR study of the fatty acid sidechain of the dihydroxyornithine residue demonstrated that optimization of the structure of this chain was found to involve a terphenyl linking unit (LY303366) (24). The homotyrosine moiety of LY303336 was phosphorylated to obtain a water soluble analog of this compound. This agent (**7**) was shown to have an expanded spectrum, rapid cidal activity, and oral activity in a mouse model of *C. albicans* (24-26).

Aureobasidins - The naturally-occuring novel cyclic depsipeptide aureobasidin A (**8**) possesses broad spectrum fungicidal activity *in vitro* and *in vivo*, though less so against

8

Aspergillus (27). Although it consists of eight L-form amino acids in a cyclic array, unlike the echinocandins it does not bear a fatty acid side chain. Recently, new congeners were isolated and characterized, several of which exhibited potency similar to aureobasidin A (28). Along with structural information from previous studies (29), this SAR data suggests that analogs of greater potency are possible. Methods for chemical synthesis of aureobasidin A were described and could be utilized for analog preparation (30). The mode of action of this orally active agent is currently under study.

Azoles - Fluconazole was an important addition to anti-fungal therapy and prophylaxis in patients at risk. Although not yet a major clinical problem, strains resistant to azoles can cause primary infection, or be selected for *in situ* during prolonged therapy with azoles (31-39). Resistant strains were isolated from AIDS patients treated with fluconazole, and since the karotypes of resistant strains from three patients attending the same hospital were identical, nosocomial spread was indicated (31). More fluconazole resistant *C. albicans* were isolated from the oropharynx of HIV positive patients with symptoms of candidiasis and history of fluconazole therapy, compared to patients not receiving fluconazole (32). In this study *in vitro* MIC testing was predictive of the clinical outcome, and azole resistance was considered a common problem in HIV positive patients (32). In a separate study, most instances of fluconazole resistance in AIDS patients with oral candidiasis still responded to therapy (33). Resistance may arise by selection of strains which have lost transport function (34), and/or which have alterations in target site susceptibility (40-42). New data on lanosterol C-14 demethylase inhibitors, and/or new derivatives were also described. Itraconazole (which has received approvals) is being considered for use in cases of fluconazole resistance (43), cryptococcal, histoplasmal, and pulmonary aspergillosis infections (44). In addition, broad spectrum triazole D-0870 showed superior activity compared to fluconazole *in vivo* against *Candida*, *Cryptococcus* and *Aspergillus* infections in normal and immunocompromised mice (45), and Ro 09-1470 showed a higher degree of selectivity for yeast P-450 compared to fluconazole (46).

Pneumocystis - Anti-*Pneumocystis carinii* agents PS-15, a folate antagonist with *in vivo* activity (47), Ro 11-8958, a trimethoprim analog active *in vivo* in combination with other drugs (48), 8-aminoquinolines, active *in vivo* (49), and topoisomerase inhibitors, active *in vitro* and *in vivo*, (50) were also described. In addition, certain lipopeptides have shown efficacy in animal models of *Pneumocystis* infection (19).

MYCOBACTERIA

Diseases - Over 20 species are documented to cause disease in humans, including *Mycobacterium tuberculosis, bovis, leprae, avium-intracellulare, kansasii, fortuitum, chelonae, scrofulaceum, xenopi , szulgai, malmoense , simiae, marinum, ulcerans* and *haemophilum*. *M. tuberculosis* kills more people world wide than any other disease ,with 8 million new cases each year, resulting in 2.6 million deaths per year. Although disease will occur in one of ten healthy individuals infected, disease in more probable, severe and lethal in immunocompromised individuals (1). *M. avium-intracellulare* also causes pulmonary disease similar to tuberculosis, with disseminated disease occurring in immunocompromised hosts, particularly AIDS patients. *M. avium* is ubiquitous in the environment, and ingestion with food or water may be an important mechanism of transmission. In addition to the increase in tuberculosis cases (especially those caused by multiply-drug-resistant [MDR] strains), and *M. avium* cases in immunocompromised hosts, an upsurge in *M. avium* infections (to 25-50% of cases) in apparently normal hosts has occurred (51). Thus, there appears to be a new opportunity taken advantage of by this organism.

The Cell Integument - Although the structure, function and synthesis of the complex mycobacterial wall structure is still poorly understood, certain key observations do reflect on the relationship of the wall to resistance to antibiotics. (52, 53). The trilayered cell wall is lipid rich and contains phospholipids, glycolipids, peptidoglycan and protein. The arabinogalactan matrix serves as a bridge between peptidoglycan and the mycolic acids, which represent 40% of the wall dry weight. X-ray diffraction measurements indicate that the mycolic acids are arranged perpendicularly to the wall plane, likely forming an asymmetric bilayer which restricts hydrophobic permeation (54). Thus, the cell wall of *M. chelonae* is several thousand fold less permeable to cephalosporins compared to the wall of *Escherchia coli* (55). The recent discovery of small amounts of low specific activity porin molecules in the mycobacterial wall (56) now appears to explain slow diffusion of small hydrophilic molecules, including some antibiotics. A better understanding of the permeability barrier will be useful in designing effective delivery strategies for anti-mycobacterial drugs.

Traditional Agents and Resistance - The traditional agents for therapy of tuberculosis and other mycobacterial infections include rifampicin, aminoglycosides, ethambutol, ethionamide, isoniazide, and pyrazinamide (57). The discovery and development of new agents, optimum treatment modalities, and strategies for dealing with resistance (58), are all essential elements of a coordinated approach to dealing with not only opportunistic mycobacterium infections in immunocompromised patients, but also with the very real problem of MDR tuberculosis which is taking advantage of opportunities created by world socioeconomics.

Rifamycins - Rifampicin inhibits RNA polymerase from sensitive mycobacterial strains (59). The gene for RNA polymerase (*rpoB*) from *M. tuberculosis* was sequenced, and information used to examine the molecular mechanism of rifampicin resistance (60). Among a group of 66 resistant isolates, 15 different mutations were detected, with only two resistant isolates showing no change in a 411 bp sequenced region. Likewise, the *rpoB* gene of *M. leprae* was the site of mutational resistance to rifampicin (61) in nine of nine resistant strains examined, with point mutations at ser-425 being the most prevalent. These and other methods may allow more rapid determination of drug susceptibility in the

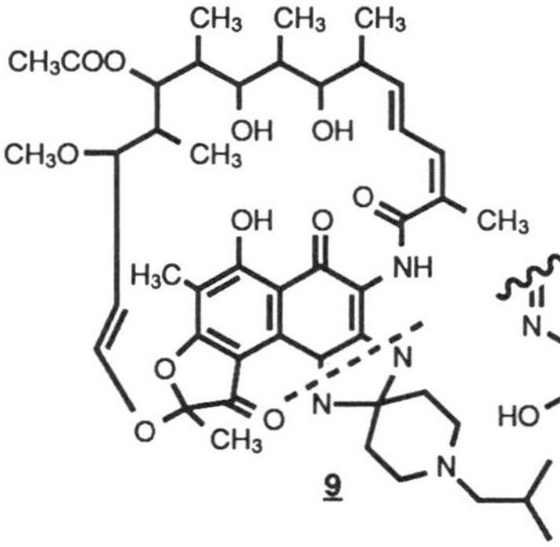

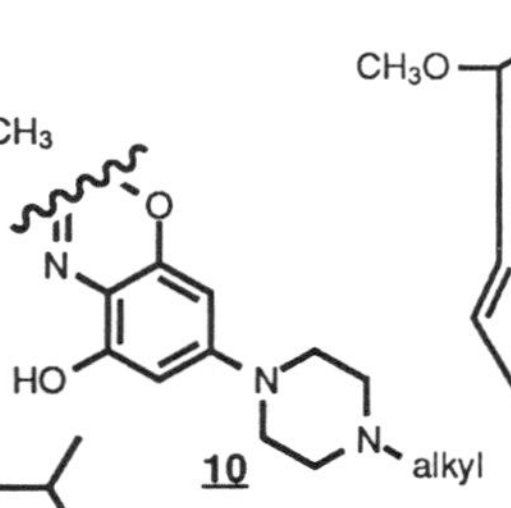

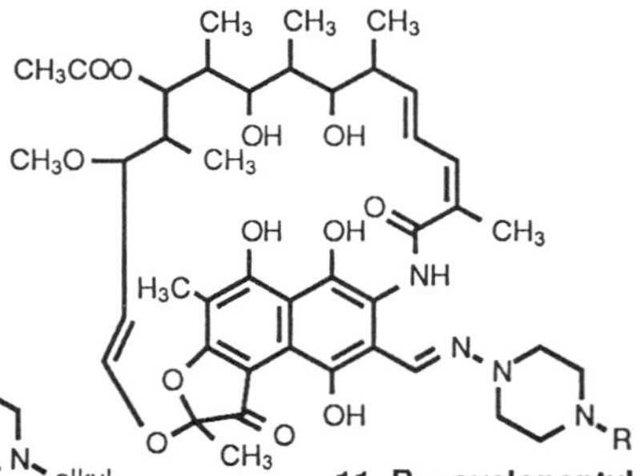

increasing background of MDR strains, thus allowing more rapid initiation of appropriate drug therapy (62-64). Several new rifampicin analogs were described and examined for activity against mycobacterial species. Rifabutin (**9**), or Mycobutin (65), was introduced for the treatment and prevention of *M. avium-intracellulare* infection in AIDS patients, and two other derivatives, rifapentine (**11**) (66), and the benzoxazinorifamycins (**10**) (67), were able to treat and prevent *M. avium* infection in animal models (68-70).

Aminoglycosides - Resistance to streptomycin at the ribosomal level was demonstrated in *M. tuberculosis*, and ribosomal protein changes were observed in some isolates (71). In addition, an A to G mutation at residue 866 of 16S rRNA gave rise to streptomycin resistance in 2 of 11 streptomycin resistant clinical isolates *M. tuberculosis* (72). Mutations at residues 491, 512, 904 (73) and 2058 (74) were also found, as well as alterations in ribosomal protein S12 (73). The existence of only a single set of rRNA genes in the slow growing mycobacterial species is a likely explanation for the relatively high frequency of selection for aminoglycoside resistance. Although aminoglycoside modifying enzymes were detected, their relationship to plasmids and resistance remains unclear (75). Thus resistance in the clinical setting appears at present to be due to selection for mutationally acquired resistance, and related to organism density and long duration of anti-mycobacterial therapies. Newer aminoglycosides (SCH 21420 and SCH 22591) exhibited good activity against several faster growing mycobacterial species (76).

Isoniazide - Although used for over thirty years, the precise mode of action and mechanism of resistance to isoniazide (**12**) are still not fully understood (58). Anti-

CONHNH₂ mycobacterial action of isoniazide (INH) involves intracellular accumulation of the drug, possible metabolic activation and antagonism of action, and susceptibility of target sites. Although synthesis of mycolic acids is specifically inhibited by INH, and not its known metabolites isonicotinic acid, 4-hydroxymethylpyridine, and isonicotinamide (77), other mechanisms may

12 be involved in anti-mycobacterial action (58). The importance of INH in anti-mycobacterial drug therapy has prompted investigation of the mechanism by which strains become resistant.

The role of *katG* encoded catalase-peroxidase activity in sensitivity and resistance to isoniazide was recently investigated in detail (78). Expression of *katG* activity is high in *M. tuberculosis*, one of the most sensitive of mycobacterial species. The *katG* gene codes for an 80 kD portion with sequence homology to other peroxidase enzymes (79), including yeast cytochrome c peroxidase for which X-ray crystal structure data are available. Many INH resistant strains show decreased catalase-peroxidase, the assumption being that *katG* is involved in uptake and/or activation of INH (58). In addition, some resistant strains may have target site mutations (e.g. in the sensitive mycolic acid synthetic factor) and work in this areas may ultimately define the critical target site, and thus lead to development of newer, more specific agents. Since INH resistant, catalase negative strains of *M. smegmatis* and *M. tuberculosis* simultaneously gained catalase activity and sensitivity to INH when transformed with the *katG* gene cloned from *M. tuberculosis*, loss of *katG* activity was demonstrated to be involved in at least one form of INH resistance (78, 80). However, the situation is further complicated by a recent study of clinical isolates of *M. tuberculosis* from New York city, where the rate of INH resistance has approached 30% (81). In this study, primers based on the published *katG* sequence were used for PCR analysis of 80 isolates (39 sensitive and 41 resistant to INH), and the results showed that resistance did not always correlate with deletion of the *katG* gene (81). Although previous work revealed that 2 of 3 high level INH resistant mutants had lost the *katG* gene (82), these combined data point out that although deletion of *katG* can lead to resistance, point mutations in *katG* and target site mutations (possibly in the mycolic acid synthetic pathway) likely contribute to the spectrum of resistance. The mechanism of action and resistance to this important anti-mycobacterial drug deserves additional study in order to define possible routes to overcoming the resistance mechanism.

Pyrazinamide - Although the precise mode of action and mechanism of resistance of pyrazinamide (PZA) (**13**) are unknown, generation of pyrazinoic acid (**14**) by mycobacterial pyrazinamidase seems to be essential for susceptibility (83, 84). Pyrazinoic acid esters were synthesized, based on the assumption that pyrazinoic acid was the active component (85). Several were active against a PZA resistant strain of *M. tuberculosis*, and certain derivatives gained activity against naturally resistant *M. bovis* and *kansasii*.

The n-propyl pyrazinoate derivative had MIC values of 3-6 µg/ml at pH 5.8, showed activity against resistant strains, and was tolerated by mice at 900 mg/kg in single dose toxicity tests. In addition, N-acylpyrazinamide derivatives were synthesized (86), and one derivative, N-palmitoyl-pyrazinamide (**15**), was active against *M. avium* when incorporated into liposomes.

13 R = NH$_2$
14 R = OH
15 R = NHCO(CH$_2$)$_{15}$CH$_3$

Mycolic acid synthesis - Mycolic acid synthesis may be the primary site of action of INH, ethambutol, and ethionamide (77). Ethambutol is used in combination therapy, and may have a primary effect on arabinogalactan synthesis (87), with secondary effects on mycolic acids metabolism, possibly due to similarities in the structure of ethambutol and trehalose monomycolate (88). Ethionamide is related to INH in structure and possibly mode of action, since it inhibited mycolic acid synthesis in mycobacterial strains (89). The existence of *in vitro* systems for measuring synthesis of mycolic acids (90, 91) represents an important tool for gaining the precise biochemical data required for drug discovery and development directed towards this critical metabolic pathway, as was attempted using anti-metabolites based on mycolic acid structure (92).

Quinolones - Several of the newer quinolones have been investigated for activity against mycobacterial infections, including ciprofloxacin in combination therapy for *M. avium* infection in AIDS (93), sparfloxacin (94), ofloxacin (94, 95), and perfloxacin (95) in human infections by *M. leprae*. An extensive study of the *in vitro* activity of 88 quinolones against *M avium* was conducted and the data analyzed so as to identify the most active and broad spectrum compounds (96, 97). Of these analogs, PD125354 (**16**) was found to be optimal and awaits further study *in vivo*. Quinolone AM-1155 (**17**) was also found to be equal to, or better than sparfloxacin and olfloxacin against various *Mycobacterium sp.* (98). A number of newer quinolones were tested for *in vitro* activity against *M. fortuitum, chelonae*

and other fast growing species (76), and while several had potent activity against *M. fortuitum* and *Mycobacterium sp.* (PD 117596, 127391, and 117558), they showed poor activity against *M. chelonae*. Further discovery and development studies with such agents may

16 R$_1$ = H$_3$C-N   N-$ R$_2$ = F
17 R$_1$ = HN   N-$ R$_2$ = OCH$_3$

improve the clinical outlook, especially in the treatment of mycobacterial strain resistant to traditional agents.

Macrolides - Among the newer macrolide antibiotics, azithromycin (99), clarithromycin (100), and roxithromycin (101-103) showed promising *in vitro* activity against *M. avium*. Clarithromycin and azithromycin showed efficacy in *in vivo* animal models, and have advanced to clinical trials. Azithromycin resolved symptoms and reduced bacteremia in an uncontrolled study of *M. avium* infection in AIDS patients (104). More clinical data are available for clarithromycin, which has recently gained approval for use in treating *M. avium* infection . Several clinical trails showed that clarithromycin was effective alone or in combination with other drugs in the treatment of *M. avium* infection in AIDS patients (105-109). Recent controlled studies examined therapy and prophylaxis, using clarithromycin as a single agent, or in combination with other traditional anti-mycobacterial agents. Clarithromycin was well tolerated in combination with other anti-mycobacterial drugs (110), and showed promising results in prophylaxis (111-113), and treatment in a study of 419 HIV-positive patents with disseminated *M. avium* infection (114). Although long term therapy with clarithromycin as a single agent was efficacious even against established *M. avium* infection, as with other anti-mycobacterial drugs single agent therapy in some cases leads to selection of resistant organisms. The spontaneous resistance frequency to azithromycin and clarithromycin was in the range of 10^{-6} to 10^{-10}, and likely involves a change in target site accessibility (115). Recently alteration at 23S rRNA position 2058 was observed in clarithromycin resistant *M. intracellulare*, which contains only a single 28S rRNA gene (116).

PROTOZOANS

Toxoplasma gondii -*T. gondii* is a protozoan which can cause disseminated disease, including encephalitis, especially in immunocompromised hosts (117, 118). As an obligate intracellular parasite *T. gongii* exists in a specialized non-acidified vacoule lined with proteinacious pores (size exclusion of about 1000 daltons) and filled with parasite encoded proteins (119). Recent advances in molecular genetic manipulation of *T. gondii* (120, 121) should assist in identification of new chemotherapuetic targets. Traditional therapy uses pyrimethamine (a dihydrofolate reductase inhibitor) plus sulfonamides (dihyropteroate reductase inhibitors); however, severe side effects can limit utility. Recently several highly selective and potent dihydrofolate reductase inhibitors were characterized, and several are scheduled for testing in animals (122). In addition, the gene for *T. gondii* dihydrofolate reductase-thymidylate synthase was cloned and sequenced, and unique structural properties of enzyme may provided useful information for the design of specific inhibitors (123). Additional antibiotics active against *T. gondii* include clindamycin (124), azithromycin (125), clarithromycin (126), roxithromycin (127) and spiramycin (128). All are effective *in vitro* and in animal models of toxoplasmosis. Although immune potentiation was considered as a mechanism of action of protein synthesis inhibitors, the presumed mechanism of action is inhibition of protein synthesis (128-130) based on *in vitro* data, and the fact that diverse structures are active; however, at present the precise site of anti-toxoplasma action is unknown. Recently, *T. gondii* mutants resistant to macrolides were isolated by mutagenesis (130). Although the *in vitro* sensitivity to macrolides was higher than the amount required for inhibition of growth in tissue culture cells, sensitivity was unchanged compared to the parental strain. Preliminary evidence indicates that cytoplasmic and mitochondrial ribosomes may not be the target site of protein synthesis inhibitors, and suggested that the apicocomplexan organelle, which contains rRNA sequences, should be considered (130). Clarithromycin plus pyrimethamine showed promising results in a small study of *Toxoplasma* encephalitis in AIDS patients (131); prophylactic trials are scheduled.

References

1.   "Emerging Infections: Microbial Threats to Health in the United States", J. Lederberg, R.E. Shope and S.C. Oaks, Eds., National Academy Press, Washington, D.C., (1992).
2.   T. Oki, M. Minoru, K. Tomatsu and K. Numata, J. Antibiotics, 42, 1756 (1989).
3.   T. Iwamoto, E. Tsjuii, M. Ezaki, A. Fujie, S. Hashimoto, M. Okuhara, M. Kohdaka and H. Imanaka, J. Antibiotics, 43, 1 (1990).
4.   J.O. Capobianco, D. Zakula, M.L. Coen and R.C. Goldman, Biochem. Biophys. Res. Comm., 190, 1037 (1993).
5.   F. Kunisch, J. Mittendorf, M. Plempel, P. Babczinski, H-C. Militzer, EP-538688,-89,-91,-92 (1993).
6.   T. Ueki, M. Oka, Y. Fukagawa and T. Oki, J. Antibiotics, 46, 465 (1992).
7.   M. Kakushima, S. Masuyoshi, M. Hirano, M. Shinoda, A. Ohta, H. Kamei, and T. Oki, Antimicorbiol. Agents Chemother. 35, 2185 (1991)
8.   T. Oki in "New Approaches for Antifungal Drugs," P.B. Fernandes, Ed., Birkhauser, Boston, MA, 1992, p. 64.
9.   K. Saitoh, Y. Sawada, K. Tomita, T. Tsuno, M. Hatori, and T. Oki, J. Antibiotics, 46, 387 (1993).

10.   T. Furumai, T. Hasegawa, M. Kakushima, K. Suzuki, H. Yamamoto, S. Yamamoto, M. Hirano, and T. Oki, J. Antibiotics, 46, 589 (1993).
11..   M. Nishio, H. Ohkuma, M. Kakushima, S-I. Ohta, S. Iimura, M. Hirano, M. Konishi, and T. Oki, J. Antibiotics, 46, 494 (1993).
12..   S. Okuyama, M. Kakushima, H. Kamachi, M. Konishi, and T. Oki, J. Antibiotics, 46, 500 (1993).
13..   S. Aburaki, H. Yamashita, T. Ohnuma, H. Kamachi, T. Moriyama, S. Masuyoshi, H. Kamei, M. Konishi, and T. Oki, J. Antibiotics, 46, 631 (1993).
14.   H. Kamachi, S. Okuyama, M. Hirano, S. Masuyoshi, M. Konishi, and T. Oki, J. Antibiotics, 46, 1246 (1993).
15.   S.Aburaki, S. Okuyama, H. Hoshi, H. Kamachi, M. Nishio, T. Hasegawa, S. Masuyoshi, S. Iimura, M. Konishi, and T. Oki, J. Antibiotics, 46, 1447 (1993);
16.   T. Furumai, H. Yamamoto, Y. Narita, T. Hasegawa, S. Aburaki, M Kakushima, and T. Oki, J. Antibiotics, 46, 1589 (1993).
17.   H. Kamachi, S. Iimura, S. Okuyama, H. Hoshi, S. Tamura, M. Shinoda, K. Saitoh, M. Konishi, and T. Oki, J. Antibiotics, 46, 1518 (1993).
18.   M. Coen, C. Lerner and R. Goldman, Microbiol. in press (1994).
19.   M. L. Hammond in "Cutaneous Antifungal Agents," J.W. Rippon and R.A. Fromtling, Eds., Marcel Dekker, New York, NY, 1993, p. 421.
20.   J.M. Balkovec, R.M. Black, G.K.Abruzzo, K. Bartizal, S. Dreikorn, and K. Nollstadt, Biorg. Med. Chem. Letters, 3, 2039 (1993).
21.   J.M. Balkovec, R.M. Black, M.L. Hammond, J.V. Heck, R.A. Zambias, G.K.Abruzzo, K. Bartizal, H. Kropp, C. Trainor, R.E. Schwartz, D.C. McFadden, K.H. Nollstadt, L.A. Pittarelli, M.A. Powles, and D.M. Schmatz, J. Med. Chem., 35, 198 (1992).
22.   R.A. Zambias, M.L. Hammond, J.V. Heck, K. Bartizal, C. Trainor, G Abruzzo, D.M. Schmatz, and K.M. Nollstadt, J. Med. Chem., 35, 2843 (1992).
23.   F.A. Bouffard, R.A. Zambias, J.F. Dropinski, J.M. Balkovec, M.L. Hammond, G.K. Abruzzo, K.F. Bartizal, J.A. Marrinan, M.B. Kurtz, D.C. McFadden, K.H. Nollstadt, M.A. Powles, and D.M. Schmatz, J. Med. Chem. 37, 222 (1994).
24.   W.Turner, M. DeBono, L. LaGrandeur, F. Burkhardt, M. Rodriquez, M. Zweifel, J. Nissen, K. Clingerman, R. Gordee, D. Zeckner, T. Parr, and J. Tang, 33rd ICAAC (1993), Abs. No. 358.
25.   D. Zeckner, T. Butler, C. Boylan, B. Boyll, Y. Lin, P. Raab, J. Schmidtke and W. Current, 33rd ICAAC (1993), Abs. No. 365.
26.   B. Pettersen, L. Green, L. Steimel, P. Barret, and W. Current, 33rd ICAAC (1993), Abs. No. 362.
27.   K. Takesako, H. Kuroda, T. Inoue, F. Haruna, Y. Yoshikawa, I. Kato, K. Uchida, T. Hiratani, and H. Yamaguchi, J. Antibiotics, 46, 1414 (1993).
28.   Y. Yoshikawa, K. Ikai, Y. Umeda, A. Ogawa, K. Takesako, I. Kato, and H. Naganawa, J. Antibiotics, 46, 1347 (1993).
29.   K. Takesako, H. Kuroda, I. Kato, T. Hiratani, K. Uchida, H. Yamaguchi, in "Recent Progress in Antifungal Chemotherapy", H. Yamaguchi, G.S. Kobayashi and H. Takahashi, Eds., Marcel Dekker Inc.., New York NY 1992, p. 502.
30.   T. Kurome, K. Inami, T. Inoue, K. Ikai, K. Takesako, I. Kato, T. Shiba, Chem. Letters, 11, 1873 (1993).
31.   P. Sandven, A. Bjorneklett, A. Maeland, and The Norweigian Study Group, Antimicrobial. Agents Chemother., 37, 2443 (1993).
32.   M.L. Cameron, W.A. Schell, S. Bruch, J.A. Bartlett, H.A.Waskin and J.R. Perfect, Antimicrobial. Agents Chemother., 37, 2449 (1993).
33.   C.A. Hitchcock, G.W. Pye, P. F. Troke, E.M. Johnson and D.W. Warnock, Antimicrobial. Agents Chemother., 37, 1962 (1993).
34.   M.A. Pfaller, J. Rhine-Chalberg, S,W, Redding, J. Smith, G. Farinacci, A.W. Fothergill, and M.G. Rinaldi, J. Clin. Microbiol., 32, 59 (1994).
35.   D.J. White, E.M. Johnson and D.W. Warnock, Genitourin. Med., 69, 112 (1993).
36.   A.Sanguineti, J.K. Carmichael and K. Campbell, Arch. Intern. Med., 153, 1122 (1993).
37.   E. Baet-Delabesse, P. Boiron, A. Carlotti and B. Dupont, J. Clin. Microbiol., 31, 2933 (1993).
38.   P. Diz, A. Ocampo, C. Miralles, I. Otero, I. Inglesia, and C. Martinez, Enferm. Infecc. Micriobiol. Clin. 11, 36 (1993).
39.   F. C. Odds, J. Antimicrobial. Chemother., 31:463 (1993).
40.   H. Vanden Bossche, P. Marichal and H. Moereels, NATO ASI Ser., Ser. H 69, 199 (1993).
41.   S. L. Kelly and D.E. Kelly, NATO ASI Ser., H 69, 215 (1993).
42.   H. Vanden Bosche. P. Marichal, J. Gorrens, D. Bellens, H. Moereels, and P. Janssen, Biochem. Soc. Trans. 18, 56 (1990).
43.   SCRIP No. 1830 , p. 31 (1993).
44.   A. M. Sugar, Curr. Clin, Top. Infect. Dis., 13, 74 (1993).
45.   H. Yamada, T. Tsuda, T. Watanabe, M. Ohashi, K. Murakame and H. Mochizuki, Antimicrob. Agents Chemother., 37, 2412 (1993).

46.   Y. Aoki, F. Yoshihara, M. Kondoh, Y. Nakamure, N. Nakayama and M. Arisawa, Antimicrob. Agents Chemother., 37, 2662 (1993).
47.   W.T. Hughes, D.P. Jacobus, C. Canfield, and J. Killmar, Antimicrob. Agents Chemother., 37, 1417 (1993).
48.   P.D. Walzer, J. Foy, P. Steele and M. White, Antimicrob. Agents Chemother., 37, 1463 (1993).
49.   S.F. Queener, M.S. Bartlett, M. Nasr and J.W. Smith, Antimicrob. Agents Chemother., 37, 2166 (1993).
50.   J.A. Fishman, S.F. Queener, R.S. Roth and M.S. Bartlett, Antimicrob. Agents Chemother., 37, 1543 (1993).
51.   J.L. Cook, Medical/Scientific Updata 10,1 (1992).
52.   N. Rastogi, Res. Microbiol., 142, 464 (1991).
53.   M.R. McNeil and P.J. Brennan, Res. Microbiol., 142, 451 (1991).
54.   H. Nikaido, S. Kim and E.Y. Rosenberg, Mol. Microbiol., 8,1025 (1993).
55.   V. Jarlier and H. Nikaido, J. Bacteriol., 172, 1418 (1990).
56.   J. Trias, V. Jalier and R. Benz, Sci. 258, 1479 (1992).
57.   P.T. Davidson and H. Q. Le, Drugs, 43, 651 (1992).
58.   N. Rastogi and H.L. David, Res. Microbiol., 144, 133 (1993).
59.   M.E. Levin and G.F. Hatfull, Mol. Microbiol., 8, 2 (1993).
60.   A. Telenti, P. Imboden, F. Marchesi, D. Lowrie, S. Cole, M. Colston, L. Matter, K. Schopfer and T. Bodmer, Lancet, 341, 647 (1993).
61.   N. Honore, and S.T. Cole, Antimicrob. Agents Chemother., 37, 414 (1993).
62.   A. Telenti, P. Imboden, F. Marchesi, T. Schmidheini and T Bodmer, Antimicrob. Agents Chemother., 37, 2054 (1993).
63.   W.R. Jacobs, R.G. Barletta, R. Udani, J. Chan, G. Kalkut, G. Sosne, T. Kieser, G.K. Sarkis, G.F. Hatfull and B.R. Bloom, Sci., 260, 819 (1993).
64.   R.C. Cooksey, J.T. Crawford, W.R. Jacobs and T.M. Shinnick, Antimicrob. Agents Chemother., 37, 1348 (1993).
65.   K.L. Rinehart, L.S. Shield, Fortschr. Chem. Org. Naturst., 33, 231 (1976).
66.   B. Cavalleri, M. Turconi, G. Tamborini, E. Occelli, G. Cietto, R. Pallanza, R. Scotti, M. Berti, G. Romano, and F. Parenti, J. Med. Chem., 33, 1470 (1990).
67.   H. Saito, H. Tomioka, K. Sato, M. Emori, T. Yamane, K. Yamashita, K. Hosoe, and T. Hidaka, Antimicrob. Agents Chemother., 35, 542 (1991).
68.   T. Yamane, T. Hashizume, K. Yamashita, K. Hosoe, T. Hidaka, K. Wantanabe, H. Kawaharada, and S. Kudoh, Chem. Pharm. Bull (Tokoyo), 41, 148 (1993).
69.   M. Emori, H. Saito, K. Sato, H. Tomioka, T. Setogawa, and T. Hidaka, Antimicrob. Agents Chemother., 37, 722 (1993).
70.   S.T. Brown, F.F. Edwards, E.M. Bernard, W. Tong and D. Armstrong, Antimicrob. Agents Chemother., 37, 398 (1993).
71.   T. Yamada, A. Nagata, Y. Ono, Y. Suzuki, and T. Yamanouchi, Antimicrob. Agents Chemother., 27, 921 (1985).
72.   J. Douglass and L.M. Steyn, J. Infect. Dis., 167, 1505 (1993).
73.   N. Honore and S.T. Cole, Antimicrob. Agents Chemother., 38, 238 (1994).
74.   A. Meier, P. Kirschner, P. Bange, U. Vogel and E.C. Bottger, Antimicrob. Agents Chemother., 38, 228 (1994).
75.   R.J. Wallace, S.I. Hull, D.G. Bobey, K.E. Price, J.M. Swenson, L.C. Steele and L. Christensen, Am. Rev. Respir. Dis. 132, 409 (1985).
76.   N. Khardori, H. Nguyen, B. Rosenbaum, K. Rolston, and G.P. Bodey, Antimicrob. Agents Chemother., 38, 134 (1994).
77.   A. Quemard, C. Lacave, and G. Laneelle, Antimicrob. Agents Chemother., 35, 1035 (1993).
78.   Y. Zang, Res. Microbiol. 144, 143 (1993).
79.   B. Heym, Y. Zhang, S. Poulet, D. Young and S.T. Cole, J. Bacteriol., 175, 4255 (1993).
80.   Y. Zhang, T. Garbe and D. Douglas, Mol. Microbiol., 8, 521 (1993).
81.   M.Y. Stoeckle, L. Guan, N. Reigler, I. Weitzman, B. Kreiswirth, J. Kornblum, F. Laraque and L.W. Riley, J. Infect. Dis., 168, 1063 (1993).
82.   Y. Zhang, B. Heym, B. Allen, D. Young and S. Cole, Nature, 358, 591 (1992).
83.   L.B. Keifets, M.A. Flory and P.J. Lindholm-Levy, Antimicrob. Agents Chemother., 33, 1252 (1989).
84.   M. Salfinger, A.J. Crowle and L.B. Reller, J. Infect. Dis., 162, 210 (1990).
85.   M.H. Cynamon and S.P. Klemens, J. Med. Chem., 35, 1212 (1992).
86.   Z.Z. Liu, X.D. Guo, L.E. Straub, G. Erdos, R.J. Prankerd, R.J. Gonzalez-Rothi and H. Schreier, Drug Des. Discov. 8, 57 (1991).
87.   G. Silve, P. Valero-Guillen, A. Quemard, M. A. Dupont, M. Daffe and G. Laneelle, Antimicrobial. Agents Chemother., 37, 1536 (1993).

88. K. Takayama, E. L. Armstrong, K. A. Kunugi, and J. O. Kilburn, Antimicrob. Agents Chemother., 16, 240 (1979).
89. A. Quemard, G. Laneelle and C. Lacave, Antimicrob. Agents Chemother., 36, 1316 (1992).
90. C. Lacave, A. Quemard and G. Laneelle, Biochim. Biophys. Acta, 1045, 58 (1990).
91. L. M. Lopez-Marin, A. Quemard, G. Laneelle, and C. Lacave, Biochim. Biophys. Acta, 1086, 22 (1991).
92. D.C. Leysen, A. Haemers, L. Blanchaert, I. Van Assche, W. Bollaert, K. Schoenmaekers and S.R. Pattyn, Framaco Sci., 42, 823 (1987).
93. J. Chiu, J. Nussbaum, S. Bozzette, J.G. Tilles, L.S. Young, J. Leedom, P.N.R. Heseltine, J.A. McCutchan, and the California Collatborative Treatment Group, Ann. Intern. Med., 113, 358 (1990).
94. G.P Chan, B.Y. Garcia-Ignacio, V.E. Chavez, J.B. Livelo, C.L. Jimenez, M.L.R. Parrilla and S.G. Franzblau, Antimicrob. Agents Chemother., 38, 61 (1994).
95. J.H. Grosset, B. Ji, C.C. Guelpa-Lauras, E.G. Perani and L.N. N'Deli, Int. J. Lepr., 58, 281 (1990).
96. G. Klopman, S. Wang, M.R. Jacobs, S. Bajaksouzian, K. Edmonds and J.J. Ellner, Antimicrob. Agents Chemother., 37, 1799 (1993).
97. G. Klopman, S. Wang, M.R. Jacobs and J.J. Ellner, Antimicrob. Agents Chemother., 37, 1807 (1993).
98. H. Tomioka and K. Sato, Antimicrob. Agents Chemother., 37, 1259 (1993).
99. C.B Inderlied, P.T. Kolonoski, M.Wu, and L.S. Young, J. Infect. Dis. 159, 994 (1989).
100. P.B. Fernandes, D.J. Hardy, D. McDaniel, C.W. Hanson and R.N. Swanson, Antimicrob. Agents Chemother., 33, 1531 (1989).
101. N. Rastogi, K.S. Goh and A. Bryskier, 2nd ICMAS (1994), Abs. No. 154.
102. N. Rastogi, K.S. Goh and A. Bryskier, 2nd ICMAS (1994), Abs. No. 299.
103. L.S. Young, L.E. Bermudez, M. Wu and C.B. Inderlied, 2nd ICMAS (1994), Abs. No. 153.
104. L.S. Young, L. Wiviott, M. Wu, P. Kolonski, R. Bolan and C.B. Inderlied, Lancet, 388, 1107 (1991).
105. B. Dautzenberg, C. Truffot, S. Legris, M.C. Meyohas, H.C. Berlie, A. Mercat, S. Chevrat and J. Grosset, Am. Rev. Respir. Med., 144, 564 (1991).
106. B. Dautzenberg, T. Saint Marc, M.C. Meyohas, M. Eliaszewitch, F. Haniez, A.M. Rogues, S. S. De Wit, L. Cotte, J.P. Chauvin and J. Grosset, Arch. Intern. Med. 153, 368 (1993).
107. L.B. Barradell, G.L. Plosker and D. McTavish, Drugs, 46, 289 (1993).
108. B. Ruf, D. Shurmann, H. Mauch, F.J. Ferenbach and H.D. Pohle, Infect. 20, 267 (1992).
109. F. de Lalla, R. Maserati, P. Scarpellini, P. Marone, R. Nicolin, F. Caccamo and R. Rifoli, Antimicrob. Agents Chemother., 36, 1567 (1992).
110. G. Notario, and D. Henry, 2nd ICMAS (1994), Abs. No. 297.
111. J.A. Sonnabend and R. Pringle Smith, 2nd ICMAS (1994), Abs. No. 303.
112. D.L. Payne, 2nd ICMAS (1994), Abs. No. 302.
113. D.R. Coulston, and C.D. Wood, 2nd ICMAS (1994), Abs. No. 298.
114. J.C. Craft and D. Henry, 2nd ICMAS (1994), Abs. No. 295.
115. L. Heifets, N. Mor and J. Vanderkolk, Antimicrob. Agents Chemother., 37, 2364 (1993).
116. A. Meier, P. Kirschner, B. Springer, V.A. Steingrube, B.A. Brown, R.J. Wallace and E.C. Bottger, Antimicrob. Agents Chemother., 37, 381 (1994).
117. K.A. Joiner and J.F. Dubremetz, Infect. Immun., 61, 1169 (1993).
118. B.J. Luft and J.S. Remington, Clin. Infect. Dis., 15, 211 (1992).
119. J.C. Schwab, C.J.M. Beckers and K.A. Joiner, Proc. Natl. Acad. Scii. USA, in press (1994).
120. R.G.K. Donald and D.S. Roos, Proc. Natl. Acad. Sic. USA, 90, 11703 (1993).
121. K. Kim, D. Soldati and J.C. Boothroyd, Sci., 262, 911 (1993).
122. L.. Chio and S.F. Queener, Antimicrob. Agents Chemother. 37, 1914 (1993).
123. D.S. Roos, J. Biol. Chem., 268, 6269 (1993).
124. R.E. McCabe and S. Oster, Drugs, 38, 973 (1989).
125. F. Derouin, R. Almadany, F. Chau, B. Rouveix and J.J. Pocidalo, Antimicrob. Agents Chemother. 36, 977 (1992).
126. F.G. Araujo, P. Prokocimer, T. Lin and J.S. Remington, Antimicrob. Agents Chemother. 36, 2454 (1992).
127. S. Romand and F. Derouin, 2nd ICMAS (1994), Abs. No. 173.
128. F.G. Araujo, R.M. Shepard and J.S. Remington, Eur. J. Clin. Microbiol. Infect. Dis. 10, 519 (1991).
129. J. Blais, C. Tardif and S. Chamberland, Antimicrob. Agents Chemother. 37, 2571 (1993).
130. J. Blais, V. Garneau and S. Chamberland, Antimicrob. Agents Chemother. 37, 1701 (1993).
131. E.R. Pfefferkorn and S.E. Borotz, Antimicrob. Agents Chemother. 38, 31 (1994).
132. J. Fernadez-Martion, C. Leport, P. Morlat, M.C. Meyohas, J.P. Chauvin and J.L. Vilde, Antimicrob. Agents Chemother. 35, 2049 (1991).

Chapter 17. *Ras* Oncogene Directed Approaches in Cancer Chemotherapy

Gary L. Bolton, Judith S. Sebolt-Leopold, and John C. Hodges
Parke-Davis Pharmaceutical Research Division
Warner-Lambert Company
Ann Arbor, Michigan 48105

Introduction - Traditional approaches to cancer management have involved cytotoxic intervention at the level of DNA replication. While cytotoxic anticancer drugs have shown limited efficacy against rapidly growing tumor cells, there remains a critical need for the development of agents targeted against more refractory, slow growing, solid tumors. The 1990's have witnessed a revolution in the understanding of signalling pathways that are important to the growth of normal and neoplastic cells. Whether a cell divides or stops dividing depends, to a great extent, on its ability to respond to membrane localized growth stimuli. Among the numerous oncogene or protooncogene encoded proteins that serve as signal transducers in the pathway from the outer membrane to the nucleus, perhaps none are as central as the Ras protein. Ras acts as a common relay point for signals from all of the growth factor receptors examined thus far (1, 2). Furthermore, single base mutations in the gene encoding the 21 kD Ras protein are found in approximately 30% of all human tumors, with the incidence as high as 50 and 90% in colon and pancreatic carcinomas, respectively (3). Comprehensive review of the function and regulation of Ras appears elsewhere (4). This chapter will be devoted to exploring the various strategies for blocking Ras function as well as reviewing progress that has been reported with anti-Ras agents.

BIOCHEMICAL FUNCTION OF RAS IN CELL SIGNALLING

Consistent with their fundamental role in cell proliferation, *ras* genes are members of a highly conserved and ubiquitous eukaryotic gene family. Mutated alleles of cellular *ras* genes were first discovered in rat sarcoma virus induced tumors (5, 6). The prevalence of activated or mutated *ras* alleles in human cancers is well documented (7, 8). All three known mammalian *ras* genes (H-, K-, and N-*ras*) encode nearly identical 21 kD proteins, commonly referred to as p21, p21^{ras} or Ras, which belong to a large superfamily of monomeric GTP-binding proteins including those from three other gene subfamilies, *rho*, *rab* and *arf* (9, 10). Collectively these proteins regulate a diverse array of cellular events ranging from cell proliferation and differentiation to cytoskeletal assembly and vesicular trafficking (11).

A distinguishing biochemical feature of members of the Ras superfamily is their ability to bind guanine nucleotides with high affinity and their ability to hydrolyze bound GTP to GDP and phosphate. Ras serves as a molecular switch for cellular growth and differentiation by cycling between an active, GTP-bound form and an inactive, GDP-bound form. The proportion of active Ras appears to be tightly regulated and is dictated by the presence of GAP (GTPase activating protein) and neurofibromin (product of NF1), which serve to accelerate the intrinsic GTPase activity of normal Ras in a negative regulatory fashion. Upon stimulation of membrane receptors, growth factor induced activation of Ras is reflected in an observed increase in cellular Ras-GTP content (12,13).

The intricacies of signal transduction can be seen in the nature of the protein-protein interactions which serve to transmit signals from growth factor receptors to Ras and subsequently from Ras to the nucleus. Protein interactions of Ras have been the subject of recent reviews (14-20). It is now believed that tyrosine kinases, e.g. EGF receptor, serve to phosphorylate an intermediary protein, the Shc protein, which binds to the SH2 domain of the Grb2 (growth factor receptor binding) protein (21). Subsequent complex formation between Shc and Grb2 allows for membrane recruitment of Sos, which stimulates guanine nucleotide dissociation (22, 23). In this way, Sos promotes GDP release from Ras, allowing it to bind GTP and thereby assume an active conformation. When activated, Ras allows further signal transduction in a process mediated by an effector protein which remains unidentified. While Ras-GAP appears to function in a negative regulatory manner, it does not appear to act alone as the effector complex for cell transformation (24). As a consequence of Ras-effector complex formation, a cascade of protein phosphorylation by cytoplasmic serine/threonine kinases is set into motion. A complex involving Ras and the Raf-1 protein, which phosphorylates and activates mitogen activated protein kinase kinase (MAPKK or MEK), is the first step in this kinase cascade (25-28). The last kinase results in the phosphorylation of several cellular proteins including the transcription factors Myc and Jun. In this manner activated Ras ultimately results in altered gene expression.

Ras does not possess a transmembrane domain; however its ability to play a pivotal role in mitogenic signalling is dependent upon localization at the inner surface of the plasma membrane. Post-translational modifications provide the lipophilicity required for membrane association. The cytoplasmic enzyme protein farnesyltransferase (PFT) recognizes the carboxyl terminus of unprocessed Ras which is characterized by a cysteine residue followed by two aliphatic amino acids and any one of several different amino acids, constituting the so-called CAAX motif. PFT catalyzes the reaction of farnesyl pyrophosphate (FPP) with the CAAX cysteine residue forming a thioether linkage (29-31). Genetically engineered Ras mutants lacking the CAAX sequence do not associate with the plasma membrane and also are not able to transform cells to malignancy (32, 33). Prior to membrane insertion of farnesylated Ras, the three terminal amino acids are trimmed by proteolytic processing (34-36). This is followed by methyl esterification at the new C-terminal cysteine residue in a step carried out by a protein methyltransferase (37, 38). Post-translational processing through prenylation, proteolysis and esterification is described in greater detail in several recent reviews (39-42). Lipid association between prenyl groups and membrane lipids may not be the only way in which prenylated proteins bind to membranes. A second hypothesis is that protein prenylation also mediates protein-protein interactions, e.g. specific interactions between membrane receptors and prenylated proteins (43). The exact nature in which farnesylation promotes membrane localization is not well understood at this time.

PFT, which is a heterodimer composed of 49 and 46 kD subunits, is ubiquitous among eukaryotic cells. Mammalian PFT is structurally and functionally similar to the yeast enzyme indicative of conservation through evolution (44). PFT shares a common prenyl pyrophosphate binding subunit (alpha subunit) with a second prenylation enzyme, protein geranygeranyltransferase type I, which unlike PFT, recognizes CAAX motifs that terminate in leucine (45). The beta subunit of PFT is believed to provide CAAX box specificity (46).

Cloning and expression of both subunits of human PFT have been reported (47, 48). A detailed kinetic analysis of this enzyme indicates a favored, but not obligate, pathway which consists of FPP binding prior to Ras binding, followed by catalysis of thioether formation as the predicted sequence *in vivo* (49).

In addition to the C-terminal CAAX motif, Ras proteins may possess additional structural features promoting membrane association. For example, K-Ras4B contains a stretch of six lysine residues upstream of the CAAX motif. Protonation of this lysine rich region is thought to provide a positively charged region that would be attracted to the negatively charged phosphate surface of membrane bilayers (50). On the other hand, H-Ras, N-Ras and K-Ras4A all contain cysteine palmitoylation sites upstream of the CAAX motif. Mutant proteins lacking these cysteine residues are no longer efficiently localized at the plasma membrane and have impaired ability to transform cells (51, 52).

STRATEGIES FOR THERAPEUTIC INTERVENTION

Potential strategies for inhibiting the function of Ras in tumors include antisense oligonucleotide approaches, inhibition of the post-translational processing enzymes, and blockade of the numerous Ras-protein interactions. Antisense Ras RNA oligonucleotides have been investigated for their ability to block the proliferative action of oncogenic Ras (53). Similarly, antisense and dominant inhibitory PFT expression plasmids have been shown to reduce colony formation in Ras-transformed cells (54). The feasibility of such antisense approaches remains to be proven through a demonstration of antitumor efficacy in animal models. Alternatively, agents that either disallow formation of Ras-GTP or restore GTPase activity to mutant Ras, e.g. by modulating Ras/GAP interaction, are also theoretical targets for therapeutic intervention. The feasibility of these approaches would depend upon a better understanding of GAP's role and a more precise knowledge of key effector molecules. Furthermore Ras activity could also be antagonized by agents that interfere with the formation of a number of Ras-protein complexes, e.g. Ras-Sos, that have recently been elucidated. Molecules that inhibit the formation of such complexes have not been reported.

Current research has been more highly focused on agents which interfere with the membrane localization of Ras (55, 56). Historically the first inhibitors of Ras membrane localization which exhibited antiproliferative effects were agents that block the biosynthetic pathway to FPP, namely, mevinolin and compactin (57). However, as inhibitors of HMG CoA reductase, these compounds also block steroid biosynthesis and hence are cytotoxic at concentrations which result in anti-Ras effects (58). The search for less toxic agents have more recently centered on inhibitors of the enzymes that catalyze the post-translational modification of Ras. Of these enzymes, PFT has been the most widely studied in terms of mechanism and inhibitors. Potential shortcomings of PFT inhibitors as anti-Ras agents include their degree of selectivity for 1) PFT relative to other prenylation enzymes, 2) Ras relative to other farnesylated substrates, and 3) oncogenic Ras relative to normal Ras. The potential for similar specificity shortcomings would also be expected for inhibitors of the protease, methyltransferase and palmitoyltransferase enzymes that complete Ras processing.

Present evidence indicates that geranylgeranyl modified proteins terminate in either CAAL, CC, CXC, or CCXX motifs and as such are processed by distinct protein geranylgeranyltransferases (PGT) (59-61). Thus by designing an agent specific for the peptide binding subunit of PFT, it should be possible to circumvent PFT/PGT specificity concerns. In addition to Ras, other proteins are substrates of PFT and would presumably be affected by a PFT inhibitor. Namely, lamin B, which is required for nuclear envelope integrity, and transducin, which is important in retinal signal transduction, are both substrates for PFT. Inhibition of their farnesylation is thus a potential toxicity concern.

Selectivity for oncogenic Ras relative to normal Ras would be a desirable property of a PFT inhibitor since Ras is expressed in virtually all tissues and plays a critical role in normal cellular proliferation. A precedent now exists for farnesylation inhibitors that appear to preferentially inhibit oncogenic Ras function (62, 63). The mechanism of this selectivity remains largely an enigma. Evidence suggesting a reduced affinity of Raf-1 for oncogenic Ras relative to normal Ras may provide a clue (64).

PROTEIN FARNESYLTRANSFERASE INHIBITORS

CAAX Analogs - Initial investigations following the isolation and purification of PFT demonstrated that the full Ras protein was not required for enzyme activity (65). The enzyme farnesylates a number of CA_1A_2X tetrapeptides, including CVIM (66-71). The preferred C-terminal (X) residues are serine, methionine, glutamine, alanine, or cysteine, which result in PFT binding. In general, A_1 and A_2 can be any aliphatic amino acid. However, substitution at the A_2 position has a critical effect on the ability of the tetrapeptide to function as a substrate. The tetrapeptide CVFM (1), containing an aromatic residue at A_2, is a potent inhibitor (IC_{50} = 25 nM) of PFT, and does not undergo farnesylation (67). A further requirement for this behavior of CVFM is a free N-terminus, in that acylation restores the ability for farnesylation (68). The liabilities of these tetrapeptides as potential drug candidates, which include their inefficient cellular uptake and rapid proteolytic degradation, have resulted in little further development. However, these compounds have served as a foundation for the design and synthesis of several classes of peptidomimetic derivatives of the CAAX sequence. A number of potent inhibitors of PFT of this type have been reported recently.

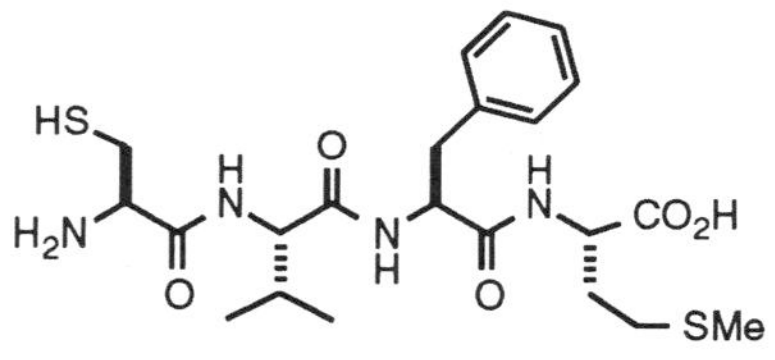

1

The pseudo-tetrapeptide L-731,735 (2) was designed as a CIIM analog in which the two N-terminal amide bonds were reduced and methionine was replaced with homoserine (62). These modifications resulted in a potent in vitro (IC_{50} =18 nM) inhibitor of PFT. The corresponding lactone derivative L-731,734 (3) was also an effective inhibitor (IC_{50} = 282 nM), and selectively inhibited both Ras processing in a transformed cell line and growth in soft agar. Systematic modification of 1 has also led to the identification of B581 (4), which showed a twofold increase in activity vs 1 in vitro (72). This analog was found to selectively inhibit the processing of farnesylated proteins rather geranylgeranylated proteins in a Ras-transformed cell line, but failed to discriminate between farnesylation of H-Ras and lamin A. Microinjection of 4 into frog oocytes also inhibited Ras-dependent maturation.

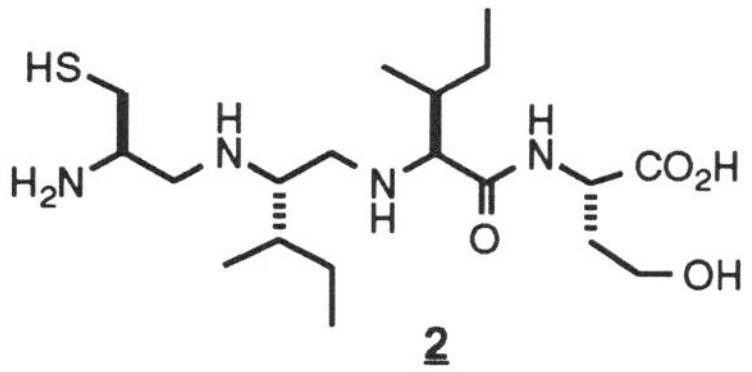

2

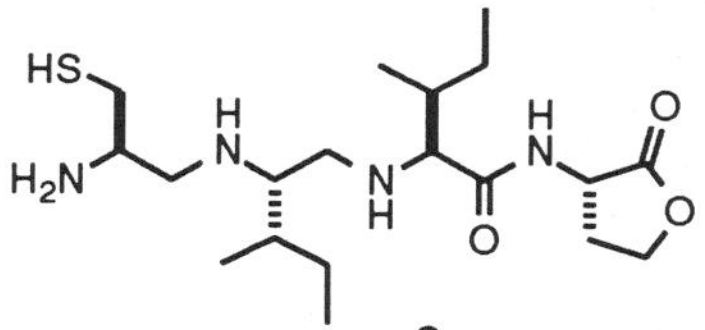

3

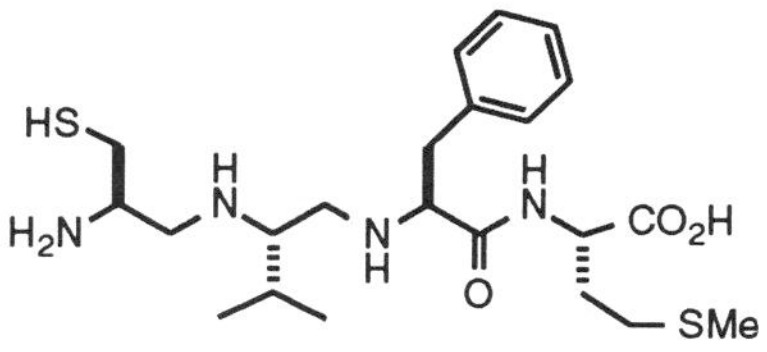

4

Others have also prepared a number of analogs of **1** to systematically examine the effect of backbone modification on activity and to further define substrate-inhibitor patterns (73). A number of modifications, including N-methylation and amide bond replacement, were investigated but these changes gave little insight to the substrate/inhibitor backbone conformation that is preferred by PFT.

Replacement of the two aliphatic residues in the CAAX motif with a benzodiazepine derived mimic of a dipeptide turn led to an extremely potent inhibitor BZA-2B (**5**) with an IC$_{50}$ of 0.85 nM (63). This compound was designed to allow both the N-terminal cysteine and the C-terminal methionine to coordinate the Zn^{2+} ion that is present at the peptide binding site (74, 75). Recent proton NMR studies with an enzyme-bound heptapeptide substrate (KTKCVFM) have shown that the C-terminal tetrapeptide portion is directly involved in binding to the enzyme via an induced type I β-turn conformation (76). Although less potent *in vitro*, the corresponding methyl ester, BZA-5B (**6**), restored a normal growth pattern to Ras-transformed cells. Incorporation of a 3-aminomethylbenzoic acid (AMBA) moiety as a replacement for the two internal aliphatic residues led to Cys-AMBA-Met (**7**), an analog containing no peptidic linkages (77). This compound was a more potent inhibitor of human PFT than CVIM, but no cellular activity was reported.

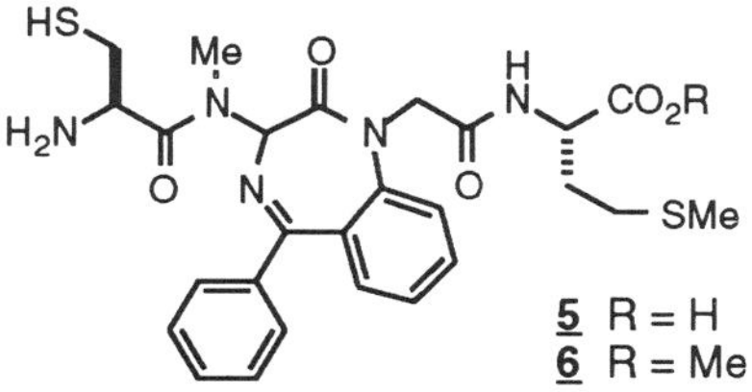

5 R = H
6 R = Me

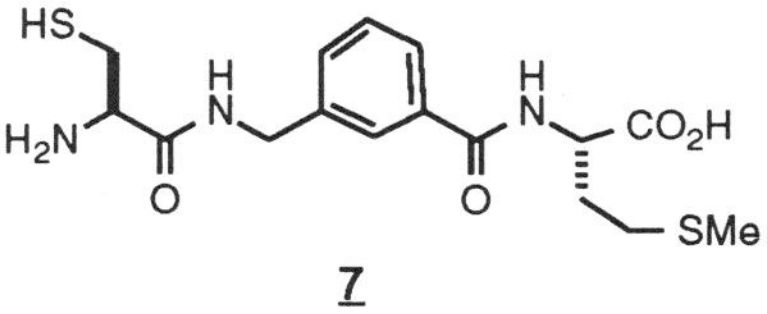

7

<u>Farnesyl Pyrophosphate Analogs</u> - Synthetic analogs of farnesyl pyrophosphate (FPP) have received less attention as potential inhibitors of PFT, perhaps because of specificity concerns due to its involvement in other biological pathways. However, two hydrolytically stable analogs of FPP have shown potent inhibition of PFT (78). Analogs **8** and **9** were found to be competitive inhibitors of FPP, with Ki values of 5 nM and 830 nM, respectively. **9** was

shown previously to be a potent inhibitor of squalene synthetase (79). A slight inhibition of Ras processing in a Ras-transformed cell line was noted with **8** at 1 uM (80). This effect, while modest, was the first demonstration that the cellular processing of Ras could be moderated by a synthetic PFT inhibitor. However, it could not be determined if higher concentrations of **8** would lead to further inhibition of Ras processing, due to the observation of cellular toxicity. Farnesylamine and other related long chain aliphatic amines have also been shown to inhibit the farnesylation and growth of Ras-transformed cells at high micromolar concentrations (81).

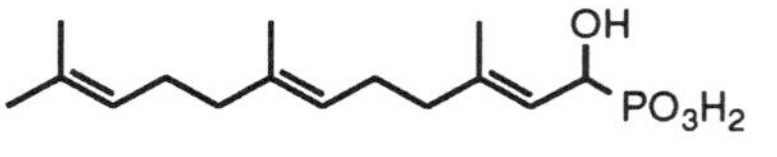

8 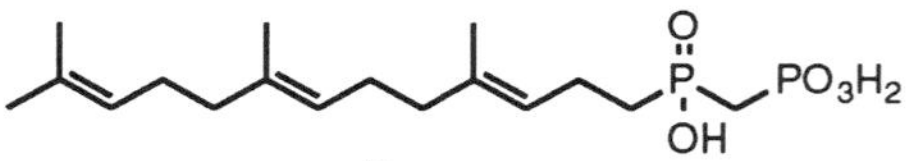   **9**

Natural Products - The use of PFT in high volume screening has led to the identification of a variety of natural products of microbial origin which inhibit PFT (82). 10'-Desmethoxy-streptonigrin (**10**), an analog of streptonigrin (**11**) was found to be a weak inhibitor (IC$_{50}$ = 21 uM) of the enzyme, and also was cytotoxic to several cell lines (83). Interestingly, streptonigrin itself was approximately threefold less potent *in vitro*.

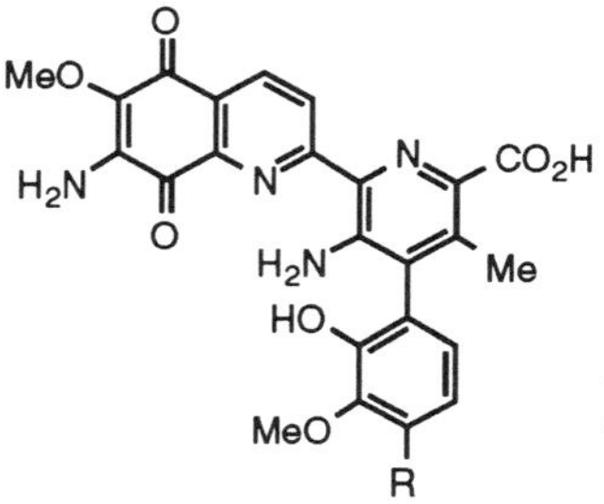

10 R = H
11 R = OMe

Other antibiotics which show activity are derivatives of the manumycin family (84). UCF1-C (**12**), also known as manumycin, was the most potent inhibitor of this class (IC$_{50}$ = 5 uM). Kinetic analysis demonstrated that **12** was a competitive inhibitor (Ki = 1.2 uM) with respect to FPP, and noncompetitive with respect to protein substrate. The reduced (dithiol) form of the fungal epipolydithiodiketopiperazine toxin, gliotoxin (**13**), was also a modest (IC$_{50}$ = 1.1 uM) inhibitor of PFT (85). This compound was found to be a noncompetitive inhibitor with respect to the Ras protein.

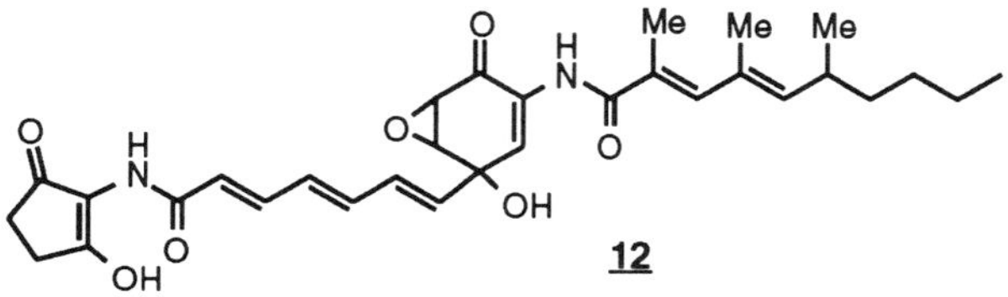

12 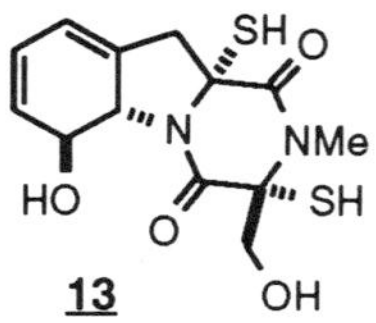   **13**

Two novel dicarboxylic acid derivatives, chaetomellic acid A (**14**) and chaetomellic acid B (**15**), were isolated from a fermentation extract and found to be potent inhibitors of PFT, with IC_{50} values of 55 nM and 185 nM, respectively (80, 86, 87). These compounds were shown to be effective mimics of FPP by kinetic analysis and by computer modeling of their overlapping steric and electrostatic regions.

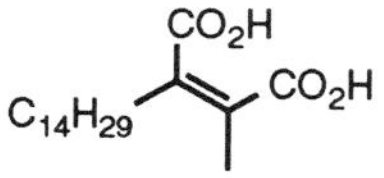

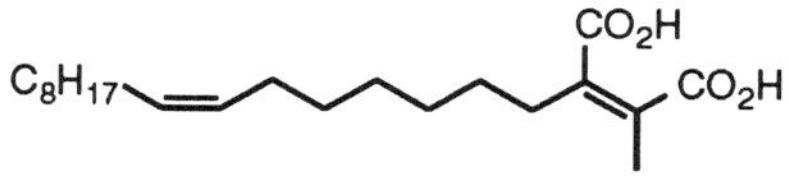

Zaragozic acid A (**16**), previously disclosed as a potent inhibitor of squalene synthetase, has also been identified as a PFT inhibitor (IC_{50} = 216 nM) (80). A synthetic analog (**17**) showed increased PFT inhibition (IC_{50} = 12 nM) and selectivity versus PGT. These compounds were competitive inhibitors with respect to FPP, and noncompetitive inhibitors with respect to the Ras protein. They had no effect on Ras processing in a Ras transformed cell line, presumably due to poor cell permeability.

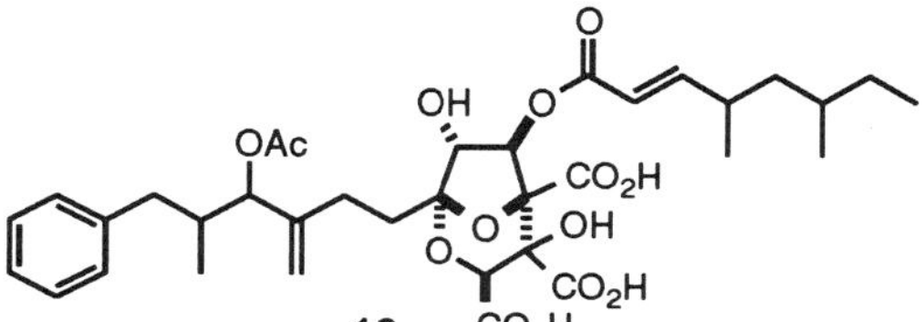

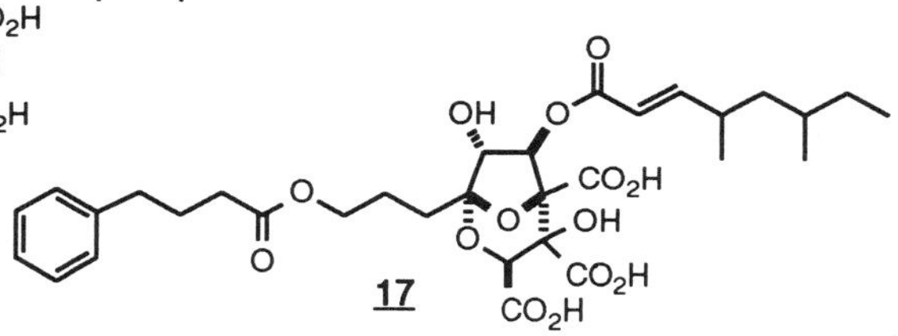

The pepticinnamins, a novel series of six related pentapeptides, were also isolated by the screening of various fermentation broths (88). Pepticinnamin E (**18**), the major component of the mixture and only member whose structure was elucidated, was found to be a good inhibitor (IC_{50} = 300 nM) of PFT (89). Limonene and related metabolites perillic acid and dihydroperillic acid, have also been demonstrated to inhibit protein isoprenylation and cell growth (90, 91). The antibiotic patulin (**19**) inhibited protein prenylation in a mouse cell line, and exhibited weak (IC_{50} = 290 uM) inhibition of PFT (92).

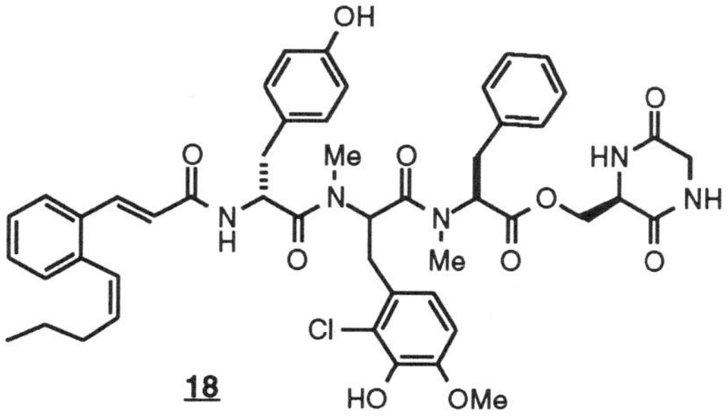

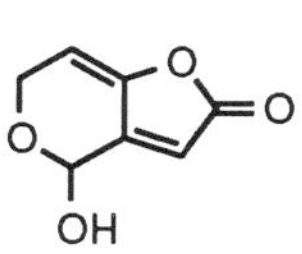

PROTEASE AND METHYLTRANSFERASE INHIBITORS

Protease Inhibitors - Compared to the rapidly growing number of recent publications surrounding PFT and its inhibitors, the literature available on the subsequent C-terminal Ras processing enzymes is small. Two microsomal enzymes that cleave AAX from a farnesyl-CAAX motif have been reported. The first is a bovine liver endoprotease which cleaves the C-terminal tripeptide from farnesylated proteins (35). The second is a rat brain carboxypeptidase which sequentially removes X, A_2 and A_1 (36). Both enzymes show high affinity for a farnesyl-CAAX motif compared to its CAAX precursor. Farnesyl-CAAX analogs with isosteric replacements for the scissile peptide bond have been shown to be inhibitors of isoprenylated protein endoprotease (93). The most potent of these are **20** (Ki = 86 nM) and **21** (Ki = 64 nM). In principle, these molecules could also be inhibitors of the carboxypeptidase although no data are currently available.

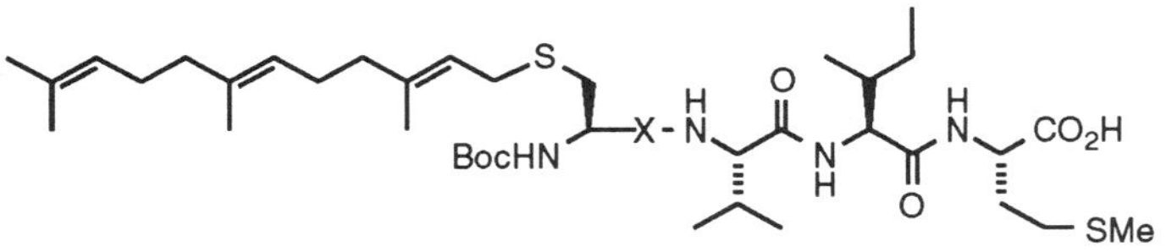

20   X = CH_2

21   X = $CH(OH)CH_2CO$

Methyltransferase Inhibitors - Microsomal enzyme preparations that catalyze the methyl esterification of farnesyl-cysteine residues located at the C-terminus of peptides have also been described (94-96). To date a number of weak inhibitors (Ki > 25 uM) have been reported including S-farnesyl mercaptopropionic acids such as MFPT (**22**) and FPA (**23**) (96).

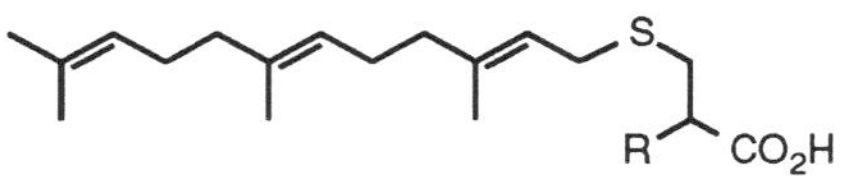

22   R = CH_3

23   R = (=O)

SUMMARY AND FUTURE DIRECTIONS

Because of the central signalling role of Ras in cell division and because of the frequency of mutant *ras* genes in human tumors, inhibition of Ras is likely to continue to be a popular target for cancer chemotherapy research. To date, the most promising strategy is to block Ras function by inhibiting its membrane localization, more specifically by preventing C-terminal processing. The bulk of current research is focused on PFT inhibitors. There are numerous CAAX analogs, FPP analogs and natural products which are inhibitors of PFT *in vitro* that provide excellent lead structures from which future drug candidates may be derived. Although several structural types have been shown to inhibit Ras processing and Ras dependant transformations in cell culture assays, the major hurdle of demonstrating the antineoplastic efficacy of a PFT inhibitor *in vivo* remains to be cleared.

References

1. J.R. Woodgett, Curr. Biol., 2, 357 (1992).
2. J. Downward, BioEssays, 14, 177 (1992).
3. J.L. Bos, Cancer Res., 49, 4682 (1989).
4. D.R. Lowy, and B.M. Willumsen, Ann. Rev. Biochem., 62, 851 (1993).
5. J.J. Harvey, Nature, 204, 1104 (1964).
6. W.H. Kirsten, and L.A. Mayer, J. Natl, Cancer Inst., 39, 311 (1967).
7. J.L. Bos, Mutat. Res., 195, 255 (1988).
8. C.J. Der, Clin. Chem., 33, 641 (1988).
9. A. Valencia, P. Chardin, A. Wittinghofer and C. Sander, Biochemistry, 30, 4637 (1991).
10. R.A. Kahn, C.J. Der and G.M. Bokoch, FASEB J., 6, 2512 (1992).
11. A. Hall, Science, 249, 635 (1990).
12. R.H. Medema, A.M.M. DeVries-Smits, G.C.M. VanderZon, J.A. Maassen, and J.L. Bos, Mol. and Cell. Biol., 13, 155 (1993).
13. J.B. Gibbs, M.S. Marshall, E.M. Scolnick, R.A.F. Dixon and U.S. Vogel, J. Biol. Chem., 265, 20437 (1990).
14. M.S. Marshall, TIBS ,18, 250 (1993).
15. P. Polakis and F. McCormick, J. Biol. Chem., 268, 9157 (1993).
16. M.S. Boguski and F. McCormick, Nature, 366, 643 (1993).
17. T. Satoh, M. Nakafuku and Y. Kaziro, J. Biol. Chem., 267, 24149 (1992).
18. T. Cartwright, A. Morgat and B. Tocque, Chim. Oggi, 10, 26 (1992).
19. M.J. Fry, G. Panayotou, G.W. Booker and M.D. Waterfield, Protein Science, 2, 1785 (1993).
20. G.M. Bokoch and C.J. Der, Faseb J., 7, 750 (1993).
21. M. Rozakis-Adcock, J. McGlade, G. Mbamalu, G. Pelicci, R. Daly, W. Li, A. Batzer, S. Thomas, J. Brugge, P.G. Pelicci, J. Schlessinger and T. Pawson, Nature, 360, 689 (1992).
22. L. Buday and J. Downward, Cell, 73, 611 (1993).
23. N.W. Gale, S. Kaplan, E.J. Lowenstein, J. Schlessinger and D. Bar-Sagi, Nature, 363, 88 (1993).
24. D.C.S Huang, C.J. Marshall and J.F. Hancock, Mol. and Cell. Biol., 13, 2420 (1993).
25. P.H. Warne, P.R. Viciana and J. Downward, Nature, 364, 352 (1993).
26. X. Zhang, J. Settleman, J.M. Kyriakis, E. Takeuchi-Suzuki, S.J. Elledge, M.S. Marshall, J.T. Bruder, V.R. Rapp and J. Avruch, Nature, 364, 308 (1993).
27. A.B.Vojtek, S.M. Hollenberg and J.A. Cooper, Cell, 74, 205 (1993).
28. B. Dickson, F. Sprenger, D. Morrison and E. Hafen, Nature, 360, 600 (1992).
29. P.J. Casey, P.A. Solski, C.J. Der and J.E. Buss, Proc. Natl. Acad. Sci., 86, 8323 (1989).
30. J. Glomset, M. Gelb and C. Farnsworth, Curr. Opin. Lipidology, 2, 118 (1991).
31. J.F. Hancock, A.I. Magee, J.E. Childs and C.J. Marshall, Cell, 57, 1167 (1989).
32. J.H. Jackson, C.G. Cochrane, J.R. Bourne, P.A. Solski, J.E. Buss and C.J. Der, Proc. Natl. Acad. Sci., 87, 3042 (1990).
33. K. Kato, A.D. Cox, M.W. Hisaka, S.M. Graham, J.E. Buss, and C.J. Der, Proc. Natl. Acad. Sci., 98, 6403 (1992).
34. M.N. Ashby, D.S. King and J. Rine, Proc. Natl. Acad. Sci., 89, 4613 (1992).
35. Y-T. Ma and R.R. Rando, Proc. Natl. Acad. Sci., 89, 6275 (1992).
36. T.N. Akopyan, Y. Couedel, A.. Beaumont, M-C. Fournie-Saluski and B.P. Roques, Biochem. Biophys. Res. Commun., 187, 1336 (1992).
37. S. Clarke, J.P. Vogel, R.J. Deschenens and J. Stock, Proc. Natl. Acad. Sci., 85, 4643 (1988).
38. D. Perez-Sala, E.W. Tan, F.J. Canada and R.R. Rando, Proc. Natl. Adad. Sci., 88, 3043 (1991).
39. W.R. Schafer and J. Rine, Annu. Rev. Genet., 30, 209 (1992).
40. C.M.H. Newman and A.I. Magee, Biochem. Biophys. Acta, 1155, 79 (1993).
41. M. Sinensky and R.J. Lutz, BioEssays, 14, 25 (1992)
42. G.M. Bokoch and C.J. Der, Faseb J., 7, 750 (1993).
43. C.J. Marshall, Science, 259, 1865 (1993).
44. R. Gomez, L.E. Goodman, S.K. Tripathy, E. O'Rourke, V. Manne and F. Tamanoi, Biochem. J., 289, 25 (1993).
45. M.C. Seabra, Y. Reiss, P.J. Casey, M.S. Brown, M.S., and J.L. Goldstein, Cell, 65, 429 (1991).
46. Y. Reiss, M.C. Seabra, S.A. Armstrong, C.A. Slaughter, J.L. Goldstein and M.S. Brown, J. Biol. Chem., 266, 10672, (1991).
47. B.T. Sheares, S.S. White, D.T. Molowa, K. Chan, V.D. Ding and R.G. Kroon, Biochemistry, 28, 8129 (1989).
48. C.A. Omer, A.M. Kral, R.E. Diehl, G.C. Prendergast, S. Powers, C.M. Allen, J.B. Gibbs and N.E. Kohl, Biochemistry,, 32, 5167 (1993).
49. D.L. Pompliano, M.D. Schaber, S.D. Mosser, C.A. Omer, J.A. Shafer and J.B. Gibbs, Biochemistry, 32, 8341 (1993).
50. J.F. Hancock, K. Cadwallader, H. Paterson and C.J. Marshall, EMBO J., 10, 4033 (1991).
51. J.F. Hancock, H. Paterson, and C.J. Marshall, Cell, 63, 133 (1990).
52. K. Kato, A.D. Cox, M.M. Hisaka, S.M. Graham, J.E. Buss and C.J. Der, Proc. Natl. Acad. Sci., 89, 6403 (1992).
53. G.D. Gray, O.M. Hernandez, H. Hebel, M. Root, J.M. Pow-Sang and E. Wickstrom, Cancer Res., 53, 577 (1993).

54. G.C. Prendergast, J.P. Davide, A. Kral, R. Diehl, J.B. Gibbs, C.A. Omer and N.E. Kohl, Cell Growth Differ., 4, 707 (1993).
55. J.B. Gibbs, Cell, 65, 1 (1991).
56. R. Khosravi-Far, A.D. Cox, K. Kato and C.J. Der, Cell. Growth and Diff., 3, 461 (1992).
57. W.A. Maltese, R. Defendini, R.A. Green, K.M. Sheridan and D.K. Donley, J. Cell Biol., 76, 1748 (1985).
58. J.E. DeClue, W.C. Vass, A.G. Papageorge, D.R. Lowy and B.M. Willumsen, Cancer Res., 51, 712 (1991).
59. F.F. Moomaw and P.J. Casey, J. Biol. Chem., 267, 17438 (1992).
60. M.C. Seabra, J.L. Goldstein, T.C. Sudhof and M.S. Brown, J. Biol. Chem., 267, 14497 (1992).
61. S. Clarke, Ann. Rev. Biochem., 61, 355 (1992).
62. N.E. Kohl, S.D. Mosser, S.J. deSolms, E.A. Giuliani, D.L. Pompliano, S.L. Graham, R.L. Smith, E.M. Scolnick, A. Oliff and J.G. Gibbs, Science, 260, 1934 (1993).
63. G.L. James, J.L. Goldstein, M.S. Brown, T.E. Rawson, T.C. Somers, R.S. McDowell, C.W. Crowley, B.K. Lucas, A.D. Levinson and J.C. Marsters, Science, 260, 1937 (1993).
64. R.E. Finney, S.M. Robbins and J.M. Bishop, Current Biology, 3, 805 (1993).
65. Y. Reiss, J.L.. Goldstein, M.C. Seabra, P.J. Casey and M.S. Brown, Cell, 62, 81 (1990).
66. Y. Reiss, S.J. Stradley, L.M. Gierasch, M.S. Brown and J.L. Goldstein, Proc. Natl. Acad. Sci., 88, 732 (1991).
67. J.L. Goldstein, M.S.Brown, S.J. Stradley, Y. Reiss and L.M. Gierasch, J. Biol. Chem., 266, 15575 (1991).
68. M.S. Brown, J.L. Goldstein, K.J. Paris, J.P. Burnier and J.C. Marsters, Jr., Proc. Natl. Acad. Sci., 89, 8313 (1992).
69. K. Kato, A.D. Cox, M.M. Hisaka, S.M. Graham., J.E. Buss and C.J. Der, Proc. Natl. Acad. Sci., 89, 6403 (1992).
70. J.B. Stimmel, R.J. Deschenes, C. Volker, J. Stock and S. Clarke, Biochem., 29, 9651 (1990).
71. S.L. Moores, M.D. Schaber, S.D. Mosser, E. Rands, M.B. O'Hara, V.M. Garsky, M.S. Marshall, D.L..Pompliano and J.B. Gibbs, J. Biol. Chem., 266, 14603 (1991).
72. A.M. Garcia, C. Rowell, K. Ackermann, J.J. Kowalczyk and M.D. Lewis, J. Biol. Chem., 268, 18415 (1993).
73. K. Leftheris, T. Kline, W. Lau, L. Mueller, V.S. Goodfellow, M.K. DeVirgilio, Y.H. Cho, C. Ricca, S. Robinson, V. Manne and C.A. Meyers, 13th American Peptide Symposium, Poster 525, Edmonton, Alberta, Canada, June 20-25, 1993.
74. Y. Reiss, M.S. Brown and J.L. Goldstein, J.Biol. Chem. 267, 6403 (1992).
75. W-J. Chen, J.F. Moomaw, L. Overton, T.A. Kost and P.J. Casey, J. Biol. Chem., 268, 9675 (1993).
76. S.J. Stradley, J. Rizo and L.M. Gierasch, Biochem., 32, 12586 (1993).
77. M. Nigam, C.-M. Seong, Y. Qian, A.D. Hamilton and S.M. Sebti, J. Biol. Chem., 268, 20695 (1993).
78. D.L. Pompliano, E. Rands, M.D. Schaber, S.D. Mosser, N.J. Anthony and J.B. Gibbs, Biochemistry, 31, 3800 (1992).
79. S.A. Biller, C. Forster, E.M. Gordon, T. Harrity, W.A. Scott and C.P. Closek, Jr., J. Med. Chem., 31, 1869 (1988).
80. J.B. Gibbs, D.L. Pompliano, S.D. Mosser, E. Rands, R.B. Lingham, S.B. Singh, E.M. Scolnick, N.E. Kohl and   A. Oliff, J. Biol. Chem., 268, 7617 (1993).
81. R. Kothapalli, N. Guthrie, A.F. Chambers and K.K. Carroll, Lipids, 28, 969 (1993).
82. F. Tamanoi, Tr. Biochem. Sci., 18, 349 (1993).
83. W.C. Liu, M. Barbacid, M. Bulgar, J.M. Clark, A.R. Crosswell, L. Dean, T.W. Doyle, P.B. Fernandes, S. Huang, V. Manne, D. M. Pirnik, J.S. Wells and E. Meyers, J. Antibiot., 45, 454 (1992).
84. M. Hara, K. Akasaka, S. Akinaga, M. Okabe, H. Nakano, R. Gomez, D. Wood, M. Uh and F. Tamanoi, Proc. Natl. Acad. Sci., 90, 2281 (1993).
85. D. Van der Pyl, J. Inokoshi, K. Shiomi, H. Yang, H. Takeshima and S. Omura, J. Antibiot., 45, 1802 (1992).
86. S.B. Singh, D.L. Zink, J.M. Liesch, M.A. Goetz, R.G. Jenkins, M. Nallin-Omstead, K.C. Silverman, G.F. Bills, R.T. Mosley, J.B. Gibbs, G. Albers-Schonberg and R.B. Lingham,Tetrahedron, 49, 5917 (1993).
87. S.B. Singh, Tetrahedron Lett., 34, 6521 (1993).
88. S. Omura, D. Van der Pyl, J. Inokoshi, Y. Takahashi and H. Takeshima, J. Antibiot., 46, 222 (1993).
89. K. Shiomi, H. Yang, J. Inokoshi, D. Van der Pyl, a. Nakagawa, H. Takeshima and S. Omura, J. Antibiot., 46, 229 (1993).
90. P.L. Crowell, S. Lin, E. Vedejs and M.N. Gould, Cancer Chemother. Pharmacol., 31, 205 (1992).
91. P.L. Crowell, R.R. Chang, Z. Ren, C.E. Elson and M.N. Gould, J. Biol. Chem., 266, 17679 (1991).
92. S. Miura, K. Hasumi and A. Endo, FEBS Lett., 318, 88 (1993).
93. Y-T. Ma, B.A. Gilbert and R.R. Rando, Biochemistry, 32, 2386 (1993).
94. C. Volker, R.A. Miller, W.R. McCleary, A. Rao, M. Poenie, J.M. Backer and J.B. Stock, J. Biol. Chem. 266, 21515 (1991).
95. C. Sobotka-Briner and D. Chelsky, J. Biol. Chem. 267, 12116 (1992).
96. B.A. Gilbert, E.W. Tan, D. Perez-Sala and R.R. Rando, J. Am. Chem. Soc., 114, 3966 (1992).

SECTION IV. IMMUNOLOGY, ENDOCRINOLOGY AND METABOLIC DISEASES

Editor: William K. Hagmann
Merck Research Laboratories, Rahway, NJ, 07065-0900

Chapter 18. Non-Immunophilin-Related Immunosuppressants

William H. Parsons
Merck Research Laboratories, Rahway, NJ 07065

Introduction - The natural product immunosuppressants cyclosporin A (CsA, Sandimmune®) (1), and FK-506 (Prograf®) (2) have revolutionized treatment in the prevention of transplant rejection and have broadened the potential transplant patient population. Perhaps equally exciting are clinical studies with CsA demonstrating the potential of immunosuppressants in the treatment of autoimmune diseases such as rheumatoid arthritis, asthma, psoriasis and dermatitis, Crohn's disease, uveitis and diabetes (3). Despite their improved clinical efficacy, CsA and FK-506 possess significant toxic side effects, in particular, nephrotoxicity and neurotoxicity which limit their potential in the treatment of autoimmune diseases in the general patient population (3). Another liability of these compound classes is their common mechanism of immunosuppression. When they are complexed with their respective immunophilins FKBP and cyclophilin, FK-506 and CsA inhibit the protein phosphatase calcineurin (4). Calcineurin has been shown to be a ubiquitous enzyme found in most tissues including kidney and brain, and it has been demonstrated that the toxicity of these compound classes is mechanism related (5). Significant efforts, therefore, are being directed toward more selective, alternate immunosuppressive strategies. Several new immunosuppressants are currently in Phase II/III clinical studies for the treatment of transplantation rejection and various autoimmune diseases. These compounds may be of intrinsic utility or may act additively or synergistically in combination with suboptimal doses of CsA or FK-506 to reduce toxicity. As well, new biological targets that may be more selective to the immune system are being identified and pursued.

IMMUNOSUPPRESSANTS CURRENTLY IN CLINICAL INVESTIGATION

Nucleotide Biosynthesis Inhibitors - Purine biosynthesis inhibitors such as azathioprine and methotrexate have been used for the prevention of transplantation rejection and the treatment of autoimmune diseases. These compounds cause non-specific antiproliferative effects, and consequently, possess inherent toxicities. Azathioprine and its metabolite 6-mercaptopurine are nonselective, inhibiting several enzymes involved with purine biosynthesis (6). Azathioprine is also mutagenic; one of its metabolites, thioguanosine, is incorporated into DNA (7). Methotrexate inhibits thymidylate synthetase, is a nonselective de novo purine synthesis inhibitor (8), and has a significant toxicity profile including hepatotoxicity, pneumonitis and bone marrow suppression (9).

It was observed that children with a deficiency in adenosine deaminase (ADA) suffer severe immunodeficiency effects. These effects result from selective depletion of T and B cells without alteration in neutrophil, erythrocyte, or platelet levels or brain function (10).

Cells have two purine biosynthesis pathways, the *de novo* and salvage pathways. ADA is an enzyme critical to the *de novo* purine biosynthesis pathway, and T and B cell proliferation is dependent on *de novo* purine biosynthesis (11). Inosine monophosphate dehydrogenase (IMPD) converts inosine monophosphate to xanthosine monophosphate and is the rate-controlling step in guanine nucleotide biosynthesis by the *de novo* purine biosynthesis pathway.

Mycophenolic acid (MPA) **1** is a selective noncompetitive inhibitor of inosine monophosphate dehydrogenase (IMPD). MPA depletes GTP and deoxyGTP pools, producing reductions in RNA and DNA synthesis, thus suppressing T and B cell proliferation (12). MPA and its orally bioavailable morpholinoethyl ester, mycophenolate mofetil (RS-61443), have demonstrated efficacy in a variety of transplantation models, and RS-61443 **2** is currently being extensively evaluated in clinical studies for the prevention and reversal of graft rejection (13).

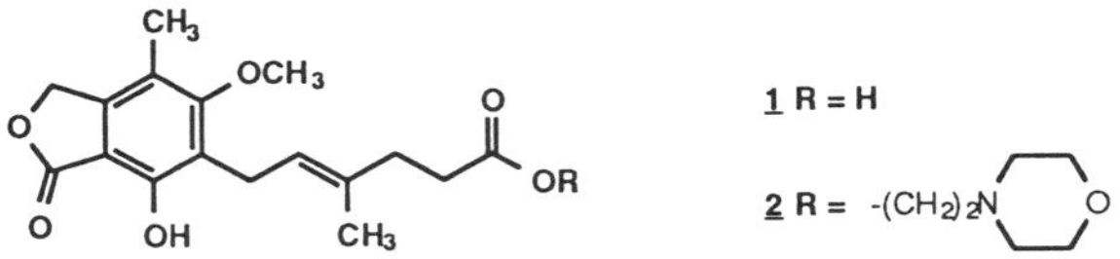

$$1 \ R = H$$
$$2 \ R = -(CH_2)_2N\!-\!O$$

To date, RS-61443 has been used for the treatment of refractory rejection in more than 150 liver (14), renal (15 - 17) and heart (18) transplant patients. RS-61443 is reported to be safe and well tolerated at oral doses as high as 3500 mg/kg/day with no evidence of nephrotoxicity, neurotoxicity or hypertension. In clinical studies evaluating combination therapy with CsA, the incidence in rejection crises are significantly reduced with RS-61443 at doses above 2000 mg/kg. In a rescue study in kidney transplant patients, RS-61443 was effective at doses greater than 2000 mg/kg/day in reversing ongoing rejection episodes (19). RS-61443 is reported to be superior to azathioprine in triple drug therapy in the treatment of early rejection episodes in renal transplantation. Potential indications for RS-61443 include CsA-sparing treatment and replacement of azathioprine in double or triple therapy (20). An NDA filing is planned for 1994 (21).

MPA may be unique among existing immunosuppressants in its potential for the treatment of chronic transplant rejection. Chronic rejection of liver, kidney and cardiac transplants is characterized by complex proliferative and obliterative arteriopathy initally of small and medium sized arteries, due to proliferation of smooth muscle cells and fibroblasts (22). CsA and FK-506 have minimal efficacy in chronic rejection models, and only toxic levels of azathioprine are effective. However, normal dose levels of RS-61443 are effective in preventing chronic rejection in several *in vivo* models (23, 24). Therefore, RS-61443 may provide adjunct treatment with CsA or FK-506.

RS-61443 is also reported to demonstrate efficacy in rheumatoid arthritis and psoriasis patients (25 - 27). Common drug-related side effects were mostly GI-related, nausea, diarrhea and cramping, along with incidences of thrombocytopenia. In a 2 year multicenter, double-blind, placebo-controlled study in 395 psoriatic patients, effective dose levels of RS-61443 in triple drug therapy were in the 1600 to 4800 mg/day range. 85 patients have been maintained on RS-61443 for 13 years. The conclusions from these studies are that with proper dosing, most patients do not experience adverse reactions, and those that do, have side effects no worse than would be expected with alternative therapies.

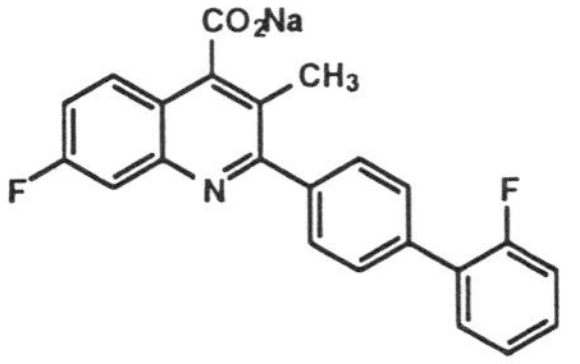

3 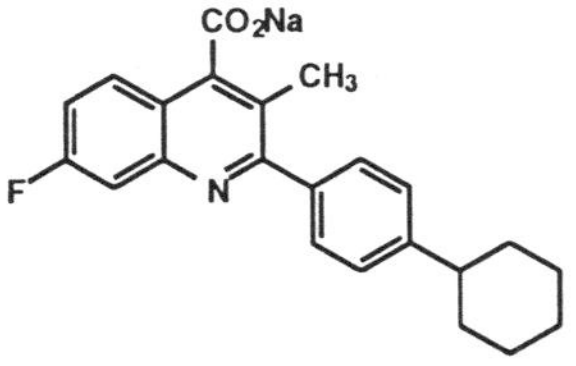   4

A second new nucleotide biosynthesis inhibitor, brequinar sodium (BQR, DUP-785) **3**, is currently in Phase II studies for the prevention of allograft rejection (28). BQR which evolved from the screening active NSC 339768 **4** inhibits *de novo* pyrimidine biosynthesis by reversible, non-competitive inhibition of dihydroorotate dehydrogenase (Ki = 16 nM), the fourth in a six enzyme cascade that produces uridine and cytidine, critical for RNA and DNA synthesis (29). BQR exhibits IC_{50} values of 66 nM in the human mixed lymphocyte reaction and 1.5 uM in inhibiting mitogen-induced proliferation of T cells (30).

Unlike, CsA and FK-506, BQR inhibits both B- and T-cell mediated responses and antibody production (31). Consequently, BQR was demonstrated to be more effective than CsA in preventing accelerated rejection of transplants in previously sensitized animals (32, 33). These observations would further suggest that BQR would be effective in xenotransplant animal models which has been confirmed (34).

Since they produce their immunosuppressive effects by distinct mechanisms, combination immunosuppression with BQR and CsA is additive if not synergistic. In a rat heterotopic cardiac transplant model, single therapy of BQR (3 mg/kg three times/week) or CsA (2.5 mg/kg/day) prolonged allograft survival 10 and 16 days, respectively. Combination therapy with the same dosing regimens were synergistic, prolonging survival 31 days (35). Synergistic activity has also been demonstrated in xenograft models and with other immunosuppressants including FK-506, RS-61443 and rapamycin.

BQR is readily water soluble and exhibits >90% oral bioavailability so it can be readily administered by the oral, i.m. or intravenous routes. Phase I clinical studies were initiated in 1991 and Phase II/III studies in transplant patients are in progress (28). Its reported toxic side effects are those expected with an antimetabolite, are dose-related, reversible and include GI distress, dermatitis, leukopenia, thrombocytopenia and bone marrow hyperplasia (36).

<u>Purine Nucleoside Phosphorylase Inhibitors</u> - In 1975, a case was reported of a child with a severe deficiency of purine nucleoside phosphorylase (PNP) resulting in significant lymphopenia and a pronounced depression of T cell responses to mitogenic and allogeneic stimuli. B cell, erythrocyte and other cell function were unaffected (37). Purine nucleoside phosphorylase catalyzes the reversible phosphorolysis of ribonucleosides and 2'-deoxyribonucleosides to the purine and ribose- or 2'-deoxyribose-1-phosphate (38). PNP produces its catabolic activity in the purine salvage pathway and its substrates are guanine, inosine and other 6-oxopurines. Inhibition of PNP in T cells blocks DNA synthesis which results in cell death. Thus, inhibition of PNP may provide a new approach to immunosuppression.

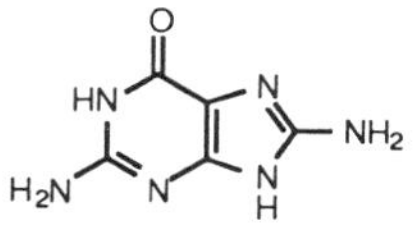

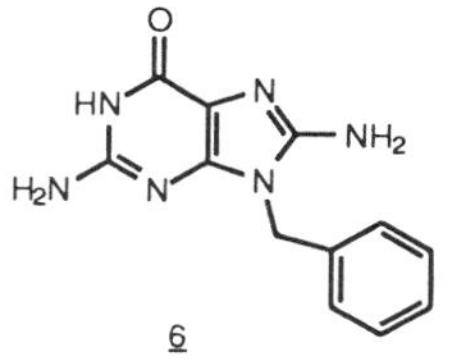

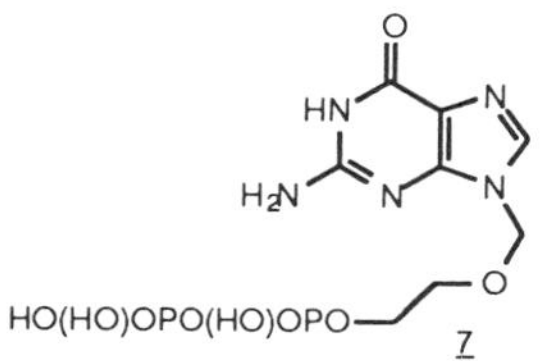

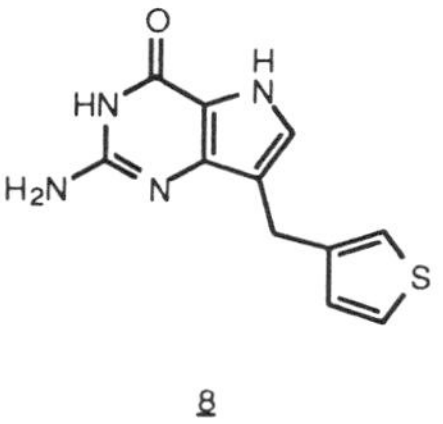

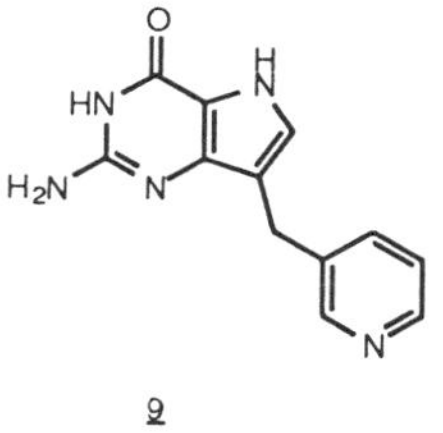

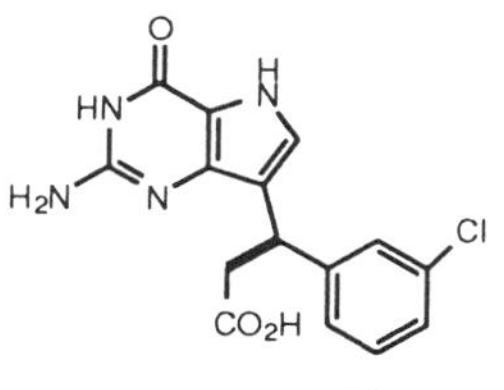

Initial attempts to develop PNP inhibitors resulted in compounds that were not sufficiently potent or cell permeable to be taken into clinical studies. Recently, a structure-based approach to design more effective PNP inhibitors proved to be more successful (39 - 41). The X-ray crystal structure of human erythrocyte PNP was determined to 2.8 Å resolution (42, 43). For design purposes, several known PNP inhibitors including 8-aminoguanine **5** (Ki = 1.2 uM), and bisubstrate inhibitors 8-amino-9-benzyl-guanine **6** (Ki =

0.2 uM) and acyclovir diphosphate $\underline{7}$ (Ki = 8.7 nM) were co-crystallized with human PNP and their X-ray crystal structures determined. Acyclovir diphosphate is a very potent PNP inhibitor, but it suffers from enzymatic and chemical lability and lacks cell permeability. Analogs derived from modelling studies based upon these early crystal structures led to a series of more potent PNP inhibitors. In an iterative process, approximately 60 analogs were prepared and the X-ray crystal structures of half of them co-crystallized with human PNP were determined.

From these studies, BCX-5 (Cl-1000) $\underline{8}$ was identified as exhibiting improved PNP inhibitory activity. BCX-5 inhibits human erythrocyte PNP with a Ki value of 67 nM and is active in *in vivo* models (40a, b). BCX-5 is reported to be in development. BCX-34 $\underline{9}$, a variant in this class, inhibits human PNP with a Ki value of 31 nM and inhibits T cell proliferation with and IC_{50} value of 1.4 uM. BCX-34 is being developed for oral, *i.v.* and topical applications, and in Phase II clinical studies, it is reported to demonstrate efficacy in the topical treatment of severe psoriasis and cutaneous T cell lymphoma (CTCL) (44).

Further modelling studies using inhibitor-PNP X-ray structures and Monte Carlo energy minimization methods led to trisubstrate inhibitors exemplified by compound $\underline{10}$ (Ki = 6 nM) and which are the most potent membrane permeable PNP inhibitors disclosed to date.

OTHER IMMUNOSUPPRESSANTS IN CLINICAL STUDIES

15-Deoxyspergualin - 15-Deoxyspergualin $\underline{11}$ (DSG, 15-dos) (45), a synthetic derivative of the antitumor and immunosuppressant spergualin $\underline{12}$ which was first isolated from *Bacillus laterosporus* (46, 47), has been recommended for approval in Japan (as Spanidin) for the treatment of kidney transplant rejection (48).

$\underline{11}$ R = H

$\underline{12}$ R = OH

The mode of immunosuppressive action of DSG, though not fully elucidated, appears to be involved with T helper cell expansion and effector cell differentiation rather than early T cell signaling events. In contrast to FK-506 and CSA, DSG has minimal affect on T cell responses to mitogen, or in the human mixed lymphocyte reaction (49). However in transplantation models, DSG is nearly as effective as FK-506 and is more potent than CsA. At the cellular level, DSG inhibits antigen processing and degranulation and presentation by macrophages and monocytes (50 - 52). At the mechanistic level, DSG binds to a member of the HSP70 family of heat shock proteins (53). This observation may lead to unveiling the specific mode by which DSG produces its immunosuppressive activity.

In the clinic, intravenous DSG has been utilized in the treatment of over 300 episodes of renal transplant rejection. The optimal dosing regimen is 3 to 5 mg/kg for 7 to 10 days. This regimen is effective in 70% of steroid-resistant rejection episodes and in 88% of episodes when DSG was used in combination with steroid therapy. Adverse reactions include facial numbness, GI distress, leukopenia, thrombocytopenia and anemia (54).

In contrast to FK-506 and CsA, DSG is effective in xenograft models (55), and initial clinical studies indicate that it may be useful in pancreatic islet xenograft therapy (56 - 58). DSG is also being evaluated for the treatment of graft vs host disease (GvHD), lupus, multiple sclerosis and nephritis. A liability of DSG is lack of oral bioavailability.

Vitamin D_3 Analogs - 1α,25-Dihydroxyvitamin D_3 (calcitriol, 1,25-$(OH)_2$-D_3) $\underline{13}$, the active form of vitamin D_3 $\underline{14}$ plays a critical role in calcium hemostasis, mineral and bone metabolism and is prescribed for treatment of hypocalcemia.

Calcitriol binds to a specific steroid-type 1,25-(OH)$_2$-D$_3$ receptor in a variety of cell types including human monocytes and T and B cells when activated by lectins such as PHA (59, 60). Calcitriol potently inhibits PHA-induced production of IL-2 by human T cells with an IC$_{50}$ value of 2 pM. Other vitamin D3 metabolites were orders of magnitude less effective (61). It also inhibits activated T cell-induced production of immunoglobulin by B cells. Calcitriol produces its immunosuppressive effects by selective inhibition of T helper (CD4) lymphocyte proliferation, but not T suppressor (CD8) lymphocytes, despite the observation that it binds to its receptors on each cell type with the same Kd value (0.1 nM). These observations suggest that the effect of calcitriol on CD4 T cells may be indirect (62, 63).

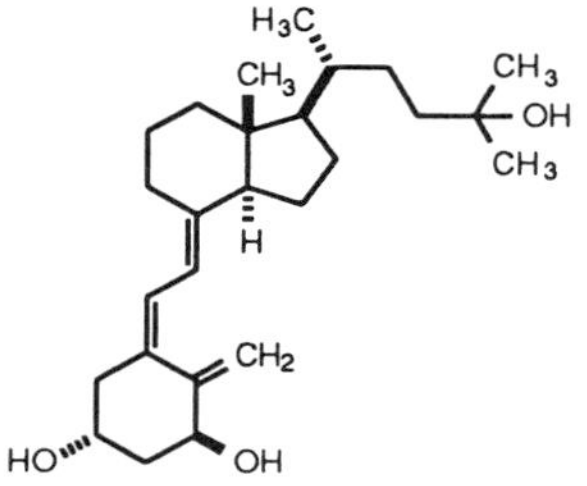

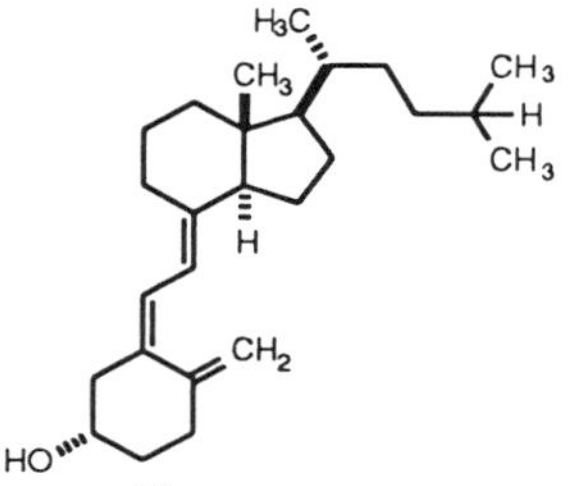

The efficacy of topical calcitriol in the treatment of psoriasis has been reported (64, 65). Patients with psoriasis that were treated topically with calcitriol ointment experienced significant reductions in lesion scores which corresponded to a decrease in PMN and T cell accumulation in the skin. Also significantly affected were epidermal proliferation and keratinization.

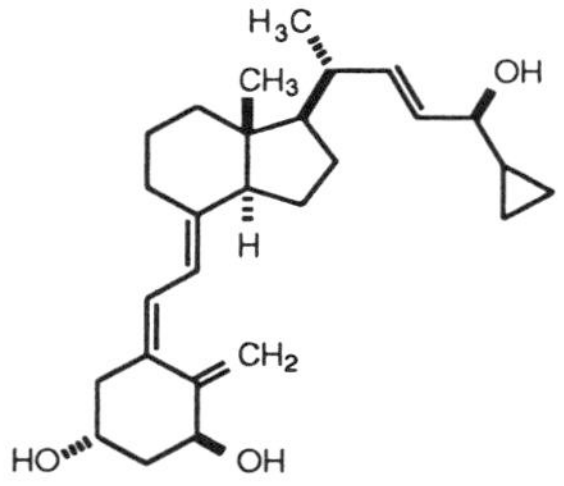

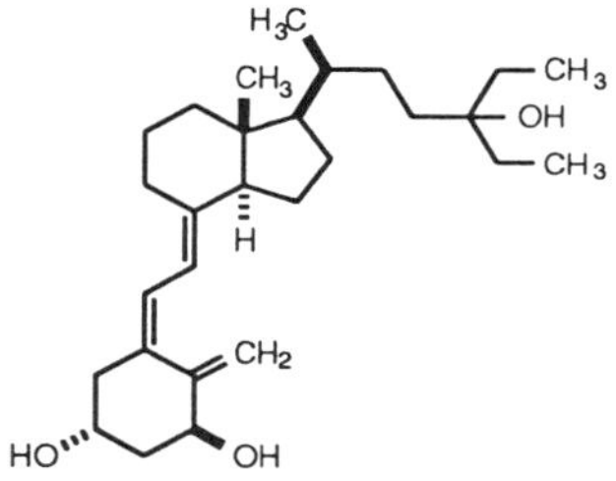

A liability of calcitriol is its calcemic activity. In animals, hypercalcemia is observed at systemic calcitriol drug levels as low as a few ug per day creating a potential risk in the topical application of calcitriol. A program to identify more specific analogs of calcitriol led to the identification of calcipotriol (MC 903) **15**, a synthetic analog wherein the C25 hydroxyl group is transposed to the C24 (R) position (66, 67). Calcipotriol is equipotent to calcitriol in binding to the 1,25(OH)$_2$D$_3$ receptor with IC$_{50}$ values in the 1 nM range. In the same cell line (human histiocytic lymphoma U937 cells), both compounds inhibited proliferation with similar IC$_{50}$ values in the 10 nM range. When their effects on calcium metabolism in rats were compared, calcipotriol was approximately 100-fold less hypercalcemic than calcitriol when given by the oral or intraperitoneal routes. Extensive studies have demonstrated the safety and effectiveness of topical calcipotriol for the treatment of psoriasis (68, 69), and it was recently approved by the FDA for the treatment of psoriasis (70).

Further structure-activity studies of over 200 vitamin D3 analogs led to the identification of significantly more potent and selective compounds (71). In particular, a class of C20-epi-vitamin D3 analogs exemplified by KH-1060 **16** exhibits superior antiproliferative effects. KH-1060 binds to the 1,25(OH)$_2$D$_3$ receptor with an IC$_{50}$ value of 13 pM and inhibits U937 cell proliferation with an IC$_{50}$ value of 1 pM. Moreover, KH-1060 inhibits production of IL-2 in PHA-activated T cells with an IC$_{50}$ value of 3×10^{-14} M compared to a value of 1 nM for calcitriol and CsA in this study. In the mixed lymphocyte reaction (MLR) KH-1060 exhibited an IC$_{50}$ value of 5×10^{-15} M compared to values of 5.2×10^{-11} M and 4.4 nM for calcitriol and CsA, respectively. Despite its increased antiproliferative efficacy, KH-1060 exhibited similar effects to calcitriol on calcium metabolism *in vivo*. More potent, safer vitamin D3 analogs may be useful intrinsically or in combination with CsA in transplantation therapy.

<u>Protein Kinase C Inhibitors</u> - The protein kinase C inhibitor kynac (L-*threo*-dihydrosphingosine) **17** is in Phase II studies for the topical treatment of psoriasis (72).

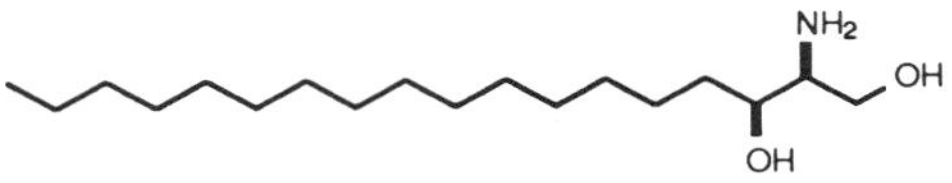

17

In Phase I trials, topically applied kynac was well tolerated, showed no skin irritation or detectable blood levels following topical administration. In the first clinical study, topical application of kynac (2%) failed to show clinical efficacy in psoriatic patients. There was a small but favorable effect on redness, scaling and lesion thickness, but it was not statistically significant.

<u>'NOVEL' IMMUNOMODULATORS AND A NEW IMMUNOSUPPRESSIVE TARGET</u>

Several interesting compounds have been recently reported to exhibit immunosuppressive activity. Listed below are a subset that, subjectively, may be more intriguing than others. There has been an explosion of publications on the significance of CD28 in T cell activation and the potential of its blockade in the treatment of autoimmune diseases.

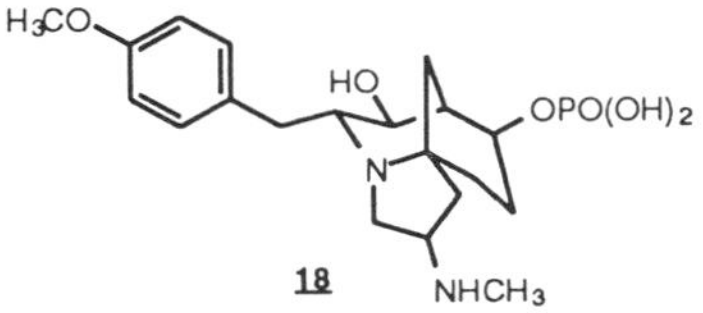

18

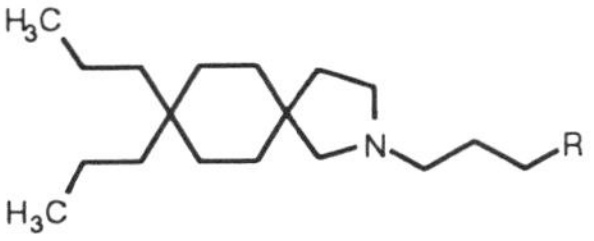

19 R=N(CH$_3$)$_2$
20 R=N(CH$_2$CH$_2$)$_2$CH$_2$

<u>WF1123A</u> **18** - This tricyclic natural product isolated from *Cladobotryum sp.* 11231, is reported to exhibit immunosuppressive properties (73). WF-11231A inhibits Con A-induced murine T cell proliferation with an IC$_{50}$ value of 60 nM. In a rodent skin allograft model, WF11231 prolonged graft survival with intraperitoneal doses of 1 to 3 mg/kg. No mechanism of action is reported.

<u>Azaspiranes</u> - Azaspirane SKF-105685 **19** is reported to be effective in rodent models of autoimmune diseases including adjuvant arthritis and experimental encephalomyelitis in rats and a lupus-like disease in MRL mice. In a model of autoimmune type I diabetes, SKF 106610 **20** was administered orally to BB rats (15 mg/kg/day) for 100 days. Drug treated rats showed a significant reduction in the incidence of diabetes compared with control rats (32% *vs* 80%). This protection was accompanied by decreased lymphocytic infiltration of the pancreatic islets (74). The mechanism by which these azaspiranes exhibit their immunosuppressive effects correlates with non-specific suppresser cell activity linked to a non-T, non-B, non-NK semi-adherent spleen cell population (75, 76). In dogs treated with SKF-105685 (1 mg/kg/day, *p.o.*) for 30 days, SKF-105685 effectively induced suppresser cell activity (defined as the ability of irradiated effector spleen cells to suppress the proliferative response of control responder spleen cells to Con A), and inhibited primary antibody responses of spleen cells to sheep erythrocytes *ex vivo* (defined as the ability of spleen cells from control or drug-treated animals to mount an antibody response to sheep red blood cells measured *ex vivo* in a hemolytic plaque-forming cell assay). Regarding generalized immunosuppression, SKF-105685 had no effect on the total number or relative abundance of white blood cell types, nor did it have any effect on the responses or IL-2 production of peripheral blood monocytes or unfractionated blood cells to pokeweed mitogen or Con A + PHA. The specific mechanism of 'selective' immunosuppression is still to be defined (77).

<u>Spiperone</u> - Spiperone **21**, a neuroleptic agent which binds to dopamine and 5-HT receptors also exhibits immunosuppressive activity (78), although its mechanism of immunosuppressive action has not been reported. Spiperone is effective at dose ranges of

1.5 to 30 mg/kg *s.c.* and topically (0.013% solution) in reducing swelling and T cell counts in the murine DH reaction. T cells have 5-HT receptors, but structurally unrelated 5-HT antagonists trazadone and mianserin and dopamine antagonist haloperidol are not effective in the murine DH reaction. A quaternary ammonium salt of spiperone is reported to exhibit reduced neuroleptic effects without altering its immunosuppressant activity.

21   **22**

Mycalamide A - Mycalamide A **22**, isolated from marine sponges (79), is cytotoxic to P388 murine leukemia cells, and exhibits general anti-proliferative activity and immunosuppressive activity. CsA and FK-506 inhibit CD4+ T cell proliferation induced by anti-CD3 Ab plus mB7CHO cells (CHO cells transfected with murine B7), but not by mB7CHO cells + PMA (via CD28). Mycalamide A, however, blocks T cell activation by both routes and is 10-fold more potent than FK-506 and 1000-fold more potent than CsA (80). The specific target by which mycalamide A produces its immunosuppressive activities is not known. However, it is known that it inhibits p21ras biosynthesis and reverts the morphology of ras-transformed NRK cells to normal morphology (81).

CD28 - After presentation of antigen to the CD3/T cell receptor complex (CD3/TCR) of T cells by antigen presenting cells (APCs) including B cells, a costimulatory signal by CD28 or CD2 is required for full T cell activation and control. CD28 is a homodimeric receptor that is constitutively expressed on the surface of 95% of CD4+ and 50 % of CD8+ T cells. The recognition of this receptor by the B cell activation antigen B7-1 (BB1), expressed on activated B cells, dendritic cells and macrophages, is critical for full T cell activation via the CD3/TCR receptor (82 - 84). B7-1 binds to CD28 with a Kd value of approximately 200 nM. CTLA4, a second higher affinity T cell receptor for B7-1 (Kd ~ 20 nM), has 28% amino acid homology to CD28 (85). In contrast to CD28, mRNA for CTLA4 is expressed on T cells only after they have been activated.

CTLA4Ig, a soluble chimeric CTLA4 - IgC fusion protein, binds potently to human and murine B7-1 with an approximate Kd value of 12 nM, blocks binding of B7-1 to CD28, inhibits T cell activation and induces T cell unresponsiveness or anergy. (86, 87). Thus, the CD28 signal transduction pathway may be a viable target for immunosuppression.

The effects of CTLA4Ig on suppressing antibody responses in a human islet to streptozotocin-treated diabetic B6 mouse xenotransplant model demonstrates its immunosuppressive properties (88). Graft survival in 3 of 7 CTLA4Ig-treated mice (10 ug for 14 consecutive days) was greater than 80 days and the average survival in the other 4 mice was 12.75 days compared to only 5.6 days in PBS-treated control mice. When the dose was increased to 50 ug every other day for 14 days, graft viability in all 12 mice was prolonged for the length of the study, over 45 days. Histological examination revealed that in contrast to controls, islets of treated mice were intact, exhibited no evidence of lymphatic infiltration and the β-cells were producing human insulin and somatostatin. CTLA4IG also demonstrated efficacy in blocking immune responses to standard immunogens keyhole limpet hemocyanin and sheep erythrocytes in mice (89).

Recent studies have identified at least two other B7 ligands, B7-2 and B7-3 (90 - 95). It would appear that B7-2, constitutively expressed, may play a more critical role early in T cell activation, followed perhaps later by events stimulated by B7-1 which is not constitutively expressed. The role of B7-3 has yet to be addressed.

Recent studies have demonstrated that activation of T cells by anti CD28 Ab leads to activation of p70 S6 kinase (96). Crosslinking the T cell receptor with anti CD3 Ab also leads to activation of P70 S6 kinase, but the activation follows a distinctly different time course than that by anti CD28 Ab. As well, this latter activation also resulted in activation of P85 rsk enzyme whereas CD28 activation does not. Activation of P70 S6 kinase by anti CD28 Ab was inhibited by rapamycin, but not by FK-506 or CsA. Therefore, P70 S6 kinase appears to be a common signaling enzyme by both the CD28 and CD3 pathways, although their kinetics are different.

The B7 - CD28 studies demonstrate the significant role the CD28 plays in T cell activation. Blockade of this co-stimulatory pathway may provide a new more selective mode of treating autoimmune-related diseases.

Summary-The dramatic advances in understanding of signal transduction processes of the immune system is unveling new approaches to immunosuppressant development. Several current immunosuppressants such as azathioprine, CsA and FK-506 were originally identified from antibacterial, antifungal or general antiproliferative screens. Critical biological steps that are more selective to the immune system (PNP, IMPD and CD28) are now being elucidated that may be viable targets for pharmaceutical intervention. Screens and medicinal chemical approaches directed toward these specific targets may provide new immunosuppressants that are more selective and therefore safer than existing drugs. Such compounds may provide novel therapeutic approaches not only for prevention of transplantation rejection, but also for treatment of broader patient populations suffering from inflammatory diseases (rheumatoid arthritis, asthma, dermatitises, nephritises) known or hypothesized to have immune bases.

References

1.   J. F. Borel, C. Feurer, H. U. Gubler, H. Stahlen, Agents Actions, 6, 468 (1976).
2.   T. Kino, H. Hatanaka, M. Hashimoto, M. Nishiyama, T. Goto, M. Okuhara, M. Kohsaka, H. Aoki, H. Imanaka, J. Antibiot., 40, 1249 (1987).
3.   W. H. Parsons, N. H. Sigal, M. J. Wyvratt, Ann. NY Acad. Sci., 685, 22 (1993).
4.   J. Liu, J. D. Farmer, W. S. Lane, J. Friedman, I. Weissman, S. L. Schreiber, Cell, 66, 807 (1991).
5.   F. J. Dumont, M. J. Staruch, S. L. Koprak, J. J. Siekierka, C. S. Lin, R. Harrison, T. Sewell, V. M. Kindt, T. R. Beattie, M. Wyvratt, N. H. Sigal, J. Exp. Med., 176, 751 (1992).
6.   G. Woldberg, Handb. Exp. Pharm., 85, 517 (1988).
7.   G. B. Elion, Science, 344, 41 (1991).
8.   J. Jolivet, K. H. Cowan, G. A. Curt, N. J. Clendenin, B. A. Chabner, New Engl. J. Med., 309, 1094 (1983).
9.   S. Hirata, T. Matsuhara, R. Saura, H. Tateishi, K. Hirohata, Arthritis Rheum., 32, 1065 (1989).
10.   E. R. Giblett, J. E. Anderson, F. Cohen, B. Polara, H. J. Meuwissen, Lancet ii, 1067 (1972).
11.   A. C. Allison, T. Hovi, R. W. E. Watts, Ciba Found Symp 48, 207 (1977).
12.   M. B. Cohen, J. Maybaum, W. Sadee, J. Biol. Chem. 256, 8713 (1981).
13.   A. C. Allison, and E. M. Eugui, Clin. Transplantation 7, 96, (1993).
14.   C. E. Freise, M. Hebert, R. W. Osorio, B. Nikolai, J. R. Lake, R. S. Kauffman, N. L. Ascher, J. P. Roberts, Transplant. Proc. 25, 1758 (1993).
15.   H. W. Solinger, M. H. Deierhoi, F. O. Belzer, A. G. Diethelm, R. S. Kauffman, Transplantation, 53, 428 (1992).
16.   M. H.. Deierhoi, H. W. Sollinger, A. G. Diethelm, F. O. Belzer, R. S. Kauffman, Transplant. Proc., 25, 693 (1993).
17.   J. R. Salaman, P. J. A. Griffin, R. W. G. Johnson, K. Kohlhaw, W. Land, R. Moore, R. Pichlmayr, R. Sells, Transplant. Proc. 25, 695 (1993).
18.   R. D. Ensley, M. R. Bristow, S. L. Olsen, D. O. Taylor, E. H. Hammond, J. B. O'Connell, D. Dunn, L. Osburn, K. W. Jones, R. S. Kauffman, W. A. Gay, D. G. Renlund, Transplantation, 56, 75 (1993).
19.   H. W. Sollinger, M. H. Deierhoi, F. O. Belzer, A. G. Diethelm, R. S. Kauffman, Transplantation 53, 428, (1992)
20.   A. C. Allison, E. M. Eugui, Clin. Transplantation, 7, 96, (1993).
21.   SCRIP 1903, 10 (1994).
22.   M. L. Foegh, Transpl. Proc., 22, 119 (1990).
23.   A. C. Allison, E. M. Eugui, Immunological Rev., 136, 5 (1993).

24.   R, E, Morris, E. G. Hoyt, M. P. Murphy, E. M. Eugui, A. C. Allison, Transpl. Proc., 22, 1659 (1990).

25.   W. W. Epinette, C. M. Parker, E. L. Jones, M. C. Greist, J. Am. Acad. Dermatol. 17, 962 (1987).

26.   C. M. Franklin, R. Goldblum, C. Robinson, Y-K. Chiang, Arthritis and Rheumatism, 36, 557 ((1993).

27.   M. H. Schiff, R. Goldblum, G. D. Fine, Arthritis and Rheumatism, 36, 5216 (1993).

28.   L. Makowka, L. S. Sher, D. V. Cramer, Immunol. Rev., 136, 51 (1993).

29.   L. Makowka, F. Chapman, D. V. Cramer, Transplant. Proc. 25, Suppl 2, 2 (1993).

30.   L. Makowka, D. V. Cramer, Transplant. Sci., 2, 50 (1992).

31.   B. D. Jaffee, E. A. Jones, S. E. Loveless, S. F. Chen, Transplant. Proc., 25, Suppl. 2, 19 (1993).

32.   C. Yasunaga, D. V. Cramer, F. A. Chapman, H. K. Wang, M. Barnett, G.-D. Wu, L. Makowka, Transplantation (in press).

33.   C. Yasunaga, D. V. Cramer, C. A. Cosenza, P. J. Tusa, F. A. Chapman, M. Barnett, G. D. Wu, B. A. Putnam, L. Makowka, Transpl. Proc., 25, Suppl. 2, 40 (1993).

34.   D. V. Cramer, F. A. Chapman, B. D. Jaffee, I. Zajac, G. Hreha-Eiras, C. Yasunaga, G.-D. Wu, L. Makowka, Transplant., 54, 403 (1992).

35.   C. A. Cosenza, D. V. Cramer, G. Eiras-Hreha, E. Cajulis, H. K. Wang, L. Makowka, Transplant., 56, 667 (1993).

36.   C. L. Arteaga, T. D. Brown, J. G. Kuhn, H. L. Shen, T. J. O'Rourke, K. Beougher, H. J. Brentzel, D. D. Von Hoff, G. R. Weiss, Cancer Res., 49, 4648 (1989).

37.   E. L. Giblett, A. J. Ammann, D. W. Wara, R. Sadman, L. K. Diamond, Lancet i, 1010 (1975).

38.   R. E. Parks, R. P. Agarwal, In The Enzymes, 3rd Ed. P. D. Boyer, Ed. 7, 483 (1972).

39.   D. Stoeckler, S. E. Ealick, C. E. Bugg, R. E. Parks, Federation Proc. 45, 2773 (1986).

40.   a. S. E. Ealick, S. Y. Babu, C. E. Bugg, M. D. Erion, W. C. Guida, J. A. Montgomery, J. A. Secrist III, Proc. Natl. Acad. Sci., 88, 11540 (1991). b. R. B. Gilbertsen, U. Josyula, J. C. Sircar, M. K. Dong, W. Wu, D. J. Wilburn, M. C. Conroy, Biochem. Pharm. 44, 996 (1992).

41.   S. E. Ealick, Y. S. Babu, C. E. Bugg, M. D. Erion, W. A. Guida, J. A. Montgomery, J. A. Secrist III, Ann. NY Acad. Sci. vol. 685, 237 (1993).

42.   W. J. Cook, S. E. Ealick, C. E. Bugg, J. D. Stoeckler, R. E. Parks, J. Biol. Chem. 256, 4079 (1981).

43.   S. E. Ealick, S. A. Rule, D. C. Carter, T. J. Greenhough, Y. S. Babu, W. J. Cook, J. Habash, J. R. Helliwell, J. D. Stoeckler, R. E. Parks, S. F. Chen, C. E. Bugg, J. Biol. Chem. 265, 1812 (1990).

44.   J. A. Montgomery, H. W. Snyder, D. A. Walsh, G. M. Walsh, Drugs of the Future, 18, 887, (1993).

45.   H. Iwasawa, S. Kondo, D. Ikeda, T. Takeuchi, H. Umezawa, J. Antibiot., 35, 1655 (1982).

46.   T. Takeuchi, H. Iinuma, S. Kunimoto, M. Takeuchi, M. Hamada, H. Naganawa, S. Kondo, H. Umezawa, J. Antibiot., 34, 1619 (1981).

47.   H. Umezawa, S. Kondo, H. Iinuma, S. Kunimoto, Y. Ikeda, H. Iwasawa, D. Ikeda, T. Takeuchi, J. Antibiot., 34, 1622 (1983).

48.   SCRIP #1884/85, 23 (1993).

49.   M. A. Tepper, S. Nadler, C. Mazzucco, C. Singh, S. L. Kelley, Ann. NY Acad. Sci., 685, 136 (1993).

50.   H. U. Schorlemmer, H. H. Sedlacek, Int. J. Immunopharmacol., 9, 559, 1987).

51.   G. Dickneite, H. U. Schorlemmer, H. H. Sedlacek, et al., Transplant. Proc., 19, 1301 (1987).

52.   S. Takahara, H. Jiang, Y. Takano, Y. Kokado, M. Ishibashi, A. Okuyama, T. Sonoda, Transplantation, 53, 914 (1992).

53.   S. B. Nadler, M. A. Tepper, B. Schacter, C. E. Mazzucca, Science, 258, 484 (1992).

54.   H. Amemiya, Ann. NY Acad. Sci., 685, 196 (1993).

55.   T. Ochiai, K. Nakajima, K. Sakamoto, N. Nagata, Y. Gunji, T. Asanao, K. Isono, T. Sakamaki, K. Hamaguchi, Transplant. Proc., 21, 829 (1989).

56.   C. G. Groth, Ann. NY Acad. Sci., 685, 193 (1993).

57.   P. F. Gores, J. S. Najarian, E. Stephanian, J. J. Lloveras, S. L . Kelly, D. E. R. Sutherland, Lancet, 341, 19 (1993).

58.   D. E. R. Sutherland, P. F. Gores, A. C. Farney, D. C. Wahoff, A. J. Matas, D. L. Dunn, R. W. G. Gruessner, J. S. Najarian, Am. J. Surgery 166, 456 (1993).

59.   M. R. Haussler, J. W. Pike, J. S. Chandler, S. C. Manolagas, L. J. Deftos, Ann. N. Y. Acad. Sci., 372, 502 (1981).

60.   D. M. Provvedini, C. D. Toukas, L. J. Deftos, S. C. Manolagas, Science 221, 1181 (1983).

61. C. D. Tsoukas, D. M. Provvedini, S. C. Manolagas, Science 224, 1438 (1984).
62. J. M. Lemire, J. S. Adams, V. Kermani-Arab, A. C. Bakke, R. Sakai, S. C. Jordan, J. I mmunol., 134, 3032 (1985).
63. D. M. Provvedini, S. C. Manolagas, J. Clin. Endocrinol. and Met., 68, 774 (1989).
64. E. M. G. J. de Jong, P. C. M. Van de Kerkhof, Br. J. Dermatol., 124, 221 (1991).
65. M. J. P. Gerritsen, H. F. C. Rulo, I. V. Vlijmen-Willems, P. E. J. Erp, P. C. M. Van de Kerkhof, Br. J. Dermatology 128, 666, (1993).
66. L. Binderup, E. Bramm, Biochem. Pharmacol., 37, 889 (1988).
67. M. J. Calverly, Tetrahedron, 43, 4609 (1987).
68. K. Kragbule, Arch. Dermatol., 125, 1647 (1989).
69. K. Kragbule, B. T. Gjertsen, D. de Hoop, T. Karlsmark, P. C. M. Van de Kerkhof, O. Larko, C. Nieboer, J. Roed-Petersen, A. Strand, G. Tikjob, Lancet, 337, 193 (1991).
70. F-D-C Reports, 56, 17 (1994)
71. L. Binderup, S. Latini, E. Binderup, C. Bretting, M. Calverley, K. Hansen, Biochem. Pharmacol., 42, 1569 (1991).
72. SCRIP 1821, 24 (1993).
73. E. Tsujii, T. Kanishi, S. Takase, M. Yamashita, S. Izume, M. Okuhara, World Patent 93/12125, (1993).
74. A. Rabinovitch, W. L. Suarez, H-Y. Qin, R. F. Power, A. M. Badger, J. of Autoimmunity 1, 39, (1993).
75. A. M. Badger, M. J. DiMartino, J. E. Talmadge, D. H. Picker, D. A. Schwartz, J. W. Dorman, C. K. Mirabelli, N. Hanna, Int. J. Immunopharmacol., 11, 839 (1989).
76. A. M. Badger, A. G. King, J. E. Talmadge, D. A. Schwartz, D. H. Picker, C. K. Mirabelli, N., Hanna, J. Autoimmunity, 3, 485 (1990).
77. J. M. Kaplan, A. M. Badger, E. V. Ruggierri, B. A. Swift, P. J. Bugelski, Int. J. Immunopharmac. 15, 113, (1993).
78. R. J. Sharpe, K. A. Arndt, S. J. Galli, P. C. Meltzer, R. K. Razdan, H. P. Sard , World Patent 93/12789 (1993).
79. N. B. Perry, J. W. Blunt, M. H. G. Munro, L. K. Pannell, J. Am. Chem. Soc., 110, 4850 (1988).
80. F. Galvin, G. J. Freeman, Z. Razi-Wolf, B. Benacerraf, L. Nadler, H. Reiser, Eur. J. Immunol., 23, 283, (1993).
81. H. Ogawara, K. Higashi, K. Uchino, N. B. Perry, Chem. Pharm. Bull (Tokyo), 39, 2152 (1991).
82. T. Hara, S. M. Fu, J. A. Hansen, J. Exp. Med., 161, 1513 (1985).
83. A. Weiss, A. B. Manger, J. Imboden, J. Immunol., 137, 819 (1986).
84. A. Aruffo, B. Seed, Proc. Natl. Acad. Sci. USA, 84, 8573 (1987).
85. G. J. Freeman, D. B. Lombard, C. D. Gimmi, S. A. Brod, K. Lee, J. C. Laning, D. A. Hafler, M. E. Dorf, G. S. Gray, H. Reiser, C. H. June, C. B. Thompson, L. M. Nadler, J. Immunol., 149, 3795 (1992).
86. P. S. Linsley, W. Brady, M. Urnes, L. S. Grosmaire, N. K. Damle, J. A. Ledbetter, J. Exp. Med., 174, 561 (1991).
87. F. A. Harding, J. G. McArthur, J. A. Gross, D. H. Raulet, J. P. Allison, Nature, 356, 607 (1992).
88. D. L. Lenschow, Y. Zeng, J. R. Thistlethwaite, A. Montag, W. Brady, M. G. Gibson, P. S. Linsley, J. A. Bluestone, Science, 257, 789 (1992)
89. P. S. Linsley, P. M. Wallace, J. Johnson, M. G. Gibson, J. L. Greene, J. A. Ledbetter, C. Singh, M. A. Tepper, Science, 257, 792 (1992).
90. K. S. Hathcock, G. Laszlo, H. B. Dickler, J. Bradshaw, P. Linsley, R. J. Hodes, Science, 262, 905 (1993).
91. G. J. Freeman, F. Borriello, R. J. Hodes, H. Reiser, K. S. Hathcock, G. Laszlo, A. J. McKnight, J. Kim, L. Du, D. B. Lombard, G. S. Gray, L. M. Nadler, A. H. Sharpe, Science, 262, 907 (1993).
92. G. J. Freeman, J. G. Gribben, V. A. Boussiotis, J. W. Ng, V. A. Restivo, L. A. Lombard, G. S. Gray, L. M. Nadler, Science, 262, 909 (1993).
93. G. J. Freeman, F. Borriello, R. J. Hodes, H. Reiser, J. G. Gribbon, J. W. Ng, J. Kim, J. M. Goldberg, K. Hathcock, G. Laszlo, L. A. Lombard, S. Wang, G. S. Gray, L. M. Nadler, A. H. Sharpe, J. Exp. Med., 178, 2185 (1993).
94. Y. Boussiotis, J. Gribben, G. Freeman, L. Nadler, Blood, 82, 188a (1993)
95. Z. Razi-Wolf, F. Galvin, G. Gray, H. Reiser, Proc. Natl. Acad. Sci. USA, 90, 11182 (1993)
96. S.-Y. Pai, V. Calvo, M. Wood, B. E. Bierer, Blood, 82, 188a (1993).

Chapter 19. Isozyme-Selective Phosphodiesterase Inhibitors as Antiasthmatic Agents

Siegfried B. Christensen and Theodore J. Torphy
SmithKline Beecham Pharmaceuticals
King of Prussia, Pennsylvania 19406

Introduction - Considerable interest has been generated in the potential utility of isozyme-selective inhibitors of cyclic nucleotide phosphodiesterase (PDE) inhibitors in the treatment of asthma and other inflammatory disorders (1,2). The scientific foundation for this interest is based upon two fundamental principles. First, inhibition of PDE activity increases the cellular content of two key second messengers, cyclic AMP (cAMP) and cyclic GMP (cGMP), thereby activating specific protein phosphorylation cascades that elicit a variety of functional responses. Increases in cAMP content suppress a broad array of functions in inflammatory and immune cells (1,3,4). Moreover, both cAMP and cGMP mediate bronchodilation (5,6). While these activities are attractive from a therapeutic standpoint, the ubiquitous distribution of cyclic nucleotides and PDEs makes the side effect profile of standard, nonselective PDE inhibitors unacceptable. This raises the second scientific principle underpinning the interest in selective PDE inhibitors as antiasthmatic and antiinflammatory agents. Specifically, it is now recognized that several distinct families of PDE isozymes exist with different tissue distributions and functional roles (1,7-9). Consequently, the side effect profile of PDE inhibitors may be improved by targeting a new generation of compounds to a specific isozyme that predominates in inflammatory cells and airway smooth muscle. As discussed in this review, considerable evidence indicates that PDE IV, PDE III and possibly PDE V are the key isozymes in these cells.

ENZYMOLOGY AND MOLECULAR BIOLOGY

PDEs inactivate cyclic nucleotides by hydrolyzing their 3'-phosphoester bonds to form 5'-nucleotide products. At least five families of isozymes are known to exist (7,10), and the recent cloning and expression of an atypical PDE hints at the existence of additional families (11). The most widely accepted PDE nomenclature is shown in Table I. Also shown in Table I are typical kinetic characteristics of enzymes in the various families as well as examples of isozyme-selective inhibitors.

TABLE I. Characteristics of phosphodiesterase isozymes[a]

Family	Isozyme	K_m (μM)		Selective Inhibitors
		cAMP	cGMP	
Iα	Ca^{2+}/CaM-stimulated	30	3	trifluoperazine, vinpocetine
Iβ	Ca^{2+}/CaM-stimulated	1	2	trifluoperazine
II	cGMP-stimulated	10	30	none
III	cGMP-inhibited	0.2	0.2	cilostamide, siguazodan
IV	cAMP-specific	3	>3000	rolipram, Ro 20-1724
V	cGMP-specific	150	1	zaprinast

[a]Nomenclature and values are from refs. 1, 7, 8, 10. Abbreviation: CaM, calmodulin.

ANNUAL REPORTS IN MEDICINAL CHEMISTRY—29

Although a detailed description of the kinetic characteristics of the various isozymes is beyond the scope of this chapter (see ref. 7-10), two general points regarding this subject warrant mention. First, while some of the isozyme families show a degree of selectivity for cAMP or cGMP, nearly all will hydrolyze either cyclic nucleotide under the appropriate conditions. Second, the catalytic activities of the various isozymes can be altered by a number of endogenous regulators (e.g., Ca^{2+}/calmodulin, cGMP, phosphorylation). Moreover, agents that increase cAMP accumulation increase the synthesis of at least one isozyme, PDE IV, probably by stimulating transcription (12).

Substantial advances have been made in understanding the genetic basis for the existence of multiple, phenotypically distinct families of PDE isozymes (10,13). Indeed, it is now believed that multiple subtypes exist within each isozyme family (10,13). For example cDNAs encoding at least four subtypes of PDE IV have been cloned and two have been expressed (14-18). While these subtypes are similar with respect to their amino acid sequence, a degree of heterology exists. For example, human PDE IV_A and PDE IV_B possess a 76% amino acid identity over the entire sequence and a 90% identity within the putative catalytic domain (16,17). In addition, alternatively spliced variants of PDE IV subtypes have been identified (19). This information, combined with data suggesting that PDE IV subtypes have differential cellular or tissue distributions (17-19), raises the possibility of designing inhibitors that are subtype specific, thereby achieving an even greater degree of cellular selectivity than that afforded by current isozyme-selective inhibitors.

There is a substantial sequence homology among members of various PDE families over an internal sequence of ~ 270 amino acids (10,13). This region contains threonines, serines and histidines that are invariant in a dozen PDE sequences, suggesting that these residues play a key role common to all PDEs, e.g., catalytic activity. Recent work on various PDE IV mutants confirms this proposal and indicates that one threonine and at least two histidine residues are critical for the expression of catalytic activity (20,21). Flanking the area of sequence homology are highly variable N-terminal and C-terminal extensions. These areas may represent regulatory domains that are targets for allosteric activators or inhibitors, or serve as substrates for phosphorylation by various protein kinases.

PDE III INHIBITORS

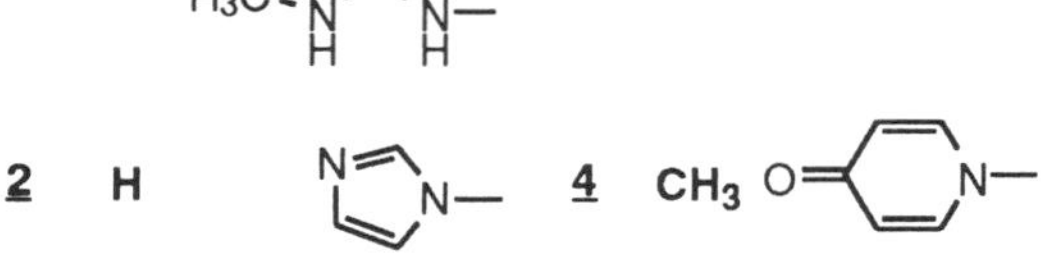

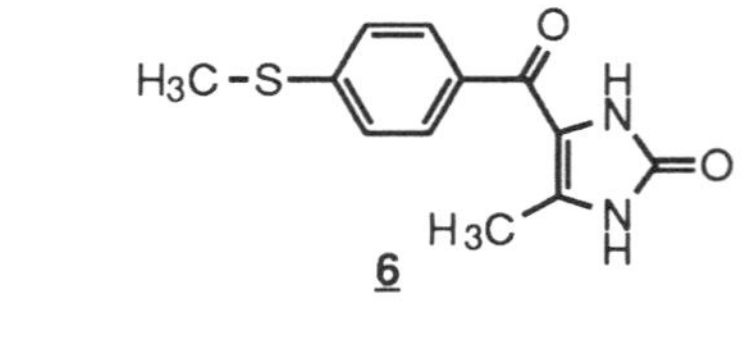

Numerous SAR studies on diverse structural classes of PDE III inhibitors developed originally as cardiotonics led to the elaboration and subsequent extension of a five-point model of the PDE III pharmacophore (22-25), though data obtained with certain classes of inhibitors have suggested the potential for multiple modes of binding to the PDE III active site (26,27). The relationship between these inhibitors and cAMP has been analyzed using computational techniques and led to the proposal that these inhibitors mimic the relatively planar *anti*-cAMP structure (23,26,28). Members of this class include siguazodan (SK&F 94836, **1**), imazodan (CI-914, **2**), CI-930 (**3**), SK&F 95654 (**4**), cilostamide (OCP 3689, **5**) and enoximone (MDL 17043, **6**) (29,30,31). Extensive data on the cardiotonic activity and the platelet anti-aggregatory activity of additional representative compounds from the various structural classes of PDE III inhibitors have been reviewed (31,32).

PDE IV INHIBITORS

The known selective PDE IV inhibitors can be divided into essentially three broad structural classes.

<u>Catechol ethers</u> - The archetypical members of this class are rolipram (ZK 62711, **7**, IC_{50} = 1 μM) and Ro 20-1724 (**8**, IC_{50} = 5μM) (1). SAR studies of rolipram using a partially purified enzyme isolated from bovine aorta and Ro 20-1724 conducted on an enzyme preparation from the hemolysate of rat erythrocytes indicated that substituent effects on the catechol oxygens of the two compounds were very similar (33,34), and a topographical model for PDE IV inhibition based upon the *R*-configuration of rolipram was derived (33). A reversal in the absolute stereochemistry of this model has been proposed based on the demonstration that the *S*-isomer of the rolipram derivative **9** is some 400-fold more potent than the *R*-isomer as an inhibitor of partially purified human monocyte PDE IV (35).

In a series of 5-(catechol ether)-2-imidazolidinones which combines structural features of both rolipram and Ro 20-1724, the primary focus of modification was the substituent on the catechol 3-oxygen, *e.g.*, **10** and **11** (36). A second series of bi-, tri- and tetracyclic hydrocarbons at the 3-alkoxy position of the same parent also was studied, with **12** and **13** among the most potent inhibitors (37). This work was extended to the tetrahydro-pyrimidinone **14**, in which the two enantiomers were essentially equipotent inhibitors of guinea-pig lung PDE IV (38).

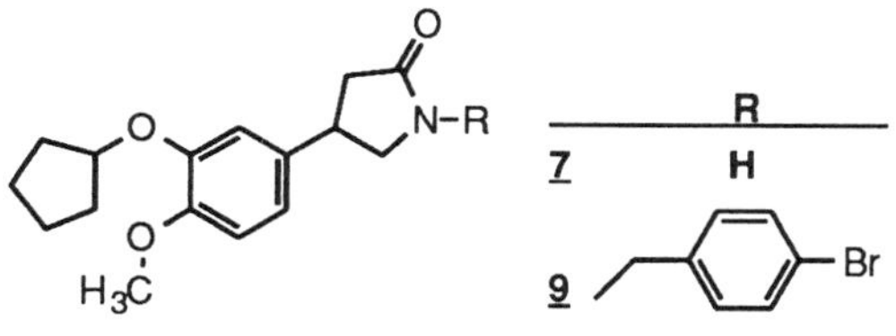

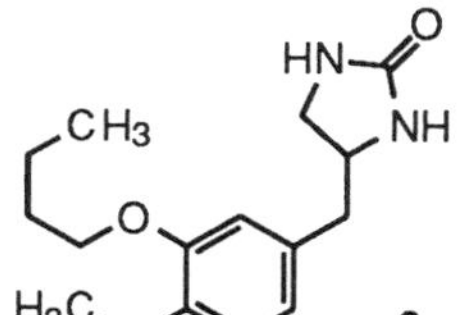

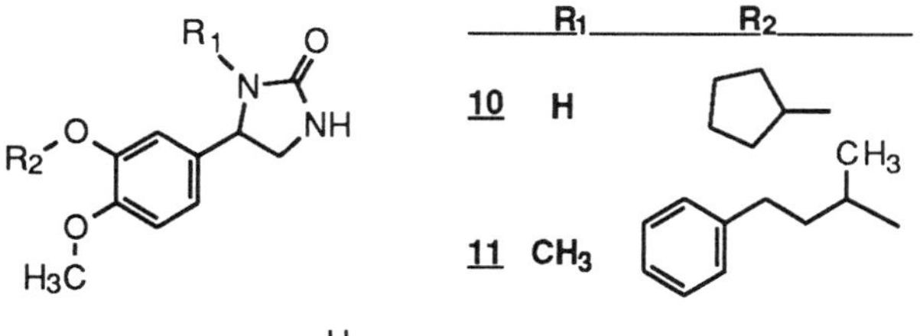

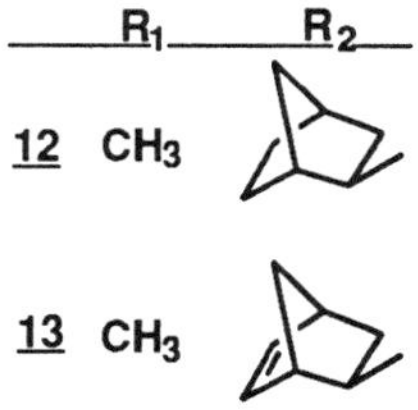

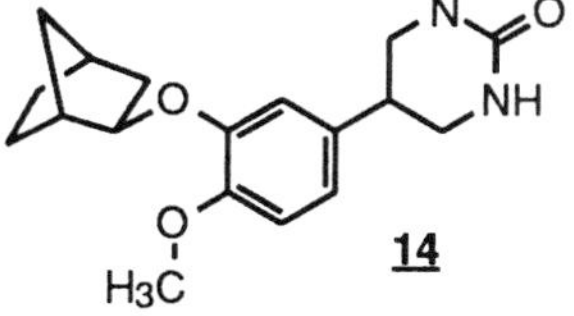

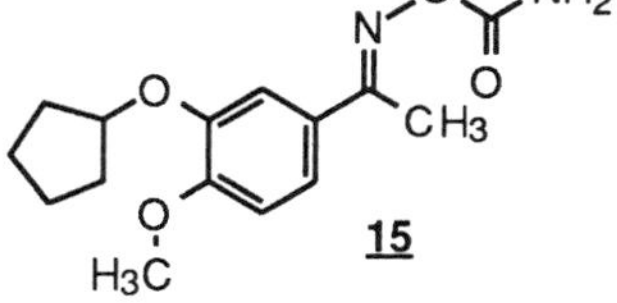

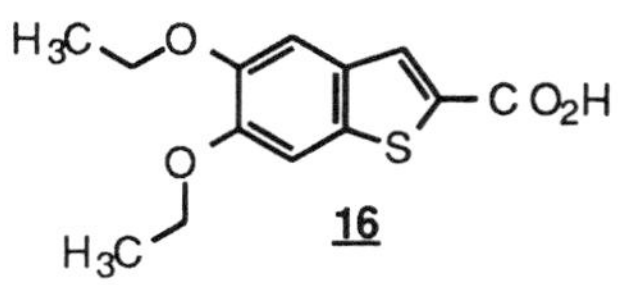

Independent modeling studies of rolipram and Ro 20-1724 led to the synthesis and evaluation of a series of oxime carbonates and carbamates (39). Of these, PDA-641 (**15**) was identified as essentially equipotent to rolipram as an inhibitor of dog trachealis PDE IV. The final member of this group, tibenelast (LY 186655, **16**) was a competitive inhibitor (K_i = 9.4 µM) of the PDE from guinea-pig peritoneal mononuclear cells (40).

Heterobicyclics and Analogs - Limited SAR studies with an enzyme mixture from rat brain cortex have been conducted in a class of quinazolinediones exemplified by nitraquazone (TVX 2706, **17**, IC_{50} = 1.9 µM) (41). Postulating an overlay of the nitro group of **17** with the lactam of rolipram, the nitro was formally replaced with an ester (**18**) while the corresponding acid and N-methyl amide produced a substantial and complete loss in potency, respectively. Although the analogous pyridopyrimidinedione **19** was less potent than **18**, replacement of the N-ethyl group in this series (*e.g.*, **20**, **21** and **22**) substantially increased potency (IC_{50} = 40-100 nM) relative to **18**. The corresponding 4-pyridyl derivative, **23**, was a very potent inhibitor (IC_{50} = 0.28 nM) of PDE IV from a human B-cell line (42). The pyridopyridazinone nucleus (**24**, **25**) also provided potent inhibitors (IC_{50} = 5 and 14 nM, respectively) of PDE IV from a human B-cell line (43). In a series of pyridine-3-carboxamide-2-ethers (**26**), effectively ring opened variants of the pyridopyrimidinediones, a variety of substituents (Z = F, Cl, OCH_3, CN, CF_3) were tolerated at the *meta* and *para* positions of the 2-aryl ether, but *ortho* substituents were not favored (44).

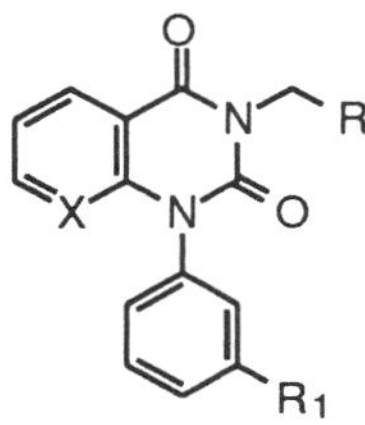

	X	R_1	R_2
17	CH	NO_2	CH_3
18	CH	CO_2CH_3	CH_3
19	N	CO_2CH_3	CH_3
20	N	CO_2CH_3	phenyl
21	N	CO_2CH_3	cyclopentyl
22	N	CO_2CH_3	norbornyl
23	N	NO_2	4-pyridyl

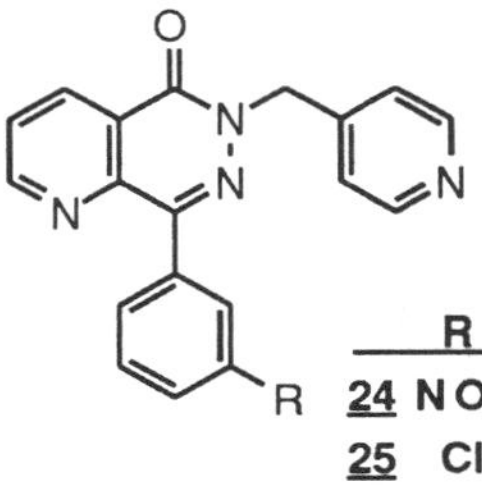

	R
24	NO_2
25	Cl

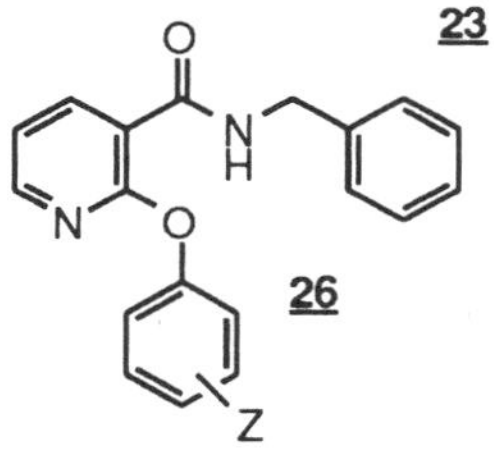

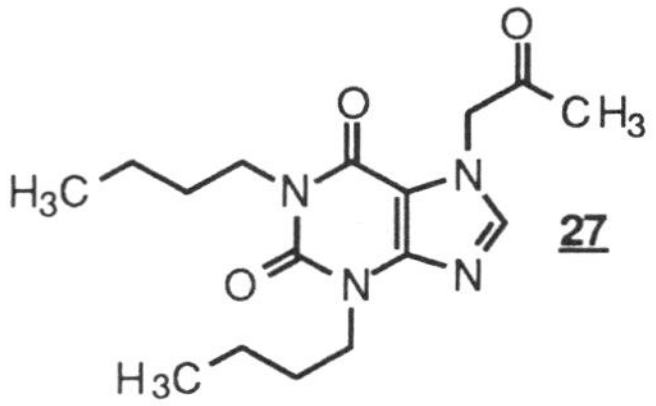

Xanthines - The xanthine denbufylline (BRL 30892, **27**), as potent and selective for PDE IV as rolipram, lacked the adenosine receptor activity common to many alkylxanthines (45). Derivatization at the N-7 or C-8 amino sites of the selective PDE IV inhibitor BRL 61063 (**28**) provided PDE IV inhibitors with varying degrees of PDE V and adenosine receptor activity (46).

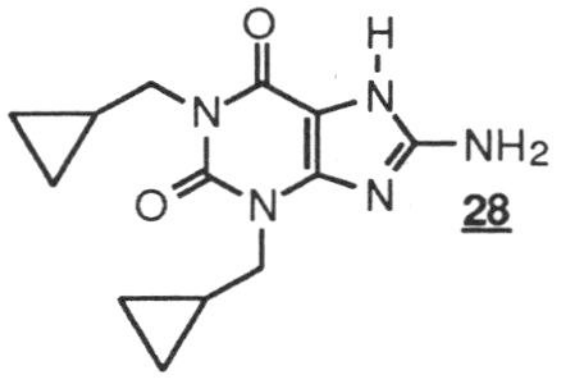

High Affinity Rolipram-binding Site - In addition to its activity as a potent (K_i = 60 nM) and selective inhibitor of PDE IV, rolipram binds saturably, stereoselectively and with high affinity (K_d = 1-2 nM for R-[-]-rolipram *vs.* 20-40 nM for S-[+]-rolipram) to preparations of brain tissue homogenate (47-50). While high affinity [^{3}H]-rolipram-binding sites do not appear to be present in all tissues that contain PDE IV (49), this activity is co-expressed with human recombinant PDE IV activity (50). Many of the known PDE IV inhibitors compete with [^{3}H]-rolipram for this binding site, though the rank order of potency of various compounds for inhibiting PDE IV catalytic activity is distinct from rolipram-binding activity (36,44,46,49,50). While the nature and function of this binding site are unclear, a strong correlation between the *in vivo* affinity of a series of selective PDE IV inhibitors for the [^{3}H]-rolipram binding site in rat and mouse forebrain with their pharmacological activity in two behavioral measures of anti-depressant activity has been observed (51).

PDE V INHIBITORS

A comprehensive review of selective PDE V inhibitors, *e.g.*, zaprinast (**29**) and SK&F 96231 (**30**), has been published (52). Since then, **31** has been identified as a potent inhibitor of PDE V which also possesses modest PDE I and II inhibitory activity (53).

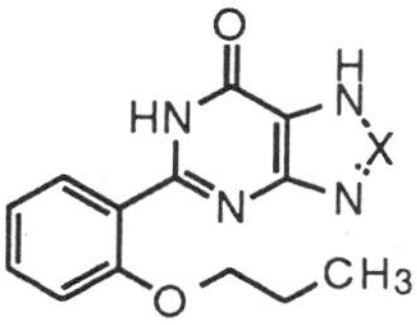

	X
29	N
30	CH

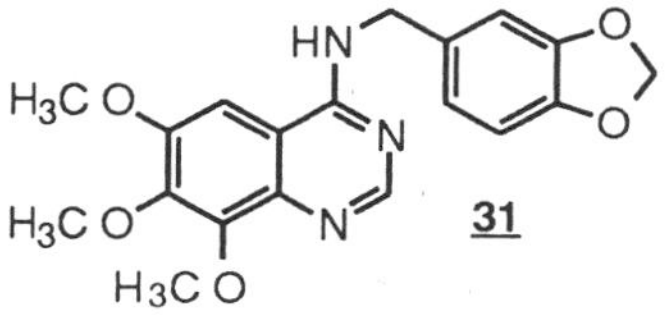

DUAL INHIBITORS

Dual PDE III/IV Inhibitors - Several PDE III/IV dual inhibitors have been identified. Structurally, each of these compounds possesses appropriately positioned functional groups within the overall flat topography required for potent PDE III inhibition (22) and the catechol ether moiety common to many PDE IV inhibitors, but no SARs for these compounds have been published. The pyridazinone zardaverine (**32**) inhibits human platelet PDE III (IC$_{50}$ = 0.58 μM) and canine trachealis PDE IV (IC$_{50}$ = 0.79 μM) (54). B9004-070 (**33**) is 6- to 13-fold more potent than **32**, and is most effective as a bronchodilator in the guinea pig when administered directly to the lung (55,56). B9004-070 and benzafentrine (AH 21-132, **34**) are unique in possessing a protonated amine moiety (at physiological pH) rather than an amide-type hydrogen as the H-bond donor required for potent PDE III inhibition (57,58). Dual PDE III/IV inhibitor profiles for EMD 54662 (**35**) and Org 30029 (**36**) also have been reported (59,60).

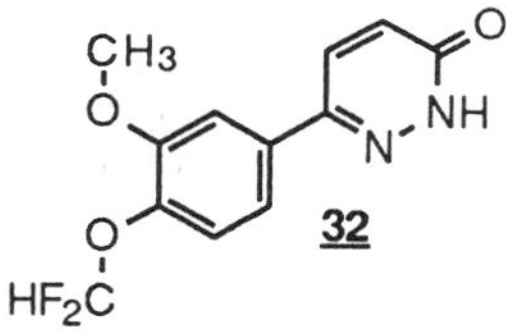

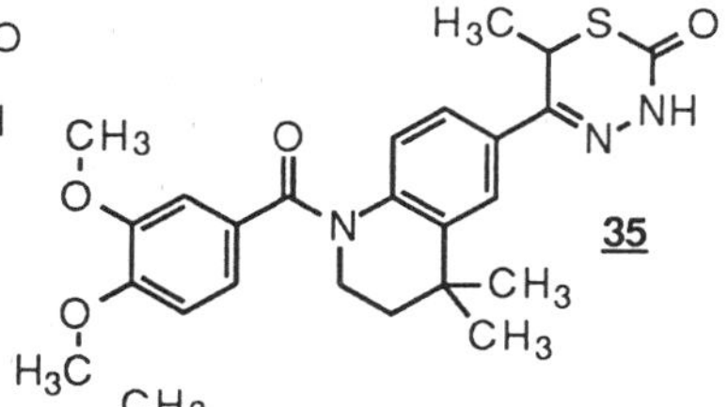

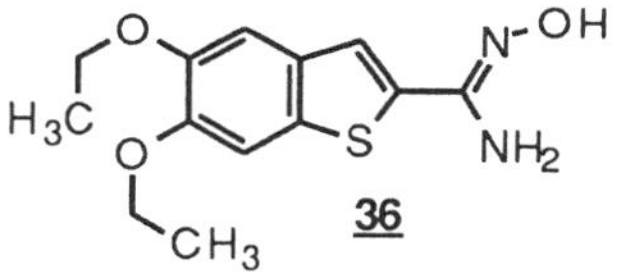

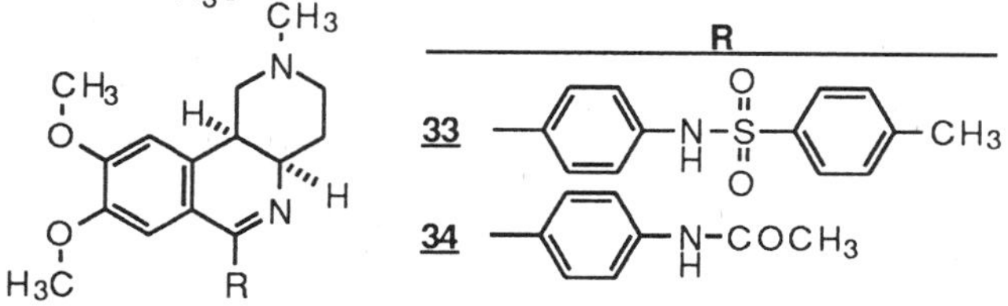

INFLAMMATORY CELLS

The PDE isozyme profile of various inflammatory and immune cells has been deduced from a combination of kinetic, pharmacologic and chromatographic techniques. In addition, the potential importance of the various PDE isozymes has been assessed using isozyme-selective inhibitors. As summarized below, PDE IV is the major cAMP PDE present in all inflammatory cell types examined thus far.

Mast Cells and Basophils - Rat peritoneal and thoracic mast cells contain PDEs II, III and V (61), whereas a mouse mast cell line (FB1) contains primarily PDE IV along with a small amount of PDE I (62). In mouse mast cells, PDE IV inhibitors increase cAMP content and reduce antigen-stimulated leukotriene (LT) and histamine release (62). The PDE isozyme profile of human mast cells has yet to be reported.

Human purified basophils contain PDEs III, IV and V (29). cAMP accumulation is increased by rolipram and the increase is associated with an inhibition of antigen-stimulated histamine and LTC_4 release (29). When used alone, neither SK&F 95654 nor zaprinast alter cyclic nucleotide content or mediator release. However, the ability of rolipram to elicit these effects is potentiated by PDE III inhibitors (29), suggesting that PDE III and PDE IV act in concert to regulate basophil cAMP content.

Eosinophils - PDE IV is the predominant isozyme in guinea-pig eosinophils (63,64). A similar profile is claimed for human eosinophils (65). Rolipram and denbufylline increase cAMP accumulation in guinea-pig eosinophils (64) and inhibit opsonized zymosan-stimulated H_2O_2 generation and basal superoxide release (64-66). Rolipram also reduces TNFα- and phorbol ester-induced adhesion of guinea-pig eosinophils to human umbilical vein endothelial cells (66). In contrast to the activity of PDE IV inhibitors, PDE III and PDE V inhibitors have little or no effect on these functions (63-66).

Neutrophils - The supernatant fraction of human neutrophils contains a predominance of PDE IV along with two additional minor activities, probably PDEs I, II or V (67-70). Consistent with the isozyme profile of neutrophils, rolipram and Ro 20-1724 as well as zardaverine reduce fMLP- and A23187-induced respiratory bursts (67,68,70). In contrast, neither zaprinast nor a series of PDE III inhibitors alters neutrophil function (67,68,70).

Monocytes and Macrophages - Like neutrophils, human monocytes contain a large amount of PDE IV (50,71). Information on additional isozymes in human monocytes or macrophages is unavailable, although analysis of PDE activity in mouse peritoneal macrophages suggests the presence of PDEs II, III, IV and V (72). Rolipram, but not inhibitors of PDEs I, III or V, is a powerful inhibitor of lipopolysaccharide (LPS)-induced TNFα production from human monocytes (73,74). Rolipram does not inhibit LPS-induced interleukin-1β production from these cells (74). Rolipram has a modest inhibitory effect on fMLP-induced superoxide generation from human monocytes (75). In contrast, Ro 20-1724 has no effect on immunologically stimulated superoxide, LTB_4, thromboxane B_2 or N-acetyl-β-D-glucosaminidase release (76).

Lymphocytes - Purified human T cells contain significant amounts of PDEs III and IV as well as a low level of a cGMP PDE activity with the characteristics of PDE V (77). An additional, unspecified activity is present with a cAMP K_m < 1.0 mM (71,77). This activity is resistant to both PDE III and PDE IV inhibitors. PDE IV inhibitors suppress the activity of human cytotoxic T Lymphocytes (78). Both Ro 20-1724 and CI-930 inhibit T cell blastogenesis (77). The effects of PDE III and PDE IV inhibitors on blastogenesis are synergistic (77), suggesting that both of these isozymes regulate cAMP content. A preliminary report indicates that rolipram, but not zaprinast or siguazodan, inhibits the production of message for interleukin-4, interleukin-5 and interferon-γ in antigen-stimulated peripheral blood mononuclear cells (30).

In Vivo Antiinflammatory Activities - The potential antiinflammatory activity of PDE IV inhibitors demonstrated in vitro has been borne out in a number of in vivo models. For example, PDE IV inhibitors have substantial inhibitory effects on platelet-activating factor (PAF)-induced microvascular leak in guinea pigs (79,80), arachidonic acid-induced ear edema and neutrophil infiltration in mice (81), LPS-induced lung damage and mortality in rats (82), and mediator-induced eosinophil infiltration into the guinea-pig conjunctiva (83).

PDE IV inhibitors also reduce allergic or inflammatory responses in animal models relevant to asthma. Antigen-induced bronchoconstriction in guinea pigs is inhibited by pretreatment with PDE IV inhibitors (84,85), primarily through the ability of these compounds to inhibit mast cell degranulation (84). Rolipram also inhibits airway hyperactivity induced by antigen (85) or PAF (86) challenge in guinea pigs. Moreover, Sephadex-induced blood eosinophilia in the rat and the associated airway hyperresponsiveness are reduced by rolipram and denbufylline (87). Finally, administration of zardaverine to rats reduces the airway hyperresponsiveness, neutrophil infiltration, elastase activity and TNFα release in response to aerosolized LPS (88).

AIRWAY SMOOTH MUSCLE

In Vitro Studies - The PDE isozyme profile of human airway smooth muscle is complex. Biochemical analyses indicate that representatives of all five isozyme families are present in both trachealis (89) and bronchus (90). The relaxant activity of isozyme-selective PDE inhibitors has been assessed in three studies (89-91). Both PDE III and PDE IV inhibitors relax tissues contracted with a variety of spasmogens, although the maximal reversal of tone attained with these agents is only 15-60% of that achieved with β-adrenoceptor agonists (89-91). PDE III and PDE IV inhibitors act synergistically, suggesting that a dual PDE III/IV inhibitor would have greater bronchodilatory activity *in vivo* than selective inhibitors of either isozyme (89), a prediction borne out in studies in the anesthetized guinea pig (92). In contrast to the activity of PDE III and PDE IV inhibitors, PDE V inhibitors do not relax human isolated airways (89). Interestingly, guinea pig tracheal relaxation by a limited series of inhibitors correlated with their ability to compete with the [^{3}H]-rolipram-binding site, not with PDE IV inhibition (93).

In Vivo Studies - Only limited information is available on the bronchodilatory activity of isozyme-selective inhibitors in whole animals. Both rolipram and PDA-641 reverse serotonin-induced bronchoconstriction in the dog when given intravenously (94). Imazodan and CI-930 also elicit bronchodilation in this model, but only at doses that produce profound cardiovascular effects (95). In the guinea pig, both PDE III and IV inhibitors reverse ongoing histamine-induced bronchoconstriction (93). However, these agents are considerably less effective against histamine-induced bronchoconstriction when given as a pretreatment, and are essentially inactive when administered prior to LTD$_4$ challenge (92,96). Consistent with the results of *in vitro* studies (92,93), the bronchodilatory actions of PDE III and IV inhibitors are at least additive *in vivo* (92).

CLINICAL EXPERIENCE

Therapeutic Activity - Clinical studies on the effects of selective PDE inhibitors on respiratory function in humans are limited. Enoximone was shown to decrease lung resistance and increase compliance in patients with decompensated chronic pulmonary disease (97). Benzafentrine was assessed using whole-body plethysmography in nonasthmatic human subjects, and evidence for bronchodilation was obtained when the compound was inhaled but not when it was administered intravenously or orally (98). Zardaverine produced modest bronchodilation in asthmatics when administered by inhalation (99), but was devoid of bronchodilator activity in patients with chronic obstructive pulmonary disease (100). A preliminary report on tibenelast showed that it slightly improved pulmonary function in asthmatics, although this effect was not statistically significant (101). In adult asthmatics, oral administration of zaprinast reduced exercise-induced bronchoconstriction but not histamine-induced bronchoconstriction (102), whereas no effects on exercise-induced bronchoconstriction were seen in a subsequent evaluation in asthmatic children (103).

<u>Side Effects</u> - Although too few clinical studies have been conducted to fully describe the side-effect profiles for the various classes of PDE inhibitors, tentative predictions can be made based upon the known pharmacology of representative isozyme-selective inhibitors as well as an understanding of the functional role of individual PDE isozymes in various tissues. For example, because of the importance of PDE III in regulating cAMP metabolism in the myocardium and in vascular smooth muscle, PDE III inhibitors increase cardiac contractility and induce vasodilation (104). A second concern with PDE III inhibitors is the potential arrhythmogenic activity of these compounds (105). In addition to the known antidepressant activity of the PDE IV inhibitor rolipram, this compound produces a variety of gastrointestinal side effects, including nausea and vomiting (106), and has been shown to produce a large but transient change in plasma osmolality (107). These adverse effects could be the result of PDE IV inhibition in the brain, gastrointestinal tract, and/or kidney. Finally, the role of cGMP as a second messenger mediating vasodilation, coupled with the importance of PDE V in regulating vascular cGMP content (108,109), suggests that inhibitors of this PDE have the potential to produce cardiovascular side effects.

<u>Conclusions and Future Directions</u> - The collective data support a potential therapeutic role for isozyme-selective PDE inhibitors in asthma, though substantive issues remain. Whether the bronchodilatory potential of a PDE III, IV or V inhibitor will provide advantages over the existing therapy offered by the β-adrenoceptor agonists has yet to be determined. Also, a combination of PDE III or V inhibitory activity along with PDE IV inhibitory activity may be required to achieve an optimal bronchodilator/antiinflammatory profile. Targeting of selective inhibitors to isozyme subtypes may achieve increased tissue selectivity and reduce side-effect potential for these compounds, but nothing along these lines has been reported. Determination of the ultimate utility of this and other (110,111) novel strategies currently being evaluated for asthma will require additional clinical studies.

References

1.   T.J. Torphy and B.J. Undem, Thorax, 46, 512 (1991).
2.   M.A. Giembycz and G. Dent, Clin.Exp.Allergy, 22, 337 (1992).
3.   H.R. Bourne, L.M. Lichtenstein, K.L. Melmon, C.S. Henney, Y. Weinstein and G.M. Shearer, Science, 184, 19 (1974).
4.   G.M. Kammer, Immunol.Today, 9, 222 (1988).
5.   T.J. Torphy, Rev.Clin.Basic Pharm., 6, 61 (1987).
6.   M.A. Giembycz and D. Raeburn, J.Auton.Pharmacol., 11, 365 (1991).
7.   J.A. Beavo, Adv.Second Messenger Phosphoprotein Res., 22, 1 (1988).
8.   W.J. Thompson, Pharmac.Ther., 51, 13 (1991).
9.   R.E. Weishaar, M.H. Cain and J.A. Bristol, J.Med.Chem., 28, 537 (1985).
10.   J.A. Beavo and D.H. Reifsnyder, Trends Pharmacol.Sci., 1, 150 (1990).
11.   T. Michaeli, T.J. Bloom, T. Martins, K. Loughney, K. Ferguson, M. Riggs, L. Rodgers, J.A. Beavo and M. Wigler, J.Biol.Chem., 268, 12925 (1993).
12.   T.J. Torphy, H.-L. Zhou and L.B. Cieslinski, J.Pharmacol.Exp.Ther., 263, 1195 (1992).
13.   M. Conti, S.L.C. Jin, L. Monaco, D.R. Repaske and J.V. Swinnen, Endocr.Rev., 12, 218 (1991).
14.   R.L. Davis, H. Takaysasu, M. Eberwine and J. Myres, Proc.Natl. Acad.Sci.USA, 86, 3604 (1989).
15.   J.V. Swinnen, D.R. Joseph and M. Conti, Proc.Natl.Acad.Sci.USA, 86, 5325 (1989).
16.   G.P. Livi, P. Kmetz, M. McHale, L. Cieslinski, G.M. Sathe, D.J. Taylor, R.L. Davis, T. Torphy and J.M. Balcarek, Mol.Cell.Biol., 10, 2678 (1990).
17.   M.M. McLaughlin, L.B. Cieslinski, M. Burman, T.J. Torphy and G.P. Livi, J.Biol.Chem., 268, 6470 (1993).
18.   G. Bolger, T. Michaeli, T. Martins, T. St. John, B. Steiner, L. Rodgers, M. Riggs, M. Wigler and K. Ferguson, Mol.Cell.Biol., 13, 6558 (1993).
19.   R. Obernolte, S. Bhakta, R. Alvarez, C. Bach, P. Zuppan, M. Mulkins, K. Jarnagin and E.R. Shelton, Gene, 129, 239 (1993).
20.   S.-L. Jin, J.V. Swinnen and M. Conti, J.Biol.Chem., 267, 18929 (1992).
21.   S. Jacobitz, M.M. McLaughlin, G.P. Livi, M.D. Ryan and T.J. Torphy, FASEB J., 8, in press (1994).
22.   J.A. Bristol, I. Sircar, W.H. Moos, D.B. Evans and R.E. Weishaar, J.Med.Chem., 27, 1099 (1984).
23.   W.H. Moos, C.C. Humblet, I. Sircar, C. Rithner, R.E. Weishaar, J.A. Bristol and A.T. McPhail, J.Med.Chem., 30, 1963 (1987).

24.   M.C. Venuti, G.H. Jones, R. Alvarez and J.J. Bruno, J.Med.Chem., 30, 303 (1987)
25.   M.C. Venuti, R.A. Stephenson, R. Alvarez, J.J. Bruno.and A.M. Strosberg, J.Med.Chem., 31, 2136 (1988).
26.   A. Davis, B.H. Warrington and J.G. Vinter, J.Computer-Aided Mol.Design, 1, 97 (1987).
27.   D.W. Robertson, N.D. Jones, J.H. Krushinski, G.D. Pollock, J.K. Swartzendruber and J.S. Hayes, J.Med.Chem., 30, 623 (1987).
28.   P.W. Erhardt, in "Cyclic nucleotide phosphodiesterases: Structure, regulation and drug action," J. Beavo and M.D. Houslay, Eds., J. Wiley and Sons, Chichester, 1990, p. 317.
29.   P.T. Peachell, B.J. Undem, R.P. Schleimer, D.W. MacGlashan, Jr., L.M. Lichtenstein and T.J. Torphy, J.Immunol., 148, 2503 (1992).
30.   D.M. Essayan, A. Kagey-Sobotka, L.M. Lichtenstein and S.K. Huang, J.Allergy Clin.Immunol., 91, 254 (1993).
31.   D.W. Robinson and D.B. Boyd, Adv.Second Messenger Phosphoprotein Res., 25, 321 (1992).
32.   N.A. Meanwell and S.M. Seiler, Drugs Future, 15, 369 (1990).
33.   M.C. Marivet, J.-J. Bourguignon, C. Lugnier, A. Mann, J.-C. Stoclet and C.-G. Wermuth, J.Med.Chem., 1989, 32, 1450 (1989).
34.   H. Sheppard and G. Wiggan, Mol.Pharmacol., 7, 111 (1971).
35.   P.W. Baures, D.S. Eggleston, K.F. Erhard, L.B. Cieslinski, T.J. Torphy and S.B. Christensen, J.Med.Chem., 36, 3274 (1993).
36.   B.K. Koe, L.A. Lebel, J.A. Nielsen, L.L. Russo, N.A. Saccomano, F.J. Vinick and I.H. Williams, Drug Dev.Res., 21, 135 (1990).
37.   N.A. Saccomano, F.J. Vinick, K. Koe, J.A. Nielson, W.M. Whalen, M. Meltz, D. Phillips, P.F. Thadieo, S. Jung, D.S. Chapin, L.A. Lebel, L.L. Russo, D.A. Helwig, J.L. Johnson, Jr., J.L. Ives and I.H. Williams, J.Med.Chem., 34, 291 (1991).
38.   N.A. Saccomano, European Patent Application, EP 0 428 302 A2 (1991).
39.   L.J. Lombardo, J.F. Bagli, J.M. Golankiewicz, R.J. Heaslip and J.K. Mudrick, Abstracts, A.C.S. 205th Natl. Meeting, Denver, CO, March 28-April 2, MEDI No. 128 (1993).
40.   P.P.K. Ho, L.Y. Wang, R.D. Towner, S.J. Hayes, D. Pollock, N. Bowling, V. Wyss and J.A. Panetta, Biochem.Pharmacol., 40, 2085 (1990).
41.   J.A. Lowe, III, R.L. Archer, D.S. Chapin, J.B. Cheng, D. Helwig, J.L. Johnson, B.K. Koe, L.A. Lebel, P.F. Moore, J.A. Nielsen, L.L. Russo and J.T. Shirley, J.Med.Chem., 34, 624 (1991).
42.   R.S. Wilhelm, R.L. Chin, B.H. Devens and R. Alvarez, International Patent Application, WO 93/19068 (1993).
43.   R.S. Wilhelm, B.E. Loe, B.H. Devens, R. Alverez and M.G. Martin, International Patent Application, WO 93/07146 (1993).
44.   F.J. Vinick, N.A. Saccomano, B.K. Koe, J.A. Nielsen, I.H. Williams, P.F. Thadieo, S. Jung, M. Meltz, J. Johnson, Jr., L.A. Lebel, L.L. Russo and D. Helwig, J.Med.Chem., 34, 86 (1991).
45.   C.D. Nicholson, S.A. Jackman and R. Wilke, Br.J.Pharmacol., 97, 899 (1989).
46.   D.R.Buckle, J.R.S. Arch, B.J. Connolly, A.E. Fenwick, K.A. Foster, K.J. Murray, S.A. Readshaw, M. Smallridge and D.G. Smith, J.Med Chem., 37, 476 (1994).
47.   H. Wachtel, J.Pharm.Pharmacol., 35, 440 (1983).
48.   J.E. Schultz and B.H. Schmidt, Naunyn-Schmiedeberg'sArch.Pharmacol., 333, 23 (1986).
49.   H.H. Schneider, R. Schmiechen, M. Brezinski and J. Seidler, Eur.J.Pharmacol., 127, 105 (1986).
50.   T.J. Torphy, J.M., Stadel, M. Burman, L.B. Cieslinski, M.M. McLaughlin, J.R. White and G.P. Livi, J.Biol.Chem., 267, 1798 (1992).
51.   R. Schmiechen, H.H. Schneider and H. Wachtel, Psychopharmacology, 102, 17 (1990).
52.   K.J. Murray, DN&P, 6, 150 (1993).
53.   B. Y. Takase, T. Saeki, M. Fujimoto and I. Saito, J.Med.Chem., 36, 3765 (1993).
54.   C. Schudt, S. Winder, B. Muller and D. Ukena, Biochem.Pharmacol., 42, 153 (1991).
55.   C. Schudt, H. Tenor, A. Wendel, M. Eltze, H. Magnussen and K.F. Rabe, Am.Rev.Respir.Dis., 147, A183 (1993).
56.   R. Beume, U. Kilian, U. Brand, D. Hafner, M. Eltze and D. Flockerzi, Am.Rev.Respir.Dis., 147, A184 (1993).
57.   K.R.F. Elliott, J.L. Berry, A.J. Bate, R.W. Foster and R.C. Small, J.Enzyme Inhibition, 4, 245 (1991).
58.   M.A. Giembycz and P.J. Barnes, Biochem.Pharmacol., 42, 663 (1991).
59.   M. Klockow and R. Jonas, Naunyn-Schmiedeberg's Arch.Pharmacol., 339, R 53 (1989).
60.   M. Shahid and C.D. Nicholson, Naunyn-Schmiedeberg's Arch.Pharmacol., 342, 698 (1990).
61.   H. Bergstrand, B. Lundqvist and A. Schurmann, Mol.Pharmacol., 14, 848 (1978).
62.   T.J. Torphy, G.P. Livi, J.M. Balcarek, J.R. White and B.J. Undem, Adv.Second Messenger Phosphoprotein Res., 25, 289 (1992).
63.   G. Dent, M.A. Giembycz, K.F. Rabe and P.J. Barnes, Br.J.Pharmacol., 103, 1339 (1991).
64.   J.E. Souness, C.M. Carter, B.K. Diocee, G.A. Hassall, L.J. Wood and N.C. Turner, Biochem.Pharmacol., 42, 937 (1991).
65.   M.A. Giembycz, Biochem.Pharmacol., 43, 2041 (1992).

66.   T.J. Torphy, M.S. Barnette, D.W.P. Hay and D.C. Underwood, Enriv.Health Perspec., 104, in press (1994).
67.   C. Schudt, S. Winder, S. Forderkunz, A. Hatzelmann and V. Ullrich, Naunyn-Schmiedeberg's Arch.Pharmacol., 344, 682 (1991).
68.   C.P. Nielson, R.E. Vestal, R.J. Sturm and R. Heaslip, J.Allergy Clin.Immunol., 86, 801 (1990).
69.   T. Engerson, J.L. Legendre and H.P. Jones, Inflammation, 10, 31 (1986).
70.   C.D. Wright, P.J. Kuipers, D. Kobylarz-Singer, L.J. Devall, B.A. Klinkefus and R.E. Weishaar, Biochem.Pharmacol., 40, 699 (1990).
71.   W.J. Thompson, C.P. Ross, W.J. Pledger and S.J. Strada, J.Biol.Chem., 251, 4922 (1976).
72.   K. Okonogi, T.W. Gettys, R.J. Uhing, W.C. Tarry, D.O. Adams and V. Prpic, J.Biol.Chem., 266, 10305 (1991).
73.   J. Semmler, H. Wachtel and S. Endres, Int.J.Immunopharmacol., 15, 409 (1993).
74.   K.L. Molnar-Kimber, L. Yonno, R.J. Heaslip and B.M. Weichman, Med.Inflamm., 1, 411 (1992).
75.   K.R.F. Elliott and E.J. Leonard, FEBS Lett., 254, 94 (1989).
76.   R.W. Fuller, G. O'Malley, A.J. Baker and J. MacDermot, Pulmon.Pharmacol., 1, 101 (1988).
77.   S.A. Robicsek, D.K. Blanchard, J.Y. Djeu, J.J. Krzanowski, A. Szentivanyi and J.B. Polson, Biochem.Pharmacol., 42, 869 (1991).
78.   M. Plaut, G. Marone and E. Gillespie, J.Immunol., 131, 2945 (1983).
79.   D. Raeburn and J.-A. Karlsson, J.Pharmacol.Exp.Therap., 267, 1147 (1993).
80.   J.L. Ortiz, J. Cortijo, J.M. Vallés, J. Bou and E.J. Morcillo, Fundam.Clin.Pharmacol., 6, 247 (1992).
81.   D.E. Griswold, E.F. Webb, J. Breton, J.R. White, M.J. DiMartino, E.F. Smith III, M.J. Slivjak and T.J. Torphy, Inflammation, 17, 333 (1993).
82.   C.R. Turner, K.M. Esser and E.B. Wheeldon, Circ.Shock, 39, 237 (1993).
83.   S.J. Newsholme and L. Schwartz, Inflammation, 17, 25 (1993).
84.   D.C. Underwood, R.R. Osborn, L.B. Novak, J.K. Matthews, S.J. Newsholme, B.J. Undem, J.M. Hand and T.J. Torphy, J.Pharmacol.Exp., 266, 306 (1993).
85.   R.E. Howell, B.D. Sickels and S.L. Woeppel, J.Pharmacol.Exp.Ther., 264, 609 (1993).
86.   D. Raeburn and S.Lewis, Eur.Respir.J., 4, 307S (1991).
87.   J.R.S. Arch, S.M. Laycock, H. Smith and B.A. Spicer, in "New Concepts in Asthma," J.P. Tarayre, B.B. Vargaftig and E. Carilla, Eds., MacMillan Press, London 1993, in press.
88.   J.C. Kips, G.F. Joos, R.A. Peleman and R.A. Pauwels, Clin.Exp.Allergy, 23, 518 (1993).
89.   T.J. Torphy, B.J. Undem, L.B. Cieslinski, M.A. Luttmann, M.K. Reeves and D.W.P. Hay, J.Pharmacol.Exp.Ther., 265, 1213 (1993).
90.   J. De Boer, A.J. Philpott, R.G.M. Van Amsterdam, M. Shahid, J. Zaagsma and D. Nicholson, Br.J.Pharmacol., 106, 1028 (1992).
91.   J. Cortijo, J. Bou, J. Beleta, I. Cardelús, J. Llenas, E. Morcillo and R.W. Gristwood, Br.J.Pharmacol., 108, 562 (1993).
92.   D.C. Underwood, C.R. Kotzer, S. Bochnowicz, R.R. Osborn, M.A. Luttmann, D.W.P. Hay and T.J. Torphy, Am.Rev.Resp.Dis., 147, A183 (1993).
93.   A.L. Harris, M.J. Connell, E.W. Ferguson, A.M. Wallace, R.J. Gordon, E.D. Pagani and P.J. Silver, J.Pharmacol.Exp.Ther., 251, 199 (1989).
94.   R.J.Heaslip, L.J. Lombardo, J.M. Golankiewicz, B.A. Ilsemann, D.Y. Evans, B.D. Sickles, J.K. Mudrick, J. Bagli and B.M. Weichman, J.Pharmacol.Exp.Ther., 268, 888 (1994).
95.   R.J.Heaslip, S.K.Buckley, B.D. Sickles and D. Grimes, J.Pharmacol.Exp.Ther., 257, 741 (1991).
96.   R.E. Howell, B.D. Sickels and S.L. Woeppel, J.Pharmacol.Exp.Ther., 264, 609 (1994).
97.   M. Leeman, P. Lejeune, C. Melot and R. Naeije, Chest, 91, 662 (1987).
98.   R.C. Small, R.W. Foster, J.L. Berry, I.D. Chapman and K.R.F. Elliott, Agents Actions, 34, S3 (1991).
99.   T. Brunnée, R. Engelstätter, V.W. Steinijans and G. Kunkel, Eur.Respir.J., 5, 982 (1992).
100.   D. Ukena, C. Rentz, C. Reiber, R. Engelslätter and G.W. Sybrecht, Am.Rev.Respir.Dis., 145, A757 (1992).
101.   E. Israel, P.N. Mathur, D. Tashkin and J.M. Drazen, Chest, 94, 71S (1988).
102.   R.M. Rudd, A.R. Gellert, P.R. Studdy and D.M. Geddes, Br.J.Dis.Chest, 77, 78 (1983).
103.   J. Reiser, Y. Yeang and J.O. Warner, Br.J. Dis.Chest, 80, 157 (1986).
104.   M.A. Wood and M.L. Hess, Am. J.Med.Sci., 297, 105 (1989).
105.   G.V. Naccarelli and R.A. Goldstein, Am.J.Cardiol., 63, 35A (1989).
106.   R. Horowski and M. Sastre-Y-Hernandez, Curr.Ther.Res., 38, 23 (1985).
107.   I. Sturgess and G.F. Searle, J.Clin.Pharmacol., 29, 369 (1990).
108.   T.M. Lincoln, Pharmacol.Ther., 41, 479 (1989).
109.   J.E. Souness, R. Brazdil, B.K. Diocee and R. Jordan, Br.J.Pharmacol., 98, 725 (1989).
110.   M.J. Sofia and S.A. Silbaugh, Ann.Rep.Med.Chem., 28, 109, (1993).
111.   A.W. Ford-Hutchinson, I.W. Rodger and T.R. Jones, DN&P, 5, 542 (1992).

Chapter 20. Human Leukocyte Elastase Inhibitors

Dennis J. Hlasta and Edward D. Pagani
Sterling Winthrop Pharmaceuticals Research Division
1250 South Collegeville Road, Collegeville, Pennsylvania 19426

Introduction - Human leukocyte elastase (HLE) is a serine protease and a basic glycoprotein with isoforms of 25-30 kD. The active site of HLE is a channel on the protein surface that has extended substrate binding sites. The catalytic triad of Ser[195], His[57], and Asp[102], which resides in the active site channel, catalyzes amide bond hydrolysis of various proteins, including elastin, the connective tissue of the lung. The S1 specificity pocket of serine proteases is largely responsible for the selectivity of this class of enzymes to cleave proteins at specific sites. HLE has a relatively small S1 pocket that is lined with hydrophobic residues, thus HLE preferentially cleaves proteins at sites with small lipophilic residues such as alanine and valine. This specificity has been exploited in the design of HLE inhibitors (1). The isoforms of HLE have identical amino acid sequences and catalytic properties, but differ based on the nature of the carbohydrate content. The different carbohydrate content of the E-1 and E-3 isoforms of elastase is proposed to direct these isoforms to secretory and lysosomal functions, respectively (2). Human elastase from polymorphonuclear neutrophils (PMN) and from purulent sputum display identical kinetics with various substrates and inhibitors suggesting that elastase from sputum is from PMNs (3). In this chapter, the term HLE will be used to indicate elastase from either PMNs or purulent sputum.

Pathophysiology of Inflammatory Disease - Elastase released from polymorphonuclear leukocytes has been proported to play a primary or contributory role in the pathogenesis of inflammatory diseases of the renal glomerulus (4), skin (5,6), and joints (7,8). HLE has also been suggested to be involved in tissue destruction associated with ischemia/reperfusion of the intestinal mucosa (9). While there is not universal agreement regarding the contribution of HLE relative to other proteases released during the inflammatory response in the pathogenesis of inflammatory diseases (10,11), there is mounting evidence that HLE plays a major role in acute and chronic inflammatory diseases of the lung (12-16).

Foreign bodies and bacterial or viral infections in the lung trigger the immune system. Among the numerous events associated with the immune cascade, neutrophils are recruited to the lung. In response to inflammatory mediators, neutrophils release the contents of intracellular granules which contain a number of proteases including HLE; these proteases are capable of proteolytic destruction of extracellular matrix proteins (17). HLE digests collagen (18), fibrinogen, and proteoglycan (7), but has been suggested to preferentially attack elastin fibers. Recently, the elastolytic property of elastase has been challenged (19). Nevertheless, indiscriminate destruction of extracellular matrix by elastase in healthy areas of the lung can account for the abnormal pulmonary mechanics and imbalance in gas exchange that is often observed in chronic inflammatory lung disease.

The activity of HLE within the interstitial space of the lung is controlled locally by endogenous inhibitors which include α_2-macroglobulin, secretory leukocyte protease inhibitor (SLPI), and α_1-protease inhibitor (α_1-PI). An imbalance in the ratio of the concentration of localized elastase to localized inhibitor, in favor of elastase, has been offered as an explanation for elastase destruction of the lung; this idea is sometimes referred to as the protease-antiprotease imbalance theory (12,20). An imbalance in the ratio could occur by an increase in the elastase load, which appears to be the case in individuals with cystic fibrosis (16) or by a reduction in the enzyme inhibitory capacity of the lung fluid. Normally, the concentrations of endogenous inhibitors are well in excess of the concentrations of both circulating and lung elastase, however, in certain circumstances the level of α_1-PI is significantly reduced. One instance is observed in individuals with genetically-linked α_1-PI deficiency in which α_1-PI levels are 10-20% of normals (21,22). The activity of α_1-PI can also be significantly reduced by oxidative inactivation of the active center Met[358] by neutrophil-generated HOCl and N-chloramines (23), by tobacco-induced release of oxidants from macrophages (24-26) or by nitrogen dioxide in tobacco smoke (27). α_1-PI is also rendered inactive by endogenous protease activity commonly seen at sites of inflammation (28,29).

Elevated elastase activity has now been associated with lung destruction in patients with cystic fibrosis (16,30), adult respiratory distress syndrome (31,32) and bronchiectasis (33). Evidence also exists for the role of elastase in the development of emphysema by individuals with genetic α_1-PI deficiency (22,34). Most evidence for the role of elastase in emphysema has been essentially extrapolated from findings in animal models in which tracheal instillation of HLE produces "emphysema-like" tissue destruction (15,35). Individuals who inhale tobacco smoke also have been found to have high levels of lung elastase activity (25), although paradoxically, not all smokers develop lung disease. Several thorough reviews discuss the role of HLE in pulmonary diseases (11,36-39).

INHIBITORS

Inhibitors of HLE can be classified based on their known or proposed mechanisms of action. A recent review defined enzyme inhibitors by mechanism and gave examples which included transition state and mechanism-based inactivators (40). Other reviews have surveyed the literature of HLE inhibitors (38,41,42). Much of the recent design of HLE inhibitors has been based on computer modeling studies of proposed enzyme-inhibitor complexes using X-ray crystal structural information from HLE or porcine pancreatic elastase (PPE), which has 40% homology with HLE. Native HLE has not been crystallized, however, enzyme-inhibitor complexes with methoxysuccinyl-Ala-Ala-Pro-AlaCH$_2$Cl (43,44) and turkey ovomucoid inhibitor (45) are reported. Structural information was available earlier with native PPE or PPE-inhibitor complexes and modeling studies have led to the design of novel and potent HLE inhibitors (46-50). A monocyclic β-lactam inhibitor, L-652,117, was proposed to cross-link the active site Ser[195] and His[57] through a "double hit" mechanism, and kinetic and SAR data supported this conclusion (51). Recently using ESI mass spectrometry, the HLE-inhibitor complex of L-652,117 was shown to form a stable acyl-enzyme with Ser[195] only (52). The inhibition reaction of PPE with methoxysuccinyl-Ala-Ala-Pro-ValCH$_2$Cl was monitored with ESI-MS. The formation of an intermediate hemiketal with Ser[195] was detected and was shown to convert to an alkylated PPE-inhibitor complex (53). Though ESI-MS studies do not give structural information, ESI-MS can be used to provide supporting evidence for proposed HLE-inhibitor complexes. The cephem inhibitors, L-658,758 and L-647,957, were proposed to inactivate HLE through similar double hit mechanisms, leading to enzyme-inhibitor complexes with the catalytic site cross-linked. ESI MS of the PPE-inhibitor complexes of cephem inhibitors, L-658,758 and L-647,957, support the proposed double hit mechanisms for these inhibitors (54). The X-ray crystal structural features of HLE and PPE inhibitor complexes have been compared (46).

Endogenous Inhibitors - The protease-antiprotease hypothesis has been used to account for the development of emphysema in α_1-protease inhibitor (α_1-PI) deficient patients and in smokers. The α_1-PI has been observed in the lung and plasma, and the deficiency or inactivation of α_1-PI by cigarette smoke has been implicated as the causative agent in emphysema. More recently, low molecular weight elastase inhibitors have been implicated as important endogenous inhibitors of HLE (55). Secretory leukocyte proteinase inhibitor, SLPI, is a 12 kD basic protein that rapidly inactivates HLE ($k_{inact}=1.1 \times 10^7$ $M^{-1}s^{-1}$) and is suggested to play a physiologic antielastase function in the upper respiratory tract, while α_1-PI is suggested to function in the lower respiratory tract (56). A recombinant truncated form of SLPI has two fold less inhibitory potency than SLPI (57). Elafin is 6 kD protein that was originally isolated from human skin (58,59), and has been identified in sputum (60). Elafin is an acid stable basic peptide composed of 57 amino acids (61). This protein is a potent, substrate-like reversible inhibitor of HLE with an inactivation rate of 3.6×10^6 $M^{-1}s^{-1}$ (59), and recombinant human elafin has a duration action of 12-18 hours when given i.t. in the HLE-induced hemorrhage model (62). Other endogenous inhibitors have been identified, but their role in the pathophysiology of disease states or in biological function is less well understood. Skin-derived antileukoproteases (SKALP) are base stable proteins isolated from psoriatic scales of human skin (63), and are potent ($K_i=20$ pM) and rapid ($k_{inact}=10^7$ $M^{-1}s^{-1}$) inhibitors. Thrombospondin is a tight-binding protein inhibitor of elastase with a k_{inact} of 5×10^6 $M^{-1}s^{-1}$, that has been isolated from human platelets (64).

Transition State Analogs - Research to discover an orally active transition state analog inhibitor has proven successful with the preparation of peptide-like derivatives that are potent HLE inhibitors, orally absorbed, and not rapidly eliminated. ICI 200,880, **1**, is the prototype trifluoromethyl ketone peptide inhibitor, and is one of a class of potent inhibitors that are active by the i.t. route in the hamster HLE-induced hemorrhage model. ICI 200,880 has a 24 hour duration of action in this model, but is rapidly eliminated by the i.v. route and is not active orally (65). New derivatives use 3-amino-2-pyridone or 5-amino-4-pyrimidone as amino acid isosteres. Compounds **2** and **3** are potent inhibitors ($K_i=39$ nM and 27 nM, respectively) and are orally active in the hamster HLE-induced hemorrhage model (active at 5 mg/kg and $ED_{50}=2.5$ mg/kg, respectively) (66,67). Since this class of

transition state analog inhibitors form a reversible covalent bond with the catalytic Ser[195], other binding interactions with HLE are necessary to slow the reactivation rate and improve inhibitory potency. Based on PPE X-ray crystal structure data and modeling experiments in the design of these analogs, inhibitors **2** and **3** bind to the HLE active site at four contact points in the enzyme-inhibitor complex **3 E-I**. A reciprocal H-bond occurs with Val[216] and the electrophilic carbonyl of the trifluoromethyl ketone forms a covalent bond with Ser[195]. The inhibitor alkoxide oxygen is hydrogen bound in the oxy-anion hole to the NH of Gly[193] and Ser[195] (48,68). The pentafluoroethyl ketone derivative, MDL 101,146 (**4**), has a $K_i=20$ nM and is orally active with 56% inhibition at 25 mg/kg in the hamster HLE-induced hemorrhage model (69). BI-RA-260, **5**, is an ICI 200,880 analog that is active in the hamster model with an ED_{50} of 3.8 µg/animal, i.t. This inhibitor has a longer duration of action than ICI 200,880 with 49% inhibition present at 72 hours (70). This use of the *N*-indanyl-glycine bioisoster for proline has been extended to a series of difluorostanone analogs (71). Tricarbonyl peptide analogs of Cbz-Ala-Ala(CO)$_3$OBn are micromolar inhibitors and are proposed to form a stable hemiketal intermediate with the HLE active site Ser^{195} (72).

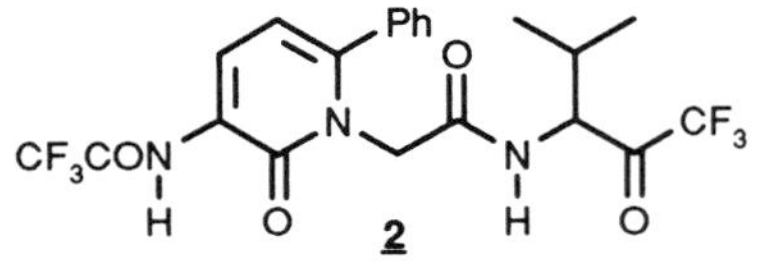

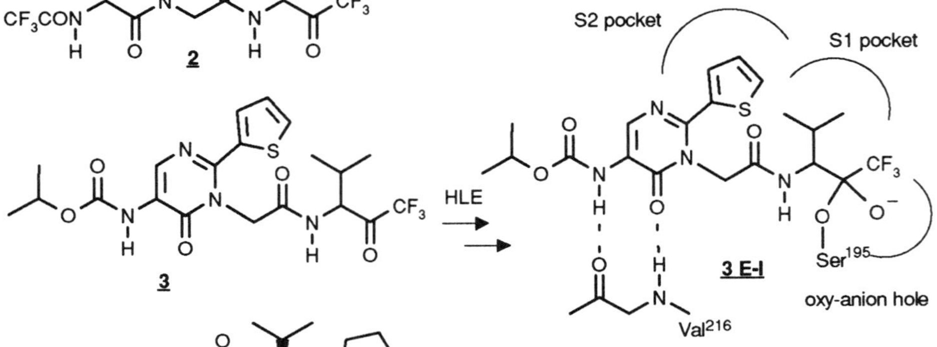

Alternate Substrates - This class of inhibitors act as substrates for HLE to acylate the active site Ser^{195}, but the hydrolysis of the acyl enzyme is sufficiently slow that inhibition occurs. CE-1037 (MDL 201,404YA, **7**) inactivates HLE with IC$_{50}$=18 nM and is active by the i.t. and i.v. route in the hamster model. A comparison of 1 hour versus 6 hour pretreatment with CE-1037 showed inhibition in the lavage fluid; however, the inhibition at 6 hours was considerably less (73,74). An X-ray crystal structure of an enzyme-inhibitor complex formed with YS 3025, **8** and PPE shows that Ser^{195} is acylated with thiophene-2-

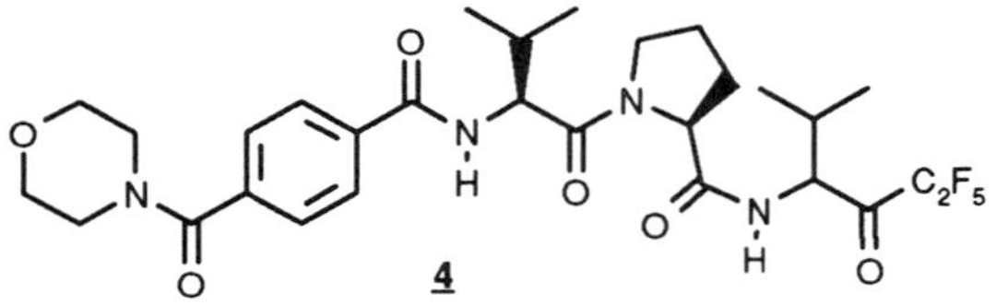

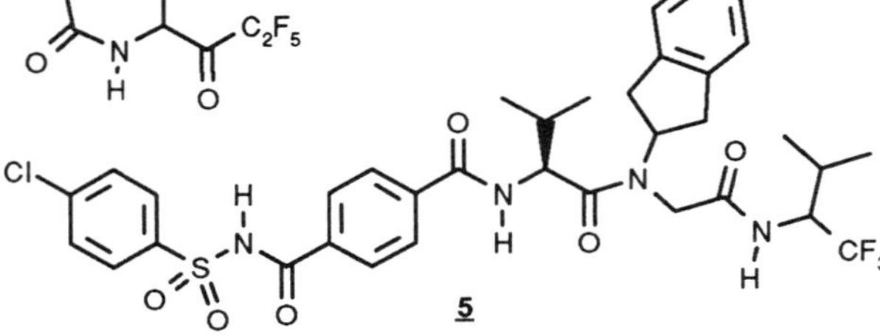

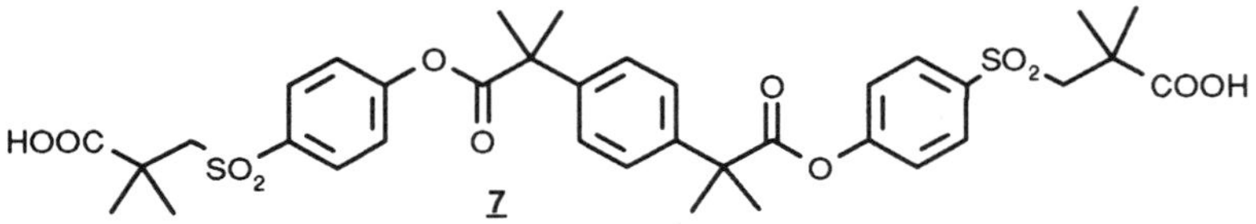

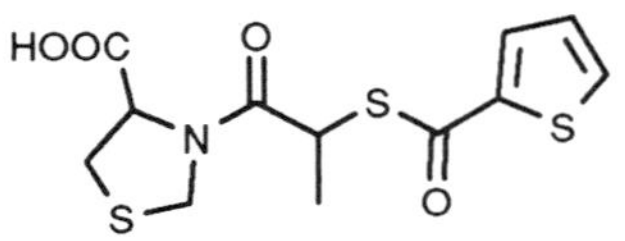

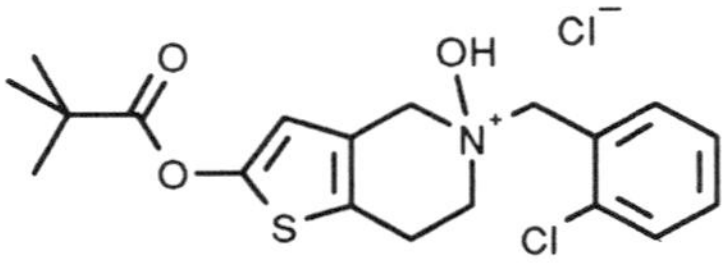

carbonyl (75). SR 26831, **9**, is a potent (IC_{50}=80 nM) and selective inhibitor of HLE that is active by i.v. and oral routes of administration in an elastase-induced hemorrhage rat model. Oral administration of 30 mg/kg of SR 16831 one hour prior to HLE insult gave 35% inhibition (76).

<u>Mechanism-Based Inhibitors</u> - This class of HLE inhibitors possess an electrophilic carbonyl that serves as the site of Ser[195] nucleophilic addition, forming an intermediate acyl enzyme which then undergoes a second reaction to generate a reactive intermediate that can react further at the active site, usually involving covalent bond formation with His[57]. The 4-chloro-3-alkoxyisocoumarin class of inhibitors can form either acyl enzyme (Ser[195] only) or double covalently bound (Ser[195] and His[57]) inhibited complexes. Both of these binding modes with PPE have been observed in crystal structures of complexes. The X-ray crystal structure of the isocoumarin inhibitor **11** complexed with PPE showed that **11** was doubly covalently bound to the enzyme. The inhibitor **11** inactivates HLE with k_{inact}=1,000 $M^{-1}s^{-1}$ and only 30% of PPE could be reactivated from the enzyme-inhibitor complex **11 E-I**, suggesting that at least 70% of the enzyme was doubly linked (77). The binding geometries of four solved PPE-isocoumarin complexes have all been different , but all have formed an acyl enzyme with Ser[195] (47).

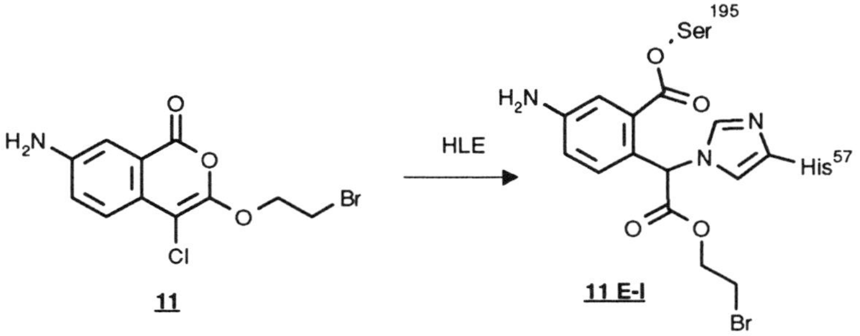

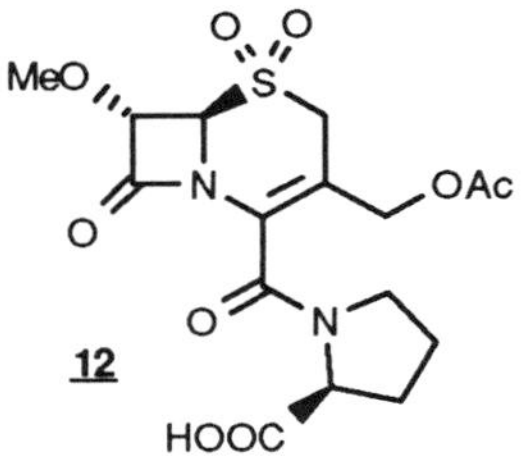

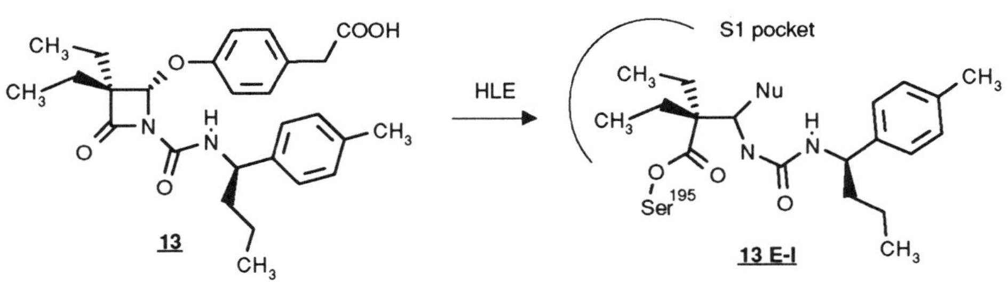

An SAR study identified L-658,758 (**12**), a selective, potent inhibitor, for clinical development by the aerosol route of administration. L-658,758 inhibits elastase induced hemorrhage by exogenous HLE in the hamster with an ED_{50}=15 µg/animal, i.t. and by endogenous elastase in an acute reverse passive Arthus reaction in the rat with an ED_{50}=55 µg/animal, i.t. (78). The penam series has also been examined, however, the biological activity was not improved over L-658,758 (79,80). L-680,833, **13**, is a potent monocyclic β-lactam inhibitor (k_{inact}=622,000 $M^{-1}s^{-1}$) that is orally bioavailable in various species with a Cmax of 34 µg/ml observed in dogs. In the hamster model, L-680,833 is orally active with an ED_{50} of 1.5 mg/kg and >50% inhibition is maintained for 13 hours (81). The SAR studies of the monocyclic β-lactams to optimize stability and oral activity focused on the effect of the C4 leaving group (82), the N-acyl substituent (83,84), stereoisomers at C3 and C4 (85), and the N-acyl α-alkyl substituent (86). L-680,833 is a specific inhibitor of human PMN elastase over various serine proteases, metalloproteases, and thiol proteases (86). Data from a mechanism study are consistent with the formation of an initial Michaelis complex

14

with HLE followed by Ser[195] acylation and loss of the C4 leaving group. This acyl enzyme partitions between enzyme regeneration through hydrolysis and formation of a stable enzyme-inhibitor complex. A double hit enzyme complex **13 E-I** where Ser[195] and His[57] are cross-linked in the active site is consistent with the kinetic data, but is not required (87). A monocyclic β-lactam **14** with an amine aqueous solubilizing group on the C4 leaving group is a potent inhibitor with k_{inact}=120,000 $M^{-1}s^{-1}$ (88).

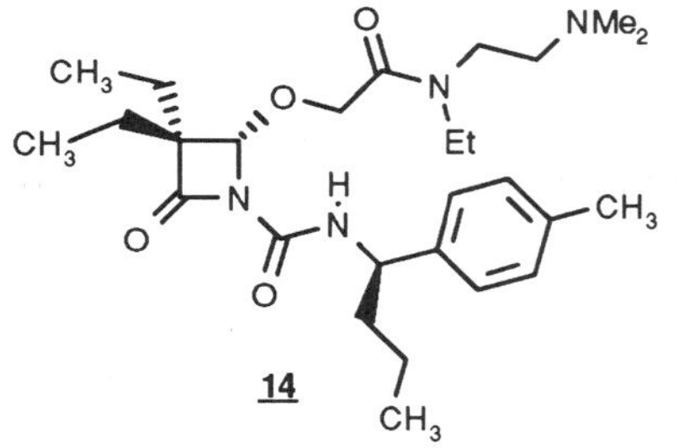

The name Lazarus Inhibitor was suggested to describe a class of benzisothiazolones with a novel mechanism of HLE inhibition. These inhibitors inactivate HLE by a suicide mechanism, but also possess a built-in mechanism for HLE to regain full activity (89,90). WIN 62225 (**15**), a prototype Lazarus inhibitor, is a potent, mechanism-based inhibitor (K_i=2 nM) that is selective for HLE over other serine proteases. Kinetic data and modeling studies support the formation of the proposed enzyme-inhibitor complex **15 E-I** (91,92). Another report describes a similar proposed mechanism for HLE inactivation by benzisothiazolones (93). More potent derivatives have been disclosed, which include the benzisothiazolones **16** and **17** with K_i of 0.007 nM and 0.013 nM, respectively. The inactivation of HLE by **16** is very rapid (k_{inact}=3X10^6 $M^{-1}s^{-1}$) (94,95).

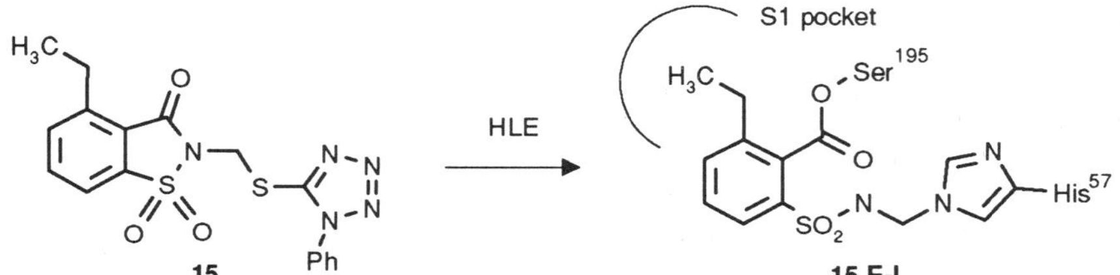

15   **15 E-I**

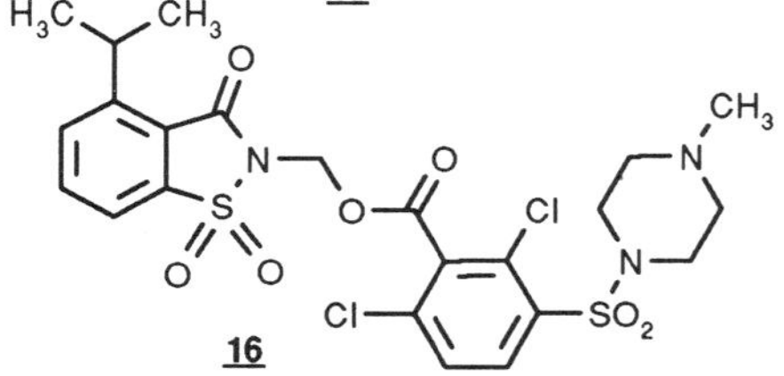

16 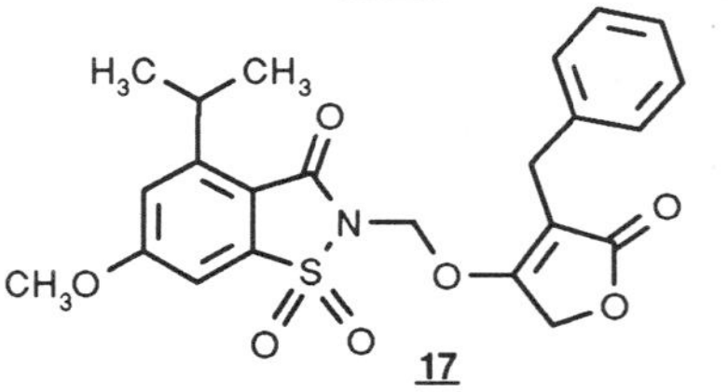   **17**

The succinimide **18** is a time-dependent, irreversible inhibitor (k_{inact}=24,000 $M^{-1}s^{-1}$) with a proposed double hit mechanism (96). After acylation of the catalytic Ser[195], the N-sulfonyloxy-amide intermediate undergoes a Lossen rearrangement to form an isocyanate which is trapped by His[57], cross-linking the catalytic site of HLE (97).

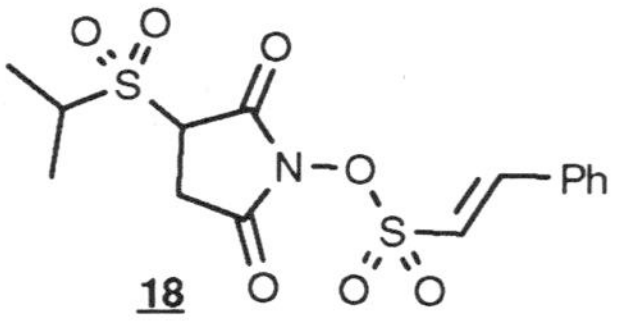

18

In Vivo Models of Elastase Inhibition - An in vivo model commonly used to assess the activity of elastase inhibitors is based on the intratracheal instillation of elastase (neutrophil or pancreatic) to small animals (13). A single instillation of elastase to the lung has been shown to cause severe tissue damage manifested by acute hemorrhage or, on a long term basis, tissue destruction and the development of emphysema-like symptoms (13,98). Models showing the effect of elastase inhibitors on the reduction of lung hemorrhage (99) or on a variety of histological (15) and mechanical (e.g., mean linear intercept of pulmonary function, FEV_1, etc.) indices of lung tissue damage/performance have been developed (100). Data derived from parental or oral administration in the lung hemorrhage model need to be interpreted cautiously. In the presence of lung hemorrhage caused by the instillation of elastase the test compound could be delivered to lung tissue by vascular diffusion or by blood entering the lung. Therefore, compounds that appear to be efficacious in the lung hemorrhage model may be inactive in the absence of hemorrhage because they do not enter the lung by vascular diffusion. Another concern with models that rely on tracheal instillation of elastase is that the initial damage to the lung caused by the instilled elastase is followed by damage caused by endogenous proteases and free radicals released during the ensuing inflammatory response. Thus efficacy or lack of efficacy by an administered compound cannot be solely ascribed to the effect on the instilled protease alone. The effect of endogenous or instilled elastase on tracheal submucosal gland secretion (101) or goblet cell metaplasia (102) has also been used to study the efficacy of elastase inhibitors (103,104). ICI 200,355 (101), α_1-PI (103), SLPI (104), and other HLE inhibitors (105) have shown efficacy for attenuation of these processes.

In Vivo Biochemical Markers of Elastase Activity - The use of biochemical markers is a valuable tool for the characterization of preclinical models of elastase-induced lung damage. They are also useful in drug discovery and development for determining elastase inhibitory effectiveness of compounds in vivo, well before changes in pulmonary function could be detected. Markers for elastase activity can be based on the measurement of elastase inhibitory capacity in bronchoalveolar lavage fluid from diseased lungs. For example, elevated elastase activity has been demonstrated in individuals with cystic fibrosis (16) or ARDS (31). Protein fragments from elastin degradation that are released into the lung and cleared to the plasma and urine have been identified as biochemical markers for lung elastase activity. These markers include fragments of elastin (106-109) and the amino acids desmosine and isodesmosine (110, 111), which are unique to the structure of elastin. These markers have been found in the urine of patients in which lung elastase activity is elevated. A cautionary note is that the source of the elastase activity and the elastin degradative products can not be unequivocally placed in the lung. Elastase has also been shown to cleave the Aα21 (Val)-Aα22 (Glu) bond on the amino-terminal region of the fibrinogen Aα–chain (112); measurement of a modified fibrinogen peptide fragment has been proposed as a way to estimate circulating elastase activity in vivo. Monitoring the levels of these or other markers of elastase activity in patients with elastase-based lung disease offers a means to investigate whether a compound is inactivating elastase during the course of compound administration. In the absence of clinical efficacy data, this type of information could be used to estimate the effective plasma concentration of a compound or predict the likelihood of seeing efficacy in treating the disease in longer term clinical studies.

Conclusions - Hypotheses on the potential effectiveness of HLE inhibitors in treatment of pulmonary diseases have yet to be proven. Approval was gained for the use α_1-PI as replacement therapy in emphysematous patients with genetic α_1-PI deficiency, since α_1-PI levels are low in these individuals. Efficacy showing improvement in pulmonary function or symptomatology has not been demonstrated. A recombinant form of SLPI is in clinical trials for cystic fibrosis (113). Increased reports of orally active HLE inhibitors suggest that clinical trials could begin in the near future.

References

1. Within this series, HLE inhibitors were reviewed in part in 1992 (A. Krantz, Ann.Rep.Med.Chem., 28, 187 (1993)
2. W. Watorek, H. van Halbeek, and J. Travis, Biol.Chem.Hoppe-Seyler, 374, 385 (1993).
3. B.G. Green, H. Weston, B.M. Ashe, J. Doherty, P. Finke, W. Hagmann, M. Lark, J. Mao, A. Maycock, V. Moore, R. Mumford, S. Shah, L. Walakovits, and W.B. Knight, Arch.Biochem.Biophys, 286, 284 (1991).
4. M.C.M. Visserts, C.C. Winterbourn, and J.S. Hunt, Biochim.Biophys. Acta., 804, 154 (1984).
5. H.L. Malech, and J.I.Gallin, N.Engl.J.Med., 317, 687 (1987).
6. P.M. Henson, and R.B. Johnston Jr, J.Clin.Invest., 79, 669 (1987).
7. M.J. Janusz, and N.S. Doherty, J.Immunol. 146, 3922 (1991).
8. A.R. Moore, H. Iwamura, J.P. Larbre, D.L. Scott, and D.A. Willoughby, Ann.Rheum.Dis., 52, 27 (1993).
9. P. Kubes, and D.N. Granger, DrugNews&Perspectives, 4, 197 (1992).
10. T.D. Tetley, Respir.Med, 86, 187 (1992).
11. T.D. Tetley, Thorax, 48, 560 (1993).
12. A. Janoff, Amer.Rev.Respir.Dis., 132, 417 (1985).
13. G.L. Snider, Am.Rev.Respir.Dis., 146, 1615 (1992).
14. J.E. Gadek, Am.J.Med., 92, 27s (1992).
15. S.M. Morris, P.J. Stone and G.L. Snider, J.Histochem.Cytochem., 41, 851 (1993).
16. K.C. Meyer, and J. Zimmerman, J. Lab. Clin. Med., 121, 654 (1993).
17. P.M. Henson, J.E. Henson, G. Kimani, D.L. Bratton, and D.W.H. Riches in "Inflammation: Basic Principles and Clinical Correlates", J.L. Gallin, I.M. Goldstein and R. Snyderman, Eds, Raven Press, N.Y., 1988, pp. 363-380.
18. R. Kittelberger, T.J. Neale, K.T. Francky, N.S. Greenhill, and G.J. Gibson, Biochim.Biophys.Acta, 1139, 295 (1992).
19. Y. Aoki, and T. Yamazaki-Hase, J.Biochem., 114, 122 (1993).
20. J.E. Gadek, G.A. Fells, R.L. Zimmerman, S.I. Rennard, and R.G. Crystal, J.Clin.Invest., 68, 889 (1981).
21. C.B. Laurell, and S. Eriksson, Scand.J.Clin.Invest., 15, 132 (1963).
22. R.G. Crystal, M.L. Brantly, R.C. Hubbard, D.T. Curiel, D.J. States, and M.D. Holmes, Chest., 95, 196 (1989).
23. S.J. Weiss, J.T. Curnutte, and S. Regiani, J.Immuno., 136, 636 (1986).
24. R.C. Hubbard, F. Ogushi, G.A. Fells, A.M. Cantin, S. Jallat, M. Courtney, and R.G. Crystal, J.Clin.Invest., 80, 1289 (1987).
25. B. Wallaert, B. Gressier, C.H. Marquette, P. Gosset, M. Remy-Jardin, J. Mizon, and A.B. Tonnel, Am.Rev.Respir.Dis., 147, 1537 (1993).
26. B. Wallaert, B. Gressier, C. Aerts, C. Mizon, C. Voisin, and J. Mizon, Am J. Respir. Cell. Moll Biol., 5, 437 (1991).
27. D.B. Hood, P. Gettins, and D.A. Johnson, Arch.Biochem.Biophys., 304, 17 (1993).
28. P.E. Desrochers and S.J. Weiss, J.Clin.Invest., 81, 1646 (1988).
29. S. Suter, and I. Chevallier, Eur.Resp.J., 4, 40 (1991).
30. C.M. O'Connor, K. Gaffney, J. Keane, A. Southey, N. Byrne, S. O'Mahoney, and M.X. Fitzgerald, Amer.Rev.Respir.Dis., 148, 1665 (1993).
31. C.T. Lee, A.M. Fein, M. Lippmann, H. Holtzman, P. Kimbel, and G. Weinbaum, N Engl J. Med., 304, 192 (1981).
32. S. Idell, R.S. Thrall, R. Maunder, T.R. martin, J. McLarty, M. Scott, and B.C. Starcher, Exp.Lung Res., 15, 739 (1989).
33. R.A. Stockley, S.L. Hill, H.M Morrison, and C.M. Starkie, Thorax, 39, 408 (1984).
34. S. Eriksson, Eur.Respir.J., 4, 1041 (1991).
35. G.L. Snider, Chest, 101, 74s (1992).
36. "Pulmonary Emphysema: The Rationale for Therapeutic Intervention," Vol. 624, G. Weinbaum, R.E. Guiles, and R.D. Krell,Eds., New York Academy of Sciences, New York, NY, 1991.
37. M. Wewers, Chest, 95, 190 (1989).
38. "Elastin and Elastases," Vol. II, L. Robert and W. Hornebeck, Eds., CRC Press, Boca Raton, FL, 1989.
39. Agents Actions Suppl., 42(Proteases, Protease Inhibitors and Protease-Derived Peptides), 1993.
40. R.B. Silverman in "The Organic Chemistry of Drug Design and Drug Action," Academic Press, San Diego, 1992, chapter 5.
41. P.D. Edwards and P.R. Bernstein, Med.Res.Rev., 14,127 (1994).
42. P.R. Bernstein, P.D. Edwards, and J.C. Williams, Progress Med.Chem., 31, 51 (1994).
43. M.A. Navia, B.M. McKeever, J.P. Springer, T.-Y. Lin, H.R. Williams, E.M. Fluder, C.P. Dorn, and K. Hoogsteen, Proc.Natl.Acad.Sci.USA, 86, 7 (1989).
44. W. An-Zhi, I. Mayr, and W. Bode, FEBSLett., 234, 367 (1988).
45. W. Bode, A.-Z. Wei, R. Huber, E. Meyer, J. Travis, and S. Neumann, EMBO J., 5, 2453 (1986).

46. W. Bode, E. Meyer, Jr., and J.C. Powers, Biochemistry, 28, 1951 (1989).
47. M.A. Hernandez, J.C. Powers, J. Glinski, J. Oleksyszyn, J. Vijayalakshmi, and E.F. Meyer, Jr., J.Med.Chem., 35, 1121 (1992).
48. P.D. Edwards, E.F. Meyer, Jr., J. Vijayalakshmi, P.A. Tuthill, D.A. Andisik, B. Gomes, and A. Strimpler, J.Am.Chem.Soc., 114, 1854 (1992).
49. M.A. Navia, B.M. McKeever, and J.P. Springer, Crystallogr.Model.MethodsMol.Des.,[Pap.Symp.] 1989, 29 (1990).
50. R.R. Plaskon, C.-M. Kam, J.E. Kerrigan, E.M. Burgess, J.C. Powers, and F.L. Suddath, Arch.Biochem.Biophys., 300, 588 (1993).
51. R.A. Firestone, P.L. Barker, J.M. Pisano, B.M. Ashe, and M.E. Dahlgren, Tetrahedron, 46, 2255 (1990).
52. W.B. Knight, K.M. Swiderek, T. Sakuma, J. Calaycay, J.E. Shively, T.D. Lee, T.R. Covey, B. Shushan, B.G. Green, R. Chabin, S. Shah, R. Mumford, T.A. Dickinson, and P.R. Griffin, Biochemistry, 32, 2031 (1993).
53. R.T. Aplin, C.V. Robinson, C.J. Schofield, and N.J. Westwood, J.Chem.Soc.,Commun., 1650 (1992).
54. R.T. Aplin, C.V. Robinson, C.J. Schofield, and N.J. Westwood, Tetrahedron, 49, 10903 (1993).
55. D. Burnett and R.A. Stockley, Am.J.Respir.Cell Mol.Biol., 8, 119 (1993).
56. C. Boudier and J.G. Bieth, Biochim.Biophys.Acta, 995, 36 (1989).
57. P. Renesto, V. Balloy,T. Kamimura, K. Masuda, A. Imazumi, and M. Chignard, Br.J.Pharmacol., 108, 1100 (1993).
58. O. Wiedow, J.-M. Schroeder, H. Gregory, J.A. Young, and E. Christophers, J.Biol.Chem., 265, 14791 (1990).
59. Q.-L. Ying and S.R. Simon, Biochemistry, 32, 1866 (1993).
60. J.-M. Sallenave, A. Silva, M.E. Marsden, and A.P. Ryle, Am.J.Respir.Cell Mol.Biol., 8, 126 (1993).
61. M. Tsunemi, H. Kato, Y. Nishiuchi, S. Kumagaye, and S. Sakakibara, Biochem. Biophys.Res.Commun., 185, 967 (1992).
62. J.C. Williams, M. Renzetti, J. Donahue, B. Lucas, A. Strimpler, J.E. Fitton, R. Anand, L.R. Farrow, B. Gomes, and R.D. Krell, Am.Rev.Respir.Dis., 145, A200 (1992).
63. J. Schalkwijk, C. de Roo, and G.J. de Jongh, Biochim.Biophys.Acta, 1096, 148 (1991).
64. P.J. Hogg, D.A. Owensby, D.F. Mosher, T.M. Misenheimer, and C.N. Chesterman, J.Biol.Chem., 268, 7139 (1993).
65. J.C. Williams, R.C. Falcone, C. Knee, R.L. Stein, A.M. Strimpler, B. Reaves, R.E. Giles, and R.D. Krell, Am.Rev.Respir.Dis., 144, 875 (1991).
66. P.R. Bernstein, A. Shaw, R.M. Thomas, D.J. Wolanin, and P. Warner, Eur.Pat.Appl.EP 509769 A2 (1992).
67. P.R. Bernstein, P.D. Edwards, R.M. Thomas, C.A. Veale, P. Warner, D.J. Wolanin, US Patent 5,254,558 (1993).
68. J.R. Damewood, Jr., C. Ceccarelli, D.W. Andisik, P.R. Bernstein, C.B. Bryant, S.W. Feeney, B.C. Gomes, R.C. Green, B.J. Kosmider, G.B. Steelman, R.M. Thomas, E.P. Vacek, C.A. Veale, P. Warner, D.J. Wolanin, and S.A. Woolson, Abstract at the 51st Annual Pittsburgh Diffraction Conference, Valley Forge, Pennsylvania, November 3-5, 1993.
69. N.P. Peet, M.R. Angelastro, and J.P. Burkhart, Eur.Pat.Appl.EP 529568 A1 (1993).
70. J.W. Skiles, V. Fuchs, C. Miao, R. Sorcek, K.G. Grozinger, S.C. Mauldin, J. Vitous, P.W. Mui, S. Jacober, G. Chow, M. Matteo, M. Skoog, S.M. Weldon, G. Possanza, J. Keirns, G. Letts, and A.S. Rosenthal, J.Med.Chem., 35, 641 (1992).
71. J.W. Skiles, C. Miao, R. Sorcek, S. Jacober, P.W. Mui, G. Chow, S.M. Weldon, G. Possanza, M. Skoog, J. Keirns, G. Letts, and A.S. Rosenthal, J.Med.Chem., 35, 4795 (1992).
72. H.H. Wasserman, D.S Ennis, P.L. Power, M,J, Ross, and B. Gomes, J.Org.Chem., 58, 4785 (1993).
73. L.M. Patton, Agents Actions Suppl., 42, 83 (1993).
74. G.P. Kirschenheuter, J. Oleksyszyn, L.W. Spruce, M. Wieczorek, T.M. Koppel, S.R. Simon, and J.C. Cheronis, Agents Actions Suppl., 42, 71 (1993).
75. K.D. Carugo, M. Rizzi, M. Fasano, M. Luisetti, C. La Rosa, P. Ascenzi, and M. Bolognesi, Biochem. Biophys.Res.Commun., 193, 32 (1993).
76. J.M. Herbert, D. Frehel, M.P. Rosso, E. Seban, C. Castet, O. Pepin, J.P. Maffrand and G. Le Fur, J.Pharmacol.Exp.Ther., 260, 809 (1992).
77. J. Vijayalakshmi, E.F. Meyer, Jr., C.-M. Kam, and J.C. Powers, Biochemistry, 30, 2175 (1991).
78. P.E. Finke, S.K. Shah, B.M. Ashe, R.G. Ball, T.J. Blacklock, R.J. Bonney, K.A. Brause, G.O. Chandler, M. Cotton, P. Davies, P.S. Dellea, C.P. Dorn, Jr., D.S. Fletcher, L.A. O'Grady, W.K. Hagmann, K.M. Hand, W.B. Knight, A.L. Maycock, R.A. Mumford, D.G. Osinga, P. Sohar, K.R. Thompson, H. Weston, and J.B. Doherty, J.Med.Chem., 35, 3731 (1992).
79. P.E. Finke, M.E. Dahlgren, H. Weston, A.L. Maycock, and J.B. Doherty, Bioorg.Med.Chem.Lett., 3, 2277 (1993).
80. K.R. Thompson, P.E. Finke, S.K. Shah, B.M. Ashe, M.E. Dahlgren, P.S. Dellea, D.S. Fletcher, K.M. Hand, A.L. Maycock, and J.B. Doherty, Bioorg.Med.Chem.Lett., 3, 2289 (1993).
81. J.B. Doherty, S.K. Shah, P.E. Finke, C.P. Dorn, Jr., W.K. Hagmann, J.J. Hale, A.L. Kissinger, K.R. Thompson, K. Brause, G.O. Chandler, W.B. Knight, A.L. Maycock, B.M. Ashe, H. Weston, P. Gale, R.A. Mumford, O.F. Andersen, H.R. Williams, T.E. Nolan, D.L. Frankenfield, D. Underwood, K.P. Vyas, P.H. Kari, M.E. Dahlgren, J. Mao, D.S. Fletcher, P.S. Dellea, K.M. Hand, D.G. Osinga, L.B. Peterson, D.T. Williams, J.M. Metzger, R.J. Bonney, J.L. Humes, S.P. Pacholok, W.A. Hanlon, E. Opas, J. Stolk, and P. Davies, Proc.Natl.Acad.Sci.USA, 90, 8727 (1993).

82.   W.K. Hagmann, A.L. Kissinger, S.K. Shah, P.E. Finke, C.P. Dorn, K.A. Brause, B.M. Ashe, H. Weston, A.L. Maycock, W.B. Knight, P.S. Dellea, D.S. Fletcher, K.M. Hand, D. Osinga, P. Davies, and J.B. Doherty, J.Med.Chem., 36, 771 (1993).
83.   W.K. Hagmann, K.R. Thompson, S.K. Shah, P.E. Finke, B.M. Ashe, H. Weston, A.L. Maycock, and J.B. Doherty, Bioorg.Med.Chem.Lett., 2, 681 (1992).
84.   S.K. Shah, C.P. Dorn, Jr., P.E. Finke, J.J. Hale, W.K. Hagmann, K.A. Brause, G.O. Chandler, A.L. Kissinger, B.M. Ashe, H. Weston, W.B. Knight, A.L. Maycock, P.S. Dellea, D.S. Fletcher, K.M. Hand, R.A. Mumford, D.J. Underwood, and J.B. Doherty, J.Med.Chem., 35, 3745 (1992).
85.   S.K. Shah, P.E. Finke, K.A. Brause, G.O. Chandler, B.M. Ashe, H. Weston, A.L. Maycock, R.A. Mumford, and J.B. Doherty, Bioorg.Med.Chem.Lett., 3, 2295 (1993).
86.   W.B. Knight, B.G. Green, R.M. Chabin, P. Gale, A.L. Maycock, H. Weston, D.W. Kuo, W.M. Westler, C.P. Dorn, P.E. Finke, W.K. Hagmann, J.J. Hale, J. Liesch, M. MacCoss, M.A. Navia, S.K. Shah, D. Underwood, and J.B. Doherty, Biochemistry, 31, 8160 (1992).
87.   R. Chabin, B.G. Green, P. Gale, A.L. Maycock, H. Weston, C.P. Dorn, P.E. Finke, W.K. Hagmann, J.J. Hale, M. MacCoss, S.K. Shah, D. Underwood, J.B. Doherty, and W.B. Knight, Biochemistry, 32, 8970 (1993).
88.   P.E. Finke, A.L. Kissinger, M. MacCoss, S.K. Shah, J.B. Doherty, W.K. Hagmann, P. Davies, J.L. Humes, E.S. Luedke, R.A. Mumford, PCT Int.Appl. WO 93 00332 A1 (1993). CA119, 49217r.
89.   R. Dunlap, N. Boaz, J. Court, C. Franke, D. Hlasta, A. Mura, Abstract at the 202nd National Meeting of the American Chemical Society, New York, New York., August 1991, MEDI 87.
90.   R.P. Dunlap, N.W. Boaz, A.J. Mura, and D.J. Hlasta, PCT Int.Appl. WO 90 13,549 (1990); CA 114:228897f.
91.   D. Hlasta, M. Bell, N. Boaz, J. Court, R. Desai, S. Eiff, C. Franke, A. Mura, C. Subramanyam, and R. Dunlap, Abstract at the 202th National Meeting of the American Chemical Society, New York, New York., August 1991, MEDI 172.
92.   D.J. Hlasta, R.C. Desai, C. Subramanyam, E.P. Lodge, R.P. Dunlap, N.W. Boaz, A.J. Mura, and L.H. Latimer, Eur.Pat.Appl.EP 542372 A1 (1993).
93.   W.C. Groutas, N. Houser-Archield, L.S. Chong, R. Venkataraman, J.B; Epp, H. Huang, and J.J. McClenahan, J.Med.Chem., 36, 3178 (1993).
94.   R.P. Dunlap, N.W. Boaz, A.J. Mura, V. Kumar, C. Subramanyam, R.C. Desai, D.J. Hlasta, M.T. Saidane, M.R. Bell, and J.J. Court, US Patent 5,128,339 (1992); CA 117, 69858n.
95.   D.J. Hlasta, J.H. Ackerman, and A.J. Mura, Eur.Pat.Appl.EP 547708 A1 (1993). CA 119, 249941j.
96.   W.C. Groutas, R. Venkataraman, M.J. Brubaker, J.B; Epp, L.S. Chong, M.A. Stanga, J.J. McClenahan, and F. Tagusagawa, Biochim.Biophys.Acta, 1164, 283 (1993).
97.   W.C. Groutas, M.J. Brubaker, M.A. Stanga, J.C. Castrisos, J.P. Crowley, and E.J. Schatz, J.Med.Chem., 32, 1607 (1989).
98.   A. Janoff, B. Sloan, G. Weinbaum, V. Damiano, R.A. Sanhaus, J. Elias, and P. Kimbel, Amer.Rev.Respir.Dis., 115, 461 (1977).
99.   D.S. Fletcher, D.G. Osinga, K.M. Hand, P.S. Dellea, B.M. Ashe, R.A. Mumford, P. Davies, W. Hagmann, P.E. Finke, J.B. Doherty, and R.J. Bonney, Am.Rev.Respir.Dis., 141, 672 (1990).
100.   P.J. Stone, J. Bryan-Rhadfi, E.C. Lucey, D.E. Ciccolella, G. Crombie, B. Faris, G.L. Snider, and C. Franzblau, Am.Rev.Respir.Dis., 144, 284 (1991).
101.   A. Schuster, I. Ueki, and J.A. Nadel, Am.J.Physiol., 262, 86 (1992).
102.   R. Breuer, T.G. Christensen, E.C. Lucey, P.J. Stone, and Snider, G.L., J.Lab.Clin.Med., 105, 635 (1985).
103.   P.J. Stone, E.C. Lucey, D.E. G.D. Virca, T.G. Christensen, R. Breuer, and Snider, G.L., Eur.Respir.J., 3, 673 (1990).
104.   E.C. Lucey, P.J. Stone, D.E. Ciccolella, R. Breuer, T.G. Christensen, R.C. Thompson, and G.L. Snider, J.Lab.Clin.Med., 115, 224 (1990).
105.   Y. Uejima, M. Kokubo, J-I. Oshida, H. Kawabata, Y. Kato, and K. Fuji, J.Pharmacol.Exp.Ther., 265, 516 (1993).
106.   G. Weinbaum, U. Kucich, P. Kimbel, S. Mette, S. Akers, and J. Rosenbloom, Ann.N.Y.Acad.Sci., 624, 147 (1991).
107.   E.E Schriver, J.M. Davidson, M.C. Sutcliffe, B.B. Swindell and G.R. Gordon, Am.Rev.Respir.Dis., 145, 762 (1992).
108.   T.J. Dillon, R.L. Walsh, R. Scicchitano, B. Eckert, E.G. Cleary and G. Mclennan, Am.Rev.Respir.Dis., 146, 1143 (1992).
109.   S. Akers, U. Kucich, M. Swartz, G. Rosen, M. Glass, J. Rosenbloom, P. Kimbel and G. Weinbaum, Am.Rev.Respir.Dis., 145, 1077 (1992).
110.   R.A. Goldstein and B.C. Starcher, J.Clin.Invest., 61, 1286 (1978).
111.   V. Pai, A. Guz, G.J. Phillips, N.T. Cooke, D.C.S. Hutchison, and T.D. Tetley, Metabolism, 40, 139 (1991).
112.   J.I. Weitz, E.K. Silverman, B. Thong, and E.J. Campbell, J. Clin. Invest., 89, 766 (1992).
113.   N.G. McElvaney, B. Doujaiji, M.J. Moan, M.R. Burnham, M.C. Wu, and R.G. Crystal, Am.Rev.Respir.Dis., 148, 1056 (1993).

Chapter 21. Interleukin-1 as a Therapeutic Target

Daniel Tracey

BASF Bioresearch Corporation, Worcester, MA 01605

Introduction - Cytokines are intercellular polypeptide messengers that regulate cell function. Among the dozens of cytokines that are now recognized, the interleukin-1 (IL-1) family has been the most intensively studied and characterized. The IL-1 family consists of two agonists, IL-1α and IL-1β, and a structurally related antagonist, IL-1ra. Unless specified otherwise, the term "IL-1" refers to IL-1α and IL-1β. Comprehensive reviews have been recently published of cytokines (1), IL-1 structure (2), IL-1 receptors (2, 3), biological activities of IL-1 (4), IL-1 signal transduction (2, 5), the role of IL-1 in disease (6), IL-1ra structure and function (7) and the modulation of IL-1 in disease (7, 8). Some of the pertinent points will be briefly summarized here as background for a discussion of the recent history of IL-1 as a therapeutic target. Volume 25 of this series contains a review (9) through 1989 of the status of pharmacological modulation of IL-1. Volume 27 contains a brief overview (10) of the status through 1992 of pharmacological modulation of many cytokines, including IL-1.

IL-1 PRODUCTION AND BIOLOGICAL ACTIVITIES

IL-1 Production - Although virtually all cell types are capable of producing IL-1, monocytes and macrophages are the major sources of these proteins. In general, IL-1 proteins are not constitutively produced, but their biosynthesis can be induced by a wide variety of agents, including microbial products, immune complexes, stress and various cytokines (4). The IL-1 agonists are synthesized from separate genes as 31 kD precursor proteins. Most of the proIL-1α molecules that are synthesized remain within the cell, including some which become plasma membrane associated and are biologically active. ProIL-1β, on the other hand, is biologically inactive and is processed by a specific intracellular proteinase, interleukin-1β converting enzyme (ICE), to generate active 17 kD IL-1β, which is released from the cell (2). The 17 kD IL-1 receptor antagonist, IL-1ra, is released by macrophages and neutrophils after the removal of a 25-amino acid signal peptide (7). The relative amounts of the IL-1 proteins synthesized varies with cell type, but human monocytes/macrophages generally produce more IL-1ra than IL-1α or IL-1β and more IL-1β than IL-1α. The cDNA's for all three proteins have been cloned and the corresponding recombinant proteins have been expressed in E. coli. Crystal structures of the proteins have been determined, revealing overall similarities in their 12-stranded β-barrel core structures, with distinct differences in the outer portions of the molecules (2, 11).

IL-1 Receptors, Biological Activities and Signal Transduction - The biological activities of IL-1 proteins on a wide variety of cell types are mediated through interaction with specific, high affinity, 80 kD, cell surface IL-1 receptors (IL-1R type I). In addition, IL-1β can bind to 65 kD cell surface IL-1 "receptors" (IL-1R type II) which do not transduce signals, but appear to be shed from the cell surface as soluble IL-1 receptors (2, 3, 12). IL-1α and IL-1β bind to the IL-

1-RI with near equal affinities and generally mediate identical activities (2-4). The spectrum of biological activities mediated by IL-1 include inflammatory, immunological, hematopoietic, cardiovascular, endocrinological and neurological activities. These activities are mediated through the stimulation by IL-1 of a wide variety of cell types, including fibroblasts, endothelial cells, chondrocytes, osteoclasts, hepatocytes, leukocytes, bone marrow cells, smooth muscle cells and cells in reproductive organs, pituitary and brain (4). The signal transduction pathways in cells that mediate these biological functions are being elucidated. Membrane phosholipid metabolism leads to intracellular kinase activation and activation of transcription factors which turn on the IL-1 responsive genes (3, 5). The biological consequences of IL-1 stimulation on various cells are as diverse as the spectrum of cells involved. Prominent among these effects are the release of cytokines, enzymes, lipid metabolites and oxygen metabolites (4).

Role of IL-1 in Disease - There is no clear evidence that IL-1 plays a homeostatic role. Plasma IL-1 is undetectable in normal individuals, except during ovulation and prolonged exercise. Rather, IL-1 appears to play a role in various pathological situations, where it mediates fever, inflammation, tissue destruction, anorexia and slow wave sleep (4, 6). Two naturally occurring mechanisms have evolved which inhibit IL-1 activities to a certain extent. These involve the production of IL-1ra and the shedding of soluble IL-1 receptors (both types I and II), which can neutralize IL-1 activities in tissues and fluids (3, 7, 8). Elevated levels of IL-1 have been seen in animal models of disease, including bacterial sepsis, endotoxic shock, inflammatory bowel diseaase, collagen-induced arthritis and insulin-dependent diabetes mellitus. Plasma IL-1β levels are elevated in patients with septic shock, rheumatoid arthritis and other diseases. Evidence of IL-1 in tissues and fluids from patients with rheumatoid arthritis, inflammatory bowel disease, asthma, periodontal disease, psoriasis, atherosclerosis and many other diseases provides further support for the hypothesis that IL-1 plays a pathogenic role in these diseases (6). More direct evidence for the role of IL-1 in certain diseases has been provided by the use of recombinant IL-1ra, soluble IL-1 receptors, anti-IL-1 antibodies or anti-IL-1RI antibodies in animal models of disease and in a limited number of clinical trials (8, 9). Significant amelioration of disease has been seen with parenteral administration of IL-1ra in murine and rabbit septic shock models, Streptococcal cell wall induced arthritis in rats, collagen-induced arthritis in mice (13) an immune complex induced colitis model in rabbits, a murine graft *vs* host disease (GVHD) model and in other models of acute and chronic inflammation (8, 9). Soluble IL-1 receptors have been used successfully in a murine cardiac allograft rejection model (2). By specifically inhibiting the biological activities of IL-1, amelioration or prevention of these diseases by IL-1ra or soluble IL-1R provides direct evidence for a role of IL-1 in their pathogenesis.

IL-1 PRODUCTION INHIBITORS

CO/LO Inhibitors and Antioxidants - Several compounds which inhibit eicosanoid metabolism via cyclooxygenase (CO) and/or lipoxygenase (LO) enzyme inhibition have been found to inhibit IL-1 production or release from monocytes or macrophages. Ironically, however, studies of a wide spectrum of CO and 5-LO inhibitors revealed no direct relationship between CO/LO inhibition and IL-1 inhibition (14,15). Several dual CO/5-LO inhibitors, including the bicyclic imidazoles SK&F 86002 (1) and SK&F 105561 (2), the active metabolite of the prodrug SK&F 105809 (3), have been reviewed in this series (9,10) and elsewhere (16). Not all the activities of these compounds can be attributed to inhibition of eicosanoid production. Equivalent inhibition of the release of IL-1β and TNFα by lipopolysaccharide (LPS)-stimulated human monocytes *in vitro* was seen with 1 (IC_{50} = 0.5 μM) and 2 (IC_{50} = 3 μM).

Some studies (17) suggest that **1** inhibits the biosynthesis of proIL-1β, whereas other studies (18) suggest that it inhibits the release of IL-1β. Imidazoles **1** and **3** have antiinflammatory activity in murine collagen induced arthritis and murine endotoxin shock models (16).

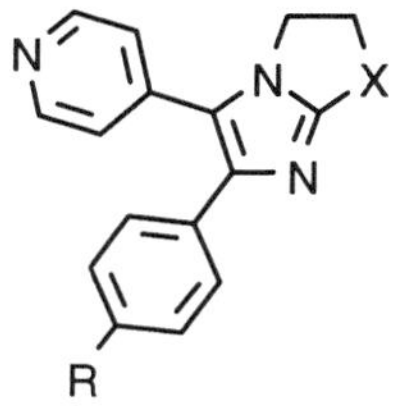

	R	X
1	F	S
2	CH₃S	CH₂
3	CH₃SO	CH₂

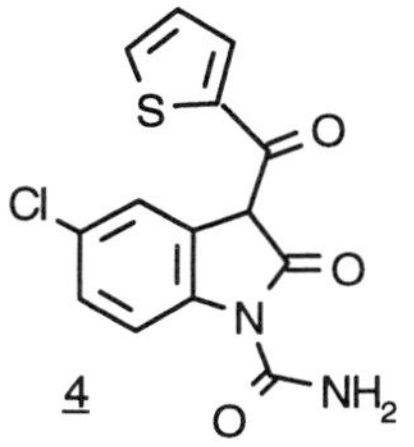

The oxindole, tenidap (**4**), is a CO (and weaker 5-LO) inhibitor that inhibits the production of IL-6, TNF and IL-1 by LPS-stimulated human peripheral blood mononuclear cells (PBMC), as previously reviewed (9). One study also showed inhibition of IL-1, but not TNF or IL-6 production, by the CO inhibitor naproxen (19), although other studies have shown little or no effect of CO inhibitors, including naproxen, on IL-1 production by human monocytes or in a human whole blood assay (15). Phase III clinical trials of **4** in rheumatoid arthritis and osteoarthritis patients have shown safety and efficacy comparable to other NSAIDs. In these trials, **4** inhibited synovial fluid IL-1 levels and the acute phase response as measured by C-reactive protein (CRP) levels, whereas naproxen had no effect on CRP (20). The role of IL-1 inhibition in the antirheumatic activity of **4** remains uncertain. Another CO inhibitor, a benzopyranone, T614 (**5**), inhibited the release of IL-1β and IL-6 from LPS-stimulated human monocytes and the human monocytic cell line THP-1 with IC_{50} values of 30 μg/ml and 7 μg/ml, respectively (21). T614 inhibited IL-1β mRNA levels in THP-1 cells, but did not inhibit intracellular proIL-1β production in human monocytes. Oral antiinflammatory activity of **5** at doses >25 mg/kg has been seen in rat adjuvant-induced arthritis and carrageenin-induced air pouch granuloma models, the latter suggesting the possible role of inhibition of cytokine synthesis in the antiinflammatory activity (21).

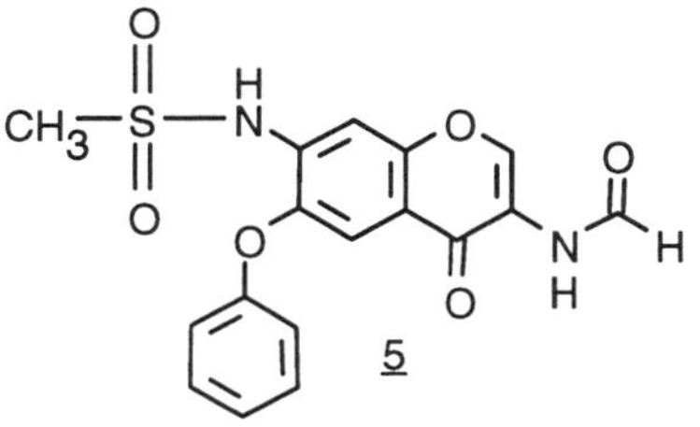

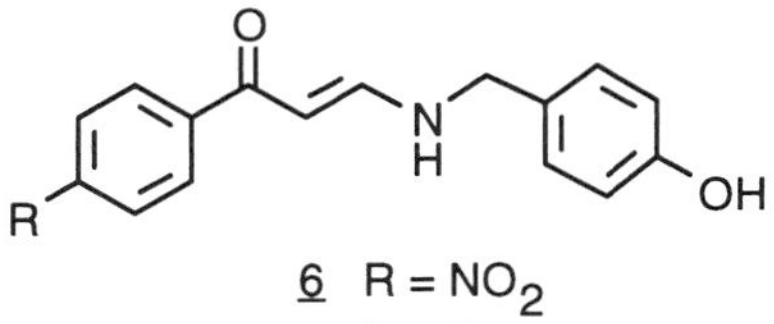

6 R = NO₂
7 R = F
8 R = Cl

Several vinylogous amides (**6**, **7**, **8**) from a large series of vinylogous hydroxamic acids (22) were found to be potent 5-LO inhibitors (IC_{50} = 2.5, 0.15, and 1.5 μM, respectively) and inhibitors of IL-1β release by LPS-stimulated human monocytes (IC_{50} = 1.7, 2.8, and 2.1 μM, respectively). These compounds are devoid of CO inhibitory activity. Vinylogous hydroxamic acids lacking the hydroxylamino moiety were uniformly inactive. There have been no reports concerning *in vivo* activities of **6**, **7**, or **8**.

No correlation was found between antioxidant activity and IL-1 production inhibition among many antioxidants tested on human monocytes (23). Several antioxidants, including α-tocopherol, probucol, BHT and mannitol, did not inhibit IL-1 production by human monocytes. The results with α-tocopherol and probucol contrast with findings of IL-1 inhibition in mouse macrophages *ex vivo* and THP-1 cells *in vitro* (9, 24). The antioxidants tetrahydropapaveroline (**9**), a tetrahydroisoquinoline derivative, and apomorphine (**10**) were potent inhibitors (IC$_{50}$ = 1.5 and 2.6 μM, respectively) of proIL-1β synthesis and IL-1β release, as well as TNFα and IL-6 release at similar concentrations, from human PBMC stimulated with LPS, zymosan, silica or pansorbin (23). In addition, **9** and **10** were active in a murine *ex vivo* assay of IL-1β release from peritoneal cells taken from mice treated orally with 100 mg/kg **9** or 50 mg/kg **10** and LPS. Apomorphine also inhibited the elevation in plasma levels of IL-1β and TNFα *in vivo* in mice following LPS injection (23).

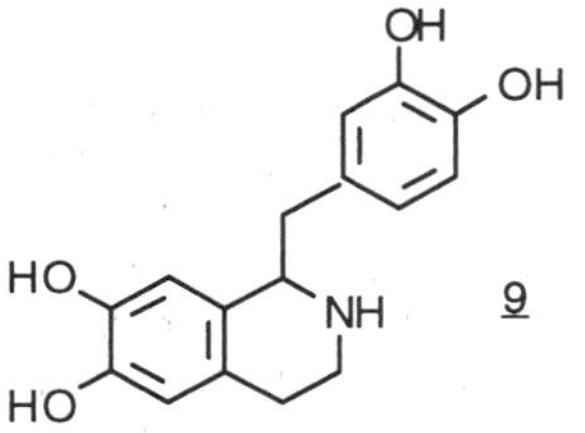

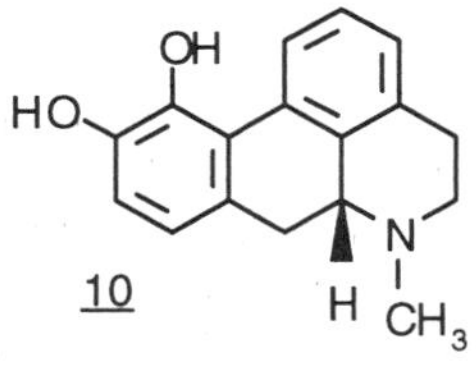

ICE Inhibitors - The recent molecular characterization of ICE and the cloning of cDNA's for human ICE (25, 26), mouse ICE (27) and rat ICE (28) has facilitated a targeted approach for the discovery of drugs which specifically inhibit IL-1β production (reviewed in 29). This unique cysteine proteinase cleaves the Asp[116]-Ala[117] bond of human proIL-1β (or the homologous Asp-Ala or Asp-Val bonds in other species' proIL-1β) to generate mature, active IL-1β. The substrate specificity of ICE has been studied using amino-terminal and carboxyl-terminal truncations of a 14-amino-acid peptide, NEAYVHDAPVRSLN, which spans the D-A cleavage site in human proIL-1β (26,30). These studies determined that Asp is absolutely required at the P1 position and small aliphatic residues are preferred at P1'. Residues beyond P1' were not required, but at least four residues N-terminal to the cleavage site were required for activity. Substitutions in P2 are well tolerated, Val is preferred at P3 and hydrophobic residues are preferred at P4. The optimal peptide substrate was found to be Ac-YVAD-methylamine. Based on this substrate sequence, two ICE inhibitors, **11** and **12**, were synthesized (26). The diazomethylketone **11** (L-707,509) irreversibly inhibited human

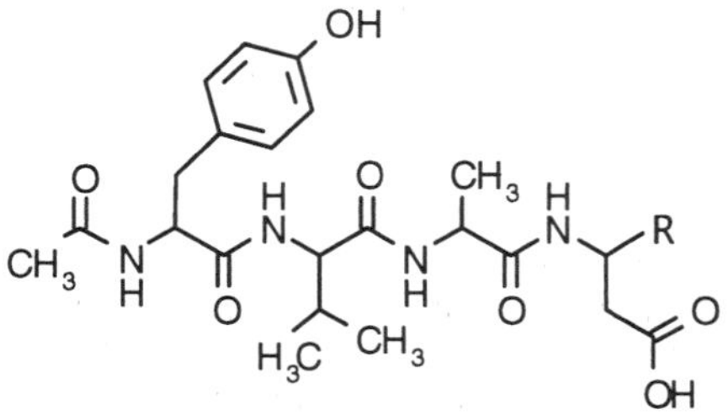

R

11	COCH$_2$N$_2$
12	CHO
13	CO(CH$_2$)$_4$C$_6$H$_5$
14	CN

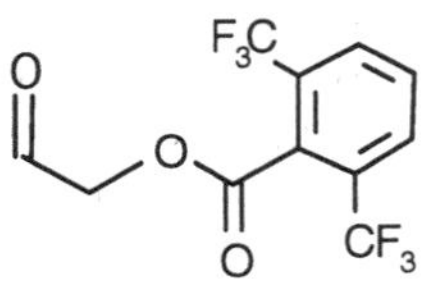

ICE, suggesting that ICE is a thiol proteinase. This was confirmed by ICE inhibition with thiol selective reagents, such as N-ethylmaleimide and iodoacetate (26, 31) and by labeling of the active site Cys[285] by [^{14}C]iodoacetate (26). The peptide aldehyde **12** (L-709,049) was found to be a potent (K_i = 0.76 nM), reversible ICE inhibitor in enzyme assays and to inhibit (IC$_{50}$ = 4 uM) IL-1β release into plasma in LPS-stimulated human whole blood (26). The diminished activity of the peptide aldehyde **12** in enzyme *vs* cellular assays probably reflects the poor penetration of the aldehyde into cells.

Additional ICE substrate analogs have been reported to inhibit ICE activity in enzyme assays. The Ac-YVAD phenylbutyl ketone **13** and the nitrile **14** are reversible ICE inhibitors with K_i values of 37 nM and 60 nM, respectively (32). Acyloxymethyl ketone derivatives of Ac-YVAD, such as **15**, irreversibly inhibit the enzyme through formation of a thiomethyl ketone with Cys[285] and have second order inactivation rates (~1 x 10^6 M^{-1}s^{-1}) that are limited by diffusion (33). Other acyloxymethyl ketones, such as **16**, have been described (34) and the free acid of **16** would be expected to inhibit ICE activity. Furthermore, **16** inhibited IL-1β release by LPS-stimulated THP-1 cells *in vitro* (IC$_{50}$ = 0.1 μM) and blocked LPS-induced fever and carrageenin-induced paw edema (ED$_{50}$ = 0.01 and 1 mg/kg, respectively) after oral administration to rats (34). Smaller acyloxymethyl ketones, such as the Asp derivative **17** and the Val-Asp derivative **18** have recently been described (35) as irreversible inactivators of ICE (k_{obs}/[I] = 7,100 and 41,000 M^{-1}s^{-1}, respectively). In addition, acyloxymethyl ketones **17** and **18** show high selectivity (>700:1 and 108:1, respectively) for ICE relative to another cysteinyl proteinase, cathepsin B (35). A macromolecular (38 kD) inhibitor of ICE, *crmA*, is encoded in cowpox virus and may be a natural antiinflammatory defense mechanism of this virus (36). The *crmA* protein inhibits ICE with very high affinity (K_i = 7 pM) and does not inhibit other proteinases (32, 36). Recently, *crmA* was found to inhibit apoptosis of neurons deprived of nerve growth factors, suggesting a possible role for ICE or ICE-like enzymes in neuronal programmed cell death (37). Moreover, overexpression of ICE in rat fibroblasts induced cell death which could be blocked by *crmA* (38).

Other IL-1 Production Inhibitors - Compounds which inhibit IL-1 production by unknown molecular mechanisms include the naphthalene propanoic acid DA-E5090 (**19**), the active form of its prodrug E5090 (**20**). DA-E5090 (**19**) and the prodrug **20** are active in rat models of acute and chronic inflammation (10, 39). DA-E5090 (**19**) inhibits IL-1 production by LPS-stimulated human monocytes and rat macrophages *in vitro* with IC$_{50}$ values of 3.0 and 1.4 μM respectively, and both **19** and **20** block granuloma formation and PGE$_2$ production in a CMC-LPS rat air pouch model at doses >25 mg/kg p.o. (40). Neither DA-E5090 (**19**), nor the prodrug **20**, inhibited cyclooxygenase activity at 300 μM (39, 40). Another potential antirheumatic compound is IX-207-887 (**21**), which selectively blocks (IC$_{50}$ = 60 μM) IL-1 release from human monocytes without blocking intracellular levels of IL-1 (10, 41). The results of a double-blind placebo-controlled clinical trial in rheumatoid arthritis patients (42) suggested that IX-207-887 is well-tolerated and effective (55% responders at 1200 mg/day **21** vs 10% in the placebo group). An amino-dithiole-one, RP 54745 (**22**), inhibits IL-1 production by LPS-stimulated murine macrophages *in vitro* (IC_{50} <1 µM) and *ex vivo* (5 or 25 mg/kg p.o.) in mice (43). Although **22** inhibits macrophage metabolism (e.g. the hexose monophosphate pathway) at µM concentrations, it inhibited IL-1α and IL-1β, but not TNF α, mRNA production.

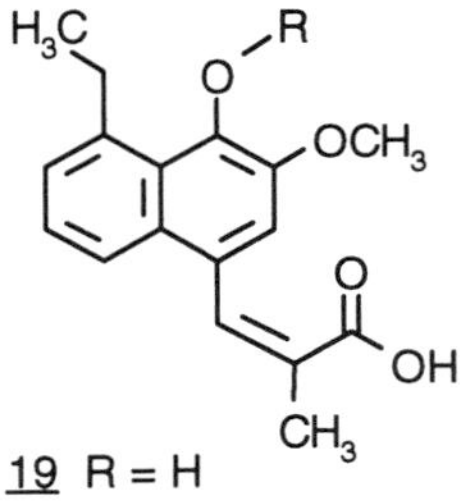

19 R = H
20 R = CH$_2$COOH

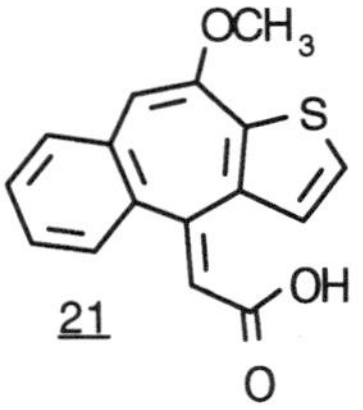

Several endogenous mediators have also been found to inhibit IL-1 production, notably the cytokines IL-4 and IL-10. Both IL-4 and IL-10 potently (IC_{50} <1 ng/ml) inhibit IL-1 α and IL-1β production at the transcriptional level and enhance IL-1ra production in human monocytes (18, 44-47). Other cytokines, such as TNFα, IL-6 and IL-8, are also inhibited by IL-4 and IL-10, but the post-receptor signalling mechanisms are unknown. Monocytes from cancer patients treated with IL-4 also produced less IL-1, and more IL-1ra, *ex vivo* than normal monocytes (48). Oxidized low density lipoprotein (LDL) has also been shown to inhibit IL-1 and IL-6 mRNA expression in LPS-stimulated murine macrophages (49).

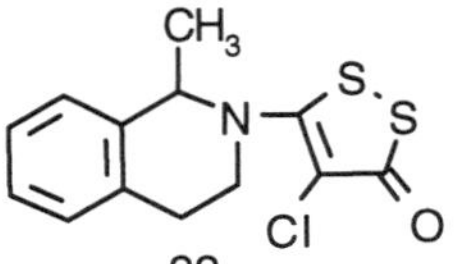

RECEPTOR LEVEL IL-1 INHIBITORS

Antibodies - Polyclonal and monoclonal antibodies to IL-1α, IL-1β and IL-1RI from various species have been used to try to neutralize IL-1 or IL-1R molecules. For example, rabbit polyclonal anti-murine IL-1β antibodies blocked neutrophil influx in a zymosan-induced murine peritonitis model (50) and reduced inflammation and cartilage destruction in a murine collagen-induced arthritis model (51). Likewise, monoclonal rat anti-mouse IL-1RI antibodies inhibit LPS-induced peritoneal neutrophil accumulation and IL-1-induced neutrophilia, acute phase response and corticosterone elevation (52, 53). However, no clinical trials with anti-human IL-1 or IL-1R antibodies have been reported.

Soluble IL-1R - Functionally, the soluble extracelluar forms of the IL-1RI and II act like neutralizing antibodies to block IL-1 function (2). Soluble IL-1RI increased heterotopic allograft survival when administered to mice undergoing heart transplantation (2). Soluble human IL-1R is in phase II clinical trials for septic shock. This protein has also been tested in a small double-blind, placebo-controlled clinical trial in allergic patients where it was co-injected intradermally with allergen at doses from 1 to 100 µg (54). The clinical late-phase allergic reaction was inhibited at all doses and there was evidence of a systemic effect at the higher doses. A pox virus protein has also been discovered which acts like soluble IL-1RII to specifically block IL-1β and reduces the immune response induced by the viral infection (55, 56).

<u>IL-1 Receptor Antagonists</u> - The IL-1ra protein has proven to be a powerful tool with which to establish a clear role for IL-1 in animal models of disease. Although IL-1ra is efficacious in many animal models of inflammation and disease, it does not inhibit antigen-specific immune responses *in vivo* (57). Furthermore, recombinant human IL-1ra (Antril) is being evaluated as a therapeutic agent in clinical trials in septic shock, rheumatoid arthritis, inflammatory bowel disease, GVHD and prospectively many other diseases (8, 9). A small phase II clinical trial in sepsis patients showed significant reduction in 28-day mortality (16% <u>vs</u> 44%) of patients treated with 133 mg/hr Antril for 72 hours, but in a larger phase III trial with 2 mg/kg/hr Antril, there was no significant difference in mortality (29% <u>vs</u> 34%). However, subgroup analysis of the phase III clinical trial data revealed that the most severely ill patients had a 22% reduction in mortality compared with the placebo (58).

Other putative IL-1 receptor antagonists have been described, including a 50 kD protein from the M20 cell line (59), but these have not been well characterized. Attempts to design or discover small molecule IL-1R antagonists have been unrewarding to date. Several structural and mutation studies with IL-1 or IL-1ra have attempted to define the receptor contact residues on the common β-barrel frameworks as a basis for the design of inhibitors (60-62). A recent study (11) comparing the mutational analysis and X-ray structures of IL-1β and IL1ra concluded that IL-1β has two receptor binding sites on opposite sides of the molecule (R11, H30, L31, Q32; K92, K93) and IL-1ra only has one of these sites (W16, Y34, L35, Q36). An alternative mechanism by which small molecules may modulate IL-1R is by down regulation of IL-1R expression on cell surfaces. Such appears to be the case with tenidap (**4**), which significantly reduced IL-1R expression on normal chondrocytes (IC_{50} = 10 μg/ml) and chondrocytes from osteoarthritis patients (IC_{50} = 22 μg/ml) in contrast to other NSAIDs and corticosteroids (63).

IL-1 SIGNAL TRANSDUCTION INHIBITORS

The prospects for therapeutic intervention at the level of IL-1 signal transduction are complicated by multiplicity of cellular targets of IL-1 and the uncertainty about whether different cell types utilize the same or different signalling pathways. Some studies implicated the involvement of G proteins and phospholipase A_2 in early membrane events following the binding of IL-1 to IL-1RI (64). Recent studies have implicated phospholipid metabolism in the early signalling events, particularly the generation of phosphatidic acid (PA) by lyso-PA acyl CoA:acyl transferase (LPAAT) and the generation of ceramide by sphingomyelinase (65, 66). Ceramide may be subsequently involved in stimulation of ceramide-activated protein kinase. A separate signalling pathway involves the activation of mitogen-activated protein (MAP)-kinase and p50 hsp kinase and the phosphorylation of proteins including the heat shock protein hsp27 (67). These early events ultimately lead to the activation of transcription factors, such as NF-κB, AP-1 and NF-IL-6, which then activate various IL-1 responsive genes (67, 68). In addition to these plasma membrane associated signalling pathways, there is evidence that complexes of IL-1/IL-1R may be internalized in cells and localize in the nuclei, suggesting the possibility that there is a nuclear signalling site for IL-1 (69).

Therefore, several molecular targets have emerged from investigations of IL-1 signalling pathways, including the enzymes LPAAT, sphingomyelinase and MAP kinase. Recent reports have described "synthetic heterocyclic compounds" (structures not available), including CT-1501R, CT-1541 and CT-2519, as LPAAT inhibitors (IC_{50} <50 μM) which block IL-1, TNF and LPS effects *in vitro* and significantly protected mice in an endotoxic shock model (65, 70). In other studies, certain serine/threonine kinase inhibitors, such as K252a

(**23**) and quercetin (**24**), selectively inhibited MAP kinase and inhibited IL-1 induced PGE_2 and IL-6 release by KB epidermoid cells *in vitro*(67). These studies showed that MAP kinase was necessary, but not sufficient, for IL-1 signal transduction in these cells.

Conclusion - Studies with the specific IL-1 antagonist, IL-1ra, *in vivo* have strengthened the evidence that IL-1 plays a significant role in the pathogenesis of inflammatory diseases. Of the many possible points of therapeutic intervention with IL-1 as the target, most of the medicinal chemistry activities have focussed on IL-1 production inhibitors. Many interesting compounds have been described which inhibit IL-1 biosynthesis, processing or release from cells. The only known molecular target for inhibition of IL-1 production is the proIL-1β processing enzyme, ICE. Many substrate analogs are now being designed and tested as ICE inhibitors. The downstream targets in the IL-1 effector cascade, namely the IL-1 receptor and the ensuing signal transduction events, have been less approachable from a medicinal chemistry point of view. However, recent information on the molecular basis of IL-1 receptor interactions and of IL-1 signal transduction pathways should also open this area to targeted drug discovery.

References

1.　J.J. Oppenheim and J. Saklatvala, Clinical Applications of Cytokines, Oxford University Press, New York, N.Y., 1993, 271.
2.　S.K. Dower, J.E. Sims, D.P. Cerretti and T.A. Bird, Chem.Immunol.,51, 33 (1992).
3.　B.M. Foxwell, K. Barrett and M. Feldmann, Clin.Exp.Immunol., 90, 161 (1992).
4.　C.A. Dinarello, Chem.Immunol.,51, 1 (1992).
5.　L.A. O'Neill, Kidney Int., 41, 546 (1992).
6.　C.A. Dinarello and S.M. Wolff, N.Engl.J.Med., 328, 106 (1993).
7.　R.C. Thompson, D.J. Dripps and S.P. Eisenberg, Int.J.Immunopharmacol., 14, 475 (1992).
8.　C.A. Dinarello, Immunol.Today, 14, 260 (1993).
9.　P.E. Bender and J.C. Lee, Ann.Rep.Med.Chem., 25, 185 (1989).
10.　K. Cooper and H. Masamune, Ann.Rep.Med.Chem., 27, 209 (1992).
11.　R.J. Evans, G.P.A. Vigers, J. Bray, T. Caffes, J.D. Childs, D.J. Dripps, R.C. Thompson, B.J. Brandhuber and S.P. Eisenberg, J.Cell.Biochem., 18A, 283 (1994).
12.　J.E. Sims, M.A. Gayle, J.L. Slack, M.R. Alderson, T.A. Bird, J.G. Giri, F. Colotta, F. Re, A.. Mantovani, K. Shanebeck, et al, Proc.Natl.Acad..Sci.USA, 90, 6155 (1993).
13.　P.H. Wooley, J.D. Whalen, D.L. Chapman, A.E. Berger, K.A. Richard, D.G. Aspar and N.D. Staite, Arth.Rheum., 36, 1305 (1993).
14.　B.A. Parkar, M.E. McCormickan and S.J. Foster, Biochem.Biophys.Res.Commun., 169, 422 (1990).

15. D.A. Hartman, S.J. Ochalski and R.P. Carlson, Agents Actions, 39, C70 (1993).

16. J.C. Lee, A.M. Badger, D.E. Griswold, D. Dunnington, A. Truneh, B. Votta, J.R. White, P.R. Young and P.E. Bender, Ann.N.Y.Acad.Sci., 696, 149 (1993).

17. P. Young, P. McDonnell, D. Dunnington, A. Hand, J. Laydon and J. Lee, Agents Actions, 39, C67 (1993).

18. J. Chin and M.J. Kostura, J.Immunol., 151, 5574 (1993).

19. J.D. Sipe, L.M. Bartle and L.D. Loose, J.Immunol., 148, 480 (1992).

20. L.D. Loose, J.D. Sipe, D.S. Kirby, A.R. Kraska, E.S. Weiner, W.R. Shanahan, M.R. Leeming, P. Farrow, C.B. Stack and N. Ting, Br.J.Rheumatol., 32, 19 (1993).

21. K. Tanaka, T. Shimotori, S. Makino, M. Eguchi, K. Asaoka, R. Kitamura and C. Yoshida, J.Pharmacobiodyn., 15, 641 (1992).

22. S.W. Wright, R.R. Harris, J.S. Kerr, A.M. Green, D.J. Pinto, E.M. Bruin, R.J. Collins, R.L. Dorow, L.R. Mantegna, S.R. Sherk, et al., J.Med.Chem., 35, 4061 (1992).

23. E.M. Eugui, B. Delustro, S. Rouhafza, R. Wilhelm and A.C. Allison, Ann.N.Y.Acad.Sci., 696, 171 (1993).

24. A.L. Akeson, C.W. Woods, L.B. Mosher, C.E. Thomas and R.L. Jackson, Atherosclerosis, 86, 261 (1991).

25. D.P. Cerretti, C.J. Kozlosky, B. Mosley, N. Nelson, N.K. Van, T.A. Greenstreet, C.J. March, S.R. Kronheim, T. Druck, L.A. Cannizzaro, et al., Science, 256, 97 (1992).

26. N.A. Thornberry, H.G. Bull, J.R. Calaycay, K.T. Chapman, A.D. Howard, M.J. Kostura, D.K. Miller, S.M. Molineaux, J.R. Weidner, J. Aunins, et al., Nature, 356, 768 (1992).

27. S.M. Molineaux, F.J. Casano, A.M. Rolando, E.P. Peterson, G. Limjuco, J. Chin, P.R. Griffin, J.R. Calaycay, G.J. Ding, T.T. Yamin, et al., Proc.Natl.Acad.Sci.USA, 90, 1809 (1993).

28. B.D. Shivers, D.A. Giegel and K.M. Keane, J.Cell.Biochem., 17B, 119 (1993).

29. D.K. Miller, J.R. Calaycay, K.T. Chapman, A.D. Howard, M.J. Kostura, S.M. Molineaux and N.A. Thornberry, Ann.N.Y.Acad.Sci., 696, 133 (1993).

30. P.R. Sleath, R.C. Hendrickson, S.R. Kronheim, C.J. March and R.A. Black, J.Biol.Chem., 265, 14526 (1990).

31. R.A. Black, S.R. Kronheim and P.R. Sleath, Febs.Lett., 247, 386 (1989).

32. N.A. Thornberry, J.Cell.Biochem., 18D, 122 (1994).

33. N.A. Thornberry, E.P. Peterson, J.J. Zhao, A.D. Howard, H.P. Griffin and K.T. Chapman, Biochem., 33, 3934 (1994).

34. R. Heng, T. Payne, L. Revesz and B. Weidmann, US Patent, WO 93/09135, (1993).

35. R.E. Dolle, D. Hoyer, C.V.C. Prasad, S.J. Schmidt, C.T. Helaszek, R.E. Miller and M.A. Ator, J.Med.Chem., 37, 563 (1994).

36. C.A. Ray, R.A. Black, S.R. Kronheim, T.A. Greenstreet, P.R. Sleath, G.S. Salvesen and D.J. Pickup, Cell, 69, 597 (1992).

37. V. Gagliardini, P.A. Fernandez, R.K. Lee, H.C. Drexler, R.J. Rotello, M.C. Fishman and J. Yuan, Science, 263, 826 (1994).

38. M. Miura, H. Zhu, R. Rotello, E.A. Hartwieg and J. Yuan, Cell, 75, 653 (1993).

39. H. Shirota, K. Chiba, M. Goto, R. Hashida and H. Ono, Agents Actions Suppl., 32, 219 (1991).

40. M. Tanaka, K. Chiba, M. Okita, T. Kaneko, K. Tagami, S. Hibi, Y. Okamoto, H. Shirota, M. Goto, H. Obaishi, et al., J.Med.Chem., 35, 4665 (1992).

41. J. Schnyder, P. Bollinger and T. Payne, Agents Actions, 30, 350 (1990).

42. M. Dougados, B. Combe, T. Beveridge, I. Bourdeix, A. Lallemand, B. Amor and J. Sany, Arth.Rheum., 35, 999 (1992).

43. F. Folliard, A. Bousseau and B. Terlain, Agents Actions, 36, 119 (1992).

44. M.J. Fenton, J.A. Buras and R.P. Donnelly, J.Immunol., 149, 1283 (1992).

45. E. Orino, S. Sone, A. Nii, T. Ogura, J.Immunol., 149, 925 (1992).

46. R. de Waal Malefyt, J. Abrams, B. Bennett, C.G. Figdor and J. de Vries, J.Exp.Med., 174, 1209 (1991).

47. P. Ralph, I. Nakoinz, J.A. Sampson, S. Fong, D. Lowe, H.Y. Min and L. Lin, J.Immunol., 148, 808 (1992).

48. H.L. Wong, G.L. Costa, M.T. Lotze and S.M. Wahl, J.Exp.Med., 177, 775 (1993).

49. L.G. Fong, T.A. Fong and A.D. Cooper, J.Lipid.Res., 32, 1899 (1991).

50. M. Perretti, E. Solito, andL. Parente, Agents Actions, 35, 71 (1992).

51. W.B. van den Berg, L.A.B. Joosten, M. Helsen and F.A.J. van de Loo, Clin.Exp.Immunol., 95, 237 (1994).

52. J.E. Gershenwald, Y.M. Fong, T. Fahey, S.E. Calvano, R. Chizzonite, P.L. Kilian, S.F. Lowry and L.L. Moldawer, Proc.Natl.Acad.Sci.USA, 87, 4966 (1990).

53. K.W. McIntyre, G.J. Stepan, K.D. Kolinsky, W.R. Benjamin, J.M. Plocinski, K.L. Kaffka, C.A. Campen, R.A. Chizzonite and P.L. Kilian, J.Exp.Med., 173, 931 (1991).

54. M.F. Mullarkey, K.M. Leiferman, M.S. Peters, I. Caro, E.R. Roux, R.K. Hanna, A.S. Rubin and C.A. Jacobs, J.Immunol., 152, 2003 (1994).

55. M.K. Spriggs, D.E. Hruby, C.R. Maliszewski, D.J. Pickup, J.E. Sims, R.M.L. Buller and J. Van Slyke, Cell, 71, (1992).

56. A. Alcami and G.L. Smith, Cell, 71, 153 (1992).

57. D.A. Faherty, V. Claudy, J.M. Plocinski, K. Kaffka, P. Kilian, R.C. Thompson and W.R. Benjamin, J.Immunol., 148, 766 (1992).

58. C.A. Dinarello, J.A. Gelfand and S.M. Wolff, J.Amer.Med.Assoc., 269, 1829 (1993).

59. V. Barak, D. Peritt, I. Flechner, P. Yanai, T. Halperin, A.J. Treves and C.A. Dinarello, Lymphokine Cytokine Res., 10, 437 (1991).

60. T.E. Labriola, C. Chandran, K.L. Kaffka, D. Biondi, B.J. Graves, M. Hatada, V.S. Madison, J. Karas, P.L. Kilian and G. Ju, Proc.Natl.Acad.Sci.USA, 88, 11182 (1991).

61. B. Veerapandian, G.L. Gilliland, R. Raag, A.L. Svensson, Y. Masui, Y. Hirai and T.L. Poulos, Proteins, 12, 10 (1992).

62. B.J. Stockman, T.A. Scahill, M. Roy, E.L. Ulrich, N.A. Strakalaitis, D.P. Brunner, A.W. Yem and M.J. Deibel, Biochem. 31, 5237 (1992).

63. J.P. Pelletier, J.M. Cloutier and P.J. Martel, Agents Actions Suppl., 39, 181 (1993).

64. L.A. O'Neill, Kidney Int., 41, 546 (1992).

65. G. Rice, S. Bursten, P. Brown, W. Tino, N. Jenkins, J. Bianco and J. Singer, J.Cell.Biochem., 17B, 107 (1993).

66. S. Mathias, A. Younes, C. Kan, I. Orlow, C. Joseph and R.N. Kolesnick, Science, 259, 519 (1993).

67. T. Bird, D. Virca, J. Slack, M. Gayle, J. Giri, H. Schule, P. de Roos, J. Sims and S. Dower, J.Cell.Biochem., 17B, 52 (1993).

68. K. Meugge, M. Vila, G.L. Gusella, T. Musso, P. Herrlich, B. Stein and S.K. Durum, Proc. Natl.Acad.Sci.USA, 90, 7054 (1993).

69. B.M. Curtis, M.B. Widmer, P. de Roos and E.E. Qwarnstrom, J.Immunol., 144, 1295 (1990).

70. R. Weeks, J. McMillan, R. G.C. and D. Lovett, J.Cell.Biochem., 18A, 300 (1994).

Chapter 22. Cell Adhesion and Carbohydrates

Daniel E. Levy, Peng Cho Tang, and John H. Musser
Glycomed, Inc. 860 Atlantic Ave. Alameda, CA 94501

<u>Introduction</u> - The interaction of cell-surface carbohydrates with protein receptors is a well accepted phenomenon. For example, cell-surface glycoconjugates found on red blood cells determine blood group specificity. These blood group epitopes are the ABO antigens (1). Recently, the discovery of selectins (2, 3), a new class of cell adhesion molecules, sparked new enthusiasm in the interaction of cell-surface carbohydrates with protein receptors. The selectins are now known to be involved in the rolling stage of the previously discovered transendothelial migration of leukocytes shown in Figure 1 (4, 5). Once out of the blood stream, leukocytes become involved in disorders ranging from autoimmune diseases (6-10) to ARDS (11). Extravasation of leukocytes also applies to the transendothelial migration of metastatic cells (12).

Figure 1

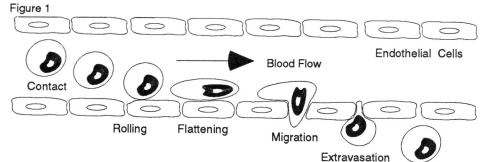

On contact with blood vessels, leukocytes roll on the endothelium (primarily selectin driven) until reaching a stationary adhesive stage (primarily integrin driven). Extravasation to sites of injury then occurs.

L-selectin (mLHR, Leu8, TQ-1, gp90MEL, CD62L, Lam-1, Lecam-1, Leccam-1), the first selectin discovered, is identified by its reaction with the monoclonal antibody MEL-14 (13). It is also a cell surface marker present on lymphocytes, neutrophils and monocytes (14-17). The events leading to platelet activation are under intense study. P-selectin (CD62, PADGEM, GMP-140) is characterized utilizing antibodies differentiating between activated and resting platelets (18, 19). This selectin, present on the surface of activated platelets, is also stored in the Weibel-Palade bodies of endothelial cells (20-23). E-selectin (ELAM-1) receptors, found on cytokine-activated endothelial cells, increase leukocyte adhesion on induction (24), and are characterized by antibodies recognizing a cytokine-inducible glycoprotein and endothelial cells at sites of inflammation. An antibody that blocks neutrophil adhesion to endothelial cells is also used for recognition of these receptors (7, 25, 26). This review focuses on the selectins, their ligands, their role in diseases and strategies towards agents directed at treating inflammatory disorders.

STRUCTURE AND FUNCTION OF THE SELECTINS AND THEIR LIGANDS

<u>Structural Features of Selectins and their Carbohydrate Ligands</u> - The functional regions of the selectin family are similar (Figure 2). Beginning with the amino terminus, each member has a lectin binding domain followed by an epidermal growth factor (EGF) region. The EGF region is followed by a number of repeating modules similar to those found in various complement binding proteins. Each selectin ends at the carboxy terminus with a transmembrane region followed by a cytoplasmic domain (27, 28). Although there is a high degree of homology among the lectin binding and EGF domains of the selectins (Figure 3), the complement binding protein-like units show little homology among the selectins.

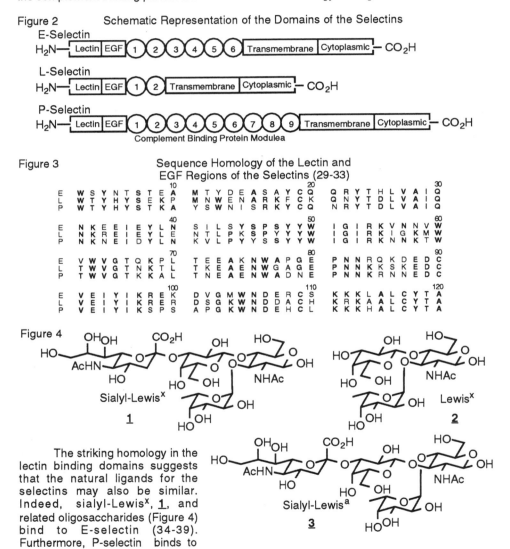

Figure 2   Schematic Representation of the Domains of the Selectins

E-Selectin
H_2N—[Lectin][EGF](1)(2)(3)(4)(5)(6)[Transmembrane][Cytoplasmic]— CO_2H

L-Selectin
H_2N—[Lectin][EGF](1)(2)[Transmembrane][Cytoplasmic]— CO_2H

P-Selectin
H_2N—[Lectin][EGF](1)(2)(3)(4)(5)(6)(7)(8)(9)[Transmembrane][Cytoplasmic]— CO_2H
          Complement Binding Protein Modulea

Figure 3   Sequence Homology of the Lectin and
          EGF Regions of the Selectins (29-33)

```
                    10                  20                        30
E  W S Y N T S T E A   M T Y D E A S A Y C Q   Q R Y T H L V A I Q
L  W T Y H Y S E K P   M N W E N A R K F C K   Q N Y T D L V A I Q
P  W T Y H Y S T K A   Y S W N I S R K Y C Q   N R Y T D L V A I Q

                    40                  50                        60
E  N K E E I E Y L N   S I L S Y S P S Y Y W   I G I R K V N N V W
L  N K R E I E Y L E   N T L P K S P Y Y Y W   I G I R K I G K M W
P  N K N E I D Y L N   K V L P Y Y S S Y Y W   I G I R K N N K T W

                    70                  80                        90
E  V W V G T Q K P L   T E E A K N W A P G E   P N N R Q K D E D C
L  T W V G T N K T L   T K E A E N W G A G E   P N N K K S K E D C
P  T W V G T K K A L   T N E A E N W A D N E   P N N K R N N E D C

                    100                 110                       120
E  V E I Y I K R E K   D V G M W N D E R C S   K K K L A L C Y T A
L  V E I Y I K R E R   D S G K W N D D A C H   K R K A A L C Y T A
P  V E I Y I K S P S   A P G K W N D E H C L   K K K H A L C Y T A
```

Figure 4

Sialyl-Lewis^x
1

Lewis^x
2

Sialyl-Lewis^a
3

The striking homology in the lectin binding domains suggests that the natural ligands for the selectins may also be similar. Indeed, sialyl-Lewis^x, **1**, and related oligosaccharides (Figure 4) bind to E-selectin (34-39). Furthermore, P-selectin binds to

Lewis[x], **2**, related carbohydrates (40, 41) and a glycoprotein receptor on human neutrophils (42). These interactions are calcium dependent and reversible. Recently, a ligand for P-selectin was cloned and shown to be a mucin-like transmembrane glycoprotein (43). This is in agreement with the observed natural ligands for L-selectin described below.

Although L-selectin is the most studied selectin, its corresponding ligand is difficult to identify due to problems in obtaining quantities of endothelial venules. Studies suggest that the ligand may contain sialic acid and fucose (44-46) with the possible addition of mannose (47). Anionic character from groups such as phosphate esters and sulfates is also postulated (44, 48). Calcium dependent interactions of L-selectin with heparin-like chains (49, 50) and sulfated glycolipids (51, 52) suggest the presence of anions in the natural ligand. Although the structure of the native ligand is unknown, L-selectin binds to sialyl-Lewis[x] when an L-selectin-IgG chimera is used in ELISA assays (53).

Two anionic glycoproteins, a major component of 50 kDa and a minor of 90 kDa are possible natural ligands of L-selectin. Their interaction with L-selectin is tissue specific and carbohydrate dependent (54, 55). The 50 kDa ligand for L-selectin can be cloned and isolated (56). Its carbohydrates are mucin-like (57, 58) and the protein itself is recognized as a mucin-like cell adhesion molecule. It is called GlyCAM-1 and will be discussed below.

Similarities to Other C-Lectins - Lectins are proteins with strong affinities for specific carbohydrate residues. Plant lectins are categorized under legumes and cereals while animal lectins are classed as S-type and C-type denoting thiol and calcium dependency, respectively (59). In addition to the selectins, C-type lectins include endocytic receptors, macrophage receptors, and molecules secreted into serum and the extracellular matrix (59). All of these lectins contain 1-8 carbohydrate recognition units, made up of approximately 120 amino acids, in addition to units that control their diverse biological functions (27, 60).

The Carbohydrate Binding Site - Although details of the selectin-ligand interaction are not well understood, some extrapolations as to the nature of the binding site can be made. Since sialic acid and some sulfated polysaccharides and lipids inhibit some selectin-mediated processes (38, 41-52), it is expected that a positive charge is present in the active site. However, as demonstrated by the interaction of sialic acid with the influenza hemaglutinin glycoprotein, a cation is not essential for this interaction (61). Essential to E- and P-selectin, as extrapolated from the NMR analysis of sialyl-Lewis[x], is a face of the protein which recognizes fucose and sialic acid when separated by approximately 10 angstroms (34, 38, 41). Monoclonal antibodies raised against E-selectin indicate that such a binding site may exist in a region of the E-selectin lectin domain near an antiparallel beta sheet (62). Furthermore, a homologous region on P-selectin is essential for binding and requires a 2',6-linked sialic acid or a sulfatide (63). Thus, the common domains on E- and P-selectins are responsible for carbohydrate recognition and cell adhesion.

Other models of the selectin-lectin binding domains are extrapolated from the crystal structure of the rat mannose binding protein (64). Homology between this protein and E/P-selectin is apparent (Figure 5). The model for the P-selectin lectin domain based on site-directed mutagenesis and neutralizing antibodies indicates that the following three amino acids, conserved among selectins, are critical for binding: Lys113, Tyr48, Tyr94 (65). A related model for E-selectin is also available (66).

The actual crystal structure of the lectin-EGF domains of E-selectin demonstrates the limited contact between the domains as well as the coordination of calcium (67). Furthermore, site mutagenesis shows the following amino acids to be critical for cell adhesion: Tyr48, Asn82, Asn83, Glu92, Tyr94, Lys111, and Lys113. Finally, the following amino acid substitutions were actually found to increase adhesion: Arg84Ala, Arg84Lys.

Figure 5   Sequence Homology of the Lectin Domains of P-Selectin,
E-Selectin and Rat Mannose Binding Protein (MBP)

```
                10                    20                    29
P    W T Y H Y S T K A Y    S W N I S R K Y C Q    N R Y T D L V A I
E    W S Y N T S T E A M    T Y D E A S A Y C Q    Q R Y T H L V A I
MBP  K F F V T N H E R M    P F S K V K A L C S    E L R G T   V A I R
                10                    20                    29
                39                    49                    57
P    Q N K N E I D Y L N    K V L P Y Y S S Y Y    W I G I R K N       N
E    Q N K E E I E Y L N    S I L S Y S P S Y Y    W I G I R K V       N
MBP  R N A E E N K A I Q    E V A K T       S A    F L G I T D E V T   E
                39                    46                    56
                66                    76                    86
P    K T W T W V   G T K    K A L T N E A E N W    A D N E P N N K R N
E    N V W V W V   G T Q    K P L T E E A K N W    A P G E P N N R Q K
MBP  G Q F M Y V T G G        L T       Y S N W    K K D E P N D H G S
                66                    72                    82
                96                   106                   116         120
P    N E D C V E I Y I K    S P S A P G K W N D    E H C L K K K H A L    C Y T A
E    D E D C V E I Y I K    R E K D V G M W N D    E R C S K K K L A L    C Y T A
MBP  G E D C V T I V D N          G L W N D        I S C Q A S H T A V    C E F P
                92                    97                   107          111
```

GlyCAM-I as a Cell Adhesion Molecule - GlyCAM-1 is a 50 kDa glycoprotein and is mucin-like as demonstrated by the structure and high mass content (approximately 70%) of its carbohydrates (68). GlyCAM-1 adhesive interactions to L-selectin are dependent on calcium, sialic acid (48), and sulfation of its carbohydrates (69). These observations agree with the requirements for L-selectin to bind to the carbohydrates of this glycoprotein. Since carbohydrate modified GlyCAM-1 recognizes L-selectin, it is classed as a cell adhesion molecule. The name arises from Glycosylation-dependent Cell Adhesion Molecule (70).

ROLE IN DISEASE

Role of Carbohydrate Interactions with Cell Adhesion Molecules - Many inflammation and immune disorders are the result of leukocytes present at the site of the relevant tissues (71). The transport of leukocytes to these tissues, illustrated in Figure 1, involves leukocyte rolling followed by adhesion to the endothelium and transendothelial migration. The first step occurs when P-selectin is expressed on the surface of endothelial cells following thrombin stimulation (72, 73). This type of rolling adhesion precedes the stationary adhesion necessary for migration of neutrophils through the endothelium (74, 75). Unlike rolling adhesion, stationary adhesion requires the stronger interactions between neutrophil integrins and endothelial intercellular adhesion molecules (5, 76). Once established, transendothelial migration can occur (77, 78). Since the later stages of cell extravasation do not involve carbohydrates, they will not be discussed further.

After neutrophils arrive at the site of injury, inflammation and immune responses begin. When these responses are misdirected, a number of diseases and disorders may occur. These may range from chronic autoimmune diseases to acute inflammation in the case of leukocytes to cancer involving metastatic cells (79-81).

Inflammatory Disease - Normal inflammation is a protective response by the immune system serving to destroy injurious materials and organisms as well as a mechanism to remodel injured tissue. However, when the immune system recognizes host tissue as foreign, an inflammatory response may result. A specific example is rheumatoid arthritis which results from the lysosomal digestion of cartilage-IgG complexes. Similar mechanisms on relevant tissues cause indications ranging from fever to psoriasis. Research on diseases such as multiple sclerosis suggest, but have not proven, autoimmune mediated mechanisms.

Reperfusion injury, which is associated with an activated inflammatory system, is the result of enhanced adhesion of leukocytes to the endothelium followed by degradation of the surrounding tissue (10, 82-86). Tumor metastasis is facilitated by platelets adhering to the malignant cells (87-90). Reperfusion injury and metastasis may be related in their dependence on the presence of P-selectin (91-93). Since the number of indications dependent on immune and inflammatory processes is significant (79-81), the remainder of this review will address potential therapies for immune-related disorders.

General Therapeutic Approaches - Understanding the mechanism of inflammation and related disorders based on selectin-induced cell adhesion may lead to the rational design of therapeutic agents. With respect to the selectins, such therapies would ideally block the initial process of leukocyte adhesion to the endothelium. These therapies may be ligand-based treatments in which small molecules, oligosaccharides, or glycomimetics are used (94). Alternately, receptor-based drugs such as protein receptors/chimeras, monoclonal antibodies, or peptide/peptidomimetic compounds are studied.

Many examples of receptor-based antagonists of inflammatory processes exist. Monoclonal antibodies raised against CD-18 and IL-8 are used in animal models for the treatment of reperfusion injuries (85, 95-100). In addition, evidence for the role of the selectins in pulmonary injuries was obtained by using antibodies raised to these receptors. Antigen-induced neutrophil influx and late-phase airway obstruction in the lungs of monkeys are blocked with antibodies to E-selectin (101). Similar results are seen in immune complex-induced neutrophil infiltration and capillary leaks in rats (102). Additionally, antibodies against P-selectin block neutrophil recruitment and vascular leaks secondary to infusion of cobra venom factor (103).

Since antibodies can be raised to specifically recognize the different selectins, these antibodies can bind to relevant sites and block the inflammatory process by inhibiting leukocyte adhesion. The recent literature is rich in the use of selectin specific antibodies as therapeutic agents (91, 102-104). The efficacy of sialylated oligosaccharides has also been demonstrated in rat models (105, 106).

As the therapeutic use of antibodies and other macromolecules is a technical challenge, most medicinally oriented research focuses on inhibitors of enzymes as well as ligand mimics. When considering adult respiratory distress syndrome (ARDS) as a therapeutic target where the mortality rate continues well above 50%, a variety of non-carbohydrate small molecules have been studied in animal models (107-109). The current research directed towards the development of carbohydrate based antagonists of ARDS and other selectin-mediated ailments will be discussed below.

GLYCOSYLTRANSFERASE INHIBITORS

Glycosyltransferases assemble polysaccharides from their monomeric units by adding, to the growing carbohydrate chain, activated nucleotide sugars (110). Since the chemistry of glycosyltransferases is well known (111-113), this section will focus on the relevance of glycosyltransferases to inflammatory diseases.

Our knowledge of the role sialyl-Lewis^x, 1, and related compounds play in the proliferation of inflammatory diseases and cancers paves the way for the development of new treatments. One such method involves the biosynthesis of sialyl-Lewis^x, itself. As illustrated in Figure 6, this tetrasaccharide is assembled from its components by aid of three glycoslytransferases (114). First, β1,4-galactosyltransferase forms the lactose scaffolding. Second, α2,3-sialyltransferase adds the sialic acid. Finally, the addition of fucose by α1,3-fucosyltransferase completes the synthesis.

Since sialic acid and fucose are important in the binding interaction of sialyl-Lewis^x to the selectins (34, 38, 41, 44-46), a reasonable conjecture is that removal of fucose from sialyl-Lewis^x will completely stop the binding of leukocytes to the endothelium. Furthermore, this may be accomplished by inhibiting the activity of α1,3-fucosyltransferase thus preventing the formation of sialyl-Lewis[x]. Further support comes from the observation that severe recurrent infections are the result of a leukocyte adhesion deficiency (114).

Figure 6   The Biosynthesis of Sialyl Lewis[x]

The glycosidic bonds are formed by the corresponding glycosyltransferases in sequence and where specified:
   a: β1,4-Galactosyltransferase
   b: α2,3-Sialyltransferase
   c: α1,3-Fucosyltransferase

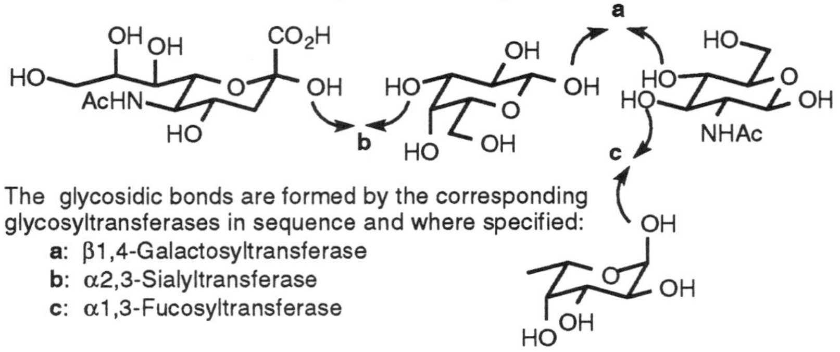

Studies directed at linking α1,3-fucosyltransferase activity to cell adhesion include cloning and transfection into COS-1 and CHO cells (35, 115). Both cell lines studied express a number of fucosylated glycoconjugates including sialyl-Lewis[x], **1**, and Lewis[x], **2**. In human umbilical vein endothelial cells (HUVEC) expressing E-selectin, transfected cell lines show adhesion while non-transfected controls show none.

Figure 7

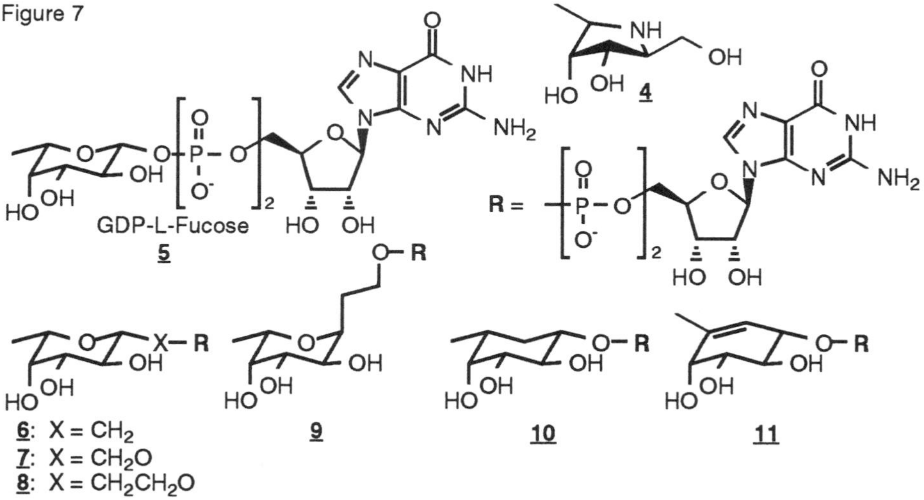

Human α1,3-fucosyltransferases are found in milk (116-118), saliva (119), and other specific cell lines (120-122). Although the cloning studies did demonstrate a role for an α1,3-fucosyltransferase in cell adhesion, they did not identify the specific fucosyltransferase responsible for the biosynthesis of sialyl-Lewis[x] on neutrophils (123, 124). Although the existence of α1,3-fucosyltransferases in plasma or as myeloid enzymes is controversial, a new α1,3-fucosyltransferase is now available via cloning (125).

Since α1,3-fucosyltransferase is linked to cell adhesion and that the activated fucose species used by this enzyme is GDP-L-fucose, **5**, a number of potential inhibitors of this enzyme are now available. These inhibitors, shown in Figure 7, are designed specifically to interfere with the cell adhesion process (126-128).

CELL ADHESION MOLECULE ANTAGONISTS

Selectin Receptor Antagonists - Because sialyl-Lewis[x], **1**, and related compounds bind to the selectins and act as anti-inflammatory agents (105, 106), research is currently focused towards designing selectin antagonists based on an understanding of the carbohydrate-selectin interaction (129-131). A number of sialyl-Lewis[x] analogs are now available (132-137). These compounds, some of which are shown in Figure 8, range from bivalent sialyl-Lewis[x] derivatives (compounds **13** and **14**) and analogues with substitutions for the sialic acid (compounds **18** and **19**) to molecules as simple as connecting sialic acid to fucose via an appropriate spacer (compound **12**) and others bearing fucose replacements and functional substitutions on the lactose (compounds **15**, **16** and **17**). These examples represent ongoing research in this rapidly emerging field.

Figure 8

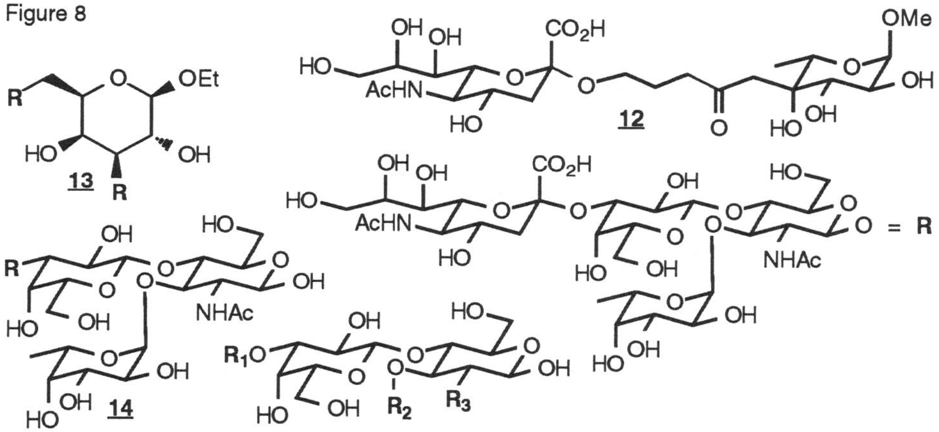

15: R_1 = Sialic Acid, R_2 = L-Fucose, R_3 = OH
16: R_1 = Sialic Acid, R_2 = $CH_2CH(OH)CH_2OH$, R_3 = NHAc
17: R_1 = Sialic Acid, R_2 = $CH_2CH(OH)CH_2OH$, R_3 = OH
18: R_1 = CH_2CO_2H, R_2 = L-Fucose, R_3 = OH
19: R_1 = $CH(CH_3)CO_2H$, R_2 = L-Fucose, R_3 = OH

CONCLUSION

The focus of this review is the mechanism of cell adhesion mediated disorders and possible treatments. Research in this area has yielded a wealth of biological information based on neutrophil-endothelial cell interactions. Furthermore, the use of carbohydrate-based drugs as novel therapeutic agents for the treatment of diseases ranging from ARDS to cancer will provide the basis for future reviews of this emerging field of glycomimetics.

REFERENCES

1. T. Feizi, Trends Biochem. Sci., _15_, 330 (1990).
2. L.M. Stoolman, Cell, _56_, 907 (1989).
3. L. Osborn, Cell, _62_, 3 (1990).
4. K.E. Arfors, C. Lundberg, L. Lindbom, P.G. Beatty and J.M. Harlan, Blood, _69_, 338 (1985).
5. M.B. Lawrence and T.A. Springer, Cell, _65_, 859 (1991).

6. H.L. Malech and J.I. Gallin, N. Eng. J. Med., 317, 687 (1987).
7. R.S. Cotran, M.A. Gimbrone Jr., M.P. Bevilacqua, D.L. Mendrick and J.S. Pober, J. Exp. Med., 164, 661 (1986).
8. J.M. Harlan, Acta. Med. Scand. Suppl., 715, 123 (1987).
9. M.I. Cybulsky, M.K.W. Chan and H.Z. Movat, Lab. Invest., 58, 365 (1988).
10. N.B. Vedder, R.K. Winn, C.L. Rice, E. Chi, K.-E. Arfors and J.M. Harlan, J. Clin. Invest., 81, 939 (1988).
11. A. Bersten and W.J. Sibbald, Critical Care Clinics, 5, 49 (1989).
12. W.L. Matis, R.M. Lavker and G.F. Murphy, J. Invest. Dermatol., 94, 492 (1990).
13. W.M. Gallatin, I.L. Weissman, and E.C. Butcher, Nature, 304, 30 (1983).
14. P.A. Gatenby, G.S. Kansas, C.Y. Xian, R.L. Evans and E.G. Engleman, J. Immunol. 129, 1997 (1982).
15. L.L. Lanier, E.G. Engleman, P. Gatenby, G.F. Babcock, N.L. Warner and L.A. Herzenberg, Immunol. Rev. 74, 143 (1983).
16. G.S. Kansas, G.S. Wood, D.M. Fishwild and E.G. Engleman, J. Immunol. 134, 1997 (1985).
17. A. Poletti, R. Manconi and P. dePaoli, Hum. Pathol. 19, 1001 (1988).
18. S.C. Hsu-Lin, C.L. Berman, B.C. Furie, D. August and B. Furie, J. Biol. Chem. 259, 9121 (1984).
19. P.E. Stenberg, R.P. McEver, M.A. Shuman, Y.V. Jacques and D.F. Bainton, J. Cell. Biol. 101, 880 (1985).
20. E. Larsen, A. Celi, G.E. Gilbert, B.C. Furie, J.K. Erban, R. Bonfanti, D.D. Wagner and B. Furie, Cell, 59, 305 (1989).
21. R. Bonfanti, B.C. Furie, B. Furie and D.D. Wagner, Blood, 73, 1109 (1989).
22. R.P. McEver, J.H. Beckstead, K.L. Moore, L. Marshall-Carlson and D.F. Bainton, J. Clin. Invest. 84, 92 (1989).
23. J.-G. Geng, M.P. Bevilacqua, K.L. Moore, T.M. McIntyre, S.M. Prescott, J.M. Kim, G.A. Bliss, G.A. Zimmerman and R.P. McEver, Nature, 343, 757 (1990).
24. M.P. Bevilacqua, J.S. Pober, M.E. Wheeler, R.S. Cotran and M.A. Gimbrone Jr., J. Clin. Invest. 76, 2003 (1985).
25. J.S. Pober, M.P. Bevilacqua, D.L. Mendrick, L.A. Lapierre, W. Fiers and M.A. Gimbrone Jr., J. Immunol. 136, 1680 (1986).
26. M.P. Bevilacqua, J.S. Pober, D.L. Mendrick, R.S. Cotran and M.A. Gimbrone Jr., Proc. Natl. Acac. Sci. USA., 84, 9238 (1987).
27. L.A. Lasky, Science, 258, 964 (1992).
28. M.P. Bevilacqua and R.M. Nelson, J. Clin. Invest., 91, 379 (1993).
29. M.P. Bevilacqua, S. Stenglin, M.A. Gimbrone Jr. and B. Seed, Science, 243, 1160 (1989).
30. M.H. Siegelman, M. van de Rijn and I.L. Weissman, Science, 243, 1165 (1989).
31. L.A. Lasky, M.S. Linger, T.A. Yednock, D. Dowbenko, C. Fennie, H. Rodriguez, T. Nguyen, S. Stachel and S.D. Rosen, Cell, 56, 1045 (1989).
32. T.F. Tedder, C.M. Isaacs, T.J. Ernst, G.D. Demetri, D.A. Adler and C.M. Disteche, J. Exp. Med., 170, 123 (1989).
33. G.I. Johnston, R.G. Cook and R.P. McEver, Cell, 56, 1033 (1989).
34. E.L. Berg, M.K. Robinson, O. Mansson, E.C. Butcher and J.L. Magnani, J. Biol. Chem., 265, 14869 (1991).
35. J.B. Lowe, L.M. Stoolman, R.P. Nair, R.D. Larsen. T.L. Berhend and R.M. Marks, Cell, 63, 475 (1990).
36. M.L. Phillips, E. Nudelman, F.C.A. Gaeta, M. Perez, A.K. Singhal, S. Hakomori and J.C. Paulson, Science, 250, 1130 (1990).
37. M. Tiemeyer, S.J. Sweidler, M. Ishihara, M. Moreland, H. Schweingruber, P. Hirtzer and B.K. Brandley, Proc. Natl. Acad. Sci. USA., 88, 1138 (1991).
38. D. Tyrrell, P. James, N. Rao, C. Foxall, S. Abbas, F. Dasgupta, M. Nashed, A. Hasegawa, M. Kiso, D. Asa, J. Kidd and B.K. Brandley, Proc. Natl. Acad. Sci. USA., 88, 10372 (1991).
39. G. Walz, A. Aruffo, W. Kolanus, M. Bevilacqua and B. Seed, Science, 250, 1132 (1990).
40. E. Larsen, T. Palabrica, S. Sajer, G.E. Gilbert, D.D. Wagner, B.C. Furie and B. Furie, Cell, 63, 467 (1990).
41. M.J. Polley, M.L. Phillips, E. Wayner, E. Nucelman, A.K. Kinghal, S. Hakomori and J.C. Paulson, Proc. Natl. Acad. Sci. USA., 88, 6224 (1991).
42. K.L. Moore, A. Varki and R.P. McEver, J. Cell Biol., 112, 491 (1991).
43. D. Sako, X.-J. Chang, K.M. Barone, G. Vachino, H.M. White, G. Shaw, G.M. Veldman, K.M. Bean, T.J. Ahern, B. Furie, D.A. Cumming and G.R. Larsen, Cell, 75, 1179 (1993).
44. Y. Imai, M.S. Singer, C. Fennie, L.A. Lasky and S.D. Rosen, J. Cell Biol., 113, 1213 (1991).
45. L.M. Stoolman and S.D. Rosen, J. Cell Biol., 96, 722 (1983).
46. D.D. True, M.S. Singer, L.A. Lasky and S.D. Rosen, J. Cell Biol., 111, 2757 (1990).
47. T.A. Yednock, L.M. Stoolman and S.D. Rosen, J. Cell Biol., 104, 713 (1987).

48.   P. Andrews, D. Milsom and W. Ford J. Cell Sci., 57, 277 (1982).
49.   K.E. Norgard-Sumnicht, N.M. Varki and A. Varki, Science, 261, 480 (1993).
50.   Y. Imai and S.D. Rosen, Glyconjugate J., 10, 34 (1993).
51.   Y. Suzuki, Y. Toda, T. Tamatani, T. Watanabe, T. Suzuki, T. Nakao, K. Murase, M. Kiso, A. Hasegawa, K. Tadano-Aritomi, I. Ishuzuka and M. Miyasaka, Biochem. Biophys. Res. Commun., 190, 426 (1993).
52.   L.K. Needham and R.L. Schnaar, Proc. Natl. Acad. Sci. USA, 90, 1359 (1993).
53.   C. Foxall, S. Watson, D. Dowbenko, C. Fennie, L.A. Lasky, M. Kiso, A. Hasegawa, D. Asa and B.K. Brandley, J. Cell Biol., 117, 895 (1992).
54.   P.R. Streeter, B.T.N. Rouse and E.C. Butcher, J. Cell Biol., 107, 1853 (1988).
55.   E. Berg, M. Robinson, R. Warnock and E.C. Butcher, J. Cell Biol., 114, 343 (1991).
56.   L.A. Lasky, M.S. Singer, D. Dowbenko, Y. Imai, W.J. Henzel, C. Grimley, C. Fennie, N. Gillett, S.R. Watson and S.D. Rosen, Cell, 69, 927 (1992).
57.   N. Jentoft, Trends Biochem. Sci., 15, 291 (1990).
58.   S.E. Harding, Adv. Carbohydr. Chem. Biochem., 47, 345 (1989).
59.   N. Sharon, Trends Biochem. Sci., 18, 221 (1993).
60.   R.C. Hughes, Curr. Opinion Struct. Biol., 2, 687 (1992).
61.   W. Weis, J.H. Brown, S. Cusak, J.C. Paulson, J.J. Skehel and D.C Wiley, Nature, 333, 426 (1988).
62.   D.V. Erbe, B.A. Wolitzky, L.G. Presta, C.R. Norton, R.J. Ramos, D.K. Burns, J.M. Rumberger, B.N.N. Rao, C. Foxall, B.K. Brandley and L.A. Lasky, J. Cell. Biol., 119, 215 (1992).
63.   D.V. Erbe, S.R. Watson, L.G. Presta, B.A. Wolitzky, C. Foxall, B.K. Brandley and L.A. Lasky, J. Cell Biol., 120, 1227 (1993).
64.   W.I. Weis, C. Drickamer and W.A. Hendrickson, Nature, 360, 127 (1992).
65.   D. Hollenbaugh, J. Bajorath, R. Stenkamp and A. Aruffo, Biochemistry, 32, 2960 (1993).
66.   A. Mills, Federation of Euro. Biochem. Soc. Letters, 319, 5 (1993).
67.   B. Graves, R.L. Crowther, C. Chandran, J.M. Rumberger, S. Li, K.-S. Huang, D.H. Presky, P.C. Familletti, B.A. Wolitzky and D.K. Burns, Nature, 367, 532 (1994).
68.   S.D. Rosen, Histochemistry, 100, 185 (1993).
69.   Y. Imai, L.A. Lasky and S.D. Rosen, Nature, 361, 555 (1993).
70.   D. Dowbenko, S.R. Watson and L.A. Lasky, J. Biolog. Chem., 268, 14399 (1993).
71.   R.M. Burch, M. Weitzberg, L. Noronha-Blob, V.C. Lowe, J.M. Bator, J. Perumattam and J.P. Sullivan, Drug News and Perspectives, 5, 331 (1992).
72.   R.P. McEver, Blood Cells, 16, 73 (1990).
73.   A. Celi, B. Furie and B.C. Furie, Proc. Soc. Exp. Biol. Med., 198, 703 (1991).
74.   M. Doré, R.J. Korthius, D.N. Granger, M.L. Entman and C.W. Smith, Blood, 82, 1308 (1993).
75.   S.M. Buttrum, R. Hatton and G.B. Nash, Blood, 82, 1165 (1993).
76.   T.K. Kishimoto, J. NIH Res., 3, 75 (1991).
77.   V.T. Marchesi, J. Exp. Physiol., 46, 115 (1961).
78.   J. Cohnheim in "Lectures on General Pathology: A Handbook for Practitioners and Students," The New Sydenham Society, London, 1889.
79.   J.I. Gallin, I.M. Goldstein and R. Snyderman in "Inflammation: Basic Principles and Clinical Correlates," J.I. Gallin, I.M. Goldstein and R. Snyderman, eds., Raven Press, New York, 1992.
80.   R.S. Cotran, V. Kumar and S.L. Robbins in "Pathologic Basis of Disease," W.B. Saunders Company, Philadelphia, 1989.
81.   J.M. Harlan and D.Y. Liu in "Adhesion: Its Role in Inflammatory Disease," W.H. Freeman and Company, New York, 1992.
82.   L.A. Hernandez, M.B. Grisham, B. Twohig, K.-E. Arfors, J.M. Harlan and D.N. Granger, Am. J. Physiol., 253, H699 (1987).
83.   N.B. Vedder, B.W. Fouty, R.K. Winn, J.M. Harlan and C.L. Rice, Surgery, 106, 509 (1989).
84.   N.B. Vedder, R.K. Winn. C.L. Rice, E. Chi, K.-E. Arfors and J.M. Harlan, Proc. Natl. Acad. Sci. USA, 81, 939 (1990).
85.   P.J. Simpson, R.F.T. III, J.C. Fantone, J.K. Mickelson, J.D. Griffin and B.R. Lucchesi, J. Clin. Invest., 81, 624 (1988).
86.   W.J. Mileski, R.K. Winn, N.B. Vedder, T.H. Pohlman, J.M. Harlan and C.L. Rice, Surgery, 108, 205 (1990).
87.   S. Karpatkin and E. Pearlstein, Ann. Intern. Med., 95, 636 (1981).
88.   G.J. Gasic, G.P. Tuszynski and E. Gorelik, Int. Rev. Exp. Pathol., 29, 173 (1986).
89.   E. Bastida and A. Ordinas, Haemostasis, 18, 29 (1988).
90.   K.V. Honn, I.M. Grossi, J. Timar, H. Chopra and J.D. Taylor in "Platelets and Cancer Metastasis, In Microcirculation and Cancer Metastasis," CRC Press, London-Oxford, 1991, p. 93.
91.   R.K. Winn, D. Liggitt, N.B. Vedder, J.C. Paulson and J.M. Harlan, J. Clin. Invest., 92, 2042 (1993).
92.   A. Aruffo, W. Kolanus, G. Walz, P. Fredman and B. Seed, Cell, 67, 35 (1991).
93.   J.P. Stone and D.D. Wagner, J. Clin. Invest., 92, 804 (1993).

94.   M.A. Nashed and J.H. Musser, Carbohydr. Chem. 250, C1 (1993).
95.   M.S. Mulligan, M.L. Jones., M.A. Balanowski, M.P. Baganoff, C.L. Deppeler, D.M. Meyers, U.S. Ryan and P.A. Ward, J. Immunol., 150, 5585 (1993).
96.   R.W. Barton, R. Rothlein, J. Ksiazek and C.J. Kennedy, Immunol., 143, 1278 (1989).
97.   M.S. Mulligan, C.W. Smith, D.C. Anderson, R.F. Todd III, M. Miyasaka, T. Tamatani, T.B. Issekutz and P.A. Ward, J. Immunol., 150, 2401 (1993).
98.   N. Sekido, N. Mukaida, A. Harada, I. Nakanishi, Y. Watanabe and K. Matsushima, Nature, 365, 654 (1993).
99.   N.B. Vedder, R.K. Winn, C.L. Rice, E.Y. Chi, K.-E. Arfors and J.M. Harlan, Proc. Natl. Acad. Sci. USA, 87, 2643 (1990).
100. L.A. Langdale, L.C. Flaherty, D. Liggitt, J.M. Harlan, C.L. Rice and R.K. Winn, J. Leukocyte Biol., 53, 511 (1993).
101. R.H. Gundel, C.D. Wegner, C.A. Torcellini, C.C. Clarke, N. Haynes, R. Rothlein, C.W. Smith and L.G. Letts, J. Clin. Invest., 88, 1407 (1991).
102. M.S. Mulligan, J. Verani, M.K. Dame, C.L. Lane, W. Smith, D.C. Anderson and P.A. Ward, J. Clin. Invest., 88, 1396 (1991)
103. M.S. Mulligan, M.J. Polley, R.J. Bayer, M.F. Nunn, J.C. Paulson and P.A. Ward, J. Clin. Invest., 90, 1600 (1992).
104. A.S. Weyrich, X.-L. Ma, D.J. Lefer, K.H. Albertine and A.M. Lefer, J. Clin. Invest., 91, 2620 (1993).
105. M.S. Mulligan, J.P. Paulson, S.D. DeFrees, Z.-L. Zheng, J.B. Lowe and P.A. Ward, Nature, 364, 149 (1993).
106. M.S. Mulligan, J.B. Lowe, R.D. Larsen, J.P. Paulson, S.D. DeFrees, Z.-L. Zheng, K. Makemura, M. Fukuda, and P.A. Ward, J. Exp. Med., 178, 623 (1993).
107. C. Metz and W.J. Sibbald, Chest, 100, 1110 (1991).
108. R.J. McDonald, Am. Rev. Respir. Dis., 144, 1347 (1991).
109. C.R. Turner, M.N. Lackey, M.F. Quinlan, L.W. Schwartz and E.B. Wheeldon, Circulatory Shock, 34, 290 (1991).
110. J.C. Paulson and K.J. Colley, J. Biol. Chem., 264, 17615 (1989).
111. R. Kornfeld and S. Kornfeld, Annu. Rev. Biochem., 54, 631 (1985).
112. J.E. Sadler, Biology of Carbohydrates, 2, 87 (1984).
113. S. Basu and M. Basu, Glycoconjugates, 3, 265 (1982).
114. J.B. Lowe, L.M. Stoolman, R.P. Nair, R.D. Larsen, T.L. Berhend and R.M. Marks, Cell, 63, 475 (1990).
115. J.B. Lowe, J. Kukowska-Latallo, R.P. Nair, R.D. Larsen, R.M. Marks, B.A. Macher, R.J. Kelly and L.K. Ernst, J. Biol. Chem., 266, 17467 (1991).
116. J.P. Prieels, D. Monnom, M. Dolmans, T.A. Beyer and R.L. Hill, J. Biol. Chem., 256, 10456 (1981).
117. P.H. Johnson and W.M. Watkins, Biochem. Soc. Trans., 10, 445 (1982).
118. S. Eppenberger-Castori, H. Lotscher and J. Finne, Glycoconjugate J., 8, 264 (1989).
119. P.H. Johnson, A.D. Yates and W.M. Watkins, Biochem. Biophys. Res. Commun., 100, 1611 (1981).
120. E.H. Holmes and S.B. Levery, Arch. Biochem. Biophys., 274, 633 (1989).
121. J.F. Kukowska-Latallo, R.D. Larsen, R.P. Nair and J.B. Lowe, Genes Dev., 4, 1288 (1990).
122. G.B. Stroup, K.R. Anumula, T.F. Kline and M.M. Caltabiano, Cancer Res., 50, 6787 (1990).
123. B.A. Macher, E.H. Holmes, S.J. Swiedler, C.L.M. Stults and C.A. Srnka, Glycobiology, 1,6 (1991).
124. T.W. Kuijpers, Blood, 81, 873 (1993).
125. K.L. Koszdin and B.R. Bowen, Biochem. Biophys. Res. Commun., 187, 152 (1992).
126. J.I. Luengo and J.G. Gleason, Tetrahedron Let., 33, 6911 (1992).
127. S. Cai, M.R. Stroud, S. Hakomori and T.J. Toyokuni, J. Org. Chem., 57, 6693 (1992).
128. C.-H. Wong, D.P. Dumas, Y. Ichikawa, K. Koseki, S.J. Danishefsky, B.W. Weston and J.B. Lowe, J. Am. Chem. Soc., 114, 7321 (1992).
129. S. Borman, C&EN News, December 7, 25 (1992).
130. K.-A. Karlsson, Trends in Pharm. Sci., 121, 265 (1991).
131. T.A. Springer and L.A. Lasky, Nature, 349, 196 (1991).
132. S.A. DeFrees, F.C.A. Gaeta, Y.-C. Lin, Y. Ichikawa and C.-H. Wong, J. Am. Chem. Soc., 115, 7549 (1993).
133. K.C. Nicolau, N.J. Bockovich and D.R. Carcanague, J. Am. Chem. Soc., 115, 8843 (1993).
134. K.C. Nicolau, C.W. Hummel and Y. Iwabuchi, J. Am. Chem. Soc., 114, 3126 (1992).
135. N.M. Allanson, A.H. Davidson and F.M. Martin, Tetrahedron Let., 34, 3945 (1993).
136. F. Dasgupta, A. Nematalla and J.H. Musser, Division of Carbohydrate Chemistry, 205th ACS National Meeting.
137. J.H. Musser, N. Rao, M.A. Nashed, F. Dasgupta, S. Abbas, A. Nematalla, V. Date, C. Foxall, D. Asa, P. James, D. Tyrrell and B.K. Brandley in "Trends in Drug Research," Vol. 20, V. Claasen, Ed., Elsevier, Amsterdam, 1993, p. 33.

Chapter 23.   Therapeutic Control of Androgen Action

Gary H. Rasmusson and Jeffrey H. Toney
Merck Research Laboratories, Rahway, NJ 07065

<u>Introduction</u> Circulating C_{19} steroids secreted from the testis, adrenal and ovary serve as the source of hormonal androgens for endocrine function. Two androgens, testosterone (T) and 5α-dihydrotestosterone (DHT) are responsible for activation of the intracellular androgen receptor (AR) which, in turn, regulates gene transcription and ultimately protein synthesis. Two major approaches have been employed to control androgen action: 1) modulation of the AR-hormone interaction and 2) inhibition of hormone production. The focus of this review are conditions which are amenable to hormonal manipulation and includes benign prostatic hyperplasia (BPH), prostatic carcinoma (PC) and skin-related problems such as acne, seborrhea, androgenic alopecia and hirsutism. Earlier reviews of these topics can be found in this series or elsewhere (general: (1-3) , BPH (4-7) , PC (8-10) , acne & seborrhea, hirsutism (11-14) , alopecia (15-17)). Studies of androgen action related to fertility or to somatic growth are not discussed in the present review.

ANDROGEN RECEPTOR

The AR is a transcription factor that responds to signaling by T and DHT and plays a major role in sexual development and sexual function in the male (18). The AR was cloned in 1988 by several laboratories and the complete sequence has been determined from sources including human, rat, and mouse (19-21). The coding sequence aligns with those of other members of the steroid/thyroid/retinoic acid receptor superfamily and includes discrete motifs for DNA and ligand binding, as well as for nuclear translocation and transcriptional activation (18). The AR is a soluble 110 kDa protein that binds strongly to both T (Kd ~ 1 nM) and to DHT (Kd ~ 0.1 nM). The binding of T or DHT to the AR is thought to alter the conformation of the protein as a first step toward regulation of gene transcription. A protease resistant fragment of the AR has been observed in the presence of DHT but not after addition of antagonists such as cyproterone acetate (CPA, **11**) or Casodex (**1**) (22). It has been suggested that an AR also exists within the plasma membrane and may play an additional role in androgen action (23, 24). Genetic data argues strongly for the presence of a single gene encoding the AR in man. Biochemical and pharmacological studies are consistent with the existence of multiple isoforms of the AR that are modified post-translationally (25). Splicing variants of the AR within the 3' untranslated region have been identified in human LNCaP prostatic carcinoma cells (26) and in rat brain tissue (27). Two mRNA isoforms of the AR have been identified in the developing larnyx of male *Xenopus laevis* (28). More recently, two isoforms of the AR have been described in a single human pedigree in which one form is truncated at the amino terminus (29). Interestingly, the truncated isoform of the AR is also expressed in normal human genital skin fibroblasts.

Mutations of the AR can result in a variety of phenotypic aberrations of male sexual development, including complete testicular feminization, Reifenstein syndrome and infertility. These mutations have been found to occur at multiple sites within the DNA and ligand binding domains of the receptor (see Scheme 1) (30-33) and can be categorized as mutations that abolish ligand binding, affect the thermolability of the receptor, or abolish DNA binding. The nature of these residues should prove to be useful for understanding structure/activity relationships of the AR in the absence of physical chemical data on the purified protein. Several somatic mutations within both the ligand binding domain and within the NH_2-terminus of the hAR have been reported for patients with PC at a frequency of 2 to 12% (34-36). Poly(CAG) sequences within the NH_2-terminus of the hAR have been found to be associated with X-linked spinal and bulbar muscular atrophy (37).

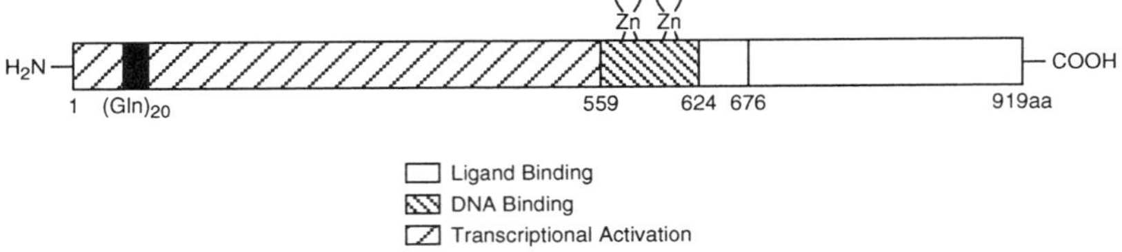

Scheme I

Antibodies are now available against peptides encoded by the AR cDNA and reveal that the AR protein is localized predominantly in the nucleus of the epithelium of the prostate and seminal vesicle (38), even in the absence of androgens (39) . There is no clear correlation between AR immunostaining and occurrence of PC, although some reports note that higher grade tumors exhibit more heterogeneity than is observed using tissues from patients with BPH or lower grade PC (40, 41). The AR has also been localized to human skin in a variety of cell types (42-44) and may be a useful target for androgen-regulated hair growth in both men and women (3, 15, 16, 42, 44-47) . AR messenger RNA has been found in a variety of tissues in rodents, including brain, kidney, liver, prostate, testes, and epididymis but is lacking in spleen (21).

Chemical Antagonists - The structure-activity relationships (SAR) describing binding to the androgen receptor have been reviewed (48-51). In general, there is a good correlation between binding to the receptor and biological potency.

Non-Steroidal - Casodex (ICI-176344, **1**), a pure androgen antagonist, given once daily has proven to be an effective monotherapy for PC but side effects may prevent it from being of general use for BPH (52-54). Flutamide (Eulexin, **2**) has been employed recently as an adjunct with luteinizing hormone releasing hormone (LHRH) agonists to achieve total androgen blockade in PC patients and seems to provide an added effect towards initial tumor regression (8, 55). The growth of prostatic cell lines such as DU145 and PC3 are inhibited by the active metabolite, hydroxyflutamide (Sch-16423, **3**) (56). In contrast, **3** promotes growth of the human PC cell line, LNCaP, in which the AR contains a point mutation within the ligand binding domain (57, 58). In this regard, relapse in PC patients receiving flutamide is sometimes repressed by withdrawal of the drug, indicating that tumor activity can become drug dependent (59, 60). Flutamide has been shown to be effective for the treatment of hirsutism in women (61, 62) and shows promise for improving symptoms in Tourette's Syndrome, a chronic neuropsychiatric disorder (63). Nilutamide (Anandron, RU 23908, **4**) in combination with orchiectomy has been employed in a large double-blinded PC clinical study and was shown to be superior to orchiectomy alone (64). New analogs (RU 59063, **6**; RU 56187, **7**) of **4** have been reported which are more active *in vitro* and *in vivo*,

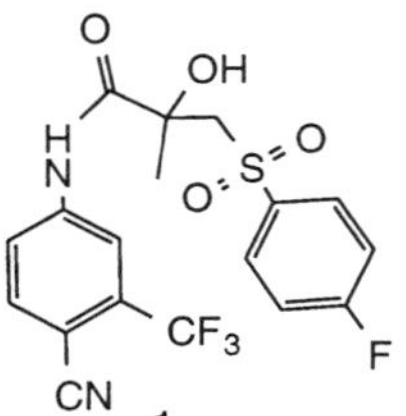

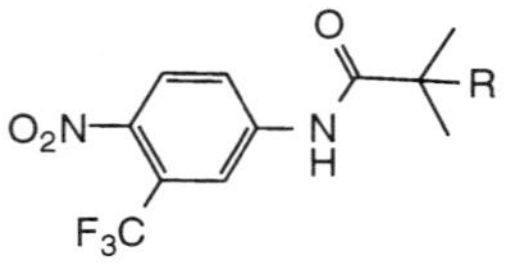

2 R = H
3 R = OH

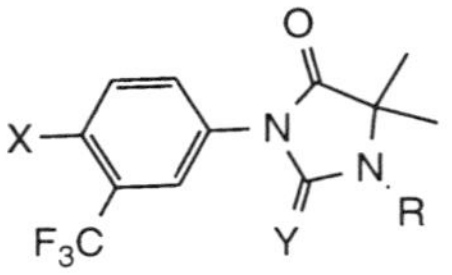

4 X = NO_2; Y = O; R = H
5 X = CN; Y = O; R = $(CH_2)_4OH$
6 X = CN; Y = S; R = $(CH_2)_4OH$
7 X = CN; Y = S; R = CH_3

respectively, than other antiandrogens (65). Another analog (RU 58841, **5**) is the most effective antagonist in selectively reducing the size of the androgen sensitive hamster flank organ (66). The locally active inocoterone (67, 68) (RU 38882, **8**) showed modest improvements in a trial for treatment of acne vulgaris. WB2838 (**9**), a weak antiandrogen, is a product of *Pseudomonas* fermentation (69). Tamandron (**10**), designed as an analog of tamoxifen, has been reported to bind weakly to the AR (70).

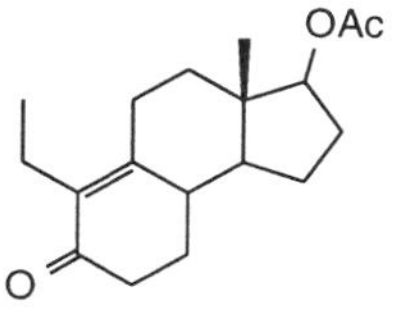

          8 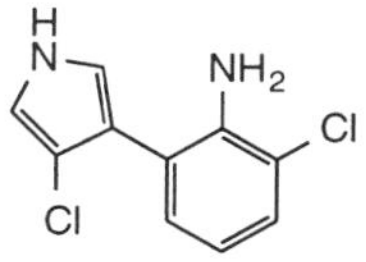   **9** 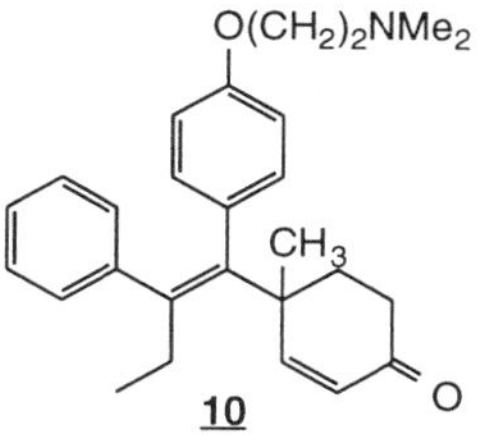   **10**

Steroidal - The classic antiandrogen CPA (Androcur, **11**) (3, 14, 53, 71-74) acts by blocking the binding of DHT to the AR, as well as by exerting progestational effects which suppress gonadotropin secretion. The growth of prostatic cell lines such as DU145 and PC3 are inhibited by **11**. However, two large double-blinded PC clinical studies have indicated that **11** was not more effective than placebo for orchiectomized patients (75, 76). CPA appears useful for the treatment of acne, since this agent decreases sebum production in humans. However, CPA is not recommended for use by males because of possible feminizing side-effects. Because of the progestational activity and potential teratogenicity, females are treated with CPA in combination with ethinyl estradiol (Diane) as a contraceptive regimen. Osaterone (TZP 4238, **13**) has been shown to be effective against the Dunning prostatic carcinoma model in the rat (77).

Combination therapy using an antiandrogen with a 5α-reductase inhibitor has been suggested as a method for reducing total androgen burden. Administration of **18** or **20** and **2** to rats resulted in a combined effect for reduced prostate growth (78, 79). Zanoterone (WIN 49596, **12**), a peripherally active antiandrogen, has been employed in combination with the 5α-reductase inhibitor finasteride (**18**) for the treatment of BPH in a dog model and seems to provide an additive effect over that of either agent alone (80). A brief trial in PC patients using the combination of **2** and **18** was found to lower the serum tumor marker enzyme, prostatic specific antigen (PSA), levels and to decrease prostate size (81).

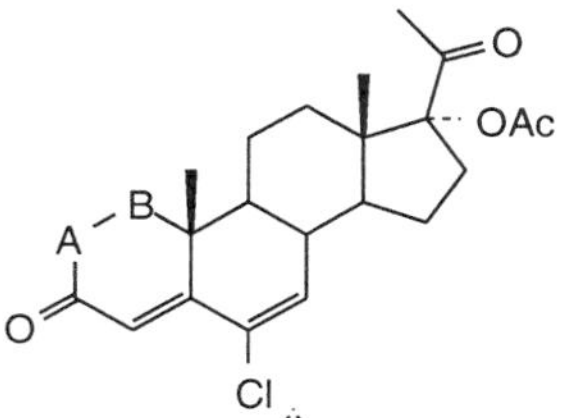

          11 A-B = ⟨△⟩

          13 A-B = O-CH$_2$ **12**

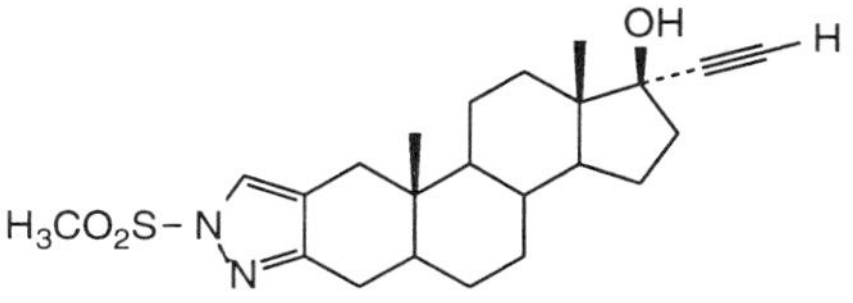

CONTROL OF ANDROGEN BIOSYNTHESIS

17α-Hydroxylase:17,20-Lyase. The C$_{19}$ androgens are derived from C$_{21}$ pregnane precursors by a cytochrome P$_{450}$ complex which oxidatively removes the acetyl side-chain at position 17. The resultant 17-oxoandrostane can provide the androgens T and DHT in subsequent metabolic steps. Selective inhibition of this cleavage should block both adrenal and testis sources of androgen, and thus limit the level of T and the hormonal metabolites

DHT and estradiol. Early studies using the lyase inhibitor ketoconazole showed promise in the treatment of PC in humans and have led to the development of liarozole (R75251, **14**), an equally effective inhibitor with a more acceptable side-effect profile (82). Reports of studies of oral liarozole in hormone refractive PC patients have been promising (83). However, circulating androgen levels were not reduced and another mechanism such as inhibition of retinoic acid metabolism may account for activity in these patients, as well as that seen in androgen insensitive PC animal models (84). Bulky esters of pyridine 3-acetates (e.g., **15**) have been claimed as selective inhibitors of the lyase (85) as well as certain 17-substituted androstenols (CB-7598, **16**, **17**) (86, 87). Antiandrogenic activity of 6-methylene progesterone (LY207320) observed in rats, previously thought to be due to 5α-reductase inhibition, is more likely a result of a block of T synthesis via the 17α-hydroxylase-17,20-lyase pathway (88).

14   **15**   **16** R = 3-pyridyl, Δ16
                            17 R = β-NHCyclopropyl

5α-Reductase - Circulating T is converted in some hormone sensitive tissues such as prostate and certain regions of skin by the enzyme 5α-reductase into the more potent DHT which is responsible for the initiation of the hormonal response (89, 90). Other endogenous 3-oxo-Δ4-steroids such as progesterone and the corticoids are also substrates for the enzyme, but their dihydro-products do not appear to be essential for expression or control of hormonal effects. Thus blockage of 5α-reductase would seem to provide a very selective method to interfere with androgen action without disturbing the desirable actions of T or of other structurally related steroid hormones. This assumption has been largely based on observations of 5α-reductase deficient individuals in whom the relative absence of DHT leads to changes in male fetal development, but to no serious adverse post-pubertal effects (91, 92).

Studies of the molecular biology of 5α-reductase have led to the discovery of isoforms of the enzyme (93). The cloning of two different human and rat (94) isoforms of 5α-reductase has led to additional studies of their properties, location and function (90). In each species the Type 1 and Type 2 enzymes differ in their in vitro pH optima (6-8.5, broad and 4.7-5.5, sharp, respectively), binding affinities for T (Type 2 > Type 1) and sensitivity to inhibitors. These membrane-associated enzymes are composed largely of hydrophobic amino acids and have similar hydropathy plots. Insertion of a four amino acid sequence from the rat enzyme for that in the human Type 1 dramatically changed the sensitivity to inhibition by **18** (95). Type 2 mRNA is expressed primarily in male sexual tissue and Type 1 in peripheral tissues in the rat (94). In regenerating rat prostate tissue both types are present, but appear in different tissue compartments (96). In humans the prostate has primarily Type 2 enzyme, the liver contains both forms, and Type 1 enzyme is found in post-pubertal non-genital skin (97, 98). Pseudohermaphrodites deficient in 5α-reductase lack only an effective Type 2 enzyme (93, 99, 100). The Type 2 enzyme thus appears to be essential for masculine development of the fetal urogenital tract and of the external phenotype. 5α-Reductase Type 2 is linked to post-pubertal hair growth in the male and to prostatic function. The Type 1 enzyme is localized primarily in the skin and may be involved in the function of the sebaceous and sweat glands and/or at times of high hormonal flux such as in the process of virilization (97, 98, 101). Interestingly, 5α-reductase 1 has been cloned independently and was shown to have sequence-specific DNA binding activity in vitro (102). However, no classical DNA binding motifs have been identified in the protein sequence. The development of selective inhibitors for each of these enzymes will help identify their individual functions and possibly lead to new and specific areas of therapeutic treatment.

<u>Inhibitors</u>   The 4-azasteroid class of 5α-reductase inhibitors has been the most thoroughly studied (103-107). One member of this series, finasteride (MK-906, L-652,932, PROSCAR®, **18**), has completed Phase III human trials for BPH and has been approved for use in this condition in over 50 countries (108, 109). Finasteride shows about 100-fold selectivity for the Type 2 (IC_{50}=4.2nM) over the Type 1 (IC_{50}=500nM) human enzyme under standard assay conditions (98). Studies using the rat prostatic and liver enzymes indicated **18** to be an equilibrium-type competitive inhibitor (110, 111). In contrast, studies using human enzymes indicate a time dependent inhibition by **18** in which the off-rate from the enzyme is so slow that the interaction can be considered irreversible (112, 113). Immunoprecipitation of the human Type 2 enzyme was blocked by **18** but not by a related inhibitor, 4-MA (L-636,028, **20**), while neither inhibitor affected the precipitation of Type 1 enzyme (114). A study of the inhibition of the 5α-reductases present in human hair follicle, foreskin and prostate by a variety of 4-azasteroids demonstrated a differential sensitivity of the enzymes in these tissues to the inhibitors (115). The azacholestane derivative **23** (L-733,692) is a selective inhibitor of the human Type 1 enzyme (116, 117).

18 R = CONHC(CH$_3$)$_3$
19 R = COCH$_2$CH(CH$_3$)$_2$

20 R = CON(C$_2$H$_5$)$_2$, R' = H
21 R = CON(iPr)CONH(iPr), R' = H
22 R = CONHCH(C$_6$H$_5$)$_2$, R' = H
23 R = C$_8$H$_{17}$, R' = CH$_3$

In rats (118, 119), dogs (120, 121) and humans (122, 123) finasteride reduces prostate size and causes a decline (70-95%) in prostatic DHT, while increasing tissue levels of T. The outcome of DHT depletion in the rat by **18** in fetal, as well as post-natal, development differs from methods providing a total androgen block (124, 125). When administered to rats from birth to puberty **18** had a negative effect on the growth of the penis, epididymus, prostate and seminal vesicles, but did not affect testis growth or spermatogenesis (126). In animal models of androgenic alopecia (the stump-tailed macaque) and seborrhea (the fuzzy rat) **18** stimulated hair growth (alone or additively with minoxidil) and reduced sebum production, respectively (127, 128). In the Dunning 3327 rat model of PC **18** lowered tumor DHT levels but was ineffective in retarding tumor progression (129). However **20**, in spite of poor control of androgen levels, was effective in this model (130). Turosteride (FCE 26073, **21**), a 5α-reductase inhibitor in Phase I studies, lowers prostatic DHT in intact rats without increasing T (131).

Three year clinical results of BPH patients treated with **18** have indicated no serious side-effects, while modest but sustained improvements in disease parameters were found in over one-half of the patients (132). Incidence of PC in the clinical trial of **18** was not significantly higher in the treatment groups relative to that found on placebo (133). A predictable reduction of PSA occurs with BPH patients treated with **18** which should not interfere in the diagnosis of PC provided the new set point concentration is determined after initiation of therapy (134). In stage D PC patients **18** reduced PSA levels, but to an extent less than that seen with complete androgen ablation (135). The safety profile of **18** has allowed the initiation of a long term (7 year) study to assess the effect of prolonged Type 2 inhibition on the incidence of PC. An efficacy trial of **18** in male androgenic alopecia has also been initiated. A ketone analog of finasteride, MK-0963 (L-654,066, **19**), has been reported to lower serum DHT levels by 44-80% after 10 days in humans at 0.1 to 25 mg/day. At 25mg/day the DHT levels remained >70% depressed for >6 days after the last dose (136).

Recently 3-oxo-Δ4-6-azasteroids have been described as potent 5α-reductase inhibitors (137). Effective inhibition of both human Type 1 and 2 enzymes was attained by compounds bearing a large lipophilic substituent attached to the 17β-carbonyl group. One of these, the benzhydrylamide **24**, had high oral bioavailability in the dog and was equal to **18** in blocking the effects of T in the castrate rat. Related 4-azasteroids bearing arylamine derived amides (**22**) at the 17-position show high potency as inhibitors of the rat prostatic enzyme (138).

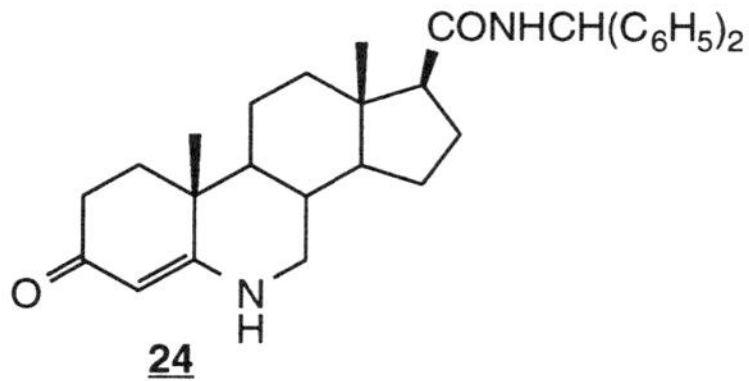

CONHCH(C$_6$H$_5$)$_2$

24

R

HO$_2$C

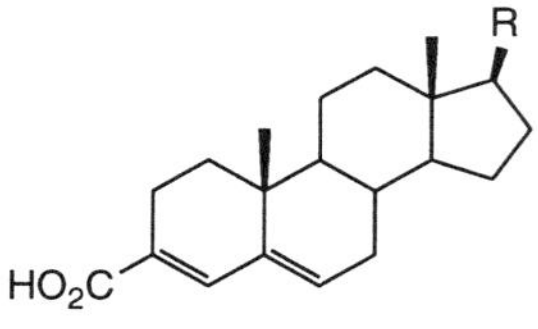

25 R = CONHC(CH$_3$)$_3$
26 R = CON(iPr)$_2$

A series of 5α-reductase inhibitors bearing an acidic function at the 3-position of an unsaturated steroid have been discovered. Designed as transition-state inhibitors, these compounds block the rat enzyme in an uncompetitive manner with respect to the substate T (111, 139). The clinical candidate, episteride (SK&F 105657, **25**), is a good inhibitor of human Type 2, primate and rat prostatic 5α-reductases (139) and of prostate growth in the rat (140). Tumor growth was suppressed by **25** in two rodent models of PC in which limiting amounts of 5α-reductase were present in the tumors, but showed no effect in a model in which the tumor levels of enzyme were relatively high (141). The orally active **25** in humans at doses up to 160mg lowers serum DHT levels to 50% without increasing serum T levels and lowers prostatic DHT ≤80% at doses up to 80mg/day (142, 143).

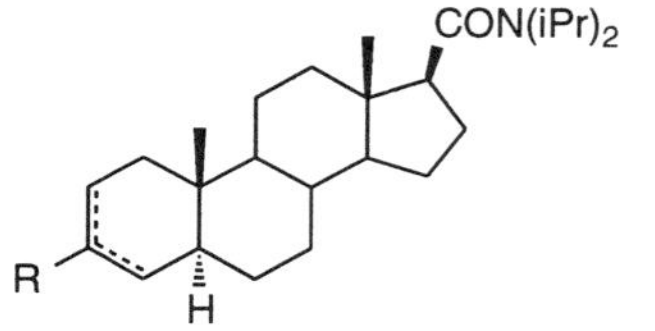

CON(iPr)$_2$

R

H

27 R = CO$_2$H, CH$_2$CO$_2$H,
         PHO$_2$H, or PO$_3$H
28 R = NO$_2$, Δ3

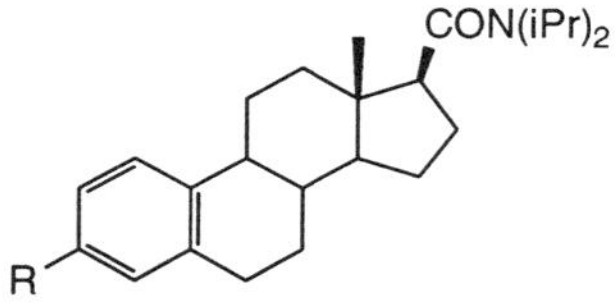

CON(iPr)$_2$

R

29 R = CO$_2$H, CH$_2$CO$_2$H,
         PHO$_2$H, PO$_3$H, orSO$_3$H

Analogs of episteride which have a trigonal atom in the ring at the 3-position bearing an acidic group, including A-ring aromatic steroids, are good inhibitors of the prostatic enzyme. Thus, Δ2- , Δ3-, Δ3,5- analogs (**27**) (144), and the estratrienes (i.e., **29**) with carboxy- (145), carboxymethyl- (146), phosphinic-, phosphonic- (139), and sulfonic- acids (147) provide potent inhibitors with apparent K$_i$'s in the low nanomolar range. Unlike the acidic compounds, the nitro analog **28** is a competitive inhibitor of the enzyme (148). This material and **26** were equally effective in lowering circulating DHT levels in monkeys.

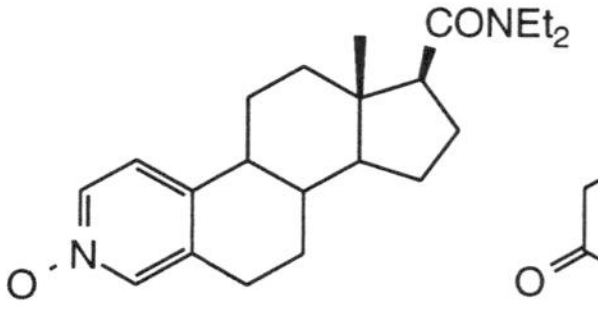

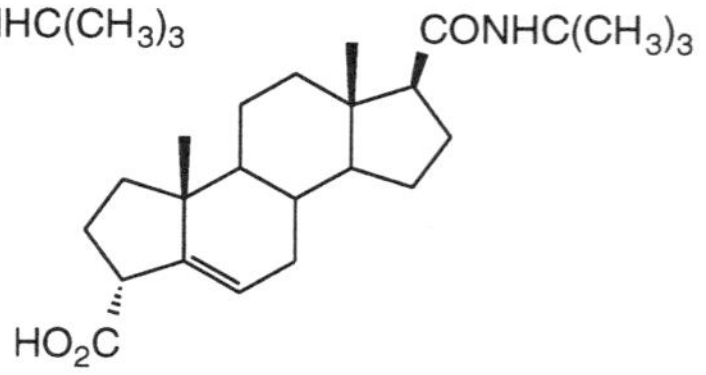

CONEt$_2$

O–N

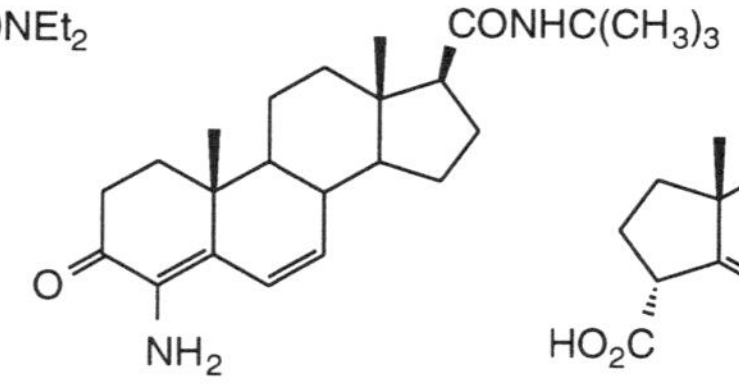

30

CONHC(CH$_3$)$_3$

O

NH$_2$

31

CONHC(CH$_3$)$_3$

HO$_2$C

32

The 3-azaestatriene **30** has recently been prepared and found to be a good (app K$_i$=31nM) Type 2 inhibitor with moderate Type 1 activity (149). Other steroidal 5α-reductase inhibitors reported in the patent literature include 4-aminodienones (**31**) (150), claimed to have low nanomolar IC_{50}s, and certain 3-carboxy-A-norsteroids (i. e., **32**) (151).

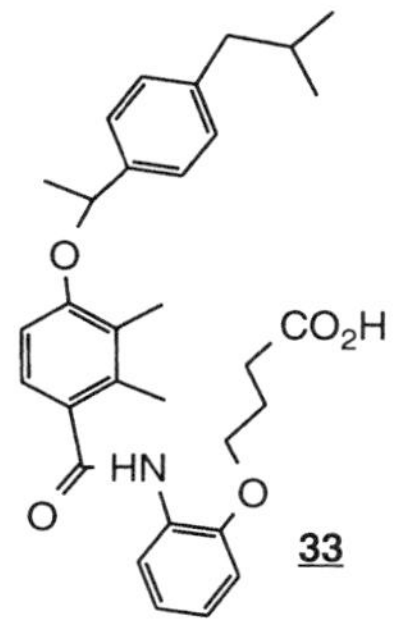

A number of non-steroidal inhibitors of 5α-reductase have been described. Some of the earliest of these were derivatives of o-aminophenoxybutyric acids including Ono 3805 (**33**) which was active in the rat (152). An isosteric series, 1,3-disubstituted benzoylindole butyric acids (**34**), have been claimed as effective inhibitors (153). One of these, FK-143, has high activity versus both human enzymes and is active in the rat (154, 155). A series of aryloxybenzoic acids, including 4-(4-biphenyloxy)benzoic acid are claimed as inhibitors of the human enzyme (156). The finding that certain polyunsaturated fatty acids (i.e., γ-linolenic acid) are good inhibitors of 5α-reductase has led to the suggestion that they may be involved in the regulation of androgen action in target cells (157). A brief SAR study of some polyhydroazaphenanthrenes was highlighted by the potent, selective human Type 1 activity of the chloroderivative **35** (158). The uncompetitive inhibitor **35** is unusual in that it is not effective against the rat enzyme.

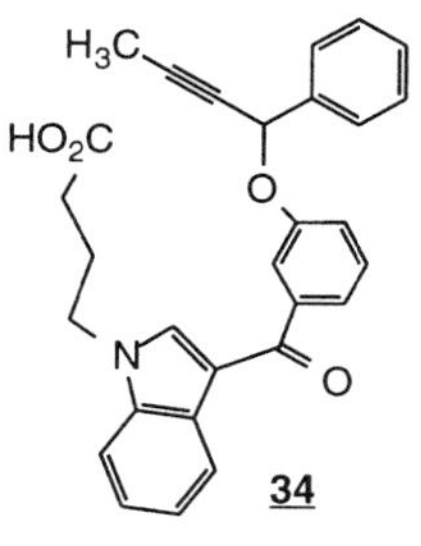

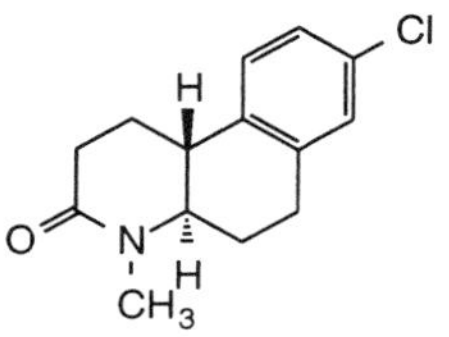

<u>LHRH Agonists/Antagonists</u> Testicular production of T is under control of pituitary gonadotropic hormone (LH & FSH) secretion. A form of chemical castration can be induced by chronic treatment using a potent synthetic analog of LHRH (GnRH) which desensitizes the pituitary to further secretion after a brief period of stimulation. These parenterally administered agents have been shown to be as effective as surgical castration for the treatment of hormone responsive PC (159). To combat the initial surge of hormone and to negate the effect of circulating adrenal hormones, antiandrogens have been coadministered with LHRH agonists in a number of studies to assess the possibility of benefit over monotherapy. Although studies are continuing, the combination treatment shows promising trends in improvement, particularly in patients with minimal disease (160). Coadministration of estrogen and progestin with the LHRH agonist decapeptyl (DP) improved efficacy and side-effects over the use of DP alone in hirsutism (161).

A LHRH antagonist has the added value of causing a direct suppression of LH without the "flare" phenomenon. Earlier problems of potent LHRH antagonists with histamine release apparently can now be overcome (162). One of these agents, Cetrorelix (SB-75), caused tumor reduction in the PC-82 implanted nude mouse (163) and provided significant decreases in androgen levels and in prostatic size in a short trial (4-6 weeks) in a small number of BPH and PC patients (164). Structural/conformational requirements for activity of LHRH agonists and antagonists have been suggested (165).

CONCLUSION

Recent years have seen a resurgence in interest in controlling the effects of androgen dependent processes in the body. The high incidence of BPH and PC in modern society has provided impetus for broadening the search for safe and specific agents to control age-related phenomena of the androgen sensitive prostate gland. New agents, such as

Casodex and finasteride appear to provide significant advances toward the treatment of PC and BPH, respectively, by offering ease of use and good side-effect profiles. The use of combined androgen ablation methods are likely to add to the length and quality of life for PC patients. Other approaches, such as the use of antisense DNA to block AR expression or blocking of the AR response elements on DNA, may lead to more specific effects, but their medical application appears to be more distant. Management of non-life threatening conditions, such as skin/hair problems, will become more common as the safety of new agents is demonstrated. New knowledge regarding the AR and 5α-reductases should lead to a fuller understanding of androgen action and possibly to new therapeutic targets.

References

1.   G. H. Rasmusson, Annual Reports in Medicinal Chemistry, 21, 179 (1986).
2.   J. P. Mallamo and P. E. Juniewicz, Annual Reports in Medicinal Chemistry, 24, 197 (1989).
3.   M. E. Sawaya and M. K. Hordinsky, Dermatologic Clinics, 11, 65 (1993).
4.   J. D. McConnell, Urol Clin N Am, 17, 661 (1990).
5.   P. C. Walsh, in Campbell's Urology P. C. Walsh, et al., Eds. (W B Saunders, Philadelphia, PA, 1992), vol. 1, pp. 1009.
6.   N. R. Banna and A. M. Rushdi, International Pharmacy Journal, 7, 101 (1993).
7.   M. Jønler, M. Riehmann, and R. C. Bruskewitz, Drugs, 47, 66 (1994).
8.   J. Geller, J Androl, 12, 364 (1991).
9.   N. J. Vogelzang and G. T. Kennealey, Cancer, 70, 966 (1992).
10. G. Alivizatos and G. O. N. Oosterhof, Anti-Cancer Drugs, 4, 301 (1993).
11. E. L. Smith and J. J. Tegeler, Annual Reports in Medicinal Chemistry, 24, 177 (1989).
12. L. F. Eichenfield and J. J. Leyden, Pediatrician, 18, 218 (1991).
13. W. J. Cunnliffe and W. W. Bottomley, Arch Dermatol, 128, 1261 (1992).
14. R. S. Jurzyk, R. L. Spielvogel, and L. I. Rose, Am Fam Phys, 45, 1803 (1992).
15. R. C. Gadwood and V. C. Fiedler, Annual Reports in Medicinal Chemistry, 24, 187 (1989).
16. D. G. Brodland and S. A. Muller, Cutis, 47, 173 (1991).
17. M. E. Sawaya and M. K. Hordinsky, in Advances in Dermatology(Moseby-Yearbook, Inc, Chicago, 1992), vol. 7, pp. 211.
18. E. M. Wilson, J. A. Simental, F. S. French, and M. Sar, Ann NY Acad Sci, 637, 56 (1991).
19. W. D. Tilley, M. Marcelli, J. D. Wilson, and M. J. McPhaul, Proc. Natl. Acad. Sci. USA, 86, 327 (1989).
20. D. B. Lubahn, D. R. Joseph, P. M. Sullivan, et al., Science, 240, 327 (1988).
21. C. Chang, J. Kokontis, and S. Liao, Science, 240, 324 (1988).
22. P. J. Kallio, O. A. Janne, and J. J. Palvimo, Endocrinology, 134, 998 (1994).
23. E. F. Konoplya and E. H. Popoff, Int J Biochem, 24, 1979 (1992).
24. A. C. Towle and P. Y. Sze, J. Steroid Biochem., 18, 135 (1983).
25. J. A. Kemppainen, M. V. Lane, M. Sar, and E. M. Wilson, J Biol Chem, 267, 968 (1992).
26. P. W. Faber, H. C. J. van Rooij, H. A. G. M. van der Korput, et al., J Biol Chem, 266, 10743 (1991).
27. R. McLachlan, B. L. Tempel, M. A. Miller, et al., Mol. Cell. Neurosci., 2, 117 (1991).
28. L. Fischer, D. Catz, and D. Kelley, Proc. Natl. Acad. Sci. USA, 90, 8254 (1993).
29. C. M. Wilson and M. J. McPhaul, Proc. Natl. Acad. Sci. USA, 91, 1234 (1994).
30. T. R. Brown, P. A. Scherer, Y.-T. Chang, et al., Eur. J. Pediatr., 152, S62 (1993).
31. M. J. McPhaul, M. Marcelli, W. D. Tilley, et al., FASEB J., 5, 2910 (1991).
32. M. J. McPhaul, M. Marcelli, S. Zoppi, et al., J. Clin. Endocrinol. Metab., 76, 17 (1993).
33. C. Sultan, S. Lumbroso, N. Poujol, et al., J. Steroid Biochem. Molec. Biol., 46, 519 (1993).
34. J. R. Newmark, D. O. Hardy, D. C. Tonb, et al., Proc. Natl. Acad. Sci. USA, 89, 6319 (1992).
35. M. P. Schoenberg, J. M. Hakimi, S. Wang, et al., Biochem. Biophys. Res. Comm., 198, 74 (1994).
36. H. Suzuki, N. Sato, Y. Watabe, et al., J. Steroid Biochem. Molec. Biol., 46, 759 (1993).
37. A. R. LaSpada, E. M. Wilson, D. B. Lubahn, et al., Nature, 352, 77 (1991).
38. N. B. West, C. Chang, S. Liao, and R. M. Brenner, J. Steroid Biochem. Molec. Biol., 37, 11 (1990).
39. Y. H. Zhuang Blauer, M., Pekki, A., Tuohimaa, P, J. Steroid Biochem. Molec. Biol., 41, 693 (1992).
40. J. A. Ruizeveld de Winter, J. Trapman, A. O. Brinkmann, et al., J Pathol, 161, 329 (1992).
41. M. V. Sadi and E. R. Barrack, Cancer, 71, 2574 (1993).
42. M. Blauer, A. Vaalasti, S.-L. Pauli, et al., J Invest Dermatol, 97, 264 (1991).
43. R. Choudhry and M. B. Hodgins, J Invest Dermatol, 98, 522 (1992).
44. R. Choudhry, M. B. Hodgins, T. H. Van der Kwast, et al., J Endocrinol, 133, 467 (1992).
45. F. Kiesewetter, A. Arai, and H. Schell, J Invest Dermatol, 101, 98 (1993).
46. T. Liang, S. Hoyer, R. Yu, et al., J Invest Dermatol, 100, 663 (1993).
47. V. A. Randall, M. J. Thornton, and A. G. Messenger, J Endocrinol, 133, 141 (1992).
48. S. Liao, T. Liang, S. Fang, et al., J. Biol. Chem., 248, 6154 (1973).

49. K. B. Chan, S. Smythe, and S. Liao, J. Steroid Biochem., 11, 1193 (1979).
50. D. J. Tindall, C. H. Chang, T. J. Lobl, and G. R. Cunningham, Pharmac. Ther., 24, 367 (1984).
51. G. H. Rasmusson, G. F. Reynolds, N. G. Steinberg, et al., J Med Chem, 29, 2298 (1986).
52. L. M. Eri and K. J. Tveter, J Urol, 150, 90 (1993).
53. U. Fuhrmann, C. Bengston, G. Repenthin, and E. Schillinger, J. Steroid Biochem. Molec. Biol., 42, 787 (1992).
54. C. J. Tyrell, The Prostate, 4, 97 (1992).
55. R. C. Benson, Prostate Supplement, 4, 85 (1992).
56. M. Bologna, P. Muzi, L. Biordi, et al., Cur Ther Res, 51, 799 (1992).
57. J. Veldscholte, C. A. Berrevoets, C. Ris-Stalpers, et al., J. Steroid Biochem. Molec. Biol., 41, 665 (1992).
58. C. Ris-Stalpers, M. C. T. Verleun-Mooijman, J. Trapman, and A. O. Brinkmann, Biochem. Biophys. Res. Commun., 196, 173 (1993).
59. W. K. Kelly and H. I. Scher, J Urol, 149, 607 (1993).
60. O. Sartor, M. Cooper, M. Weinberger, et al., J Nat Can Inst, 86, 222 (1994).
61. F. Fruzzetti, D. De Lorenzo, C. Ricci, and P. Fioretti, Fertil Steril, 60, 806 (1993).
62. L. Cusan, R. R. Tremblay, A. Dupont, et al., Fertil Steril, 61, 281 (1994).
63. B. S. Peterson, J. F. Leckman, L. Scahill, et al., J. Clin. Psychopharm 14, 131 (1994).
64. R. A. Janknegt, C. C. Abbou, R. Bartoletti, et al., J Urol, 149, 77 (1993).
65. G. Teusch, F. Goubet, T. Battmann, et al., J Steroid Biochem Molec Biol, 48, 111 (1994).
66. T. Battmann, A. Bonfils, C. Branche, et al., J. Steroid Biochem. Molec. Biol., 48, 55 (1994).
67. D. P. Lookingbill, B. B. Abrams, C. N. Ellis, et al., Arch Dermatol, 128, 1197 (1992).
68. S. Puri, B. Abrams, R. Cherill, and D. Trembley, Clin Pharmacol Ther, 53, 154 (1993).
69. Y. Hori, Y. Abe, H. Nakajima, et al., The Journal of Antibiotics, 46, 1327 (1993).
70. G. A. Potter and R. McCague, J. Chem. Soc., Chem. Commun., , 635 (1992).
71. M. M. Bouton, D. Lecaque, J. Secchi, and C. Tournemine, J. Invest. Derm., 86, 163 (1986).
72. F. H. de Jong, P. J. Reuvers, J. Bolt-de Vries, et al., J Steroid Biochem Molec Biol, 42, 49 (1992).
73. P. E. Pochi, Drug Development Research, 13, 157 (1988).
74. J. Steinsapir, G. Mora, and T. G. Muldoon, Biochem Biophys Acta, 1094, 103 (1991).
75. T. Jorgensen, K. J. Tveter, S.-2. Group, and L. H. Jorgensen, Eur. Urol., 24, 466 (1993).
76. M. R. G. Robinson, Cancer, 72, 3855 (1993).
77. T. Ichikawa, S. Akimoto, and J. Shimazaki, Endocr J, 40, 425 (1993).
78. C. Labrie, C. Trudel, S. Li, et al., Endocrinology, 129, 566 (1991).
79. N. E. Fleshner and J. Trachtenberg, J Urol, 148, 1928 (1992).
80. P. E. Juniewicz, S. J. Hoekstra, B. M. Lemp, et al., Endocrinology, 133, 904 (1993).
81. N. Fleshner and J. Trachtenberg, J Urol, 149, 258 (1993).
82. H. Vanden Bossche, G. Willemsens, D. Bellens, et al., Biochem Soc Trans, 18, 10 (1990).
83. J. Trachtenberg and A. Toledo, J Urol, 145, 317 (1991).
84. G. A. Dijkman, R. J. A. van Moorselaar, R. van Ginckel, et al., J Urol, 151, 217 (1994).
85. S. E. Barrie, M. Jarman, R. McCague, et al., GB 2253851, British Technology Group, (1992).
86. M. R. Angelastro and T. R. Blohm, US 4,966,898, Merrell Dow Pharmaceuticals Inc, (1990).
87. S. E. Barrie, G. A. Potter, M. Jarman, and M. Dowsett, Br J Cancer, 67, 75 (1993).
88. B. L. Neubauer, K. L. Best, T. R. Blohm, et al., Prostate, 23, 181 (1993).
89. E. D. Lephart, Mol Cell Neurosci, 4, 473 (1993).
90. D. W. Russell and J. D. Wilson, Ann Rev Biochem, 63, 25 (1994).
91. J. D. Wilson, J. E. Griffin, and D. W. Russell, Endocr Rev, 14, 577 (1993).
92. J. Imperato-McGinley and T. Gautier, Trends in Genetics, 2, 130 (1986).
93. S. Andersson, D. Berman M, E. P. Jenkins, and D. W. Russell, Nature, 354, 159 (1991).
94. K. Normington and D. W. Russell, J Biol Chem, 267, 19548 (1992).
95. A. E. Thigpen and D. W. Russell, J Biol Chem, 267, 8577 (1992).
96. D. M. Berman and D. W. Russell, Proc Natl Acad Sci U S A, 90, 9359 (1993).
97. A. E. Thigpen, R. I. Silver, J. M. Guileyardo, et al., J Clin Invest, 92, 903 (1993).
98. G. Harris, B. Azzolina, W. Baginsky, et al., Proc Natl Acad Sci U S A, 89, 10787 (1992).
99. W. C. Wigley, J. S. Prihoda, I. Mowszowicz, et al., Biochemistry, 33, 1265 (1994).
100. A. E. Thigpen, D. L. Davis, T. Gautier, et al., N Engl J Med, 327, 1216 (1992).
101. V. Luu-The, Y. Sugimoto, L. Puy, et al., J Invest Dermatol, 102, 221 (1994).
102. K. Gaston and M. Fried, Nucleic Acids Research, 20, 6297 (1992).
103. G. H. Rasmusson, in Pharmacology and clinical uses of inhibitors of hormone secretion and action B. J. A. Furr, A. E. Wakeling, Eds. (Baillière Tindall, London, 1987) pp. 308.
104. H. J. Smith, in Design of Enzyme Inhibitors as Drugs M. Sandler, H. J. Smith, Eds. (Oxford University Press, Oxford, 1988), vol. 2, pp. 779.
105. B. W. Metcalf, M. A. Levy, and D. A. Holt, Trends Pharmacol Sci, 10, 491 (1989).
106. J. S. Tenover, in Steroid hormones: Synthesis, metabolism, and action in health and disease J. F. Strauss III, Eds. (W B Saunders Company, Philadelphia, PA, 1991), vol. 20, pp. 893.

107. J. R. Brooks, G. S. Harris, and G. H. Rasmusson, in *Design of Enzyme Inhibitors as Drugs* M. Sandler, H. J. Smith, Eds. (Oxford University Press, Oxford, 1994), vol. 2, pp. 495.
108. S. L. Sudduth and M. J. Koronkowski, Pharmcother, 13, 309 (1993).
109. R. S. Rittmaster, N Engl J Med, 330, 120 (1994).
110. T. Liang, M. A. Cascieri, A. H. Cheung, et al., Endocrinology, 117, 571 (1985).
111. M. A. Levy, M. Brandt, J. R. Heys, et al., Biochemistry, 29, 2815 (1990).
112. B. Faller, D. Farley, and H. Nick, Biochemistry, 32, 5705 (1993).
113. G. Tian, J. D. Stuart, M. L. Moss, et al., Biochemistry, 33, 2291 (1994).
114. A. E. Thigpen, K. M. Cala, and D. W. Russell, J Biol Chem, 268, 17404 (1993).
115. T. N. Mellin, R. D. Busch, and G. H. Rasmusson, J Steroid Biochem Molec Biol, 44, 121 (1993).
116. R. K. Bakshi, G. H. Rasmusson, R. L. Tolman, et al., WO 9323419, Merck & Co, Inc, (1993).
117. R. K. Bakshi, G. F. Patel, G. H. Rasmusson, et al., Abst 207th ACS National Meeting, , Medi 8 (1994).
118. T. C. Shao, A. Kong, P. Marafelia, and G. R. Cunningham, J Androl, 14, 79 (1993).
119. J. R. Brooks, C. Berman, R. L. Primka, et al., Steroids, 47, 1 (1986).
120. J. R. Brooks, C. Berman, D. Garnes, et al., The Prostate, 9, 65 (1986).
121. S. M. Cohen, K. H. Taber, P. F. Malatesta, et al., Magnetic Resonance in Medicine, 21, 55 (1991).
122. J. Geller and B. S. Sionit, J Cell Biochem, 16H Supplement, 109 (1992).
123. J. D. McConnell, J. D. Wilson, F. W. George, et al., J Clin Endocrin Metab, 74, 505 (1992).
124. J. R. Spencer, T. Torrado, R. S. Sanchez, et al., Endocrinology, 129, 741 (1991).
125. J. Imperato-McGinley, R. S. Sanchez, J. R. Spencer, et al., Endocrinology, 131, 1149 (1992).
126. F. W. George, L. Johnson, and J. D. Wilson, Endocrinology, 125, 2434 (1989).
127. A. R. Diani, M. J. Mulholland, K. L. Shull, et al., J Clin Endocrinol Metab, 74, 345 (1992).
128. S. Douglas, S. Packard, S. Kurata, and H. Uno, Clin Res, 40, 731 (1992).
129. J. R. Brooks, C. Berman, H. Nguyen, et al., The Prostate, 18, 215 (1991).
130. A. A. Geldof, M. F. A. Meulenbroek, I. Dijkstra, et al., J Cancer Res Clin Oncol, 118, 50 (1992).
131. E. di Salle, D. Guidici, G. Briatico, et al., J Steroid Biochem Molec Biol, 46, 549 (1993).
132. E. Stoner and Study Group, Urology, 43, (1994).
133. E. Stoner and Study Group, Arch Intern Med, 154, 83 (1994).
134. H. A. Guess, J. F. Heyse, and G. J. Gormley, Prostate, 22, 31 (1993).
135. J. C. Presti, W. R. Fair, G. Andriole, et al., J Urol, 148, 1201 (1992).
136. J. Schwartz, O. Laskin, S. Schneider, et al., Clin Pharmacol Ther, 53, 231 (1993).
137. S. V. Frye, C. D. Haffner, P. R. Maloney, et al., J Med Chem, 36, 4313 (1993).
138. K. Kojima, H. Kurata, H. Horikoshi, and T. Hamada, EP 0484094, Sankyo Company Ltd, (1992).
139. M. A. Levy, B. W. Metcalf, M. Brandt, et al., Bioorg Chem, 19, 245 (1991).
140. J. C. Lamb, H. English, P. L. Levandoski, et al., Endocrinology, 130, 685 (1992).
141. J. C. Lamb, M. A. Levy, R. K. Johnson, and J. T. Isaacs, Prostate, 21, 15 (1992).
142. P. Audet, H. Nurcombe, Y. Lamb, et al., Clin Pharmacol Ther, 53, 231 (1993).
143. R. E. Johnsonbaugh, B. R. Cohen, E. M. McCormick, et al., J Urol, 149, 432 (1993).
144. D. A. Holt, M. A. Levy, H.-J. Oh, et al., J Med Chem, 33, 943 (1990).
145. D. A. Holt, M. A. Levy, D. L. Ladd, et al., J Med Chem, 33, 937 (1990).
146. D. A. Holt, H.-J. Oh, L. W. Rozamus, et al., Bioorg Med Chem Let, 3, 1735 (1993).
147. D. A. Holt, H.-J. Oh, M. A. Levy, and B. Metcalf, Steroids, 56, 4 (1991).
148. D. A. Holt, M. A. Levy, H.-K. Yen, et al., Bioorg Med Chem Let, 1, 27 (1991).
149. C. Haffner, Tetrahedron Letters, 35, 1349 (1994).
150. P. M. Weintraub, EP 469547, Merrell Dow Pharmaceuticals Inc, (1992).
151. G. A. Flynn, P. Bey, and T. R. Blohm, EP 0435321, Merrell Dow Pharmaceutical Inc., (1991).
152. O. Takahashi, K. Imai, K. Watanabe, et al., Hinyokika Kiyo (Japan), 38, 30 (1992).
153. S. Okada, K. Sawada, A. Kuroda, et al., WO 9316996, Fujisawa Pharmaceutical Co Ltd, (1993).
154. Bulletin:, Pharma Japan, Issue 1367, 16 (1993).
155. J. Hirosumi, O. Nakayama, T. Fagan, et al., Abstracts of the XIIth International Congress of Pharmacology, Montreal, Canada (1994).
156. H. Hara, S. Igarashi, T. Kimura, et al., WO 9324442, Yamanouchi Pharmaceutical Co Ltd, (1993).
157. T. Liang and S. Liao, Biochem J, 285, 557 (1992).
158. C. D. Jones, J. E. Audia, D. E. Lawhorn, et al., J Med Chem, 36, 421 (1993).
159. A. V. Schally, A. M. Comaru-Schally, and D. Gonzolez-Barcena, Biomedicine & Pharmacotherapy, 46, 465 (1992).
160. L. Denis and G. P. Murphy, Cancer, 72, 3888 (1993).
161. E. Carmina, A. Janni, and R. A. Lobo, J Clin Endocrin Metab, 78, 126 (1994).
162. M. J. Karten, in *Modes of Acton of GnRH and GnRH Analogs* W. F. Crowley, M. P. Conn, Eds. (Springer: New York, NY, 1992) pp. 275.
163. T. W. Redding, A. V. Schally, S. Radulovic, et al., Cancer Res, 52, 2538 (1992).
164. D. Gonzalez-Barcena, M. Vadillo-Buenfil, F. Gomez-Orta, et al., Prostate, 24, 84 (1994).
165. G. V. Nikiforovich and G. R. Marshall, Int J Peptide Protein Res, 42, 181 (1993).

SECTION V.   TOPICS IN BIOLOGY

Editor: John C. Lee, SmithKline Beecham Pharmaceuticals
King of Prussia, PA   19406

Chapter 24. Transcription Factor NF-κB: An Emerging Regulator of Inflammation

Anthony M. Manning and Donald C. Anderson
Adhesion Biology, The Upjohn Company, Kalamazoo, MI

Introduction - Since its identification as an inducible activator of transcription from both the intronic κ light chain and HIV-1 enhancers, the nuclear transcription factor NF-κB has been implicated in the regulation of an increasing number of gene products contributing to the pathogenesis of clinical disorders. Emerging evidence indicates that NF-κB may be involved in the onset of multiple forms of vascular pathobiology, including inflammation. Fundamental to this concept is the observation that multiple proinflammatory genes of endothelial cells have functional NF-κB sites, and that diverse inflammatory mediators are capable of stimulating signal transduction pathways leading to the activation and function of NF-κB in these cells. Thus, this pleiotropic transcription factor is uniquely situated to coordinate the expression of numerous endothelial products (adhesion molecules, cytokines, growth factors and coagulation factors), which cooperate to promote inflammatory responses. Based on these unique properties, NF-κB may represent an attractive molecular target in the development of novel anti-inflammatory therapies.

This chapter will address the hypothesis that antagonism or regulation of NF-κB in vascular endothelial cells modulates their proinflammatory functions. Most importantly, the effects of NF-κB regulation on leukocyte adhesion and functional activation will be considered within the context of modulating leukocyte sequestration and inflammatory injury in vivo. Descriptions of our current understanding of the molecular biology of NF-κB and its mechanism of activation in endothelial or other cell types will be provided as a basis for understanding the relative activities (as well as a mechanism of action) of several classes of anti-inflammatory pharmacologic agents. This chapter will also examine potential applications of NF-κB antagonists in other types of clinical disorders and the potential advantages or disadvantages of such approaches, and consider key questions deserving further investigation.

ROLE OF NF-κB IN LEUKOCYTE-ENDOTHELIAL ADHESION

The recruitment of blood leukocytes to sites of inflammation involves a well coordinated and dynamic sequence of events in which several endothelial cell adhesion molecules (CAMs) and chemotactic cytokines play an active role (1 and references therein). Several lines of investigation in vitro predict a multistep model involving: 1. initial low affinity selectin molecule-dependent "vascular rolling" contacts of neutrophils or monocytes (termed "margination"), 2. leukocyte activation by chemoattractants secreted by cytokine stimulated endothelium (e.g. IL-8, MCP-1) and 3. a transition to β_2-integrin-dependent high-affinity leukocyte adhesion to the endothelial determinant intercellular adherence molecule-1 (ICAM-1) and other CAMs, and subsequent transendothelial migration. This multistep model for in vivo leukocyte sequestrants at the site of inflammation has been well documented using an intravital microscope to examine the microcirculation (2), and has included observations of the characteristics of adhesion and transmigration of leukocytes genetically deficient in β_2 integrins (3) or synthesis of SLe^x, a ligand for selectin adhesion molecules (4).

Endothelial Genes and Leukocyte Adhesion - Central to this mechanism of tissue leukocyte localization is the induction of several endothelial CAMs and endothelially derived cytokines elicited by diverse inflammatory mediators. These include E-selectin (CD62E), a counterreceptor for undefined ligands on neutrophils and other granulocytes, monocytes, and T cell subsets (5). E-selectin is not constitutively expressed, but it is rapidly induced by TNFα, lymphotoxin, IL-1β, and LPS on cultured endothelium, and is markedly upregulated ion venules and capillary endothelium in inflamed tissues (6,7). Vascular cell adhesion molecule-1 (VCAM-1) is another cytokine-inducible endothelial CAM that is recognized by VLA-4 (α4P1) or $\alpha_4\beta_7$ on lymphocytes, monocytes, and eosinophils. VCAM-1 is not constitutively expressed, but is rapidly induced by TNFα, IL-4, IL-5 and other mediators (8,9). ICAM-1 (CD54) is a cognate ligand for the β_2 integrins on neutrophils, monocytes, and lymphocytes, where it plays an obligatory role in transendothelial migration (10). This endothelial CAM is constitutively expressed, but is also induced on cytokine (IL-1, TNFα) stimulated endothelium (10). Endothelial cells express or secrete several chemotactic cytokines capable of mediating the selective recruitment of leukocyte populations (11). Interleukin-8 (IL-8), a member of the C-X-C family of chemokines, is elicited in cytokine activated endothelium. This molecule is a potent chemotactic factor capable of upregulating β_2-dependent neutrophil adhesion and transmigration (12). Monocyte chemoattractant protein 1 (MCP-1), a representative member of the C-C family of chemokines, is also induced in endothelium by inflammatory stimuli (13). This molecule selectively activates and promotes the accumulation of monocytes/macrophages and T lymphocytes at sites of inflammation (14, 15).

Cytokine-induced cell surface expression of E-selectin, VCAM-1 and ICAM-1 and the enhanced secretion of IL-8, MCP-1 and other chemokines is regulated at the transcriptional level in endothelial cells (reviewed in 16). Structural analysis of the genes encoding these proteins reveals the presence of putative binding sites for the nuclear transcription factor NF-κB in the 5' flanking sequences (17-21). Deletion analysis and site-directed mutagenesis studies have revealed that cytokine induced activation of transcription from these genes requires the presence of functional NF-κB binding sites (18, 19, 22-24). Endothelial cells express the p50 and RelA components of NF-κB system (25). NF-κB-like factors are activated in cytokine-stimulated endothelial cells, and NF-κB binds specifically to the κB sites present in the E-selectin, VCAM-1, ICAM-1, and IL-8 genes (18, 22-25). The transfection of primary cultures of endothelial cells with constructs encoding RelA resulted in the activation of E-selectin and VCAM-1 transcriptional activity (25, 26). Endothelial cells express a cytoplasmic inhibitor of NF-κB activity, IκBα, and overexpression of recombinant IκB inhibits E-selectin and VCAM-1 transcriptional activity induced by TNFα and by overexpression of p65 (25, 26). These studies suggest that the nuclear transcription factor NF-κB plays a central role in the activation of genes whose products promote the adhesion and extravasation of leukocytes at sites of inflammation (Figure 1).

MOLECULAR BIOLOGY OF NF-κB

NF-κB was first identified as the nuclear factor that binds a decameric DNA sequence within the intronic immunoglobulin kappa light chain enhancer (27). The binding of NF-κB to this DNA sequence is responsible for the inducible activity of the enhancer element present in the immunoglobulin gene. Recent reviews have covered different aspects of NF-κB, including its physiology (28-30), biochemistry (31, 32), and molecular biology (33, 34) as well as its relationship to v-rel (35, 36) and the *Drosophila melanogaster* factor dorsal (37). The role of NF-κB in the regulation of the expression of genes involved in T and B cells and in macrophage/monocyte immune functions has also been reviewed in detail (27).

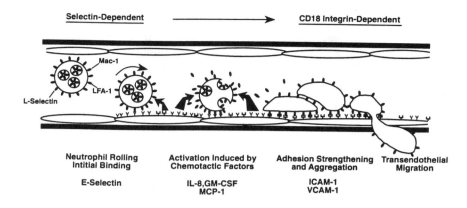

Selectin-Dependent ⟶ **CD18 Integrin-Dependent**

Mac-1

LFA-1

L-Selectin

| Neutrophil Rolling Intitial Binding | Activation Induced by Chemotactic Factors | Adhesion Strengthening and Aggregation | Transendothelial Migration |

E-Selectin　　　　　IL-8,GM-CSF　　　　　ICAM-1
　　　　　　　　　　　　　MCP-1　　　　　　VCAM-1

Figure 1. NF-κB-regulated endothelial ligands and chemotactic cytokines involved in the multi-step process of leukocyte adhesion and migration.

NF-κB is a member of the Rel family of proteins, a novel family of transcription factors sharing a common structural motif for DNA binding and dimerization (33, 38). The rel family of transcription factors can be defined as a group of proteins that share sequence homology over a 300 amino acid region termed the NF-κB/ rel domain (NRD). The rel family comprises important regulatory proteins from a wide variety of species in addition to the mammalian NF-κB, including the *Drosophila* morphogen dorsal, the avian oncogene *v-rel*, and the cellular proto-oncogene *c-rel* (30). Several other NF-κB-related polypeptides have been identified in mammalian cells, including RelB, p52, and its precursor protein p100 (also known as lyt10) (27, 34). These proteins can exist either as homo- or heterodimers, each with a specific affinity for different decameric binding sites fitting the κB motif (30).

NF-κB is found in the cytoplasm of non-stimulated cells, where it is associated with an inhibitor protein called IκB (40). Five distinct IκB species have been identified in vertebrates (41). These proteins can bind selectively to the individual NF-κB and rel-related polypeptides (42). IκBα binds selectively to NF-κB heterodimers and inhibits the DNA binding activity of NF-κB by preventing nuclear uptake of NF-κB complexes (39). Antibodies specific for the NLS of p50 and RelA can no longer bind in the presence of IκB, suggesting a direct physical contact between IκB and the NLS (43). Mutation of the RelA NLS abolishes IκB binding *in vitro* and prevents RelA nuclear localization when overexpressed in cells, suggesting that IκB prevents translocation of NF-κB to the nucleus in non-stimulated cells by binding to the NLS (43).

MECHANISM OF ACTIVATION OF NF-κB

NF-κB activation represents the terminal step in a signal transduction pathway leading from the cell-surface to the nucleus. A wide variety of extracellular stimuli trigger the activation of NF-κB, including pro-inflammatory cytokines, bacterial and viral products, and even physical or oxidative stress (27, 30 and references therein). A general pathway of NF-κB activation has been proposed based upon the cumulative study of NF-κB activation in various transformed cell lines (27). These studies have suggested a role for phosphorylation, redox signals, and proteolysis in the activation of NF-κB.

Phosphorylation - NF-κB activity is still susceptible to inhibition by IκB in cells stimulated with the PKC activator PMA whereas IκB activity is not detectable (40). PKC and other kinases can activate NF-κB under cell-free conditions, apparently by direct phosphorylation and release of IκB (44, 45). TNFα and IL-1β-induced loss of IκB from the cytoplasm of cultured cells is preceded by IκB phosphorylation (46). In the same experiments, however, no such phosphorylation is observed following stimulation with PMA. The importance of IκB phosphorylation in NF-κB activation remains unclear. Further support for a role of phosphorylation in the signal transduction events leading to NF-κB activation has come from reports that the kinase inhibitor H7 can block PMA-induced NF-κB activation and that the phosphatase inhibitor okadaic acid can induce NF-κB activation in intact cells (47, 48). Recent studies have demonstrated a role for the tyrosine kinases, Ras and Raf-1 and for a specific isoform of PKC in the activation of NF-κB DNA-binding activity and κB site-dependent transcription (49, 50). A role for ceramide as a second messenger in the TNFα-induced activation of NF-κB has been revealed by studies demonstrating that the binding of TNFα to the 55kD TNF receptor leads to the activation of an acidic sphingomyelinase (SMase) by 1,2 diacylglycerol produced by a phosphatidylcholine-specific phospholipase C (51). It is hypothesized that ceramide released from sphingomyelin by this SMase controls the activity of specific kinases and phosphatases.

Intracellular Proteolysis - The activation of NF-κB DNA-binding activity in transformed cell lines, primary human T lymphocytes, and endothelial cells is associated with the loss of IκB protein from the cytoplasm and the translocation of NF-κB to the nucleus (27, 25). The complete and rapid loss of IκB protein from the cytoplasm of stimulated cells has suggested that specific proteolysis of IκB is the critical step in NF-κB activation. Indeed, alkylating and non-alkylating substrate analogues of chymotrypsin-like serine proteases are potent inhibitors of IκB degradation and NF-κB activation in intact cells (52, 53). The half-life of IκB protein was approximately 2 hours in protein synthesis-arrested pre-B cells, whereas in PMA-stimulated cells it is reduced dramatically to only 1.5 minutes, leading to the suggestion that a proteolytic system, or even a specific protease, might be activated following cellular stimulation, leading to the rapid degradation of almost all of the IκB protein in the cytoplasm (52). It is unclear whether the IκB is disassociated from the NF-κB:IκB complex before proteolytic degradation.

Redox Signalling - Reactive oxygen intermediates (ROIs) have been recognized as important inducers of gene expression (54). Many of the agents that activate NF-κB, such as TNFα, LPS, T and B cell-activating stimuli and UV irradiation, also cause an increase in the cellular production of ROIs (55). ROIs have been directly implicated as second messengers in the activation of NF-κB, based upon the ability of a variety of antioxidants to inhibit NF-κB activation in intact cells (58). These include N-acetyl-L-cysteine (a precursor of glutathione), vitamin E derivatives, 2-mercaptoethanol, dithiocarbamates, butylated hydroxyanisol and chelators of iron and copper ions. The chemical diversity of these inhibitors suggest that they inhibit NF-κB activity as a result of their ability to scavenge free radicals, not because of a specific structure-related activity. In addition, hydrogen peroxide or depletion of glutathione levels rapidly activates NF-κB in some T cell lines (55). The antioxidant PDTC has been shown to inhibit degradation of IκB, suggesting that ROIs may play a role in the activation of IκB proteolysis (52). Redox changes play a role in the activation of serine proteases in general, through the modification of residues required for the binding of inhibitory proteins belonging to the serpin family of serine protease inhibitors (57). It has been reported that the cytotoxic and NF-κB-activating potential of TNFα is dependent upon ROI produced by mitochondria (58). NF-κB itself is sensitive to redox variations *in vitro*, being inactivated by oxidation. Site-directed mutagenesis studies have demonstrated that a conserved motif containing a cysteine residue is critical for both optimal DNA binding and redox regulation of the related v-rel protein (59). This suggests that ROIs may play a further role in the DNA-binding activity of NF-κB *in vivo*.

NF-κB ACTIVATION AND ENDOTHELIAL GENE EXPRESSION

An understanding of the mechanism of activation of NF-κB in transformed cell lines has led to an assessment of the effects of inhibitors of NF-κB activation on the expression of cell adhesion molecules and chemotactic cytokines in primary cultured human umbilical vein endothelial cells. These studies have provided a pharmacologic confirmation of the critical role that NF-κB plays in the expression of endothelial cell adhesion molecules and chemotactic cytokines and the promotion of leukocyte adhesion, activation, and transendothelial migration.

Antioxidants - The antioxidants N-acetyl-cysteine (NAC), pyrrolidone dithiocarbamate (PDTC) and bunaprolast (U-66858; **1**) inhibits the TNFα-, IL1β-, and LPS-induced activation of NF-κB in primary cultures of human umbilical vein endothelial cells (60, 61). Treatment of endothelial cells with these inhibitors results in a dose-dependent inhibition of E-selectin, VCAM-1, and ICAM-1 gene expression (60). These antioxidants also inhibit the cytokine-induced expression of the IL-8, IL-6, and GM-CSF genes (62). Neutrophil and eosinophil adhesion and transendothelial migration through cytokine-activated endothelial cell monolayers is significantly inhibited when the endothelium is pretreated with concentrations of antioxidants that inhibit NF-κB activation (60). **1** represents the most potent antioxidant inhibitor of endothelial NF-κB activation and NF-κB-mediated gene expression, with an IC_{50} for E-selectin and VCAM-1 expression of approximately 3μM (60).

1

Kinase Inhibitors - Protein kinase C and cAMP play a part in the cytokine-induced expression of E-selectin, VCAM-1, and ICAM-1. Whereas the PKC inhibitors calphostin C and staurosporine inhibit cytokine-induced VCAM-1 cell surface expression, this effect is not mediated by inhibition of activation of NF-κB (63). The lack of effect of PKC inhibition on NF-κB activation was further corroborated by the inability of these inhibitors to block cytokine-induced E-selectin expression. However, pretreatment of endothelial cells with H7, another inhibitor of PKC (but one which may inhibit other kinases) inhibited TNFα, IL-1 and LPS induction of ICAM-1 and VCAM-1 (64). Agonists of PKC, including PMA, 12-deoxyphorbol 13-phenylacetate 20-acetate (dPPA), and thymealatoxin, induced NF-κB binding activity and surface expression of E-selectin and VCAM-1 in endothelial cells (65). The β1-selective PKC agonist, dPPA, induced NF-κB DNA-binding activity and adhesion molecule surface expression to the same extent as PMA and thymealatoxin, indicating that the activation of the β1 isozyme of PKC may be sufficient for induction of expression of E-selectin and VCAM-1. One interpretation of these observations is that PKC activity, inhibited by staurosporine or H7, plays a role in the induction of VCAM-1 expression, while kinases other than PKC, but also inhibited by H7, play an important role in the expression of E-selectin and ICAM-1. Pretreatment of endothelial cells with a combination of forskolin, an inducer of adenylate cyclase, and isobutyl methylxanthine, a phosphodiesterase inhibitor, has been shown to inhibit the TNFα-induced upregulation of E-selectin and VCAM-1 mRNA, but not of ICAM-1 mRNA (66). Further examination of the effects of selective inhibitors of kinase and phosphatase activity is necessary to resolve the role that these components play in the activation of NF-κB and adhesion molecule expression in endothelial cells.

Miscellaneous - Several other pharmacologic inhibitors of endothelial cell adhesion molecule expression have been reported, although their effects upon NF-κB activation have not yet been examined. Treatment of endothelial cells with PD 144795 **2** significantly inhibits the induction of E-selectin, VCAM-1 and ICAM-1 cell surface expression and IL-8 secretion following TNFα or PMA stimulation (IC_{50} average 3μM) (67). In addition, at doses of 0.6 - 2.0 mg/animal **2** significantly inhibits neutrophil influx into the peritoneal cavity of mice challenged 2 hours earlier with thioglycollate (68). At doses of 10-30 mg/kg **2** also inhibited neutrophil accumulation into the pleural cavity in a rat reverse passive arthus model. A series of substituted pyridines (for example **3**) inhibit E-selectin and/or ICAM-1 expression on TNFα-activated endothelial cells (69). Compounds of this class have anti-inflammatory, antipyretic, analgesic, anti-allergy and immunomodulatory activities. Further analysis of the intracellular mechanism of action of **2** and **3** should determine what effect they have on the NF-κB system in endothelial cells.

2   **3**

INHIBITORS OF NF-κB AS ANTI-INFLAMMATORY THERAPEUTICS

The inhibition of the activity of endothelial transcription factors represents a novel approach to the antagonism of leukocyte-endothelial cell adhesion, through inhibition of gene expression. Our understanding of NF-κB biology suggests several strategies for modulating the activity of this transcription factor in endothelial cells (Figure 2). The distinct mechanism of activation of NF-κB and our developing understanding of the signal transduction cascade that leads to the activation of NF-κB in endothelial cells suggests that it is feasible to develop small molecule inhibitors of NF-κB activation that would be potent inhibitors of E-selectin, VCAM-1, ICAM-1, IL-8, and MCP-1 gene expression (A in Figure 2). These inhibitors might represent redox-regulators such as antioxidants or inhibitors of proteolytic degradation of IκB, which would prevent the release of IκB from the cytoplasmic NF-κB complex, effectively inhibiting the translocation of NF-κB to the nucleus. Many of the components of this NF-κB activation cascade are still unresolved. The elucidation of new components of this signal transduction cascade will offer new opportunities to develop specific inhibitors of NF-κB activation.

It may be possible to interfere with the function of NF-κB following activation (B in Figure 2). When short, double-stranded oligonucleotides encoding the NF-κB binding site sequences are introduced into endothelial cells, they inhibit the activation of VCAM-1 transcription (26). These transcription factor decoys (TFDs) do not inhibit activation of NF-κB, but simply induce binding of activated NF-κB within the cytoplasm, effectively blocking NF-κB translocation to the nucleus. It may be possible to develop effective therapeutics based on this TFD concept.

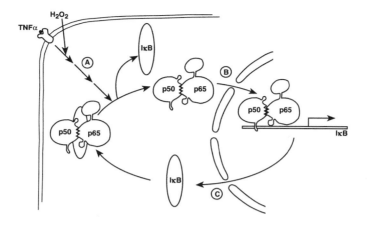

Figure 2. Regulation of NF-κB activation by the cytoplasmic inhibitor, IκB, and points of possible antagonism of NF-κB function.

It may be possible to develop small molecular weight compounds that can interact selectively with the NLS of NF-κB, preventing nuclear translocation of activated NF-B, mimicking the function of IκB. Alternatively, an understanding of the molecular processes involved in nuclear uptake of NF-κB may lead to the development of inhibitors of nuclear uptake. Activated NF-κB induces the transcription of its own inhibitor, IκB, and so acts as a negative feedback regulator of its own activity (27). It may be possible to selectively induce activation of the IκB gene, or to introduce IκB directly into cells in order to overcome the activation pathway that results in the degradation of IκB (C in Figure 2). There are other strategies for regulating endothelial gene expression. Anti-sense oligonucleotide technology has been utilized to inhibit IL-1β-induced ICAM-1 expression in human umbilical vein endothelial cells, and have shown potential in murine models of acute inflammation (70). The effective delivery of oligonucleotide inhibitors in humans represents a challenge for the development of stable and diffusible therapeutic entities (71).

Small molecule inhibitors of NF-κB activation would inhibit the expression of multiple cell adhesion receptors and chemotactic cytokines which regulate leukocyte adhesion and transendothelial migration, and thus should possess broad anti-inflammatory properties. The pharmacologic and immunologic inhibition of leukocyte-endothelial cell adhesion has become the central focus of research directed at attenuating granulocyte-mediated tissue injury in a variety of pathologic disorders of importance to the critical care physician (72). These disorders include ischemia/ reperfusion injury, hemorrhagic and septic shock, allograft rejection, frostbite injury, bacterial meningitis, acute airway inflammation and the pulmonary complications induced by cardiopulmonary bypass. An elevated expression of endothelial cell adhesion molecules has been consistently observed in association with many of these disorders, demonstrating that endothelial activation is a common component of acute inflammatory disorders (73). Cell adhesion mechanisms appear to contribute to chronic inflammatory disorders of major clinical significance, including asthma, arthritis, and even atherosclerosis (74-76). Small molecule inhibitors of NF-κB activation would be selective for sites of endothelial activation and would be rapidly delivered to endothelium lining the vasculature. These inhibitors would be functional within the cytoplasm, and would not require effective delivery to the nucleus. The effective suppression of leukocyte-endothelial cell adhesion *in vivo* may require antagonism of multiple adhesive mechanisms. Inhibition of NF-κB activation represents a single molecular mechanism by which the expression of multiple genes involved in leukocyte sequestration could be suppressed.

CHALLENGES FOR THE FUTURE

Molecular genetic and pharmacologic studies in cell culture systems have led to the paradigm of NF-κB as a critical regulator of inflammatory genes, as presented herein. The validation of this paradigm in man will be an important step in the realization of NF-κB as a novel molecular target for the development of anti-inflammatory therapeutics. The presence of activated NF-κB within the nucleus of endothelial cells in inflammatory disorders of man has not yet been demonstrated. Such results will be required in order to validate the paradigm of NF-κB regulation of endothelial gene expression that has been developed from molecular genetic and pharmacologic studies in cell culture systems. The activation of NF-κB DNA binding activity *in vivo* has been reported in the liver of animals following ischemia and reperfusion (77), and after animals were fed an atherogenic diet (78), supporting the contention that NF-κB activation may play a role in acute and chronic inflammatory pathologies.

NF-κB is important in the transcriptional activation of a variety of cellular and viral genes in addition to endothelial cell adhesion molecules and chemotactic cytokines (27). All of these genes have been associated with inflammatory pathology *in vivo*, raising the possibility that NF-κB-mediated transactivation may be a common event in the initiation of multiple pathologic processes. Inhibitors of NF-κB activation may therefore have broad applications as novel therapeutics in human disease. Evidence exists that antioxidant inhibitors of NF-κB may attenuate replication of the human immunodeficiency virus (HIV) in host cells (79). Inhibition of NF-κB function by p65 anti-sense technology modulates the expression of surface adhesion molecules of certain tumor-derived cell lines and inhibits tumor formation in mice (80). These preliminary findings of anti-viral or anti-tumor activity by NF-κB antagonists may lead to still other potential clinical applications.

It is important to elucidate the signal transduction events that lead to the activation of NF-κB in response to pro-inflammatory stimuli in order to develop highly specific inhibitors of NF-κB function. The tissue and molecular specificities of NF-κB antagonists must be defined, especially in light of the broad range of genes regulated, at least in part, by NF-κB activation. Possible synergistic combinations of other transcription factors together with NF-κB in inflammatory processes should be considered in developing potent therapeutic reagents. These challenges are faced by all signal transduction research. Recent successes in the analysis of signal transduction mechanisms regulating gene expression and the transcription factors that act as components of this system hold out the hope that much more will be achieved in the near future.

References

1.   T.A. Springer, Cell 76, 301 (1994).
2.   J.M. Harlan, R.K. Winn, N.B. Vedder, C.M. Doerschuk and C.L. Rice in "Adhesion: Its Role In Inflammation," J.M. Harlan and D.Y. Liu, Ed., W.H Freeman and Co., New York, 1992, p117.
3.   D.C. Anderson and T.A. Springer, Am. Rev. Med., 38, 175 (1987).
4.   U.H. von Adrian, E.M. Berger, L. Ramezani, J.D. Chambers, H.D. Ochs, J.M. Harlan, J.C.

Paulson, A. Etzioni and K.-E. Arfors, J. Clin. Invest., <u>91</u>, 2893 (1993).

5.   M.P. Bevilacqua, S. Stengelin, M.A. Gimbrone and B. Seed, Science <u>243</u>, 1160 (1989).

6.   J. Whelan, P. Ghersa, R.H. van Huijsduijnen, J. Gray, G. Chandra, F. Talbot, Nuc. Acids. Res., <u>19</u>, 2645 (1991)

7.   R.S. Cotran, M.A. Gimbrone, M.P. Bevilacqua, D.L. Mendrick and J.S. Pober. J. Exp. Med., <u>164</u>, 661 (1986).

8.   M.J. Elices, L. Osborn, Y. Takada, C. Crouse, S. Luhowshyj, M.E. Hemmler and R.R. Lobb, Cell <u>60</u>, 577 (1990).

9.   A.S. Neish, A.J. Williams, H.J. Palmer, M.Z. Whitley and T. Collins, J. Exp. Med., <u>176</u>, 1583 (1992).

10.   J.S. Pober, M.A. Gimbrone, L.A. Lapierre, D.L. Mendrick, W. Fiers, R. Rothlein and T.A. Springer, J. Immunol., <u>137</u>, 1893 (1986).

11.   J.J. Oppenheim, C.O. Zachariae, N. Mukaida and K. Matsushima, Ann. Rev. Immunol., <u>9</u>, 617 (1991).

12.   A.R. Huber, S.L. Kunkel, R.F. Todd and S.J. Weiss, Science, <u>58</u>, 227 (1991).

13.   B.J. Rollins, T. Yoshimura, E.J. Leonard and J.S. Pober, Am. J. Pathol., <u>82</u>, 5375 (1990).

14.   E.A. Robinson, T. Yoshimura, E.J. Leonard, S. Tanaka, P.R. Griffin, J. Shabanowitz, D.F. Hunt and E. Appella, Proc. Natl. Acad. Sci. USA, <u>86</u>, 1850 (1989).

15.   M.W. Carr, S.J. Roght, E. Luther, S.S. Rose and T.A. Springer, Proc. Natl. Acad. Sci. USA, in press (1994).

16.   T. Collins, Lab. Invest., <u>68</u>, 499 (1993).

17.   M.I. Cybulsky, J.W.U. Fries, A.J. Williams, P. Sultan, R. Eddy, M. Byers, T. Shows, M.A. Gimbrone, Jr. and T. Collins, Proc. Natl. Acad. Sci. USA, <u>88</u>, 7859 (1991).

18.   M.F. Iademarco, J.J. McQuillan, G.D. Rosen and D.C. Dean, J. Biol. Chem., <u>267</u>, 16323 (1992).

19.   K. Degitz, L. Lian-Jie and S.W. Caughman, J. Biol. Chem., <u>266</u>, 14024 (1991).

20.   T. Collins, A. Williams, G.I. Johnston, J. Kimm, R. Eddy, T. Shows, M.A. Gimbrone, Jr. and M.P. Bevilacqua, J. Biol. Chem. <u>266</u>, 2466 (1991).

21.   K.F. Montgomery, L. Osborn, C. Hession, R. Tizard, D. Goff, C. Vassallo, P.I. Tarr, K. Bomsztyk, R. Lobb, J.M. Harlan and T.H. Pohlman, Proc. Natl. Acad. Sci. USA, <u>88</u>, 6523 (1991).

22.   G. Voraberger, R. Schafer and C. Strotowa, J. Immunol., <u>147</u>, 2777 (1991).

23.   R.H. van Huijsduijnen, J. Whelan, R. Pescini, M. Becker-Andre, A-M. Schenk and J.F. DeLamarter, J. Biol. Chem., <u>267</u>, 22385 (1992).

24.   T. Collins, M.Z. Whitley, H. Palmer, A.S. Neish and A.J. Williams, FASEB J., <u>7</u>, A198 (1992).

25.   M.A. Read, M.Z. Whitley, A.J. Williams and T. Collins, J. Exp. Med., <u>179</u>, 503 (1994).

26.   H.B. Shi, A. Agranoff, E.G. Nabel, K. Leung, A.S. Neish, T. Collins and G. Nabel, Mol. Cell. Biol. <u>13</u>, 6283 (1993).

27.   P.A. Baeuerle and T. Henkel, Annu. Rev. Immunol., <u>12</u>, 141 (1994).

28.   M.J. Lenardo and D. Baltimore, Cell, <u>58</u>, 227 (1989).

29.   P.A. Baeuerle and D. Baltimore in "Molecular Aspects of Cellular Regulation: Hormonal Control Regulation of Gene Transcription", P. Cohen and J.G. Foulkes, Eds., Elsevier/North Holland Biomed., Amsterdam, 1991, pp. 409.

30.   M. Grilli, J-S. Jason and M.J. Leonardo, Int. Rev. Cytol., <u>143</u>, 1 (1993).

31.   P.A. Baeuerle, Biochim. Biophys. Acta, <u>1072</u>, 63 (1991).

32.   S. Grimm and P.A. Baeuerle, Biochem. J., <u>290</u>, 297 (1993).

33.   V. Blank, P. Kourilsky and A. Israel, Trends Biochem. Sci., <u>17</u>, 135 (1992).

34.   G.P. Nolan and D. Baltimore, Curr. Opin. Genet. Dev., <u>2</u>, 135 (1992).

35.   T.D. Gilmore, Trends Genet., <u>7</u>, 313 (1991).

36.   H.J. Bose, Biochim. Biophys. Acta, <u>1114</u>, 1 (1992).

37.   S. Govind and R. Steward, Trends Gen. <u>7</u>, 119 (1991).

38.   G.J. Nabel and I.M. Verma, Genes & Dev., <u>7</u>, 2063 (1993).

39.   A.A. Beg, S.M. Ruben, R.I. Scheinman, S. Haskill, C.A. Rosen and A.J. Baldwin, Genes Dev. <u>6</u>, 1899 (1992).

40.   P.A. Baeuerle and D. Baltimore, Science, <u>242</u>, 540 (1988).

41.   T.D. Gilmore and P.J. Morin, Trends in Genetics, <u>9</u>, 427 (1993).

42.   A.A. Beg and A.S. Baldwin, Jr., Genes & Dev., <u>7</u>, 2064 (1993).

43.   U. Zabel, T. Henkel, M.S. Silva and P.A. Baeuerle, EMBO J., <u>12</u>, 201 (1993).

44. S. Ghosh and D. Baltimore, Nature, 344, 678 (1990).
45. E. Link, L.D. Kerr, R. Schreck, U. Zabel, I. Verma and P.A. Baeuerle, J. Biol. Chem., 267, 239 (1992).
46. A.A. Beg, T.S. Finco, P.V. Nantermet and A.S. Baldwin, Jr., Mol. Cell. Biol., 13, 3301 (1993).
47. S.W. Tong-Starksen, P.A. Luciw and B.M. Peterlin, J. Immunol., 142, 702 (1989).
48. C. Thevenin, S-J., Kim, P. Rieckmann, H. Fujiki, M.A. Norcross, M.B. Sporn and J.H. Kehrl, New Biologist, 2, 793 (1991).
49. T.S. Finco and A.S. Baldwin, J. Biol. Chem. 268, 17676 (1993).
50. S. Li and J.M. Sedivy, Proc. Natl. Acad. Sci. USA, 90, 9247 (1993).
51. S. Schütze, K. Pothoff, T. Machleidt, D. Bercovic, K. Wiegmann and M. Krönke, Cell, 71, 765 (1992).
52. T. Henkel, T. Machleidt, I. Alkalay, M. Krönke, Y. Ben-Neriah and P.A. Baeuerle, Nature, 365, 182 (1993).
53. P.J. Chiao, S. Miyamoto and I.M. Verma, Proc. Natl. Acad. Sci. USA, 91, 28 (1994).
54. B. Demple and C.F. Amabile-Cuevas, Cell, 67, 837 (1991).
55. R. Schreck and P.A. Baeuerle, Trends Cell Biol., 9, 39 (1991).
56. R. Schreck, B. Meier, D.N. Männel, W. Dröge and P.A. Baeuerle, J. Exp. Med., 175, 1181 (1992).
57. E. Remold-O'Donnell, J.C. Nixon and R.M. Rose, J. Exp. Med., 169, 1071 (1989).
58. K. Schulze-Osthoff, A.C. Bakker, B. Vanhaesebroeck, R. Beyaert, W.A. Jacob and W. Fiers, J. Biol. Chem., 267, 5317 (1992).
59. M.B. Toiedano and W.J. leonhard, Proc. Natl. Acad. Sci. USA, 88, 4328 (1991).
60. A.M. Manning, C.C. Chen, A.R. McNab, C.L. Rosenbloom and D.C. Anderson, J. Cell. Biochem., 18B, 298 (1994).
61. N. Marui, M.K. Offermann, R. Swerlick, C. Kunsch, C.A. Rosen, M. Ahmad, R.W. Alexander and R.M. Medford, J. Clin. Invest., 92, 1866 (1993).
62. C.C. Chen, C.L. Rosenbloom, T.P. Nguyen, A.R. McNab, D.C. Anderson and A.M. Manning, J. Cell. Biochem., 18B, 321 (1994).
63. T.A. Deisher, T.L. Haddiz, K.F. Montgomery, T.H. Pohlman, K. Kaushansky and J.M. Harlan, FEBS, 331, 285 (1993).
64. P. Mattila, M.L. Majuri, P.S. Mattila and R. Renkonen, Scand. J. Immunol., 36, 159 (1992).
65. T.A. Deisher, T.T. Sato, T.H. Polhman and J.M. Harlan, Biochem. Biophys. Res. Comm., 193, 1283 (1993).
66. P. Ghersa, R. Hooft van Huijsduijnen, J. Whelan, Y. Cambet and J.F. Delamarter, J. Cell. Biochem., 18B, 278 (1994).
67. S.K. Sanders, J.B. Marine and S.M. Crean, J. Cell. Biochem., 18B, 298 (1994).
68. D.J. Sohrier, M.E. Lesch, C.D. Wright, G.C. Okonkwo, M.P. Finkel and K. Imre, J. Cell. Biochem., 18B, 298 (1994).
69. D.H. Boschelli and C.D. Wright, Curr. Opin. Invest. Drugs, 2, 723 (1993).
70. M.Y. Chiang, H. Chan, M.A. Zounes, S. Freier, W.F. Lima and C.F. Bennett, J. Biol. Chem., 266, 18162 (1991).
71. C.K. Mirabelli, C.F. Bennett, K. Anderson and S.T. Crooke, Anticancer Drug Design, 6, 647 (1991).
72. R.J. Korthuis, D.C. Anderson and D.N. Granger, J. Crit. Care, in press (1994).
73. T. M. Carlos and J. M. Harlan, Blood, in press (1994)
74. C.D. Wegner, R. Rothlein and R.H. Gundel, Agents and Actions Suppl., 34, 529 (1991).
75. J.S. Grober, B.L. Bowen, H. Ebling, B. Athey, C.B. Thompson, D.A. Fox and L.M. Stoolman, J. Clin. Invest., 91, 2609 (1993).
76. R. Ross, Nature, 362, 801 (1993).
77. S.L. Liu, T. Yao, A.M. Diehl, S. Degli Esposti and M.A. Zern, Hepatology, 18, 148A (1993).
78. F. Liao, A. Andalibi, F.C. deBeer, A.M. Fogelman, A.J. Lusis, J. Clin. Invest. 76: 1709 (1990).
79. M. Roederer, S.W. Ela, F.J.T. Staal, L.A. Herzenberg and L.A. Herzenberg, AIDS Res. Human Retroviruses, 8, 209 (1992).
80. K.A. Higgins, J. R. Perez, T.A. Coleman, K. Dorshkind, W.A. McComas, U.M. Sarmiento, C.A. Rosen and R. Narayanan, Proc. Natl. Acad. Sci. USA, 90, 9901 (1993).

Chapter 25. Translational Control of Gene Expression

Lee Gehrke, Louane E. Hann, Roger L. Kaspar
Division of Health Sciences and Technology
Massachusetts Institute of Technology
Cambridge, MA 02139

Introduction - Significant new developments have stimulated renewed interest in translation-level gene regulatory mechanisms. Among the new findings, investigators have demonstrated that overexpression of protein synthesis initiation factor 4E (eIF-4E) transforms cells, and that injection of these cells into nude mice causes tumors. These data indicate that translational control mechanisms should be considered in the genetic basis for cancers. In other studies of viral and cellular mRNA translation, ribosomes have been observed to initiate protein synthesis not only by the 5' scanning mechanism, but also by binding internally. These findings suggest that the scanning model for initiation of protein synthesis does not account for all initiation events. Specific RNA-protein interactions are at the core of translational control mechanisms, and a great deal has been learned about the specific determinants that permit RNA and protein molecules to bind specifically and form ribonucleoprotein complexes. One of the most exciting new developments in translational control is the emerging role of 3' untranslated region. Although the 3' terminus was previously believed to be a relatively benign stretch of non-coding nucleotides, it is now clear that the 3' end is essential for mRNA localization, stability, protein binding, and translational control.

INITIATION FACTOR eIF-4E, PROTEIN KINASES, AND CANCER

Overexpression of eIF-4E leads to a Transformed Phenotype - Translational regulation of gene expression occurs at the levels of protein synthesis initiation, polypeptide chain elongation and probably termination, but among these, the initiation step is thought to be the major regulatory site. Protein synthesis requires protein initiation factors that facilitate a number of reactions or interactions, including ribosome-mRNA recognition and unwinding of secondary structure (reviewed in 1-3). The dominant model for initiation of translation is that the recognition of the 5' end of a messenger RNA is facilitated by the m7GpppG cap structure in the presence of initiation factor complex eIF-4F. Factor eIF-4F has three subunits (4): a 220 kilodalton subunit (also called eIF-4 gamma), a 50 kilodalton subunit (referred to as eIF-4A), and a 25 kilodalton subunit (called eIF-4E). Factor eIF-4E, which is thought to interact directly with the m7GpppG cap structure, is found in limiting concentrations in the cell, and its activity is regulated positively by phosphorylation at serine-53 (5).

To test the potential involvement of a protein synthesis initiation factor in the mitogenic signal transduction pathway, factor eIF-4E was overexpressed in NIH 3T3 or Rat 2 fibroblast cells (6). While cells that overexpressed eIF-4E lost anchorage-dependent growth properties and induced tumors when injected into nude mice, there were no morphological changes or tumor formation when a non-phosphorylatable mutant eIF-4E protein was expressed (6). Subsequent studies demonstrated that phosphorylation of eIF-4E is increased in Src-transformed cell lines, suggesting that eIF-4E is a downstream target of a phosphorylation cascade activated by tyrosine-specific protein kinases (reviewed in 7).

The mitogenic and oncogenic properties of eIF-4E correlate with the activation of the Ras signaling transduction pathway.   Rat embryo fibroblast cells overexpressing eIF-4E have increased levels of active GTP-bound Ras; moreover, an anti-Ras antibody inhibited the mitogenic activity of eIF-4E (8). Nerve growth factor-induced phosphorylation of eIF-4E in PC12 cells can be blocked by the expression of a dominant inhibitory *ras* mutant. These data indicate that the mitogenic activities of eIF-4E are mediated through Ras activation.

These recent findings define eIF-4E as a proto-oncogene and suggest that maintenance of growth rate control depends in part upon translation-level regulation. The mechanism of eIF-4E-induced cell transformation is not understood. Although stable secondary structure in the 5' untranslated region of mRNAs diminishes translational efficiency (9), cells overexpressing eIF-4E seem to translate structured mRNAs efficiently (10). It has been suggested that increased eIF-4E activity due to phosphorylation or protein overexpression could enhance the translational expression of oncogenes (6), thereby leading to the malignant transformation.

PKR May Be a Tumor Suppressor- Initiation factor eIF-2 plays a critical role in protein synthesis. eIF-2 is involved in binding GTP·Met-tRNA to the 40S ribosomal subunit, forming the 43S preinitiation complex (reviewed in 1). Like eIF-4E, the activity of eIF-2 is regulated by phosphorylation; however, phosphorylation of the alpha subunit of eIF-2 severely attenuates, rather than stimulates, its activity. Two protein kinases that phosphorylate eIF-2 alpha at Ser-51 are the double-stranded RNA-dependent protein kinase (formerly named dsI or DAI; now referred to as PKR) and the heme-regulated protein kinase (formerly named HCR, now known as HRI) (reviewed in 11). Both kinases have been molecularly cloned (12, 13) and detailed biochemical characterization of the enzymes is in progress (reviewed in 14, 15).

PKR is an interferon-induced cAMP-independent protein kinase that is activated by double-stranded RNA. In the presence of low concentrations of double-stranded RNA, PKR is activated by autophosphorylation and then specifically phosphorylates eIF-2 alpha. The phosphorylated form of eIF-2 alpha (*i.e.* eIF-2 alpha·P) forms a very stable complex with the guanine nucleotide exchange factor, eIF-2B (*i.e.* eIF-2(alpha·P)·eIF-2B). The formation of the eIF-2(alpha·P)·eIF-2B complex sequesters eIF-2B, resulting in severe inhibition of protein synthesis because functional eIF-2B is required for the exchange of tightly-bound GDP for GTP in the eIF-2·GDP complex (reviewed in 16).

Infection of cells by RNA viruses often activates PKR. The subsequent inhibition of cellular protein synthesis may be a host cell defense, causing slower rates of virus replication. A minimum of eighty-five base pairs of double-stranded RNA is required for optimal activation of PKR (17); additionally, *in vitro* experiments demonstrate that PKR can be activated by the structured TAR region of HIV RNA (18, 19).

Following the reports demonstrating that overexpression of eIF-4E caused malignant transformation of fibroblast cells (cited above), similar experiments were carried out by overexpressing PKR in NIH 3T3 cells. Although injection of mice with cells overexpressing the wild-type PKR enzyme failed to elicit tumors, injection of cells producing an inactive mutant form of PKR gave rise to large tumors in the inoculated mice (20, 21).   These surprising results suggest, therefore, that PKR may have a tumor suppressor activity.

INTERNAL INITIATION OF PROTEIN SYNTHESIS

<u>More Than Scanning</u> - The scanning hypothesis for initiation of protein synthesis states that the 40S ribosomal subunit recognizes the 5' end of mRNA at the cap site and then 'scans' down the 5' untranslated leader sequence until it encounters the AUG initiation codon and subsequently binds the 60S ribosomal subunit (22, 23). The fact that insertion of stable secondary structure in the 5' UTR diminishes translational efficiency (9, 24) supports the scanning hypothesis; moreover, mRNAs lacking stable secondary structure translate with enhanced efficiency both *in vitro* (25) and *in vivo* (26). Recent data generated using viruses and mammalian cells strongly indicate, however, that protein synthesis can be initiated by a second mechanism: internal initiation.

The study of poliovirus has been instrumental in characterizing internal initiation of protein synthesis. Infection of cells by poliovirus is accompanied by proteolytic cleavage of the p220 subunit of eIF-4F (27) and results in the selective translation of the uncapped poliovirus messenger RNA (reviewed in 28). The poliovirus RNA has a very long 5' UTR (>700 nucleotides) that contains a number of AUG initiation codons, some of which are out of frame with the known protein coding regions. The question of how the correct AUG is selected for translation led to experiments that defined internal initiation.

Bistronic messenger RNA constructs have been used to demonstrate that the poliovirus RNA 5' UTR contains a *cis*-acting element that enables ribosomes to initiate translation in downstream regions of the mRNA without recognizing the 5' end (29). The poliovirus RNA 5' UTR was inserted between two coding regions, where the termination codon of the first open reading frame was followed by a highly stable hairpin to prevent reinitiation by terminating ribosomes. Thus, ribosomes that initiated translation at the 5' end were unable to translate the second cistron. Translation of the second open reading frame was dependent upon the poliovirus RNA 5' UTR. The mRNA element within the poliovirus RNA 5' UTR that was required for internal initiation was called the "ribosome landing pad", now referred to as internal ribosome entry site (IRES). Presumably, the IRES permits the poliovirus RNA to translate in the absence of intact cap binding protein (eIF-4F) while translation of the host mRNA templates is severely compromised in the absence of eIF-4F.

Internal initiation has been best characterized for two groups of picornaviruses, represented first by the cardioviruses (encephalomyocarditis virus) and aphthoviruses (foot and mouth disease virus), and secondly by poliovirus and rhinovirus (30). Internal initiation is apparently not, however, restricted to RNA viruses. Translation of the immunoglobulin heavy chain binding protein mRNA, a cellular template, has been shown to initiate internally (31), as has translation of the *Drosophila* antennapedia mRNA (32). A further variation of the theme has been reported for sendai virus, which seems to initiate by a cap-dependent mechanism, but the initiated ribosome seems to hop to the initiation site without scanning the 5' leader (33). As a practical application of internal initiation phenomena, retroviral vectors containing an IRES have been generated to permit translation of a second cistron (34).

The mechanism of internal initiation is not understood. One hypothesis is that the untranslated leader sequences contain a structurally complex region that facilitates ribosome binding (29). The minimal poliovirus IRES is about 450 nucleotides in length, containing regions that are predicted to be both structured and unstructured (35). Proteins have been reported to bind specifically to the IRES (36-39). These proteins may interact with the IRES and also with the ribosome to guide it to the internal AUG. Additional evidence suggesting a role for cellular factors comes from *in vitro* translation experiments. Poliovirus RNA translates poorly in reticulocyte lysates (40) but efficiently in HeLa cell extracts (28); however, supplementing reticulocyte lysate translation reactions with HeLa cell extracts enhances poliovirus RNA translation (40, 41). These results are consistent with the notion that cellular factors mediate internal initiation.

RNA -PROTEIN INTERACTIONS

<u>Informosomes and Ribonucleoprotein Particles</u> - Messenger RNAs are not naked cytoplasmic polynucleotide chains; rather, they bind cellular proteins to form ribonucleoprotein particles (42, 43). During the past few years, significant progress has been made toward understanding the sequence and structural features of RNAs and proteins that permit them to form specific RNA-protein complexes. The field is moving closer to elucidating high resolution structures of a larger number of RNAs and RNA-protein complexes. In the following paragraphs, several examples are described to illustrate current progress in characterizing specific RNA-protein or RNA-peptide interactions. In general, much less is known about RNA-protein interactions than DNA-protein interactions, and there is minimal information about specific protein binding to messenger RNAs. The examples cited below do not necessarily relate directly to translational control mechanisms, but rather demonstrate that progress is being made toward future analysis of the very large multi-subunit protein synthesis initiation factors and their complexes with messenger RNA.

<u>Translational Regulation and Iron Metabolism</u> - One of the most elegant examples of translational control of gene expression is the regulation of ferritin mRNA translation (reviewed in 44). Ferritin is an iron storage protein, and the 5' untranslated region of the ferritin mRNA contains a hairpin structure known as the iron-responsive element (IRE). The IRE is specifically bound by the IRE-binding protein (IRE-BP), and bound IRE-BP blocks the translation of ferritin mRNA. The understanding of ferritin regulation is significant not only for its relevance to translational regulation, but also for providing insight into the problem of how cells sense and respond to changes in environmental conditions. When circulating iron concentrations are low, the IRE-BP binds to the IRE, blocking ferritin mRNA translation. However, when iron concentrations increase, the IRE-BP is unable to bind the 5' end of ferritin mRNA and translation is unimpeded. What controls the reversible ability of the IRE-BP to bind RNA is not completely understood, but seems to relate to the redox state and the formation of an iron-sulfur complex with the protein (45-47).

<u>Classification of RNA Binding Proteins</u> - Molecular cloning, coupled with in vitro transcription of RNAs from cloned cDNAs, has enabled investigators to compare the sequences and biochemical characteristics of proteins that bind RNA. It is now clear that related sequence motifs unite many RNA binding proteins into classes. The first motif that was described is the RNA binding domain (RBD; now referred to as the RNA recognition motif, or RRM) that was identified in poly(A) binding protein (PABP) by Dreyfuss and colleagues (48). Initiation factor eIF-4B also contains an RRM (49). The PKR kinase described previously in this chapter has a double-stranded RNA binding motif that has also been identified in the Drosophila staufen protein (50) and the TAR binding protein (51). In the example of transcription factor IIIA (TFIIIA), which binds both DNA and RNA specifically, it appears that different sets of zinc fingers specify DNA or RNA binding (52). To date, at least nine classes of RNA binding proteins have been described (reviewed in 53).

<u>RNA-Peptide Interactions</u> - The arginine-rich class of RNA binding proteins (54) is a significant group that includes the Tat and Rev proteins of human immunodeficiency virus (HIV). An important aspect of the Tat/Rev analysis is that peptides--some fewer than twenty amino acids in length--have been shown to have nearly the same binding affinity and biological activity as the full-length proteins. Frankel and colleagues demonstrated that a single arginine residue surrounded by an environment of basic amino acids is sufficient for specific Tar RNA binding (55). Circular dichroism studies have shown that RNA-peptide interactions can alter RNA structure (56, 57). Significantly, the RNA binding site for a Rev peptide has been shown by RNAse protection analysis to be the same as that using the full-length protein (58). The peptide data demonstrate the compartmentalization of functionally significant domains in some proteins; moreover, the potential for generating high resolution structural data is increased by the use of small peptides, as compared to the full-length protein.

Structural Data - Far less is known about the high resolution structural details of RNA and RNA-protein complexes than the DNA counterparts (reviewed in 59, 60). One of the major challenges of research in this area is to understand the contacts between RNA and proteins that bind specifically. The most obvious differences between duplex RNA (A-form helix) and DNA (B-form helix) are that the major groove of DNA is wide and deep, but the major groove of RNA is narrow and deep, with a corresponding shallow but relatively wide minor groove (60). Unlike the DNA major groove, which has a relatively rich variety of functional groups available for protein interactions, the RNA major groove is too narrow to permit penetration by protein alpha-helices (61). Recent data suggest that distortions in the RNA structure such as bulge loops or non-Watson-Crick base pairing are necessary to widen the major groove and thereby expose contact areas (62, 63).

One of the most important determinants of specific RNA-protein interactions may be the inherent, sequence-dependent ability of an RNA structure to be deformed (61). An emerging theme from experimental analysis is that structural perturbations accompany specific RNA-protein interactions. The best-understood examples of RNA structure and RNA-protein complexes are transfer RNAs and transfer RNA-aminoacyl synthetase complexes (64-66). When the phosphate backbone of tRNA^{Gln} complexed to glutaminyl-tRNA synthetase is compared with that of uncomplexed yeast tRNA^{Phe}, two major differences are found (66). First, a base pair between nucleotides 1 and 72 is disrupted, and second, the anticodon loop opens out, allowing the bases of the anticodon to interact directly with the synthetase protein. Similarly, the binding of an arginine analog to HIV Tar RNA leads to significant changes in the structure of a bulge loop and the formation of a base triple interaction that stabilizes binding (67). Alterations in RNA conformation induced by peptide binding can be measured by circular dichroism (CD) (56, 68); however, CD data do not give many clues about the molecular details (*i.e.* changes in base pairing; altered base stacking) that underlie the overall changes in conformation. The significance of "deformability" may be in providing an element of specificity to RNA-protein interactions. In other words, a lower free energy cost for assuming a structure favorable for binding a protein or peptide would facilitate a specific RNA-protein interaction, while a high free energy cost would not.

There are, to date, no high resolution structure data for protein synthesis initiation factors, but the crystal structure of the N-terminal RNA binding domain of the U1 A small ribonuclear protein has been described (69, 70), as has the structure of protein synthesis elongation factor Tu (EF-Tu) (71). The RNA binding domain of the U1A protein is similar to the RRM found in eIF-4B (49); it contains two alpha-helices and a four-stranded ß-sheet. Loops between the ß strands and the alpha-helices form a pair of basic 'jaws' that may interact with RNA. The crystal structure of EF-Tu, a GTP binding protein, has been determined at 1.7 Å resolution (71). When GDP is exchanged for GTP, EF-Tu undergoes a conformational switch, and as a result, a cleft required for aminoacyl-tRNA binding is exposed. Despite the wealth of data provided by the biophysical analyses, however, progress toward understanding the functional significance of RNA-protein complexes will continue to require a combination of biochemical and structural studies.

In vitro Selection of Random RNAs - A recent technique, *in vitro* selection of random RNAs, is providing new insights into RNA-ligand interactions. The *in vitro* selection technique gained wide attention after reports appeared from two laboratories at about the same time (72, 73). The basic technique centers on the following steps: First, chemical DNA synthesis is used to produce a library of random sequence oligomers, each of which is preceded by the bacteriophage T7 polymerase promoter. Second, T7 RNA polymerase is added, resulting in the synthesis of a pool of RNAs containing random sequences, representing a theoretical complexity of up to 4^n different molecules, where n equals the number of nucleotides in the oligomer. Ligand (for example, an RNA binding protein), is then added and RNA-protein complexes are selected by a number of techniques including nitrocellulose filter binding or polyacrylamide gels. The selected RNAs are then extracted, converted to DNA using reverse transcriptase, amplified using polymerase chain reaction, re-transcribed into RNA, and then subjected to additional rounds of selection. After a number of iterations, the selected sequences are cloned and the nucleotide sequence is determined. Patterns in the selected RNAs illustrate sequences and/or structures that are important for binding the ligand.

In vitro selection has provided important new insights on the functional characterization of RNA-protein complexes. The method been used to isolate new ribozymes (74) as well as new ligands that bind the HIV Rev protein (75). The translational regulation of bacteriophage R17 coat protein expression has been studied, resulting in the identification of an RNA substrate that binds the coat protein with greater affinity than wild-type RNA (76). Single-stranded DNA molecules that bind thrombin were isolated (77), and this work raises the possibility that single-stranded nucleic acids may have potential utility as antagonist molecules. A three-dimensional structure for a selected thrombin-binding DNA was reported recently (78).

The protein equivalent of in vitro selection of random nucleic acid sequences is represented by the use of random peptide libraries that are expressed by bacteriophages (79, 80); (reviewed in 81). An important application of peptide libraries has been to identify antibody epitopes (79); however, the technology has been used recently to characterize nucleic acid binding peptides (82, 83).

Random RNA and random peptide libraries offer a nearly unbiased approach for defining the determinants required for specific intermolecular interactions. An added benefit is that there is good potential for identifying ligands that have greater affinity for the substrate than the known molecule. These data may be useful for the design of antagonists or other new therapeutic compounds.

A NEW IDENTITY FOR THE 3' UNTRANSLATED REGION

Until recently, the 3' untranslated region of messenger RNAs was thought to be a relatively benign stretch of nucleotides that were significant only in terms of influencing mRNA half-life. Several recent mini-reviews (84-87) address a range of important functions for the 3' UTR in cell differentiation, RNA localization, translational control and even tumor suppression.

Translational Control, Sex, and Development - Developmental biologists noted more than ten years ago that recruitment of maternal messenger RNA into translating polysomes was coupled with increases in the length of poly(A) tail (88); moreover, genetic evidence later suggested that translation initiation and the function of poly(A) binding protein were somehow related (89). The role of the 3' poly(A) tail in translation is still not clear, but experimental evidence generated using the free-living soil nematode C. elegans, strongly suggests that the 3' UTR is used as an important regulatory locus.

The sexual identity of C. elegans seems to be controlled at the translational level by a mechanism that involves a perfect direct repeat found in the 3' UTR of the tra -2 messenger RNA (90), which encodes a putative RNA binding protein (91). Mutations in the 3' direct repeat elements lead to increased tra -2 activity and association of the mRNA with larger polysomes. In addition, an RNA binding activity specific for the direct repeats was found in cell extracts. The data suggest that a protein factor may bind the direct repeats in the 3' end of the tra- 2 mRNA and inhibit translation by an unknown mechanism.

Possible Regulation by RNA-RNA Interactions - A role for 3' untranslated regions has recently been proposed to explain the temporal pattern of C. elegans cell lineage. A temporal gradient of the Lin-14 protein specifies the normal temporal sequence of cell lineages in this animal (92). lin-14 loss of function mutations cause premature execution of cell lineage patterns, while lin-14 gain of function mutations results in repeated patterns of cell lineage appropriate for an earlier developmental stage (93). Genetic analysis defined a second regulatory locus referred to as *lin* - 4; however, the negative regulator *lin -4* gene product does not seem to encode a protein (94). Unexpectedly, the functional products of the *lin*- 4 gene product are small (fewer than 65 nucleotides) RNAs; moreover, the *lin* -4 RNAs have antisense complementarity to repeated sequences found in the 3' UTR of *lin* -14 mRNA. A speculative model to explain the regulation of *lin* -14 mRNA translation states that the antisense *lin*-4 RNAs bind to the *lin-14* 3' UTR repeats and somehow block translation (in a manner perhaps analogous to the regulation proposed for the *tra* RNA described above). At this time, there is no direct evidence of pairing between the *lin-14* 3' UTR and small *lin* -4 RNAs.

Cytokine mRNA Translational Control - In 1986, a conserved AUUUA nucleotide sequence motif was described (95) with reference to mRNAs that encode inflammatory mediators, and the functional significance of this sequence element was also associated with RNA stability (96). A number of groups have described proteins that bind AUUUA elements (97-99), and the molecular cloning of one such factor has been reported (100). In addition to a possible role in regulating RNA stability, the AUUUA elements may also be important in regulating translation (101, 102); however, the mechanism is undefined.

Translational regulation of human interleukin-1ß (IL-1ß) mRNA expression has been reported in peripheral blood mononuclear cells stimulated by C5a, a component of the complement cascade (103, 104). In this system, IL-1ß mRNA is present, but neither intracellular proIL-1ß protein nor released mature IL-1ß is detectable. Despite the apparent absence of protein, the IL-1ß mRNA sediments with large polysomes, indicating that the translational block is downstream of the initiation step of protein synthesis (105). Based upon the fact that the IL-1ß mRNA has AUUUA elements that have been associated with a translational blockade (101), the regulatory mechanism may involve 3' UTR sequences.

Pseudoknots and mRNA Translation - The term 'pseudoknot' describes a structural motif that forms when the unpaired loop bases of a hairpin stem-loop hydrogen bond with unpaired bases upstream or downstream of the stem (106). Pseudoknots are important in translational control processes (reviewed in 107). Tang and Draper described a complex double pseudoknot structure that is recognized by a translational repressor (108). Pseudoknots are also found at the 3' end of some viral RNAs that lack a poly(A) tail, and recent evidence suggests that the 3' end of these RNAs may be functionally equivalent to a poly(A) tail (109).

Potential Long range Interactions Between 5' and 3' UTRs - Recent experimental data indicate that efficient translation of the naturally uncapped satellite tobacco necrosis virus RNA requires both the 5' and 3' UTR, leading to the suggestion that there may be long-range interactions that facilitate translation. Deletion of 5' or 3' UTR nucleotides diminished the *in vitro* translational efficiency of STNV RNA, but the efficiency was restored by adding a 5' m7GpppG cap structure (110). These data suggest that both the 5' and 3' UTRs of STNV RNA are essential for cap-independent translation. In the case of STNV RNA, the 3' UTR sequences required for efficient translation do not overlap with pseudoknot structures; however, two areas of potential base pairing between 5' and 3' ends have been proposed (111). The proposed mechanisms for translational enhancement by interacting 5' and 3' UTRs include the possibility that unfavorable secondary structures within the 5' leader could be masked or that the proximity of ends could facilitate reinitiation of ribosomes by bringing terminating ribosomes into close proximity with the initiation codon (111). In related findings, it has been proposed that ribosomes bypass regions of mRNAs and perhaps even shunt in *trans* between two separate RNA molecules (112).

RNA Localization - In the early 1980s, investigators described specific cytoplasmic localization of messenger RNAs (113) (reviewed in 114). Messenger RNA localization is apparently dependent upon nucleotide sequence elements that are found in the 3' UTR (reviewed in 114, 115). Specific cytolocalization of actin mRNA has been documented (116), and although it is likely that there is an interaction with cytoskeletal elements (117, 118), the details have not been clearly elucidated. The localization and transport of messenger RNAs may also have a translational control component because mRNA translation seems to be repressed during RNA transport, only to be reactivated upon delivery to the correct site (114, 119).

3' UTRs, Cell Differentiation, and Tumor Suppression - Using a genetic screen to identify regulators of cell growth and differentiation, Rastinejad and Blau found that the 3' untranslated regions of muscle structural genes inhibited cell division and promoted muscle cell differentiation (120). In a subsequent study, constitutive expression of a 200-nucleotide fragment of the alpha-tropomyosin 3' UTR suppressed anchorage-independent growth and tumor formation by a nondifferentiating mutant myogenic cell line (121). Control 3' UTR sequences had no effect, suggesting that RNA from the untranslated region of an mRNA can function as a tumor suppressor. These findings, coupled with the *lin*-14/*lin* 4 data (discussed above) suggest a new role for RNAs as "riboregulators" (121). The mechanism is, again, unknown, but the UTR may bind growth factors or regulators of cell growth and thereby limit access to their targets.

PERSPECTIVES

During the past few years, striking discoveries have been reported in the field of translational control, and these findings will significantly impact our overall understanding of regulated gene expression. The experimental evidence for many of the phenomena described in this chapter is compelling; however, there are significant gaps in knowledge because many of the mechanisms have not been elucidated. Significant data can be anticipated in several areas. As a result of recent progress in obtaining RNA crystals for X-ray diffraction analysis (122) and in the analysis of nucleic acids and nucleic acid-protein complexes by NMR (123), it is likely that our understanding of the structural details of RNA-protein or RNA-peptide complexes will increase substantially in the next few years. At the same time, however, the problem of understanding the higher order structure of large RNAs will continue to be imposing because of technical difficulties associated with experimental analysis and theoretical predictions for large molecules. Many laboratories are overexpressing proteins that are components of the translational machinery, and it is likely that additional examples oncoproteins will be described. To date, there have been few reports of the use of random nucleic acid or peptide libraries towards understanding translational regulation of gene expression; however, this is likely to change in the near future. Much of translational regulation of gene expression centers on RNA structure, RNA-protein and perhaps RNA-RNA interactions--subjects that are well suited for random sequence approaches.

Translation-level regulation of gene expression has often been considered to be the poor cousin of transcriptional control in terms of its significance for regulating gene expression. New data now demonstrate the flexibility and elegance of translational control mechanisms where the structural diversity of RNA molecules provides specific binding sites for proteins and where messenger RNAs are shuttled through the cytoplasm in silent form and somehow activated when they reach their destination. There is much to be learned about translational control, and the future will likely provide surprising answers to long-standing questions.

REFERENCES

1.   J.W.B. Hershey, Ann. Rev. Biochem., 60, 717 (1991).
2.   W.C. Merrick, Microbiol Rev, 56, 291 (1992).
3.   R.E. Rhoads, J Biol Chem, 268, 3017 (1993).
4.   J.A. Grifo, S.M. Tahara, M.A. Morgan, A.J. Shatkin and W.C. Merrick, J. Biol. Chem., 258, 5804 (1983).
5.   W. Rychlik, M.A. Russ and R.E. Rhoads, J. Biol. Chem., 262, 10434 (1987).
6.   A. Lazaris-Karatzas, K.S. Montine and N. Sonenberg, Nature, 345, 544 (1990).
7.   R.M. Frederickson and N. Sonenberg in "Translational Regulation of Gene Expression 2," Vol. J. Ilan, Ed., Plenum, New York, 1993, p. 143.

8.   A. Lazaris-Karatzas, M.R. Smith, R.M. Frederickson, M.L. Jaramillo, Y.L. Liu, H.F. Kung and N. Sonenberg, Gene Develop, 6, 1631 (1992).
9.   J. Pelletier and N. Sonenberg, Cell, 40, 515 (1985).
10.   A.E. Koromilas, A. Lazaris-Karatzas and N. Sonenberg, EMBO J, 11, 4153 (1992).
11.   C.E. Samuel, J. Biol. Chem., 268, 7603 (1993).
12.   J.-J. Chen, M.S. Throop, L. Gehrke, I. Kuo, J.K. Pal, M. Brodsky and I.M. London, Proc. Natl. Acad. Sci. USA, 88, 7729 (1991).
13.   E. Meurs, K. Chong, J. Galabru, N.S.B. Thomas, I.M. Kerr, B.R.G. Williams and A.G. Hovanessian, Cell, 62, 379 (1990).
14.   A.G. Hovanessian in "Translational Regulation of Gene Expression 2," Vol. J. Ilan, Ed., Plenum, New York, 1993, p. 163.
15.   J.-J. Chen in "Translational Regulation of Gene Expression 2," Vol. J. Ilan, Ed., Plenum, New York, 1993, p. 349.
16.   I.M. London, D.H. Levin, R.L. Matts, N.S.B. Thomas, R. Petryshyn and J.J. Chen, The Enzymes, XVII, 359 (1987).
17.   L. Manche, S.R. Green, C. Schmedt and M.B. Mathews, Mol Cell Biol, 12, 5238 (1992).
18.   I. Edery, R. Petryshyn and N. Sonenberg, Cell, 56, 303 (1988).
19.   D.N. SenGupta and R.H. Silverman, Nucl. Acids Res., 17, 969 (1989).
20.   E.F. Meurs, J. Galabru, G.N. Barber, M.G. Katze and A.G. Hovanessian, Proc Natl Acad Sci USA, 90, 232 (1993).
21.   A.W. Koromilas, S. Roy, G.N. Barber, M.G. Katze and N. Sonenberg, Science, 257, 1685 (1992).
22.   M. Kozak, Cell, 15, 1109 (1978).
23.   M. Kozak, Cell, 22, 7 (1980).
24.   M. Kozak, Proc. Natl. Acad. Sci. USA, 83, 2850 (1986).
25.   S.A. Jobling and L. Gehrke, Nature, 325, 622 (1987).
26.   D.R. Gallie, D.E. Sleat, J.W. Watts, P.C. Turner and T.M.A. Wilson, Nucl. Acids Res., 15, 8693 (1987).
27.   D. Etchison, S.C. Milburn, I. Edery, N. Sonenberg and J.W.B. Hershey, J. Biol. Chem., 257, 14806 (1982).
28.   N. Sonenberg, Adv. Virus Res., 33, 175 (1987).
29.   J. Pelletier and N. Sonenberg, Nature, 334, 320 (1988).
30.   R.J. Jackson, Nature, 353, 14 (1991).
31.   D.G. Macejak and P. Sarnow, Nature, 353, 90 (1991).
32.   S.K. Oh, M.P. Scott and P. Sarnow, Gene Develop, 6, 1643 (1992).
33.   J. Curran and D. Kolakofsky, 7, 2869 (1988).
34.   M.A. Adam, N. Ramesh, A.D. Miller and W.R.A. Osborne, J Virol, 65, 4985 (1991).
35.   R. Nicholson, J. Pelletier, S.Y. Le and N. Sonenberg, J Virol, 65, 5886 (1991).
36.   J.R. Gebhard and E. Ehrenfeld, J Virol, 66, 3101 (1992).
37.   N. Luz and E. Beck, J Virol, 65, 6486 (1991).
38.   S.K. Jang and E. Wimmer, Gene Develop, 4, 1560 (1990).
39.   K. Meerovitch, J. Pelletier and N. Sonenberg, Genes & Devel., 3, 1026 (1989).
40.   A.J. Dorner, B.L. Semler, R.J. Jackson, R. Hanecak, E. Duprey and E. Wimmer, J. Virol., 50, 507 (1984).
41.   B.A. Phillips and A. Emmert, Virology, 148, 255 (1986).
42.   A.S. Spirin in "Current topics in developmental biology," Vol. 1, A. A. Moscona and A. Monroy, Ed., Academic Press, New York, 1966, p. 1.
43.   R.E. Swiderski and J.D. Richter, Devel. Biol., 128, 349 (1988).
44.   R.D. Klausner, T.A. Rouault and J.B. Harford, Cell, 72, 19 (1993).
45.   R.D. Klausner and J.B. Harford, Science, 246, 870 (1989).
46.   D.J. Haile, T.A. Rouault, J.B. Harford, M.C. Kennedy, G.A. Blondin, H. Beinert and R.D. Klausner, Proc Natl Acad Sci USA, 89, 11735 (1992).
47.   C.C. Philpott, D. Haile, T.A. Rouault and R.D. Klausner, J Biol Chem, 268, 17655 (1993).
48.   S.A. Adam, T. Nakagawa, M.S. Swanson, T.K. Woodruff and G. Dreyfuss, Mol. Cell. Biol., 6, 2932 (1986).
49.   S.C. Milburn, J.W.B. Hershey, M.V. Davies, K. Kelleher and R.J. Kaufman, EMBO J, 9, 2783 (1990).
50.   D. St. Johnston, N.H. Brown, J.G. Gall and M. Jantsch, Proc Natl Acad Sci USA, 89, 10979 (1992).
51.   A. Gatignol, C. Buckler and K.T. Jeang, Mol Cell Biol, 13, 2193 (1993).
52.   K.R. Clemens, V. Wolf, S.J. McBryant, P. Zhang, X. Liao, P.E. Wright and J.M. Gottesfeld, Science, 260, 530 (1993).
53.   I.W. Mattaj, Cell, 73, 837 (1993).
54.   D. Lazinski, E. Grzadzielska and A. Das, Cell, 59, 207 (1989).
55.   B.J. Calnan, S. Biancalana, D. Hudson and A.D. Frankel, Genes & Dev., 5, 201 (1991).
56.   M. Baer, F. Houser, L.S. Loesch-Fries and L. Gehrke, EMBO J., 13, 727 (1994).
57.   R.Y. Tan and A.D. Frankel, Biochemistry, 31, 10288 (1992).
58.   J. Kjems, B.J. Calnan, A.D. Frankel and P.A. Sharp, EMBO J., 11, 1119 (1991).
59.   J.A. Jaeger, J. Santalucia and I. Tinoco, Annual Rev Biochem, 62, 255 (1993).
60.   T.A. Steitz in "The RNA World," Vol. R. F. Gesteland and J. F. Atkins, Ed., Cold Spring Harbor Laboratory Press, Cold Spring Harbor, 1993, p. 219.
61.   T.A. Steitz, Q. Rev. Biophys., 23, 205 (1990).
62.   K.M. Weeks and D.M. Crothers, Biochemistry, 31, 10281 (1992).
63.   J.L. Battiste, R.Y. Tan, A.D. Frankel and J.R. Williamson, Biochemistry, 33, 2741 (1994).
64.   M. Ruff, S. Krishnaswamy, M. Boeglin, A. Poterszman, A. Mitschler, A. Podjarny, B. Rees, J.C. Thierry and D. Moras, Science, 252, 1682 (1991).
65.   J.J. Perona, R.N. Swanson, M.A. Rould, T.A. Steitz and D. Soll, Science, 246, 1152 (1989).
66.   M.A. Rould, J.J. Perona, D. Soll and T.A. Steitz, Science, 246, 1135 (1989).
67.   J.D. Puglisi, R. Tan, B.J. Calnan, A.D. Frankel and J.R. Williamson, Science, 257, 76 (1992).
68.   T.J. Daly, J.R. Rusche, T.E. Maione and A.D. Frankel, Biochemistry, 29, 9791 (1990).

69.   P.R. Evans, C. Oubridge, T.H. Jessen, J. Li, C.H. Teo, C. Pritchard, N. Ito and K. Nagai, Biochem Soc Trans, 21, 605 (1993).
70.   K. Nagai, C. Oubridge, T.H. Jessen, J. Li and P.R. Evans, Nature, 348, 515 (1990).
71.   H. Berchtold, L. Reshetnikova, C.O.A. Reiser, N.K. Schirmer, M. Sprinzl and R. Hilgenfeld, Nature, 365, 126 (1993).
72.   A.D. Ellington and J.W. Szostak, Nature, 346, 818 (1990).
73.   C. Tuerk and L. Gold, Science, 249, 505 (1990).
74.   D.P. Bartel and J.W. Szostak, Science, 261, 1411 (1993).
75.   K.B. Jensen, L. Green, S. Macdougalwaugh and C. Tuerk, J Mol Biol, 235, 237 (1994).
76.   D. Schneider, C. Tuerk and L. Gold, J Mol Biol, 228, 862 (1992).
77.   L.C. Bock, L.C. Griffin, J.A. Latham, E.H. Vermass and J.J. Toole, Nature, 355, 564 (1992).
78.   P. Schultze, R.F. Macaya and J. Feigon, J Mol Biol, 235, 1532 (1994).
79.   S.F. Parmley and G.P. Smith, Gene, 73, 305 (1988).
80.   S.E. Cwirla, E.A. Peters, R.W. Barrett and W.J. Dower, Proc. Natl. Acad. Sci. USA, 87, 6378 (1990).
81.   J.K. Scott, Trends Biochem. Sci., 17, 241 (1992).
82.   E.J. Rebar and C.O. Pabo, Science, 263, 671 (1994).
83.   A.C. Jamieson, S.-H. Kim and J.A. Wells, Biochemistry, in press, (1994).
84.   M. Wickens, Nature, 363, 305 (1993).
85.   R.J. Jackson, Cell, 74, 9 (1993).
86.   M. Wickens and K. Takayama, Nature, 367, 17 (1994).
87.   N. Standart, Sem. Dev. Biol., 3, 367 (1993).
88.   E. Rosenthal, T. Tansey and J. Ruderman, J. Mol. Biol., 166, 309 (1983).
89.   A.B. Sachs and R.W. Davis, Cell, 58, 857 (1989).
90.   E.B. Goodwin, P.G. Okkema, T.C. Evans and J. Kimble, Cell, 75, 329 (1993).
91.   H. Amrein, M. Gorman and R. Nothiger, Cell, 55, 1025 (1988).
92.   G. Ruvkun and J. Giusto, Nature, 338, 857 (1989).
93.   V. Ambros and H.R. Horvitz, Gene Dev., 1, 389 (1987).
94.   B. Wightman, I. Ha and G. Ruvkun, Cell, 75, 855 (1993).
95.   G. Shaw and R. Kamen, Cell, 46, 659 (1986).
96.   S. Savantbhonsale and D.W. Cleveland, Gene Develop, 6, 1927 (1992).
97.   J.S. Malter, Science, 246, 664 (1989).
98.   P.R. Bohjanen, B. Petryniak, C.H. June, C.B. Thompson and T. Lindsten, Mol Cell Biol, 11, 3288 (1991).
99.   N.B.K. Raj and P.M. Pitha, FASEB J, 7, 702 (1993).
100.   W. Zhang, B.J. Wagner, K. Ehrenman, A.W. Schaefer, C.T. Demaria, D. Crater, K. Dehaven, L. Long and G. Brewer, Mol Cell Biol, 13, 7652 (1993).
101.   V. Kruys, O. Marinx, G. Shaw, J. Deschamps and G. Huez, Science, 245, 852 (1989).
102.   V. Kruys, M. Wathelet, P. Poupart, R. Contreras, W. Fiers, W. Content and G. Huez, Proc. Natl. Acad. Sci. USA, 84, 6030 (1987).
103.   C.A. Dinarello and R. Schindler, Cytokines and Lipocortins in Inflammation and Differentiation, 349, 195 (1990).
104.   T. Geiger, C. Rordorf, N. Galakatos, B. Seligmann, R. Henn, J. Lazdins, F. Erard and K. Vosbeck, Lymphokine Cytokine Res, 11, 55 (1992).
105.   R.L. Kaspar and L. Gehrke, J. Immunol., in press, (1994).
106.   C.W.A. Pleij, K. Rietveld and L. Bosch, Nucl. Acids Res., 13, 1717 (1985).
107.   P. Schimmel, Cell, 58, 9 (1989).
108.   C.K. Tang and D.E. Draper, Cell, 57, 531 (1989).
109.   D.R. Gallie and V. Walbot, Genes & Devel., 4, 1149 (1990).
110.   R.T. Timmer, L.A. Benkowski, D. Schodin, S.R. Lax, A.M. Metz, J.M. Ravel and K.S. Browning, J Biol Chem, 268, 9504 (1993).
111.   X. Danthinne, J. Seurinck, F. Meulewaeter, M. Vanmontagu and M. Cornelissen, Mol Cell Biol, 13, 3340 (1993).
112.   J. Fütterer, Z. Kiss-Laszlo and T. Hohn, Cell, 73, 789 (1993).
113.   W.R. Jeffery, C.R. Tomlinson and R.D. Brodeur, Dev. Biol., 99, 408 (1983).
114.   J.E. Wilhelm and R.D. Vale, J Cell Biol, 123, 269 (1993).
115.   R.H. Singer, J Cell Biochem, 52, 125 (1993).
116.   E.H. Kislauskis, Z.F. Li, R.H. Singer and K.L. Taneja, J Cell Biol, 123, 165 (1993).
117.   D.L. Gard, Dev. Biol., 143, 346 (1991).
118.   W.E. Theurkauf, S. Smiley, M.L. Wong and B.M. Alberts, Development, 115, 923 (1992).
119.   T.C. Evans, S.L. Crittenden, V. Kodoyianni and J. Kimble, Cell, 77, 183 (1994).
120.   F. Rastinejad and H.M. Blau, Cell, 72, 903 (1993).
121.   F. Rastinejad, M.J. Conboy, T.A. Rando and H.M. Blau, Cell, 75, 1107 (1993).
122.   J.A. Doudna, C. Grosshans, A. Gooding and C.E. Kundrot, Proc. Natl. Acad. Sci. USA, 90, 7829 (1993).
123.   G. Wagner, S.G. Hyberts and T.F. Havel, Ann. Rev. Biophys. Biomolec. Structure, 21, 167 (1992).

Chapter 26. Protein Kinases and Phosphatases: Structural Biology and Synthetic Inhibitors

Kenneth J Murray and William J Coates, SmithKline Beecham Pharmaceuticals, The Frythe, Welwyn, Herts Al6 9AR, UK.

Introduction - The reversible covalent phosphorylation of proteins was discovered in 1955 and now represents a well established mechanism for the regulation of biological function (1). A large number of proteins are known to exist in phospho- and dephospho- forms; the exact effect of phosphorylation varies from protein to protein, but changes in enzyme kinetics and ligand binding are common responses. Ultimately the level of phosphoryation of any protein is determined by the activity of two classes of enzyme, the protein kinases, which catalyse the transfer of the terminal phosphoryl moiety of ATP (or very occasionally GTP) to the acceptor amino acid, and the phosphoprotein phosphatases which remove the phosphate. It is common for a protein to be acted on by a number of protein kinases either at different sites or at the same amino acid residue (2). Largely as a result of molecular biology, the number of known protein kinases and phosphatases is rapidly increasing (3,4), and information regarding their cellular function is also being generated quickly. Two features of protein kinases relevant to medicinal chemistry, the crystal structure and synthetic inhibitor molecules are reviewed, and a brief overview of phosphatases is presented.

PROTEIN KINASE STRUCTURE

Classification - In eukaryotes, the vast majority of protein kinases form an acid stable phosphate ester with the three hydoxyamino acids serine, threonine, and tyrosine. These protein kinases are the subject of this review. However, acid labile phosphoramidates are also known to occur on the basic amino acids histidine, lysine, and arginine (2,5,6). There are well characterised protein kinases that form a mixed phosphate-carboxylate acid anhydride with aspartyl residues that play a role in bacterial signal transduction; initially autophosphoryation on histidine residues occurs, followed by transfer of the phosphoryl group to an aspartyl residue (7). Traditionally, the protein kinases acting on serine, threonine, and tyrosine have been split into two families; the serine/ threonine PrKs (S/T-PrKs) and the tyrosine protein kinases (Y-PrKs), and this nomenclature will be followed here. However, while the homology between the catalytic domains of the two families does form a basis for this classification, (8,9) it is not absolute. For example, there are S/T-PrKs that do not phosphorylate both amino acids (2), and "dual specificity" protein kinases that phosphorylate all three residues have recently been described (10).

Catalytic Domain - Both S/T-PrKs and Y-PrKs share a common catalytic domain of 250-300 amino acids, although the rest of the molecule shows a wide variation in primary structure.

The catalytic domain consists of 11 highly conserved motifs (or subdomains) interspersed with regions of sequence that show reduced conservation (see Fig. 1). However, S/T-PrKs show specific differences from Y-PrKs in sequences located in two of the subdomains (VIb and VIII). Alignment of the catalytic site sequences allows PrKs to be grouped in 10 families of S/T-PrKs and 11 families of Y-PrKs; members of a family tend to share either a common mode of activation or substrate specificity (8,9).

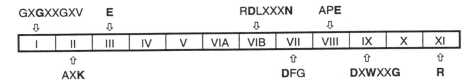

Fig. 1. The domain structure of the catalytic core of protein kinases.

The position of invariant (shown in bold) and highly conserved residues is indicated. For further details see (2, 8, 9). The one-letter amino acid code is used and X represents any amino acid.

Crystal Structure of Protein Kinases - To date, the crystal structure of three protein kinases has been determined. Most studied is the catalytic subunit of the cAMP-dependent protein kinase (cA-PrK), which is active in its monomeric state; physiological inhibition occurs by binding to the regulatory subunit. The cA-PrK catalytic subunit has a bilobal structure, with the carboxy terminus forming the larger lobe that interacts with the peptide substrate; the residues involved in catalysis are also found in this lobe (11,12). ATP binds to the smaller amino terminal lobe with the γ-phosphate positioned for transfer, allowing catalysis to take place in the cleft between the two lobes. It is thought that binding of the substrates results in closure of the cleft (13-15). In the smaller lobe, the glycine rich loop (subdomain I) provides backbone amides that hydrogen bond to the β-phosphate. Ion pairing occurs between the conserved lysine (II) and the α- and β-phosphates; an ionic interaction between the lysine and glutamate (III) serves to fix the lysine in place (13-15). Mutation of this lysine is known to lead to loss of catalytic activity in a number of protein kinases, including pp56[lck], although in this case the binding of ATP itself was unaffected (16).

Two loops of the large lobe, lying at the surface of the cleft where catalysis is thought to take place, contain the conserved residues. The aspartate in subdomain VIb acts as the catalytic base and is held in place by the asparigine in this subdomain. The Mg^{2+} ion that interacts with the α- and β- phosphates is chelated by the aspartate in subdomain IX. The remaining conserved residues [glutamate (VIII), aspartate (IX) and arginine (XI)] do not appear to interact directly with either of the substrates, but presumably are essential to maintaining the structure of the lobe. Three further residues that are not totally conserved also play a role in this lobe. The lysine in subdomain VIb is conserved in all S/T-PrKs and interacts with the γ- phosphate of ATP. The highly conserved arginine (VIb) interacts with the phospho-threonine between subdomains VII and VIII. The phosphorylation of this threonine is unusual in that it occurs soon after translation and is turned over extremely slowly;

therefore, the enzyme normally contains this phospho-threonine both in the cell and when purified. The kinase has now been purified with and without phosphate at this residue, and the results show that the phospho-form has a considerably lower Km for both substrates (17). The crystal structure has also revealed interactions with the peptide inhibitor (11,12,14,18), helped to define a new conserved motif (19) and provided a model for the Y-PrKs (20).

The other two protein kinases have been crystallised in catalytically inactive conformations. Comparison of their structure with that of the cA-PrK catalytic subunit described above has suggested molecular mechanisms that regulate the activity of these kinases. The mitogen activated protein (MAP) kinases form an expanding family of protein kinases that are activated in cells in response to growth factors and many other agents (21-25). The distinguishing feature of MAP kinases is that to obtain full activity they require phosphorylation on both the threonine and tyrosine residues found in the motif TXY (where X is usually E), located between subdomains VII and VIII; note that this position is analogous to the stable phosphorylation site found in the cA-PrK catalytic subunit. In cells MAP kinase is part of a protein kinase cascade and is activated by a dual specificity upstream kinase (termed MAP kinase kinase or MEK) that phosphorylates MAP kinase on both residues; in turn, MAP kinase kinase is activated by further upstream kinases including raf, mos, and an activity termed MAP kinase kinase kinase (23).

A member of the MAP kinase family, extracellular regulated kinase-2 (ERK2) has been crystallised in its inactive de-phospho form (26) and from the derived structure it appears that both steric interference with protein substrate binding and misalignment of key catalytic residues may contribute to the lack of activity. It is proposed that this form of ERK2 is locked in an open conformation with which ATP can bind but without forming the Michaelis-like complex seen in cA-PrK catalytic subunit; for example, the close interactions of subdomain I with the α- and β- phosphates are not present (26). In ERK2 the loop containing the TEY motif is in a position similar to the residues in the cA-PrK catalytic subunit that interact with the peptide inhibitor; it has, therefore, been suggested that this region may sterically block the binding of protein substrates, which is relieved on phosphorylation of the enzyme (26).

Cyclin-dependent protein kinases (CDKs) play a major role in the control of the eukaryotic cell cycle (27,28). CDKs have a molecular weight of approximately 34,000, thus containing barely more than the protein kinase catalytic core and are regulated by two mechanisms. The first step is the interaction between CDK and one of the cyclins, a family of cell regulated proteins, resulting in a complex that has some kinase activity (27,28). However, as with MAP kinase, the action of another protein kinase is required for full activation of CDKs. In the case of CDK2 one such upstream kinase has been identified as p40[MO15] (29), which phosphorylates CDK2 on a single threonine residue. Like ERK2, the crystal structure of the inactive CDK2 has been solved (30), and similar, but not identical mechanisms operate to suppress kinase activity. Again, the loop between subdomains VII and VIII (which contains the phosphorylated threonine) appears to be in a position to block protein substrate binding, whereas the ATP is bound in a conformation that does not allow transfer to the attacking hydroxyl (12,30).

PROTEIN KINASE INHIBITORS

Protein kinase C - The protein kinase C (PrKC) family currently consists of 12 proteins denoted by Greek letters (31-36). Individual isoforms are well conserved over a number of mammalian species, suggesting that there is a physiological reason for this apparent plethora of PrKCs, and recent reports provide at least a partial explanation for their existence (31). Studies of the individual isolated enzymes have revealed three classes: (i) the "conventional" Ca^{2+}-dependent, diacylglycerol(DAG)/phorbol ester(PE) activated PrKCs (α, β, γ), (ii) the "novel" Ca^{2+}-independent DAG/PE activated forms (δ,ε,ζ, η,) and (iii) the "atypical" PrKCs (ι, λ, μ),which are not activated by DAG or PEs (31). The properties of the last group have not yet been fully characterised. The individual PrKCs also show differences in their specificity with respect to protein substrates and models to explain this (37,38) and other features of the active site (39,40) have been proposed. In addition to differences in their regulatory and catalytic properties, the individual PrKCs also show a varying tissue distribution and subcellular localisation (31); the latter may be due to intracellular receptors for PrKC, which have been shown to be homologous to the G-protein β-subunit (41). Taken together, these observations provide a theoretical basis for PrKC isoenzymes having distinct physiological roles, and there is now evidence for this from intact cell studies (31).

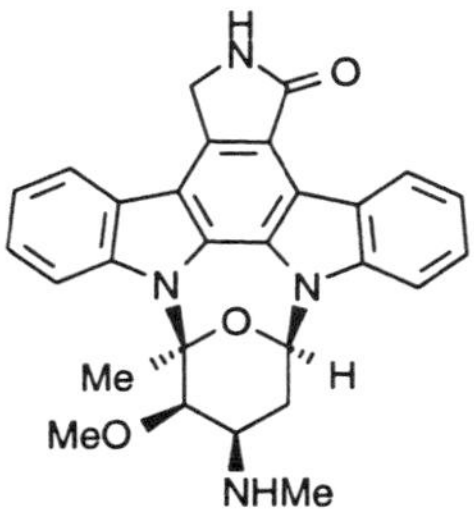

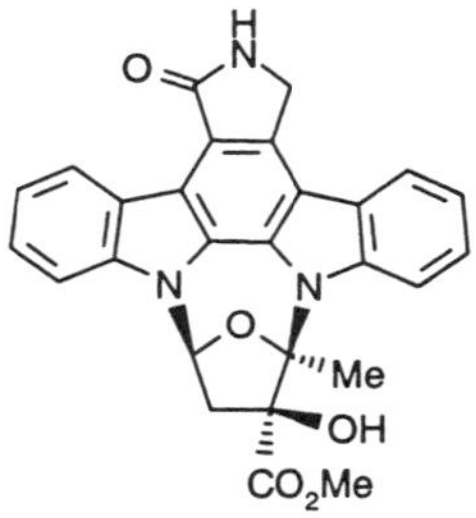

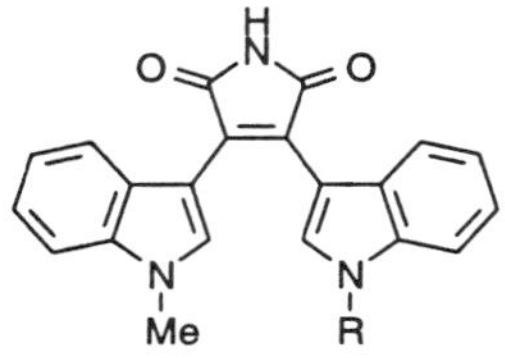

R=
3, Me
4, Pr
5, H$_2$NPr
6, H$_2$NNH=CSPr
7, H$_2$NO$_2$NN=CNHPr

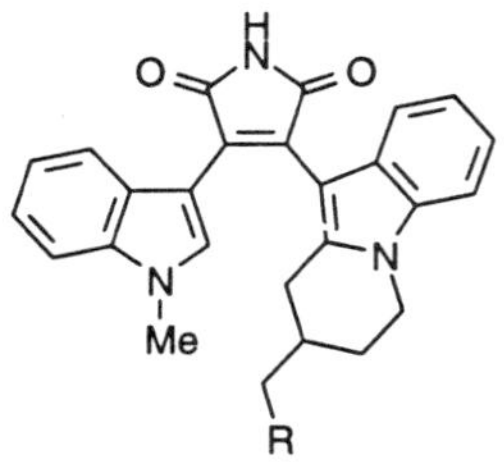

8 R=NH2; **9** R=NMe2 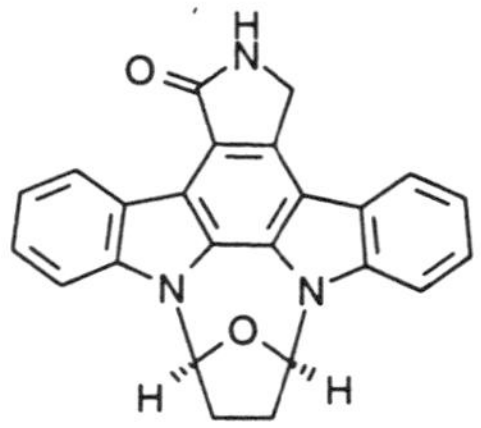   **10** 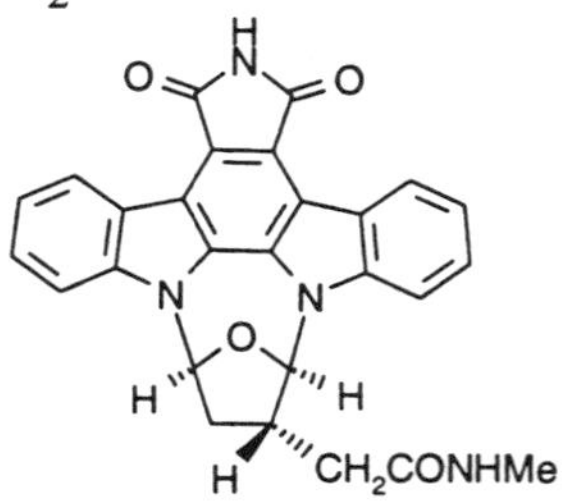   **11**

Inhibitors of PrKC with improved selectivity have been designed starting from the non selective microbial products staurosporine (**1**, see (42) for absolute stereochemistry) and K-252a (**2**). Thus, bisindolylmaleimides (BIMs, e.g., **3**, **4**) retain some of the potency against

PrKC of staurosporine but are more selective with respect to cA-PrK (43). With a few exceptions (e.g., 1-naphthyl, 2-chlorophenyl) replacement of one indole ring by an alternative arene resulted in loss of activity. Inhibition by BIMs was greatly increased by introducing a basic centre that could interact at the putative methylamino binding site of staurosporine, as in compounds **5** - **7**. These compounds inhibit PrKC in cells, but the less active amine **5** showed a better pharmacokinetic profile (44). Using **5** as a lead, conformational restriction of the amino function gave, for example **8** and **9**, which have potency comparable to staurosporine but with improved selectivity; compound **9** inhibited inflammatory responses in two animal models (45). The S enantiomer of **8** was more potent than the R as a PrKC inhibitor. Using a different approach (46), it has been shown that the core structure of K-252a (**10**) is a more potent PrKC inhibitor than K-252a itself; from a series of derivatives the amide **11** is the most potent. The furan ring is tolerant of substitution by neutral, acidic, and basic groups, which allows scope for modification of overall physicochemical properties. An unexpected rearrangement reaction was used (47) to synthesise a novel potent PrKC inhibitor **12**. PrKC inhibition of the staurosporine aglycone has been confirmed, and it is reported that substitution of one indole NH by a polar alkyl group increased potency; the cyanoethyl compound Gö 6976 (**13**) is a potent inhibitor of PrKC with >100 fold selectivity with respect to a number of other kinases (48). The potency and selectivity of a series of staurosporine derivatives has been reported (49). The PrKC isoenzyme selectivity of some of these compounds has been described; **4-10** showed little discrimination between the isoforms (50), whereas **13** inhibited conventional PrKCs but not the novel isoforms (51).

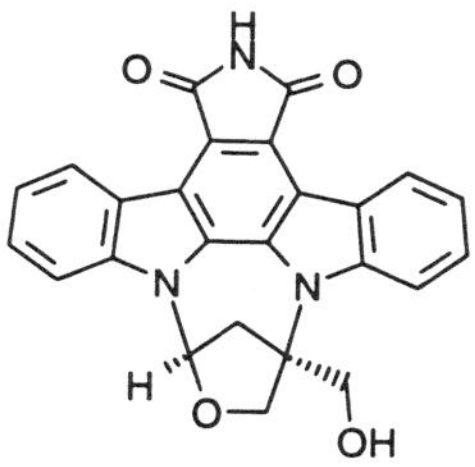

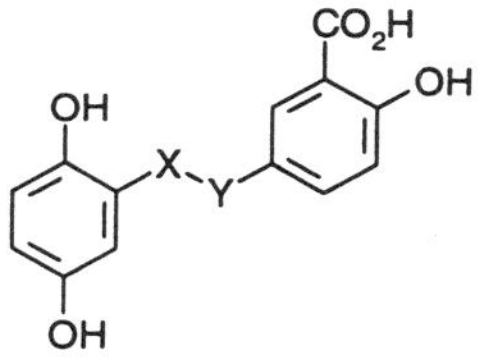

XY =
14, CH=CH
15, CH$_2$CH$_2$
16, CH$_2$NH
17, CH=N

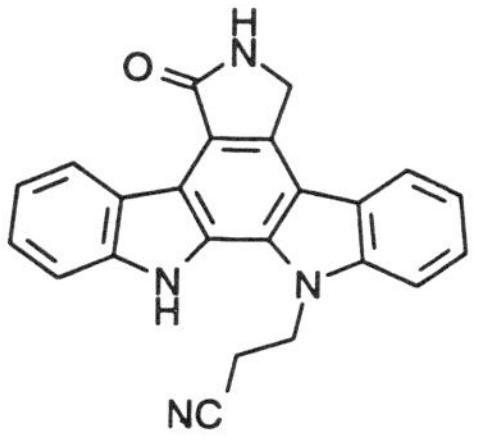

12   **13**

Tyrosine Kinase Inhibitors - Tyrosine protein kinases (Y-PrKs) and their inhibitors are reviewed in (52, 53). The salicyl stilbene derivative **14**, an intermediate in the synthesis of the carbon analogue **15** of the lavendustin A pharmacophore (**16**), shows remarkable potency, especially compared to the imine **17** (54). The importance of the hydroquinone and salicyl residues of **14** has been confirmed (55), and either different sites of binding or binding of a 'cis' conformation of **14** has been suggested to explain the marked difference in activity between **14** and **17**. Lack of cell penetration by **14** is consistent with its poor activity in intact cells compared with the much greater activity of the methyl ester despite the lower potency of the latter against the enzyme. Rigid analogues of **14**, e.g. **18**, are inactive (56). SAR of a series of indolinethione alkanoic acid derivatives **19** is difficult to delineate in part because of thione/disulphide interconversion (57). The more active compounds appear to be the disulphides of the propanoic acids which are competitive with neither ATP nor substrate. All of the active compounds inhibit epidermal growth factor (EGF) stimulated cell growth, but again acids and esters show an inverse relationship for enzyme and whole cell activity consistent with poor cell penetration by the free acids. A further series of tyrphostins (**20**) have been described (58) in which a substituted thiomethyl group effectively replaces one of the hydroxy groups. A range of thiol derivatives have been investigated, but none of them were found to be better than benzylthiomethyl (**20**, R = PhCH$_2$). The differences seen between inhibition of EGFR and HER 1-2 autophosphorylation do not correlate with the antiproliferative activity of these compounds. While breaking the biindolyl bond of the staurosporine-like aglycone **21** gives PrKC inhibitory BIMs, dissection of the indole 3,4 bonds gives dianilinophthalimides, e.g., **22**, which is a potent and selective inhibitor of EGF receptor kinase (59,60). Even trivial structural modifications lead to loss of activity in this series; the asymmetric propeller conformation of **22** is invoked to explain the contrast in activity between BIMs and (**21**).

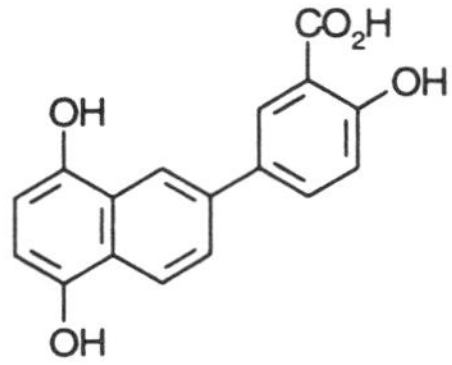

18 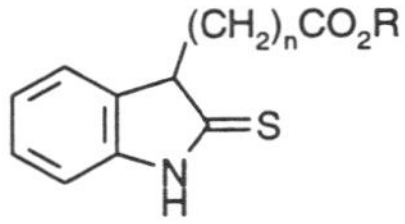   **19** 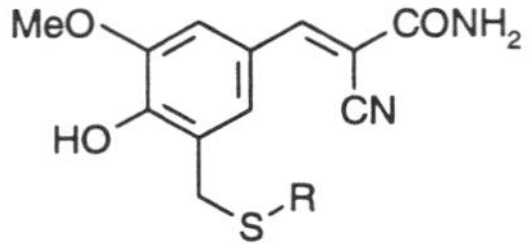   **20**

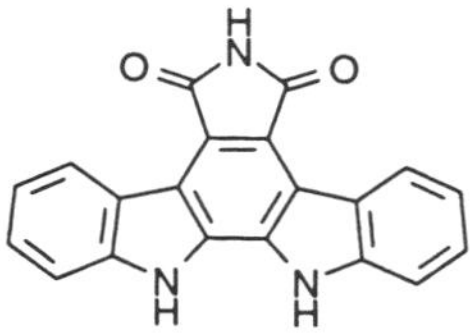

21 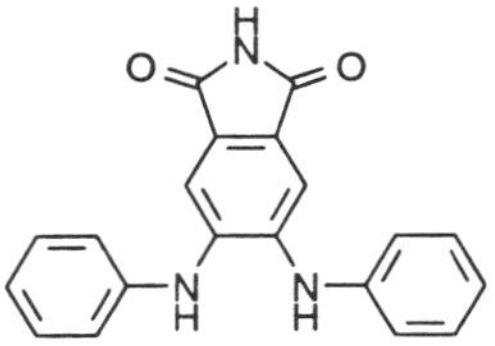   **22**

PHOSPHOPROTEIN PHOSPHATASES

Phosphoserine/threonine Phosphatases (PS/TPs) - The PS/TPs have been classified into four groups, PP1, PP2A, PP2B and PP2C, according to their substrate specifity, metal ion dependence and sensitivity to inhibition (61). PP1 and PP2A are both active independent of metal ions and are potently inhibited by microcystin LR; the protein inhibitors 1 and 2 act only on PP1, whereas okadaic acid is a more potent inhibitor of PP2A (62). PP1 is found as a complex in which the catalytic subunit (PP1C) is bound to regulatory targeting proteins that direct the subcellular localisation and modify the phosphatase activity (63). The protein inhibitor 2 has been reported to be required for the correct folding of PP1C, and may also act as an intracellular store for the catalytic subunit (64). Tyrosine phosphorylation inactivates PP2A and occurs in cells (65). The activity of PP2B, also known as calcineurin, is dependent on the Ca^{2+}-calmodulin complex and is only weakly inhibited by microcystin, okadaic acid, and the protein inhibitors. However, T cell PP2B is inhibited by the immunosuppressants cyclosporin and FK506 when these agents are bound to their respective immunophilins (66).

The final class, PP2C, is Mg^{2+}-stimulated but insensitive to any of the inhibitors described. The alignment of the protein sequences of 44 PS/TPs reveals a 280 residue catalytic core containing a number of invariant positions, which is common to members of the PP1, PP2A, and PP2B families, but not to PP2C. It is predicted that this core, which shows similarity to E. coli diadenosine tetraphophatase, consists of two domains: one being largely α-helical, and the other has a single β-sheet surrounded by α-helices (67). Co-crystals of microcystin LR and PP1γ have been reported (68).

Phosphotyrosine Phosphatases (PTPs) - Over 40 PTPs have been characterised (69-72); the majority are cytoplasmic enzymes while the remaining, of which CD45 (73) is an example, are transmembrane molecules with receptor-like domains. There is a 230 amino acid catalytic domain common to all PTPs and, in contrast to protein kinases, there is no homology with PS/TPs (70,71). In addition, various PTPs contain other well defined domains (see below) which in the majority of transmembrane enzymes includes a second, tandem catalytic domain. The consensus motif (I/V)HCXAGXXR(S/T)G is found in the catalytic domain, with the essential cysteinyl residue forming a thiol-phosphate intermediate during general acid-general base catalysis (70,74). The crystal structure of a cytoplasmic PTP (the truncated form of PTP1B) has been solved, revealing a single domain structure with the eleven-residue motif forming a catalytic site located at the base of a shallow cleft. Tungstate, a potent inhibitor of PTP1B, was used as a substrate analogue to indicate the phosphotyrosine binding site, and it has been suggested that the active site cleft is too deep to allow penetration of the smaller phosphoserine or phosphothreonine. The surface around the cleft is open and could allow a number of forms of protein-protein interaction, which could explain the lack of selectivity of PTPs for particular phosphotyrosine-containing proteins (69).

Surrounding the catalytic core of PTPs there are a range of domains that appear to be important for subcellular localisation and subsequent regulation of enzymatic activity (75). The full-length PTP1B contains a carboxy-terminal sequence that directs it to the endoplasmic reticulum; on agonist stimulation in platelets this is cleaved giving a cytoplasmic PTP1B with increased activity (76). The interactions of proteins containing src homology 2 (SH2) domains with the phosphotyrosines found in activated growth factor receptors (GFRs) is a well established form of intracellular signalling (77), and a number of PTPs (e.g., SHP, SYP) are known to contain such domains. SHP and SYP can bind to a number of activated GFRs, and although their exact role in the complex is not clear, it appears they can act to potentiate rather than antagonise the effects of tyrosine kinase activation (72,75). For example, the corkscrew protein (the Drosophila homologue of SYP) acts as a positive regulator of the Y-PrK Torso (78). Two mechanisms have been proposed to explain this: 1. the PTP activates a member of the Src Y-PrK family by removing an inhibitory phosphate (72); and 2. SYP acts as an adapter molecule that allows the ras-activating complex GRB2-Sos to bind to activated platelet-derived GFR (79). Recent observations indicate that the SH2 domains may also play a direct role in regulating the catalytic activity of SYP (80-83). In addition to membrane targeting signals, PTPs are also known with nuclear localisation and cytoskeletal association domains and it has been suggested that the correct intracellular localisation of PTPs is important in restricting their substrate specificity (75).

Dual Specificity Phosphatases (DSPs) - DSPs are named after their ability to dephosphorylate phosphoserine and/or phosphothreonine in addition to phosphotyrosine in protein substrates; this group has also been termed VH1 phosphatases after the vaccina virus H1 enzyme (70), and similar DSPs have been found in other viruses (84). DSPs are members of the PTP family, as their catalytic domains contain the HCXAGXXR motif (70): like PTPs (85) they are inhibited by vanadate but not by agents that inhibit PS/TPs. Several members of this family have now been characterised; all of the genes are induced by growth factors and similar agents (86). Three DSPs can dephosphorylate and inactivate active MAPK by removing phosphate from both threonine and tyrosine residues. The growth factor -inducible immediate-early gene 3CH134 (also cloned as erp (87)) of mouse fibroblasts encodes a DSP that shows a 15-fold selectivity for MAPK over artificial substrates (88), whereas the human homologues CL100 (89) and HVH1 (90) only effectively dephosphorylated MAPK. However, there is no correlation between the distribution of CL100 and MAPK in the brain, suggesting that it may have further physiological substrates (91). PAC-1, a mitogen activated gene closely related to CL100, has been cloned from human T cells, although the substrate specificity of the gene product has not been reported (92). A mouse DSP, cdc25M2 (93), is reported to dephosphorylate CDK and a further enzyme, termed KAP, can also interact with CDKs (94).

References

1. L.N. Johnson and D. Barford, Annu. Rev. Biophys. Biomol. Struct., 22, 199 (1993).
2. D.P. Leader, Pharmac. Ther., 59, 343 (1993).
3. T. Hunter, Cell, 50, 823 (1987).
4. N. Perrimon, Curr. Op. Cell Biol., 6, 260 (1994).
5. B.T. Wakim and G.D. Aswad, J. Biol. Chem., 269, 2722 (1994).
6. K. Motojima and S. Goto, J. Biol. Chem., 269, 9030 (1994)
7. R.V. Swanson and M.I. Simon, Curr. Biol., 4, 234 (1994).
8. S.K. Hanks, A.M. Quinn and T. Hunter, Science, 241, 42 (1988).
9. S.K. Hanks, Curr. Op. Struct. Biol., 1, 369 (1991).
10. E. Douville, P. Duncan., N. Abraham and J.C. Bell, Cancer Metastasis Revs., 13, 1 (1994)
11. S.S. Taylor, D.R. Knighton, J. Zheng, J.M. Sowadski, C.S. Gibbs and M.J. Zoller, Trends Biochem. Sci., 18, 84 (1993).
12. D.O. Morgan and H.L. De Bondt, Curr. Op. Cell Biol., 6, 239 (1994).
13. D. Bossemeyer, R.A. Engh, V. Kinzel, H. Ponstingl and R. Huber, EMBO J., 12, 849 (1993).
14. J. Zheng, D.R. Knighton, L.F. Ten Eyck, R. Karlsson, N-H Xuong, S.S. Taylor and J.M. Sowadski, Biochemistry, 32, 2154 (1993).
15. G.A. Olah, R.D. Mitchell, T.R. Sosnik, D.A. Walsh and J. Trewhella, Bichemistry, 32, 3649 (1993).
16. A.C. Carrea, K. Alexandrov and T.M. Roberts, Proc. Natl. Acad. Sci. USA, 90, 442 (1993).
17. R.A. Steinberg, R.D. Cauthron, M.M. Symcox and H. Shuntoh, Mol. Cell Biol., 13, 2332 (1993).
18. W. Wen and S.S. Taylor, J. Biol. Chem., 269, 8423 (1994).
19. M. Vernon, E. Radzio-Andzelm, I. Tsigelny, L.F. Ten Eyck and S.S. Taylor, Proc. Natl. Acad. Sci. USA, 90, 10618 (1993).
20. D.R. Knighton, D.L. Cadena, J. Zheng, L.F. Ten Eyck, S.S. Taylor, J.M. Sowadski and G.N. Gill, Proc. Natl. Acad. Sci. USA, 90, 5001 (1993).
21. E. Nishida and Y. Gotoh, Trends Biochem. Sci., 18, 128 (1993).

22. P. Cohen, Biochem. Soc. Trans, 21, 555 (1993).

23. R. J. Davis, J. Biol. Chem., 268, 14553 (1993).

24. J. Blenis, Proc. Natl. Acad. Sci. USA, 90, 5889, (1993).

25. G.L. Johnson and R.V. Vaillancourt, Curr. Op. Cell Biol., 6, 230 (1994).

26. F. Zhang, A. Strand, D. Robbins, M.H. Cobb and E.J. Goldsmith, Nature, 367, 704 (1994).

27. J. Pines, Curr. Biol., 3, 544 (1993).

28. J. Pines, Trends Biochem. Sci.,, 19, 143 (1994).

29. M.J. Solomon, J.W. Harper and J. Shuttleworth, EMBO J., 12, 3133 (1993).

30. H.L. De Bondt, J. Rosenblatt, J. Jancarik, H.D. Morgan and S-H. Kim, Nature, 363, 595 (1993).

31. L.V. Dekker and P.J. Parker, Trends Biochem. Sci., 19, 73 (1994).

32. S.E. Wilkinson and T.J. Hallam, Trends Pharmacol. Sci., 15, 53 (1994).

33. F-J. Johannes, J. Prestle, P. Oberhagemann and K. Pfizenmaier, J. Biol. Chem., 269, 6140 (1994).

34. L.A. Selbie, C. Schmitz-Peiffer, Y. Sheng and T.J. Biden, J. Biol. Chem., 268, 24296 (1993).

35. A. Basu, Pharmac. Ther., 59, 257 (1993).

36. H. Hug and T.F. Sarre, Biochem. J., 291, 329 (1993).

37. L.V.Dekker, P. McIntyre and P.J. Parker, J. Biol. Chem., 268, 19498 (1993).

38. L.V.Dekker, P. McIntyre and P.J. Parker, FEBS Lett., 329, 129 (1993).

39. J.W. Orr and A.C. Newton, J. Biol. Chem., 269, 8383 (1994).

40. Y-G. Kwon, M. Mendelow and D.S. Lawrence, J. Biol. Chem., 269, 4389 (1994).

41. D. Ron, C-H. Chen, J. Caldwell, L. Jamieson, E. Orr and D. Mochly-Rosen, Proc. Natl. Acad. Sci. USA, 91, 839 (1994).

42. N. Funato, H. Takayanagi, Y. Konda, Y. Toda, Y. Harigaya, Y. Iwai and S. Omura, Tetrahedron Letts, 35, 1251 (1994).

43. P.D. Davis, C.H. Hill, G. Lawton, J.S. Nixon, S.E. Wilkinson, S.A. Hurst, E. Keech and S.E. Turner, J. Med. Chem., 35, 177 (1992).

44. P.D. Davis, L.H. Elliot, W. Harris, C.H. Hill, S.A. Hurst, E. Keech, M.K.H. Kumar, G. Lawton, J.S. Nixon and S.E. Wilkinson, J. Med. Chem., 35, 994 (1992).

45. R.A. Bit, P.D. Davis, L.H. Elliot, W. Harris, C.H. Hill, E. Keech, H. Kumar, G. Lawton, A. Maw,   J.S. Nixon, D.R. Vesey, J. Wadsworth and S.E. Wilkinson, J. Med. Chem., 36, 21 (1993).

46. S.W. McCombie, P.W. Bishop, D. Carr, E. Domek, M.P. Kirkup, P. Kirschmeier, S-I. Lin, J. Petrin, K. Rosinski, B.B. Shankar and O. Wilson, Biorg. Med. Chem. Letts., 3, 1537 (1993).

47. B.B. Shankar, S.W. McCombie, M.P. Kirkup, A.Q. Viet, M.S. Puar and A.K. Ganguly,   Tetrahedron Letts, 34, 5685 (1993).

48. J. Kleinschroth, J. Hartenstein, and C. Schachtele, Biorg. Med. Chem. Letts., 3, 1959 (1993).

49. G. Caravatti, T. Meyer, A. Fredenhagen, U. Trinks, H. Mett and D. Fabbro, Biorg. Med. Chem. Letts., 4, 399 (1994).

50. S.E. Wilkinson, P.J. Parker and J.S. Nixon, Biochem. J., 294, 335 (1993).

51. G. Martiny-Baron, M.G. Kazanietz, H. Mischak, P.M. Blumberg, G. Kochs, H. Hug, D. Marme and C. Schachtele, J.Biol. Chem., 268, 9194 (1993).

52. W.J. Fantl, D.E. Johnson and L.T. Williams, Annu. Rev. Biochem., 62, 453 (1993).

53. E. M. Dobrusin and D.W. Fry, Annu. Rep. Med. Chem., 27, 169 (1992).

54. M.S. Smyth, I. Stefanova, F. Hartmann, I.D. Horak, N. Osherov, A. Levitzki and T.R. Burke Jr., J. Med Chem., 36, 3010 (1993).

55. H. Chen, J. Boiziau, F. Parker, R. Maroun, B. Tocque, B.P. Roques and C. Garbay-Jauregiberry, J. Med Chem., 36, 4094 (1993).

56. M.S. Smyth, I. Stefanova, I.D. Horak and T.R. Burke Jr., J. Med Chem., 36, 3015 (1993).

57. A.M. Thompson, G.W. Rewcastle, M. Tercel, E.M. Dobrusin, D.W. Fry, A.J. Kraker and W.A. Denny, J. Med Chem., 36, 2459 (1993).

58. A. Gazit, N. Osherov, I. Posner, A. Bar-Sinai, and A. Levitzki, J. Med Chem., 36, 3556 (1993).

59. U. Trinks, E. Buchdunger, P. Furet, W. Kump, H. Mett, T. Meyer, M. Muller, U. Regenass, G. Rihs, N. Lydon and P. Traxler, J. Med Chem., 37, 1015 (1994).

60. E. Buchdunger, U. Trinks, H. Mett, U. Regenass, M. Muller, T. Meyer, E. McGlynn, L.A. Pinna, P. Traxler and N.B. Lydon, Proc. Natl. Acad. Sci. USA, 91, 2334 (1994).

61. M.C. Mumby and G. Walter, Physiol. Revs., 73, 673 (1993).

62. K. Sasaki, M. Murata, T. Yasumoto, G. Mieskes and A. Takai., Biochem. J., 298, 259 (1994).

63. M.J. Hubbard and P. Cohen, Trends Biochem. Sci., 18, 172 (1993).

64. D.R. Alessi, A.J. Street, P.Cohen and P.T.W. Cohen, Eur. J. Biochem., 213, 1055 (1993).

65. J. Chen, S. Parsons and D.L. Brautigan., J. Biol. Chem., 269, 7957 (1994).

66. J. Liu, Trends Pharmacol. Sci., 14, 182 (1993).

67. G.J. Barton, P.T.W. Cohen and D. Barford, Eur. J. Biochem., 220, 225 (1994).

68. D. Barford and J.C. Keller, J. Mol. Biol., 235, 763 (1994).

69. D. Barford, A.J. Flint and N.K. Tonks., Science, 263, 1397 (1994).

70. Z-Y. Zhang and J.E. Dixon, Adv. Enzymol., 68, 1 (1994).

71. K.M. Walton and J.E. Dixon, Annu. Rev. Biochem., 62, 101 (1993).

72. S.M. Brady-Kalnay and N.K. Tonks, Trends Cell Biol., 4, 73 (1994).

73. M.L. Thomas, Curr. Biol., 6, 247 (1994).

74. Z-Y. Zhang, Y. Wang and J.E. Dixon, Proc. Natl. Acad. Sci. USA., 91, 1624 (1994).

75. L.J. Mauro and J.E. Dixon, Trends Biochem., 19, 151 (1994).

76. J.V. Frangioni, A. Oda, M. Smith, E.W. Salzman and B.G. Neel, EMBO J., 12, 4843 (1993).

77. T. Pawson and J. Schlessinger, Curr. Biol. 3, 434 (1993).

78. L.A. Perkins, I. Larsen and N. Perrimon, Cell, 70, 225 (1992).

79. W. Li, R. Nishimura, A. Kashisian, A.G. Batzer, W.J.H. Kim, J.A. Cooper and J. Schlessinger, Mol. Cell. Biol. 14, 509 (1994).

80. R.J. Lechleider, R.M. Freeman and B.G. Neel, J. Biol. Chem., 268, 6593 (1993).

81. U. Dechert, M. Adam, K.W. Harder, I. Clark-Lewis and F. Jirik, J. Biol. Chem., 269, 5602 (1994).

82. H. Maegawa, S. Ugi, M. Adachi, Y. Hinoda, R. Kikkawa, A. Yachi, Y. Shigeta and A. Kashiwagi, Biochem. Biophys. Res. Commun., 199, 780 (1994).

83. Z. Zhao, R. Larocque, W-T. Ho, E.H. Fischer and S-H Shen, J. Biol. Chem., 269, 8780 (1994).

84. D.J. Hakes, K.J. Martell, W.G. Zhao, R.F. Massung, J.J. Esposito and J.E. Dixon, Proc. Natl. Acad. Sci. USA, 91, 1731 (1994).

85. B.I. Posner, R. Faure, J.W. Burgess, A.P. Bevan, D. Lachance, G. Zhang-Sun, I.G. Fantus, J.B. Ng, D.A. Hall, B.S. Lum and A. Shaver, J. Biol. Chem., 269, 4596 (1994).

86. A.R. Nebreda, Trends Biochem. Sci., 19, 1 (1994).

87. T. Noguchi, R. Metz, M-G. Mattei, D. Carrasco and R. Bravo, Mol. Cell. Biol., 13, 5195 (1993).

88. C.H. Charles, H. Sun, L.F. Lau and N.K. Tonks, Proc. Natl. Acad. Sci. USA, 90, 5292 (1993).

89. D.R. Alessi, C. Smythe and S.M. Keyse, Oncogene, 8, 2015 (1994).

90. C-F. Zheng and K-L. Guan, J. Biol. Chem., 268, 16116 (1993).

91. S.P Kwak, D.J. Hakes, K.J. Martell and J.E. Dixon, J. Biol. Chem., 269, 3596 (1994).

92. P.J. Rohan, P. Davis, C.A. Moskuluk, M. Kearns, H. Krutzsch, U. Siebenlist and K. Kelly, Science, 259, 1763 (1993).

93. B. Sebastian, A.Kakizuka and T.Hunter, Proc. Natl. Acad. Sci. USA, 91, 3521 (1994).

94. G.J Hannon, D. Casso and D. Beach, Proc. Natl. Acad. Sci. USA, 91, 1731 (1994).

Chapter 27. Transgenic and Gene Targeting Technology in Drug Discovery

Mark E. Swanson, David S. Grass, Vincent B. Ciofalo*
DNX Biotherapeutics Inc., Princeton NJ 08540
*Pharmakon USA, Waverly PA 18471

Introduction- The fast and efficient development of new pharmaceuticals requires the availability of high throughput biological test systems to identify and characterize potentially therapeutic agents prior to human clinical trials. Historically, lead compounds have often been identified by random screening of chemical or fermentation product libraries. Peptide, oligonucleotide, carbohydrate, and combinatorial chemical libraries have now been added to the list of novel structure sources from which to identify new chemical leads. Ideally, for rapid and economical development of new pharmaceuticals the biological test systems used to identify and characterize new lead compounds should mimic as closely as possible the human condition being targeted.

Biological test systems employed in drug discovery include in vitro assays, cell culture based assays, organ perfusion and whole animal models. In vitro assay systems often have the advantage of low cost and high throughput screening, but lack the biological processes present in a whole cell system, hence limiting their usefulness. Cell culture and organ perfusion systems attempt to bridge the gap between in vitro and in vivo models. The availability of human cell lines is a step toward simulating a human condition but lacks in vivo complexity. In vivo models have the greatest advantage in providing the metabolic, physiologic, and perhaps pathologic complexity absent in cell culture systems, but they are generally more costly to perform and can be limited by their lack of similarity to conditions in human cells. The development of technologies to alter the germ line of some experimental animals by pronuclear microinjection of DNA fragments (gene addition) or homologous recombination (gene targeting) in embryonic stem (ES) cells can provide an unlimited supply of new animal models that more closely resemble conditions in human cells, opening the way for rapid and accurate evaluation of new pharmaceuticals.

During early stages of drug development, transgenic animals can provide a critical tool for identifying novel sites for drug intervention through improved understanding of biological processes, identification of rate limiting components and in vivo hypothesis testing. With this information in hand, transgenic animals can be used to provide new in vivo models that produce human targeted proteins, mimic human biological processes and model human pathological conditions. This chapter will provide a brief overview of transgenic technology and provide examples of how this technology can be used to provide improved animal models for drug discovery.

TRANSGENIC ANIMAL PRODUCTION BY GENE ADDITION

Current transgenic animal technologies allow exogenous genes to be introduced into an animal genome by gene addition and endogenous genes to be specifically modified by gene targeting. Pronuclear microinjection is the most common method used to produce transgenic animals by gene addition. Successful microinjection and integration of a DNA fragment into a transgenic mouse was first reported in 1980, however, expression of this gene was not detected (1). Shortly thereafter, several groups reported functional expression of gene fragments in transgenic mice (2-5). Since this time, hundreds of scientific reports have been published describing transgenic mice produced by pronuclear microinjection. While pronuclear microinjection has become the method of choice in gene addition experiments, infection of preimplantation embryos with retroviral vectors (6) and sperm-mediated transfer of DNA have also been reported (7). Some success has also been achieved by injecting chromosome fragments into mouse embryos (8). Pronuclear microinjection has been used to produce transgenic mice, rats, rabbits, chickens, goats, sheep, cows, pigs and fish.

To produce transgenic mice by pronuclear microinjection, single cell fertilized embryos are collected from donor females. Approximately 100-200 copies of a DNA fragment are injected into an embryo shortly after fertilization, when the individual male and female pronuclei are visible but prior to fusing of the two pronuclei. Usually the larger, more visible, male pronuclei is injected. The injected embryo is then returned to the reproductive tract of a pseudopregnant recipient female and developed to term. In mice, usually about 20-25% of the offspring contain detectable levels of the injected fragment. The presence of the transgene fragment is most often determined by analysis of DNA isolated from tail biopsies of the surviving pups. The transgene integration-positive animals are referred to as transgenic founders. When the germ cells of the founder contain the transgene (generally about 90% of the time) the transgene can be transmitted to its progeny and transgenic lines can be established.

Every transgenic founder produced by gene transfer is a unique animal. Integration of the transgene fragment is a random event and therefore each founder contains the transgene at a different chromosomal locus. Generally, the transgene fragment is present in head to tail arrays at a single chromosomal site. The number of copies of a transgene fragment at the site of integration is also unpredictable. The level of transgene expression obtained, and the stage and site specific nature of that expression, is determined by the properties of the transgene fragment, the number of transgene fragments integrated, and the influence of surrounding sequences at the site of integration. Transgene fragments can be envisioned as consisting of two types of sequence information: "structural" sequences that encode the amino acid sequence of the protein to be produced and "regulatory" sequences that mediate when, where, and to what extent the protein will be expressed. Regulatory sequences can be located near the site of initiation of RNA synthesis or many kilobase pairs upstream or downstream of the RNA initiation site. Regulatory sequences include sequences that control processing and stability of the RNA transcript which can also affect the overall level of protein expression obtained.

Large DNA fragments containing the maximum amount of sequence that flank the structural sequences are most likely to contain all the regulatory sequences necessary for efficient gene expression in a spatially normal manner and may be less influenced by sequences juxtaposed by integration in the genome. Regulatory sequences from one gene can also be fused to a heterologous gene. Thus, a recombinant gene fragment can be used to direct the expression of a gene product at levels and in cells where it is not normally expressed.

GENE MODIFICATION BY HOMOLOGOUS RECOMBINATION IN ES CELLS

Gene targeting by homologous recombination in ES cells provides the potential for modifying endogenous genes in defined ways and examining the consequences in vivo. The key to producing mice harboring the modified endogenous gene is the availability of pluripotent ES cells that can contribute to all the cell lineages of a developing mouse embryo and the ability of exogenous DNA fragments to recombine with endogenous targeted genes in these cells. ES cells were first isolated from mouse blastocyts in 1981 (9,10). Subsequently, it was demonstrated that ES cells could contribute to the germline of a mouse following injection into a blastocyst when a chimera capable of transmitting ES cell derived genome to its offspring was produced (11). The first mice containing a gene mutated by homologous recombination in ES cells was reported in 1987 when mice deficient in HPRT were produced (12). Subsequently, alternative selection procedures including positive and negative selection methods (13) and screening procedures such as the polymerase chain reaction (14) have facilitated the identification of homologous recombination events for non-selectable genes.

ES cells are derived from the inner cell mass of a mouse blastocyst and can be maintained, replicated, and modified in culture. It is critical that ES cells retain their potential to contribute to somatic and germ cells following injection into a blastocyst in order to allow the production of a chimeric mouse and eventually a transgenic offspring. To retain this potential, it is important ES cells are cultured and maintained properly (see 15). A targeting vector containing the desired gene modification can be introduced to ES cells using calcium transfection, electroporation, microinjection or other methods. The frequency of homologous recombination between the vector and its endogenous target is much lower than the frequency of random integration of the targeting vector. Therefore, a target vector must include features designed to select for these events. Selection and screening identifies the ES cells containing these rare homologous recombination events. ES cells containing the modified endogenous gene are then cloned and propagated. The modified ES cells are injected into the blastocyst of a preimplantation mouse embryo which is transferred into the uterus of a recipient host to complete development. The resulting mouse is a chimera, derived from both donor stem cells and the host blastocyst. If the donor cells are derived from a pigmented strain and the host cells from an albino strain, chimeric offspring will have a mosaic coat pattern. To produce a line of mice containing the modified gene, the chimeric mouse must successfully transmit the altered gene to its offspring. Genomic screening of the progeny is therefore performed to identify mice with the modified gene. Mice homozygous for the mutation can then be produced by interbreeding.

Gene addition and gene targeting are powerful technologies with distinct applications. Gene addition is a powerful tool for rapidly mapping cis acting DNA sequences responsible for gene regulation. Multiple gene promoter mutations, constructed in vitro, can be rapidly tested by gene addition for their ability to target expression of the transgene. Gene addition is also useful for determining whether overexpression or ectopic expression of a wild type gene product will cause an abnormal phenotype in a transgenic animal and perhaps mimic a human condition. Gene addition is limited however, to characterization of "dominant" mutations because of the presence of the "wild type" endogenous gene. Gene targeting technology has a different set of applications, including the ability to create null mutations. Null mutations have been useful in elucidating the function of several gene products during embryonic development and adulthood. However, homozygous null mutations often fail to exhibit an observable phenotype, perhaps due to redundancy of gene function. Structural mutations can also be generated using this technology. If a particular gene defect is known to be associated with a disease phenotype, a targeted mutation can be utilized to create a mouse model of that particular disease. These mutations need not be dominant mutations, as the newly created mutant allele can be bred to homozygousity in these animals. Expression of the mutant allele in these cases will be determined by the same regulatory sequences that control expression of the wild type gene.

UTILITY OF TRANSGENIC ANIMALS IN DRUG DISCOVERY

Providing Human Pharmacological Targets- Transgenic animals have become a valuable tool in elucidating the role of serum lipoproteins and lipoprotein modifying enzymes in cholesterol metabolism and atherosclerosis. In the past decade, most of the protein components of lipoproteins have been identified and their corresponding genes have been cloned and characterized. These genes have been used to produce transgenic animals to elucidate the way that these genes are regulated and to examine the consequences of overexpression of these lipoproteins. A recent review summarizes the many applications of transgenics to the field of lipoprotein metabolism and atherosclerosis (16). Many of these new strains of mice lend themselves to use in drug discovery efforts.

In some cases, a human transgene product is itself a target for pharmacological intervention. The level of apolipoprotein B100 (apo B), the major protein constituent of low density lipoprotein (LDL), is directly proportional to the risk of developing atherosclerosis and coronary heart disease. A potentially important pharmacolocigical approach to controlling heart disease is to reduce apo B and LDL-cholesterol. Pharmacologically, apo B levels could be reduced by numerous mechanisms including inhibition of gene expression, inhibition of protein synthesis or secretion and increasing apo B clearance. Unlike humans, mice have low levels of apo B and LDL-cholesterol and have high levels of high density lipoprotein (HDL)-cholesterol, limiting their usefulness for modeling human lipoprotein metabolism. However, apo B transgenic mice were recently produced with levels of human apo B100 similar to humans (17). These mice, which were produced with a gene fragment containing the entire human apo B gene along with several kilobase pairs of upstream and downstream flanking sequence also have a significant increase in LDL-cholesterol. Since these mice contain the human apo B gene, they provide a target for inhibition of human apo B gene expression, and may be an especially important model for pharmaceutical approaches, such as antisense or triplex inhibition, that require the presence of human gene sequences. In addition, since the apo B in these mice is primarily human, these mice may become a preferred model for targeting apo B containing lipoproteins.

Cholesteryl ester transfer protein (CETP) transgenic mice are another example of a transgenic model providing a human target for pharmacological inhibition. CETP is a plasma enzyme that exchanges cholesterol ester and triglyceride between lipoproteins. The presence of CETP activity leads to an increase in LDL-cholesterol and a decrease in HDL-cholesterol. Since this transfer of cholesterol facilitates an atherogenic lipoprotein profile, CETP is believed to be atherogenic and is a pharmacological target for inhibition. Convenient small animal models, such as the mouse and rat, lack detectable CETP plasma activity and are therefore useless as models for CETP inhibition. Several laboratories have developed CETP transgenic mice that contain appreciable CETP plasma activity (18,19). Depending upon the gene promoter used to direct CETP gene expression, different levels of plasma CETP activity were achieved. Regardless, the presence of CETP activity in these mice results in a reduction of HDL-cholesterol as compared to non-transgenic controls. Since plasma CETP activity is inversely correlated with HDL cholesterol levels the efficacy of compounds that inhibit CETP can be measured easily by determining HDL-cholesterol. Of course, mice transgenic for the human CETP gene express the human protein, the ultimate target of a pharmaceutical agent.

Providing Human-like Pathological Phenotypes- A second, potentially valuable application of transgenic technology to drug discovery, is the ability to provide a consistent genetically programmed phenotype similar to a human pathological condition. Nowhere is this more relevant than in the study of Alzheimer's Disease (AD). Extracellular deposition of amyloid plaque is a hallmark of AD, and genetic data linking Familial Alzheimer's Disease (FAD) to the gene encoding the amyloid precursor protein (APP) supports its role in the development of AD. Despite the identification of this critical target for pharmacological intervention, only aged non-human primates develop beta-amyloid deposits similar to those observed in human AD. The lack of small animal models of AD is a critical shortcoming in the identification and characterization of agents that inhibit amyloid deposition.

Numerous reports have recently been published describing transgenic mice that express either the entire human amyloid precursor protein (APP) or portions of the APP protein (20). Unfortunately, these attempts have fallen short of providing a mouse model of human AD. One approach has been to express the entire native human APP in transgenic mice, hoping that overexpression of APP would lead to amyloid plaque deposition. This approach has been limited by the level of human APP expression obtained, despite the numerous gene promoter fragments used to drive its expression. In cases in which the level of APP has been well characterized, overexpression of total APP has not exceeded more than two fold that of endogenous mouse APP. To date, only one report has described extra-cellular deposition of beta amyloid epitopes in APP transgenic mice (21), but these results have not been reproduced by others who have achieved a similar level of APP overexpression (22-24). While these transgenic animals have not yet provided a model with plaque deposition, they may provide an in vivo model of APP processing. Pharmacological approaches to reduce amyloid production by interfering with APP processing could eventually prove useful in the treatment of AD. Furthermore, many laboratories are now using APP gene fragments containing FAD mutations (25) in another attempt to create an animal model containing amyloid plaque.

As discussed above, the "normal" laboratory mouse is a poor model of human lipoprotein metabolism. Mice circulate most of their cholesterol in "anti-atherogenic" HDL and are extremely resistant to the development of human-like atherogenic lesions. Certain strains of mice have been shown to develop atherosclerotic lesions but only following a prolonged diet high in cholesterol, fat and cholic acid (26). To facilitate the identification and development of new anti-atherogenic pharmaceuticals it would be desirable to have a small animal model that more consistently develops atherosclerosis in a short time period. Recently several transgenic mouse models have been engineered that develop atherosclerotic lesions. Homozygous apolipoprotein E (apo E) "knock-out" mice have been produced that develop foam cell lesions in the aorta within 3 months on a normal chow diet (27,28) These mice, which lack any apo E, the major protein constituent of very low density lipoprotein responsible for clearance of VLDL by the liver, have extremely elevated cholesterol levels. CETP transgenic mice, discussed above, also develop atherosclerotic lesions following an 18 week diet high in fat and cholesterol (29). In addition, transgenic mice expressing human apolipoprotein (a), a large glycoprotein linked to LDL by disulfide bonding, also develop features of human atherosclerosis (30). However, in these mice, the human apo (a) is found free in plasma and not present as lipoprotein (a). While the relevance to human atherosclerosis must be demonstrated in these models, undoubtedly, transgenic animal technology will provide small animal models of atherosclerosis that will be beneficial to the drug development community.

Models of Human Viral Infection-Transgenic technology can also be very beneficial in creating mouse models of human viral infection. Often, human viruses do not infect laboratory animals due to the lack of a human specific viral receptor. However, gene addition provides a mechanism to overcome this limitation by introducing a human viral receptor, once it is identified and cloned. Recent work relevant to creating a mouse model for HIV-1 infection was performed independently by several groups (31-33). These groups were able to achieve tissue specific expression of the human CD4 molecule, known to be part of the receptor for the HIV-1 virus (34). It is unlikely that these particular mice will be useful as a model for HIV-1 infection since experiments have shown that the presence of human CD4 on a variety of murine cell lines has not rendered these cell lines infectable with HIV (the virus can bind the cells but not internalize)(34-35). These mice might become useful in the future, however, if the molecule(s) required for internalization of the virus is identified. Transgenic mice expressing both the human CD4 gene and the "new" gene products required for HIV entry can then be established and tested for susceptibility to HIV-1 infection.

Another approach used to generate models of viral infection is to express individual viral proteins that may facilitate disease progression in transgenic mice. This approach is usually attempted as a tool to determine the role of the particular protein during the course of the viral infection. As an example of this, mice expressing the HIV-1 *nef* gene have been produced (36). The functional role for nef, a 27 kDa membrane associated myristoylated protein, has not yet been determined. Mice in which *nef* expression was directed by the HIV-1 LTR exhibited expression in the skin. A large percentage of these animals spontaneously developed skin lesions resembling papillomas, which had hyperkeratosis, epidermal proliferation and follicular hyperplasia. This information should be useful in helping to define a role for nef in HIV-1 infection and may lead to the use of these animals as a model for evaluating nef antagonists.

The human papillomavirus type 16 (HPV-16) is thought to be involved in most cervical cancers. More specifically, the E6 and E7 HPV-16 oncogenes, which have been shown to associate with the p53 tumor suppresser gene product (37) and the retinoblastoma susceptibility gene product Rb (38), have been implicated in the etiology of this disease (39). In a variation of the above approach, transgenic mice were produced in which the expression of the E6 and E7 genes were targeted to the mouse ocular lens by using the αA crystalline promoter (40). The ocular lens was chosen for a number of reasons. First, it is an epithelial tissue similar to the epidermis, a site of papilloma virus infection. Second, both tissues are composed of a single layer of undifferentiated cells, with the remaining layers containing cells at various stages of differentiation. In addition, the lens is not essential for the viability of the animal, avoiding the possibility of lethality. The consequence of expression of E6 and E7 in the ocular lens was bilateral microphthalmia and cataracts. This was caused by lens fiber cell differentiation and induction of cell proliferation. These histological properties are similar to the hyperproliferation of basal epithelial cells and the impairment of epithelial cell differentiation seen in HPV-positive cervical dysplasias. In addition, one line of mice expressing the highest levels of the transgene had a high incidence of lens tumors. This is consistent with the supposed role of these oncogenes in cervical carcinoma. Thus, although expression was specifically targeted to the ocular lens rather than to the cervical epithelia, these mice provide a model for the role and possible inhibition of viral oncogenes in human cancer.

Utility of Null Mutations in Drug discovery-The p53 gene is the most commonly detected genetic lesion in human cancers and is normally thought to play a role in the regulation of the cell cycle. Recently mice were created in which a null mutation of the p53 gene was introduced by homologous recombination in murine embryonic stem cells (41). They showed that although p53 is thought to be involved in the cell cycle, it is not required for proper embryonic development. Mice homozygous for the mutation were susceptible to lymphomas, sarcomas and a variety of other tumors before the age of 6 months. Hemizygous mice took longer to develop tumors. These mice may be useful for testing substances suspected of being carcinogenic.

The gene targeting technique has been used to understand the role of many molecules believed to be associated with lymphocyte development and function. This approach has been useful in dissecting the role of the interaction between the CD8 molecule on immature T cells and the MHC class I molecules of antigen presenting cells on T cell differentiation and development. The role of the CD4-MHC class II molecule interaction has also been elucidated using this technique. In addition, mutations in the T cell receptor α and β genes have shed light on the differentiation of immature T cells. Defects in several X-linked immunodeficiencies have recently been discovered and characterized on a molecular level. For example, the IL-2R was implicated in X-linked severe combined immunodeficiency (SCID) (42). This knowledge could lead to the use of better mouse models using gene targeting technology. The gene targeting technique as it has been applied to studying the immune system was recently reviewed (43).

Other uses of transgenic animals in drug discovery-There are two other utilizations of transgenic animals, that, although beyond the scope of this review, warrant being mentioned here. Transgenic animals have tremendous potential for employment as models for gene therapy. For example, a gene therapy approach was taken to correct a defect of the cystic fibrosis transmenbrane regulator (CFTR) gene in transgenic mice in which the CFTR gene was disrupted by gene targeting (44). The CFTR gene product is a cyclic-AMP-regulated chloride channel (45,46). Disruptions of this gene are responsible for cystic fibrosis (47). Permanent lung damage often occurs because the ion transport abnormalities in the lungs of people with cystic fibrosis lead to accumulation of mucus and bacterial colonization. Mice with the null mutation of the CFTR gene (48), which exhibited some phenotypic characteristics of human cystic fibrosis, including intestinal obstruction, and ion transport abnormalities in the lung, were used to test a gene therapy protocol in which a wild type CFTR gene expression plasmid was delivered into the lungs via liposome mediated transfection. The results of these experiments indicated that the ion conductance defects found in the trachea of these mice could be corrected and suggest that gene therapy for the pulmonary aspects of human cystic fibrosis may be feasible.

Another application of transgenic technology is to produce pharmaceuticals. Transgenic farm animals can produce large quantities of biologically active protein for human use. Large quantities of human globin to be used as a potential blood substitute have been produced in swine by the gene addition technique (49). Human protein C, a molecule involved in blood clotting, has been produced in the milk of transgenic swine (50). In addition, transgenic sheep and goats have been produced that express human alpha-1 antitrypsin (51) and human tissue-type plasminogen activator (52) respectively in their milk. Recently, by using both gene addition and gene targeting technology, mice were produced that when immunized, produce "human" antibodies (53). The genes encoding the human heavy and light chain loci were inserted and murine antibody production was disrupted. This should enable the development of therapeutic human monoclonal antibodies that have the advantage of specificity for the antigen without being immunogenic in humans.

ADVANTAGES OF TRANSGENIC MODELS FOR USE IN DRUG DISCOVERY

The use of transgenic animals could prove to be a remarkable technical advance for drug development. These transgenic models could provide distinct advantages in identifying disease treatments that have eluded the more classical screening approaches (e.g., broad screen profile, receptor binding assay, computer molecular design, enzyme inhibition, recombinant DNA technology, etc.). This approach could also allow the pharmaceutical and biotechnology industries to move lead compounds from animals to human clinical trials more efficiently and cost effectively. Validated transgenic animal models containing the genetic material that simulate human pathophysiological disease states are more relevant models for evaluating lead compounds obtained from high output bioassay screens. Ameliorating symptoms or altering biochemical events in such models could reduce the need for more exhaustive, costly, and time consuming tests associated with the more classical screening approaches. Moreover, efficacy testing of lead compounds in transgenic models could significantly reduce the number of animals on study and yet be more predicative of success in the clinic.

References

1. J.W. Gordon, G.A. Scangos, D.J. Plotkin, J.A. Barbosa and F.H. Ruddle, Proc. Natl. Acad. Sci. USA, **77**, 7380 (1980).
2. R.L. Brinster, H.Y. Chen, M.E. Trumbauer, A.W. Senear, R. Warren and R.D. Palmiter, Cell, **27**, 223 (1981).
3. F. Costantini and E. Lacy, Nature, **294**, 92 (1981).
4. T.E. Wagner, P.C. Hoppe, J.D. Jollick, D.R. Scholl, R.L. Hodinka and J.B. Gault, Proc. Natl. Acad. Sci. USA, **78** 6376 (1981).
5. J.W. Gordon and F.H. Ruddle, Science, **214** 1244 (1981).
6. R. Jaenisch, Proc. Natl. Acad. Sci. USA, **73**, 1260 (1976).
7. M. Lavitrano, A. Camaioni, V.M. Fazio, S. Dolci, M.G. Farace and C. Spadafora Cell, **57**, 717 (1989).
8. J. Richa and C.W. Lo, Science, **245**, 175 (1989).
9. M.J. Evans and M.H. Kaufman, Nature, **292**, 154 (1981).
10. G.R. Martin, Proc. Natl. Acad. Sci. USA, **78**, 7634 (1981).
11. A. Bradley, M.H. Kaufman, M.J. Evans and E.J. Robertson, Nature, **309**, 255 (1984).
12. M.R. Kuehn, A. Bradley, E.J. Robertson and M.J. Evans, Nature, **326**, 295 (1987).
13. S.L. Mansour, K.R. Thomas and M.R. Capecchi, Nature, **336**, 348 (1988).
14. H.S. Kim and O. Smithies, Nucleic Acids Research, **16**, 8886 (1988).
15. E.J. Robertson in "Teratocarcinomas and Embryonic Stem Cells A Practical Approach", E.J. Robertson, Ed. IRL Press, Washington DC, 1987, p.71.
16. J.L. Breslow, Proc. Natl. Acad. Sci. USA, **90**, 8314 (1993).
17. M. F. Linton, R. V. Farese, Jr., G. Chiesa, D.S. Grass, P. Chin, R. E. Hammer, H. H. Hobbs and S. G. Young, J. Clin. Invest., **92**, 3029 (1993).
18. L. B. Agellon, A. Walsh, T.Hayek, P. Moulin, X. C. Jiang, S. A. Shelanski, J. L. Breslow and A.R. Tall, J. Biol.Chem., **266**, 10796 (1991).
19. K.R. Marotti, C.K. Castle, R.W. Murray, E.F. Rehberg, H.G. Polites and G.W. Melchior, Arterioscler. and Thromb., **12**, 736 (1992).
20. R.W. Scott, D.S. Howland, B.D. Greenberg, M.J. Savage and M.E. Swanson in "Strategies in Transgenic Animal Science", G.M. Monastersky and J.M. Robl, Ed., American Society for Microbiology, in press (1994).
21. D. Quon, Y. Wang, R. Catalano, J.M. Scardina, K. Murakami and B. Cordell, Nature, **352**, 239 (1991).
22. B.T. Lamb, S.S. Sisodia, A.M. Lawler, H.H. Slunt, C.A. Kitt, W.G. Kearns, P.L. Pearson, D.L. Price and J.D. Gearhart, Nature Genetics, **5**, 22 (1993).
23. B.E. Pearson and T.K. Choi, Proc. Natl. Acad. Sci. USA, **90**, 10578 (1993).
24. J.D. Buxbaum, J.L. Christensen, A.A. Ruefli, P. Greengard and J.F. Loring, Biochem. Biophys. Res. Commun., **197**, 639 (1993).
25. R.F. Clark and A.M. Goate, Arch. Neurol., **50**, 1162 (1993).
26. P.M. Nishina, J. Verstuyft and B. Paigen, J. Lipid Res., **31**, 859 (1990).

27. A.S. Plump, J.D. Smith, T. Hayek, K. Aalto-Setala, A. Walsh, J.G. Verstuyft, E.M. Rubin, and J.L. Breslow, Cell, **71**, 343 (1992).

28. S.H. Zhang, R.L. Reddick, J.A. Piedrahita and N. Maeda, Science, **258**, 468 (1992).

29. K.R. Marotti, C.K. Castle, T.P. Boyle, A.H. Lin, R.W. Murray and G.W. Melchior, Nature, **364**, 73 (1993).

30. R.M. Lawn, D.P. Wade, R.E. Hammer, G. Chiesa, J.G. Verstuyft and E. M. Rubin, Nature, **360**, 670 (1992).

31. N. Killeen, S. Sawada and D.R. Littman, EMBO J., **12**, 1547(1993).

32. F.P. Gillespie, L. Doros, J. Vitale, C. Blackwell, J. Gosselin, B.W. Snyder, S.C., Wadsworth, Mol. Cell. Biol., **13**, 2952 (1993).

33. M.D. Blum, G.T. Wong, K.M. Higgins, M.T. Sunshine and E. Lacy, J. Exp. Med., **177**, 1343 (1993).

34. P.J. Maddon, A.G. Dalgleish, J.S. McDougal, P.R. Chapham, R.A. Weiss and R. Axel, Cell, **47**, 333(1986).

35. P. Lores, V. Boucher, C. MacKay, M. Pla, H. Von Boehmer, J. Jami, F. Barre-Sinoussi and J.-C Weill, Res. Hum. Retroviruses, **8**, 2063 (1992).

36. P. Dickie, F. Ramsdell, A.L. Notkins, S. Venkatesan, Virology, **197**, 431 (1993).

37. B.A. Werness, A.J. Levine, P.M. Howley, Science, **248**, 76 (1990).

38. N. Dyson, P.M. Howley, K. Munger and E. Harlow, Science, **243**, 934 (1989).

39. H. zur Hausen, Virology, **184**, 9 (1991).

40. A.E. Griep, R. Herber, S. Jeon, J.K. Lohse, R.R. Dubielzig, P.F. Lambert, J. Virol., **67**, 1373 (1993).

41. L.A. Donehower, M. Harvey, B.L. Slagle, M.J. McArthur, C.A. Montgomery, Jr., J.S. Butel, A. Bradley, Nature, **356**, 215 (1992).

42. M. Noguchi, H. Yi, H.M. Rosenblatt, A.H. Filipovich, S. Adelstein, W.S. Modi, O.W. McBride, W.J. Leonard, Cell, **73**, 147 (1993).

43. R.S. Yeung, J. Penninger, T.W. Mak, Curr Opin Immunol, **5**, 585 (1993).

44. S.C. Hyde, D.R. Gill, C.F. Higgins and A.E.O. Trezise, L.J. MacVinish, A.W. Cuthbert, R. Ratcliff, M.J. Evans and W.H. Colledge, Nature, **362**, 250 (1993).

45. M.P. Anderson, R.J. Gregory, S. Thompson, D.W. Souza, S. Paul, R.C. Mulligan, A.E. Smith and M.J. Welsh, Science, **253**, 202 (1991).

46. C.E. Bear, C. Li, N. Kartner, R.J. Bridges, T.J. Jensen, M. Ramjeesingh and J.R. Riordan, Cell, **68**, 809 (1992).

47. J.R. Riordan, J.M. Rommens, B-S. Kerem, N. Alon, R. Rozmahel, Z. Grzelczak, J. Zielenski, S. Lok, N. Plavsic, J-L. Chou, M.L. Drumm, M.C. Iannuzzi, F.S. Collins and L-C. Tsui, Science, **245**, 1066 (1989).

48. W.H. Colledge, R. Ratcliff, D. Foster, R. Williamson and M.J. Evans, The Lancet, **340**, 680 (1992).

49. M.E. Swanson, M.J. Martin, J.K. O'Donnell, K. Hoover, W. Lago, V. Huntress, C.T. Parsons, C.A. Pinkert, S. Pilder and J.S. Logan, Bio/Technology, **10**, 557 (1992).

50. W.H. Velander, J.L. Johnson, R.L. Page, C.G. Russell, A. Subramanian, T.D. Wilkins, F.C. Gwazdauskas, C. Pittius, W.N. Drohan, Proc. Natl. Acad. Sci. USA, **89**, 12003 (1992).

51. A.S. Carver, M.A. Dalrymple, G. Wright, D.S., Cottom, D.B. Reeves, Y.H. Gibson, J.L. Keenan, J.D. Barrass, A.R. Scott, A. Colman and I. Garner, Bio/Technology, **11**, 1263 (1993).

52. K.M. Ebert, J.P. Selgrath, P. DiTullio, J. Denman, T.E. Smith, M.A. Memon, J.E. Schindler, G.M. Monastersky, J.A. Vitale and K. Gordon, Bio/Technology, **9**, 835 (1991).

53. S.L. Morrison, Nature, **368**, 812 (1994).

Chapter 28. Emerging Therapies in Osteoporosis

Gideon A. Rodan
Merck Research Laboratories
West Point, PA 19486

<u>Introduction</u> -   The recent major advance in osteoporosis is the ability to recognize the disease prior to the occurrence of fractures by measuring bone mineral density (BMD). It is now well-established that the risk of fractures increases with the decrease in bone mass (1). This has led to the following definition: Osteoporosis is a systemic skeletal disease characterized by low bone mass and microarchitectural deterioration of bone tissue, with a consequent increase in bone fragility and susceptibility to fracture (2). Bone mass increases during childhood and adolescence, plateaus at "peak bone mass" between the ages of 20 and 30, and declines thereafter (3). Peak bone mass is lower in women than in men and decreases faster during the first few years after menopause, putting women at a greater risk of fractures. It was estimated that in the United States caucasian women who reach the age of 50 have a 50 percent chance of experiencing an osteoporotic fracture during the rest of their life (4). The earliest fractures are in the wrist (5), followed by spine fractures and eventually hip fractures, which increase exponentially with age (6). In the U.S., 25 million women have a BMD two standard deviations below mean peak bone mass. Each year there are 1.5 million osteoporotic fractures, including 250,000 hip fractures and 550,000 spine fractures, causing significant suffering and cost (4). Increased mortality due to hip fractures is about 15% (6). Sixty percent of patients with hip fractures do not regain full function and 15-35% are committed to nursing homes. Osteoporosis also occurs in men, albeit later in life. About 30% of hip fractures are in men. Osteoporosis is thus a significant public health problem which, in the absence of effective intervention, will worsen due to demographic changes and a secular trend showing an increase in fracture incidence for people of the same age (7).

There is an increasing tendency and need to recognize osteoporosis by measuring BMD with single or dual beam photon absorptiometry (DPA) or dual energy x-ray absorptiometry (DXA), which has an accuracy of 2-3% and a precision of about 1% (8). These methods measure the attenuation of x-rays or gamma-rays while crossing the bones of the spine, hip, or radius before they reach a detector. Less expensive methods using ultrasound or traditional x-rays are under development (9,10), and it is likely that affordable ways for measuring bone mass and diagnosing osteoporosis, defined by the World Health Organization as a bone mass 2-2.5 S.D. below peak bone mass, will be widely available in developed countries. It should be mentioned that demographic projections indicate that the greatest increase in the incidence of osteoporosis in the next 40 years will take place in Asia (11). Other diagnostic procedures include biochemical tests of blood and urine, which estimate bone turnover. Combining the measurement of bone mass at menopause with biochemical estimates of bone loss was shown to predict bone mass 12 years later (12).

PATHOPHYSIOLOGY OF OSTEOPOROSIS

Bone mass in an individual reflects the "peak bone mass", reached at the age of about 20, and the rate of bone loss after the age of 35-40 (12). Peak bone mass is determined genetically (13), and is influenced by the availability of calcium, nutrition, and exercise during development (14). There may be a very modest decline in bone mass between the ages of about 35 and 40 (3), but a very rapid decline of about 2% per year can occur in the spine in women following menopause (3). This decline is due to estrogen deficiency, is also induced by ovariectomy or amenorrhea, and can be prevented by estrogen replacement (15,16). Bone mass continues to decline with age, and can reach 50% of peak bone mass in the ninth decade (17). It also declines in men and is lower in men with delayed puberty (18), suggesting that testosterone plays a similar role to estrogen in the maintenance of the skeleton. Bone loss occurs because bone packets are continuously destroyed (resorbed) by osteoclasts and rebuilt (formed) by osteoblasts. During growth, formation exceeds resorption; during young adulthood, the two processes are in balance; after 40, bone resorption exceeds formation. Modern therapeutic interventions target the bone cells, which are briefly reviewed.

Bone loss is most often the result of increased osteoclastic bone resorption. Osteoclasts are large, hemopoietically-derived, multinucleated cells formed by the fusion of precursors related to monocyte/macrophages. Osteoclasts can be generated *in vitro* from mouse, rat or rabbit bone marrow (19). Agents that stimulate bone resorption *in vivo,* such as parathyroid hormone (PTH), $1,25(OH)_2D_3$, interleukin-1 (IL-1), interleukin-6 (IL-6), tumor necrosis factor-α, prostaglandin E (PGE), and several colony-stimulating factors, increase osteoclast formation in these assays. It has been suggested that the increase in bone resorption caused by estrogen deficiency is due to higher levels of IL-1 and IL-6 in bone (20,21).

Current knowledge suggests that bone resorption by the mature osteoclast (22) proceeds as follows. Part of the osteoclast membrane comes into contact with mineralized bone. The bone surface may have been prepared for resorption by lining cells, which digest a layer of protective matrix, leaving the mineral denuded. The osteoclast attaches firmly via an impermeable "sealing zone," which defines a round area of about 500-1000 μm^2, which will be excavated. The osteoclast membrane facing the resorption surface becomes highly convoluted, forming the so-called "ruffled border," probably through fusion with lysosomal vesicles. This membrane is enriched in a vacuolar H^+-ATPase, which acidifies the resorption space (22). The acid dissolves the mineral and lysosomal enzymes, including cathepsin B, L, and D, and possibly tripeptidyl peptidase, are proposed to digest the matrix (23), which consists of 90% type 1 collagen. Recently, collagenase (24) and gelatinase (25) were also shown to be present in osteoclasts and bone. The osteoclast is endowed with several pH regulating systems, including carbonic anhydrase-2 (26) and a chloride bicarbonate exchanger (22). Inhibitors of the proton pump (Bafilomycin A) (27), carbonic anhydrase-2 (sulfonamides) (28), chloride bicarbonate exchange (22), lysosomal metallo-proteinases (29) and collagenase (TIMP) (30) inhibit bone resorption. Bone resorption assays include the release of ^{45}Ca from long bones (28) or calvaria of rat pups; the release of calcium from mouse calvaria (30); the excavation of pits in bovine or dentin slices (31) by osteoclasts isolated from rats, rabbits, or chicks, or by osteoclasts formed in culture (19). Other structures described in osteoclasts, that may play a role in bone resorption include c-src, since mice that do not express c-src are osteopetrotic (32); the vitronectin receptor $\alpha v\beta 3$ (33); a calcium "sensor" that increases intracellular calcium in response to high extracellular calcium (31); an inward rectifying K^+ channel; a transient and a Ca^{++}-dependent K^+ outward channel (34); nitrous oxide synthase; and sodium/potassium ATPase. With the possible exception of the vacuolar ATPase (35,36), none of these structures are unique to osteoclasts. Only calcitonin acts directly on the osteoclast at physiological concentrations and rapidly suppresses osteoclast activity (31). Calcitonin is released from the thyroid C-cells in response to elevations in serum calcium, but it does not seem to play a role in calcium or bone homeostasis in the adult.

The bone forming cells, the osteoblasts, are of mesenchymal origin and in mature bone originate from committed precursors present in the periosteum or bone marrow/endosteum (37). During bone formation, the osteoblast precursors undergo several cell divisions and start expressing in a stepwise fashion the bone-specific genes needed for matrix formation, its mineralization and for the regulation of bone tunover. Many hormones, cytokines, and growth factors regulate osteoblast proliferation and maturation *in vitro*, and some of them have similar effects *in vivo*. The *in vitro* bone formation assays, which are not as reliable as resorption assays, include: proliferation of osteoblastic cells; regulation of phenotype-related genes in these cells, including alkaline phosphatase, type 1 collagen, osteocalcin, etc. (37); formation of mineralized nodules in primary cultures of calvaria (38) or bone marrow-derived cells (39); bone formation in folded periosteum (40); and collagen synthesis in calvaria explants (41).

The following factors stimulate bone formation *in vivo*: transforming growth factor β, when injected locally next to the periosteum in calvaria (42) or long bones (43); bone morphogenetic proteins, related structurally to TGFβ, when injected into muscle (44), skin (45), or locally into bone (46); PGE, when injected locally (47) or systemically (48); and PTH, injected systemically (49). The data are less clear for IGFs and FGFs, for which both positive effects and no effect have been reported (50-52). Receptors for TGFβ (53), PTH (54), and PGE (55) have recently been cloned. The target cells for the osteogenic effects of PGE and PTH may be different, since PGE has a pronounced effect on periosteal bone formation (48), whereas PTH produces a greater effect in cancellous bone (49). Non-physiological agents reported to stimulate bone formation *in vivo* include fluoride, which has been used extensively in osteoporosis therapy, especially in Europe (56), and experimental agents including aluminum (57) and zeolite (58).

CURRENT THERAPIES

In addition to estrogen deficiency, osteoporosis can be caused by hyperparathyroidism (59), hyperthyroidism (60), renal disease, glucocorticoid treatment or immunosuppressive therapy following organ transplants (61). These are referred to as secondary osteoporosis. In the aged who are institutionalized, osteoporosis can be caused by vitamin D deficiency (62), which may produce secondary hyperparathyroidism. An adequate calcium intake or calcium supplements are now an integral part of the management of osteoporosis (63,64) with all other therapies. For so-called secondary osteoporosis, the obvious therapy is removing the primary cause, when possible, with medical or surgical treatment. When glucocorticoids are given for life-threatening conditions, co-administration of vitamin D offers partial protection of the skeleton (65).

For postmenopausal osteoporosis, the only treatment so far shown unequivocally to reduce both bone loss and the incidence of fractures is estrogen replacement therapy (66). Equine conjugated estrogens, 0.625 mg/day, or an equivalent amount of other formulations, causes a moderate increase in bone mass and prevents bone loss. Recent studies show that 7-10 years of treatment are required to protect the skeleton (67), indicating that cessation of estrogen treatment increases bone loss. Recent studies have shown that elderly patients, many years after menopause, can also benefit from estrogen treatment (68). In women with uteri, progestogens are usually prescribed along with estrogen to reduce the risk of uterine cancer. The compliance with estrogen/progestogen replacement therapy is low due to unpleasant side effects, such as bleeding or the continuation of menses (69), tension headaches, and a possible increase in the risk of cancer (70). Contraindications for estrogen replacement therapy include breast cancer or a family history of breast cancer, thrombotic events, liver disease, or fibroids in the uterus.

Calcitonin (CT) is a 32 amino acid hormone that suppresses osteoclast activity. Salmon, eel, and human CTs have been developed for osteoporosis treatment. The formulations include intramuscular injection (the only form approved in the U.S.) (71), intranasal CT spray (72), and suppositories (73). The recommended dosage is 50-400 units per day, and various regimens are used to administer it: daily, every other day, and alternating periods of treatment and non-treatment (74,75). A double-blind, placebo-controlled study with continuous CT administration of intranasal spray showed that 200 IU/day for two years increased bone mineral content in the spine by 3% and reduced the number of patients with new vertebral fractures (76). In most CT studies, the effect on fractures could not be assessed because of statistical power or patient selection (77). A recent randomized, placebo-controlled study using 100 IU salmon CT suppositories, six times per week, showed no effect on bone mass and turnover (78). No other treatments are approved for osteoporosis in the U.S.

Bisphosphonates, which are approved in other countries, are analogues of pyrophosphate in which carbon is substituted for oxygen in the P - O - P structure and the carbon side chains R1 and R2 form various derivatives. Etidronate (Didronel, R1 = OH and R2 = CH_3), which is approved in several countries, increased BMD of the spine by 5% after two years of treatment, but its effect on fractures remains inconclusive (79), probably because there have not been enough patients treated in this way.

Anabolic steroids, especially nandrolene decanoate, have been used, particularly in older patients (80). Their use decreased recently because of side effects, such as virilization and undesirable changes in the lipid profile.

EMERGING THERAPIES

Estrogen Analogues. - The mode of action of estrogen on bone is not firmly established. It has been suggested that estrogen acts locally via stromal or osteoblast lineage cells to inhibit osteoclastic bone resorption (20). Tamoxifen, which prevents bone loss in estrogen-deficient rats and possibly in women (81), seems to be an agonist in bone and an antagonist in the breast and other tissues. Since no differences in receptors have been found, the different effects may be due to target cell and gene dependent action of the estrogen receptor (82), as explained by the roles that its TAF1 and TAF2 domains play in transcription activation. This raises the possibility that there may be estrogen ligands with tissue-specific effects. Raloxifen is such a compound. It was initially developed as an anti-estrogen, and was recently reported to partially protect bone loss caused by ovariectomy in rats (83). It also reduces cholesterol level in rats and has no estrogen-like effects on the uterus in rats or in women (84,85).

Bisphosphonates - Several compounds of this family are currently in clinical trials for the treatment of osteoporosis. Their mechanism of action at the molecular level is not known. The P - C - P structure reduces absorption from the gut to about 1-2%, and limits penetration into cells, which explains the lack of metabolism, low toxicity, tissue-specificity, and favorable pharmacokinetics, with 50% of the absorbed dose being taken up by bone within 4-6 hours and the balance being excreted in the urine (85). The pharmacologically active bisphosphonates inhibit osteoclastic bone resorption. Some concentrate on bone resorption surfaces (86) and inhibit osteoclast activity, reflected in the disappearance of the osteoclast ruffled border (86,87). Histologically, this results in reduced bone resorption and reduced bone turnover (88), which at the tissue level are effects similar to those of estrogen. The following bisphosphonates are under development for osteoporosis (89).

Alendronate (1-hydroxy-4-aminobutylidene bisphosphonate monosodium salt trihydrate) is 1000 times more potent than etidronate. Alendronate prevents bone loss caused by ovariectomy in baboons and rats and increases bone mass and bone strength, in ovariectomized animals relative to vehicle-treated (87). Given orally to osteoporotic women at 5 or 10 mg in addition to 500 mg calcium supplements for 18 months, alendronate increases BMD in the spine by about 7% and in the hip by about 4% relative to baseline (89). No noticeable side effects were reported at these doses. Large-scale studies, powered to detect a 32% reduction in spine or all other fractures, are currently in progress.

Tiludronate (chloro-4-phenyl-thiomethylene bisphosphonate) is about 10 times more potent than etidronate. Given orally for six months at 200, 400 or 800 mg/d, it increases BMD by 2% above baseline and 8% above placebo. Six months after cessation of treatment, patients had no further bone loss (91). Placebo-treated patients continued to lose bone. In rats and monkeys (papio papio), treatment increased BMD and bone strength (92). Clinical trials with a fracture endpoint are currently in progress.

Risedronate [2-(3-pyridinyl)-2-hydroxyethylidene-1,1-bisphosphonate disodium] is about 1000-5000 fold more potent than etidronate (93), and was shown to inhibit bone loss in ovariecotmized rats (94). It reduced bone metastases in an animal model (95) and inhibited bone resorption in hyperparathyroid patients (96). Risedronate is being developed for the prevention of osteoporotic fractures.

YM175 [disodium dihydrogen-(cycloheptylamino)-methylene- bisphosphonate monohydrate] has a potency and characteristics that are similar to alendronate and risedronate. It inhibits tumor osteolysis in animals (97). This compound is in early clinical development in Japan.

Pamidronate (3-amino-1-hydroxypropylidene-1,1-bisphosphonate) is about 100-fold more potent than etidronate (93), is approved in most countries for the treatment of hypercalcemia of malignancy, and has been the subject of many preclinical and clinical (98) studies. Given daily for four years, it produces a sustained increase in BMD (99), but it is apparently not being developed for osteoporosis, possibly because it can cause gastrointestinal irritation.

EB-1053 [disodium 1-hydroxy-3-(1-pyrrolidinyl)-propylidene-1,1-bisphosphonate] is about 1000 times more potent than etidronate and inhibits bone resorption in vitro and in vivo (100). No human studies were reported on this compound.

BM21.0955 [1-hydroxy-3-(methylpentylamino) propylidene bisphosphonate] is 10,000-fold more potent than etidronate (92). It reduces cancer-associated hypercalcemia (101) and prevents bone loss in ovariectomized dogs (102). We have no knowledge of its development for osteoporosis.

Flavinoids - Flavinoids have been approved for the treatment of osteoporosis in a small number of countries . Their mode of action is not well established. It has been suggested that they are weak estrogen agonists. The approved compound is ipriflavone, which is a mixture of several isomers. In studies on a small number of postmenopausal (103) and oophorectomized (104) patients, ipriflavone increased BMD in the spine but not the hip.

Other Antiresorptive Therapies - Several approaches for inhibiting osteoclastic activity are being investigated preclinically by academic and industrial laboratories, and could become treatments in the future.

The H^+ vacuolar ATPase is a multimeric ubiquitous structure, which may have unique features in osteoclasts (35). The enzyme H^+/K^+-ATPase was reported to be present in osteoclasts on the basis of crossreactivity with an antibody raised against the stomach enzyme (104). H^+/K^+ ATPase antagonists were shown to inhibit bone resorption in vitro (106); however, omeprazole produced only minor effects in vivo (107).

Integrins are a family of heterodimeric membrane receptors containing an α-subunit of 1200-1800 amino acids and β-subunits of about 800 amino acids (108). The osteoclast α_v/β_3 integrin, which was reviewed in last year's Annual Reports, is a potential target for inhibition of bone resorption. The α_v/β_3 ligands echistatin (109) and kistrin (110) were shown to inhibit bone resorption in vivo.

Ca^{++} receptors presumably mediate the increase in intracellular calcium in response to an elevation in extracellular calcium in avian and mammalian osteoclasts. This causes inhibition of osteoclast activity in vitro and could be a therapeutic target (111).

The recently cloned parathyroid gland calcium receptor is a seven-membrane spanning-receptor (112), which signals via G proteins and inhibits PTH secretion. Inhibitory or stimulatory ligands of this receptor are being developed to control hyperparathyroidism or stimulate bone formation (see below).

Several protease enzyme inhibitors have been proposed to hydrolyze the matrix in bone resorption, including the lysosomal enzymes, cathepsin B, L, and D (23), as well as the mammalian collagenase (24). Their inhibitors were shown to block bone resorption, but there is no evidence far that these structures are osteoclast specific.

Stimulators of Bone Formation

Bone morphogenetic proteins (BMPs) and the structurally related transforming growth factor β (TGFβ) are among the few natural substances that stimulate bone formation *in vivo* when injected into animals or humans (43). They belong to a large family of dimeric proteins of about 25 kDa that are synthesized as larger precursors, which are activated by cleavage (113). These factors are involved in embryonic development and are found in many tissues. Several BMPs are currently being developed for local application to stimulate fracture repair. Both TGFβ and BMPs have been shown to stimulate the proliferation and/or differentiation of osteoblastic cells *in vitro* (114). The obvious difficulties in using TGFβ or BMPs for the treatment of osteoporosis relate to their effect on other target organs and the administration of relatively large peptides. TGFβ, for example, acts on virtually all cell types, and TGFβ1 was shown to regulate the immune response (115). An attractive proposition would be to promote the production or activation of these compounds in the tissue as a way of stimulating bone formation.

Parathyroid hormone (PTH), administered intermittently, was shown many years ago to stimulate trabecular bone formation in rats. It was recently shown that in ovariectomized rats PTH treatment produced more bone than estrogen replacement (116). Intermittent administration may be important for producing the osteogenic effect (117). It is not known which cells respond to the anabolic effect of PTH, and could thus be used to screen for PTH analogues. The anabolic response to PTH has been histomorphometrically dissociated from its effect on bone resorption in rats (117). PTH studies conducted in humans support its action as a bone forming agent (118), and when given in conjunction with estrogen to osteoporotic women, PTH increased the amount of cancellous bone (119). The mid-molecule fragments of PTH produce various effects in bone-derived cells *in vitro* (120); however, no receptor has been identified for this action. Attempts are under way to develop PTH or PTH analogues as bone-forming agents.

Growth hormone (GH) levels decrease with age, and several cross-sectional studies have shown a correlation between GH deficiency and a decrease in bone mineral content, as well as a positive correlation between IGF1 levels and bone mineral content (121). IGF1, which is produced in the liver in response to GH, stimulates *in vitro* collagen synthesis and proliferation of osteoblastic cells (122). In limited studies in patients, GH increased bone density (123). It is not clear if the effects of growth hormone on the skeleton are all mediated by IGF1, or if GH may also act directly on the skeleton. IGFs are abundant in human bone (IGF2 > IGF1). IGF expression in bone cells is stimulated by parathyroid hormone and prostaglandin E2, which are potent stimulators of bone formation *in vivo*. However, the administration of IGF1 to rats did not produce consistent osteogenic effects (124). This may be the result of the complex relationship of IGF with IGF binding proteins (IGFBP), which are locally produced in response to IGF1 itself, GH as well as other agents. There are at least six IGFBPs, some of which were reported to enhance IGF1 action on target cells (IGFBP3), and some to inhibit it (IGFBP4). IGFBP3 was shown to be decreased in osteoporosis (125) and it has been proposed that IGFBPs can be used to stimulate bone formation, rather than IGF. IGF and many other growth and differentiation factors act via tyrosine kinase receptors. Their action is balanced by a large family of tyrosine phosphatases (TPT). Fluoride stimulation of bone formation may be the result of its inhibition of TPT. Indeed, inhibitors of TPTs, such as vanadate, can stimulate DNA synthesis in osteoblastic cells in culture, but this action is not specific to bone cells. Patents are now pending for the use of tyrosine phosphatase inhibitors to stimulate bone formation, and early clinical trials may be under way. The *in vivo* effects of these compounds in animals have not been reported.

Conclusion - This brief review indicates that several osteoporosis therapies are currently under development, and shows that many additional approaches are explored at the preclinical level to identify new targets. One of the main challenges is the development of bone-forming agents.

References

1.   P.D. Ross, H.K. Genant, J.W. Davis, P.D. Miller and R.D. Wasnich, Osteoporosis Int., 3, 120 (1993).
2.   Conference Report, Am. J. Med., 94, 646 (1993).
3.   J.M. Pouilles, F. Tremollieres and C. Ribot, Calcif. Tissue Int., 52, 340 (1993).
4.   R.L. Riggs and L.J. Melton III, New Engl. J. Med., 327(9), 620 (1992).
5.   J.L. Kelsey, W.S. Browner, D.G. Seeley, M.C. Nevitt and S.R. Cummings, Am. J. Epidemiology, 135, 477 (1992).
6.   C. Cooper, E.J. Atkinson, S.J. Jacobsen, W.M. O'Fallon and L.J. Melton III, Am. J. Epidemiology, 137, 1001 (1993).
7.   C. Cooper, E.J. Atkinson, M. Kotowicz, W.M. O'Fallon and L.J. Melton III, Calcif. Tissue Int., 51, 100 (1992).
8.   M. Jergas and H.K. Genant, Arthritis & Rheumatism, 36, 1649 (1993).
9.   A.M. Schott, D. Hans, E. Sornay-Rendu, P.D. Delmas and P.J. Meunier, Osteoporosis Int., 3, 249 (1993).
10.   C.M. Schnitzler, D.G.K. Pitchford, E.M. Willis and K.A. Gear, Osteoporosis Intl., 3, 293 (1993).
11.   C. Cooper, G. Campion and L.J. Melton III, Osteoporosis Int., 2, 285 (1992).
12.   M.A. Hansen, K. Overgaard, G.J. Riis and C. Christiansen, BMJ, 303, 961 (1991).
13.   N.A. Morrison, J.C. Qi, A. Tokita, P.J. Kelly, L. Crofts, T.V. Nguyen, P.N. Sambrook and J.A. Eisman, Nature, 367, 284 (1994).
14.   V. Matkovic, J. Rheumatol., 19, 54 (1992).
15.   D.W. Dempster and R. Lindsay, The Lancet, 341, 797 (1993).
16.   E.G. Lufkin, H.W. Wahner, W.M. O'Fallon, S.F. Hodgson, M.A. Kotowicz, A.W. Lane, H.L. Judd, R.H. Caplan and B.L. Riggs, Annals Int. Med., 117, 1 (1992).
17.   L. Mosekilde, Acta Obstet. Gynecol. Scand., 72, 409 (1993).
18.   J.S. Finkelstein, R.M. Neer, B.M.K. Biller, J.D. Crawford and A. Klibanski, New Engl. J. Med., 326, 600 (1992).
19.   T. Suda, N. Takahashi and T.J. Martin, Endocrine Rev., 13, 66 (1992).
20.   R.L. Jilka, G. Hangoc, G. Girasole, G. Passeri, D.C. Williams, J.S. Abrams, B. Boyce, H. Broxmeyer and S.C. Manolagas, Science, 257, 88 (1992).
21.   R. Pacifici, J.L. Vannice, L. Rifas and R.B. Kimble, J. Clin. Endocrinol. Metab., 77, 1135 (1993).
22.   H.C. Blair, P.H. Schlesinger, F.P. Ross and S.L. Teitelbaum, Clin. Orthopaed. Related Res., 294, 7 (1993).
23.   A.E. Page, A.R. Hayman, L.M.B. Andersson, T.J. Chambers and M.J. Warburton, Int. J. Biochem., 25, 545 (1993).
24.   J.M. Delaisse, Y. Eeckhout, L. Neff, Ch. Francois-Gillet, P. Henriet, Y. Su, G. Vaes and R. Baron, J. Cell Sci., 106, 1071 (1993).
25.   A.-M. Bollen and D.R. Eyre, Conn. Tissue Res., 29, 223 (1993).
26.   T. Laitala and K. Väänänen, J. Bone Miner. Res., 8, 119 (1993).
27.   K. Sundquist, P. Lakkakorpi, B. Wallmark and K. Väänänen, Biochem. Biophys. Res. Commun., 168, 309 (1990).

28. L.G. Raisz, H.A. Simmons, W.J. Thompson, K.L. Shepard, P.S. Anderson and G.A. Rodan, Endocrinology, 122, 1083 (1988).
29. G. Vaes, Clin. Orthop. Rel. Res., 231, 239 (1988).
30. P.A. Hill, J.J. Reynolds and M.C. Meikle, Biochim. Biophys. Acta, 1177, 71 (1993).
31. M. Zaidi, M. Pazianas, V.S. Shankar, B.E. Bax, C.M.R. Bax, P.J.R. Bevis, C. Stevens, C.L.-H. Huang, D.R. Blake, B.S. Moonga and A.S.M.T. Alam, Exper. Physiol., 78, 721 (1993).
32. C. Lowe, T. Yoneda, B.F. Boyce, H. Chen, G.R. Mundy and P. Soriano, Proc. Natl. Acad. Sci. USA, 90, 4485 (1993).
33. S. Nesbitt, A. Nesbit, M. Helfrich and M. Horton, J. Biol. Chem., 268, 16737 (1993).
34. A.F. Weidema, J.H. Ravesloot, G. Panyi, P.J. Nijweide and D.L. Ypey, Biochim. Biophys. Acta, 1149, 63 (1993).
35. D. Chatterjee, M. Chakraborty, M. Leit, L. Neff, S. Jamsa-Kellokumpu, R. Fuchs and R. Baron, Proc. Natl. Acad. Sci. USA, 89, 6257 (1992).
36. B. van Hille, H. Richener, D.B. Evans, J.R. Green and G. Bilbe, J. Biol. Chem., 268, 7075 (1993).
37. G.S. Stein and J.B. Lian, Endocrine Rev., 14, 424 (1993).
38. A.M. Flanagan and T.J. Chambers, Endocrinology, 130, 443 (1992).
39. S.-L. Cheng, J.W. Yang, L. Rifas, S.-F. Zhang and L.V. Avioli, Endocrinology, 134, 277 (1994).
40. H.C. Tenenbaum and J.N.M. Heersche, Calcif. Tissue Int., 38, 262 (1986).
41. J.M. Hock and E. Canalis, Endocrinology, 134, 1423 (1994).
42. M. Noda and J.J. Camilliere, Endocrinology, 124, 2991 (1989).
43. M.E. Joyce, A.B. Roberts, M.B. Sporn and M.E. Bolander, J. Cell Biol., 110, 2195 (1990).
44. U. Ripamonti, S. Ma, N.S. Cunningham, L. Yeates and A.H. Reddi, Matrix, 12, 369 (1992).
45. T.K. Sampath, J.C. Maliakal, P.V. Hauschka, W.K. Jones, H. Sasak, R.F. Tucker, K.H. White, J.E. Coughlin, M.M. Tucker, R.H.L. Pang, C. Corbett, E. Ozkaynak, H. Oppermann and D.C. Rueger, J. Biol. Chem., 267, 20352 (1992).
46. U. Ripamonti, S. Ma, N.S. Cunningham, L. Yeates and A.H. Reddi, Matrix, 12, 369 (1992).
47. R.-S. Yang, T.-K. Liu and S.-Y. Lin-Shiau, Calcif. Tissue Int., 52, 57 (1993).
48. H.Z. Ke, W.S.S. Jee, Q.Q. Zeng, M. Li and B.Y. Lin, Bone & Miner., 21, 189 (1993).
49. D.W. Dempster, F. Cosman, M. Parisien, V. Shen and R. Lindsay, Endocrine Rev., 14, 690 (1993).
50. M. Machwate, E. Zerath, X. Holy, P. Pastoureau and P.J. Marie, Endocrinology, 134, 1031 (1994).
51. K.J. Ibbotson, C.M. Orcutt, S.M. D'Souza, C.L. Paddock, J.A. Arthur, M.L. Jankowsky and R.W. Boyce, J. Bone Miner. Res., 7, 425 (1992).
52. H. Mayahara, T. Ito, H. Nagai, H. Miyajima, R. Tsukuda, S. Taketomi, J. Mizoguchi and K. Kato, Growth Factors, 9, 73 (1993).
53. C.H. Bassing, J.M. Yingling, D.J. Howe, T. Wang, W.W. He, M.L. Gustafson, P. Shah, P.K. Donahoe and X.-F. Wang, Science, 263, 87 (1994).
54. H. Juppner, A.B. Abou-Samra, M.W. Freeman, X.F. Kong, E. Schipani, J. Richards, L.F. Kolakowski, J. Hock and J.T. Potts Jr., Science, 254, 1024 (1991).
55. C.D. Funk, L. Furci, G.A. FitzGerald, R. Grygorczyk, C. Rochette, M.A. Bayne, M. Abramovitz, M. Adam and K.M. Metters, J. Biol. Chem., 268, 26767 (1993).
56. J. Dequeker and K. Declerck, Schweiz. Med. Wochenschr., 123, 2228 (1993).
57. L.D. Quarles, G. Murphy, J.B. Vogler and M.K. Drezner, J. Bone Miner. Res., 5, 625 (1990).
58. M.K. Drezner, T. Nesbitt and L.D. Quarles, Proceedings 1993, Fourth International Symposium on Osteoporosis and Consensus Development Conference, Hong Kong, March 27-April 2, 1993, pp. 114.
59. C.L. Benhamou, D. Chappard, J.B. Gauvain, M. Popelier, Ch. Roux, G. Picaper and C. Alexandre, Clin. Rheumatol., 10, 144 (1991).
60. A. Schoutens, E. Laurent, E. Markowicz, J. Lisart and V. De Maertelaer, Calcif. Tissue Int., 49, 95 (1991).
61. G.M. Rich, G.H. Mudge, G.L. Laffel and M.S. LeBoff, J. Heart Lung Transplant, 11, 950 (1992).
62. M.C. Chapuy, M.E. Arlot, F. Duboeuf, J. Brun, B. Crouzet, S. Arnaud, P.D. Delmas and P.J. Meunier, N. Engl. J. Med., 327, 1637 (1992).

63. I.R. Reid, R.W. Ames, M.C. Evans, G.D. Gamble and S.J. Sharpe, N. Engl. J. Med., 328, 460 (1993).

64. J.F. Aloia, A. Vaswani, J.K. Yeh, P.L. Ross, E. Flaster and F.A. Dilmanian, Ann. Intern. Med., 120, 97 (1994).

65. P. Sambrook, J. Birmingham, P. Kelly, S. Kempler, T. Nguyen, M. Stat, N. Pocock and J. Eisman, N. Engl. J. Med., 328, 1747 (1993).

66. R. Lindsay, Am. J. Med., 95, 5A-37S (1993).

67. D.T. Felson, Y. Zhang, M.T. Hannan, D.P. Kiel, P.W.F. Wilson and J.J. Andersen, N. Engl. J. Med., 329, 1141 (1993).

68. E.G. Lufkin, H.W. Wahner, W.M. O'Fallon, S.F. Hodgson, M.A. Kotowicz, A.W. Lane, H.L. Judd, R.H. Caplan and B.L. Riggs, Ann. Intern. Med., 117, 1 (1992).

69. B. Ettinger, J.V. Selby, J.T. Citron, V.M. Ettinger and D. Zhang, Maturitas, 17, 197 (1993).

70. L.A. Brinton and C. Schairer, Epidemiol. Rev., 15, 66 (1993).

71. R. Civitelli, S. Gonneli, F. Zacchei, S. Bigazzi, A. Vattimo, L.V. Avioli and C. Gennari, J. Clin. Invest., 82, 1268 (1988).

72. J.Y. Reginster, D. Denis, R. Deroisy, M.P. Lecart, M. De Longueville, B. Zegels, N. Sarlet, P. Noirfalisse and P. Franchimont, J. Bone Miner. Res., 9, 69 (1994).

73. S. Gonnelli, D. Agnusdei, A. Camporeale, R. Palmieri and C. Gennari, Curr. Therap. Res., 54, 458 (1993).

74. J.-Y. Reginster, Am J. Med., 95 (Suppl. 5A), 44S (1993).

75. K. Overgaard, M.A. Hansen, V.-A.H. Nielsen, B.J. Riis and C. Christiansen, Am J. Med., 89, 1 (1990).

76. K. Overgaard, M.A. Hansen, S.B. Jensen and C. Christiansen, BMJ, 305, 556 (1992).

77. P. Burckhardt and B. Burnand, Osteoporosis Int., 3, 24 (1993).

78. G. Kollerup, A.P. Hermann, K. Brixen, B.E. Lindblad, L. Mosekilde and O.H. Sorensen, Calcif. Tissue Int., 54, 12 (1994).

79. S.T. Harris, N.B. Watts, R.D. Jackson, H.K. Genant, R.D. Wasnich, P. Ross, P.D. Miller, A.A. Licata, and C.H. Chesnutt III, Am. J. Med., 95, 557 (1993).

80. M. Passeri, M. Pedrazzoni, G. Pioli, L. Butturini, A.H.C. Ruys and M.G.G. Cortenraad, Maturitas, 17, 211 (1993).

81. A.J. Neal, K. Evans and P.J. Hoskin, Eur. J. Cancer, 29A, 1971 (1993).

82. M.T. Tzukerman, A. Esty, D. Santiso-Mere, P. Danielian, M.G. Parker, R.B. Stein, J.W. Pike and D.P. McDonnell, Mol. Endocrinol., 8, 21 (1994).

83. L.J. Black, M. Sato, E.R. Rowley, D.E. Magee, A. Bekele, D.C. Williams, G.J. Cullinan, R. Bendele, R.F. Kauffman, W.R. Bensch, C.A. Frolik, J.D. Termine and H.U. Bryant, J. Clin. Invest., 93, 63 (1994).

84. M.W. Draper, D.E. Flowers, W.J. Huster and J.A. Neild, Proceedings 1993, Fourth International Symposium on Osteoporosis and Consensus Development Conference, Hong Kong, March 23-April 2, 1993, pp. 119.

85. J.H. Lin, D.E. Duggan, I.W. Chen and R.L. Ellsworth, Drug Metab. Disp. Biol. Fate Chem., 19, 926 (1991).

86. M. Sato, W. Grasser, N. Endo, R. Akins, H. Simmons, D.D. Thompson, E. Golub and G.A. Rodan, J. Clin. Invest., 88, 2095 (1991).

87. C.M.T. Plasmans, P.H.K. Jap, W. Kuypers and T.J.J. Slooff, Calcif. Tissue Intl, 32, 247 (1980).

88. R. Balena, B.C. Toolan, M. Shea, A. Markatos, E.R. Myers, S.C. Lee, E.E. Opas, J.G. Seedor, H. Klein, D. Frankenfield, H. Quartuccio, C. Fioravanti, J. Clair, E. Brown, W.C. Hayes and G.A. Rodan, J. Clin. Invest., 92, 2577 (1993).

89. H. Fleisch, Osteoporosis Int., Suppl. 2, S15 (1993).

90. A.C. Santora, N.H. Bell, C.H. Chesnut, K. Ensrud, H.K. Genant, R. Grimm, S.T. Harris, M.R. McClung, F.R. Singer, J.L. Stock, R.A. Yood, P.D. Delmas, S. Pryor-Tillotson and J. Weinberg, J. Bone Miner. Res., 8(Suppl. 1), S135 (1993).

91. J.Y.L. Reginster, Bone, 13, 351 (1992).

92.   P. Geusens, J. Nijs, G. Van Der Perre, R. Van Audekercke, G. Lowet, S. Goovaerts, A. Barbier, F. Lacheretz, B. Remandet, Y. Jiang and J. Dequeker, J. Bone Miner. Res., 7, 599 (1992).

93.   S.M. Ott, J. Bone Miner. Res., 8 (Suppl. 2), S597 (1993).

94.   T.J. Wronski, L.M. Dann, H. Qi and C.-F. Yen, Calcif. Tissue Int., 53, 210 (1993).

95.   D.G. Hall and G. Stoica, J. Bone Miner. Res., 9, 221 (1994).

96.   C.A. Reasner, M.D. Stone, D.J. Hosking, A. Ballah and G.R. Mundy, J. Clin. Endocrinol. Metab., 77, 1067 (1993).

97.   R. Nemoto, Y. Nishijima, K. Uchida and K. Koiso, Br. J. Cancer, 67, 893 (1993).

98.   R.D. Devlin, R.W. Retallack, A.J. Fenton, V. Grill, D.H. Gutteridge, G.N. Kent, R.L. Prince and G.K. Worth, J. Bone Miner. Res., 9, 81 (1994).

99.   S.E. Papapoulos, Am. J. Med., 95 (Suppl. 5A), 48S (1993).

100. G. Van Der Pluijm, L. Binderup, E. Bramm, L. Van Der Wee-Pals, H. De Groot, E. Binderup, C. Löwik and S. Papapoulos, J. Bone Miner. Res., 7, 981 (1992).

101. C. Wüster, K.H. Schöter, D. Thiébaud, Ch. Manegold, D. Krahl, M.F. Clemens, M. Ghielmini, Ph. Jaeger and S.H. Scharla, Bone and Mineral, 22, 77 (1993).

102. M.-C. Monier-Faugere, R.M. Friedler, F. Bauss and H.H. Malluche, J. Bone Miner. Res., 8, 1345 (1993).

103. D. Agnusdei, A. Camporeale, F. Zacchei, C. Gennari, M.C. Baroni, D. Costi, M. Giondi, M. Passeri, A. Ciacca, C. Sbrenna, E. Falsettini and A. Ventura, Curr. Therap. Res., 51, 82 (1992).

104. M. Gambacciani, A. Spinetti, B. Cappagli, F. Taponeco, R. Felipetto, D. Parrini, N. Cappelli and P. Fioretti, J. Endocrinol. Invest., 16, 333 (1993).

105. R. Baron, L. Neff, D. Louvard and P.J. Courtoy, J. Cell Biol., 101, 2210 (1985).

106. R. Sarges, A. Gallagher, T.J. Chambers and L.-A. Yeh, J. Med. Chem., 36, 2828 (1993).

107. K. Mizunashi, Y. Furukawa, K. Katano and K. Abe, Calcif. Tissue Int., 53, 21 (1993).

108. R.O. Hynes, Cell, 69, 11 (1992).

109. J.E. Fisher, M.P. Caulfield, M. Sato, H.A. Quartuccio, R.J. Gould, V.M. Garsky, G.A. Rodan and M. Rosenblatt, Endocrinology, 132, 1411 (1993).

110. K.L. KIng, J.J. D'Anza, S. Bodary, R. Pitti, M. Siegel, R.A. Lazarus, M.S. Dennis, R.G. Hammonds Jr. and S.C. Kukreja, J. Bone Miner. Res., 9, 381 (1994).

111. M. Zaidi, A.S.M.T. Alam, C.L.-H. Huang, M. Pazianas, C.M.R. Bax, B.E. Bax, B.S. Moonga, P.J.R. Bevis and V.S. Shankar, Cell Calcium, 14, 271 (1993).

112. E.M. Brown, G. Gamba, D. Riccardi, M. Lombardi, R. Butters, O. Kifor, A. Sun, M.A. Hediger, J. Lytton and S.C. Hebert, Nature, 366, 575 (1993).

113. A.B. Roberts and M.B. Sporn, Growth Factors, 8, 1 (1993).

114. I. Asahina, T.K. Sampath, I. Nishimura and P.V. Hauschka, J. Cell Biol., 123, 921 (1993).

115. M.M. Shull, I. Ormsby, A.B. Kier, S. Pawlowski, R.J. Diebold, M. Yin, R. Allen, C. Sidman, G. Proetzel, D. Calvin, N. Annunziata and T. Doetschman, Nature, 359, 693 (1992).

116. T.J. Wronski, C.-F. Yen, H. Qi and L.M. Dann, Endocrinology, 132, 823 (1993).

117. J.M. Hock and I. Gera, J. Bone Miner. Res., 7, 65 (1992).

118. A.B. Hodsman, L.J. Fraher, T. Ostbye, J.D. Adachi and B.M. Steer, J. Clin. Invest., 91, 1138 (1993).

119. J.N. Bradbeer, M.E. Arlot, P.J. Meunier and J. Reeve, Clin. Endocrinol., 37, 282 (1992).

120. C. Nakamoto, H. Baba, M. Fukase, K. Nakajima, T. Kimura, S. Sakakibara, T. Fujita and K. Chihara, Acta Endocrinol., 128, 367 (1993).

121. A.G. Johansson, E. Lindh and S. Ljunghall, J. Int. Med., 234, 553 (1993).

122. C. Schmid, J. Int. Med., 234, 535 (1993).

123. M. Thorén, M. Soop, M. Degerblad and M. Sääf, Acta Endocrinol., 128, 41 (1993).

124. J.H. Tobias, J.W.M. Chow and T.J. Chambers, Endocrinology, 131, 2387 (1992).

125. C. Wüster, W.F. Blum, S. Schlemilch, M.B. Ranke and R. Ziegler, J. Int. Med., 234, 249 (1993).

126. P.A. Maher, J. Cell. Physiol., 151, 549 (1992).

SECTION VI.   TOPICS IN DRUG DESIGN AND DISCOVERY

Michael C. Venuti
Parnassus Pharmaceuticals, Inc.
Alameda, CA 94502

Chapter 29.   Adenylate Cyclase Subtypes as Molecular Drug Targets

James F. Kerwin, Jr.
Abbott Laboratories
Abbott Park, IL 60064

Introduction - Adenylate cyclase (AC, E.C. 4.6.1.1.) catalyzes the transformation of ATP into cyclic AMP (cAMP). Enzymatic activity is generally quantified by direct measurement of radiolabeled ($[^3H]$ or $[^{32}P]$) product (1). Receptor-mediated AC regulation has been among the most-studied of the signal transduction pathways. While the heterogeneity of G-proteins and their subunits has added complexity to AC-coupled pathways, recent evidence supports a substantial amount of molecular diversity of the effector protein, AC (2,3). This array of diversity coupled with emerging differential localizations and multiple regulatory pathways offers a complex yet potentially exploitable molecular target for drug discovery.

Adenylate Cyclase (AC) - The presence of different isoforms of AC was first demonstrated by the separation of calmodulin (CaM)-sensitive and CaM-insensitive ACs using CaM-sepharose affinity chromatography (4). Antibodies were later developed to distinguish between different AC isoforms (5). There are eight known mammalian ACs; six of these have been isolated and two other partial sequences exist (Table 1). Partial sequences of two human brain ACs have been determined (6,7), as well as three partial sequences of human ACs from erythroleukemia cell line (HEL) and four from HEK293 cells (8). The eight ACs are grouped into five families by homology (Figure 1), Type II (containing Types II, IV, and VII) and Type V (containing Types V and VI) families, and Types I, III, and VIII constituting individual representatives. Families were determined by comparing a 57 amino acid fragment from the central cytoplasmic loop among the different isoforms of AC in the rat (9). HEK293 cells express AC Types II, III, and VI, while HEL cells express AC Types III and VI (8). Purification of Type I AC from a bovine brain cDNA library produced a 120-kDa AC stimulated by G_s and Ca^{2+}/CaM (10). Using sequence data from this AC, full or partial sequences encoding an additional seven ACs have been identified (11-17). ACs are approximately 120-130 kDa and fit a common topology on the basis of hydropathy plots. ACs are membrane glycoproteins with 12 transmembrane spanning domains, a topology which is similar to ion channels and transporters (10). Both the N- and C-terminal domains are predicted to be cytoplasmic and two large cytoplasmic domains, a 350-amino acid loop between transmembrane domains 6 and 7, and a 250-300-amino acid domain at the C-terminal tail, contain the catalytic core (Figure 2). Expression of truncated forms of the Type I AC support the necessity of both of these domains for catalytic activity (18,19). The ends of these domains closest to the membrane mediate interactions with G-proteins. Half-maximal stimulation of Type I AC was found at 50 nM for Ca^{2+} and 20 nM for CaM (20), with maximal stimulation at 120 nM Ca^{2+} and 67 nM CaM (21). In vivo, Ca^{2+} increases regulate Type I AC (22). Type I AC is concentrated in the granule cells of the cerebellum, dentate gyrus, neocortex, and olfactory

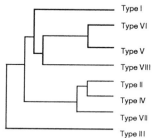

Figure 1. Homology of known mammalian ACs with relative homologies indicated by branching.

Type I
Type VI
Type V
Type VIII
Type II
Type IV
Type VII
Type III

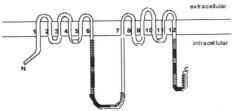

Figure 2. Predicted structure of mammalian ACs with N-terminal and C-terminal ends indicated by N and C respectively. Dark regions (>80%) and speckled regions (60-80%) indicate similarity over large stretches of amino acids.

system (23). Within the retina, the photoreceptor cells were particularly high in AC mRNA content, suggesting a role for cAMP-dependent events in photoadaptation. Type II AC, which is less abundant than Type I, is found in the granule cells of the cerebellum, CA-1, CA-4 regions of hippocampus and granule cells of the dentate gyrus (24). In situ hybridization of Type I and II AC mRNA in rat brain indicate clear differences in localization (25). In the paraventricular nucleus, substantia nigra, amygdala, habenular nuclei, and hypothalmus Type II predominates while in the hippocampus, entorhinal cortex and

neocortex, both are found, but with qualitatively different expression. Type III AC was first found in the olfactory neurons (12). Later, mRNA was detected in spinal cord, adrenal medulla and cortex, heart atrium, aorta, lung, as well as other brain regions (26). The localization of Type I AC and Type I deletion mutants in Drosophila melanogaster suggest a role in learning and memory presumably via linkage of Ca^{2+} and cAMP regulatory systems (18,27). Type III AC is regulated by Ca^{2+}/CaM but, unlike Type I, is not stimulated by Ca^{2+}/CaM in the absence of other effectors (20). Half-maximal stimulation of Type III AC occurs at 1 μM CaM and 5 μM Ca^{2+} in the presence of effectors such as forskolin (Fsk) and guanylyl-5'-imidodiphosphate (GppNHp). Type III AC may allow for Ca^{2+} amplification of cAMP signals generated by other ACs. The existence of cAMP-gated ion channels in olfactory sensory neurons suggests that cAMP signals are amplified by increases in intracellular Ca^{2+}, stimulating Type III AC (28). Type VI AC is inhibited by low concentrations of Ca^{2+} independently of CaM (15,29). Examples of such regulation include AC from pituitary, somatotrophs (30), cardiac sarcolemma (31), platelets (32), and C6-2B glioma (33).

Table 1. Isoenzymes of Adenylate Cyclase

AC Isoform	Amino acid length Source	Regulated by	mRNA (kB)	M_r (kDa)	Location
I	1,134 (10) bovine brain cDNA	Ca^{2+}/CaM βγ dimer	11.5	120	brain, retina adrenal medula
II	1,090 (13) rat olfactory cDNA	G_i, βγ dimer PKC	4.1	119	brain, lung olfactory epithelium
III	1,144 (14) rat olfactory (34)	Ca^{2+}/CaM	4.7	125	brain, lung, heart retina, adrenal medulla
IV	1,064 (11) rat testis cDNA	G_i, βγ dimer	3.5	117	brain, kidney, lung liver, heart, testis
V	1,223 (16,35) rat striatum, canine heart	Ca^{2+}	5.7	119	brain, kidney, lung liver, heart, testis
VI	1,165 (15) NCB-20 cells canine cardiac cDNA	G_i, Ca^{2+} PKA	6.1	132	heart, kidney, brain, testis intestine, liver, lung skeletal muscle
VII					brain, heart
VIII					brain

Type I AC was also cloned from a human fetal brain cDNA library and has 87% nucleotide sequence and 92% amino acid identity with bovine AC (36). This AC was mapped to the proximal portion of the short arm of chromosome 7. Two partial AC sequences isolated from human brain map to chromosome 8 (8q24.2) and to chromosome 5 (5p15.3) (7).

<u>Cofactors, Substrates and Inhibitors of AC</u> - Rat brain localization of mRNA for CaM, AC, and guanylate cyclase (GC) was compared to AC and GC activities in cortex, cerebellum, olfactory bulb, hippocampus, and striatum (37). While AC basal activities in these brain regions were equivalent (0.22-0.32 nmol/mg protein/min), differences existed between Fsk and Ca^{2+}/CaM stimulation. While Fsk increased basal activity in most brain regions by 10-fold, striatum levels increased 23-fold. In contrast the effect of Ca^{2+}/CaM was least in striatum (2.7 fold) and more pronounced in cerebellum (4.7-fold), cortex (3.8 fold), and hippocampus (3.6 fold). <u>In situ</u> hybridization confirmed higher CaM concentrations in cerebellum, pyramidal cells of the hippocampus, granule cells of the dentate gyrus, and cerebral cortex. CaM was abundant in striatum, suggesting that ACs in the striatum may not be under CaM regulation.

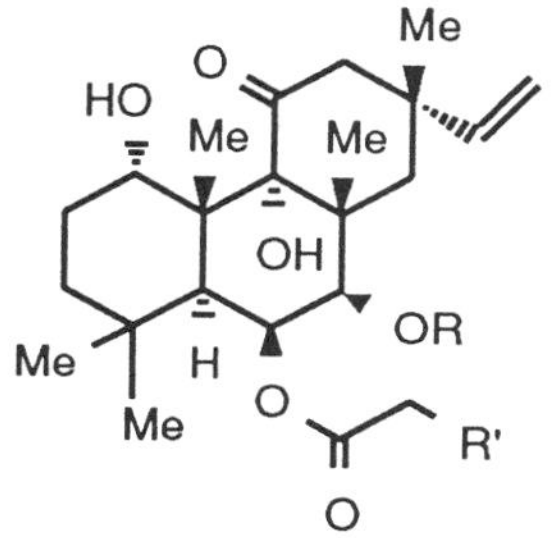

1 R = Ac, R' = CH$_2$NMe$_2$
2 R = H, R' = N(CH$_2$)$_5$

Fsk is a direct activator of AC (38). Recently, a water soluble derivative, NKH477 (**1**), was reported (39). Another derivative, HL 706 (**2**), is a vasodilator and produced a dose related increase in left ventricular dP/dt when administered orally (40). Other analogs differentiate AC and glucose transporters (41-43). MDL 12,330A (**3**) inhibits the nocturnal increase in serotonin N-acetyltransferase (NAT) activity in chick pineal gland and amphibian retina (44). The increase in NAT is mimicked by Fsk and cAMP analogs suggesting that AC activity was required for the regulation of NAT. Furthermore **3** had no effect on the stimulation of NAT by 8-bromo cAMP, consistent with **3** being a direct AC inhibitor. Adenosine and its P-site analogs inhibit Types I, V, and VI AC; other isoforms have not been examined (2).

<u>G-Protein Coupling to AC</u> - Many G-protein receptor systems are coupled to AC. Examples of stimulatory-coupled receptors are amylin, calcitonin, calcitonin-gene-related neuropeptide (CGRP), adrenocorticotropin, gonadotropin, adenosine (A$_{2A}$ and A$_{2B}$), PGE$_1$, dopamine (D$_1$), vasopressin, peptide histidine-isoleucine (PHI), secretin, endothelin (ET$_A$), serotonin (5-HT$_4$), pituitary adenylate cyclase activating polypeptide (PACAP), and β-adrenoceptors. Examples of inhibitory-coupled receptors include adenosine (A$_1$ and A$_3$), galanin, GABA-B, opioid, somatostatin, muscarinic, α$_2$-adrenoceptors, dopamine (D$_2$), serotonin (5-HT$_1$), and endothelin (ET$_B$). AC is regulated by two homologous guanine nucleotide binding regulatory proteins G$_s$ and G$_i$ (3). Both are composed of α, β, and γ subunits; the G$_s$ is composed of 45, 35, and 10 kDa and G$_i$ comprises 41, 35, and 10 kDa subunits. Currently 16 α subunits (350 - 380 amino acids long), 4 β subunits (340 amino

3

acids long), and 5 γ subunits (68-75 amino acids long) are known to comprise the heterotrimeric G-proteins. G-proteins are GTPases and stimulated receptors catalyze the release of GDP from G$_\alpha$ subunits. Activation of G$_s$ was presumed to result in dissociation of the α subunit from the βγ dimer. However, recent evidence suggests the entire G-protein complex may remain fully intact after activation (45). At present 6 of the known α subunits are involved with direct

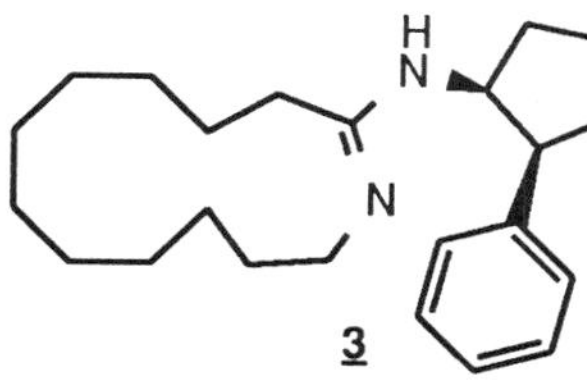

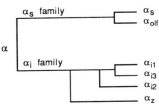

Figure 3. G_α subunits that are involved with the regulation of adenylate cyclase.

control of AC activity, α_s and α_{olf} - are activators of AC; α_{i1}, α_{i2}, α_{i3}, and α_z - are inhibitory (Figure 3) (46). In addition, α_s and α_{olf} may also stimulate L-type calcium channels, potassium-activated calcium channels, and Na^+-H^+ exchange (47); α_i units may be involved with activation of inwardly rectifying potassium channels and activation of ATP-inhibited potassium channels (46). The α_i units are subject to ADP ribosylation by pertussis toxin; the α_s, by cholera toxin. The eight ACs are all stimulated by α_s, however certain ACs are regulated by $\beta\gamma$ dimers (48,49). Type I AC is inhibited by $\beta\gamma$ dimer while Type II and IV AC are stimulated. Inhibition by $\beta\gamma$ dimer is independent of α_s, but requires that the AC be stimulated by α_s (50). An endogenous dopamine receptor that mediates pertussis toxin-sensitive inhibition of AC and an exogenous stimulatory receptor for Type II AC which was inhibited by dopamine were co-expressed (51). Dopamine application inhibited AC in the absence of the stimulatory receptor but enhanced cAMP formation in the presence of Type II AC. This is consistent if the endogenous receptor generated an α_i and a $\beta\gamma$ dimer that is neutral or inhibitory to the endogenous AC but stimulatory to the exogenous Type II AC. The exogenous AC activity was sensitive to pertussis toxin and also to co-transfected α_t of transducin, which scavenges $\beta\gamma$ dimers. AC stimulation by $\beta\gamma$ is particularly important because non-AC-coupled G-proteins or even G_i coupled AC receptors may activate Type II and IV AC via liberation of $\beta\gamma$ dimers (52). For example, β-adrenergic responses in rat hippocampus were potentiated by G_i-coupled receptor stimulation (53). Most $\beta\gamma$ regulation of AC appears to be associated with CNS, endocrine, exocrine, and epithelial cells. A stretch of amino acids (236-356) within α_s is responsible for AC activation (54). Mutation within four clusters of residues within this 121-residue region prevented AC activation and were also found in homologous sequences of α_{i2}. Modeling of G_α based on the structure of p21^{ras} suggests that effector activating regions form a surface on the membrane side of the protein that accommodates GTP (54). Mutations within G-protein coupled receptors at the C-terminal of the third cytoplasmic loop have led to constitutively active mutants, the β_2-receptor linked to AC, for example. Constitutive activity was rationalized by an allosteric transition of the receptor-G protein complex to a conformation with increased affinity for agonists and propensity to adopt the active conformation in the absence of agonist (55).

Desensitization and Sensitization of AC - Desensitization of hormonally regulated AC can be characterized as homologous or heterologous (56). Homologous desensitization can result from uncoupling of the receptor protein from the G-protein complex and sequestration/internalization of the protein. Uncoupling can be mediated by phosphorylation either by protein kinase A (PKA) or β-adrenergic receptor kinase (β-ARK) (57,58). Type V and VI AC contain a predicted PKA phosphorylation site in the central cytoplasmic loop and evidence suggests that this is a component of heterologous desensitization of AC (59). Heterologous desensitization is defined by a general loss of stimulation of AC by several hormones or activators. Underlying mechanisms of heterologous desensitization are not understood. However, functional changes in G-proteins are involved in some cases (58). D_1 dopamine receptors in opossum kidney cells down-regulate with chronic agonist exposure both in terms of receptor number and functional response. Co-expression of a cAMP phosphodiesterase resulted in clones with decreased AC responses to agonist stimulation but maintained receptor number (60). Chronic exposure of NG-108-15 cells to prostanoid agonists resulted in a concurrent down-regulation of prostanoid receptor and the $G\alpha_S$ subunit (61). Multiple pathways of heterologous desensitization may operate differentially. $G\alpha_S$ can also serve as a substrate for tyrosine kinases. Insulin, which activates a tyrosine kinase, down-regulates β_1-adrenoceptor coupled AC in SK-N-MC cells (62). cAMP regulates tyrosine kinases as in myocytes where it is responsible for p140^{trk} induction (63). Crosstalk between AC and tyrosine kinases remains to be fully elucidated. cAMP also regulates β-adrenoceptor coupled AC by inhibiting receptor recruitment to the plasma membrane, leading to a down-regulation of β-adrenoceptors and decreased AC activity (64). AC present in olfactory neurons is highly regulated by desensitization. Odorants elicit a rapid increase in cAMP evidenced in 20 seconds. Longer exposure times, 2-5 min, result in lower cAMP levels which trend towards baseline. Interestingly, if an odorant is reapplied shortly thereafter (~ 1 min), a potentiated response results implying a cellular "memory" (65). Multiple second messenger systems may be utilized allowing differentiation of odorants as well as of varying concentrations (65). Sensitization of AC-coupled receptor systems may also occur in response to loss of stimulus, either by conditioning or denervation (66,67). In some cases, sensitization is mediated by increased levels of G_α and G_β (68), or of receptor (69).

Other Regulatory Mechanisms - Ca^{2+}-stimulated (Type I), Ca^{2+}-insensitive (Type II), and Ca^{2+}-inhibited (Type VI) AC were transiently expressed in the HEK293 cell line (70). Phorbol esters increased cAMP synthesis by Type II AC greater than 9-fold over basal, with no effect on other ACs. This stimulatory effect was dose-dependent and synergistic with Fsk stimulation. Type II AC is insensitive to regulation by $[Ca^{2+}]_i$, but can be regulated by PKC (71). Type I and III ACs expressed in HEK293 cells demonstrated an increased responsiveness to Fsk stimulation when PKC was activated by TPA (72). This effect was not due to elevated intracellular Ca^{2+} and suggests that PKC is directly altering the responsiveness of ACs to Fsk. A major complication of acute myocardial ischemia is the occurrence of tachyarrhythmias, which may be due in part to transient increases in cellular AC activity (73). In myocardial ischemia, functionally active β-adrenoceptors are increased and at short times sensitization of AC results. At longer times down-regulation of G-proteins, particularly G_S, leads to desensitization. Short term sensitization of AC is mediated by PKC (74). Another pathway for AC sensitization is loss of inhibition coupling through G_i-coupled systems (73,75). PKC phosphorylation of AC could account for loss of inhibition by $G\alpha_{i2}$ (76,77). Thus several modes are available to regulate AC activity during acute myocardial ischemia. The importance of AC in vascular disorders is highlighted by the finding that spontaneously hypertensive rats (SHR), though having no differences in β-adrenoceptor number in myocardial tissue, possess an increased efficiency of AC-coupling (78). This effect apparently compensates for a lowered basal activity of AC (79). The necessity for selectivity in therapeutic targeting is supported by the finding that AC stimulation prevents pulmonary capillary injury in ischemia-reperfusion models (80,81).

In C6-2B cells Types VI and III AC predominate. Inhibition of AC by substance K (SK) ($t_{1/2}$ = 35 sec) correlated with an increase in $[Ca^{2+}]_i$ ($t_{1/2}$ = 25 sec). Inhibition of AC ($t_{1/2}$ = 1.6 min) by thapsigargin, a selective microsomal Ca^{2+}-ATPase inhibitor (33), also correlated with an increase in intracellular Ca^{2+} ($t_{1/2}$ = 1.5 min). Activation of transfected SK receptor led to Ca^{2+} release through PI hydrolysis, but PKC was not involved with AC inhibition (82).

A truncated form of AC (Type V-α) containing six transmembrane domains and a cytoplasmic loop was cloned from a canine cardiac cDNA library (19). This half-form of AC is generated by a polyadenylation signal located within an intronic sequence. Enzymatic activity is reconstituted by heterodimerization, another mechanism for regulation of AC.

Nuclear hormonal receptor-mediated regulation of AC activity results from changes which affect coupling efficiency with G-protein units (83) or increased activity of the catalytic unit (84). Testosterone down-regulates G_i and G_s (85) and cytokines enhance basal and stimulated AC activity (86). Cytokine's effects can be inhibited by AC-activation (87).

In S. cerevisiae and S. pombe, cyclase activity is further modulated by AC associated proteins (CAP). In S. cerevisiae, CAP mediates cellular responses of the ras/cAMP pathway (88). A human cDNA encoding a CAP-related 475-amino-acid peptide has been identified and co-expression of human CAP protein in S. cerevisiae substituted for CAP-defective phenotypes. AC modulatory effector proteins also exist in rat (89).

A bicarbonate-stimulated AC-activity exists in ocular ciliary processes. Membrane-bound bicarbonate-sensitive AC in fluid-transporting tissues may be part of a cellular HCO_3^- autoregulatory mechanism acting via cAMP-control of membrane ion transporters (90,91).

Pharmacological Roles for AC - Dopamine can activate or inhibit ACs. Dopamine receptor-coupled ACs are highly concentrated in the corpus striatum and limbic structures of the CNS. Northern blot analysis of various brain regions and peripheral tissues detected Type V in striatum, with small amounts in heart and kidney (92). Only Type V AC is localized to corpus striatum and other limbic structures (35). G_{olf} is the predominant G-protein in striatum suggesting that striatal dopamine neurotransmission is uniquely coupled via G_{olf} and Type V AC (35). Type I AC is important in neuroplasticity. A Drosophila melanogaster learning-deficient mutant, rutabaga, is deficient in Type I AC which maps within the X chromosome at the rut locus (93). Studies in Aplysia have found a role for CaM-sensitive AC in associative learning interactions between conditioned and unconditioned stimuli (29). AC activation enhanced the field excitatory post synaptic potential (EPSP) evoked by stimulation of Schaffer/commissural afferents in rat hippocampus (94), supporting a role for cAMP in the modulation of long term potentiation (LTP).

Two AC genes have been isolated from Dictyostelium. One gene product, ACA, resembles mammalian ACs while the other, ACG, resembles mammalian membrane bound GCs (95). ACA is required for aggregation; ACG is normally expressed during germination. Dictyostelium possess cAMP receptors which require $G_{\alpha 2}$ that activate AC (96). One of these receptors (cAR1), but not $G_{\alpha 2}$, is required for the normal adaptive response of AC activity suggesting that AC is coupled to an unidentified G-protein. Mechanisms by which cyclic nucleotides regulate nuclear events are of interest for anti-infective approaches and understanding of human physiology. cAMP regulates cellular circadian functions, meiosis, and mitosis (97). In S. pombe, a glucose-induced AC activity was found encoded by the git2 gene whose function is to provide cAMP to repress transcription of the fbl1 gene (98).

Therapeutic Directions for AC Inhibition - Transcription factors are modulated by intracellular signal transduction pathways, in particular AC. Transcriptional response of cAMP-inducible genes is achieved by interplay of nuclear transcription factors such as CREB and CREM (99,100). Since prokaryotes utilize cAMP-responsive transcription, AC represents a potential therapeutic target (101). Trypanosoma brucei differentiate synchronously from pleomorphic bloodstream forms to procyclic forms. AC activity occurs 6-10 hours after differentiation, immediately preceding cell division, and again at 20-40 hours, when cells emerging from the first division begin proliferation (102). The second phase of activity does not appear to be associated with the bloodstream-specific AC, ESAG 4, which is shed at 9 hours. Three other ACs, GRESAG 4.1, 4.2, and 4.3, may also control life cycle and/or proliferation (103). Adenylate cyclase toxin (ACT) is a competent AC produced and secreted by Bordetella pertussis (104) which hemolyzes erythrocytes (105). The ability to increase intracellular cAMP in macrophages alters macrophage killing following uptake and thereby increases bacterial survival (105). ACT deficient mutants lack the ability to initiate infection indicating ACT as a potential therapeutic target. In Chlamydomonas, AC plays a role in gametogenesis (106). In man, cAMP transduces positive and negative signals on cell growth and differentiation which are transmitted by RI and RII (cAMP-binding regulatory subunits of cAMP dependent protein kinases). The growth stimulatory RI and inhibitory RII are in equilibirium to maintain normal cell function and loss of balance can cause malignancy and other pathology (107). Activation of AC inhibits mitogen induction of glycolytic isozymes in human peripheral T cells (108). Fsk added with mitogen, or 12-24 hours after mitogen, inhibits DNA synthesis and causes cells to accumulate in the G_0 or early G_1 phase (108,109). T-cell regulation produced by different AC-coupled receptor systems such as β-adrenoceptors and PGE_2 is distinct (110). PGE_2 receptor stimulation provides cAMP accumulation, which is greater and longer lasting than that of β-adrenoceptor stimulation, resulting in a greater functional inhibition. In tumor lymphoid cells, β-adrenoceptor loss or a loss of coupling efficiency may contribute to proliferation (111,112). Neutrophil phagocytic functions are inhibited by cAMP elevation (113). AC regulates growth in human gastric cancer cell lines, and AC inhibition has been advanced as an anticancer target (114), suggesting a viable antiproliferative approach to control of tumors that are non-responsive to hormonal regulation. In cultured neurons, abnormal cAMP elevations result in altered G-protein expression and susceptibility to proliferation (115).

Type V AC coupled via G_{olf} mediates dopamine neurotransmission in striatum (35). The relative absence of Type V AC from peripheral tissues suggests the possibility of neurological and psychiatric therapeutic agents based on Type V AC modulation (35,116). D_1 receptor AC activity facilitates ischemia-induced CNS damage (117).

Sustained synthesis of proopiomelanocortin (POMC) from rat anterior pituitary evoked during stress is paralleled by enhanced AC activity (118). Type II AC mRNA was found to be significantly increased in the pituitary and frontal cortex while G-proteins and Type I AC mRNA levels were unaffected. Physiological mechanisms regulating the hypothalamic pituitary axis in chronic stress are of interest and a major focus is glucocorticoid and CRH receptors. Since cAMP alters POMC gene expression, it is hypothesized that AC may be an important component in chronic stress response. Altered responsiveness to CRH, an AC coupled system, may explain compensation for a diminished number of CRH receptors during chronic stress, which results in higher sensitivity than in non-stressed controls (119). Chronic stress attenuates the AC response to CaM in cortex (120). Adrenalectomy abolished a diurnal sensitivity of hippocampal AC to CaM-stimulation. Since corticosterone treatment did not block the effect of adrenalectomy, other adrenal secretions are involved in regulating hippocampal AC, supporting the idea of adrenal regulation of information processing and LTP mechanisms in the hippocampus (121). Since hippocampal LTP appears to require AC activity (122), the presence of Type II AC in pituitary presents a viable alternative target for stress-related disorders.

Homologous desensitization of myocardial AC-coupled β-adrenoceptors during septic insult is hypothesized to disrupt signal transduction pathways mobilizing cardiac reserve. Thus effectively a larger burden is placed on remaining myocardial tissue until cardiac failure results (123). A study of AC responsiveness in SHR concluded that an increase in $G_{\alpha i}$ expression may play a pathophysiological role in terminal heart failure and in hypertrophic cardiomyopathy (124). Similar studies in human myocardial tissue support this finding (125) and AC regulation may be contributory in idiopathic dilated cardiomyopathy (126), ischemic dilated cardiomyopathy (127), and primary pulmonary hypertension (128).

Other potential AC-targeted endocrine and metabolic disorders exist. Alterations in G_i and G_s expression may contribute to insulin-resistant diabetes (129). CGRP and amylin exert effects on inhibition of bone resorption via quiescence (Q effect), which is AC-mediated (130). Elevation of intracellular cAMP causes excessive growth hormone secretion and somatotroph proliferation in acromegalic patients (131,132). Elevated cAMP regulates net collagen content in hepatocytes by signaling intracellular collagen degradation (133).

<u>Future Directions</u> - The heterogeneity of ACs will continue to unfold as new tools allow further differentiation and definition of subtype-specific pharmacology. As focus shifts from receptor-based pharmacotherapy to the modulation of transduction pathways, increased research on AC may stimulate novel approaches to therapeutics. Current evidence clearly suggests a role for ACs in disease processes, although in some cases the role of AC may be unclear and yet to be elucidated. For example, in fragile X syndrome, the ability of platelet-derived AC to regulate cAMP levels is impaired in some patients (134), and this is thought to result from a defect in AC regulation or activity. Lithium, a current treatment for bipolar affective disorder, modulates AC and PI systems; whether therapeutic effects could be derived from selective AC inhibitors is an open question (135).

References

1. S.P. Brooks, M.J. Mancebo, C.P. Holden and K.B. Storey, Anal.Biochem., <u>210</u>, 419 (1993).
2. R. Iyengar, FASEB J., <u>7</u>, 768 (1993).
3. J.P. Pieroni, O. Jacobowitz, J. Chen and R. Iyengar, Curr.Opin.Neurobiol., <u>3</u>, 345 (1993).
4. K.R. Westcott, D.C. LaPorte and D.R. Storm, Proc.Natl.Acad.Sci.USA, <u>76</u>, 204 (1979).
5. S. Mollner and T. Pfeuffer, Eur.J.Biochem., <u>171</u>, 265 (1988).
6. J. Parma, D. Stengel, M.-H. Gannage, M. Poyard, R. Barouki and J. Hanoune, Biochem.Biophys.Res. Commun., <u>179</u>, 455 (1991).
7. D. Stengel, J. Parma, M.H. Gannage, N. Roeckel, M.G. Mattei, R. Barouki and J. Hanoune, Hum.Genet., <u>90</u>, 126 (1992).
8. K. Hellevuo, M. Yoshimura, M. Kao, P.L. Hoffman, D.M. Cooper and B. Tabakoff, Biochem.Biophys. Res.Commun., <u>192</u>, 311 (1993).
9. J. Krupinski, T.C. Lehman, C.D. Frankenfield, J.C. Zwaagstra and P.A. Watson, J.Biol.Chem., <u>267</u>, 24858 (1992).
10. J. Krupinski, F. Coussen, H.A. Bakalyar, W.-J. Tang, P.G. Feinstein, K. Orth, C. Slaughter, R.R. Reed and A.G. Gilman, Science, <u>244</u>, 1558 (1989).
11. B. Gao and A.G. Gilman, Proc.Natl.Acad.Sci.USA, <u>88</u>, 10178 (1991).
12. H.A. Bakalyar and R.R. Reed, Science, <u>250</u>, 1403 (1990).
13. P.G. Feinstein, A. Schrader, H.A. Bakalyar, W.J. Tang, J. Krupinski, A.G. Gilman and R.R. Reed, Proc.Natl.Acad.Sci.USA, <u>88</u>, 10173 (1991).
14. Y. Ishikawa, S. Katsushika, L. Chen, N.J. Halnon, J. Kawabe and C.J. Homcy, J.Biol.Chem., <u>267</u>, 13553 (1992).
15. M. Yoshimura and D.M. Cooper, Proc.Natl.Acad.Sci.USA, <u>89</u>, 6716 (1992).
16. S. Katsushika, L. Chen, J. Kawabe, R. Nilakantan, N.J. Halnon, C.J. Homcy and Y. Ishikawa, Proc.Natl.Acad.Sci.USA, <u>89</u>, 8774 (1992).
17. R.T. Premont, J. Chen, H. Ma, M. Ponnapalli and R. Iyengar, Proc.Natl.Acad.Sci.USA, <u>89</u>, 9809 (1992).
18. L.R. Levin, P.L. Han, P.M. Hwang, P.G. Feinstein, R.L. Davis and R.R. Reed, Cell, <u>68</u>, 479 (1992).
19. S. Katsushika, J. Kawabe, C.J. Homcy and Y. Ishikawa, J.Biol.Chem., <u>268</u>, 2273 (1993).
20. E.J. Choi, Z. Xia and D.R. Storm, Biochemistry, <u>31</u>, 6492 (1992).
21. M. Gnegy, N. Muirhead, J.-M. Roberts-Lewis and G. Treisman, J.Neurosci., <u>4</u>, 2712 (1984).
22. E.J. Choi, S.T. Wong, T.R. Hinds and D.R. Storm, J.Biol.Chem., <u>267</u>, 12440 (1992).
23. Z. Xia, E.J. Choi, F. Wang, C. Blazynski and D.R. Storm, J.Neurochem., <u>60</u>, 305 (1993).
24. T. Furuyama, S. Inagaki and H. Takagi, Brain.Res.Mol.Brain.Res., <u>19</u>, 165 (1993).
25. N. Mons, M. Yoshimura and D.M. Cooper, Synapse, <u>14</u>, 51 (1993).
26. Z. Xia, E.J. Choi, F. Wang and D.R. Storm, Neurosci.Lett., <u>144</u>, 169 (1992).
27. Z. Xia, D.R. Cheryl, K.M. Merchant, D.M. Dorsa and D.R. Storm, Neuron, <u>6</u>, 431 (1991).
28. T. Kurahashi and K.W. Yau, Nature, <u>363</u>, 71 (1993).
29. V. Yovell, E.R. Kandel, Y. Dudai and T.W. Abrams, J.Neurochem., <u>59</u>, 1736 (1992).
30. N. Narayanan, B. Lussier, M. French, B. Moor and J. Kraicer, Endocrinology, <u>124</u>, 485 (1989).
31. R.A. Colvin, J.A. Oibo and R.A. Allen, Cell Calcium, <u>12</u>, 19 (1991).
32. K.K. Caldwell, C.L. Boyajian and D.M. Cooper, Cell Calcium, <u>13</u>, 107 (1992).
33. M.A. Debernardi, R. Munshi, M. Yoshimura, D.M. Cooper and G. Brooker, Biochem.J., <u>293</u>, 325 (1993).
34. B.P. Menco, R.C. Bruch, B. Dau and W. Danho, Neuron, <u>8</u>, 441 (1992).

35. C.E. Glatt and S.H. Snyder, Nature, 361, 536 (1993).
36. E.C. Villacres, Z. Xia, L.H. Bookbinder, S. Edelhoff, C.M. Disteche and D.R. Storm, Genomics, 16, 473 (1993).
37. I. Matsuoka, G. Giuili, M. Poyard, D. Stengel, J. Parma, G. Guellaen and J. Hanoune, J.Neurosci., 12, 3350 (1992).
38. A. Laurenza, J.D. Robbins and K.B. Seamon, Mol.Pharmacol., 41, 360 (1992).
39. M. Hosono, T. Takahira, A. Fujita, R. Fujihara, O. Ishizuka, T. Tatee and K. Nakamura, J.Cardiovasc.Pharmacol., 19, 625 (1992).
40. R. Rajagopalan, A.V. Ghate, P. Subbarayan, W. Linz and B.A. Schoelkens, Arzneimittelforschung, 43, 313 (1993).
41. N.M. Appel, J.D. Robbins, E.B. De Souza and K.B. Seamon, J.Pharmacol.Exp.Ther., 263, 1415 (1992).
42. J.D. Robbins, N.M. Appel, A. Laurenza, I.A. Simpson, E.B. De Souza and K.B. Seamon, Brain Res., 581, 148 (1992).
43. E.M. Sutkowski, F. Maher, A. Laurenza, I.A. Simpson and K.B. Seamon, Biochemistry, 32, 2415 (1993).
44. J.H. Boatright, J. Gan, B.J. Butler, G. Avendano and P.M. Iuvone, FASEB J., 5, A1499 (1991).
45. A. Bar-Sinai, I. Marbach, R.G. Shorr and A. Levitzki, Eur.J.Biochem., 207, 703 (1992).
46. L. Birnbaumer, Cell, 71, 1069 (1992).
47. D.L. Barber, M.B. Ganz, P.B. Bongiorno and C.D. Strader, Mol.Pharmacol., 41, 1056 (1992).
48. K. Enomoto, K. Hayama, M. Takano and T. Asakawa, Jpn.J.Pharmacol., 62, 103 (1993).
49. R. Taussig, L.M. Quarmby and A.G. Gilman, J.Biol.Chem., 268, 9 (1993).
50. W.-J. Tang and A.G. Gilman, Science, 254, 1500 (1991).
51. A.D. Federman, B.R. Conklin, K.A. Schrader, R.R. Reed and H.R. Bourne, Nature, 356, 159 (1992).
52. K.D. Lustig, B.R. Conklin, P. Herzmark, R. Taussig and H.R. Bourne, J.Biol.Chem., 268, 13900 (1993).
53. R. Andrade, Neuron, 10, 83 (1993).
54. C.H. Berlot and H.R. Bourne, Cell, 68, 911 (1992).
55. R.J. Lefkowitz, S. Cotecchia, P. Samama and T. Costa, Trends Pharmacol.Sci., 14, 303 (1993).
56. L. Rampello, Funct.Neurol., 7, 97 (1992).
57. J.L. Arriza, T.M. Dawson, R.B. Simerly, L.J. Martin, M.G. Caron, S.H. Snyder and R.J. Lefkowitz, J.Neurosci., 12, 4045 (1992).
58. R.T. Premont and R. Iyengar, Endocrinology, 125, 1151 (1989).
59. R.T. Premont, O. Jacobowitz and R. Iyengar, Endocrinology, 131, 2774 (1992).
60. M.D. Bates, C.L. Olsen, B.N. Becker, F.J. Albers, J.P. Middleton, J.G. Mulheron, S.L. Jin, M. Conti and J.R. Raymond, J.Biol.Chem., 268, 14757 (1993).
61. E.J. Adie, I. Mullaney, F.R. McKenzie and G. Milligan, Biochem.J., 285, 529 (1992).
62. S.W. Bahouth and S. Lopez, Life Sci., 51, PL271 (1992).
63. P.B. Ehrhard, U. Ganter, J. Bauer and U. Otten, Proc.Natl.Acad.Sci.USA, 90, 5423 (1993).
64. D. Sandnes, F.W. Jacobsen, M. Refsnes and T. Christoffersen, Eur.J.Pharmacol., 246, 163 (1993).
65. G.V. Ronnett, H. Cho, L.D. Hester, S.F. Wood and S.H. Snyder, J.Neurosci., 13, 1751 (1993).
66. W.W. Fleming and D.A. Taylor, J.Neural Transm.Suppl., 34, 179 (1991).
67. H.K. Hammond, D.A. Roth, C.E. Ford, G.W. Stamnas, M.G. Ziegler and C. Ennis, Circulation, 85, 666 (1992).
68. T. Babila and D.C. Klein, J.Neurochem., 59, 1356 (1992).
69. B. Mouillac, M. Caron, H. Bonin, M. Dennis and M. Bouvier, J.Biol.Chem., 267, 21733 (1992).
70. M. Yoshimura and D.M. Cooper, J.Biol.Chem., 268, 4604 (1993).
71. O. Jacobowitz, J. Chen, R.T. Premont and R. Iyengar, J.Biol.Chem., 268, 3829 (1993).
72. E.J. Choi, S.T. Wong, A.H. Dittman and D.R. Storm, Biochemistry, 32, 1891 (1993).
73. B. Rauch and F. Niroomand, Eur.Heart J., 12, 76 (1991).
74. R.H. Strasser, R. Braun-Dullaeus, H. Walendzik and R. Marquetant, Circ.Res., 70, 1304 (1992).
75. F. Niroomand, M. Bangert, T. Beyer and B. Rauch, J.Mol.Cell.Cardiol., 24, 471 (1992).
76. J. Chen and R. Iyengar, J.Biol.Chem., 268, 12253 (1993).
77. C. Gallego, S. Gupta, S. Winitz, B. Eisfelder and G. Johnson, Proc.Natl.Acad.Sci.USA, 89, 9695 (1992).
78. F. Ohsuzu, S. Katsushika, S. Maie, M. Akanuma, S. Yanagida, N. Sakata, H. Ishida, N. Aosaki and H. Nakamura, Jpn.Circ.J., 56, 301 (1992).
79. C.J. Clark, G. Milligan, A.R. McLellan and J.M. Connell, Hypertension, 21, 204 (1993).
80. W.K. Adkins, J. Barnard, S. May, A.. Seibert, J. Haynes and A.E. Taylor, J.Appl.Physiol., 72, 492 (1992).
81. J. Haynes Jr., J. Robinson, L. Saunders, A.E. Taylor and S.J. Strada, Am.J.Physiol., 262, H511 (1992).
82. M.A. Debernardi, R. Munshi and G. Brooker, Mol.Pharmacol., 43, 451 (1993).
83. B. Feve, B. Baude, S. Krief, A. Strosberg, J. Pairault and L.J. Emorine, J.Biol.Chem., 267, 15909 (1992).
84. A. McLellan, S. Tawil, F. Lyall, G. Milligan, J. Connell and C. Kenyon, J.Mol.Endocrinol., 9, 237 (1992).
85. M.N. Dieudonne, R. Pecquery, J.P. Dausse and Y. Giudicelli, Biochim.Biophys.Acta, 1176, 123 (1993).
86. R.J. Anderson and R. Breckon, Kidney Int., 42, 559 (1992).
87. G. De Sarro and G. Nistico, Int.J.Neurosci., 59, 67 (1991).

88.   H. Matviw, G. Yu and D. Young, Mol.Cell.Biol., 12, 5033 (1992).
89.   A. Zelicof, J. Gatica and J.E. Gerst, J.Biol.Chem., 268, 13448 (1993).
90.   M.L. Toews, L.J. Arneson-Rotert and S.A. Liewer, J.Pharmacol.Exp.Ther., 264, 1211 (1993).
91.   T.W. Mittag, W.B. Guo and K. Kobayashi, Am.J.Physiol., 264, F1060 (1993).
92.   J.P. Pieroni, D. Miller, R.T. Premont and R. Iyengar, Nature, 363, 679 (1993).
93.   M.S. Livingston, P.P. Sziber and W.G. Quinn, Cell, 37, 205 (1984).
94.   L.E. Chavez-Noriega and C.F. Stevens, Brain Res., 574, 85 (1992).
95.   G.S. Pitt, N. Milona, J. Borleis, K.C. Lin, R.R. Reed and P.N. Devreotes, Cell, 69, 305 (1992).
96.   M. Pupillo, R. Insall, G.S. Pitt and P.N. Devreotes, Mol.Biol.Cell, 3, 1229 (1992).
97.   L. Edmunds Jr., I.A. Carre, C. Tamponnet and J. Tong, Chronobiol.Int., 9, 180 (1992).
98.   S.M. Byrne and C.S. Hoffman, J.Cell.Sci., 105, 1095 (1993).
99.   E. Borrelli, J.P. Montmayeur, N.S. Foulkes and P. Sassone-Corsi, Crit.Rev.Oncog., 3, 321 (1992).
100.   N.S. Foulkes, B. Mellstrom, E. Benusiglio and P. Sassone-Corsi, Nature, 355, 80 (1992).
101.   J.L. Botsford and J.G. Harman, Microbiol.Rev., 56, 100 (1992).
102.   S. Rolin, P. Paindavoine, J. Hanocq-Quertier, F. Hanocq, Y. Claes, D. Le Ray, P. Overath and E. Pays, Mol.Biochem.Parasitol., 61, 115 (1993).
103.   P. Paindavoine, S. Rolin, S. Van Assel, M. Geuskens, J.C. Jauniaux, C. Dinsart, G. Huet and E. Pays, Mol.Cell.Biol., 12, 1218 (1992).
104.   M.K. Gross, D.C. Au, A.L. Smith and D.R. Storm, Proc.Natl.Acad.Sci.USA, 89, 4898 (1992).
105.   H.R. Masure, Microb.Pathog., 14, 253 (1993).
106.   Y. Zhang and W.J. Snell, J.Biol.Chem., 268, 1786 (1993).
107.   Y.S. Cho-Chung, Semin.Cancer Biol., 3, 361 (1992).
108.   S. Marjanovic, P. Wollberg, S. Skog, T. Heiden and B.D. Nelson, Arch.Biochem.Biophys., 302, 398 (1993).
109.   S. Valitutti, M. Dessing and A. Lanzavecchia, Eur.J.Immunol., 23, 790 (1993).
110.   M.M. Bartik, W.H. Brooks and T.L. Roszman, Cell.Immunol., 148, 408 (1993).
111.   G.A. Cremaschi, P. Fisher and F. Boege, Immunopharmacology, 22, 195 (1991).
112.   C. D'Amico, M. Crescimanno, M.G. Armata, V. Leonardi, M. Palazzoadriano and N. D'Alessandro, Anticancer Res., 12, 2253 (1992).
113.   C. Bengis-Garber and N. Gruener, Cell.Signal, 4, 247 (1992).
114.   M. Piontek, K.-J. Hengles, R. Porschen and G. Strohmeyer, J.Cancer Res.Clin.Oncol., 119, 697 (1993).
115.   P.S. Eriksson, M. Nilsson, B. Carlsson, O.G. Isaksson, L. Ronnback and E. Hansson, Neurosci.Lett., 135, 28 (1992).
116.   J.B. Thilo and P.W. Burnet, Psychopharmacol.Bull., 28, 477 (1992).
117.   R. Prado, R. Busto and M.Y. Globus, J.Neurochem., 59, 1581 (1992).
118.   A.C. Morrill, D. Wolfgang, M.A. Levine and G.S. Wand, Life Sci., 53, 1719 (1993).
119.   E. Young and H. Akil, Psychoneuroendocrinology, 13, 317 (1988).
120.   M.N. Gannon, R.E. Brinton, R.R. Sakai and B.S. McEwen, J.Neuroendocrinol., 3, 37 (1991).
121.   D.M. Diamond, M.C. Bennet, D.A. Engstromn, M. Fleshner and G.M. Rose, Brain Res., 492, 356 (1989).
122.   T.V. Dunwiddie, M. Taylor, L.R. Heginbotham and W.R. Proctor, J.Neurosci., 12, 506 (1992).
123.   S.B. Jones and F.D. Romano, Circ.Shock, 30, 51 (1990).
124.   M. Bohm, P. Gierschik, A. Knorr, K. Larisch, K. Weismann and E. Erdmann, J.Hypertens., 10, 1115 (1992).
125.   M. Bohm, P. Gierschik and E. Erdmann, Basic Res.Cardiol., 87, 37 (1992).
126.   C. Kawai, Cardiovasc.Drugs Ther., 6, 7 (1992).
127.   L.X. Fu, Q.M. Liang, F. Waagstein, J. Hoebeke, C. Sylven, E. Jansson, P. Sotonyi and A. Hjalmarson, Cardiovasc.Res., 26, 950 (1992).
128.   A.M. Feldman, Circulation, 87, IV27 (1993).
129.   T.M. Palmer, P.V. Taberner and M.D. Houslay, Cell.Signal, 4, 365 (1992).
130.   A.S. Alam, C.M. Bax, V.S. Shankar, B.E. Bax, P.J. Bevis, C.L. Huang, B.S. Moonga, M. Pazianas and M. Zaidi, J.Endocrinol., 136, 7 (1993).
131.   E.F. Adams, S. Brockmeier, E. Friedmann, M. Roth, M. Buchfelder and R. Fahlbusch, Neurosurgery, 33, 198 (1993).
132.   G. Faglia, Acta Endocrinol., 129, 1 (1993).
133.   K.I. Andrabi, N. Kaul, S. Mudassar, J.B. Dilawari and N.K. Ganguly, Mol.Cell.Biochem., 109, 89 (1992).
134.   E. Berry-Kravis and P.R. Huttenlocher, Ann.Neurol., 31, 22 (1992).
135.   M.I. Masana, J.A. Bitran, J.K. Hsiao and W.Z. Potter, J.Neurochem., 59, 200 (1992).

Chapter 30. Recent Advances in Antisense Technology

John S. Kiely
Isis Pharmaceuticals, Inc.
Carlsbad, CA 92008

Introduction - The treatment of disease by modulation of biological processes can be undertaken at the level of the gene, the mRNA, or the protein. Since the 1970's, proteins associated with disease pathology have been the major focus of drug discovery efforts. More recently however, the goal of creating specific drugs to act directly on the mRNA is now the focus of significant research and development efforts (1,2). This paradigm is a method of affecting the expression of a gene by interfering with the translation of the gene's mRNA into a functional protein. This exploits the susceptibility of formally single stranded mRNA to undergo sequence-specific high-affinity binding to a complementary oligonucleotide sequence, called the antisense sequence, via Watson-Crick hydrogen bonding. By binding to the mRNA, the antisense oligomer can by any one of several mechanisms interfere with and arrest cytosolic translation of the mRNA into protein (Figure 1). Each mechanism assumes that the antisense oligomer has found and bound to the target mRNA (3). The first is translational (ribosomal) blockade where the antisense strand hybridizes to the sense strand and prevents the ribosome from completely reading the mRNA message, resulting in synthesis of truncated nonfunctional protein. The second is binding of a DNA-like antisense oligonucleotide to the mRNA and activation of RNase H, an enzyme that specifically cleaves the RNA strand of a RNA/DNA duplex. The cleavage by RNase H of the RNA strand results in destruction of the message and arrest of protein production. Thirdly, within the 5'-untranslated region (5'-UTR) of the mRNA, failure to initiate translation can be achieved by competition between the ribosome and the antisense strand for binding to the 5'-UTR. In this competition, binding of the antisense strand to the 5'-UTR can also lead to activation of RNase H and cleavage of the mRNA, preventing binding of the ribosome and its translation of the message into protein. Finally, by any of the previous mechanisms the appearance of fully mature mRNA in the cytosol can be prevented at the level of RNA transcription, splicing, processing or nucleus to cytoplasm transport. During 1993 antisense technology has been reviewed in several different forums (1,4-6). Recent reviews on the synthesis of phosphodiester oligonucleotides have appeared (7-9).

OLIGONUCLEOTIDE STRUCTURAL MODIFICATIONS

Medicinal chemistry efforts directed towards antisense mechanisms have been focused on modifying oligonucleotides to achieve stability towards degradation and to increase binding affinity. Natural phosphodiester oligomers are too rapidly degraded by both endo- and exo- nucleases to be used as drugs directly in cellular or whole animal systems, except at very high doses. Recent efforts have been directed at creating nuclease-resistant high-affinity oligonucleotide analogs by a variety of approaches.

Backbone Modified Oligomers - The phosphodiester backbone, **1a**, consists of the phosphate linkage having a 3' to 5' linkage with the sugar ring, as shown in Figure 2 (10). The simplest backbone modification incorporated to date, and the most studied, is replacement of one phosphate oxygen with a sulfur to produce a phosphorothioate oligonucleotide, **1b** (11-13). This modification imparts increased stability towards degradation by nucleases, but concomitantly decreases binding affinity. The synthesis of phosphodiester and phosphorothioate oligonucleotides is reasonably routine, and is usually carried out on automated synthesizers (9). Sequences consisting of a mixture of phosphodiester and phosphorothioate linkages have been investigated. Increasing sulfur content lowers the duplex thermal melting temperature (Tm) relative to that of the natural phosphodiester (14). In addition, phosphorothioates contribute to oligomer stability toward nucleases dependent on position and total sulfur content.

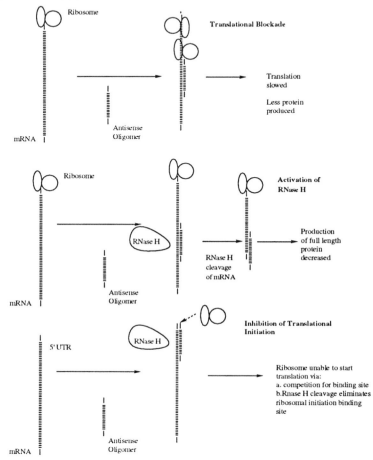

Figure 1. Antisense Modes of Action (Adapted from reference 3) __

Oxidation of the phosphorus from P(III) to P(V) with sulfur produces a chiral center. Phosphorothioate oligonucleotide sequences are in general made and evaluated as diastereomeric mixtures. The probable need to devise stereospecific routes to phosphorothioate oligomers has been recognized and some progress has been made (15,16).

Peptide nucleic acids (PNAs, see Figure 2) are the first successful substitute for the sugar-phosphate backbone that have displayed hybridization equal to or better than natural DNA or RNA (17-19). The replacement of the backbone with a N-(2-

aminoethyl)glycine unit is a significant departure from previous modifications (10). PNAs can bind to a target oligonucleotide sequence in two orientations, either parallel, where the PNA amino terminus is aligned with the DNA 5'-end, or antiparallel, where the PNA amino terminus aligns with the 3'-end of the DNA. The PNAs display strong sequence-dependent specific duplex binding in an antiparallel manner when hybridized to a mixed purine -pyrimidine DNA or RNA sequence (20). As measured by thermal Tm studies, this hybridization imparts a +1.0-1.2 °C per modification increase in Tm over DNA. The alternative parallel binding mode was less stable but was still as stable as DNA-DNA or DNA-RNA duplexes. Quite uniquely, for homopyrimidine PNAs binding to homopurine DNA or RNA, a complex that takes the form of a $(PNA)_2DNA$ triplex is observed (21). When homopyrimidine PNA is targeted to duplex DNA containing a homopurine target, the $(PNA)_2DNA$ complex still forms by displacing the homopyrimidine DNA stretch from the DNA duplex (19,22,23). The displaced homopyrimidine DNA segment is then susceptible to enzymatic (24,25) and chemical (19) degradation, while the DNA homopurine strand is protected under the same cleavage conditions. Molecular modeling studies for duplex and triplexed PNAs with both DNA and RNA have proposed a 'cyclic' seven-membered inter-residue hydrogen bond that serves to preorganize the PNA strand and stabilize the duplex (26,27).

In vitro inhibition of RNA translation has been demonstrated using a PNA in a rabbit reticulocyte lysate (28). In tsa 8 cells, translation of RNA from a microinjected plasmid was inhibited by a coinjected homopyrimidine PNA. Additional examples of PNA's ability to inhibit oligonucleotide processing enzymes *in vitro* were also recently disclosed (29,30). What has yet to be documented is the cellular activity and *in vivo* activity of PNAs or PNA analogs. These data will be the key to determining the real potential of this new and promising oligomer as an antisense drug.

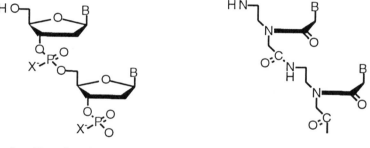

1a, X = O, phosphodiester
1b, X = S, phosphorothioate

peptide nucleic acid (PNA)

Figure 2. Comparison of Backbone Modifications

Alternative backbone modifications, including carbonates, carbamates, sulfur[II, IV, and VI], formacetal, oximes and imines have been reviewed recently (10,31). These modifications are usually incorporated into oligomers as dimers flanked by phosphodiester linkages and do not represent fully modified backbones. Continuing efforts to extend this modification strategy (Figure 3) has lead to investigation of the methylene(dimethylhydrazo) linkage, which has been shown to bind in a sequence-specific manner with a small loss of duplex stability (ΔTm of -0.7 °C/modification), but with improved resistance to degradation by fetal calf serum nucleases (32). The 3'-thioformacetal motif has improved stability compared to the parent phosphodiester, with a reported ΔTm of +0.8 °C/modification (33). In comparison, the formacetal linkage decreases stability by -0.7 °C/modification. The 3'-thioformacetal shows good mismatch specificity, and triplex stability.

The use of amide linkages as an alternative backbone has been adapted to the dimer approach to preparing modified oligomers. Of the possible isomeric amide linkages, which could replace the normal four atom backbone in a natural oligonucleotide, five were recently disclosed (34). Two of these have been described more fully; namely, Amide 2 (35) and Amide 1 (36). The Amide 2 isomer destabilized the duplexes by >-2.0 °C/modification, while Amide 1 destabilized the duplex by -1.6 °C/modification. Recent reports have appeared that the Amide 3 isomer did not alter the duplex stability against an RNA target, but was somewhat destabilizing against DNA (37-39). Reports have yet to appear on the remaining isomeric structures. Additional backbones have been reported but do not improve duplex stability (40,41) or no stability data is yet available (42).

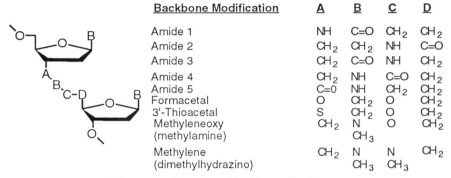

Backbone Modification	A	B	C	D
Amide 1	NH	C=O	CH$_2$	CH$_2$
Amide 2	CH$_2$	CH$_2$	NH	C=O
Amide 3	CH$_2$	C=O	NH	CH$_2$
Amide 4	CH$_2$	NH	C=O	CH$_2$
Amide 5	C=O	NH	CH$_2$	CH$_2$
Formacetal	O	CH$_2$	O	CH$_2$
3'-Thioacetal	S	CH$_2$	O	CH$_2$
Methyleneoxy (methylamine)	CH$_2$	N CH$_3$	O	CH$_2$
Methylene (dimethylhydrazino)	CH$_2$	N CH$_3$	N CH$_3$	CH$_2$

Figure 3: Phosphate Backbone Replacements

Sugar Modified Oligomers - Another area of recent significant effort and progress is the preparation and evaluation of 2'-modified oligonucleotides. To assess the benefit of alkylation of the RNA 2'-hydroxyl moiety, 2'-O-alkyl adenosine monomers were prepared by direct alkylation of unprotected adenosine and incorporated as point modifications in DNA oligomers (43). The comparison of the 2'-O-alkyl series (**2a-i**) demonstrated that alkyl substituents smaller than propyl were stabilizing relative to phosphodiester DNA. The 2'-O-propyl substituent displayed the same stability as phosphodiester DNA and any substituent larger and/or longer than propyl was destabilizing. These data lead to the conclusion that electronegative substitution at the 2' position alters the duplex to more A-form like structure and that 2' bulk is destabilizing likely due to steric crowding in the minor groove. An independent study comparing 2'-O-methyl to 2'-O-allyl confirmed these data (44).

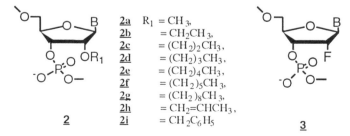

2a	R$_1$ = CH$_3$,
2b	= CH$_2$CH$_3$,
2c	= (CH$_2$)$_2$CH$_3$,
2d	= (CH$_2$)$_3$CH$_3$,
2e	= (CH$_2$)$_4$CH$_3$,
2f	= (CH$_2$)$_5$CH$_3$,
2g	= (CH$_2$)$_8$CH$_3$,
2h	= CH$_2$=CHCH$_3$,
2i	= CH$_2$C$_6$H$_5$

The 2'-fluoro-2'-deoxy modification, **3**, was also found to be highly stabilizing with a ΔTm equal to approximately +1.3 °C/modification (43). Using all four 2'-fluoro modified nucleotides demonstrated that uniformly modified oligomers were stabilized by +1.5

°C/modification (45). While highly stabilizing towards duplex formation, this modification was not measurably more resistant to nuclease degradation, and did not support RNase H cleavage of the RNA strand within a 2'-fluoro DNA/RNA duplex (45). The corresponding 2'-fluoro phosphorothioate sequences stabilized the oligomer against nuclease degradation and increased duplex stability to above that of the wild type diester, but RNase H activity was still lacking within these doubly modified sequences (45).

A comparison of duplex stability, nuclease degradation rates and cellular anti-HIV activity of a series of oligomers encompassing both α- and β-anomeric nucleotides as phosphodiesters and phosphorothioates and the corresponding methyl phosphonates demonstrated duplex stability data was condition dependent (44).

These researchers concluded that DNA as a model for target RNA is not appropriate in assessing potential antisense oligomers. The suggestion was advanced that thermoelution measurements against longer RNA sequences should be employed to confirm stability data from length-matched target sequences. In this comparison of duplex stability, α-DNA was slightly less stable than β-DNA under low salt and thermoelution conditions, but more stable under high salt conditions. Nuclease degradation studies showed that α-DNA is more stable than β-DNA. The α-DNA oligomers degraded to a shorter oligomer which was significantly more stable than the parent, a result which remains unexplained. The anti-HIV activity of the α-DNA and β-DNA sequences was non-specific.

Modified Base Oligomers - Modification of the nucleobases is an alternative means of preparing novel monomers for incorporation into antisense oligomers. The preparation of C5-(propyn-1-yl)uracil and -cytosine, **4**, and their incorporation into phosphorothioate oligomers was reported (46-48). In a 20-mer having 18 modified C5-propynyl pyrimidines showed a +1.5-2.0 °C/modification increase in Tm. This propyne-modified sequence showed length- and sequence-specific inhibition of an intracellular plasmid carrying the SV40 large T antigen when plasmid and oligomer were delivered by microinjection. The expression of the gene was inhibited for 48 hours by an intracellular concentration of drug of 0.5 µM. The corresponding control oligomers, including the 2'-O-allyl phosphorothioate analog with or without C5-propyne did not inhibit expression of this gene, indicating the mechanism of action was manifested through an RNase H-directed mechanism. Activity was also observed

4	**5**	**6**
X = O, NH	X = O, R = CH$_3$	X = O, NH
R= C≡CCH$_3$, F, Br, I	X = NH, R = H	

when oligomer delivery was by microinjection or by addition to the cellular media in the presence of lipofectin, a cationic lipid known to increase cell permeability. Preparation and testing of 6-aza analogs of thymidine and cytosine, **5**, revealed that these modifications destabilized DNA/RNA duplexes by -1.4 °C/modification and -3.1 °C/modification, respectively (49). Both modifications enhanced oligomer nuclease stability and 6-aza thymidine supported RNase H cleavage of a modified DNA/RNA duplex. This study also evaluated 5-fluoro, 5-bromo, and 5-iodouridines, **4**, along with 5-bromocytidine, **4**. In the uridine series the halogen substitution had no effect on duplex stability. Likewise, 5-bromo cytidine had a +1.2 °C/modification stabilization of the duplex, supported RNase H activity and imparted only a slight (2X) increase in nuclease stability.

Purine nucleobases have also been modified. Attachment of a 3-(1H-imidazol-yl)propyl fragment to the N2 position of guanine or 2-amino adenosine, **6**, gave modified bases that enhanced duplex binding affinity by about +0.5 °C/modification against RNA (50). Additionally, these modifications support RNase H activity even when a modified guanine was placed on the antisense strand at the primary site of RNase H cleavage. Substitution of three modified nucleotides at the 3' end of an oligomer enhanced nuclease stability .

OLIGOMER BIOLOGICAL ACTIVITY

In Vitro Activity - The understanding of the mechanism(s) by which unmodified and modified oligonucleotides are taken up by cells and their cellular distribution is of obvious importance to understanding and assessing *in vitro* activity of antisense oligomers (51-53). Comparison of phosphodiester, phosphorothioate, mixed diester-thioate and methyl phosphonate oligomers in cellular association and uptake demonstrated that phosphorothioates had the highest degree of cellular association. Competition studies also demonstrated that the same cellular oligonucleotide binding sites are involved in diester and thioate uptake and that as little as 10 mole percent of phosphorothioate would block phosphodiester uptake. After oligomer internalization, fluorescent microscopy did not reveal nuclear staining (54). It was reported that uptake of phosphorothioate oligomers decreased as cellular density in cultured cells increased (55). These data, if confirmed in additional cell lines, will have significant impact on *in vitro* testing strategies.

For *in vitro* cellular studies, the use of cationic lipid preparations to complex the oligonucleotides provides increased cellular uptake (8-16X) and, more significantly, alters the intracellular distribution to provide nuclear localization of the oligomer (56,57). With cationic lipid, a 1000X increase in antisense activity of a sequence targeted to the AUG translation initiation codon for ICAM-1 in human umbilical vein endothelial cells was observed (57,58).

In vitro activity of phosphorothioate oligonucleotides has been reviewed (1,52). Recent reports include inhibition of cytomegalovirus major immediate early (IE-2) gene by an antisense oligomer that was 30-fold more potent than ganciclovir (59). The E2 gene of papillomavirus (HPV) was targeted with a series of phosphorothioate oligomers from which one targeted to the AUG codon was found to possess an IC_{50} of 5-7 µM (60). In both oral and cervical cancer cell lines which contained HPV-18 DNA, antisense oligomers to the E6 and E7 genes significantly inhibited growth when dosed at 1-5 µM (61). Antisense phosphorothioates directed against the REV-responsive element inhibited Rev activity and HIV-1 replication (62,63).

An antisense phosphorothioate that was selective for a mutant H-ras mRNA point mutation at codon 12 was chosen for studies incorporating 2'-modifications into the sequence (64) . In the follow-up study, the activity of the modified oligomers was correlated to their hybridization affinity and the absolute need for an unmodified "gap" of 2'-deoxynucleotides in the middle of the oligomer in order to support an RNase H mechanism of action (65). The modifications investigated had a rank order following the trend 2'-O-pentyl < 2'-O-propyl < 2'-O-methyl < 2'-fluoro-2'-deoxy wherein the 2'-fluoro-2'-deoxy oligomer had an

IC_{50} of 10 nM. The position within the oligomer and the width of the gap were also found to be important to the antisense potency of the oligomer, with a seven base 2'-deoxy gap seemingly optimal.

The 5'-termini of many mRNAs carry a 7-methyl guanosine 5'-linked to the mRNA 5'-end by a triphosphate. This unique terminus has been investigated as a site to target antisense oligomers to modulate protein binding (66). It was discovered that antisense oligomers with one, two, or three additional bases 3' to the natural end of the sequence had effects on eucaryotic initiation factor-4E binding to an ICAM-1 mRNA. Extended antisense oligomers with a guanosine as the first additional base inhibited protein binding, while use of adenosine, thymidine, or cytosine as the first extra base unexpectedly enhanced protein binding by as much as 200%.

In Vivo Activity - Reports of _in vivo_ effects of antisense oligomers continue to appear and provide encouraging data indicating that systemic effects are measurable and efficacy can be achieved. The potential for _ex vivo_ uses of oligonucleotide antisense molecules has been demonstrated with oligomer targeted against c-_myb_ proto-oncogene in human malignant hematopoietic cells (67). In _scid_ mice with established human leukemia, treatment with antisense phosphorothioate oligomers to c-_myb_ resulted in a dose-dependent increase in survival (3-8 times greater than control) and a lessening of clinical manifestations of the leukemia (68). In two separate studies, mice transplanted with fibrosarcomas were treated after establishment of the tumors with antisense oligomers targeted to the p65 protein of NF-kB. These experiments showed complete regression of the tumors and no reoccurrence up to 5 months (69,70). Antisense oligomers to the p50 protein of NF-kB were not as potent. A phosphorothioate antisense oligomer targeted to the 3' UTR of p120, a proliferation-associated nucleolar antigen, inhibited tumor growth in a dose and sequence specific manner in nude mice injected i.p. with LOX ascites tumor cells. The effect was enhanced significantly by using a cationic lipid, DOTMA (71).

The dose-dependent inhibition of viral replication in ducks infected with duck hepatitis B virus has been reported using a phosphorothioate oligonucleotide targeted to the viral pre-S/S region (72). The animals were shown to be free of viral markers in serum and liver cells. The effect was specific, in that neither a scrambled nor the sense sequence showed activity. Additionally, infection was prevented by administering antisense drug prior to infection and continuing for 12 days.

By localized delivery in rats, antisense c-_myb_ oligomers have also been utilized to inhibit vascular smooth muscle cell (VSMC) accumulation after injury by angioplasty (73). This effect was limited to the area treated by the local application of drug, but the effect persisted. Mismatch or sense oligomers did not show the desired inhibition. This result was confirmed by a complementary study that simultaneously employed two antisense oligomers directed at proliferating-cell nuclear antigen and at cdc2 kinase to inhibit VSMC accumulation (74).

An antisense phosphorothioate targeted to the region just after the translation initiation codon of the NMDA-R1 has been employed _in vivo_ to inhibit infarct size caused by occlusion of the middle cerebral artery (75). This sequence-specific effect suggests a non-RNase H mechanism, since the mRNA concentration was not decreased while the NMDA receptor density was decreased. An oligonucleotide 18-mer targeted to the neuropeptide Y1 receptor diminished receptor density in cultured rat cortical neurons (76). When injected into the lateral cerebral ventricle of rats, the antisense oligomer showed a marked increase in anxiogenic-like effect and a decrease in NPY1 receptor density after sacrifice.

The distribution of ^{14}C- and of ^{35}S-radiolabeled phosphorothioates in mice after a single i.v. injection was reported to be wide spread and disappeared in a multiphase manner

(77,78). Binding to serum proteins (78) and degradation of the drug was observed. Either 3' or 5' capping of the oligomer with a propanediol phosphate moiety slowed the rate of degradation. In pregnant mice, i.p. injection of phosphorothioates at different stages of embryo development revealed no toxicity or teratogenic effects (79).

Clinical Trials - Clinical trials of antisense drugs have only recently begun and few data are available. Trials are underway for treatment of human papillomavirus induced genital warts (80), cytomegalovirus caused retinitis in acquired immune deficiency syndrome patients (81), AIDS infection (82), and acute myelogenous leukemia (83). Initial results from the Phase 1 study indicate no major toxicities in a small number of patients with acute myelogenous leukemia (84).

Target Site Selection - In searching for an mRNA sequence that will be susceptible to an antisense effect, efforts have been made to take into consideration the secondary structure of mRNA. The use of computational methods to predict single-stranded regions has been reported (85,86). In an RNase H-based assay targeted to rabbit β-globin (85) and in an evaluation of HIV antisense sequences (86), the computational predictions yielded active sequences but failed for a translational arrest-based assay (85). A model of mRNA-protein kinetics was reported as an aid to deciding the best antisense mode of action to employ depending on the target mRNA and protein lifetimes (3).

Conclusions - The antisense drug paradigm is just now advancing from the stage of concept validation to the level of refinement of technology. Animal models have demonstrated that antisense regulation of biological processes is possible. Medicinal chemistry efforts are demonstrating that much can be done to alter the structure, activities, and properties of antisense oligomers to obtain real drug molecules. A recent report listing one set of criteria that antisense drugs should meet to be viable requires the antisense agent to: 1) be synthesizable in quantity, 2) be stable *in vivo*, 3) be widely distributed in the body, 4) possess a good cellular half-life, 5) bind effectively to the targeted sequence and achieve reduction in the corresponding protein, and 6) not interact in a non-sequence specific manner with other macromolecules (87). Little has been published on the scaleup of antisense oligomers (88), but the fact some agents are entering clinical trails effectively argues this is an attainable goal. Although significant progress has been made on points 2 through 5 adherence to the requirement prohibiting non-sequence specific interactions is neither reasonable nor attainable. When one considers that antisense drug are chemical entities as are all biological macromolecules and that chemical compounds do interact with one another, it is inescapable that some non-specific interactions will occur. However, the promise of antisense drugs relies on the potential that the problems of subtype specificity which plague many small molecule drugs will be significantly reduced.

References

1.   S.T. Crooke and B. Lebleu, Antisense Research and Applications. S.T. Crooke, and B. Lebleu, Eds., CRC Press Inc., Boca Raton, 1993, p. 1.
2.   M.D. Matteucci and N. Bischofberger, Annu.Rep.Med.Chem., 26, 287 (1991).
3.   M. Ramanathan, R.D. MacGregor and C.A. Hunt, Antisense Res.Dev., 3, 3 (1993).
4.   J.P. Leonetti, G. Degols, P. Clarenc, N. Mechti and B. Lebleu, Prog.Nucl.Acid Res.Mol.Biol., 44, 143 (1993).
5.   J.F. Milligan, M.D. Matteucci and J.C. Martin, J.Med.Chem., 36, 1923 (1993).
6.   A. Frauendorf and J.W. Engels, in "Studies in Natural Products Chemistry" Atta-ur-Rahman, Eds., Elsevier Science Publishers B. V., Amsterdam, 1993, vol. 13, p. 257.
7.   S.L. Beaucage and R. Iyer, Tetrahedron, 49, 6123 (1993).
8.   S.L. Beaucage and R.P. Iyer, Tetrahedron, 49, 1925 (1993).
9.   S.L. Beaucage and R.P. Iyer, Tetrahedron, 48, 2223 (1992).
10.   R.S. Varma, SynLett, 1993, 621 (1993).
11.   J.S. Cohen, in "Antisense Research and Application" S.T. Crooke, and B. Lebleu, Eds., CRC Press, Boca Raton, 1993 p. 205.

12. J.S. Cohen, L. Neckers, C. Stein, S.L. Loke and K. Shinozuka, US Patent 5,276,019, (1994).
13. J.S. Cohen, L. Neckers, C. Stein, S.L. Loke and K. Shinozuka, US Patent 5,264,423, (1993).
14. M.K. Ghosh, K. Ghosh and J.S. Cohen, Anticancer Drug Design, 8, 15 (1993).
15. A. Suska, A. Grajkowski, A. Wilk, B. Uznanski, J. Blaszczyk, M. Wieczorek and W.J. Stec, Pure Appl.Chem., 65, 707 (1993).
16. Z.J. Lesnikowski, Bioorg.Chem., 21, 127 (1993).
17. M. Egholm, O. Buchardt, P.E. Nielsen and R.H. Berg, J.Amer.Chem.Soc., 114, 1895 (1992).
18. M. Egholm, P.E. Nielsen, R.H. Buchardt and R.H. Berg, J.Amer.Chem.Soc., 114, 9677 (1992).
19. P.E. Nielsen, M. Egholm, R.H. Berg and O. Buchardt, Science, 254, 1497 (1991).
20. M. Egholm, O. Buchardt, L. Christensen, C. Behrens, S.M. Freier, D.A. Driver, R.H. Berg, S.K. Kim, B. Norden and P.E. Nielsen, Nature, 365, 566 (1993).
21. S.K. Kim, P.E. Nielsen, M. Egholm, O. Buchardt, R.H. Berg and B. Norden, J.Amer.Chem.Soc., 115, 6477 (1993).
22. D.Y. Cherny, B.P. Belotserkovskii, M.D. Frank-Kamenetskii, M. Egholm, O. Buchardt, R.H. Berg and P.E. Nielsen, Proc.Natl.Acad.Sci.USA, 90, 1667 (1993).
23. N.J. Peffer, J.C. Hanvey, J.E. Bisi, S.A. Thomson, C.F. Hassman, S.A. Noble and L.E. Babiss, Proc.Natl.Acad.Sci.USA, 90, 10648 (1993).
24. P.E. Nielsen, M. Egholm, R.H. Berg and O. Buchardt, Anti-Cancer Drug Design, 8, 53 (1993).
25. V.D. Demidov, M.D. Frank-Kamenetskii, M. Egholm, O. Buchardt and P.E. Nielsen, Nucleic Acids Res., 21, 2103 (1993).
26. O. Almarsson and T.C. Bruice, Proc.Natl.Acad.Sci.USA, 90, 9542 (1993).
27. O. Almarsson, T.C. Bruice, J. Kerr and R.N. Zuckermann, Proc.Natl.Acad.Sci.USA, 90, 7518 (1993).
28. J.C. Hanvey, N.J. Peffer, J.E. Bisi, S.A. Thomson, R. Cadilla, J.A. Josey, D.J. Ricca, C.F. Hassman, M.A. Bonham, K.G. Au, S.G. Carter, D.A. Bruckenstein, A.L. Boyd, S.A. Noble and L.E. Babiss., Science, 258, 1481 (1992).
29. H. Orum, P.E. Nielsen, M. Egholm, R.H. Berg, O. Buchardt and C. Stanley, Nucleic Acids Res., 21, 5332 (1993).
30. P.E. Nielsen, M. Egholm, R.H. Berg and O. Buchardt, Nucleic Acids Res., 21, 197 (1993).
31. Y.S. Sanghvi and P.D. Cook, in "Nucleosides and Nucleotides as Antitumor and Antiviral Agents" C.K. Chu and D.C. Baker, Eds., Plenum Press, Inc., New York, 1993, p. 311.
32. Y.S. Sanghvi, J.-J. Vasseur, F. Debart and P.D. Cook, Collect.Czech.Chem.Commun., 58, 158 (1993).
33. R.J. Jones, K.-Y. Lin, J.F. Milligan, S. Wadwani and M.D. Matteucci, J.Org.Chem., 58, 2983 (1993).
34. A. De Mesmaeker, J. LeBreton, A. Waldner and P.D. Cook, Int.Patent WO 92/20,823, (1992).
35. A. De MesMaeker, J. LeBreton, A. Waldner, V. Fritsch, R.M. Wolf and S.M. Freier, Synlett, 1993, 733 (1993).
36. J. LeBreton, A. De Mesmaeker, A. Waldner, V. Fritsch, R.M. Wolf and S.M. Freier, Tetrahedron Lett., 34, 6383 (1993).
37. A. De Mesmaeker, J. Lebreton, A. Waldner, P. Hoffman, R.M. Wolf and S.M. Freier, Angew.Chem.Int.Ed.Engl., 33, 226 (1994).
38. I. Idziak, G. Just, M.J. Damha and P.A. Giannaris, Tetrahedron Lett., 34, 5417 (1993).
39. J. Lebreton, A. Waldner, C. Lesueur and A. De Mesmaeker, Synlett, 137 (1994).
40. A. De Mesmaeker, A. Waldner, Y.S. Sanghvi and J. Lebreton, Bioorg.Med.Chem.Lett., 4, 395 (1994).
41. A. Waldner, A. De Mesmaeker, J. Lebreton, V. Fritsch and R.M. Wolf, Synlett, 57 (1994).
42. K. Teng and P.D. Cook, J.Org.Chem., 59, 278 (1994).
43. E.A. Lesnik, C.J. Guinosso, A.M. Kawasaki, H. Sasmor, M. Zounes, L.L. Cummins, D.J. Ecker, P.D. Cook and S.M. Freier, Biochemistry, 32, 7832 (1993).
44. F. Morvan, H. Porumb, G. Degols, I. Lefebvre, A. Pompon, B.S. Sproat, B. Rayner, C. Malvy, B. Lebleu and J.-L. Imbach, J.Med.Chem., 36, 280 (1993).
45. A.M. Kawasaki, M.D. Casper, S.M. Freier, E.A. Lesnik, M.C. Zounes, L.L. Cummins, C. Gonzalez and P.D. Cook, J.Med.Chem., 36, 831 (1993).
46. B.C. Froehler, S. Wadwani, T.J. Terhorst and S.J. Gerrard, Tetrahedron Lett., 33, 5307 (1992.).
47. B.C. Froehler, R.J. Jones, X. Cao and T.J. Terhorst, Tetrahedron Lett., 34, 1003 (1993).
48. R.W. Wagner, M.D. Matteucci, J.G. Lewis, A.J. Gutierrez, C. Moulds and B.C. Froehler, Science, 260, 1510 (1993).
49. Y.S. Sanghvi, G.D. Hoke, S.M. Freier, M.C. Zounes, C. Gonzalez, L.L. Cummins, H. Sasmor and P.D. Cook, Nucleic Acids Res., 21, 3197 (1993).
50. K.S. Ramasamy, M. Zounes, C. Gonzalez, S.M. Freier, E.A. Lesnik, L.L. Cummins, R.H. Griffey, B.P. Monia and P.D. Cook, Tetrahedron Lett., 35, 215 (1994).
51. J.-P. Clarenc, B. Lebleu and J.-P. Leonetti, J.Biol.Chem., 268, 5600 (1993).
52. R.M. Crooke, in "Antisense Research and Applications" S.T. Crooke, and B. Lebleu, Eds., CRC Press, Baco Raton, 1993, p471.

53.   L.M. Neckers, in "Antisense Research and Applications" S.T. Crooke, and B. Lebleu, Eds., CRC Press, Boca Raton, 1993 p. 451.
54.   Q. Zhao, S. Matson, C.J. Herrera, E. Fisher, H. Yu and A.M. Krieg, Antisense Res.Dev., $\underline{3}$, 53 (1993).
55.   T. Iwanaga and P.C. Ferriola, Biochem.Biophys.Res.Commun., $\underline{191}$, 1152 (1993).
56.   J. Felgner, F. Bennett and P.L. Felgner, Method: A Companion to Methods in Enzymology, $\underline{5}$, 67 (1993).
57.   C.F. Bennett, M.-Y. Chiang, H. Chan, J. Shoemaker and C.K. Mirabelli, Mol.Pharmacol., $\underline{41}$, 1023 (1992).
58.   C.F. Bennett, M.-y. Chiang, H. Chan and S. Grimm, J.Liposome Res., $\underline{3}$, 85 (1993).
59.   R.F. Azad, V.B. Driver, K. Tanaka, R.M. Crooke and K.P. Anderson, Antimicrobial Agents Chemother., $\underline{37}$, 1945 (1993).
60.   L.M. Cowsert, M.C. Fox, G. Zon and C.K. Mirabelli, Antimicrobial Agents Chemother., $\underline{37}$, 171 (1993).
61.   C. Steele, L.M. Cowsert and E.J. Shillitoe, Cancer Res., $\underline{53}$, 2330 (1993).
62.   J. Lisziewicz, D. Sun, V. Metelev, P. Zamecnik, R.C. Gallo and S. Agrawal, Proc.Natl.Acad.Sci.USA, $\underline{90}$, 3860 (1993).
63.   G. Li, J. Lisziewicz, D. Sun, G. Zon, S. Daeffler, F. Wong-Staal, R.C. Gallo and M.E. Klotman, J.Virol., $\underline{67}$, 6882 (1993).
64.   B.P. Monia, J.F. Johnston, D.J. Eckers, M.A. Zounes, W.F. Lima and S.M. Freier, J.Biol.Chem., $\underline{267}$, 19954 (1992).
65.   B.P. Monia, E.A. Lesnik, C. Gonzalez, W.F. Lima, D. McGee, C.J. Guinosso, A.M. Kawasaki, P.D. Cook and S. Freier, J.Biol.Chem., $\underline{268}$, 14514 (1993).
66.   B.F. Baker, L. Miraglia and C.H. Hagedorn, J.Biol.Chem., $\underline{267}$, 11495 (1992).
67.   M.Z. Ratajczak, J. Hijiya, L. Catani, K. DeReil, S.M. Luger, P. McGlave and A.M. Gewirtz, Blood, $\underline{79}$, 1956 (1992).
68.   M.Z. Ratajczak, J.A. Kant, S.M. Luger, J. Hijiya, J. Zhang, G. Zon and A.M. Gewirtz, Proc.Natl.Acad.Sci.USA, $\underline{89}$, 11823 (1992).
69.   K.A. Higgins, J.R. Perez, T.A. Coleman, K. Dorshkind, W.A. McComas, U.M. Sariento, C.A. Rosen and R. Narayanan, Proc. Natl. Acad. Sci. USA, $\underline{90}$, 9901 (1993).
70.   I. Kitajima, T. Shinohara, J. Bilakovics, D.A. Brown, X. Xu and M. Nerenberg, Science, $\underline{258}$, 1792 (1992).
71.   L. Perlaky, Y. Saijo, R.K. Busch, C.F. Bennett, C.K. Mirabelli, S.T. Crooke and H. Busch, Anti-Cancer Drug Design, $\underline{8}$, 3 (1993).
72.   W.-B. Offensperger, S. Offensperger, E. Walter, K. Teubner, G. Igloi, H.E. Blum and W. Gerok, EMBO J., $\underline{12}$, 1257 (1993).
73.   M. Simons, E.R. Edelman, J.-L. DeKeyser, R. Langer and R.D. Rosenberg, Nature, $\underline{359}$, 67 (1992).
74.   R. Morishita, G.H. Gibbons, K.E. Ellison, M. Nakajima, L. Zhang, Y. Kaneda, T. Ogihara and V.J. Dzau, Proc.Natl.Acad.Sci.USA, $\underline{90}$, 8474 (1993).
75.   C. Wahlestedt, E. Golanov, S. Yamamoto, F. Yee, H. Ericson, H. Yoo, C.E. Inturrisi and D.J. Reis, Nature, $\underline{363}$, 260 (1993).
76.   C. Wahlestedt, E.M. Pich, G.F. Koob, F. Yee and M. Heilig, Science, $\underline{259}$, 528 (1993).
77.   J.-Y.T. J. Temsamani T. Padmapriya, M. Kubert, S. Agrawal, Antisense Res.Dev., $\underline{3}$, 277 (1993).
78.   P.A. Cossum, H. Sasmor, D. Dellinger, L. Truong, L. Cummins, S.R. Owens, P.M. Markham, J.P. Shea and S.T. Crooke, J. Pharmacol.Exp.Ther., $\underline{267}$, 1181 (1993).
79.   M.F. Gaudette, G.H. Hampikan, V. Metelev, S. Agrawal and W.R. Crain, Antisense Res.Dev., $\underline{3}$, 391 (1993).
80.   Cancer Weekly, July 13, 1993, .
81.   BioCentury, Dec 20, 1993, p. B7.
82.   Genetic Eng. News, Oct 15, 1993, p. 1.
83.   Cancer Researcher Weekly, Jun 7, 1993, p. 7.
84.   E. Bayever, P.I. Iversen, M.R. Bishop, J.G. Sharp, H.K. Tewary, M. Arneson, S.J. Pirruccello, R.W. Ruddon, A. Kessinger, G. Zon and J.O. Armitage, Antisense Res.Devel., $\underline{3}$, 383 (1993).
85.   J.W. Jaroszewski, J.-L. . Syi, M. Ghosh, K. Ghosh and J.S. Cohen, Antisense Res.Dev., $\underline{3}$, 339 (1993).
86.   G. Sczakiel, M. Homann and K. Rittner, Antisense Res.Dev., $\underline{3}$, 45 (1993).
87.   C.A. Stein and Y.-C. Cheng, Science, $\underline{261}$, 1004 (1993).
88.   F.X. Montserrat, A. Grandas, R. Eritja and E. Pedroso, Tetraherdon, $\underline{50}$, 2617 (1994).

Chapter 31. *In Vitro* Approaches for the Prediction of Human Drug Metabolism

John O. Miners, Maurice E. Veronese and Donald J. Birkett
Flinders Medical Centre, Bedford Park, SA 5042, Australia

Introduction - Early knowledge of the potential human metabolism and pharmacokinetics of new drug candidates is of great importance at several stages of drug development. The design, interpretation and relevance to humans of animal toxicology studies is critically dependent on interspecies comparisons of routes of metabolism and pharmaco/toxico-kinetics. In early Phase I studies, prediction of human drug clearance and clearance pathways aids in dose-ranging, choice of sampling schedules and analysis for potential metabolites. In Phase II development, drug interaction studies can be targeted to those likely to be relevant and clinically important with major savings in development time and resources. Knowledge that the metabolism of a new drug is likely to be subject to a genetic polymorphism (such as the poor debrisoquine or poor mephenytoin metaboliser phenotypes) also can guide decisions at several stages in development. At a very early stage the drug may be dropped or an analogue not subject to the polymorphism chosen as a better development candidate. In early Phase I studies, volunteer subjects can be phenotyped or genotyped so that affected individuals can be excluded to reduce variability, or observed more closely for adverse effects. At later stages in clinical development, knowledge of pharmacogenetic influences can help explain population variability in pharmacokinetics, therapeutic response and occurrence of adverse effects. This review describes recent advances in the development of *in vitro* methods to predict human drug metabolism and pharmacokinetics before *in vivo* studies, animal or human, are carried out.

PREDICTION OF *IN VIVO* PHARMACOKINETICS FROM *IN VITRO* ENZYME KINETICS

For drugs subject to linear pharmacokinetics the *in vitro* correlate of *in vivo* intrinsic clearance (CL_{int}) is given by the sum of the V_{max}/K_m terms for each metabolic pathway. As V_{max} is usually expressed per mg of microsomal protein, potentially the microsomal activity could be scaled up to activity for a whole liver and *in vivo* CL_{int} estimated. The relative values of V_{max}/K_m for different pathways give an index of the relative contributions of the pathways to the overall elimination of the drug *in vivo*. Little systematic investigation of such an approach has been reported in the literature.

In vitro enzyme kinetics of a reaction for a single metabolic pathway are often biphasic indicating the involvement of multiple isoforms with differing affinities for the drug substrate. In this case, the velocity of enzyme reaction is given by the sum of the activities of each of the isoforms. Usually only the high affinity activity is relevant to *in vivo* substrate (drug) concentrations. The relative contributions of high and low affinity activities at any substrate concentration can be calculated by substitution in the Michaelis-Menten equation. It is important in carrying out correlation or inhibitor studies (see later) to choose a substrate concentration at which the relevant activity (usually the high affinity activity) dominates (1-3). Similar considerations apply to inhibitor enzyme kinetics. As with substrate probes, inhibitor probes may lose isoform selectivity at high concentrations. Also the inhibitor concentration giving 50% inhibition (IC_{50}) varies with the substrate concentration and K_i is a better parameter to define interaction of an inhibitor probe with a particular enzyme.

Finally, in using cDNA expressed isoforms to define the isoform profile of a drug substrate, it is important to show similarity of kinetic constants with the expressed isoform and human liver microsomes. Demonstration of activity at high substrate concentrations with expressed isoforms may be due to low affinity activities not relevant to the *in vivo* situation.

DEFINITION OF METABOLIC PATHWAYS *IN VITRO*

In vitro (microsomal) reaction systems normally use conditions such that there is <10% (often <1%) conversion of drug substrate to primary metabolic products to maintain linearity of reaction rate with time and protein concentration. Under these conditions, further biotransformation of metabolites usually does not occur. The primary metabolic pathways can then be defined by HPLC separation and collection of metabolite peaks provides compound for metabolite identification by methods such as mass spectrometry. Secondary metabolism of the primary metabolites can similarly be defined by using the primary metabolites as substrates in further *in vitro* incubations. A recent example of this approach is the use of *in vitro* methods to define the primary and secondary metabolism of omeprazole and to identify the cytochrome P450 isoforms carrying out the various reactions (3,4).

GENE SUPERFAMILIES OF DRUG METABOLISING ENZYMES

The great diversity of structures that act as substrates for xenobiotic metabolising enzymes is due to the presence of multiple forms (isoforms) which have differing but often overlapping substrate selectivity. This review will concentrate on the cytochrome P450 (CYP) gene superfamily of enzymes which is the best characterised and quantitatively the most important of the xenobiotic metabolising enzymes. The CYP superfamily is subdivided into gene families and gene subfamilies on the basis of the degree of homology in amino acid sequence between the isoforms. A total of 36 CYP gene families have been described but only a few are relevant to drug metabolism in humans (viz. gene families 1,2 and 3). Nelson *et al* (5) summarises CYP genes characterised to the end of 1992. In general, there is greater sequence homology across species within gene families and subfamilies than there is across gene families and subfamilies within the same species. Unfortunately, this interspecies homology in amino acid sequence is not accompanied by a similar degree of homology in functional characteristics (substrate selectivity) frequently making animal models of metabolism poor predictors of human metabolism.

Individual CYP isoforms have differing substrate and inhibitor specificities and differ in terms of *in vivo* regulation (induction, repression, inhibition, tissue-specific expression, polymorphism) (6,7). Given the differing regulation of CYP isoforms, it is apparent that prediction of drug interactions and other pharmacogenetic and environmental influences on the metabolism of a new drug requires identification of the specific CYP isoforms responsible for its metabolism. The concept of the CYP "isoform profile" of a drug substrate (8,9) is useful, that is the quantitative contributions made by specific CYP isoforms to the overall metabolism, and therefore elimination, of a drug substrate. Changes in activity of one isoform are likely to have a clinically important influence on the pharmacokinetics of a drug only if the drug is metabolised solely or mainly (>50%) by that isoform. Recent advances in approaches to defining *in vitro* the human CYP isoforms profile of drug substrates are detailed below. Also of importance is the availability of isoform-selective probe drugs which can safely be administered to humans to investigate further the *in vivo* regulation of specific isoforms. This also is discussed.

SUBSTRATE AND INHIBITOR PROBES FOR CYTOCHROME P450 ISOFORMS *IN VITRO* AND *IN VIVO*

A variety of approaches have been developed for the identification of the human enzymes involved in the metabolism of any given drug (or non-drug xenobiotic) *in vitro* (9-11). These include: (i) investigation of correlations between the rates of metabolism of the drug and immunoreactive CYP isoform contents or prototypic isoform-specific activities in a "panel" of human liver microsomes; (ii) comparison of the kinetics of metabolism of the drug by human liver microsomes and cDNA-expressed or purified isoforms; (iii) competitive inhibition of the metabolism of isoform specific substrates by the drug, using both human liver microsomes and expressed isoforms, and; (iv) characterisation of the effects of known isoform specific xenobiotic or antibody inhibitors on the metabolism of the drug by human liver microsomes. There are limitations associated with the interpretation of certain of these procedures in isolation (9). For example, competitive inhibitors are not invariably alternate substrates and correlations between activities may be observed when compounds are metabolised by separate enzymes. However, a combination of the approaches allows identification of the enzyme(s) responsible for the metabolism of a xenobiotic with a high degree of certainty.

Substrate probes are required for investigation of the effects of genetic, physiological and environmental factors on isoform activity *in vivo*. If a drug is to be used as a substrate probe *in vivo*, confirmation that measurement of metabolic activity *in vivo* does indeed reflect isoform activity is desirable. *In vivo* isoform-specific inhibitors are extremely useful in this regard.

It is important that the index of metabolism measured *in vivo* is appropriate. Theoretically, the closest measure of enzyme activity is partial metabolic clearance of unbound drug, which equates to intrinsic clearance (V_{max}/K_m). More often, however, simpler but indirect measures of metabolic activity are determined. These include the CO_2 breath test (for demethylation reactions), urine metabolite excretion and the urinary metabolic ratio (MR). Measurement of fractional urinary metabolite excretion is the least desirable index of intrinsic clearance since the excretion of any given metabolite may be influenced by alterations in clearances along other pathways, either renal or metabolic (12). While of use for the detection of polymorphisms of drug metabolism, the MR (which is usually expressed as the ratio of unchanged drug to metabolite in urine) is dependent not only on enzyme activity but also renal clearance of unchanged drug. Thus, for lipophilic compounds changes in urine flow (and subsequently tubular reabsorption) will influence the MR and this confounding issue must be taken into account in the interpretation of metabolism studies which utilise the MR (13,14).

CYP1A   In the past compounds such as phenacetin, 7-ethoxycoumarin and 7-ethoxyresorufin have found widespread use as hepatic CYP1A2 probes *in vitro*. While evidence for the use of phenacetin as a specific CYP1A probe is now compelling (1), the broad isoform specificity of 7-ethoxycoumarin and 7-ethoxyresorufin is less well defined. Indeed, 7-ethoxycoumarin may also be metabolised by CYP2A6 (15). More recently, attention has focussed on the use of caffeine (1,3,7-trimethylxanthine) as a CYP1A substrate probe, given the potential for the use of this compound *in vivo*. The primary metabolic pathways of caffeine in humans are N1-, N3- and N7-demethylation to form theobromine, paraxanthine and theophylline, respectively, and 8-hydroxylation to form trimethyluric acid (2,16). Of these pathways, paraxanthine formation dominates at the substrate concentrations normally associated with dietary caffeine intake. Using a combination of *in vitro* approaches it has been demonstrated that CYP1A2 is solely responsible for the **high affinity** components of the three caffeine N-demethylations, and under appropriate experimental conditions caffeine N3-demethylation may serve as a measure of human hepatic CYP1A2 (2,16,17). Another reaction applied recently for the investigation of human CYP1A2 is the 6-hydroxylation of R-warfarin (18), and it is now apparent that R,S-warfarin may serve as a multi-isoform substrate probe *in vitro* (see later sections). Comparative kinetic and inhibition studies using cDNA expressed enzymes have shown that most, if not all, CYP1A substrate probes are non-specific in their recognition of the two isoforms comprising this subfamily (1). Although of possible significance to the measurement of CYP1A activity in extrahepatic tissues, this CYP1A non-specificity of substrates is unlikely to be of major importance to the measurement of hepatic CYP1A2 given the negligible expression of CYP1A1 in uninduced liver (19).

Since the CYP1A2 catalysed N-demethylations account for the majority of caffeine metabolism (20), plasma clearance of caffeine is assumed to reflect human hepatic CYP1A2 activity *in vivo*. Once formed *in vivo*, the primary monodemethylated biotransformation products undergo further extensive biotransformation to form a range of methyl-xanthines and -urates, as well ring-opened uracil derivatives. A number of MRs utilising various caffeine metabolites have been considered as markers of CYP1A2 activity *in vivo*. Of these, the ratio of (1-methylxanthine + 1-methyluric acid + 6-amino-5-[N-formylmethylamino]-3-dimethluric acid) to 1,7-dimethyluric acid excreted in urine following caffeine ingestion has been utilised most widely for the assessment of human CYP1A2 activity (21). This ratio correlates highly with caffeine plasma clearance ($r^2 = 0.82$) but less convincingly with the caffeine breath test ($r^2 = 0.54$) (21). The denominator chosen for the "caffeine MR" is 1,7-dimethyluric acid, rather than caffeine itself. Use of the urine excretion of this compound minimised variability in the MR, (21), presumably by eliminating the confounding affect of urine flow rate on caffeine renal clearance (13). Other versions of the caffeine MR utilised recently are the ratios of (paraxanthine) or (paraxanthine + 1,7-dimethyluric acid) to unchanged caffeine (22); respective r^2 values compared to the rate constant for the conversion of caffeine to paraxanthine (i.e. 1,7-dimethylxanthine) were, however, only 0.29 and 0.54.

Like CYP1A substrate probes, the majority of commonly used CYP1A inhibitors, including ellipticine and α-naphthoflavone, also affect both CYP1A1 and CYP1A2 (1). Furafylline, which has been shown previously to be a specific CYP1A inhibitor (18,23), appears to be the only xenobiotic inhibitor probe capable of differentiating CYP1A1 and CYP1A2 activities. Under conditions which favour mechanism-based inhibition, furafylline exhibits more than 100-fold specificity towards CYP1A2 (24). Thus, the use of furafylline should allow the separation of CYP 1A1- and 1A2-catalysed activities where co-expression occurs. Furafylline also has the capacity to abolish CYP1A2 catalysed reactions in humans *in vivo*. Apart from known affects on caffeine metabolism, coadministration of furafylline was shown recently to completely inhibit phenacetin O-deethylation and the CYP1A2 mediated components of the metabolism of dietary heterocyclic amines in healthy volunteers (25). It is noteworthy, however, that furafylline appears currently not to have been approved for general use in humans. Other compounds shown recently to inhibit human hepatic CYP1A2 *in vitro* include fluvoxamine (26) and derivatives of 4-oxoquinolone-3-carboxylic acid (27). However, the specificity of fluvoxamine is questionable (28) and further studies are necessary to define the broad specificity of the quinolones as inhibitors of human CYP.

CYP2A6 - Microsomal xenobiotic- and immuno-inhibition data are consistent with coumarin 7-hydroxylation being catalysed almost exclusively by CYP2A6 (29,30). Apart from serving as an *in vitro* CYP2A6 substrate, this compound represents a potentially useful inhibitor probe for the screening of human liver microsomal CYP2A6 mediated reactions given its high affinity (apparent K_m ~0.5μM) for the enzyme (30). The urinary excretion of 7-hydroxycoumarin following the oral administration of coumarin (2 or 5mg) has been investigated in two groups (comprising 64 and 110 healthy subjects) (31,32). Both studies reported marked interindividual variability in 7-hydroxycoumarin excretion. Although it might be assumed that marked variability in CYP2A6 activity occurs *in vivo* on the basis of the demonstrated involvement of CYP2A6 in the 7-hydroxylation of coumarin *in vitro*, the relationship between the urinary excretion of 7-hydroxycoumarin and enzyme activity was not considered in either study. Whether 7-hydroxycoumarin excretion represents a reliable measure of human CYP2A6 activity *in vivo* awaits further investigation.

CYP2B6 - There are no currently accepted specific substrate or inhibitor probes for CYP2B6. Recent data suggests, however, that CYP2B6 is the enzyme responsible for the high affinity component of human liver microsomal cyclophosphamide 4-hydroxylation (the pathway responsible for phosphoramide mustard generation)(33). The utility of this pathway for assessment of *in vitro* CYP2B6 activity awaits further investigation. There is an unsubstantiated report that orphenadrine may be a selective inhibitor of CYP2B6 (33).

CYP2C8 - Although cDNA-expressed CYP2C8 has the capacity to oxidise a number of drugs, the contribution of this constitutively expressed enzyme to human hepatic xenobiotic metabolism seems more limited (34). A minor role for CYP2C8 in the metabolism of carbamazepine and the endogenous substrate retinoic acid has been suggested (35,36), but neither can usefully serve as a substrate probe for the human liver enzyme.

CYP2C9/10 - Tolbutamide and sulphaphenazole are established CYP2C9/10 substrate and inhibitor probes, respectively, and recent reports have confirmed the CYP2C9/10 specificity of both compounds (34,37,38). Combinations of cDNA-expression, sulphaphenazole inhibition, and tolbutamide competitive inhibition and activity correlation studies have been used to link phenytoin hydroxylation (34,37,38), (S)-warfarin 7-hydroxylation (39) and torasemide tolylhydroxylation (40) to human hepatic CYP2C9/10. Tolbutamide, phenytoin, torasemide and S-warfarin (7-hydroxylation pathway) may therefore all be used as CYP2C9/10 substrate probes *in vitro*, but the higher rates of reaction associated with the hydroxylations of tolbutamide and torasemide have advantages in terms of metabolite analysis.

As *in vitro*, the CYP2C9 inhibitor sulphaphenazole has been used to demonstrate that the hydroxylation of tolbutamide *in vivo* is also catalysed essentially exclusively by this isoform. Coadministration of sulphaphenazole and tolbutamide to healthy volunteers reduced tolbutamide plasma clearance by more than 80% (41). The same study also showed that the tolbutamide MR is a reasonably good predictor of tolbutamide plasma clearance ($r^2 = 0.72$) and that the MR can distinguish tolbutamide hydroxylase activity in normal subjects and those converted to model phenotypically "poor" metabolises with sulphaphenazole. Phenytoin is known to be metabolised by CYP2C9/10 *in vitro* (see previously) and a comparative pharmacokinetic study has demonstrated that the metabolism of phenytoin and tolbutamide are co-regulated *in vivo* (42). Thus, phenytoin may also serve as an alternate CYP2C9/10 substrate probe *in vivo*. The 6- and 7-hydroxylations of S-warfarin have also been used recently to assess CYP2C9/10 activity *in vivo* (43). As indicated earlier, a number of CYP isoforms are known to catalyse specifically the various hydroxylations of R,S-warfarin *in vitro* and it is possible that this compound may be useable as a multi-isoform probe *in vivo*.

S-Mephenytoin hydroxylase (SMPH) - While it is generally accepted that the polymorphic SMPH is a member of the CYP2C subfamily, the precise identity of the enzyme remains unknown. The current favoured candidate enzyme is CYP2C19 (44). In addition to the previously characterised S-mephenytoin (4-hydroxylation) and hexobarbitone (3'-hydroxylation), more recently investigated SMPH substrates *in vitro* include diazepam (45,46), omeprazole (3,47), proguanil (48) and R-warfarin (8-hydroxylation) (49). Whereas SMPH is responsible solely for the high affinity component of omeprazole hydroxylation (3), CYP3A isoforms contribute to a variable extent to the conversion of proguanil to cycloguanil (48) and to diazepam N-demethylation and 3-hydroxylation (45). At least *in vitro*, the use of these latter two compounds as SMPH probes would seem inappropriate. Further studies are also necessary to confirm the specificity of SMPH for R-warfarin 8-hydroxylation. Of the various compounds known to interact with SMPH, S-mephenytoin remains the preferred inhibitor probe given the sole involvement of SMPH in its metabolism.

The procedure used most frequently for assessment of SMPH activity *in vivo*, and especially for assignment of SMPH phenotype, is measurement of the ratio of S- to R-mephenytoin excreted in urine following administration of racemic drug (reviewed in ref. 50). However, problems are known to occur with the reproducibility of this method (51,52) and the procedure appears to be relatively insensitive for the separation of activities within the two phenotypes. The activation of the antimalarial drug proguanil to cycloguanil has been shown recently to cosegregate with the polymorphic hydroxylation of S-mephenytoin *in vivo*, but interestingly no correlation was apparent across all subjects when proguanil activation and S-mephenytoin hydroxylation were compared (53,54). The reason for this latter observation is not clear, but may relate to the variability inherent in the MR if conditions are not carefully controlled (see earlier discussion). Furthermore, CYP3A isoforms are known to contribute to proguanil activation *in vitro* (48) and this may also confound the *in vivo* assessment of SMPH activity. The known sole involvement of the SMPH in omeprazole hydroxylation, both *in vitro* and *in vivo* (3,47,55), suggests that measurement of this pathway, either as an MR or partial metabolic clearance, may provide a convenient and more precise estimate of enzyme activity *in vivo*.

CYP2D6 - Compounds utilised as CYP2D6 substrate probes *in vitro* and *in vivo* have been reviewed elsewhere (56,57) and include bufurolol, codeine, debrisoquine, dextromethorphan, metoprolol and sparteine. With the exception of bufurolol, all have been used for assignment of CYP2D6 phenotype *in vivo*, usually by application of an MR. The codeine, dextromethorphan and metoprolol MRs have been validated by comparison with the appropriate partial metabolic clearance and debrisoquine/sparteine phenotype (see for example ref. 58). Although not a substrate for CYP2D6, quinidine is a potent and highly selective competitive inhibitor of this enzyme which has found widespread use for the identification of CYP2D6-catalysed xenobiotic oxidations *in vitro* and *in vivo* (56). *In vivo*, a single quinidine dose of 50-100mg is generally sufficient to convert extensive metabolisers to activities associated with the CYP2D6 poor metaboliser phenotype (56,57). Examples of this approach may be found in recent publications (59,60).

CYP2E1 - With the increasing awareness of the toxicological importance of CYP2E1 (61), there has been considerable interest in the development of substrate and inhibitor probes for the human enzyme. Microsomal kinetic, inhibition and correlation approaches, with corroborative evidence from catalytic studies with purified- or cDNA expressed-enzyme, have shown conclusively that chlorzoxazone 6-hydroxylation and 4-nitrophenol hydroxylation (to form 4-nitrocatechol) are performed exclusively by human CYP2E1 (62,63). Measurement of both activities is experimentally convenient. Furthermore, chlorzoxazone and 4-nitrophenol have advantages over other putative CYP2E1 markers, such aniline and N-nitrosodmethylamine, in terms of specificity and safety of handling. Diethyldithiocarbamate acts as a mechanism-based inhibitor of CYP2E1. At diethyldithiocarbamate concentrations $<50\mu M$ inhibition seems to be selective for CYP2E1 (62,64), but at higher concentrations non-specific inhibition of human hepatic CYP occurs (16,64).

It has been reported recently that pretreatment of human subjects with a single dose of disulfiram (which is reduced *in vivo* to diethyldithiocarbamate) inhibits the formation clearance of 6-hydroxychlorzoxazone by more than 90% (65). Together with the specificity of those compounds *in vitro*, these data suggest chlorzoxazone and single-dose disulfiram pretreatment may be used for the investigation of CYP2E1 activities *in vivo*. The inhibitor specificity of the single dose disulfiram treatment warrants further investigation, however, since chronic treatment with this drug is known to inhibit CYP-mediated reactions in a non-specific manner. Since no chlorzaxazone is excreted unchanged in urine (65), use of a chlorzoxazone MR for assessment of *in vivo* CYP2E1 activity is not possible.

CYP3A - A number of xenobiotics metabolised by CYP3A3/4 have been utilised in the past as human hepatic substrate probes. These include aldrin, erythromycin, lignocaine, midazolam and nifedipine (reviewed in ref. 56). More recent studies have shown that dapsone N-oxidation (66), caffeine 8-hydroxylation to form trimethyluric acid (2,16), and terfenadine N-dealkylation and hydroxylation (67) are catalysed by human hepatic CYP3A3/4. Under appropriate conditions, all of these reactions are readily measurable in human liver microsomes and thus the three compounds represent potential CYP3A3/4 substrate probes. In addition to the well documented use of triacetyloleoandomycin as an irreversible, specific inhibitor of CYP3A (56), the 17α-acetylenic steroid gestodene has been demonstrated to be a specific mechanism-based inhibitor of CYP3A4 (68). Thus, preincubation of human liver microsomes with either compound leads to selective loss of CYP3A activity. Another compound of use in identifying CYP3A catalysed reactions *in vitro* is α-naphthoflavone. While α-naphthoflavone inhibits CYP1A2 at low concentration ($<10\mu M$), higher concentrations of the compound in incubations of human liver microsomes leads to activation of most CYP3A activities, presumably via allosteric modification (16,36,66,69).

There is direct evidence *in vitro* and indirect evidence *in vivo* supporting the involvement of CYP3A in the oxidative dehydrogenation of nifedipine, the dealkylation of erythromycin and lignocaine, and the 6β-hydroxylation of cortisol and all of these reactions have been applied to the investigation of CYP3A activity *in vivo* (56). However, there are difficulties, either technical or theoretical, associated with the use of these substrates (see refs 66 and 70 for a discussion) and this lead to a recent investigation of the use of the urinary excretion of N-hydroxydapsone as a marker of CYP3A activity *in vivo* (70). Comparison of urinary N-hydroxydapsone excretion, erythromycin demethylation and cortisol 6β-hydroxylation surprisingly demonstrated no statistically significant correlations among any of the measurements ($r = 0.13$ to -0.12, $p > 0.5$). It was suggested that different routes of administration and varying contributions of extrahepatic metabolism accounted for the lack of a significant relationship (70). Consistent with this observation, only a weak relationship ($r^2 = 0.35$) has been reported between the erythromycin breath test and 6β-hydroxycortisol excretion in 47 subjects (71). Despite the fact that the erythromycin breath test is considered the current benchmark for measurement of CYP3A *in vivo*, only a modest relationship ($r^2 = 0.56$) was observed between the erythromycin breath test and hepatic CYP3A content in a group of patients with liver disease (72). Taken together, it would appear that none of the currently available *in vivo* procedures are optimal for measurement of CYP3A.

APPLICATION OF cDNA-EXPRESSED ISOFORMS

A number of heterologous expression systems have been used to express isoforms of drug metabolizing enzymes. Mammalian cell systems which transiently express recombinant proteins include transformed African green monkey kidney (COS) cells (73) and vaccinia virus mediated expression in human HepG2 cells (74). Stable expression has been achieved in V79 Chinese hamster cells (75) and human lymphoblastoid cells (76). The latter are available commercially (Gentest Corporation, Woburn, MA, USA). The mammalian cells provide a suitable membrane environment and coenzymes (e.g. NADPH-cytochrome P450 reductase) for efficient CYP catalytic function, but expression levels are low making assay sensitivity a problem and generally precluding measurement of functional holoenzyme by CO difference spectrophotometry.

Higher levels of expression are possible in yeast cell systems (77). However, expression is variable between CYP isoforms and activities may be low unless the assay system is supplemented with NADPH-cytochrome P450 reductase and cytochrome b5. Yeast systems engineered to overexpress these co-enzymes have been described recently (77). High CYP isoform expression levels are achieved in the baculovirus expression systems which uses insect cells as the expression host (78). The use of prokaryotic E coli expression systems generally requires modification of the coding sequence. The cells lack an internal membrane structure and post-translational modification processes which may cause incorrect protein folding and the formation of insoluble inclusion bodies. Nevertheless, successful expression of functional CYP isoforms in E coli has been achieved (79). Both the baculovirus and E coli systems require supplementation with, or co-expression of, reductase to achieve optimal activity (80).

PREDICTIVE SUBSTRATE STRUCTURE-ACTIVITY MODELLING

Molecular modelling techniques have been used recently to derive substrate and inhibitor structural features directing metabolism selectively to CYP2D6, CYP3A and CYP2C9 (81-85). Some models also incorporate information on the crystal structure of the bacterial isoform CYP101 (83-84). It is likely that such methods will become more widely applied as further selective substrates and inhibitors for individual CYP isoforms are identified and the tertiary structure of mammalian isoforms becomes known.

Summary - With the exception of CYP 2B6 and 2C8, substrate and inhibitor probes are available for the investigation of all known drug metabolising human CYP isoform activities *in vitro* (see Table). Utilising combinations of kinetic and inhibitor techniques with human liver microsomes and cDNA-expressed isoforms, these probes enable linkage of the metabolism of any given compound to a CYP isoform(s). This potentially allows prediction of factors likely to influence metabolism of the compound prior to its administration to humans. Although substrate and inhibitor probes are also available for the investigation of most CYP isoform activities *in vivo*, methods for assessment of isoform activity *in vivo* are frequently less satisfactory than those used *in vitro*. For example, while methods may be adequate for characterising phenotypes of polymorphically expressed isoforms, they may be less reliable for differentiating activities within a phenotype. While this review has focussed on probes for CYP, it is now recognised that UDP-glucuronosyltransferase (86) and sulphotransferase (87) also exist as gene superfamilies. Since few UDP-glucuronosyltransferase or sulphotransferase isoforms metabolising drugs have been characterised to date, the absolute specificity of probes utilised *in vitro* or *in vivo* remains unknown. Methods are, however, available for assessment of N-acetyltransferase (NAT2) (21,88) and xanthine oxidase (21) activities *in vitro* and/or *in vivo*.

Table. In vitro substrate and inhibitor probes for human CYP

Isoform	Substrates	Inhibitors
CYP1A1	Caffeine, phenacetin	α-Naphthoflavone
CYP1A2	Caffeine, phenacetin, R-warfarin	Furafylline, α-naphthoflavone
CYP2A6	Coumarin	
CYP2C9/10	Tolbutamide, torasemide, S-warfarin, phenytoin	Sulphaphenazole
SMPH	S-Mephenytoin, omeprazole	
CYP2D6	Debrisoquine, codeine, dextomethorphan, sparteine	Quinidine
CYP2E1	Chlorzoxazone, 4-nitrophenol	Diethyldithiocarbamate
CYP3A3/4	Caffeine, dapsone, erythromycin, lignocaine, midazolam	Gestodene, triacetyloleandomycin

REFERENCES

1   W. Tassaneeyakul, D.J. Birkett, M.E. Veronese, M.E. McManus, R.H. Tukey, H.V. Gelboin and J.O. Miners, J.Pharmacol.Exp.Ther., 265, 401 (1993).
2   W. Tassaneeyakul, Z. Mohamed, D.J. Birkett, M.E. McManus, M.E. Veronese, R.H. Tukey, F.J. Gonzalez and J.O. Miners, Pharmacogenetics, 2, 173 (1992).
3   T. Andersson, J.O. Miners, D.L. Rees, W. Tassaneeyakul, M.E. Veronese, U.A. Meyer and D.J. Birkett, Br.J.Clin.Pharmacol., 36, 521 (1993).
4   T. Andersson, J.O. Miners, M.E. Veronese and D.J. Birkett, Br.J.Clin.Pharmacol., in press (1994).
5   D.R. Nelson, T. Kamataki, D.J. Waxman, F.P. Guengerich, R.W. Estabrook, R. Feyereisen, F.J. Gonzalez, M.J. Coon, I.C. Gunsalus, O. Gotoh, K. Okuda and D.W. Nebert, DNA Cell Biol., 12, 1 (1993).
6   F.J. Gonzalez, Trends Pharmacol.Sci., 13, 346 (1992).
7   F.J. Gonzalez, Pharmacol.Ther., 45, 1 (1990).
8   D.J. Birkett, R.A. Robson, M.E. McManus and J.O. Miners, in "Toxicological and Immunological Aspects of Drug Metabolism and Environmental Chemicals", R.W. Estabrook, F. Oesch and A. L. de Wech, Eds, Symposia Medica Hoechst, 1988, p 261.
9   D.J. Birkett, M.E. Veronese, P.I. Mackenzie and J.O. Miners, Trends Pharmacol.Sci., 14, 292 (1993).
10 F.J. Gonzalez, Anesthesiology, 77, 41 (1992).
11 F.P. Guengerich, FASEB J, 6, 745 (1992).
12 J.O. Miners and D.J. Birkett, Pharmacogenetics, 3, 58 (1993).

13 D.J. Birkett and J.O. Miners, Br.J.Clin.Pharmacol., 31, 405 (1991).
14 J.O. Miners, N.J. Osborne, A.L. Tonkin and D.J. Birkett, Br.J.Clin.Pharmacol., 34, 359 (1992).
15 C-H. Yun, T. Shimada and F.P. Guengerich, Mol.Pharmacol, 40, 679 (1991).
16 W. Tassaneeyakul, D.J. Birkett, M.E. McManus, M.E. Veronese, T. Andersson, R.H. Tukey and J.O. Miners, Biochem.Pharmacol. (in press).
17 U. Fuhr, J. Doehmer, N. Battula, C. Wolfed, C. Kidla, Y. Keita and C. Staib, Biochem.Pharmacol., 43, 225 (1992).
18 K.L. Kunze and W.F. Trager, Chem.Res.Toxicol., 6, 649 (1993).
19 R.A. McKinnon, P.M. Hall, L. Quattrochi, R.H. Tukey and M.E. McManus, Hepatology, 14, 848 (1991)
20 A. Lelo, J.O. Miners, R.A. Robson and D.J. Birkett, Br.J.Clin.Pharmacol, 22, 183 (1986).
21 W. Kalow and B-K. Tang, Clin.Pharmacol.Ther., 53, 503 (1993).
22 M.A. Butler, N.P. Lang, J.F. Young, N.E. Caporaso, P. Vineis, R.B. Hayes, C.H. Teitel, M.F. Lawsen and F.F. Kadlubar, Pharmacogenetics, 2, 116 (1992).
23 D. Sesardic, A. Boobis, B. Murray, S. Murray, J.D. Segura, R. Torre and D.S. Davies, Br.J.Clin. Pharmacol, 29, 651 (1990).
24 W. Tassaneeyakul, PhD Thesis, Flinders University of South Australia, (1994).
25 A.R. Boobis, A.M. Lynch, S. Murray, R. de la Torre, A.Solans, M. Farre, J. Segura, N.J. Gooderham and D.S. Davies, Cancer Res., 54, 89 (1994).
26 K. Brosen, E. Skjelbo, B.B. Rasmussen, H.E. Poulson and S. Loft, Biochem.Pharmacol., 45, 1211 (1993).
27 U. Fuhr, G. Strobl, F. Mandut, E-M. Anders, F. Sorgel, D.T.W Chu, A.G. Pernet, G. Mahr, F. Sanz and A.H. Staib, Mol.Pharmacol., 43, 191 (1993).
28 E. Skjelbo and K. Brosen, Br.J.Clin.Pharmacol., 34, 256 (1992).
29 J. Maenpaa, H. Sigusch, H. Raunio, T. Syngelma, P. Vuorela, H. Vuorela and O. Pelkonen, Biochem.Pharmacol., 45, 1035 (1993).
30 R. Pearce, D. Greenway and A. Parkinson, Arch.Biochem.Biophys., 298, 211 (1992).
31 S. Cholerton, M.E. Idle, A. Vas, F.J. Gonzalez and J.F. Idle. J.Chromatogr., 575, 325 (1992).
32 A. Rautio, H. Kraul, A. Kojo, E. Salmela and O. Pelkonen, Pharmacogenetics, 2, 227 (1992).
33 T.K.H. Chang, G.F. Weber, C.L. Crespi and D.J. Waxman, Cancer Res., 53, 5629 (1993).
34 M.E. Veronese, C.J. Doecke, P.I. Mackenzie, M.E. McManus, J.O. Miners, D.L.P. Rees, R. Gasser, U.A. Meyer and D.J. Birkett, Biochem.J., 289, 533 (1993).
35 M-A. Leo, J.M. Lasker, J.L. Raucy, C-L. Kim, M. Black and C.S. Lieber, Arch.Biochem.Biophys., 269, 305 (1989).
36 B.M. Kerr, K.E. Thummel, C.M. Johnson, S.M. Klein, D.L. Kroetz, F.J. Gonzalez and R.H. Levy, Biochem.Pharmacol., in press (1994).
37 M.E. Veronese, P.I. Mackenzie, C.J. Doecke, M.E. McManus, J.O. Miners and D.J. Birkett, Biochem.Biophys.Res.Commun., 175, 1112 (1991).
38 C.J. Doecke, M.E. Veronese, S.M. Pond, J.O. Miners, D.J. Birkett, L. Sansom and M.E. McManus, Br.J.Clin.Pharmaol., 31, 125 (1991).
39 A.E. Rettie, K.R. Korzekwa, K.L. Kunze, R.F. Lawrence, A.C. Eddy, T. Aoyama, H.V. Gelboin, F.J. Gonzalez and W. Trager, Chem.Res.Toxicol., 5, 54 (1992).
40 J.O. Miners, D.L.P. Rees, M.E. Veronese and D.J. Birkett, Clin.Exp.Pharmacol.Physiol., suppl 1, 49 (1993).
41 M.E. Veronese, J.O. Miners, D. Randles, D. Gregov and D.J. Birkett, Clin.Pharmacol.Ther., 47, 403 (1990).
42 W. Tassaneeyakul, M.E. Veronese, D.J. Birkett, C.J. Doecke, M.E. McManus, L.N. Sansom and J.O. Miners, Br.J.Clin.Pharmacol., 34, 494 (1992).
43 T.A. O'Sullivan, J-P. Wang, J.D. Unadkat, S.M.H. Al-Haket, W.F. Trager, A.L. Smith, S. McNamara and M.L. Aitken, Clin.Pharmacol.Ther., 54, 323 (1993).
44 S.A. Wrighton, J.C. Stevens, G.W. Becker and M. Vandenbranden, Arch.Biochem.Biophys., 306, 2400 (1993).
45 T. Andersson, J.O. Miners, M.E. Veronese and D.J. Birkett, Br.J.Clin.Pharmacol., in press (1994).
46 T.V. Beischlag, W. Kalow, W.A. Mahon and T. Inaba, Xenobiotica, 22, 559 (1992).
47 K. Chiba, K. Kobayashi, K. Manebe, M. Tani, T. Kamataki and T. Ishizaki, J.Pharmacol.Exp.Ther., 296, 52 (1993).
48 D.J. Birkett, D.L.P. Rees, T. Andersson, J.O. Miners, F.J. Gonzalez and M.E. Veronese, Br.J.Clin. Pharmacol., in press (1994).
49 L.S. Kaminsky, S.M.F. de Morais, M.B. Faletto, D.A. Dunbar and J.A. Goldstein, Mol.Pharmacol., 43, 234 (1993).
50 G.R. Wilkinson, F.P. Guengerich and R.A. Branch, Pharmacol.Ther., 43, 53 (1989).
51 G. Tybring and L. Bertilsson, Pharmacogenetics, 2, 241 (1992).
52 Y. Zhang, R.A. Bloun, P.J. McNamara, J. Steinmetz and P.J. Wedlund, Br.J.Clin.Pharmacol., 31,

350 (1991).

53 C.Funck-Bretano, O. Bosco, E. Jacqz-Aigrain, A. Keundjian and P. Jaillon, Clin.Pharmacol.Ther., 51, 507 (1992).

54 S.A. Ward, N.A. Helsby, E. Skjelbo, K. Brosen, L.F. Gram and A.M. Breckenridge, Br.J.Clin. Pharmacol., 31, 689 (1991).

55 T. Andersson, C-G. Regardh, Y-C. Lou, Y. Zhang, M-L. Dahl and L. Bertilsson, Pharmacogenetics, 2, 25 (1992).

56 S.A. Wrighton and J.C. Stevens, Crit.Rev.Toxicol., 22, 1 (1992).

57 S. Cholerton, A.K. Daley and J.R. Idle, Trends Pharmacol.Sci., 13, 434 (1992).

58 Z.R. Chen, A.A. Somogyi, G. Reynolds and F. Bochner, Br.J.Clin.Pharmacol., 31, 381 (1991).

59 N. Feifel, K. Kucher, L. Fuchs, M. Jedrychowski, E. Schmidt, P.R. Bieck and C.H. Gleiter, Eur.J. Clin.Phamacol., 45, 265 (1993).

60 D. Young, K.K. Midha, M.J. Fossler, E.M. Hawes, J.W. Hubbard, G. McKay and E.D. Korchiniski,Eur.J.Clin.Pharmacol., 44, 433 (1993).

61 D.R. Koop, FASEB J., 6, 724 (1992).

62 W. Tassaneeyakul, M.E. Veronese, D.J. Birkett, F.J. Gonzalez and J.O. Miners, Biochem. Pharmacol.,46, 1975 (1993).

63 R. Peter, R. Bocker, P.H. Beaune, M. Iwasaki, F.P. Guengerich and C.S. Yang, Chem.Res.Toxicol., 3, 556 (1990).

64 F.P. Guengerich, D-H. Kim and M. Iwasaki, Chem.Res.Toxicol., 4, 168 (1991).

65 E.D. Kharasch, K.E. Thummel, J. Mhyre and J.H. Lillibridge, Clin.Pharmacol.Ther., 53, 643 (1993).

66 C.M. Fleming, R.A. Branch, G.R. Wilkinson and F.P. Guengerich, Mol.Pharmacol., 41, 975 (1992).

67 C-H. Yun, A. Okerholm and F.P. Guengerich, Drug Metab.Disp., 21, 403 (1993).

68 F.P. Guengerich, Chem.Res.Toxicol., 3, 363 (1990).

69 C-H. Yun, T. Shimada and F.P. Guengerich, Cancer Res., 52, 1868 (1992).

70 M.T. Kinirons, D. O'Shea, T.E. Downing, A.T. Fitzwilliam, L. Joellenbeck, J.D. Groopman, G.R. Wilkinson and A.J.J. Wood, Clin.Pharmacol.Ther., 54, 621 (1993).

71 P.B. Watkins, D.K. Turgeon, P. Saenger, K.S. Lown, J.C. Kolars, T. Hamilton, K. Fishman, P.S. Guzelian and J.J. Voorheis, Clin.Pharmacol.Ther., 52, 265 (1992).

72 K.S. Lown, J. Kolars, D.K. Turgeon, R. Merion, S.A. Wrighton and P.B. Watkins, Clin.Pharmacol. Ther., 51, 229 (1992).

73 B.J. Clark and M.R. Waterman, Methods Enzymol., 206, 100 (1991).

74 F.J. Gonzalez, Toxicology, 82, 77 (1993).

75 J. Doehmer, Toxicology, 82, 105 (1993).

76 C.L. Crespi, R. Lagenbach and B.W. Penman, Toxicology, 82, 89 (1993).

77 J.P. Renaud, M.A. Peyronneau, P. Urban, G. Truan, C. Cullin, D. Pompon, R. Beaune and D. Mansuy, Toxicology, 82, 39 (1993).

78 F.J. Gonzalez, S. Kimura, S. Tamura and H.V. Gelboin, Methods Enzymol., 206, 93 (1991).

79 M.R. Waterman, Biochem.Soc.Trans., 21, 1081 (1993).

80 M.S. Shet, C.W. Fisher, P.L. Holmans and R.W. Estabrook, Proc.Natl.Acad.Sci., 90, 11748 (1993).

81 D.A. Smith and B.C. Jones, Biochem.Pharmacol., 44, 2089 (1992).

82 L.Koymans, N.P.E.Vermeulen, S. van Acker, J.M. Koppele, J.J.P. Heykants, K. Larrijsen, W. Meuldermans and G.M. Donné-Op den Kelder, Chem.Res.Toxicol., 5, 211 (1992).

83 K.R. Korzekwa and J.P. Jones, Pharmacogenetics, 3, 1 (1993).

84 S.A. Islam, C.R. Wolf, M.S. Lennard and M.J.E. Sternberg, Carcinogenesis, 12, 2211 (1991).

85 G.R. Strobl, S. von Kruedener, J. Stökigt, F.P. Guengerich and T. Wolff, J.Med.Chem., 36, 1136 (1993).

86 J.O. Miners and P.I. Mackenzie, Pharmacol.Ther., 51, 347 (1991).

87 M.E. Veronese, W. Burgess, X. Zhu and M.E. McManus, Biochem.J., in press (1994).

88 D. M. Grant, M. Blum and U.A. Meyer, Xenobiotica, 22, 1073 (1992).

Chapter 32. Humanized Monoclonal Antibodies

Leonard Presta

Genentech, Inc., 460 Point San Bruno Blvd.
South San Francisco, CA 94080

Introduction - Almost two decades ago, the development of monoclonal hybridoma technology — the ability to make cultured cells produce antibodies of predefined specificity — promised a new weapon in the arsenal of molecules able to combat disease (1). Antibodies against a specific antigen (or target molecule) could be generated, incorporated into a hybridoma for production, and then used in diagnosis or therapy. However, only rodent monoclonal antibodies could be made due to technological limitations. In most clinical applications, these rodent monoclonal antibodies exhibit properties which severly limit their utility. First, they may induce an immunogenic response in humans, referred to as HAMA (human anti-mouse antibodies), when the human immune system recognizes them as foreign substances. Second, the therapeutic efficacy of the rodent antibody may also be reduced because a rodent antibody is cleared from serum more rapidly than human ones. Third, in humans, rodent antibodies generally exhibit only weak recruitment of effector functions, such as antibody-dependent cell-mediated cytotoxicity and complement fixation, which may be requisite for the function of the antibody. Use of a human antibody would circumvent these three problems.

With advances in molecular biology and mammalian tissue culture, it became possible to obtain useful amounts of any antibody, thereby surmounting the limitations of hybridoma technology and use of only rodent antibodies. But one problem still existed — how does one obtain *human* antibodies against a particular human antigen? Even if one could ethically use a human subject as the biological factory, the human immune system, in general, does not produce antibodies against human proteins. Two techiques have been developed to address this problem. Earliest of these was the construction of chimeric antibodies in which entire antigen-binding domains from a rodent antibody are substituted for those of a human antibody (Fig. 1) (2). In many cases, these molecules still exhibited some, albeit reduced, HAMA. The next step was generation of humanized antibodies in which a significantly reduced number of rodent residues are incorporated into a human antibody such that the humanized antibody has the same binding characteristics and specificity as the original rodent antibody (3-5). In order to understand the technique of humanization and the difference between a chimeric and humanized antibody, the structure of antibodies must be appreciated.

Antibody Structure - An antibody consists of four peptide chains — two identical light chains and two identical heavy chains — which form a 'Y' shape. The functions of the antibody reside in different domains. Antigen binding occurs at the ends of the arms of the 'Y', each arm being referred to as a Fab or antigen binding fragment. Hence each antibody can bind two antigen molecules. The effector functions reside in the base of the 'Y', referred to as the Fc portion. Each light chain consists of one variable domain and one constant domain; each heavy chain consists of one variable domain and three constant domains (Fig. 1). The 'constant' notation refers to the fact that for a particular species and immunoglobulin class (e.g. IgG, IgE, IgA) the amino acid sequence for the constant domain is, for our purposes, invariant. The 'variable' notation refers to the fact that certain portions of variable domains differ extensively in sequence among antibodies and are used in the binding and specificity of each particular antibody for its particular antigen. These portions are referred to as complementarity-determining regions (CDRs). Both light and heavy chain variable domains contain three CDRs to give a total of six CDRs involved in binding antigen.

Each CDR is a contiguous sequence of amino acids which form a loop connecting two β-strands of the framework. Note though that the amino acid sequence of the non-CDR portion of the 'variable' domain is relatively invariant and is used to categorize these domains into subgroups (6).

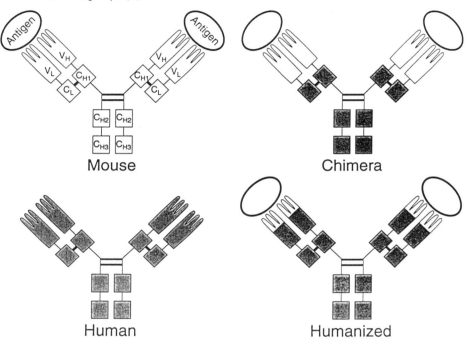

Figure 1. Chimeric and Humanized Antibodies. CDRs are represented by 'fingers' on variable domains of the light chain (V_L) and heavy chain (V_H). Note that only the mouse, chimera and humanized antibodies bind antigen. Disulfide bonds between the two heavy chains and between the light and heavy chains are denoted by thicker, dark lines.

In one type of chimeric antibody all of the variable domains from a rodent antibody are fused onto the constant domains of a human antibody (Fig. 1). This type of chimeric antibody is comprised of four rodent domains and eight human domains and consequently retains approximately 33% rodent residues. In the corresponding humanized antibody, only the six rodent CDRs replace the six human CDRs (Fig. 1). The resulting antibody contains only about 5-10% rodent residues and is still chimeric in that it has residues derived from different species. From the perspective of the human immune system, the humanized antibody is the more 'human' of the two types as it contains fewer rodent residues. Moreover one must appreciate that the amino acids in CDRs are highly variable and this variance is independent of species. The nature of the CDRs is dependent upon the target, whereas the framework is species dependent. Hence even though the humanized antibody still retains about 5-10% rodent residues, these residues would also be variant in human as well as rodent and thus the humanized antibody could well be very similar to a human antibody directed against the same site on the same target as the original rodent antibody.

Humanized Antibody Construction - At first glance constructing a humanized antibody is rather simple and straightforward. One chooses a human antibody, clips off the six human CDRs and replaces them with their structurally analogous rodent CDRs. Indeed some humanized antibodies have approached this simplicity in method (7,8). However, there are both theoretical and technical difficulties. Among the former is the exact definition of a CDR, i.e. exactly which residues comprise the CDRs and which comprise the framework. Two definitions have been proposed. The first relies solely on analyses of the amino acid sequences of a multitude of antibodies (6). The variance of amino acid types at each residue position of antibodies was evaluated and those positions which exhibited a relatively large variance were included in a CDR as long as they were contiguous to other positions with high variance. Residue positions showing relatively low variance were termed framework. The second definition was derived from structures of antibodies (9). When the two definitions are compared some striking differences are apparent. While the sequence-based CDRs are generally larger than, and encompass, the structure-based CDRs, a notable discrepancy occurs in the definition of CDR-H1 (i.e. the first CDR loop in the heavy chain sequence) where the two definitions overlap by only two residues. Generally, the sequence-based definition of CDRs has been used when an antibody is humanized.

The technical difficulties fall into three categories. First, the choice of constant domains must be made. To date almost all antibodies humanized have been those of the IgG class. In humans there are four recognized subclasses, denoted IgG1, IgG2, IgG3 and IgG4, the classification of which is based on the sequence of their constant domains. Each subclass also exhibits differences in effector, or biological, functions mediated by the constant domains. The intended therapeutic use of the antibody must be considered when choosing the IgG subclass, e.g. is complement fixation an advantage or disadvantage for the particular application?

Second, the choice of human variable domains, both light and heavy, must be made. In the homology or 'best-fit' method, the sequence of the rodent variable domain is screened against the entire library of known human variable domain sequences. The human sequence which is closest to that of the rodent is then accepted as the human framework (8,10). Another method uses a particular framework derived from the *consensus* sequence of all human antibodies of a particular subgroup of light or heavy chains. The same framework may be used for several different humanized antibodies (11,12). Remembering that the purpose of humanization is to trick the human immune system into recognizing the humanized antibody as one of its own, is the 'best-fit' or the consensus framework better for this trickery? Unfortunately there are very limited *in vivo* data on the reaction of the human immune system to humanized antibodies and no comparative study has been performed on the two types. In the final analysis both may function well with regard to acceptance by the human immune system, with perhaps an occasional aberration.

Third is the most imposing technical difficulty -— designing the humanized antibody such that it has the same binding affinity and specificity as the original rodent antibody. When the rodent CDRs are simply grafted onto a human framework the resultant humanized antibody may exhibit signficiantly reduced binding compared to the original rodent antibody (11,12). This reduction in binding may be caused by certain buried framework residues. Each CDR is a loop and these loops may be anchored to the framework not only covalently at their ends but through noncovalent interactions with the sidechains of buried framework residues. For example, in CDR-L1 the sidechain at position 29 (residue numbering is according to ref. 6) is usually hydrophobic, buried and contacts hydrophobic sidechains at framework positions 2, 25, 33, 71 and 90 (13). If an incorrect amino acid is chosen for any of these framework positions the packing of CDR-L1 against the protein might be altered, resulting in an incorrect presentation of the exposed CDR-L1 sidechains and consequent reduction in binding to antigen. Such a case may occur when the sidechains at position 29 in the rodent CDR and positions 2, 25, 33, 71 and 90 in the rodent framework differ from those in the human antibody. In order to retain proper binding the offending framework residues must be altered from human to murine so that the CDR packs against the protein properly. Alternatively, the buried sidechain in the CDR itself, such as at position 29 in CDR-L1, might be changed from murine to human to evaluate whether proper CDR conformation can be attained. This regimen must be evaluated for each of the six CDRs because if just one CDR has an improper conformation it may reduce the binding of the humanized antibody to the point that it is useless as a therapeutic replacement for the rodent antibody. While buried framework residues can affect the binding of the humanized antibody through an indirect mechanism, i.e. by influencing the conformation of a CDR, occassionally an exposed framework residue is involved in direct binding to the antigen and must be included in the humanized form (11,14).

While performing this regimen for six CDRs may seem a daunting task, especially since crystal structures of the rodent and human antibodies are usually not available, nature has provided some guidance. Even though CDRs vary widely in sequence and to a lesser extent in size, analysis of crystal structures of Fab fragments shows that for each CDR (except CDR-H3) only a limited number of conformations may be utilized (13). These canonical conformations are dependent on size and sequence of the CDR. Hence if one inspects the sequence and size of the particular rodent CDR the canonical conformation can then be assigned. This provides information as to which CDR residues will be exposed, which will be buried and which framework residues may play a role in stabilizing the CDR conformation.

How has this technique fared regarding the ability of a humanized antibody to reproduce the binding affinity of its parent rodent antibody? The first humanization involved transferring only the CDRs of the heavy chain from a murine to human antibody. When expressed with the murine light chain, the humanized heavy chain exhibited binding less than 2-fold down from the parent murine (15). Two years later another two humanizations were published in which both the light and heavy chain CDRs were transferred. In the first, a humanized anti-lysozyme antibody, binding was reduced by 10-fold (16). The second was more interesting since it was potentially clinically relevant, underscored the importance of a single buried residue for maintaining proper CDR conformation, and noted the difference between the sequence-based and structure-based CDR definitions (17). Directed against the CAMPATH-1 antigen on human lymphocytes and monocytes, the initial humanized version was 40-fold down in binding. Changing the human Ser27 to murine Phe27 in the heavy chain resulted in a version which exhibited only a 3-fold reduction in affinity. Was this a framework or a CDR residue? According to the structure-based definition of CDRs used to design the humanized anti-CAMPATH-1 antibody, heavy chain position 27 was part of the framework while according to the sequence-based definition position 27 was part of the CDR. Regardless of the definition, the marked improvement in binding when the murine Phe was substituted for the human Ser emphasized the need to pay attention to framework, as well as CDR, sequence in future humanizations.

Since 1988 the technique has developed and become increasingly utilized for clinically relevant antibodies. The 'best-fit' method, used first in 1989 (10), has remained the more popular method for designing the sequence of the humanized antibody than the later consensus method (11). More recently, in the design of an anti-IgE humanized antibody, both methods were evaluated (14). The study concluded that although the consensus sequence showed the best overall binding, no clear advantage in binding was evident for the consensus antibody versus the 'best-fit' antibody. Another humanization of an anti-IgE antibody, utilizing only the consensus method, was reported earlier that year (12).

Both of the anti-IgE humanizations used another recent technology in the design of the antibody — molecular modeling. As the database of antibody crystal structures grows (which consists predominanly of structures of Fab fragments), information accumulates on the three-dimensional structure of variable domains and the relationship between CDRs and framework. Since the structure of the rodent antibody of interest will usually not be available, a computer graphical model based on crystal structures can be generated and used to determine which framework residues might influence CDR conformation. Many of the humanizations published of late include use of molecular modeling or crystal structure information (7,12,14,18,19). Only recently, however, have crystal structures of humanized antibodies been determined and compared to the molecular model used to design the humanized antibody (20,21). In this case, the model correctly predicted the conformation of the framework and five of the CDR loops. The conformation of the sixth CDR in the model differed from that in the crystal structure, but this same CDR also varied in conformation among the crystal structures of three closely related humanized variants (21).

Though some humanizations have been successful without any alteration of the chosen human framework (7,8), other studies have found it necessary to alter some human framework residues to murine in order to attain binding equivalent to the parent murine antibody. In both of the aforementioned anti-IgE humanizations, murine framework residues were important (12,14). In humanization of the Mikβ1 antibody, eighteen framework changes from the 'best-fit' sequence were incorporated (18). Another version of Mikβ1 with just six changes bound only about 2-fold less well than the version with eighteen changes. However the latter was superior in biological activity and it was noted that the bioactivities of these two versions could not be predicted solely from their binding affinities to the target molecule, in this case the IL2 receptor β-chain. Such a lack of correlation between binding affinity and bioactivity had been noted previously for another humanized antibody (11). The importance of a particular framework residue may vary from antibody to antibody. For example, among six humanizations reported very recently, two retained the human amino acid at heavy chain position 71 (7,8), three changed to the murine amino acid (14,18,19) and in the last the murine and human consensus residues at position were identical and no change was required (12). This emphasizes the importance of considering each murine antibody as idiosyncratic, i.e. each antibody may possess its own unique CDR-framework interactions which are sensitive to changes in amino acid constitution. As more humanizations are performed and antibody crystal structures become increasing available, perhaps nature's rules for determining these characteristic packing interactions may become more evident so that they can be ascertained during the initial design of the humanized antibody. Some effort in this area has already been accomplished (13,22,23).

While buried framework residues have been shown to be important, less attention has been focused on exposed (or semi-exposed) framework residues. Crystal structures of Fab-antigen complexes have shown that on occassion a framework residue will be involved in direct binding to the antigen (24). The first mention of an exposed framework residue important for retention of binding and bioactivity in a humanized antibody was for the anti-p185HER2 antibody (11). The murine Arg at position 66 in the light chain, predominantly Gly in human kappa light chains, was shown to improve binding by 4-fold and improve bioactivity. More recently, having the murine amino acids at light chain residues 1 and 3 were shown to enhance binding (14). Though the involvement of exposed framework residues in binding antigen may not be common, they deserve consideration when attempting to further improve the binding of a less than competent humanized antibody.

The focus of humanized antibodies has been on the variable domains since these bind antigen, but some studies have begun to concentrate on the rest of the humanized molecule, namely the Fc portion which possesses the effector functions. While some humanized antibodies have exhibited the ability to effect antibody-dependent cell-mediated cytolysis (11,18,25,26) others have not (8). In other studies the differences between the various classes of IgG have been evaluated (27) and, recently, the removal of conserved carbohydrate in the Fc has been investigated (28). As humanizations become more commonplace (or are replaced by human antibodies) and clinical data become available, the necessity for understanding and tailoring the Fc for a desired effector function will increase.

Therapeutic Use of Chimeric and Humanized Antibodies - By the mid-1980's, it was recognized that rodent antibodies used as therapeutics would be problematic in that they may elicit a HAMA response, are more rapidly cleared from serum and less efficacious at eliciting effector functions (29). The HAMA response was also categorized into at least two types of anti-antibodies: those directed against the CDRs, referred to as anti-idiotypic antibodies, and those directed against the remainder of the molecule, referred to as non-anti-idiotypic antibodies (30). Initially chimeric antibodies were thought be the solution. In one interesting study the immunogenicity of chimeric antibodies was evaluated by introducing human antibodies and human/murine chimeric antibodies into mice (31). The murine response to fully human antibodies was robust, with 90% of the anti-human antibodies directed against the constant domains and 10% directed against the variable domains. The chimeric antibodies also elicited a response directed against the human variable domains, whereas fully murine antibodies gave no response or only an anti-allotypic response. The latter arise from variation in the sequence of constant (and sometimes variable) domains among individuals and is akin to blood group antigens (A, B, AB, O). In humans, chimeric antibodies have given mixed results. For example, the first chimeric antibody in a clinical trial, directed against a surface antigen on adenocarcinoma cells of gastrointestinal origin, elicited a detectable but modest HAMA response in only 1 of 10 patients (32). In contrast, the same investigators later reported that a different chimeric antibody, directed against a different surface antigen on adenocarcinoma cells, elicited a HAMA response in 7 of 12 patients after only one infusion (33). For both antibodies the HAMA response was directed against the murine variable region.

The first humanized antibody used in humans, CAMPATH-1H, induced remission in two patients with non-Hodgkin lymphoma and showed no detectable HAMA (34). CAMPATH-1H was evaluated a few years later for rheumatoid arthritis (35). In this study, 7 of 8 patients showed significant clinical benefit and all 8 patients exhibited no HAMA after one course of treatment. However upon retreatment, 3 of 4 patients did show a response. Of these, two had only an anti-idiotypic response while the third had both an anti-idiotypic and anti-allotypic response. Another antibody which was among the earliest humanized, the anti-Tac antibody (10), has been shown to be less immunogenic in cynomolgus monkeys than the parent murine antibody (36). As expected the response in the cynomolgus monkeys to the humanized antibody was anti-idiotypic while that against the murine was directed against the rest of the immunoglobulin. These anti-idiotypic antibodies have recently been shown to be directed primarily against combinations of CDRs H1, H2 and L3 rather than against a single CDR (37). In a similar study, murine and humanized OKT4A antibodies were evaluated in cynomolgus monkeys (38). All 17 monkeys which received murine OKT4A showed a response which was both idiotypic and non-idiotypic. While all 8 monkeys which received the humanized OKT4A also showed a response, it was only anti-idiotypic. In addition, higher serum levels of the humanized antibody were maintained for a longer period following treatment than were serum levels of the murine OKT4A.

Finally, during the past year a study was published in which mice injected with murine antibodies were used as a model for estimating potential patient sensitization against humanized antibodies (39). As in a similar earlier study (31), anti-idiotypic and anti-allotypic antibodies were elicited. That both types of anti-antibodies have been found underscores the need to determine the biological and therapeutic consequences of each. Anti-idiotypic antibodies directed against the CDRs may prove to be a necessary evil when humanized antibodies are used as therapeutics since the CDRs are the part of the immunoglobulin which bind antigen and therefore cannot be altered. Anti-idiotypic antibodies could present a problem in that they would block the function of the humanized antibody after a repeated application, thereby requiring an increased dose and/or significantly reducing the efficacy of the humanized antibody. If this is the *only* problem caused by anti-idiotypic antibodies, then one possible solution would be to have several bioactive humanized antibodies which bind the same antigen but have different CDRs. After a course of therapy with the first, the second (or even third) could then be used if a chronic therapy was required. Anti-allotypic antibodies, directed against idiosyncratic sequences of an individual, will be more problematic. Chronic therapy may require matching of a recipient's allotype prior to therapy (39). If the allotype of the recipient differs from that of the available humanized antibody then treatment might have to be abandoned or, alternatively, the constant portion of the humanized antibody would have to be redesigned for the patient's allotype. While such a 'tailored-to-fit' antibody is technically feasible, it would undoubtedly significantly increase the cost of treatment. Since no studies have been published on the biological consequences of either anti-idiotypic or anti-allotypic antibodies in humans, these questions await further clinical trial.

Conclusion - Although several successful humanizations of antibodies have been published, clinical data at this point in time are scarce due to the relatively recent development of the technique. What is available supports the contention that antibodies hold great promise as therapeutics (3,4). Most likely, however, humanization as a technique will be replaced by novel methods of generating *human* antibodies. Two such technologies are the production of human antibody repertoires in transgenic mice (40) and from phage display libraries (41,42). Though it is destined for replacement, humanization of antibodies has stimulated research on the structure-function of antibodies and provided an important intermediary step in utilization of monoclonal antibodies as therapeutics.

References

1. G. Kohler and C. Milstein, Nature, 256, 52 (1975).
2. S.L. Morrison, M.J. Johnson, L.A. Herzenberg and V.T. Oi, Proc.Natl.Acad.Sci.USA, 81, 6851 (1984).
3. G. Winter and W.J. Harris, Trends Pharmacol.Sci., 14, 139 (1993).
4. L.K. Jolliffe, Intl.Rev.Immunol., 10, 241 (1993).
5. L.G. Presta, Curr.Opinion Struct.Biol., 2, 593 (1992).
6. E.A. Kabat, T.T. Wu, H.M. Perry, K.S. Gottesman and C. Froeller, "Sequences of Proteins of Immunological Interest", Vol. 1, 5th Ed., NIH Publ. No. 91-3242 (1991).
7. M.E. Verhoeyen, J.A. Saunders, M.R. Price, J.D. Marugg, S. Briggs, E.L. Broderick, S.J. Eida, A.T.A. Mooren and R.A. Badley, Immunology, 78, 364 (1993).
8. M.J. Sims, D.G. Hassal, S. Brett, W. Rowan, M.J. Lockyer, A. Angel, A.P. Lewis, G. Hale, H. Waldmann and S.J. Crowe, J.Immunol., 151, 2296 (1993).
9. C. Chothia and A.M. Lesk, J.Mol.Biol., 196, 901 (1987).
10. C. Queen, W.P. Schneider, H.E. Selick, P.W. Payne, N.F. Landolfi, J.F. Duncan, N.M. Avdalovic, M. Levitt, R.P. Junghans and T.A. Waldmann, Proc.Natl.Acad.Sci.USA, 86, 10029 (1989).
11. P. Carter, L. Presta, C.M. Gorman, J.B.B. Ridgway, D. Henner, W.L.T. Wong, A.M. Rowland, C. Kotts, M.E. Carver and H.M. Shepard, Proc.Natl.Acad.Sci.USA, 89, 4285 (1992).
12. L.G. Presta, S.J. Lahr, R.L. Shields, J.P. Porter, C.M. Gorman, B.M. Fendly and P.M. Jardieu, J.Immunol., 151, 2623 (1993).
13. C. Chothia, A.M. Lesk, A. Tramontano, M. Levitt, S.J. Smith-Gill, G. Air, S. Sheriff, E.A. Padlan, D. Davies, W.R.Tulip, P.M. Colman, S. Spinelli, P.M. Alzari and R.J. Poljak, Nature, 342, 877 (1989).
14. F. Kolbinger, J. Saldanha, N. Hardman and M.M. Bendig, Protein Eng., 6, 971 (1993).
15. P.T. Jones, P.H. Dear, J. Foote, M.S. Neuberger and G. Winter, Nature, 321, 522 (1986).
16. M. Verhoeyen, C. Milstein and G. Winter, Science, 239, 1534 (1988).
17. L. Reichmann, M. Clark, H. Waldmann and G. Winter, Nature, 332, 323 (1988).
18. J. Hakimi, V.C. Ha, P. Lin, E. Campbell, M.K. Gately, M. Tsudo, P.W. Payne, T.A. Waldmann, A.J. Grant, W.-H. Tsien and W.P. Schneider, J.Immunol., 151, 1075 (1993).
19. K. Sato, M. Tsuchiya, J. Saldanha, Y. Koishihara, Y. Ohsugi, T. Kishimoto and M.M. Bendig, Cancer Res., 53, 851 (1993).
20. C. Eigenbrot, M. Randal, L. Presta and A.A. Kossiakoff, J.Mol.Biol., 229, 969 (1993).
21. C. Eigenbrot, T. Gonzalez, J. Mayeda, P. Carter, W. Werther, T. Hotaling, J. Fox and J. Kessler, Proteins, 18, 49 (1994).
22. J. Foote and G. Winter, J.Mol.Biol., 224, 487 (1992).
23. A. Tramontano, C. Chothia and A.M. Lesk, J.Mol.Biol., 215, 175 (1990).

24. I.S. Mian, A.R. Bradwell and A.J. Olson, J.Mol.Biol., 217, 133 (1991).
25. R.P. Junghans, T.A. Waldmann, N.F. Landolfi, N.M. Avdalovic, W.P. Schneider and C. Queen, Cancer Res., 50, 1495 (1989).
26. P.C. Caron, M.S. Co, M.K. Bull, N.M. Avdalovic, C. Queen and D.A. Scheinberg, Cancer Res., 52, 6761 (1992).
27. J. Greenwood, M. Clark and H. Waldmann, Eur. J.Immunol., 23, 1098 (1993).
28. S. Bolt, E. Routledge, I. Lloyd, L. Chatenoud, H. Pope, S.D. Gorman and H. Waldmann, Eur.J.Immunol., 23, 403 (1993).
29. R.J. Benjamin, S.P. Cobbold, M.R. Clark and H. Waldmann, J.Exp.Med., 163, 1539 (1986).
30. G.J. Jaffers, T.C. Fuller, A.B. Cosimi, P.S. Russell, H.J. Winn and R.B. Colvin, Transplantation, 41, 572 (1986).
31. M. Brueggemann, G. Winter, H. Waldmann and M.S. Neuberger, J.Exp.Med., 170, 2153 (1989).
32. A.F. LoBuglio, R.H. Wheeler, J. Trang, A. Haynes, K. Rogers, E.B. Harvey, L. Sun, J. Ghrayeb and M.B. Khazaeli, Proc.Natl.Acad.Sci.USA, 86, 4220 (1989).
33. M.B. Khazaeli, M.N. Saleh, T.P. Liu, R.F. Meredith, R.H. Wheeler, T.S. Baker, D. King, D. Secher, L. Allen, K. Rogers, D. Colcher, J. Schlom, D. Shochat and A.F. LoBuglio, Cancer Res., 51, 5461 (1991).
34. G. Hale, M.R. Clark, R. Marcus, G. Winter, M.J.S. Dyer, J.M. Philips, L. Reichmann and H. Waldmann, Lancet,2, 1394 (1988).
35. J.D. Isaacs, R.A. Watts, B.L. Hazleman, G. Hale, M.T. Keogan, S.P. Cobbold and H. Waldmann, Lancet, 340, 748 (1992).
36. J. Hakimi, R. Chizzonite, D.R. Luke, P.C. Familletti, P. Bailon, J.A. Kondas, R.S. Pilson, P. Lin, D.V. Weber, C. Spence, L.J. Mondini, W.-H. Tsien, J.L. Levin, V.H. Gallati, L. Korn, T.A. Wakdmann, C. Queen and W.R. Benjamin, J.Immunol., 147, 1352 (1991).
37. W.P. Schneider, S.M. Glaser, J.A. Kondas and J. Hakimi, J.Immunol., 150, 3086 (1993).
38. F.L. Delmonico, A.B. Cosimi, T. Kawai, D. Cavender, W.H. Lee, L.K. Jolliffe and R.W. Knowles, Transplantation, 55, 722 (1993).
39. E. Kremmer, J. Mysliwietz, R. Lederer and S. Thierfelder, Eur.J.Immunol., 23, 1017 (1993).
40. M. Bruggemann, N.P. Davies and I.R. Rosewell, Year in Immunol., 7, 33 (1993).
41. H. R. Hoogenboom and G. Winter, J.Mol.Biol., 227, 381 (1992).
42. J.D. Marks, H.R. Hoogenboom, T.P. Bonnert, J. McCafferty, A.D. Griffiths and G. Winter, J.Mol.Biol., 222, 581 (1991).

Chapter 33. Ethnobotany As A Source For New Drugs

Michael S. Tempesta and Steven R. King
Shaman Pharmaceuticals, Inc.
South San Francisco, CA 94080

Introduction - Plants have been utilized as a primary source of medicines for millennia, with the earliest recorded uses found in Babylon circa 1770 B. C. in the Code of Hammurabi and in ancient Egypt circa 1550 B. C. (1). Medicinal plants from that period, believed to have utility even in the afterlife of the pharaohs, have been recovered from the Giza pyramids, and can be found on display in a dark corner of the Cairo Museum. In fact, Neanderthal man may have had beliefs of a similar nature, albeit much earlier (2).

Many of our modern therapeutic agents have been derived from medicinal plants (3). In fact, over one-hundred and twenty pharmaceutical products currently in use are plant-derived, many with established ethnobotanical uses (4-6). Historically, even until the 1970's, there was a strong interest within the pharmaceutical industry in utilizing plants as sources of new pharmaceuticals. Breakthroughs in molecular biology and genetic engineering in the late 1970's and early 1980's, tied to advances in synthetic and computational chemistry offered great promise for generating a very large number of novel target-based compounds for evaluation as potential drugs. This served to significantly diminish plant natural product chemistry efforts underway in most U.S. pharmaceutical companies, as they built large synthetic chemistry efforts to provide an estimated 3 million compounds for high-throughput screens derived from molecular biology (7). The industrial emphasis on rational design and synthetic chemical random screening prompted downsizing of many plant natural product chemistry and pharmacognosy programs in the U.S. (8), lead by federal granting agencies and research institutions. Subsequently, most university programs in the plant natural product chemistry and pharmacognosy areas began to atrophy in the 80's (9) and this trend has continued until the present.

The initial promise of rational drug design was perhaps too optimistic, although recent successes in a number of therapeutic areas have certainly validated the approach. However, the need for new chemical entities is still as strong as ever, even as the research pipelines in major pharmaceutical companies become thinner as evidenced by the limited number of new chemical entities patented and developed in the U.S. each year. Perhaps the most significant therapeutic area of immediate need and concern is in infectious diseases, with drug resistance, particularly in the antibacterial area, emerging as the potential scourge of the 90's and beyond. The U. S. pharmaceutical industry is again recognizing the potential chemical diversity found in plants, and are renewing their efforts in their discovery programs to include plant sources. Both Merck's INBIO program in Costa Rica and Pfizer's more recent collaboration with The New York Botanical Gardens serve as recent highly-publicized examples, but many other companies are also involved in significant ethnobotanical discovery efforts.

THE SCIENCE OF ETHNOBOTANY

Ethnobotany is the study of how human beings have utilized plants for a wide diversity of primary survival and esthetic purposes. The area covers historical as well as present usage. People use plants for food, shelter, medicine, clothing, hunting, religious as well as other purposes. The study of how native peoples use plants therapeutically is that subdiscipline of ethnobotany of particular interest to current drug discovery programs. In this case, ethnopharmacologists are specifically looking at the pharmacological actions of plants, with emphasis on those plants used in medicine, religious or sacred activities (10).

Ethnobotanists possess training in diverse fields depending on the specific individual, but are usually biologists with a botanical training who have acquired some training in anthropology, linguistics or other social sciences. Anthropologists looking at the cultural context in which the plants are handled are another important aspect of ethnobotany.

A unified approach to this complex aspect of drug discovery involves coupling broadly-trained ethnobotanists with Western-trained physicians who are disease specialists to help diagnose specific pathologies treated by traditional healers or shamans. The physicians ideally would also be widely trained in the areas of anthropology or the relevant social sciences, with a genuine receptivity to the distinctly unique world views in the healing systems practiced by indigenous peoples. These physicians can be referred to as specialists in the area of ethnobiomedicine, a new discipline bridging ethnomedicine and contemporary biomedical research. One of the critical factors required to implement this approach is the creation of reciprocal relationships with indigenous people, their communities as well as national government agencies and national scientific institutions (11,12). A combination of the expertise of ethnobotanists, physicians and local indigenous healers has significantly enhanced our ability to identify and collect plants and data with a high degree of confidence in their potential as therapeutic agents in humans. This methodology is undergoing evolution through our field research expeditions, which has yielded a significant number of compounds under active development at this time.

Historical Utility of Ethnobotany - Ethnobotany and indigenous knowledge has had a profound impact on the development of modern medicine. However, it has not been highlighted in the education of medical professionals or the general public. Numerous ethnobotanically-derived therapeutic agents are utilized daily throughout the world to treat a wide diversity of patients and health conditions. Among the more well known examples are the skeletal muscle relaxant *d*-tubocuraine (**1**), (derived from the active principle of *Curarae*), utilized in surgical procedures throughout the world on a routine basis, and quinine-based anti-malarial drugs. The successful malarial therapy literally changed history by protecting and treating foreign and local peoples working in developing regions of West and East Africa, as well as many areas of South East Asia, where malaria previously had a devastating impact on local populations and visitors from other countries. These quinine-based anti-malarial drugs were developed by isolating the active compound of the "fever bark" (from *Cinchona*) utilized by the indigenous people in the Andean forest of Ecuador, Peru and Bolivia.

There have been numerous other drugs developed (3,4). One of the more recent examples of a compound derived from ethnobotanical knowledge is NDHG (nor-dihydroguaiaretic acid, **2**), marketed as Actinex since 1988. This drug is utilized to treat multiple actinic keratoses, and is extracted from the plant *Larrea tridentata,* commonly known as the creosote bush. Extracts of the creosote bush have been utilized by numerous indigenous groups in the Southwestern U. S. and Northern Mexico (13). The Pima Indians, in particular, have utilized creosote both orally and topically for a number of medical conditions. The Pima Indians have used a decoction of this species to treat several skin conditions including skin infections and sores.

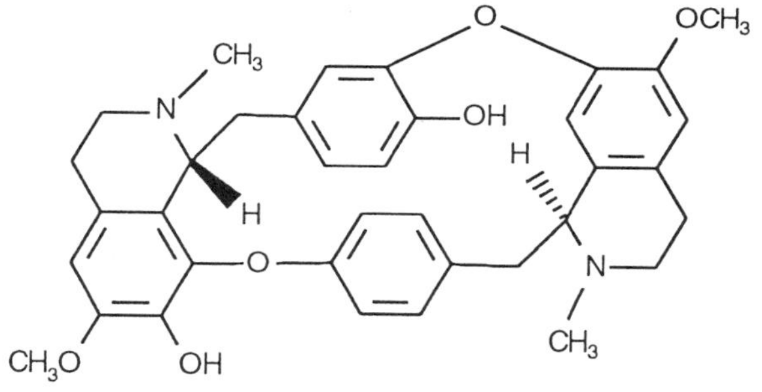

1

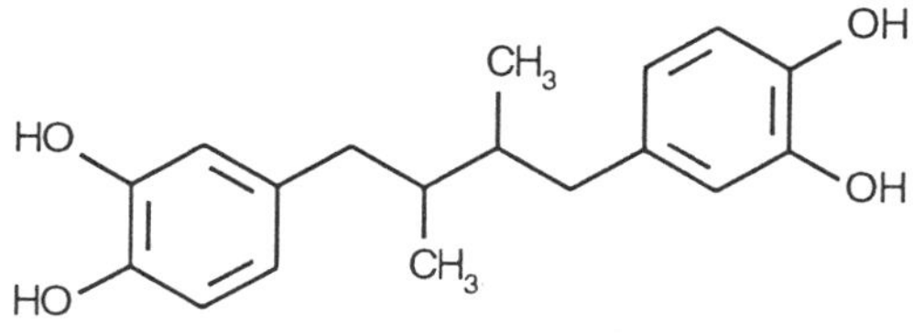

2

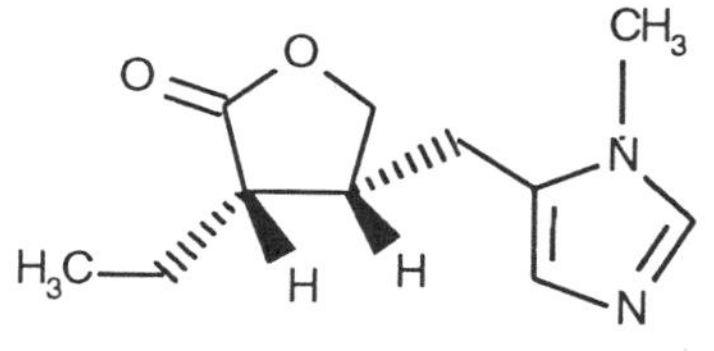

3

Another example of a drug currently in clinical trials derived from ethnobotanical knowledge is Salagen. The active compound, pilocarpine (**3**), is derived from *Pilocarpus jaborandi.* This species has multiple uses by native people in Brazil. Drug derived from this species is on the market to treat glaucoma. Currently, the drug is in clinical trials with an indication to treat erostoma (dry mouth syndrome) induced by chemotherapy. The species name of the plant "jaborandi" is a Tupi Indian word which means "slobber-mouth plant", and decoctions taken orally induce salivation. Indigenous knowledge continues to prove valuable in directing drug discovery efforts to yield new therapeutic agents.

Ethnobotanical Field Discovery Process - The initial phase in the ethnobotanical discovery process of new therapeutic agents involves an interdisciplinary field research program studying indigenous peoples in tropical forests. In order to prioritize plant collections, a detailed knowledge of the local and indigenous people who live in and around tropical forests is required. Prior to any research expedition a regional study on the epidemiology, traditional medicine, culture and ecology of the people and environment in which they live is prepared;. detailed information on the plants known to be utilized in any given area is obtained by searching a number of international databases (*e.g.,* NAPRALERT, MEDLINE, BIOSIS, CA, AGRICOLA) to give ethnomedical, biological and phytochemical information. Data from international and national remote area hospitals and treatment programs that work with local and native peoples are also gathered. The above information is synthesized and integrated into the field research program teams consisting of an ethnobotanist and a Western trained physician.

In the field, ethnobotanists and physicians utilize a focused data collection strategy. The goal of this method is to identify and collect plants with the highest probability of activity in our therapeutic areas. This integrated system is aimed at maximizing the probability of discovery of new drug candidates. The ethnobotanist and physician present case presentations of diseases described both verbally and visually as identified in epidemiological reports generated as described above. These brief case descriptions of individual diseases are presented to shamans and the local healers. Photographs of diseases with readily visible clinical manifestations (*i.e.,* fungal or viral skin infections) are utilized with the case presentations. Interviewing the informants is conducted very carefully, by presenting cases *without* using medical terminology for diseases. Terms such as "herpes", "hepatitis" or "parasites" are not utilized by the interviewing team, nor would they necessarily be understood by the local healers. The focus is on common signs and symptoms which are easily recognized. A translator for the local language is usually necessary to conduct this phase, and translators for Swahili, Quichua, Waorani and many other languages have been utilized without difficulty, as the case symptoms are designed to be simple but highly specific. Once a healer has recognized and described the same or similar disease state, the botanical treatment for that condition is recorded in detail (including preparation and posology). If several independent and reliable shamans describe a similar treatment for a disease, the plant is collected.

Additional plant collections and observations are made when the Western trained physician provides health care to the local people (12). The patient and the local healer are asked about the types plants, if any, used to treat the disease state (i.e. wounds, fevers or other symptomatology) by the treating physician. Often, use of antiinfective plants prior to the arrival of the ethnobotanist/physician team are described. Those plants that appear interesting are also collected for later analysis. The plant collecting process involves standard methodology of field botanists. This includes the preparation of multiple plant voucher specimens, which are deposited in the host country as well as in various U. S. herbaria. General bulk field collecting methods are well described by one of the National Cancer Institute's collecting scientists (14).

One of the critical components to successfully applying the ethnobotanical process is the development of strong reciprocal relationships with indigenous peoples and national governments (15-18). Case studies detailing approaches to building reciprocal relationships have been published in separate publications (11,19).

Ethobotanical Drug Development - Modern ethnobotanical drug discovery begins with targeting specific plants which have been used for medicinal purposes by native people for various infectious diseases (specifically viral and fungal infections), central nervous system disorders and complications of diabetes. After collection, the plants are then processed, and tested in first tier bioassays, which include state-of-the-art *in vitro* (*e.g.,* HPLC-guided assays) and *in vivo* (*e.g.,* transgenic animal) screens relevant to the therapeutic areas, and which are analogous to the historical approaches utilized (20,21). The most promising initial leads are then directly fractionated using the relevant *in vitro* activity as a guide to obtain pure compounds in milligram amounts. Those natural products meeting *in vitro* criteria, when compared against the best available therapeutics for comparison, are structurally characterized and subjected to confirmatory biological testing.

The selection to proceed to animal testing for determination of safety and efficacy is primarily based on structural novelty associated with confirmatory bioactivity. These criteria cover both the expected patentability and the likely mechanism of action. Compounds that are able to lead to patent protection and have positive confirmatory bioassay results are then scaled up to provide gram quantities for animal testing. Metabolism, pharmacokinetics, medicinal chemistry and formulation are brought in at this stage in order to insure the compounds (or the compound class) are given a reasonable chance of having satisfactory *in vivo* activity. Candidates whose *in vivo* activities compare favorably with the best available marketed therapeutics are significantly scaled up (either via isolation, semisynthesis of total synthesis) to provided hundreds of grams and proceed to detailed formulation, toxicology, ADME, supporting efficacy and pharmacological studies. Those candidate compounds with successful toxicological profiles are then subjected to manufacturing and QA/QC evaluation to provide GMP materials for supporting the IND application.

In the U. S. pharmaceutical industry, only one in ten IND's actually proceeds to NDA submission and eventual FDA approval. The failure of nine in ten IND's accounts for the bulk of the estimated cost of drug development of approximately $250-300 MM per approved drug. The preclinical development costs are a small fraction of these costs, and an obvious efficiency would be to have a significantly higher percentage of IND candidates making it to the market. The ethnobotanical approach starts with bioactivity in humans, knowing that in the end, specific and nontoxic efficacy in the target species will have to be demonstrated. The historical evidence, however, may give ethnobotanists a significant advantage in proving efficacy and safety in the clinic, and lead to an approval rate of much better than the U. S. pharmaceutical industry average of one in ten drugs currently in the clinic.

References

1.   B.B. Simpson, in "Economic Botany: Plants in Our World", M.J. Martin and B. Benjamin, eds., p 351 (1986).
2.   R. Solecki and I.V. Shanidar, Science, 190, 180 (1975).
3.   A.H. Marderosian and L.Liberti, Natural product medicine - a scientific guide to foods, drugs and cosmetics, G. F. Stickley Co., Philadelphia, PA (1988).
4.   N.R. Farnsworth, O. Akerele and A.S. Bingel, Bull.World Health Org. 63, 965 (1985).
5.   N.R. Farnsworth, in "Biodiversity", E.O.Wilson, ed. pp 83-97, National Academy Press, Washington, DC (1988).
6.   D.D. Soejarto and N.R. Farnsworth, Perspect.Biol.Med. 32,244 (1989).
7.   A.L. Harvey, in "Drugs from Natural Products - Pharmaceuticals and Agrochemicals" Series in Pharmaceutical Technology, Ellis Horwood, Ltd., (1993).
8.   G.A. Cordell, Am.Druggist, 96 (1987).
9.   G.M. Hocking, Am. Pharmacy, NS28, 12, 8 (1988).
10.   P.A.G.M. de Smet and L. Rivier, J. Ethnopharmacol. 25, 127 (1989).
11.   S.R. King, "Establishing Reciprocity: Biodiversity, Conservation and New Models for Cooperation Between Forest-Dwelling Peoples and the Pharmaceutical Industry. Intellectual Property Rights for Indigenous Peoples, A Sourcebook", T. Greaves, ed., The Society for Applied Anthropology,. Oklahoma City, OK, pp. 69-82, (1994).
12.   S.R. King and T.J. Carlson, "Biological Diversity, Indigenous Knowledge, Drug Discovery and Intellectual Property Rights: Creating Reciprocity and Maintaining Relationships," Proceedings of NSF sponsored workshop on Indigenous People and Intellectual Property Rights, Lake Tahoe, California, (1994).
13.   P. Train, J. R. Henrichs and W. A. Archer, "Medicinal uses of plants by Indian tribes of Nevada." Quaterman Publications, Lawrence, MA, pp 64-65 (1957).
14.   D. Soejarto, in A.D. Kinghorn and M.F.Balandrin (eds.) "Human medicinal agents from plants." American Chemical Society (ACS Symposium Series 534) Washington, DC, pp 96-111 (1993).
15.   B. M. Boom, in "Ethnobiology: Implications and Applications," Proceedings of the First Congress of Ethnobiology, Vol. 2, Part F. (D. Posey et al., eds.). Belém, Pará: Museu Paraense Emilio Goeldi, pp. 147-153 (1990).
16.   A. B. Cunningham, Ethics, Ethnobiological Research, and Biodiversity. WWF/UNESCO/KEW "People and Plants" Initiative, Gland, Switzerland, (1993).
17.   E. Elisabetsky, Cultural Survival Quarterly 15, 9 (1991).
18.   J. R. Axt, M. L. Corn, M. Lee and D. Ackerman, Biotechnology, Indigenous Peoples and Intellectual Property Rights. Congressional Research Service, The Library of Congress, (1993).
19.   S. R. King, in "Sustainable Harvest and Marketing of Rainforest Products," M. Plotkin and L. Familore, eds., Washington, D.C. Island Press, pp.231-239 (1992).
20.   A.B. Svendsen and J.J.C. Scheffer, Pharmaceut. Weekblad Sci.Ed. 4, 93 (1982).
21.   M. Hamburger and K. Hostettmann, Phytochemistry, 30, 3864 (1991).

SECTION VII.   TRENDS AND PERSPECTIVES

Editor: James A. Bristol
Parke-Davis Pharmaceutical Research
Division of Warner-Lambert Company, Ann Arbor, MI 48105

Chapter 34.   To Market, To Market - 1993

Xue-Min Cheng
Parke-Davis Pharmaceutical Research
Division of Warner-Lambert Company, Ann Arbor, MI 48105

New chemical entities (NCEs), including several genetically engineered molecular entities, introduced for human therapeutic use into the world market for the first time during 1993 totaled 43 (1). This is well above the 36 introduced in both 1992 (2) and 1991 (3) and the 37 NCEs in 1990 (4).

Once again, Japan was the leader for the new launches in 1993 with 16 NCE introductions. Italy had a remarkable record of 9 new drugs holding the second place and the United States was the third with 6 NCEs. Germany and the United Kingdom tied in the forth place, each with 4 NCEs followed by France and Canada, both with 3 new launches. Within the 43 NCEs, 7 of them were launched simultaneously in more than one country. Following the trend for many years, Japan and the United States were the originators for the majority of the NCEs, with 15 and 10 respectively, totaling approximately 60% of all new launches. Italy followed with 5, and Switzerland and Germany with 3 each.

Eight new antiinfectives, including 3 new fluoroquinolones and an antiviral, stood as the leader in therapeutic categories in 1993 followed closely by 7 CNS drugs and 6 antineoplastic agents. Several new drugs in 1993 reflected real advances in treatment for certain serious diseases. Highlights include the long-awaited introduction of tacrine as the first therapeutic for Alzheimer s disease; interferon β-1b for multiple sclerosis; and taxol for metastatic ovarian cancer. Novel NCEs of special interests also include acetorphan, the first enkephalinase inhibitor for diarrhea; formestane, the first specific aromatase inhibitor for breast cancer; miltefosine, the first phospholipid-based anticancer; risperidone, the first of a new class of antipsychotic; structurally complicated FK 506 (tacrolimus) as a potent immunosuppressive substance; and tamsulosin hydrochloride as the first α-1A adrenoceptor antagonist for dysuria associated with benign prostatic hypertrophy. Among the 43 NCEs, besides 5 biological products, 23 synthetic substances have more than 1 asymmetric center (including those that are derived from partial synthesis of natural products). Of those, 17 (74%) are used in enantiomerically pure form compared with 25% for a selection of 866 representative synthetic drugs introduced prior to 1985 (5). This indicates a clear trend towards the development of optically pure agents within the pharmaceutical industry.

In 1993, 25 new molecular entities were approved in the United States (6, 7, 8), similar to the 26 approvals in 1992 (2) but considerably fewer than the record number of 31 in 1991. Among these, 11 were given the priority classification by the FDA. Three of them are CNS agents and only 1 for AIDS related conditions, compared with 4 in 1992. Five of the 25 new approvals, cladribine, felbamate, interferon β-1b, taxol, tacrine, along with histrelin which was approved in previous year, were marketed in the US as the first worldwide introduction.

Acetorphan (Antidiarrheal) (9-11)

Country of Origin: **France**
Originator: **Bioprojet**
First Introduction: **France**
Introduced by: **Bioprojet**
Trade Name: **Tiorfan**
CAS Registry No.: **81110-73-8**
Molecular Weight: **385.48**
Dosage Form: **100 mg Capsule**

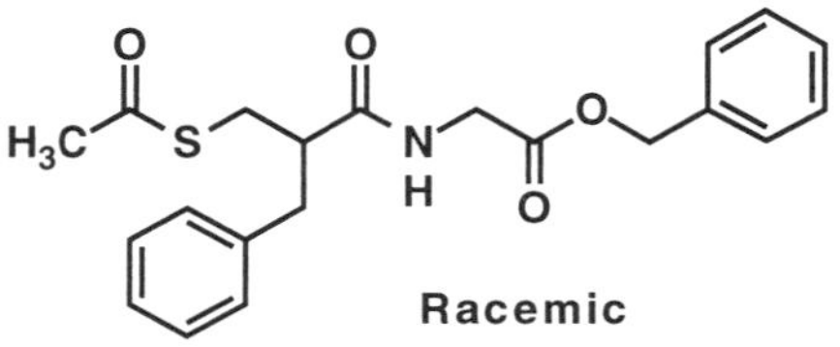

Racemic

Acetorphan is an orally-active enkephalinase inhibitor introduced in 1993 for the treatment of acute diarrhea in adults. It displays antidiarrheal activity via a purely antisecretory mechanism without inhibiting intestinal transit. Acetorphan acts as rapidly as loperamide, a classical mu opiate receptor agonist that exhibits its antidiarrheal effect via an antitransit mechanism, without producing constipation or other side effects. The novel mechanism of action may render acetorphan potentially useful in the management of infectious diarrhea.

Alendronate Sodium (Osteoporosis) (12-15)

Country of Origin: **Italy**
Originator: **Istituto Gentili**
First Introduction: **Italy**
Introduced by: **Merck; Sigma Tau; Neopharmed**
Trade Name: **Alendros**
CAS Registry No.: **129318-43-0**
Molecular Weight: **271.08**
Dosage Form: **5 mg Tablet**

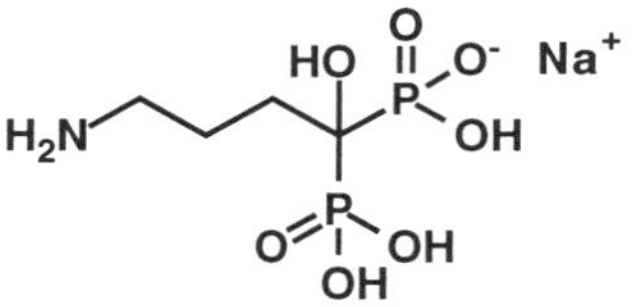

Alendronate sodium is the fourth bisphosphonate to reach the market for the treatment of postmenopausal osteoporosis. Bisphosphonates are potent inhibitors of bone resorption. They reduce pain and complications due to bone metastases, are effective in Paget's disease and in increasing bone mineral density. These agents bind tightly to hydroxyapatite crystals and are retained on bone resorption surfaces. Local release of the bisphosphonates occurs by acidification during the process of the bone resorption to impair the osteoclasts' ability to resorb bone. Alendronate is more potent than other bisphosphonates such as clodronate, pamidronate, and etidronate and is reported to have no deleterious effects on bone. It has also been shown to reduce hypercalcemia in cancer patients.

Amtolmetin Guacil (Antiinflammatory) (16-18)

Country of Origin: **Italy**
Originator: **Sigma Tau**
First Introduction: **Italy**
Introduced by: **Sigma Tau; Medosan**
Trade Name: **Artromed**
CAS Registry No.: **87344-06-7**
Molecular Weight: **420.47**
Dosage Form: **600 mg Tablet**

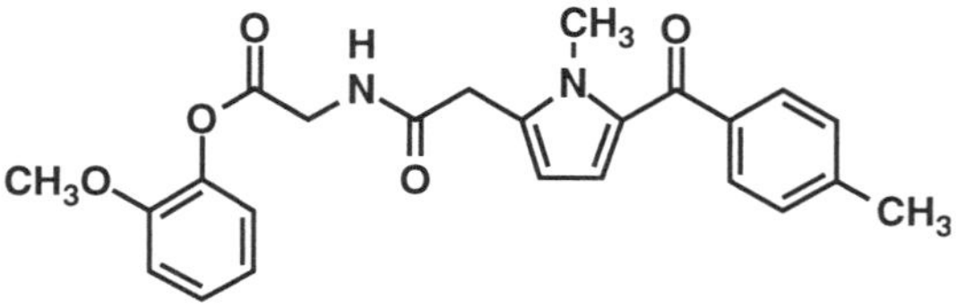

Amtolmetin guacil is an orally active non-steroidal antiinflammatory drug (NSAID) introduced for the treatment of osteoarthritis, rheumatoid arthritis and post-operative pain. Amtolmetin guacil is reported to elicit a more rapid and improved analgesic action than other agents such as paracetamol. In models of adjuvant arthritis and rheumatic diseases, amtolmetin guacil is more efficacious than existing drugs such as naproxen, indomethacin and piroxicam. As a non-acidic prodrug of tolmetin, it has similar *in vivo* activity to the parent but with minor ulcerogenic action, lower acute toxicity, and excellent biological and gastric tolerability.

Aniracetam (Cognition Enhancer) (19-21)

Country of Origin:	**Switzerland**
Originator:	**Roche**
First Introduction:	**Italy; Japan**
Introduced by:	**Roche; Toyama; Menarini;**
	Biomedica Foscama
Trade Name:	**Draganon; Sarpul;**
	Ampamet; Reset
CAS Registry No.:	**72432-10-1**
Molecular Weight:	**219.24**
Dosage Form:	**100; 200; 750 mg Tablet**
	1500 mg Sachet

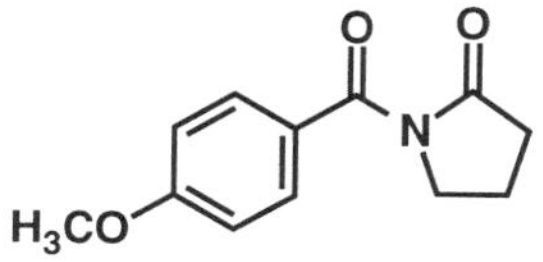

Aniracetam is a non-NMDA receptor agent useful for the treatment of cognitive impairment. In a clinical study, it showed good to very good improvement in 60% of patients with cerebral insufficiency. Statistically significant improvement in the psychobehavioral parameters has been observed in a group of patients with senile dementia of the Alzheimer's type. Aniracetam is well tolerated and efficacious in improving velocity and accuracy of the saccades, complex reaction time, and other aspects of performance compared with placebo group in hypoxic volunteers. Its nootropic property has been attributed to the cholinergic blockade.

Brodimoprim (Antibiotic) (22)

Country of Origin:	**Switzerland**
Originator:	**Helsinn**
First Introduction:	**Italy**
Introduced by:	**Fisons; Master Pharma**
Trade Name:	**Hyprim; Unitrim**
CAS Registry No.:	**56518-41-3**
Molecular Weight:	**339.19**
Dosage Form:	**100 mg Tablet;**
	1% Oral susp

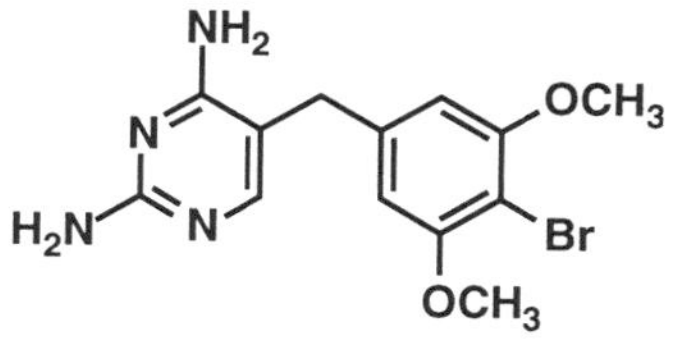

Brodimoprim, a dihydrofolate reductase inhibitor, was introduced for the treatment of common infections of upper respiratory tract such as sinusitis and exacerbation of chronic bronchitis. It has a broad spectrum of antimicrobial activities against most Gram-positive and -negative pathogens including some bacterial strains that are resistant to amoxicillin, a widely used antibiotic for the indicated infections. Brodimoprim also has a favorable pharmacokinetic profile exemplified by its long half-life and good tissue penetration.

Cabergoline   (Antiprolactin) (23-26)

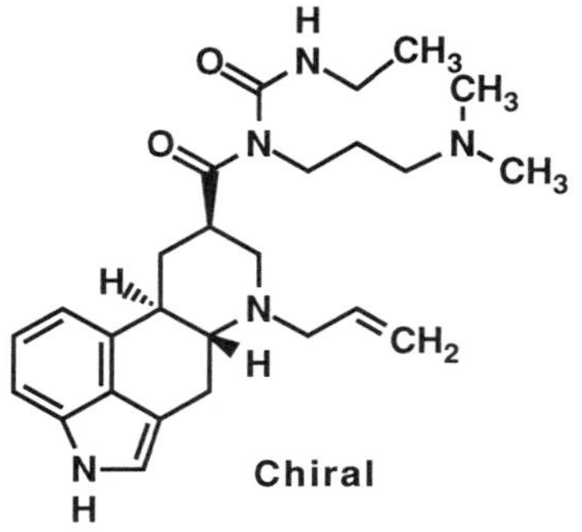

Country of Origin:	**Sweden**
Originator:	**Kabi Pharmacia**
First Introduction:	**Belgium**
Introduced by:	**Kabi Pharmacia**
Trade Name:	**Dostinex**
CAS Registry No.:	**81409-90-7**
Molecular Weight:	**451.62**
Dosage Form:	**0.5 mg Tablet**

Chiral

Cabergoline is a potent, selective, and long-lasting dopamine D_2 receptor agonist launched in 1993 in Belgium as a prolactin inhibitor. A single 1 mg dose of cabergoline effectively prevents puerperal lactation for up to 14 days, remarkably superior to other drugs that require a daily regimen. It has a low rate of rebound breast activity and good tolerability. Cabergoline is also in clinical trials for Parkinson's disease, breast cancer, and hyperprolactinaemia.

Cefepime   (Antibiotic) (27-30)

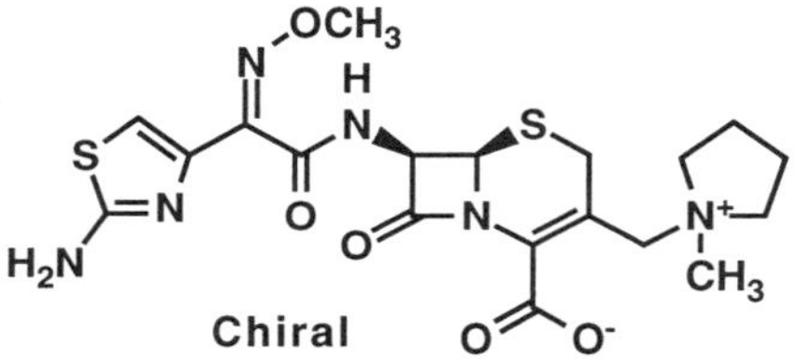

Country of Origin:	**U.S.A.**
Originator:	**Bristol-Myers Squibb**
First Introduction:	**Sweden; France**
Introduced by:	**Bristol-Myers Squibb**
Trade Name:	**Maxipime; Axepim**
CAS Registry No.:	**88040-23-7**
Molecular Weight:	**480.56**
Dosage Form:	**0.5; 1; 2 g Vial**

Chiral

Cefepime is a new fourth-generation parenteral cephalosporine antibiotic launched in 1993 in Sweden and France. Cefepime has broad spectrum antimicrobial activity against *Staphylococcus, Streptococcus, Pseudomonas*, and the *Enterobacteriaceae*, including many bacterial isolates that are resistant to commonly used ceftazidime and cefotaxime. Its efficacy has been demonstrated in the treatment of lower respiratory tract infections especially pneumonia, intra-abdominal and urinary tract infections, skin and soft tissue infections, chronic osteomyelitis and in prophylaxis of biliary tract and prostate infections. It is well tolerated by patients and is reported to exhibit no significant drug interactions.

Cinolazepam   (Hypnotic) (31, 32)

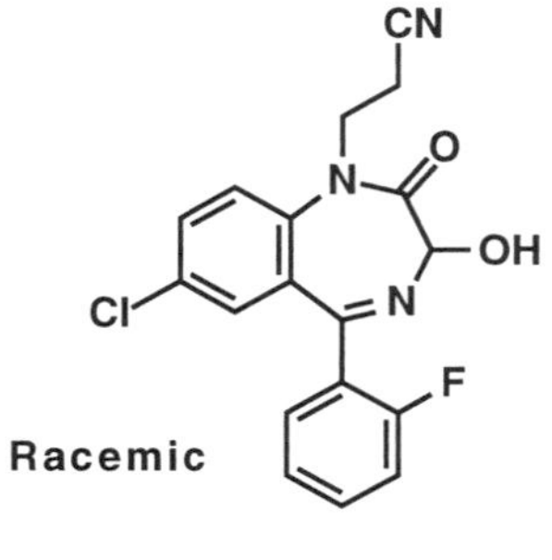

Country of Origin:	**Austria**
Originator:	**Gerot**
First Introduction:	**Austria**
Introduced by:	**Gerot**
Trade Name:	**Gerodorm**
CAS Registry No.:	**75696-02-5**
Molecular Weight:	**357.77**
Dosage Form:	**40 mg Tablet**

Racemic

Cinolazepam is a benzodiazepine hypnotic introduced first in Austria for the treatment of sleep disturbance due to somatic or environmental factors. It has a half-life of 9 hr and effectively restores disturbed sleep to normal but does not significantly affect normal sleep patterns. This compound is well tolerated and has been reported to have no hang-over side effect after sleep.

Cladribine (Antineoplastic) (33-36)

Country of Origin:	**U.S.A.**
Originator:	**Johnson & Johnson**
First Introduction:	**U.S.A.**
Introduced by:	**Ortho Biotech**
Trade Name:	**Leustatin**
CAS Registry No.:	**4291-63-8**
Molecular Weight:	**285.69**
Dosage Form:	**1 mg/ml Inj.**

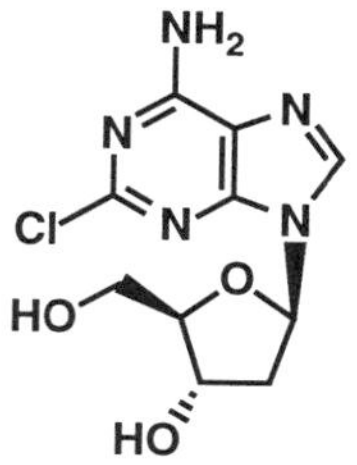

Cladribine, an adenosine deaminase inhibitor, was introduced in the United States as a single intravenous treatment for hairy cell leukemia. The incorporation of a chlorine atom at the 2-position of deoxyadenosine renders cladribine more resistant to enzymatic attack by adenosine deaminase, resulting in a more prolonged cytotoxic effect. Cladribine efficiently crosses lymphocyte and monocyte cell membranes and is metabolized in cells to the biologically active triphosphate, which inhibits DNA synthesis. While most antineoplastic drugs are active primarily against dividing cells, cladribine destroys both resting and proliferating cells. Its potential uses in the treatment of autoimmune hemolytic anemia, multiple sclerosis, chronic lymphocytic leukemia and various lymphomas have also been evaluated.

Cytarabine Ocfosfate (Antineoplastic) (37, 38)

Country of Origin:	**Japan**
Originator:	**Yamasa**
First Introduction:	**Japan**
Introduced by:	**Yamasa; Nippon Kayaku**
Trade Name:	**Starasid**
CAS Registry No.:	**65093-40-5**
Molecular Weight:	**615.68**
Dosage Form:	**50; 100 mg Capsule**

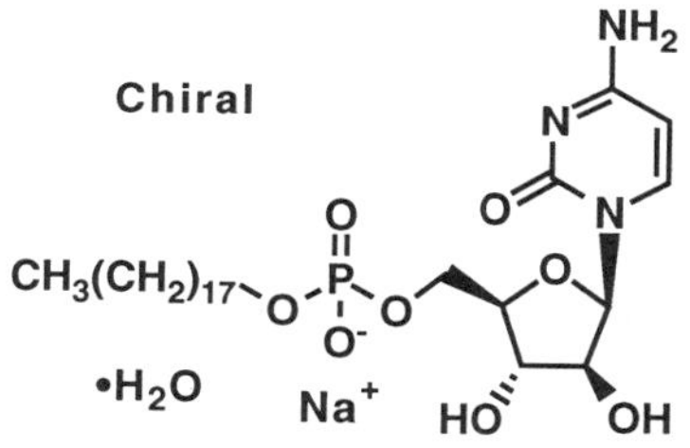

Cytarabine ocfosfate is the first orally active stearate derivative of cytarabine introduced for the treatment of certain types of malignancies including myelodysplastic syndromes that eventually evolve into acute leukemia. Cytarabine is one of the most effective drugs for treating acute myeloid leukemia but suffers from poor bioavailability. As a mono-phosphorylated analog of cytarabine, the title compound exhibits potent antitumor activity and slowly converts to cytarabine in the liver, releasing cytarabine into the bloodstream over a prolonged period. Cytarabine ocfosfate is effective by both parenteral and oral administration and, therefore, has an advantage over the parenteral form of cytarabine or its analogs. It should be noted that cytarabine ocfosfate has major side effects such as myelosuppression and gastrointestinal toxicity.

Dirithromycin   (Antibiotic) (39-41)

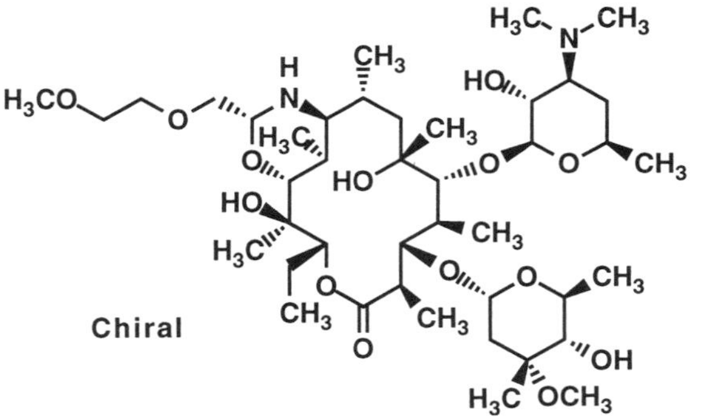

Country of Origin:	**Germany**
Originator:	**Boehringer Ingelheim**
First Introduction:	**Spain**
Introduced by:	**Lilly**
Trade Name:	**Nortron**
CAS Registry No.:	**62013-04-1**
Molecular Weight:	**835.09**
Dosage Form:	**250 mg Tablet**

Chiral

Dirithromycin, an orally active oxazine derivative of erythromycin, is a second-generation macrolide antibiotic. It was first introduced in Spain for the treatment of respiratory tract, skin and soft tissue infections. It is safe and effective in treating acute bronchitis, acute bacterial exacerbation of chronic bronchitis, pneumonia, pharyngitis, and tonsillitis. Dirithromycin has been reported to have a similar antimicrobial spectrum and potency to erythromycin. However, compared to other macrolide antibiotics, dirithromycin has improved oral bioavailability, higher tissue permeability and longer duration of action. The once-daily therapy regiment is advantageous in terms of patient compliance and is a good alternative for patients who are unable to take penicillin.

Ecabet Sodium   (Antiulcerative) (42-44)

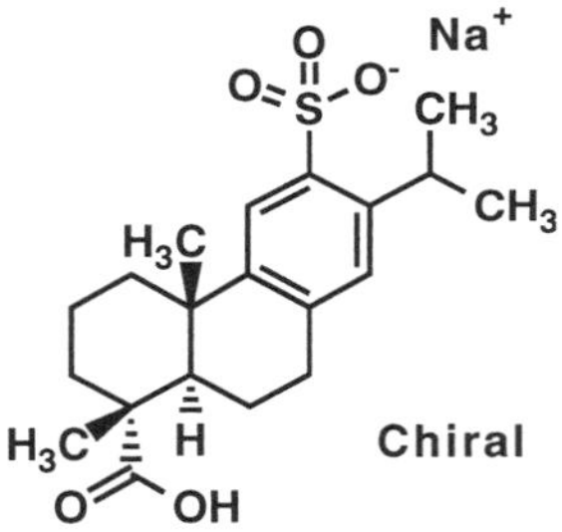

Country of Origin:	**Japan**
Originator:	**Tanabe**
First Introduction:	**Japan**
Introduced by:	**Tanabe; Nippon Boehringer Ingelheim**
Trade Name:	**Gastrom**
CAS Registry No.:	**86408-72-2**
Molecular Weight:	**402.48**
Dosage Form:	**66.7% Granule**

Chiral

Ecabet sodium is a new antiulcer agent launched in Japan for the treatment of gastric ulcer. It has both mucosal-protective and tissue-repairing properties. In rats, it effectively prevents gastrointestinal lesion formation induced experimentally by EtOH, HCl, NaOH, and boiling water. The mechanism of action has been suggested to involve inhibition of pepsin activity in gastric juice by precipitating pepsin in the form of a complex and inactivating pepsinogen.

Emedastine Difumarate   (Antiallergic and Antiasthmatic) (45, 46)

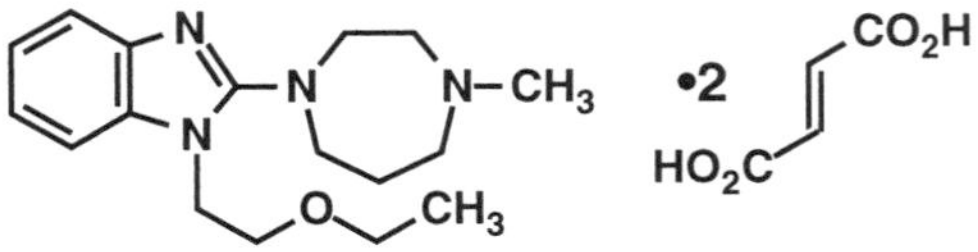

Country of Origin:	**Japan**
Originator:	**Kanebo**
First Introduction:	**Japan**
Introduced by:	**Kanebo; Kowa**
Trade Name:	**Daren; Remicut**
CAS Registry No.:	**87233-62-3**
Molecular Weight:	**534.57**
Dosage Form:	**1; 2 mg Capsule**

Emedastine difumarate, a potent H_1-receptor antagonist, was launched in Japan for the treatment of allergic rhinitis and urticaria. Emedastine exerts its antiallergic effect via inhibition of substance P-induced histamine release. It has been demonstrated both *in vitro* and *in vivo* that this effect is mediated by the inhibition of Ca^{2+} release from extracellular stores and of Ca^{2+} influx into mast cells. In a clinical trial with bronchial asthma, emedastine improved asthmatic symptoms in 55.3% of patients.

Felbamate (Antiepileptic) (47-50)

Country of Origin:	**U.S.A.**
Originator:	**Carter-Wallace**
First Introduction:	**U.S.A.**
Introduced by:	**Carter-Wallace**
Trade Name:	**Felbatol**
CAS Registry No.:	**25451-15-4**
Molecular Weight:	**238.24**
Dosage Form:	**400; 600 mg Tablet; 600 mg/5ml Susp.**

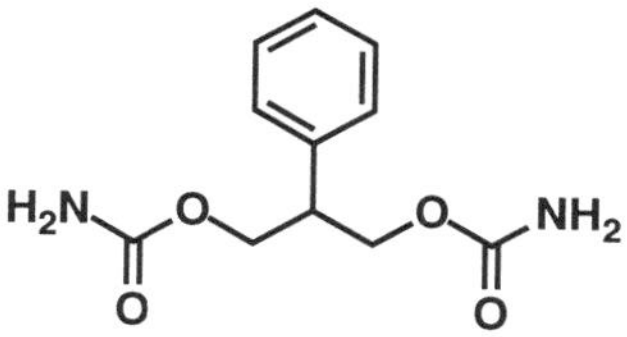

Felbamate, characterized by its low toxicity and wide margin of safety, is efficacious in treating refractory patients with generalized tonic-clonic and complex partial seizures as monotherapy and adjunctive therapy. It has also been demonstrated to have a neuroprotective effect in cerebral ischemia and hypoxia. It has been suggested that the mechanism of its anticonvulsant activity is possibly through an interaction with the strychnine-insensitive receptor site on the NMDA receptor complex.

Formestane (Antineoplastic) (51, 52)

Country of Origin:	**Switzerland**
Originator:	**Ciba-Geigy**
First Introduction:	**United Kingdom**
Introduced by:	**Ciba-Geigy**
Trade Name:	**Lentaron**
CAS Registry No.:	**566-48-3**
Molecular Weight:	**302.42**
Dosage Form:	**250 mg i.m. Depot**

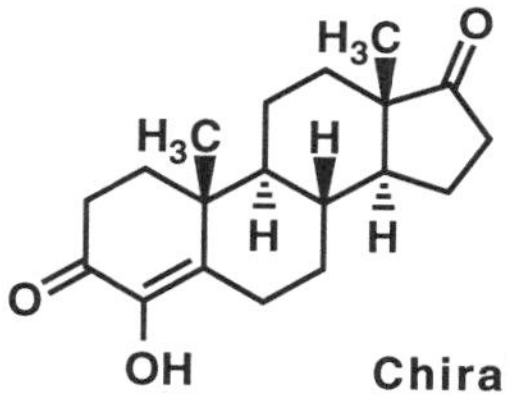

Formestane is a potent aromatase inhibitor launched in the UK as a second-line endocrine treatment for breast cancer. As a synthetic derivative of androstanedione, the natural substrate for the biosynthesis of estrogen by the enzyme aromatase, formastane selectively inhibits aromatase and binds to its steroid receptor site to cause a rapid and sustained fall in circulating estrogen level and, therefore, inhibits tumor growth. In patients with existing bulky primary tumors, formastane effectively reduces the size of the tumors. Formastane has apparent tolerability advantages and less side effects than other agents such as aminoglutethimide.

Gabapentin (Antiepileptic) (53-56)

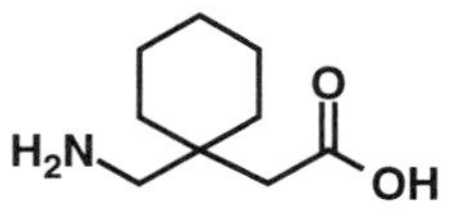

Country of Origin:	**U.S.A.**
Originator:	**Warner-Lambert**
First Introduction:	**United Kingdom**
Introduced by:	**Warner-Lambert**
Trade Name:	**Neurontin**
CAS Registry No.:	**60142-96-3**
Molecular Weight:	**171.24**
Dosage Form:	**100; 300; 400 mg Capsule**

Gabapentin was introduced in 1993 in the UK and early 1994 in the USA as an adjunctive therapy in the treatment of refractory partial seizures and secondarily generalized tonic-clonic seizures. Although being a lipophilic analog of the neurotransmitter GABA, gabapentin appears to exert its anticonvulsive function by a GABA receptor independent mechanism, possibly involving the *L*-system amino acid transporter protein. Gabapentin easily crosses the blood brain barrier and exhibits a favorable pharmacokinetic profile with high tolerability. It does not interfere with the metabolism of other concomitant administered antiepileptic drugs, thus having a low potential for drug interactions. Studies are currently underway for the use of gabapentin as mono-therapy for the treatment of various seizures.

Glucagon, rDNA (Hypoglycemia) (57-59)

Country of Origin:	**Denmark**	Introduced by:	**Novo-Nordisk**
Originator:	**Novo-Nordisk**	Trade Name:	**GlucaGen**
First Introduction:	**Denmark;**	CAS Registry No.:	**9007-92-5**
	Netherlands;	Dosage Form:	**1 mg/ml Vial**
	Switzerland		

GlucaGen, a recombinant DNA product, was developed as a nasal formulation of glucagon for the treatment of insulin-induced hypoglycemia. Glucagon is widely used by insulin dependent diabetic patients as an emergency treatment for severe hypoglycemic reactions. Use of genetically engineered glucagon by both intranasal and intramuscular methods has a similar effect on hepatic glucose production rate as natural glucagon from pancreatic extraction.

Histrelin (Precocious Puberty) (60, 61)

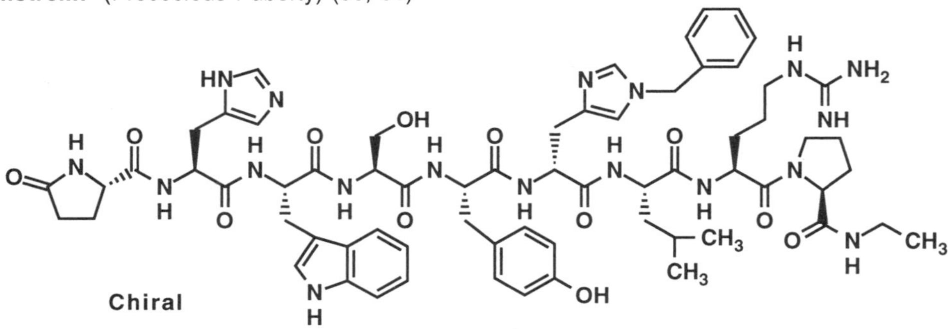

Chiral

Country of Origin:	**U.S.A.**	Trade Name:	**Supprelin**
Originator:	**Johnson & Johnson**	CAS Registry No.:	**76712-82-8**
First Introduction:	**U.S.A.**	Molecular Weight:	**1323.53**
Introduced by:	**Roberts**	Dosage Form:	**120 mcg/0.6ml Vial**

Histrelin, a synthetic gonadotrophin-releasing hormone (GnRH) agonist, was introduced as a first-line therapy for the treatment of central precocious puberty. When administered over a prolonged period, it suppresses the release of gonadotrophin, inhibits ovarian and testicular steroidogenesis, and prevents sexual maturation. Superior to the traditional therapy with progestational agents, histrelin appears to have decelerating effects on skeletal maturation allowing more statural growth and significantly increased final adult height. Histrelin has also been used to treat other disorders including endometriosis, polycystic ovarian disease, uterine leiomyomas, severe premenstrual syndrome and to prevent acute intermittent porphyria.

Imidapril Hydrochloride (Antihypertensive) (62-64)

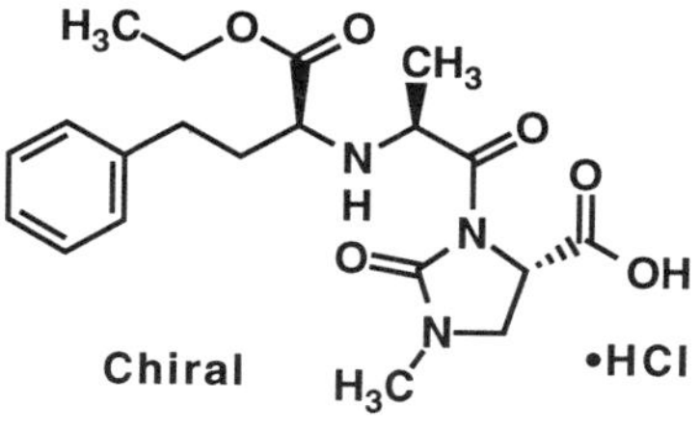

Country of Origin:	**Japan**
Originator:	**Tanabe**
First Introduction:	**Japan**
Introduced by:	**Tanabe**
Trade Name:	**Tanatril; Novarok**
CAS Registry No.:	**89371-37-9**
Molecular Weight:	**441.91**
Dosage Form:	**2.5; 5; 10 mg Tablet**

Imidapril is a new ACE inhibitor launched in Japan as a once-daily therapy for hypertension. Its active metabolite has more potent inhibitory effects on all tissue ACEs from spontaneously hypertensive rats (SHRs) and Wistar Kyoto rats (WKYs) than many other ACE inhibitors including enalapril and captopril. It has a very long duration of action. In stroke-prone spontaneously hypertensive rats (SHRSP), imidapril is effective in stroke prevention at a considerably lower dose than enalapril. This stroke preventive effect has been suggested to be primarily by an indirect mechanism related to the amelioration of kidney dysfunction.

Interferon β-1b (Multiple Sclerosis) (65-67)

Country of Origin:	**U.S.A.**	Introduced by:	**Berlex**
Originator:	**Chiron**	Trade Name:	**Betaseron**
First Introduction:	**U.S.A.**	Dosage Form:	**0.3 mg Vial**

Interferon β-1b, a stable analog of human interferon-β produced by recombinant DNA technique, is the first biotechnology product to be licensed under the FDA's accelerated approval regulation. It is the first drug available for the treatment of relapsing remitting multiple sclerosis (MS). The mechanism of action of interferon β-1b in MS remains unknown. It has shown immunoregulatory properties including its ability to decrease T-cell proliferation, to block the synthesis of interferon γ that may be involved in MS, to inhibit release of other cytokines that damage oligodendrocytes, and to increase T suppresser-cell activity. Interferon β-1b has also demonstrated potential in melanoma, renal cell carcinoma, genital warts, hepatitis, rhinovirus, and several types of malignant tumors.

Levofloxacin   (Antibiotic) (68-71)

Country of Origin:	**Japan**
Originator:	**Daiichi**
First Introduction:	**Japan**
Introduced by:	**Daiichi**
Trade Name:	**Floxacin; Cravit**
CAS Registry No.:	**100986-85-4**
Molecular Weight:	**361.38**
Dosage Form:	**100 mg Tablet**
	100 mg/g Granule

Levofloxacin, the optically active *S*- isomer of the fluoroquinolone antibiotic ofloxacin, is two to four times more potent than ofloxacin with reportedly less side effects in treating infections of the lower respiratory and urinary tract, prostate infections and sexually transmitted diseases. It has broad and potent antibacterial activity over common Gram-positive and -negative aerobic pathogens and obligate anaerobes. Different from the cephem antibiotics, levofloxacin is unique in its marked selectivity against members of the family *Enterobacteriaceae* and its negligible effect on predominant anaerobes. Levofloxacin also exhibits satisfactory antimicrobial effects in surgical infections and it may be used for treatment of gastrointestinal infections such as traveler's diarrhea associated with the pathogenic *Enterobacteriaceae*.

Miltefosine   (Topical Antineoplastic) (72-75)

Country of Origin:	**Germany**
Originator:	**Asta Medica**
First Introduction:	**Germany**
Introduced by:	**Asta Medica**
Trade Name:	**Miltex**
CAS Registry No.:	**58066-85-6**
Molecular Weight:	**407.58**
Dosage Form:	**60 mg/ml Oily soln**

Miltefosin, representing the prototype of a new phospholipid structure, was introduced for the palliative treatment of skin metastases in patients with breast cancer. It is highly active against the human leukemia tumor cells xenograft in nude mice, leading to growth inhibition and regression of large established tumors. Its mode of antitumor activity is not mediated by the host immune system but by its pharmacological effects at the level of the cancer cell membrane, distinctly different from that of the classical cytostatic drugs which interact with cell proliferation at the level of DNA replication. Protein kinase C inhibition has been suggested as a possible mechanism.

Nadifloxacin   (Topical Antibiotic) (76-78)

Country of Origin:	**Japan**
Originator:	**Otsuka**
First Introduction:	**Japan**
Introduced by:	**Otsuka**
Trade Name:	**Acuatim**
CAS Registry No.:	**124858-35-1**
Molecular Weight:	**360.39**
Dosage Form:	**1% Cream**

Racemic

Nadifloxacin, one of the three new fluoroquinolone antibiotics launched in 1993 is indicated for topical treatment of acne vulgaris and other skin infections. Nadifloxacin has a potent and broad spectrum of activity against aerobic Gram-positive and -negative bacteria and against anaerobic bacteria. It produces significant improvement in patients with *Propionibacterium acnes* infection and does not appear to cause cross-resistance to other antibiotic agents. Its potent antimicrobial activity has also been demonstrated in the *Pseudomonas aeruginosa* burn wound infection model in mice. Nadifloxacin exerts its antibiotic activity by inhibiting the formation of supercoiled DNA by DNA gyrase.

Neltenexine (Cystic Fibrosis) (79)

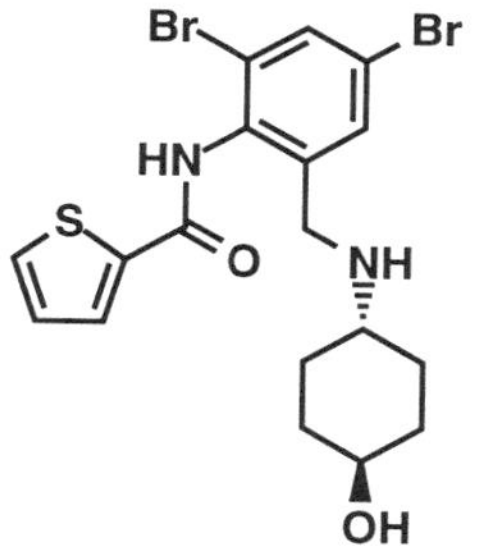

Country of Origin:	**Italy**
Originator:	**Pulitzer**
First Introduction:	**Italy**
Introduced by:	**Pulitzer; IBI;**
	Maggioni Winthrop
Trade Name:	**Tenoxol; Alveoten; MUCO 4**
CAS Registry No.:	**99453-84-6**
Molecular Weight:	**488.24**
Dosage Form:	**37 mg Tablet and oral susp.**

Neltenexine is a mucolytic launched the first time in Italy for the treatment of cystic fibrosis. It is also effective for the treatment of acute and chronic bronchopulmonary diseases.

Neticonazole Hydrochloride (Topical Antifungal) (80, 81)

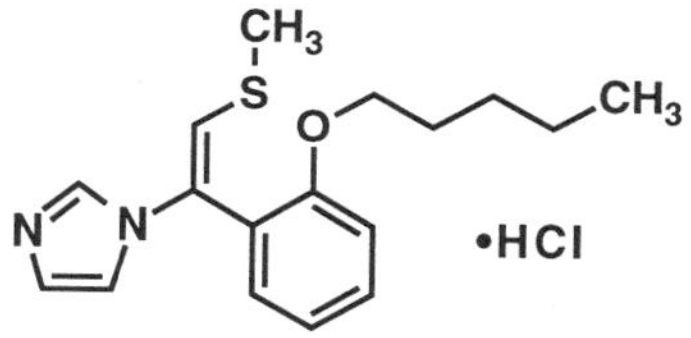

Country of Origin:	**Japan**
Originator:	**SS Pharmaceutical**
First Introduction:	**Japan**
Introduced by:	**SS Pharmaceutical**
Trade Name:	**Atolant**
CAS Registry No.:	**130773-02-3**
Molecular Weight:	**338.90**
Dosage Form:	**1% Cream and solution**

Neticonazole hydrochloride is a novel imidazole antifungal agent introduced for the treatment of infections caused by *Candida* species. One percent neticonazole cream once-daily application has been reported to be effective in treating patients with tinea pedis, tinea corporis, tinea cruris, intertrigo-type candidiasis, erosio interdigitalis and tinea versicolor. Compared with other agents of the azole class, neticonazole has the most potent activity against a variety of fungi, including the dermatophytes and pathogenic yeasts. It also has activity against Gram-positive, but not Gram-negative bacteria.

Paclitaxel (Antineoplastic) (82-86)

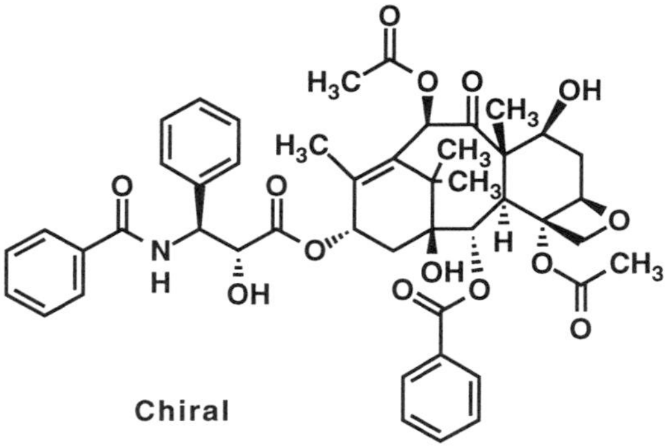

Chiral

Country of Origin:	**U.S.A.**	Trade Name:	**Taxol**
Originator:	**NIH**	CAS Registry No.:	**33069-62-4**
First Introduction:	**U.S.A.; Canada; Sweden**	Molecular Weight:	**853.93**
Introduced by:	**Bristol-Myers Squibb**	Dosage Form:	**6 mg/ml Inj.**

Paclitaxel, a natural product isolated from the bark of the Pacific yew, is effective in treating refractory metastatic ovarian cancer. Unlike any other antineoplastic agents, paclitaxel appears to have several possible mechanisms of action, including an antimicrotubule action through the promotion of tubulin polymerization and stabilization of microtubules, thereby, halting mitosis and promoting cell death. The supply of paclitaxel is limited by its low natural abundance and currently it is being manufactured by a semi-synthetic route from deacetylbaccatin III that is isolated from the needles of the yew tree. Recent completion of two total syntheses of taxol conquered the structural complexity of the title compound and may be useful in obtaining certain closely related analogs, some of which have been found to have antitumor activity. Paclitaxel has potential uses in the treatment of metastatic breast cancer, lung cancer, head and neck cancer, and malignant melanoma.

Parnaparin Sodium (Anticoagulant) (87-89)

Country of Origin:	**Italy**	Trade Name:	**Fluxum**
Originator:	**Opocrin**	CAS Registry No.:	**9005-49-6**
First Introduction:	**Italy**	Molecular Weight:	**~4.5 kD**
Introduced by:	**Alfa**	Dosage Form:	**3200 Anti-XaU/0.3ml Inj.**
	Wassermann		**4250 Anti-XaU/0.4ml Inj.**

Parnaparin sodium is a low molecular weight heparin obtained from bovine mucosal heparin by chemical depolymerization. It has more potent antithrombotic and profibrinolytic activity than heparin evidenced by its higher activity in inhibiting factor Xa and in reducing plasma activity of platelet activator inhibitor. It is effective in improving the venous blood outflow of lower limbs in deep vein thrombosis (DVT) patients in addition to preventing DVT following orthopaedic surgery, reportedly without causing bleeding complications. Parnaparin has also shown efficacy in inflammatory occlusive complications of postphlebitic syndrome and in acute myocardial infarction.

Pidotimod (Immunostimulant) (90-93)

Country of Origin:	**Italy**
Originator:	**Poli**
First Introduction:	**Italy**
Introduced by:	**Poli; Fidia; Max Pharma; Boehringer Mannheim**
Trade Name:	**Polimod; Pigitil; Axil; Onaka**
CAS Registry No.:	**121808-62-6**
Molecular Weight:	**244.27**
Dosage Form:	**200; 400 mg Oral vial; 400 mg Tablet; 800 mg Sachet**

Pidotimod, a dipeptide immunomodulating agent, has been introduced as immunostimulant therapy in patients with cell-mediated immunosuppression during respiratory or urinary tract infections. Its mode of action on the immune system is by activation of T-lymphocytes via interleukin (IL)-2 stimulation and macrophages via activation of superoxide dismutase and chemotaxis. In mice, treatment with pidotimod causes significant increase in the natural killer cell activity that may play an important role in immunosurveillance against tumors and in physiological homeostasis. Pidotimod can reverse the immunosuppression caused by surgical stress and has antiinflammatory, antioxidant, and antiaging properties. It has also been reported to provide protection from bacterial infections in mice.

Porfimer Sodium (Antineoplastic Adjuvant) (94-96)

Country of Origin:	**Canada**	Introduced by:	**Quadra Logic; Cyanamid**
Originator:	**Quadra Logic**		
First Introduction:	**Canada**	Trade Name:	**Photofrin**
		CAS Registry No.:	**87806-31-3**

Porfimer sodium, a polyporphyrin derivative, is a photosensitizer introduced for the treatment of refractory lung, bladder, and esophageal cancers. Porfimer sodium treatment with photodynamic therapy inhibits or stops recurrence of superficial bladder cancer tumors. In patients with inoperable early stage lung cancer, use of porfimer sodium prior to argon dye laser treatment produced complete response in >60% of carcinomas. It also has potential for the treatment of cervical dysplasia, atherosclerosis, and psoriasis.

Pramiracetam Sulfate (Cognition Enhancer) (97-99)

Country of Origin:	**U.S.A.**
Originator:	**Warner-Lambert**
First Introduction:	**Italy**
Introduced by:	**Lusofarmaco; Firma; Boehringer Mannheim**
Trade Name:	**Remen; Neupramir; Pramistar**
CAS Registry No.:	**72869-16-0**
Molecular Weight:	**367.46**
Dosage Form:	**600 mg Tablet**

Pramiracetam sulfate, together with aniracetam are two nootropics of the piracetam family introduced in 1993 for the treatment of attention and memory disorders resulting from degenerative or vascular disorders in the elderly. In patients with mild to moderate primary degenerative dementia, treatment with 75-300 mg/day of pramiracetam reduced depression, anxiety, sleep disturbance, and hostility with no observable adverse reactions. Significant improvement of performance on measures of memory, especially delayed recall, have been observed for pramiracetam treated patients with brain injury.

Reviparin Sodium (Anticoagulant) (100-102)

Country of Origin:	**Germany**	Trade Name:	**Clivarin**
Originator:	**Knoll**	CAS Registry No.:	**9041-08-1**
First Introduction:	**Germany**	Molecular Weight:	**~3.9 kD**
Introduced by:	**Knoll**	Dosage Form:	**0.25 ml Ampule**

Reviparin sodium, a second-generation low molecular weight heparin produced from porcine mucosal heparin, has been introduced for the prevention of deep vein thrombosis and pulmonary embolism following surgery. It has sustained activity and increased bioavailability over unfractioned heparin. More noticeably, reviparin shows a pronounced inhibitory effect, both *in vitro* and *in vivo*, on smooth muscle cell proliferation which plays a predominant role in restenosis following angioplasty. Indeed, a lower incidence of restenosis without major bleeding complications has been reported for reviparin treated patients who successfully underwent percutaneous transluminal coronary angioplasty.

Risperidone (Neuroleptic) (103-106)

Country of Origin:	**U.S.A.**
Originator:	**Janssen**
First Introduction:	**Canada;** **United Kingdom**
Introduced by:	**Janssen; Organon**
Trade Name:	**Risperdal**
CAS Registry No.:	**106266-06-2**
Molecular Weight:	**410.50**
Dosage Form:	**1; 2; 3; 4 mg Tablet**

Risperidone is a novel antipsychotic introduced for the treatment of acute and chronic schizophrenia. It has a balanced serotonin 5-HT$_2$ and dopamine D$_2$ receptor antagonist activity. While the anti-D$_2$ activity may relate to the antipsychotic potency of neuroleptic agents, an antidepressive efficacy of substances with anti-5-HT$_2$ activity has been suggested. Risperidone, therefore, has therapeutic action on both positive and negative symptoms of schizophrenia and produces significantly fewer side effects especially extrapyramidal symptoms compared with commonly used pure D$_2$ antagonist antipsychotics. It also has potential for management of alcohol withdrawal and cocaine addiction.

Sarpogrelate Hydrochloride (Platelet Antiaggregant) (107-110)

Country of Origin:	**Japan**
Originator:	**Mitsubishi Kasei**
First Introduction:	**Japan**
Introduced by:	**Tokyo Tanabe**
Trade Name:	**Anplag**
CAS Registry No.:	**135159-51-2**
Molecular Weight:	**465.98**
Dosage Form:	**50; 100 mg Tablet**

Racemic   **•HCl**

Sarpogrelate hydrochloride, a potent and selective serotonin 5-HT$_2$ receptor antagonist, was launched in Japan as an antithrombotic. It exhibits inhibition of *ex vivo* platelet aggregation stimulated by serotonin in combination with collagen and suppression of blood vessel constriction mediated by 5-HT$_2$ *in vitro*. Its antithrombotic effects have been demonstrated in several *in vivo* experimental models including reduction of the mortality rate in acute pulmonary thromboembolic disease, arterial thrombosis, and peripheral obstructive disease. Sarpogrelate has been shown to be especially useful as an antiplatelet agent for patients with Type 2 diabetes mellitus, in whom 5-HT$_2$ mediated amplification of collagen-induced platelet aggregation is significantly increased.

Sorivudine (Antiviral) (111-114)

Country of Origin:	**Japan**
Originator:	**Yamasa Shoyu**
First Introduction:	**Japan**
Introduced by:	**Yamasa Shoyu;**
	Nippon Shoji
Trade Name:	**Usevir**
CAS Registry No.:	**77181-69-2**
Molecular Weight:	**349.14**
Dosage Form:	**50 mg Tablet**

Sorivudine is an orally-active antiviral nucleoside analog. It has potent and selective activity against herpes simplex virus type-1 (HSV-1) and varicella-zoster virus (VZV). In a cutaneous model infection of mice with HSV-1, oral administration of sorivudine at dose of 20 mg/kg resulted in a significant increase in the survival rate. Five percent sorivudine cream is also useful in the topical treatment of cutaneous HSV-1 infection, including immunocompromised patients. It has been suggested that its antiviral activity may be mediated by inhibition of DNA synthesis in virus infected cells.

Since the metabolite of sorivudine inhibits the enzyme for pyrimidine catabolism, caution should be taken not to use sorivudine in combination with fluorouracil drugs in order to avoid hematological disorders.

Sparfloxacin (Antibiotic) (115-118)

Country of Origin:	**Japan**
Originator:	**Dainippon**
First Introduction:	**Japan**
Introduced by:	**Dainippon**
Trade Name:	**Spara**
CAS Registry No.:	**110871-86-8**
Molecular Weight:	**392.41**
Dosage Form:	**100; 150 mg Tablet**

Sparfloxacin is the most potent fluoroquinolone antibiotic introduced for the treatment of community acquired infections. It has superior and broad *in vitro* activity against members of family *Enterobacteriaceae* and anaerobic bacteria, some of which are resistant to β-lactam antibiotics or to aminoglycosides. In patients with surgical infections, sparfloxacin shows excellent activity against resistant pathogens. It is effective in treating patients with bladder irritability and is reported to have potential in the treatment of leprosy and *Mycobacterium tuberculosis* in mice. Favorable pharmacokinetic properties, good intracellular penetration and a lack of transferable resistance have been reported.

Tacalcitol (Topical Antipsoriatic) (119-121)

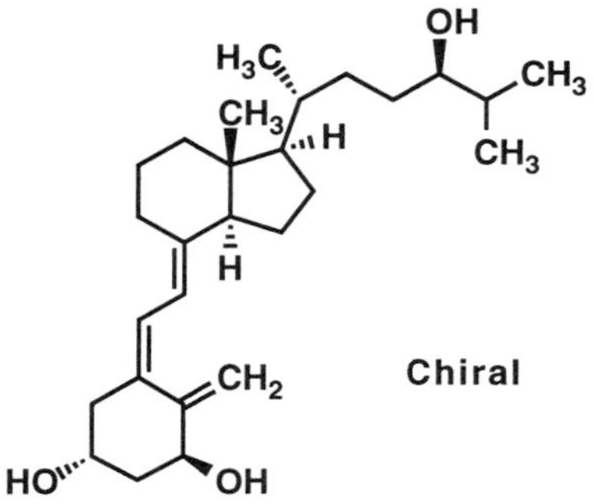

Country of Origin:	**Japan**
Originator:	**Teijin**
First Introduction:	**Japan**
Introduced by:	**Teijin; Fujisawa**
Trade Name:	**Bonalfa**
CAS Registry No.:	**57333-96-7**
Molecular Weight:	**416.64**
Dosage Form:	**2 mcg/g Ointment**

Chiral

Tacalcitol, the second vitamin D analog developed as a topical antipsoriatic, was introduced for keratosis, psoriasis, ichthyosis, pityriasis rubra pilaris and palmoplantar pustulosis and keratoderma. It shows improvement in appearance of skin lesions for psoriasis patients without significant side effects. In patients with psoriasis vulgaris, it effectively inhibits keratinocyte growth. Some *in vitro* studies have shown that tacalcitol inhibits interleukin-1-α-induced granulocyte macrophage colony-stimulating-factor mRNA expression of the human dermal microvascular endothelial cells, which may be involved in mediation of angioproliferation in skin inflammation.

Tacrine Hydrochloride (Treatment for Alzheimer's Disease) (122-125)

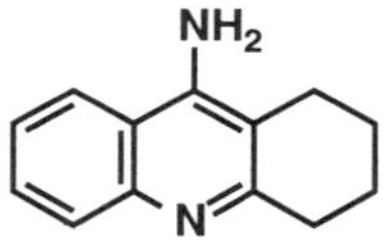

Country of Origin:	**U.S.A.**
Originator:	**NIH**
First Introduction:	**U.S.A.**
Introduced by:	**Warner-Lambert**
Trade Name:	**Cognex**
CAS Registry No.:	**1684-40-8**
Molecular Weight:	**234.73**
Dosage Form:	**10; 20; 30; 40 mg Capsule**

•HCl

Tacrine is the first therapeutic launched specifically for the treatment of Alzheimer's disease which affects four million people in the US alone. Clinically significant improvement in cognition has been demonstrated in the patients with Alzheimer's disease. Although the cause of Alzheimer's disease is not understood, degeneration of cholinergic neurons is thought to be a primary factor in the development and progression of the disease. Tacrine is a reversible acetylcholinesterase inhibitor that presumably acts centrally by elevating the acetylcholine level in the cerebral cortex and by slowing degradation of acetylcholine from intact cholinergic neurons. While tacrine improves the symptoms of Alzheimer's disease, the underlining cause of the dementing process is not altered. It has also been suggested that the beneficial effects of tacrine may be due to its multiple effects on several neurotransmitter systems.

Tacrolimus (Immunosuppressant) (126-129)

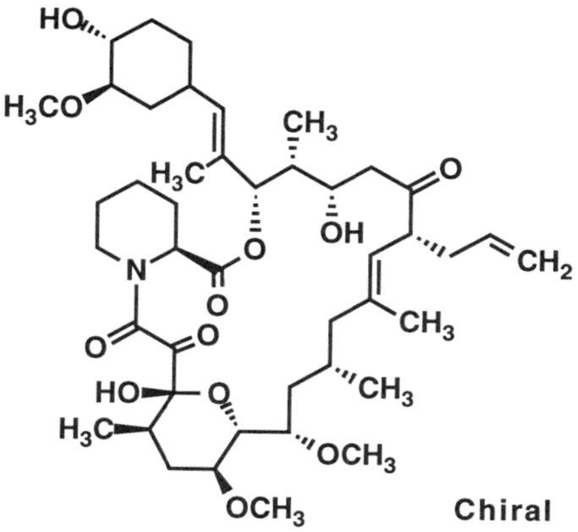

Country of Origin: **Japan**
Originator: **Fujisawa**
First Introduction: **Japan**
Introduced by: **Fujisawa**
Trade Name: **Prograf**
CAS Registry No.: **104987-11-3**
Molecular Weight: **804.04**
Dosage Form: **1 mg Capsule; 5 mg Inj.**

Chiral

Tacrolimus, isolated from the microorganism *Streptomyces tsukubaensis*, is a macrolide immunosuppressant developed by Fujisawa for organ transplantation. It displays similar but more potent immunosuppressive activity than cyclosporin. It inhibits both cell mediated and humoral immune responses. In animal models of organ transplantation, tacrolimus has been shown to prolong survival of hepatic, renal, cardiac, small intestine, pancreatic and skin allografts and to reverse cardiac and renal allograft rejection. It has been used effectively in humans as rescue or primary immunosuppressant therapy in liver or kidney transplantation. Compared to cyclosporin, tacrolimus causes reduced incidence of infectious complications and of hypertension and hypercholesterolemia for the allograft recipients. In common with cyclosporin, tacrolimus binds with high affinity to a family of cytoplasmic immunosuppressant binding proteins, the immunophilins. This tight complex is proposed as the biologically active moiety that interacts with intracellular molecules involved in signal transduction. It inhibits phosphatase activity of calcineurin, an action that may impair the generation and/or activation of nuclear transcription factors required for lymphokine (particularly interleukin-2) gene expression. Tacrolimus has also been reported to have potential in multiple sclerosis, psoriasis, rheumatoid arthritis and uveitis associated with Behcet's disease.

Tamsulosin Hydrochloride (Antiprostatic Hypertrophy) (130-132)

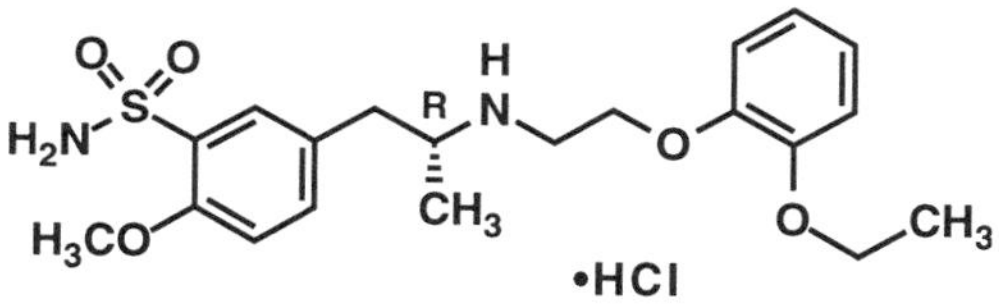

•HCl

Country of Origin: **Japan**
Originator: **Yamanouchi**
First Introduction: **Japan**
Introduced by: **Yamanouchi**

Trade Name: **Harnal**
CAS Registry No.: **106133-20-4**
Molecular Weight: **444.98**
Dosage Form: **0.1; 0.2 mg Capsule**

Tamsulosin hydrochloride is the first in the class of potent and selective α-1A adrenoceptor antagonists introduced for the treatment of dysuria associated with benign prostatic hypertrophy (BPH). In a clinical study, significant improvement in irritative and obstructive symptoms has been reported for tamsulosin treated patients with BPH. *In vitro* studies in human penile erectile tissue and vas deferens indicated that tamsulosin may be of use in treating male sexual dysfunction.

Torasemide (Diuretic) (133-136)

Country of Origin:	**Norway**
Originator:	**Hafslund Nycomed**
First Introduction:	**Germany; Italy**
Introduced by:	**Boehringer Mannheim**
Trade Name:	**Unat; Toradiur**
CAS Registry No.:	**56211-40-6**
Molecular Weight:	**348.42**
Dosage Form:	**2.5; 5; 10; 200 mg Tablet; 10 mg/2ml Ampule for inj.**

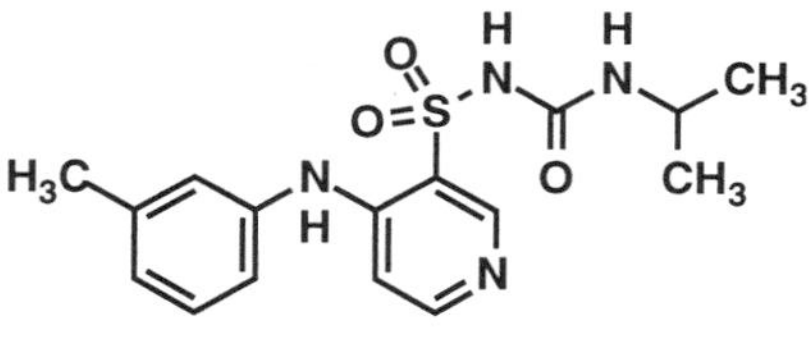

Torasemide is a novel loop diuretic launched in 1993 after a 12-year gap from the last diuretic introduction. It is indicated for the treatment of hypertension and edema associated with chronic congestive heart failure, renal disease and hepatic cirrhosis. Torasemide exerts its major diuretic activity on the thick ascending limb of the Henle's loop to promote rapid and marked excretion of water, Na^+, Cl^-, and to a lesser extent, K^+ and Ca^{2+}. Compared with other loop diuretics such as furosemide, torasemide has a stronger antihypertensive action, a higher bioavailability, a longer duration of action that is independent of the renal function, and has no side effects such as paradoxical antidiuresis. The mechanism of its vasodilating effect has been suggested to result from, at least in part, the competitive antagonism of the thromboxane A_2 receptor.

Trandolapril (Antihypertensive) (137-141)

Country of Origin:	**France**
Originator:	**Roussel Uclaf**
First Introduction:	**France; Germany; United Kingdom**
Introduced by:	**Roussel Uclaf; Knoll**
Trade Name:	**Odrik; Udrik; Gopten**
CAS Registry No.:	**87679-37-6**
Molecular Weight:	**430.55**
Dosage Form:	**0.5; 1; 2 mg Capsule**

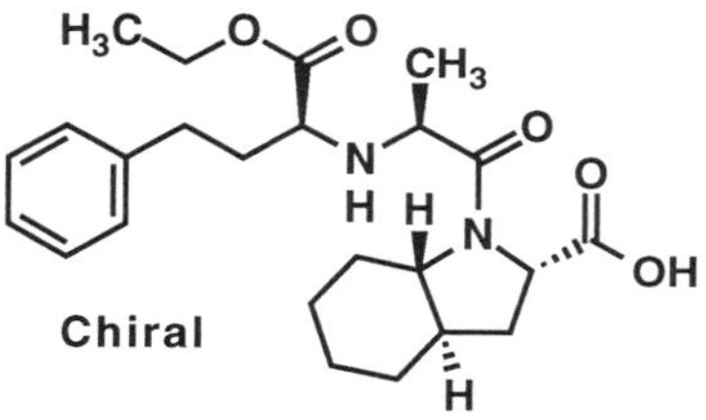

Chiral

Trandolapril is a new ACE inhibitor that is rapidly hydrolyzed, mainly in the liver, to its biologically active form, trandolaprilat. Compared with all other ACE inhibitors, trandolaprilat is reported to have the highest lipophilicity and the most prolonged ACE inhibitory activity. In hypertensive patients, trandolapril at a dose of 2 mg reduces blood pressure consistently throughout the 24 hour period after intake, making it one of the best once a day antihypertensive drugs. It has also been demonstrated to inhibit aortic atherosclerosis in the hyperlipidemic rabbit.

Tretinoin Tocoferil (Antiulcerative) (142-144)

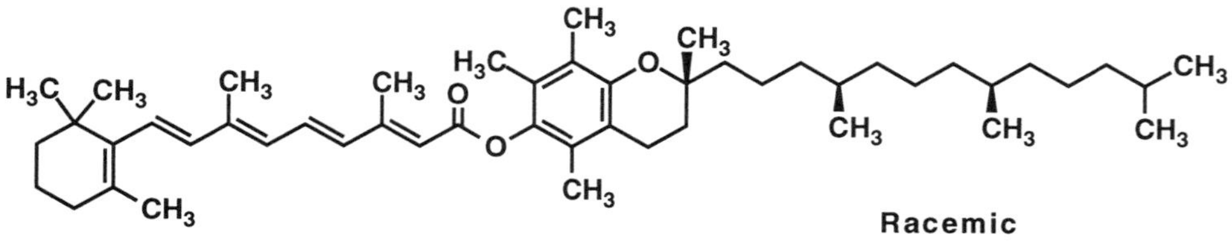

Racemic

Country of Origin:	**Japan**	Trade Name:	**Olcenon**
Originator:	**Nisshin Flour Milling**	CAS Registry No.:	**40516-48-1**
First Introduction:	**Japan**	Molecular Weight:	**713.15**
Introduced by:	**Nisshin Flour Milling; Lederle**	Dosage Form:	**0.25% Ointment**

Tretinoin tocoferil, the α-tocopherol ester of all-*trans*-retinoic acid, was launched for the treatment of bedsores and skin ulcers such as burn, leg, and diabetic ulcers. It is also an orally active agent for gastric ulcer treatment. Tretinoin tocoferil represents a new class of antiulcer drugs which act by directly promoting tissue repair, distinguished from common drugs that exert an indirect function by suppression of acid-secretion and mucosa protection. The mode of its action has been proposed to relate to its ability to stimulate DNA synthesis of growth-arrested human skin fibroblasts by promoting expression of the epidermal growth factor receptor.

Zaltoprofen (Antiinflammatory) (145-147)

Country of Origin:	**Japan**
Originator:	**Nippon Chemiphar**
First Introduction:	**Japan**
Introduced by:	**Nippon Chemiphar; Zeria**
Trade Name:	**Soleton; Peon**
CAS Registry No.:	**89482-00-8**
Molecular Weight:	**298.36**
Dosage Form:	**800 mg Tablet**

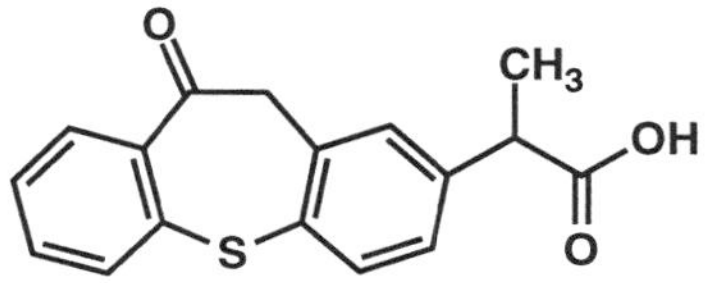

Racemic

Zaltoprofen is a potent non-steroidal antiinflammatory drug (NSAID) with analgesic activity. In rats and mice, zaltoprofen is reported to be equipotent or superior to other NSAIDs in bradykinin-induced pain, acetic acid-induced writhing, carrageenan-induced hyperalgesia, and in several other experimental models of analgesia. It acts by selectively suppressing the production of prostaglandins at the inflammatory site and not in other organs such as stomach and kidney, therefore, has remarkably low gastric side effects that are associated with conventional antiinflammatory agents.

References

1. The material in this chapter is based on the combined information from the following sources:
    a. Scrip Magazine, January, 1994.
    b. Pharmaprojects.
    c. IMSworld Publication.
    d. J.R. Prous, DN&P, 7, 26 (1994).
2. J.D. Strupczewski and D.B. Ellis, Annu. Rep. Med. Chem., 28, 325 (1993).
3. J.D. Strupczewski and D.B. Ellis, Annu. Rep. Med. Chem., 27, 321 (1992).
4. J.D. Strupczewski, D.B. Ellis, and R.C. Allen, Annu. Rep. Med. Chem., 26, 297 (1991).
5. H.J. Roth and A. Kleemann in Pharmaceutical Chenistry, Vol. 1, Halsted Press, New York, N.Y., 1988, p17.
6. F-D-C Reports, 6 (January 4, 1994).
7. R.F. Borne and M.C. Vinson, Drug Topics, 40 (February 7, 1994).
8. D.A. Hussar, American Pharmacy, NS34, 24 (1994).
9. J. Roge, P. Baumer, H. Berard, J.C. Schwartz, and J.M. Lecomte, Scand. J. Gastroenterol. 28, 352 (1993).
10. P. Baumer, E. Danquechin Dorval, J. Bertrand, J.M. Vetel, J.C. Schwartz, and J.M. Lecomte, Gut. 33, 753 (1992).
11. J.F. Bergmann, S. Chaussade, D. Couturier, P. Baumer, J.C. Schwartz, and J.M. Lecomte, Aliment. Pharmacol. Ther. 6, 305 (1992).
12. G.A. Rodan and R. Balena, Ann. Med. 25, 373 (1993).
13. B.L. Riggs and L.J. Melton, N. Engl. J. Med., 327, 620 (1992).
14. M. Sato, W. Grasser, N. Endo, R. Akins, H. Simmons, D.D. Thompson, E. Golub, and G.A. Rodan, J. Clin. Invest. 88, 2095 (1991).
15. S.R. Nussbaum, R.P. Warrell, Jr., R. Rude, J. Glusman, J.P. Bilezikian, A.F. Stewart, M. Stepanavage, J.F. Sacco, S.D. Averbuch, and B.J. Gertz, J. Clin. Oncol. 11, 1618 (1993).
16. J.R. Prous, ed., Drugs Future, 17, 969 (1993).
17. A. Caruso, V.M.C. Cutuli, E. de Bernardis, G. Attaguile, and M. Amico-Roxas, Drugs Exp. Clin. Res., 18, 481 (1992).
18. E. Arrigoni-Martelli, Drugs Exp. Clin. Res., 16, 63 (1990).
19. J.R. Prous, ed., Drugs Future, 18, 257 (1993).
20. U. Senin, G. Abate, C. Fieschi, G. Gori, A. Guala, G. Marini, C. Villardita, and L. Parnetti, Eur. Neuropsychopharmacol. 1, 511 (1991).
21. K. Wesnes, R. Anand, P. Simpson, and L. Christmas, J. Psychopharmacol., 4, 219 (1990).
22. R. Nyffenegger, D. Riebenfeld, and A. Macciocchi, Clin. Ther., 13, 589 (1991).
23. J.R. Prous, ed., Drugs Future, 17, 837 (1992).
24. R. Rolland, G. Piscitelli, C. Ferrari, and A. Petroccione, Br. Med. J., 302, 1367 (1991).
25. A. Caballero-Gordo, N. Lopez-Nazareno, M. Calderay, J.L. Caballero, E. Macheno, and D. Sghedoni, J. Reprod. Med., 36, 717 (1991).
26. E. Frans, R. Dom, and M. Demedts, Eur. Resp. J., 5, 263 (1992).
27. J.R. Prous, ed., Drugs Future, 17, 947 (1992).
28. M.P. Okamoto, R.K. Nakahiro, A. Chin, and A. Bedikian, Clin. Pharmacokinet., 25, 88 (1993).
29. F.P.V. Maesen, B.I. Davies, W. Geraedts, and R. Costongs, Eur. Resp. J., 4, Suppl. 14, 244S (1991).
30. L.O. Gentry and G.G. Rodriguez, Antimicrob. Agents. Chemother., 35, 2371 (1991).
31. J.R. Prous, ed., Drugs Future, 18, 166 (1993).
32. *Cinolazepam*. Annu Drug Data Rep., 14, 948 (1992).

33.   H.M. Bryson and E.M. Sorkin, Drugs, _46_, 872 (1993).
34.   A. Saven and L.D. Piro, Cancer Invest., _11_, 559 (1993).
35.   E.H. Estey, R. Kurzrock, H.M. Kantarjian, S.M. O'Brien, K.B. McCredie, M. Beran, C. Koller, M.J. Keating, C. Hirsch-Ginsberg, Y.O. Huh, S. Stass, and E.J. Freireich, Blood, _79_, 882 (1992).
36.   E. Beutler, Lancet, _340_, 952 (1992).
37.   J.R. Prous, ed., Drugs Future, _17_, 1037 (1992).
38.   R. Ohno, N. Tatsumi, M. Hirano, K. Imai, H. Mizoguchi, T. Nakamura, M. Kosaka, K. Takatsuki, T. Yamaya, K. Toyama, T. Yoshida, T. Masaoka, S. Hashimoto, T. Ohshima, I. Kimura, K. Yamada, K. Kimura, Oncology, _48_, 451 (1991).
39.   J.R. Prous, ed., Drugs Future, _18_, 170 (1993).
40.   O. Muller and K. Wettich, J. Antimicrob. Chemother., _31_, Suppl. C, 97 (1993).
41.   G.D. Sides and P.M. Conforti, J. Antimicrob. Chemother., _31_, Suppl. C, 175 (1993).
42.   J.R. Prous, ed., Drugs Future, _17_, 1041 (1992).
43.   Y. Ito, S. Nakamura, Y. Onoda, Y. Sugawara, and O. Takaiti, Jpn. J. Pharmacol. _62_, 169 (1993).
44.   Y. Ito, Y. Onoda, S. Nakamura, K. Tagawa, T. Fukushima, Y. Sugawara, and O. Takaiti, Jpn. J. Pharmacol. _62_, 175 (1993).
45.   J.R. Prous, ed., Drugs Future, _18_, 475 (1993).
46.   T. Saito, A. Hagihara, N. Igarashi, N. Matsuda, A. Yamashita, and K. Ito, Jpn. J. Pharmacol. _62_, 137 (1993).
47.   J.R. Prous, ed., Drugs Future, _17_, 1042 (1992).
48.   K.J. Palmer and D. McTavish, Drugs, _45_, 1041 (1993).
49.   R.T. McCabe, C.G. Wasterlain, N. Kucharczyk, R.D. Sofia, and J.R. Vogel, J. Pharmacol. Exp. Ther., _264_, 1248 (1993).
50.   H.S. White, H.H. Wolf, E.A. Swinyard, G.A. Skeen, and R.D. Sofia, Epilepsia, _33_, 564 (1992).
51.   J.R. Prous, ed., Drugs Future, _18_, 599 (1993).
52.   L.R. Wiseman and D. McTavish, Drugs, _45_, 66 (1993).
53.   J.R. Prous, ed., Drugs Future, _18_, 572 (1993).
54.   K.L. Goa and E.M. Sorkin, Drugs, _46_, 409 (1993).
55.   A. Handforth and D.M. Treiman, Epilepsia, _34_, Suppl. 6, 109 (1993).
56.   M.W. Pierce, H. Anhut, and W. Sauermann, Epilepsia, _34_, Suppl. 2, 181 (1993).
57.   A. Hvidberg, S. Jorgensen, and J. Hilsted, Br. J. Clin. Pharmacol., _34_, 547 (1992).
58.   J. Hilsted, A. Hvidberg, and R. Djurup, Diabetologia, _35_, Suppl. 1, A184 (1992).
59.   C.V. Pollack, Jr., J. Emerg. Med., _11_, 195 (1993)..
60.   J.R. Prous, ed., Drugs Future, _17_, 847 (1992).
61.   L.B. Barradell and D. McTavish, Drugs, _45_, 570 (1993).
62.   J.R. Prous, ed., Drugs Future, _18_, 661 (1993).
63.   N. Ogiku, H. Sumikawa, Y. Hashimoto, and R. Ishida, Stroke, _24_, 245 (1993).
64.   N. Ogiku, H. Sumikawa, S. Minamide, and R. Ishida, Jpn. J. Pharmacol. _61_, 69 (1993).
65.   M. Abramowicz, ed., Med. Lett. Drugs Ther., _35_, 61 (1993).
66.   The IFNB Multiple Sclerosis Study Group, Neurology, _43_, 655 (1993).
67.   H.S. Panitch, Drugs, _44_, 946 (1992).
68.   J.R. Prous, ed., Drugs Future, _18_, 676 (1993).
69.   M.A. Pfaller, A.L. Barry, and P.C. Fuchs, J. Clin. Microbiol., _31_, 1924 (1993).
70.   Y. Inagaki, R. Nakaya, T. Chida, and S. Hashimoto, Jpn. J. Antibiot. _45_, 241 (1992).
71.   K. Morimoto, H. Kinoshita, S. Nakatani, K. Sakai, M. Fujimoto, K. Ohno, T. Ueda, K. Ohmori, O. Yamazaki, and S. Doi, Jpn. J. Antibiot. _45_, 258 (1992).
72.   J.R. Prous, ed., Drugs Future, _17_, 1050 (1992).

73. C.H. Simon, A.S.T. Planting, J.H.M. Schellens, G. Stoter, and J. Verweij, Neth. J. Med., 42, A76 (1993).

74. C.C. Geilen, R. Haase, K. Buchner, T. Wieder, F. Hucho, and W. Reutter, Eur. J. Cancer, 27, 1650 (1991).

75. P. Hilgard, E. Kampherm, L. Nolan, J. Pohl, and T. Reissmann, J. Cancer Res. Clin. Oncol., 117, 403 (1991).

76. J.R. Prous, ed., Drugs Future, 18, 666 (1993).

77. I. Kurokawa, J. Invest. Dermatol., 101, 471 (1993).

78. K. Vogt, J. Hermann, U. Blume, H. Gollnick, H. Hahn, U.F. Haustein, and C.E. Orfanos, Eur. J. Clin. Microbiol. Infect. Dis., 11, 943 (1992).

79. R. Aquilina, F. Bergero, P. Nocetti, and C. De Michelis, Arch. Med. Interna., 38, 157 (1986).

80. J.R. Prous, ed., Drugs Future, 18, 324 (1993).

81. K. Maebashi, T. Itoyama, K. Uchida, H. Yamaguchi, T. Asaoka, and A. Iwasa, Jpn. J. Antibiot., 46, 896 (1993).

82. J.R. Prous, ed., Drugs Future, 18, 92 (1993).

83. C.D. Runowicz, P.H. Wiernik, A.I. Einzig, G.L. Goldberg, and S.B. Horwitz, Cancer, 71, Suppl. 4, 1591 (1993).

84. R.E. Gregory and A.F. DeLisa, Clin. Pharm., 12, 401 (1993).

85. R.A. Holton, C. Somoza, H.-B. Kim, F. Liang, R.J. Biediger, P.D. Boatman, M. Shindo, C.C. Smith, S. Kim, H. Nadizadeh, Y. Suzuki, C. Tao, P. Vu, S. Tang, P. Zhang, K.K. Martin, L.N. Gentile, and J.H. Liu, J. Am. Chem. Soc., 116, 1597, 1599 (1994).

86. K.C. Nocolaou, Z. Yang, J.J. Liu, H. Ueno, P.G. Nantermet, R.K. Guy, C.F. Claiborne, J. Renaud, E.A. Couladouros, K. Paulyannan, E.J. Sorensen, Nature, 367, 630 (1994).

87. A. Tedoldi, F. Botticella, and M.R. Maloberti, Clin. Trials Meta-Analys., 28, 215 (1993).

88. A.M. Laguardia and G.C. Caroli, Curr. Med. Res. Opin., 12, 584 (1992).

89. G. Melandri, A. Branzi, F. Semprini, V. Cervi, and B. Magnani, Thromb. Res., 66, 141 (1992).

90. J.R. Prous, ed., Drugs Future, 17, 1142 (1992).

91. A. Pugliese, A. Biglino, C. Uslenghi, L. Marinelli, B. Forno, and R. Girardello, Int. J. Immunother. 8, 212 (1992).

92. A. Auteri, A.L. Pasqui, G. Gotti, F. Bruni, M. Saletti, M. Di Renzo, G. Bova, G, Borlini, S. Gori, G. Fanetti, G. Campoccia, D. Maggiore, and R. Girardello, Int. J. Immunother. 9, 95 (1993).

93. G. Migliorati, L. D Adamio, G. Coppi, I. Nicoletti, and C. Riccardi, Immunopharmacol. Immunotoxicol., 14, 737 (1992).

94. K. Furuse, M. Fukuoka, H. Kato, T. Horai, K. Kubota, N. Kodama, Y. Kusunoki, N. Takifuji, T. Okunaka, C, Konaka, H. Wada, and Y. Hayata, J. Clin. Oncol., 11, 1852 (1993).

95. J.A. Chapman, Y. Tadir, B.J. Tromberg, K. Yu, A. Manetta, C.H. Sun, and M.W. Berns, Am. J. Obstet. Gynecol., 168, 685 (1993).

96. M. Leroy, C. Sari, and A. Bisson, Proc. Am. Soc. Clin. Oncol., 12, 332 (1993).

97. J.R. Prous, ed., Drugs Future, 17, 852 (1992).

98. J.J. Claus, C. Ludwig, E. Mohr, M. Giuffra, J. Blin, and T.N. Chase, Neurology, 41, 570 (1991).

99. A. McLean, Jr., D.D. Cardenas, D. Burgess, and E. Gamzu, Brain Inj., 5, 375 (1991).

100. K.M. Schmid, M. Preisack, W. Voelker, M. Sujatta, and K.R. Karsch, Semin. Thromb. Hemost., 19, Suppl. 1, 155 (1993).

101. W. Jeske, B. Lojewski, J.M. Walenga, D. Hoppensteadt, A. Ahsan, and J. Fareed, Semin. Thromb. Hemost., 19, Suppl. 1, 229 (1993).

102. V.V. Kakkar, B. Boneu, A.T. Cohen, and M. Suijata, Thromb. Haemostasis, 69, 651 (1993).

103.   J.R. Prous, ed., Drugs Future, 17, 145 (1992).
104.   P.A.J. Janssen, A. Schotte, A.A.H.P. Megens, and J.E. Leysen, Pharmacopsychiatry, 26, 165 (1993).
105.   S. Nyberg, L. Farde, L. Eriksson, C. Halldin, and B. Eriksson, Psychopharmacology, 110, 265 (1993).
106.   A. Hillert, W. Maier, H. Wetzel, and O. Bekert, Pharmacopsychiatry, 25, 213 (1992).
107.   J.R. Prous, ed., Drugs Future, 17, 1093 (1992).
108.   H. Hara, M. Osakabe, A. Kitajima, Y. Tamao, and R. Kikumoto, Thromb. Haemostasis, 65, 415 (1991).
109.   H. Hara, A. Kitajima, H. Shimada, and Y. Tamao, Thromb. Haemostasis, 66, 484 (1991).
110.   M.H. Pietrasek, Y. Takada, A. Taminato, T. Yoshima, I. Watanabe, and A. Takada, Thromb. Res., 70, 131 (1993).
111.   J.R. Prous, ed., Drugs Future, 18, 676 (1993).
112.   H. Machida and Y. Watanabe, Microbiol. Immunol., 35, 139 (1991).
113.   H. Machida, K. Ijichi, and J. Takezawa, Antiviral Res., 17, 133 (1992).
114.   K. Ijichi, N. Ashida, S. Varia, and H. Machida, Antiviral Res., 21, 47 (1993).
115.   J.R. Prous, ed., Drugs Future, 18, 490 (1993).
116.   T. Watanabe, Y. Akieda, T. Suzuki, K. Itokawa, E. Yamaji, and I. Nakayama, Drugs, 45, Suppl. 3, 388 (1993).
117.   P. Richard and L. Gutmann, J. Antimicrob. Chemother., 30, 739 (1992).
118.   R.N. Jones, M.S. Barrett, M.E. Erwin, B.M. Briggs, D.M. Johnson, Diagn. Microbiol. Infect. Dis., 14, 319 (1991).
119.   J.R. Prous, ed., Drugs Future, 18, 190 (1993).
120.   B. Farkas, T. Fujimura, T. Tone, H. Eto, M. Masuzawa, and F. Otani, J. Invest. Dermatol., 101, 490 (1993).
121.   M. Tsuchimoto, N. Saito, O. Ushijima, N. Okada, I. Kaneda, Y. Okamiya, K. Aoki, K. Sunakawa, K. Hoshina, K. Inoue, I. Nagata, H. Horiuchi, K. Komoriya, T. Takeshita, and T. Naruchi, Pharmacometrics, 39, 1 (1990).
122.   J.R. Prous, ed., Drugs Future, 17, 1060 (1992).
123.   N.J. Owens, Hosp. Formul. 28, 670 (1993).
124.   H.A. Berman and K. Leonard, Mol. Pharmacol., 41, 412 (1992).
125.   A. Adem, Acta Neurol. Scand. 85, Suppl. 139, 69 (1992).
126.   J.R. Prous, ed., Drugs Future, 17, 732 (1992).
127.   D.H. Peters, A. Fitton, G.L. Plosker, and D. Faulds, Drugs, 46, 746 (1993).
128.   J. Kunz and M.N. Hall, Trends Biochem. Sci., 18, 334 (1993).
129.   S.L. Schreiber and G.R. Crabtree, Immunol. Today, 13, 136 (1992).
130.   J.R. Prous, ed., Drugs Future, 18, 395 (1993).
131.   K. Kawabe, A. Ueno, Y. Takimoto, Y. Aso, and H. Kato, J. Urol. 144, 908 (1990).
132.   F. Holmquist, H. Hedlund, and K.E. Andersson, Eur. J. Pharmacol., 186, 87 (1990).
133.   J.R. Prous, ed., Drugs Future, 18, 294 (1993).
134.   H.A. Friedel and M.M. Buckley, Drugs, 41, 81 (1991).
135.   J. Kindler, Cardiovasc. Drugs Ther., 7, Suppl. 1, 75 (1993).
136.   C.P. Caputo and R.J. Cody, Clin. Res., 41, 642A (1993).
137.   J.R. Prous, ed., Drugs Future, 17, 760 (1992).
138.   H. Conen and H.R. Brunner, Am. Heart J., 125, 1525 (1993).
139.   B. Guller, J. Hall, and R.L. Reeves, Am. Heart J., 125, 1536 (1993).
140.   A.V. Chobanian, C.C. Haudenschild, C. Nickerson, and S. Hope, Hypertension, 20, 473 (1992).
141.   L.N.C. Duc and H.R. Brunner, Am. J. Cardiol. 70, 27D (1992).
142.   K. Sakyo, N. Nakaya, M. Mori, H. Hamada, S. Ogawa, and K. Nishiki, Jpn. J. Pharmacol., 58, Suppl. 1, 132P (1992).

143.   Y. Masukawa, Y. Urano, H. Hamada, K. Shirogane, K. Nishiki, and H. Takagi,
        Jpn. J. Pharmacol., 55, Suppl. 1, 85P (1991).
144.   K. Sakyo, H. Hamada, Y. Masukawa, K. Nishiki, Y. Furukawa, and N. Ohtsuka, Jpn.
        J. Pharmacol., 55, Suppl. 1, 386P (1991).
145.   J.R. Prous, ed., Drugs Future, 18, 401 (1993).
146.   T. Ishizaki, Y. Horai, H. Echizen, K. Kubota, K. Chiba, and M. Kusaka, Drug Invest.,
        3, 1 (1991).
147.   A. Ito and Y. Mori, Res. Commun. Chem. Pathol. Pharmacol., 70, 131 (1990).

GENERIC NAME	INDICATION	YEAR INTRODUCED	ARMC VOL., PAGE	
acarbose	antidiabetic	1990	26,	297
aceclofenac	antiinflammatory	1992	28,	325
acetohydroxamic acid	hypoammonuric	1983	19,	313
acetorphan	antidiarrheal	1993	29,	332
acipimox	hypolipidemic	1985	21,	323
acitretin	antipsoriatic	1989	25,	309
acrivastine	antihistamine	1988	24,	295
adamantanium bromide	antiseptic	1984	20,	315
adrafinil	psychostimulant	1986	22,	315
AF-2259	antiinflammatory	1987	23,	325
afloqualone	muscle relaxant	1983	19,	313
alacepril	antihypertensive	1988	24,	296
alclometasone dipropionate	topical antiinflammatory	1985	21,	323
alendronate sodium	osteoporosis	1993	29,	332
alfentanil HCl	analgesic	1983	19,	314
alfuzosin HCl	antihypertensive	1988	24,	296
alglucerase	enzyme	1991	27,	321
alminoprofen	analgesic	1983	19,	314
alpha-1 antitrypsin	protease inhibitor	1988	24,	297
alpidem	anxiolytic	1991	27,	322
alpiropride	antimigraine	1988	24,	296
alteplase	thrombolytic	1987	23,	326
amfenac sodium	antiinflammatory	1986	22,	315
aminoprofen	topical antiinflammatory	1990	26,	298
amisulpride	antipsychotic	1986	22,	316
amlexanox	antiasthmatic	1987	23,	327
amlodipine besylate	antihypertensive	1990	26,	298
amorolfine hydrochloride	topical antifungal	1991	27,	322
amosulalol	antihypertensive	1988	24,	297
amrinone	cardiotonic	1983	19,	314
amsacrine	antineoplastic	1987	23,	327
amtolmetin guacil	antiinflammatory	1993	29,	332
aniracetam	cognition enhancer	1993	29,	333
APD	calcium regulator	1987	23,	326
apraclonidine HCl	antiglaucoma	1988	24,	297
APSAC	thrombolytic	1987	23,	326
arbekacin	antibiotic	1990	26,	298
argatroban	antithromobotic	1990	26,	299
arotinolol HCl	antihypertensive	1986	22,	316
artemisinin	antimalarial	1987	23,	327
aspoxicillin	antibiotic	1987	23,	328
astemizole	antihistamine	1983	19,	314
astromycin sulfate	antibiotic	1985	21,	324
atovaquone	antiparasitic	1992	28,	326
auranofin	chrysotherapeutic	1983	19,	314
azelaic acid	antiacne	1989	25,	310
azelastine HCl	antihistamine	1986	22,	316
azithromycin	antibiotic	1988	24,	298
azosemide	diuretic	1986	22,	316

GENERIC NAME	INDICATION	YEAR INTRODUCED	ARMC VOL., PAGE	
aztreonam	antibiotic	1984	20,	315
bambuterol	bronchodilator	1990	26,	299
barnidipine hydrochloride	antihypertensive	1992	28,	326
beclobrate	hypolipidemic	1986	22,	317
befunolol HCl	antiglaucoma	1983	19,	315
benazepril hydrochloride	antihypertensive	1990	26,	299
benexate HCl	antiulcer	1987	23,	328
benidipine hydrochloride	antihypertensive	1991	27,	322
beraprost sodium	platelet aggreg. inhibitor	1992	28,	326
betaxolol HCl	antihypertensive	1983	19,	315
bevantolol HCl	antihypertensive	1987	23,	328
bifemelane HCl	nootropic	1987	23,	329
binfonazole	hypnotic	1983	19,	315
binifibrate	hypolipidemic	1986	22,	317
bisantrene hydrochloride	antineoplastic	1990	26,	300
bisoprolol fumarate	antihypertensive	1986	22,	317
bopindolol	antihypertensive	1985	21,	324
brodimoprin	antibiotic	1993	29,	333
brotizolam	hypnotic	1983	19,	315
brovincamine fumarate	cerebral vasodilator	1986	22,	317
bucillamine	immunomodulator	1987	23,	329
bucladesine sodium	cardiostimulant	1984	20,	316
budralazine	antihypertensive	1983	19,	315
bunazosin HCl	antihypertensive	1985	21,	324
bupropion HCl	antidepressant	1989	25,	310
buserelin acetate	hormone	1984	20,	316
buspirone HCl	anxiolytic	1985	21,	324
butenafine hydrochloride	topical antifungal	1992	28,	327
butibufen	antiinflammatory	1992	28,	327
butoconazole	topical antifungal	1986	22,	318
butoctamide	hypnotic	1984	20,	316
butyl flufenamate	topical antiinflammatory	1983	19,	316
cabergoline	antiprolactin	1993	29,	334
cadexomer iodine	wound healing agent	1983	19,	316
cadralazine	hypertensive	1988	24,	298
calcipotriol	antipsoriatic	1991	27,	323
camostat mesylate	antineoplastic	1985	21,	325
carboplatin	antibiotic	1986	22,	318
carumonam	antibiotic	1988	24,	298
carvedilol	antihypertensive	1991	27,	323
cefbuperazone sodium	antibiotic	1985	21,	325
cefdinir	antibiotic	1991	27,	323
cefepime	antibiotic	1993	29,	334
cefetamet pivoxil hydrochloride	antibiotic	1992	28,	327
cefixime	antibiotic	1987	23,	329
cefmenoxime HCl	antibiotic	1983	19,	316
cefminox sodium	antibiotic	1987	23,	330
cefodizime sodium	antibiotic	1990	26,	300

GENERIC NAME	INDICATION	YEAR INTRODUCED	ARMC VOL., PAGE	
cefonicid sodium	antibiotic	1984	20,	316
ceforanide	antibiotic	1984	20,	317
cefotetan disodium	antibiotic	1984	20,	317
cefotiam hexetil hydrochloride	antibiotic	1991	27,	324
cefpimizole	antibiotic	1987	23,	330
cefpiramide sodium	antibiotic	1985	21,	325
cefpirome sulfate	antibiotic	1992	28,	328
cefpodoxime proxetil	antibiotic	1989	25,	310
cefprozil	antibiotic	1992	28,	328
ceftazidime	antibiotic	1983	19,	316
cefteram pivoxil	antibiotic	1987	23,	330
ceftibuten	antibiotic	1992	28,	329
cefuroxime axetil	antibiotic	1987	23,	331
cefuzonam sodium	antibiotic	1987	23,	331
celiprolol HCl	antihypertensive	1983	19,	317
centchroman	antiestrogen	1991	27,	324
centoxin	immunomodulator	1991	27,	325
cetirizine HCl	antihistamine	1987	23,	331
chenodiol	anticholelithogenic	1983	19,	317
choline alfoscerate	nootropic	1990	26,	300
cibenzoline	antiarrhythmic	1985	21,	325
cicletanine	antihypertensive	1988	24,	299
cilazapril	antihypertensive	1990	26,	301
cilostazol	antithrombotic	1988	24,	299
cimetropium bromide	antispasmodic	1985	21,	326
cinitapride	gastroprokinetic	1990	26,	301
cinolazepam	hypnotic	1993	29,	334
ciprofibrate	hypolipidemic	1985	21,	326
ciprofloxacin	antibacterial	1986	22,	318
cisapride	gastroprokinetic	1988	24,	299
citalopram	antidepressant	1989	25,	311
cladribine	antineoplastic	1993	29,	335
clarithromycin	antibiotic	1990	26,	302
clobenoside	vasoprotective	1988	24,	300
cloconazole HCl	topical antifungal	1986	22,	318
clodronate disodium	calcium regulator	1986	22,	319
cloricromen	antithrombotic	1991	27,	325
clospipramine hydrochloride	neuroleptic	1991	27,	325
cyclosporine	immunosuppressant	1983	19,	317
cytarabine ocfosfate	antineoplastic	1993	29,	335
dapiprazole HCl	antiglaucoma	1987	23,	332
defibrotide	antithrombotic	1986	22,	319
deflazacort	antiinflammatory	1986	22,	319
delapril	antihypertensive	1989	25,	311
denopamine	cardiostimulant	1988	24,	300
deprodone propionate	topical antiinflammatory	1992	28,	329
desflurane	anesthetic	1992	28,	329
dexrazoxane	cardioprotective	1992	28,	330

GENERIC NAME	INDICATION	YEAR INTRODUCED	ARMC VOL., PAGE	
dezocine	analgesic	1991	27,	326
diacerein	antirheumatic	1985	21,	326
didanosine	antiviral	1991	27,	326
dilevalol	antihypertensive	1989	25,	311
dirithromycin	antibiotic	1993	29,	336
disodium pamidronate	calcium regulator	1989	25,	312
divistyramine	hypocholesterolemic	1984	20,	317
dopexamine	cardiostimulant	1989	25,	312
doxacurium chloride	muscle relaxant	1991	27,	326
doxazosin mesylate	antihypertensive	1988	24,	300
doxefazepam	hypnotic	1985	21,	326
doxifluridine	antineoplastic	1987	23,	332
doxofylline	bronchodilator	1985	21,	327
dronabinol	antinauseant	1986	22,	319
droxicam	antiinflammatory	1990	26,	302
droxidopa	antiparkinsonian	1989	25,	312
ebastine	antihistamine	1990	26	302
ecabet sodium	antiulcerative	1993	29,	336
emedastine difumarate	antiallergic and antiasthmatic	1993	29,	336
emorfazone	analgesic	1984	20,	317
enalapril maleate	antihypertensive	1984	20,	317
enalaprilat	antihypertensive	1987	23,	332
encainide HCl	antiarrhythmic	1987	23,	333
enocitabine	antineoplastic	1983	19,	318
enoxacin	antibacterial	1986	22,	320
enoxaparin	antithrombotic	1987	23,	333
enoximone	cardiostimulant	1988	24,	301
enprostil	antiulcer	1985	21,	327
epalrestat	antidiabetic	1992	28,	330
eperisone HCl	muscle relaxant	1983	19,	318
epidermal growth factor	wound healing agent	1987	23,	333
epirubicin HCl	antineoplastic	1984	20,	318
epoprostenol sodium	platelet aggreg. inhib.	1983	19,	318
eptazocine HBr	analgesic	1987	23,	334
erythromycin acistrate	antibiotic	1988	24,	301
erythropoietin	hematopoetic	1988	24,	301
esmolol HCl	antiarrhythmic	1987	23,	334
ethyl icosapentate	antithrombotic	1990	26,	303
etizolam	anxiolytic	1984	20,	318
etodolac	antiinflammatory	1985	21,	327
exifone	nootropic	1988	24,	302
factor VIII	hemostatic	1992	28,	330
famotidine	antiulcer	1985	21,	327
felbamate	antiepileptic	1993	29,	337
felbinac	topical antiinflammatory	1986	22,	320
felodipine	antihypertensive	1988	24,	302
fenbuprol	choleretic	1983	19,	318
fenticonazole nitrate	antifungal	1987	23,	334
filgrastim	immunostimulant	1991	27,	327

GENERIC NAME	INDICATION	YEAR INTRODUCED	ARMC VOL., PAGE	
finasteride	5α-reductase inhibitor	1992	28,	331
fisalamine	intestinal antiinflammatory	1984	20,	318
fleroxacin	antibacterial	1992	28,	331
flomoxef sodium	antibiotic	1988	24,	302
flosequinan	cardiostimulant	1992	28,	331
fluconazole	antifungal	1988	24,	303
fludarabine phosphate	antineoplastic	1991	27,	327
flumazenil	benzodiazepine antag.	1987	23,	335
flunoxaprofen	antiinflammatory	1987	23,	335
fluoxetine HCl	antidepressant	1986	22,	320
flupirtine maleate	analgesic	1985	21,	328
flutamide	antineoplastic	1983	19,	318
flutazolam	anxiolytic	1984	20,	318
fluticasone propionate	antiinflammatory	1990	26,	303
flutoprazepam	anxiolytic	1986	22,	320
flutropium bromide	antitussive	1988	24,	303
fluvoxamine maleate	antidepressant	1983	19,	319
formestane	antineoplastic	1993	29,	337
formoterol fumarate	bronchodilator	1986	22,	321
foscarnet sodium	antiviral	1989	25,	313
fosfosal	analgesic	1984	20,	319
fosinopril sodium	antihypertensive	1991	27,	328
fotemustine	antineoplastic	1989	25,	313
gabapentin	antiepileptic	1993	29,	338
gallium nitrate	calcium regulator	1991	27,	328
gallopamil HCl	antianginal	1983	19,	319
ganciclovir	antiviral	1988	24,	303
gemeprost	abortifacient	1983	19,	319
gestodene	progestogen	1987	23,	335
gestrinone	antiprogestogen	1986	22,	321
glucagon, rDNA	hypoglycemia	1993	29,	338
goserelin	hormone	1987	23,	336
granisetron hydrochloride	antiemetic	1991	27,	329
guanadrel sulfate	antihypertensive	1983	19,	319
halobetasol propionate	topical antiinflammatory	1991	27,	329
halofantrine	antimalarial	1988	24,	304
halometasone	topical antiinflammatory	1983	19,	320
histrelin	precocious puberty	1993	29,	338
hydrocortisone aceponate	topical antiinflammatory	1988	24,	304
hydrocortisone butyrate	topical antiinflammatory	1983	19,	320
ibopamine HCl	cardiostimulant	1984	20,	319
ibudilast	antiasthmatic	1989	25,	313
idarubicin hydrochloride	antineoplastic	1990	26,	303
idebenone	nootropic	1986	22,	321
iloprost	platelet aggreg. inhibitor	1992	28,	332
imidapril HCl	antihypertensive	1993	29,	339
imipenem/cilastatin	antibiotic	1985	21,	328
indalpine	antidepressant	1983	19,	320

GENERIC NAME	INDICATION	YEAR INTRODUCED	ARMC VOL., PAGE	
indeloxazine HCl	nootropic	1988	24,	304
indobufen	antithrombotic	1984	20,	319
interferon, β-1b	multiple sclerosis	1993	29,	339
interferon, gamma	antiinflammatory	1989	25,	314
interferon, gamma-1α	antineoplastic	1992	28,	332
interferon gamma-1b	immunostimulant	1991	27,	329
interleukin-2	antineoplastic	1989	25,	314
ipriflavone	calcium regulator	1989	25,	314
irsogladine	antiulcer	1989	25,	315
isepamicin	antibiotic	1988	24,	305
isofezolac	antiinflammatory	1984	20,	319
isoxicam	antiinflammatory	1983	19,	320
isradipine	antihypertensive	1989	25,	315
itraconazole	antifungal	1988	24,	305
ivermectin	antiparasitic	1987	23,	336
ketanserin	antihypertensive	1985	21,	328
ketorolac tromethamine	analgesic	1990	26,	304
lacidipine	antihypertensive	1991	27,	330
lamotrigine	anticonvulsant	1990	26,	304
lansoprazole	antiulcer	1992	28,	332
lenampicillin HCl	antibiotic	1987	23,	336
lentinan	immunostimulant	1986	22,	322
leuprolide acetate	hormone	1984	20,	319
levacecarnine HCl	nootropic	1986	22,	322
levobunolol HCl	antiglaucoma	1985	21,	328
levocabastine hydrochloride	antihistamine	1991	27,	330
levodropropizine	antitussive	1988	24,	305
levofloxacin	antibiotic	1993	29,	340
lidamidine HCl	antiperistaltic	1984	20,	320
limaprost	antithrombotic	1988	24,	306
lisinopril	antihypertensive	1987	23,	337
lobenzarit sodium	antiinflammatory	1986	22,	322
lodoxamide tromethamine	antiallergic ophthalmic	1992	28,	333
lomefloxacin	antibiotic	1989	25,	315
lonidamine	antineoplastic	1987	23,	337
loprazolam mesylate	hypnotic	1983	19,	321
loracarbef	antibiotic	1992	28,	333
loratadine	antihistamine	1988	24,	306
lovastatin	hypocholesterolemic	1987	23,	337
loxoprofen sodium	antiinflammatory	1986	22,	322
mabuterol HCl	bronchodilator	1986	22,	323
malotilate	hepatroprotective	1985	21,	329
manidipine hydrochloride	antihypertensive	1990	26,	304
masoprocol	topical antineoplastic	1992	28,	333
medifoxamine fumarate	antidepressant	1986	22,	323
mefloquine HCl	antimalarial	1985	21,	329
meglutol	hypolipidemic	1983	19,	321
melinamide	hypocholesterolemic	1984	20,	320

GENERIC NAME	INDICATION	YEAR INTRODUCED	ARMC VOL., PAGE	
mepixanox	analeptic	1984	20,	320
meptazinol HCl	analgesic	1983	19,	321
metaclazepam	anxiolytic	1987	23,	338
metapramine	antidepressant	1984	20,	320
mexazolam	anxiolytic	1984	20,	321
mifepristone	abortifacient	1988	24,	306
milrinone	cardiostimulant	1989	25,	316
miltefosine	topical antineoplastic	1993	29,	340
miokamycin	antibiotic	1985	21,	329
misoprostol	antiulcer	1985	21,	329
mivacurium chloride	muscle relaxant	1992	28,	334
mitoxantrone HCl	antineoplastic	1984	20,	321
mizoribine	immunosuppressant	1984	20,	321
moclobemide	antidepressant	1990	26,	305
mometasone furoate	topical antiinflammatory	1987	23,	338
moricizine hydrochloride	antiarrhythmic	1990	26,	305
moxonidine	antihypertensive	1991	27,	330
mupirocin	topical antibiotic	1985	21,	330
muromonab-CD3	immunosuppressant	1986	22,	323
muzolimine	diuretic	1983	19,	321
nabumetone	antiinflammatory	1985	21,	330
nadifloxacin	topical antibiotic	1993	29,	340
nafamostat mesylate	protease inhibitor	1986	22,	323
nafarelin acetate	hormone	1990	26,	306
naftifine HCl	antifungal	1984	20,	321
naltrexone HCl	narcotic antagonist	1984	20,	322
nedocromil sodium	antiallergic	1986	22,	324
neltenexine	cystic fibrosis	1993	29,	341
nemonapride	neuroleptic	1991	27,	331
neticonazole HCl	topical antifungal	1993	29,	341
nicorandil	coronary vasodilator	1984	20,	322
nilutamide	antineoplastic	1987	23,	338
nilvadipine	antihypertensive	1989	25,	316
nimesulide	antiinflammatory	1985	21,	330
nimodipine	cerebral vasodilator	1985	21,	330
nipradilol	antihypertensive	1988	24,	307
nisoldipine	antihypertensive	1990	26,	306
nitrefazole	alcohol deterrent	1983	19,	322
nitrendipine	hypertensive	1985	21,	331
nizatidine	antiulcer	1987	23,	339
nizofenzone fumarate	nootropic	1988	24,	307
nomegestrol acetate	progestogen	1986	22,	324
norfloxacin	antibacterial	1983	19,	322
norgestimate	progestogen	1986	22,	324
octreotide	antisecretory	1988	24,	307
ofloxacin	antibacterial	1985	21,	331
omeprazole	antiulcer	1988	24,	308
ondansetron hydrochloride	antiemetic	1990	26,	306
ornoprostil	antiulcer	1987	23,	339

GENERIC NAME	INDICATION	YEAR INTRODUCED	ARMC VOL., PAGE	
osalazine sodium	intestinal antinflamm.	1986	22,	324
oxaprozin	antiinflammatory	1983	19,	322
oxcarbazepine	anticonvulsant	1990	26,	307
oxiconazole nitrate	antifungal	1983	19,	322
oxiracetam	nootropic	1987	23,	339
oxitropium bromide	bronchodilator	1983	19,	323
ozagrel sodium	antithrombotic	1988	24,	308
paclitaxal	antineoplastic	1993	29,	342
parnaparin sodium	anticoagulant	1993	29,	342
paroxetine	antidepressant	1991	27,	331
pefloxacin mesylate	antibacterial	1985	21,	331
pegademase bovine	immunostimulant	1990	26,	307
pemirolast potassium	antiasthmatic	1991	27,	331
pentostatin	antineoplastic	1992	28,	334
pergolide mesylate	antiparkinsonian	1988	24,	308
perindopril	antihypertensive	1988	24,	309
picotamide	antithrombotic	1987	23,	340
pidotimod	immunostimulant	1993	29,	343
piketoprofen	topical antiinflammatory	1984	20,	322
pilsicainide hydrochloride	antiarrhythmic	1991	27,	332
pimaprofen	topical antiinflammatory	1984	20,	322
pinacidil	antihypertensive	1987	23,	340
pirarubicin	antineoplastic	1988	24,	309
piroxicam cinnamate	antiinflammatory	1988	24,	309
plaunotol	antiulcer	1987	23,	340
porfimer sodium	antineoplastic adjuvant	1993	29,	343
pramiracetam H_2SO_4	cognition enhancer	1993	29,	343
pravastatin	antilipidemic	1989	25,	316
prednicarbate	topical antiinflammatory	1986	22,	325
progabide	anticonvulsant	1985	21,	331
promegestrone	progestogen	1983	19,	323
propacetamol HCl	analgesic	1986	22,	325
propentofylline propionate	cerebral vasodilator	1988	24,	310
propiverine hydrochloride	urologic	1992	28,	335
propofol	anesthetic	1986	22,	325
quazepam	hypnotic	1985	21,	332
quinapril	antihypertensive	1989	25,	317
quinfamide	amebicide	1984	20,	322
ramipril	antihypertensive	1989	25,	317
ranimustine	antineoplastic	1987	23,	341
rebamipide	antiulcer	1990	26,	308
remoxipride hydrochloride	antipsychotic	1990	26,	308
repirinast	antiallergic	1987	23,	341
reviparin sodium	anticoagulant	1993	29,	344
rifabutin	antibacterial	1992	28,	335
rifapentine	antibacterial	1988	24,	310
rifaximin	antibiotic	1985	21,	332
rifaximin	antibiotic	1987	23,	341

GENERIC NAME	INDICATION	YEAR INTRODUCED	ARMC VOL., PAGE	
rilmazafone	hypnotic	1989	25,	317
rilmenidine	antihypertensive	1988	24,	310
rimantadine HCl	antiviral	1987	23,	342
risperidone	neuroleptic	1993	29,	344
rokitamycin	antibiotic	1986	22,	325
romurtide	immunostimulant	1991	27,	332
ronafibrate	hypolipidemic	1986	22,	326
rosaprostol	antiulcer	1985	21,	332
roxatidine acetate HCl	antiulcer	1986	22,	326
roxithromycin	antiulcer	1987	23,	342
rufloxacin hydrochloride	antibacterial	1992	28,	335
RV-11	antibiotic	1989	25,	318
salmeterol hydroxynaphthoate	bronchodilator	1990	26,	308
sapropterin hydrochloride	hyperphenylalaninemia	1992	28,	336
sargramostim	immunostimulant	1991	27,	332
sarpogrelate HCl	platelet antiaggregant	1993	29,	344
schizophyllan	immunostimulant	1985	22,	326
sertaconazole nitrate	topical antifungal	1992	28,	336
setastine HCl	antihistamine	1987	23,	342
setiptiline	antidepressant	1989	25,	318
setraline hydrochloride	antidepressant	1990	26,	309
sevoflurane	anesthetic	1990	26,	309
simvastatin	hypocholesterolemic	1988	24,	311
sodium cellulose PO4	hypocalciuric	1983	19,	323
sofalcone	antiulcer	1984	20,	323
somatropin	hormone	1987	23,	343
sorivudine	antiviral	1993	29,	345
sparfloxacin	antibiotic	1993	29,	345
spizofurone	antiulcer	1987	23,	343
succimer	chelator	1991	27,	333
sufentanil	analgesic	1983	19,	323
sulbactam sodium	B-lactamase inhibitor	1986	22,	326
sulconizole nitrate	topical antifungal	1985	21,	332
sultamycillin tosylate	antibiotic	1987	23,	343
sumatriptan succinate	antimigraine	1991	27,	333
suprofen	analgesic	1983	19,	324
surfactant TA	respiratory surfactant	1987	23,	344
tacalcitol	topical antipsoriatic	1993	29,	346
tacrine HCl	Alzheimer's disease	1993	29,	346
tacrolimus	immunosuppressant	1993	29,	347
tamsulosin HCl	antiprostatic hypertrophy	1993	29,	347
tazobactam sodium	β-lactamase inhibitor	1992	28,	336
tazanolast	antiallergic	1990	26,	309
teicoplanin	antibacterial	1988	24,	311
telmesteine	mucolytic	1992	28,	337
temafloxacin hydrochloride	antibacterial	1991	27,	334
temocillin disodium	antibiotic	1984	20,	323

GENERIC NAME	INDICATION	YEAR INTRODUCED	ARMC VOL., PAGE	
tenoxicam	antiinflammatory	1987	23,	344
teprenone	antiulcer	1984	20,	323
terazosin HCl	antihypertensive	1984	20,	323
terbinafine hydrochloride	antifungal	1991	27,	334
terconazole	antifungal	1983	19,	324
tertatolol HCl	antihypertensive	1987	23,	344
thymopentin	immunomodulator	1985	21,	333
tiamenidine HCl	antihypertensive	1988	24,	311
tianeptine sodium	antidepressant	1983	19,	324
tibolone	anabolic	1988	24,	312
tilisolol hydrochloride	antihypertensive	1992	28,	337
timiperone	neuroleptic	1984	20,	323
tinazoline	nasal decongestant	1988	24,	312
tioconazole	antifungal	1983	19,	324
tiopronin	urolithiasis	1989	25,	318
tiquizium bromide	antispasmodic	1984	20,	324
tiracizine hydrochloride	antiarrhythmic	1990	26,	310
tiropramide HCl	antispasmodic	1983	19,	324
tizanidine	muscle relaxant	1984	20,	324
toloxatone	antidepressant	1984	20,	324
tolrestat	antidiabetic	1989	25,	319
torasemide	diuretic	1993	29,	348
toremifene	antineoplastic	1989	25,	319
tosufloxacin tosylate	antibacterial	1990	26,	310
trandolapril	antihypertensive	1993	29,	348
tretinoin tocoferil	antiulcer	1993	29,	348
trientine HCl	chelator	1986	22,	327
trimazosin HCl	antihypertensive	1985	21,	333
tropisetron	antiemetic	1992	28,	337
troxipide	antiulcer	1986	22,	327
ubenimex	immunostimulant	1987	23,	345
vesnarinone	cardiostimulant	1990	26,	310
vigabatrin	anticonvulsant	1989	25,	319
vinorelbine	antineoplastic	1989	25,	320
xamoterol fumarate	cardiotonic	1988	24,	312
zalcitabine	antiviral	1992	28,	338
zaltoprofen	antiinflammatory	1993	29,	349
zidovudine	antiviral	1987	23,	345
zolpidem hemitartrate	hypnotic	1988	24,	313
zonisamide	anticonvulsant	1989	25,	320
zopiclone	hypnotic	1986	22,	327
zuclopenthixol acetate	antipsychotic	1987	23,	345

GENERIC NAME	INDICATION	YEAR INTRODUCED	ARMC VOL., PAGE
gemeprost	ABORTIFACIENT	1983	19, 319
mifepristone		1988	24, 306
nitrefazole	ALCOHOL DETERRENT	1983	19, 322
tacrine HCl	ALZHEIMER'S DISEASE	1993	29, 346
quinfamide	AMEBICIDE	1984	20, 322
tibolone	ANABOLIC	1988	24, 312
mepixanox	ANALEPTIC	1984	20, 320
alfentanil HCl	ANALGESIC	1983	19, 314
alminoprofen		1983	19, 314
dezocine		1991	27, 326
emorfazone		1984	20, 317
eptazocine HBr		1987	23, 334
flupirtine maleate		1985	21, 328
fosfosal		1984	20, 319
ketorolac tromethamine		1990	26, 304
meptazinol HCl		1983	19, 321
propacetamol HCl		1986	22, 325
sufentanil		1983	19, 323
suprofen		1983	19, 324
desflurane	ANESTHETIC	1992	28, 329
propofol		1986	22, 325
sevoflurane		1990	26, 309
azelaic acid	ANTIACNE	1989	25, 310
emedastine difumarate	ANTIALLERGIC	1993	29, 336
nedocromil sodium		1986	22, 324
repirinast		1987	23, 341
tazanolast		1990	26, 309
lodoxamide tromethamine	ANTIALLERGIC OPHTHALMIC	1992	28, 333
gallopamil HCl	ANTIANGINAL	1983	19, 319
cibenzoline	ANTIARRHYTHMIC	1985	21, 325
encainide HCl		1987	23, 333
esmolol HCl		1987	23, 334
moricizine hydrochloride		1990	26, 305
pilsicainide hydrochloride		1991	27, 332
tiracizine hydrochloride		1990	26, 310

GENERIC NAME	INDICATION	YEAR INTRODUCED	ARMC VOL., PAGE
amlexanox	ANTIASTHMATIC	1987	23, 327
emedastine difumarate		1993	29, 336
ibudilast		1989	25, 313
pemirolast potassium		1991	27, 331
ciprofloxacin	ANTIBACTERIAL	1986	22, 318
enoxacin		1986	22, 320
fleroxacin		1992	28, 331
norfloxacin		1983	19, 322
ofloxacin		1985	21, 331
pefloxacin mesylate		1985	21, 331
rifabutin		1992	28, 335
rifapentine		1988	24, 310
rufloxacin hydrochloride		1992	28, 335
teicoplanin		1988	24, 311
temafloxacin hydrochloride		1991	27, 334
tosufloxacin tosylate		1990	26, 310
arbekacin	ANTIBIOTIC	1990	26, 298
aspoxicillin		1987	23, 328
astromycin sulfate		1985	21, 324
azithromycin		1988	24, 298
aztreonam		1984	20, 315
brodimoprin		1993	29, 333
carboplatin		1986	22, 318
carumonam		1988	24, 298
cefbuperazone sodium		1985	21, 325
cefdinir		1991	27, 323
cefepime		1993	29, 334
cefetamet pivoxil hydrochloride		1992	28, 327
cefixime		1987	23, 329
cefmenoxime HCl		1983	19, 316
cefminox sodium		1987	23, 330
cefodizime sodium		1990	26, 300
cefonicid sodium		1984	20, 316
ceforanide		1984	20, 317
cefotetan disodium		1984	20, 317
cefotiam hexetil hydrochloride		1991	27, 324
cefpimizole		1987	23, 330
cefpiramide sodium		1985	21, 325
cefpirome sulfate		1992	28, 328
cefpodoxime proxetil		1989	25, 310
cefprozil		1992	28, 328
ceftazidime		1983	19, 316
cefteram pivoxil		1987	23, 330
ceftibuten		1992	28, 329
cefuroxime axetil		1987	23, 331
cefuzonam sodium		1987	23, 331
clarithromycin		1990	26, 302

GENERIC NAME	INDICATION	YEAR INTRODUCED	ARMC VOL., PAGE	
dirithromycin		1993	29,	336
erythromycin acistrate		1988	24,	301
flomoxef sodium		1988	24,	302
imipenem/cilastatin		1985	21,	328
isepamicin		1988	24,	305
lenampicillin HCl		1987	23,	336
levofloxacin		1993	29,	340
lomefloxacin		1989	25,	315
loracarbef		1992	28,	333
miokamycin		1985	21,	329
rifaximin		1985	21,	332
rifaximin		1987	23,	341
rokitamycin		1986	22,	325
RV-11		1989	25,	318
sparfloxacin		1993	29,	345
sultamycillin tosylate		1987	23,	343
temocillin disodium		1984	20,	323
mupirocin	ANTIBIOTIC, TOPICAL	1985	21,	330
nadifloxacin		1993	29,	340
chenodiol	ANTICHOLELITHOGENIC	1983	19,	317
parnaparin sodium	ANTICOAGULANT	1993	29,	342
reviparin sodium		1993	29,	344
lamotrigine	ANTICONVULSANT	1990	26,	304
oxcarbazepine		1990	26,	307
progabide		1985	21,	331
vigabatrin		1989	25,	319
zonisamide		1989	25,	320
bupropion HCl	ANTIDEPRESSANT	1989	25,	310
citalopram		1989	25,	311
fluoxetine HCl		1986	22,	320
fluvoxamine maleate		1983	19,	319
indalpine		1983	19,	320
medifoxamine fumarate		1986	22,	323
metapramine		1984	20,	320
moclobemide		1990	26,	305
paroxetine		1991	27,	331
setiptiline		1989	25,	318
sertraline hydrochloride		1990	26,	309
tianeptine sodium		1983	19,	324
toloxatone		1984	20,	324
acarbose	ANTIDIABETIC	1990	26,	297
epalrestat		1992	28,	330
tolrestat		1989	25,	319

GENERIC NAME	INDICATION	YEAR INTRODUCED	ARMC VOL., PAGE
acetorphan	ANTIDIARRHEAL	1993	29, 332
granisetron hydrochloride	ANTIEMETIC	1991	27, 329
ondansetron hydrochloride		1990	26, 306
tropisetron		1992	28, 337
felbamate	ANTIEPILEPTIC	1993	29, 337
gabapentin		1993	29, 338
centchroman	ANTIESTROGEN	1991	27, 324
fenticonazole nitrate	ANTIFUNGAL	1987	23, 334
fluconazole		1988	24, 303
itraconazole		1988	24, 305
naftifine HCl		1984	20, 321
oxiconazole nitrate		1983	19, 322
terbinafine hydrochloride		1991	27, 334
terconazole		1983	19, 324
tioconazole		1983	19, 324
amorolfine hydrochloride	ANTIFUNGAL, TOPICAL	1991	27, 322
butenafine hydrochloride		1992	28, 327
butoconazole		1986	22, 318
cloconazole HCl		1986	22, 318
neticonazole HCl		1993	29, 341
sertaconazole nitrate		1992	28, 336
sulconizole nitrate		1985	21, 332
apraclonidine HCl	ANTIGLAUCOMA	1988	24, 297
befunolol HCl		1983	19, 315
dapiprazole HCl		1987	23, 332
levobunolol HCl		1985	21, 328
acrivastine	ANTIHISTAMINE	1988	24, 295
astemizole		1983	19, 314
azelastine HCl		1986	22, 316
ebastine		1990	26, 302
cetirizine HCl		1987	23, 331
levocabastine hydrochloride		1991	27, 330
loratadine		1988	24, 306
setastine HCl		1987	23, 342
alacepril	ANTIHYPERTENSIVE	1988	24, 296
alfuzosin HCl		1988	24, 296
amlodipine besylate		1990	26, 298
amosulalol		1988	24, 297
arotinolol HCl		1986	22, 316
barnidipine hydrochloride		1992	28, 326
benazepril hydrochloride		1990	26, 299

GENERIC NAME	INDICATION	YEAR INTRODUCED	ARMC VOL., PAGE
benidipine hydrochloride		1991	27, 322
betaxolol HCl		1983	19, 315
bevantolol HCl		1987	23, 328
bisoprolol fumarate		1986	22, 317
bopindolol		1985	21, 324
budralazine		1983	19, 315
bunazosin HCl		1985	21, 324
carvedilol		1991	27, 323
celiprolol HCl		1983	19, 317
cicletanine		1988	24, 299
cilazapril		1990	26, 301
delapril		1989	25, 311
dilevalol		1989	25, 311
doxazosin mesylate		1988	24, 300
enalapril maleate		1984	20, 317
enalaprilat		1987	23, 332
felodipine		1988	24, 302
fosinopril sodium		1991	27, 328
guanadrel sulfate		1983	19, 319
imidapril HCl		1993	29, 339
isradipine		1989	25, 315
ketanserin		1985	21, 328
lacidipine		1991	27, 330
lisinopril		1987	23, 337
manidipine hydrochloride		1990	26, 304
moxonidine		1991	27, 330
nilvadipine		1989	25, 316
nipradilol		1988	24, 307
nisoldipine		1990	26, 306
perindopril		1988	24, 309
pinacidil		1987	23, 340
quinapril		1989	25, 317
ramipril		1989	25, 317
rilmenidine		1988	24, 310
terazosin HCl		1984	20, 323
tertatolol HCl		1987	23, 344
tiamenidine HCl		1988	24, 311
tilisolol hydrochloride		1992	28, 337
trandolapril		1993	29, 348
trimazosin HCl		1985	21, 333
aceclofenac	ANTIINFLAMMATORY	1992	28, 325
AF-2259		1987	23, 325
amfenac sodium		1986	22, 315
amtolmetin guacil		1993	29, 332
butibufen		1992	28, 327
deflazacort		1986	22, 319
droxicam		1990	26, 302
etodolac		1985	21, 327

GENERIC NAME	INDICATION	YEAR INTRODUCED	ARMC VOL., PAGE
flunoxaprofen		1987	23, 335
fluticasone propionate		1990	26, 303
interferon, gamma		1989	25, 314
isofezolac		1984	20, 319
isoxicam		1983	19, 320
lobenzarit sodium		1986	22, 322
loxoprofen sodium		1986	22, 322
nabumetone		1985	21, 330
nimesulide		1985	21, 330
oxaprozin		1983	19, 322
piroxicam cinnamate		1988	24, 309
tenoxicam		1987	23, 344
zaltoprofen		1993	29, 349
fisalamine	ANTIINFLAMMATORY,	1984	20, 318
osalazine sodium	INTESTINAL	1986	22, 324
alclometasone dipropionate	ANTIINFLAMMATORY,	1985	21, 323
aminoprofen	TOPICAL	1990	26, 298
butyl flufenamate		1983	19, 316
deprodone propionate		1992	28, 329
felbinac		1986	22, 320
halobetasol propionate		1991	27, 329
halometasone		1983	19, 320
hydrocortisone aceponate		1988	24, 304
hydrocortisone butyrate propionate		1983	19, 320
mometasone furoate		1987	23, 338
piketoprofen		1984	20, 322
pimaprofen		1984	20, 322
prednicarbate		1986	22, 325
pravastatin	ANTILIPIDEMIC	1989	25, 316
artemisinin	ANTIMALARIAL	1987	23, 327
halofantrine		1988	24, 304
mefloquine HCl		1985	21, 329
alpiropride	ANTIMIGRAINE	1988	24, 296
sumatriptan succinate		1991	27, 333
dronabinol	ANTINAUSEANT	1986	22, 319
amsacrine	ANTINEOPLASTIC	1987	23, 327
bisantrene hydrochloride		1990	26, 300
camostat mesylate		1985	21, 325
cladribine		1993	29, 335
cytarabine ocfosfate		1993	29, 335
doxifluridine		1987	23, 332

GENERIC NAME	INDICATION	YEAR INTRODUCED	ARMC VOL., PAGE
enocitabine		1983	19, 318
epirubicin HCl		1984	20, 318
fludarabine phosphate		1991	27, 327
flutamide		1983	19, 318
formestane		1993	29, 337
fotemustine		1989	25, 313
idarubicin hydrochloride		1990	26, 303
interferon gamma-1α		1992	28, 332
interleukin-2		1989	25, 314
lonidamine		1987	23, 337
mitoxantrone HCl		1984	20, 321
nilutamide		1987	23, 338
paclitaxal		1993	29, 342
pentostatin		1992	28, 334
pirarubicin		1988	24, 309
ranimustine		1987	23, 341
toremifene		1989	25, 319
vinorelbine		1989	25, 320
porfimer sodium	ANTINEOPLASTIC ADJUVANT	1993	29, 343
masoprocol	ANTINEOPLASTIC, TOPICAL	1992	28, 333
miltefosine		1993	29, 340
atovaquone	ANTIPARASITIC	1992	28, 326
ivermectin		1987	23, 336
droxidopa	ANTIPARKINSONIAN	1989	25, 312
pergolide mesylate		1988	24, 308
lidamidine HCl	ANTIPERISTALTIC	1984	20, 320
gestrinone	ANTIPROGESTOGEN	1986	22, 321
cabergoline	ANTIPROLACTIN	1993	29, 334
tamsulosin HCl	ANTIPROSTATIC HYPERTROPHY	1993	29, 347
acitretin	ANTIPSORIATIC	1989	25, 309
calcipotriol		1991	27, 323
tacalcitol	ANTIPSORIATIC, TOPICAL	1993	29, 346
amisulpride	ANTIPSYCHOTIC	1986	22, 316
remoxipride hydrochloride		1990	26, 308
zuclopenthixol acetate		1987	23, 345

GENERIC NAME	INDICATION	YEAR INTRODUCED	ARMC VOL., PAGE
diacerein	ANTIRHEUMATIC	1985	21, 326
octreotide	ANTISECRETORY	1988	24, 307
adamantanium bromide	ANTISEPTIC	1984	20, 315
cimetropium bromide	ANTISPASMODIC	1985	21, 326
tiquizium bromide		1984	20, 324
tiropramide HCl		1983	19, 324
argatroban	ANTITHROMBOTIC	1990	26, 299
defibrotide		1986	22, 319
cilostazol		1988	24, 299
cloricromen		1991	27, 325
enoxaparin		1987	23, 333
ethyl icosapentate		1990	26, 303
ozagrel sodium		1988	24, 308
indobufen		1984	20, 319
picotamide		1987	23, 340
limaprost		1988	24, 306
flutropium bromide	ANTITUSSIVE	1988	24, 303
levodropropizine		1988	24, 305
benexate HCl	ANTIULCER	1987	23, 328
ecabet sodium		1993	29, 336
enprostil		1985	21, 327
famotidine		1985	21, 327
irsogladine		1989	25, 315
lansoprazole		1992	28, 332
misoprostol		1985	21, 329
nizatidine		1987	23, 339
omeprazole		1988	24, 308
ornoprostil		1987	23, 339
plaunotol		1987	23, 340
rebamipide		1990	26, 308
rosaprostol		1985	21, 332
roxatidine acetate HCl		1986	22, 326
roxithromycin		1987	23, 342
sofalcone		1984	20, 323
spizofurone		1987	23, 343
teprenone		1984	20, 323
tretinoin tocoferil		1993	29, 348
troxipide		1986	22, 327
didanosine	ANTIVIRAL	1991	27, 326
foscarnet sodium		1989	25, 313
ganciclovir		1988	24, 303
rimantadine HCl		1987	23, 342

GENERIC NAME	INDICATION	YEAR INTRODUCED	ARMC VOL., PAGE
sorivudine		1993	29, 345
zalcitabine		1992	28, 338
zidovudine		1987	23, 345
alpidem	ANXIOLYTIC	1991	27, 322
buspirone HCl		1985	21, 324
etizolam		1984	20, 318
flutazolam		1984	20, 318
flutoprazepam		1986	22, 320
metaclazepam		1987	23, 338
mexazolam		1984	20, 321
flumazenil	BENZODIAZEPINE ANTAG.	1987	23, 335
bambuterol	BRONCHODILATOR	1990	26, 299
doxofylline		1985	21, 327
formoterol fumarate		1986	22, 321
mabuterol HCl		1986	22, 323
oxitropium bromide		1983	19, 323
salmeterol hydroxynaphthoate		1990	26, 308
APD	CALCIUM REGULATOR	1987	23, 326
clodronate disodium		1986	22, 319
disodium pamidronate		1989	25, 312
gallium nitrate		1991	27, 328
ipriflavone		1989	25, 314
dexrazoxane	CARDIOPROTECTIVE	1992	28, 330
bucladesine sodium	CARDIOSTIMULANT	1984	20, 316
denopamine		1988	24, 300
dopexamine		1989	25, 312
enoximone		1988	24, 301
flosequinan		1992	28, 331
ibopamine HCl		1984	20, 319
milrinone		1989	25, 316
vesnarinone		1990	26, 310
amrinone	CARDIOTONIC	1983	19, 314
xamoterol fumarate		1988	24, 312
brovincamine fumarate	CEREBRAL VASODILATOR	1986	22, 317
nimodipine		1985	21, 330
propentofylline		1988	24, 310
succimer	CHELATOR	1991	27, 333
trientine HCl		1986	22, 327
fenbuprol	CHOLERETIC	1983	19, 318

GENERIC NAME	INDICATION	YEAR INTRODUCED	ARMC VOL., PAGE
auranofin	CHRYSOTHERAPEUTIC	1983	19, 314
aniracetam	COGNITION ENHANCER	1993	29, 333
pramiracetam H_2SO_4		1993	29, 343
nicorandil	CORONARY VASODILATOR	1984	20, 322
neltenexine	CYSTIC FIBROSIS	1993	29, 341
azosemide	DIURETIC	1986	22, 316
muzolimine		1983	19, 321
torasemide		1993	29, 348
alglucerase	ENZYME	1991	27, 321
cinitapride	GASTROPROKINETIC	1990	26, 301
cisapride		1988	24, 299
erythropoietin	HEMATOPOETIC	1988	24, 301
factor VIII	HEMOSTATIC	1992	28, 330
malotilate	HEPATROPROTECTIVE	1985	21, 329
buserelin acetate	HORMONE	1984	20, 316
goserelin		1987	23, 336
leuprolide acetate		1984	20, 319
nafarelin acetate		1990	26, 306
somatropin		1987	23, 343
sapropterin hydrochloride	HYPERPHENYLALANINEMIA	1992	28, 336
cadralazine	HYPERTENSIVE	1988	24, 298
nitrendipine		1985	21, 331
binfonazole	HYPNOTIC	1983	19, 315
brotizolam		1983	19, 315
butoctamide		1984	20, 316
cinolazepam		1993	29, 334
doxefazepam		1985	21, 326
loprazolam mesylate		1983	19, 321
quazepam		1985	21, 332
rilmazafone		1989	25, 317
zolpidem hemitartrate		1988	24, 313
zopiclone		1986	22, 327
acetohydroxamic acid	HYPOAMMONURIC	1983	19, 313

GENERIC NAME	INDICATION	YEAR INTRODUCED	ARMC VOL., PAGE	
sodium cellulose PO4	HYPOCALCIURIC	1983	19,	323
divistyramine	HYPOCHOLESTEROLEMIC	1984	20,	317
lovastatin		1987	23,	337
melinamide		1984	20,	320
simvastatin		1988	24,	311
glucagon, rDNA	HYPOGLYCEMIA	1993	29,	338
acipimox	HYPOLIPIDEMIC	1985	21,	323
beclobrate		1986	22,	317
binifibrate		1986	22,	317
ciprofibrate		1985	21,	326
meglutol		1983	19,	321
ronafibrate		1986	22,	326
bucillamine	IMMUNOMODULATOR	1987	23,	329
centoxin		1991	27,	325
thymopentin		1985	21,	333
filgrastim	IMMUNOSTIMULANT	1991	27,	327
interferon gamma-1b		1991	27,	329
lentinan		1986	22,	322
pegademase bovine		1990	26,	307
pidotimod		1993	29,	343
romurtide		1991	27,	332
sargramostim		1991	27,	332
schizophyllan		1985	22,	326
ubenimex		1987	23,	345
cyclosporine	IMMUNOSUPPRESSANT	1983	19,	317
mizoribine		1984	20,	321
muromonab-CD3		1986	22,	323
tacrolimus		1993	29,	347
sulbactam sodium	β-LACTAMASE INHIBITOR	1986	22,	326
tazobactam sodium		1992	28,	336
telmesteine	MUCOLYTIC	1992	28,	337
interferon β-1B	MULTIPLE SCLEROSIS	1993	29,	339
afloqualone	MUSCLE RELAXANT	1983	19,	313
doxacurium chloride		1991	27,	326
eperisone HCl		1983	19,	318
mivacurium chloride		1992	28,	334
tizanidine		1984	20,	324

GENERIC NAME	INDICATION	YEAR INTRODUCED	ARMC VOL., PAGE
naltrexone HCl	NARCOTIC ANTAGONIST	1984	20, 322
tinazoline	NASAL DECONGESTANT	1988	24, 312
clospipramine hydrochloride	NEUROLEPTIC	1991	27, 325
nemonapride		1991	27, 331
risperidone		1993	29, 344
timiperone		1984	20, 323
bifemelane HCl	NOOTROPIC	1987	23, 329
choline alfoscerate		1990	26, 300
exifone		1988	24, 302
idebenone		1986	22, 321
indeloxazine HCl		1988	24, 304
levacecarnine HCl		1986	22, 322
nizofenzone fumarate		1988	24, 307
oxiracetam		1987	23, 339
alendronate sodium	OSTEOPOROSIS	1993	29, 332
beraprost sodium	PLATELET AGGREG. INHIBITOR	1992	28, 326
epoprostenol sodium		1983	19, 318
iloprost		1992	28, 332
sarpogrelate HCl	PLATELET ANTIAGGREGANT	1993	29, 344
histrelin	PRECOCIOUS PUBERTY	1993	29, 338
gestodene	PROGESTOGEN	1987	23, 335
nomegestrol acetate		1986	22, 324
norgestimate		1986	22, 324
promegestrone		1983	19, 323
alpha-1 antitrypsin	PROTEASE INHIBITOR	1988	24, 297
nafamostat mesylate		1986	22, 323
adrafinil	PSYCHOSTIMULANT	1986	22, 315
finasteride	5α-REDUCTASE INHIBITOR	1992	28, 331
surfactant TA	RESPIRATORY SURFACTANT	1987	23, 344
APSAC	THROMBOLYTIC	1987	23, 326
alteplase		1987	23, 326
tiopronin	UROLITHIASIS	1989	25, 318

GENERIC NAME	INDICATION	YEAR INTRODUCED	ARMC VOL., PAGE	
propiverine hydrochloride	UROLOGIC	1992	28,	335
clobenoside	VASOPROTECTIVE	1988	24,	300
cadexomer iodine	WOUND HEALING AGENT	1983	19,	316
epidermal growth factor		1987	23,	333